# L.E.J. Brouwer – Topologist, Intuitionist, Philosopher

Dirk van Dalen

# L.E.J. Brouwer – Topologist, Intuitionist, Philosopher

## How Mathematics Is Rooted in Life

Springer

Dirk van Dalen
Department of Philosophy
Utrecht University
Utrecht, Netherlands

Whilst we have made considerable efforts to contact all holders of copyright material contained in this book we have failed to locate some of them. Should holders wish to contact the Publisher, we will make every effort to come to some arrangement with them.

ISBN 978-1-4471-4615-5 ISBN 978-1-4471-4616-2 (eBook)
DOI 10.1007/978-1-4471-4616-2
Springer London Heidelberg New York Dordrecht

Library of Congress Control Number: 2012954496

AMS Subject Classification: 01A70, 01A55, 01A60

Based on a previous edition of the two-volume work:
Mystic, Geometer, and Intuitionist: The Life of L. E. J. Brouwer by Dirk van Dalen


*Front cover image*: Hanna Elkan/MAI, Brouwer Archive/ ©

Printed on acid-free paper

Springer is part of Springer Science+Business Media (www.springer.com)

*In loving memory of our daughter Tineke*
*8 July 1958–14 October 2009*

# Preface

> Faire abstraction du monde d'objets (ce qui est necessaire pour travailler dans les mathématiques intuitionistes) n'est possible qu'en éprouvant la vie comme un rêve.
>
> Brouwer

Among the leading scientists of all times, Brouwer occupies a somewhat unorthodox position that merits a closer look. A genius is usually supposed to be continually involved in brilliant and illustrious activities. Mozart, for example, was said to the embodiment of music, his mind at all times and places emerged in creation and reflection. In mathematics, Euler would be the perfect example—always investigating, creating, publishing, until a ripe old age. Brouwer belonged to different class of genius; gifted with a deep intuition, he had an unparalleled access to the secrets and intricacies of mathematics and other subjects, but the manifestations of his genius were rather the eruptions of a proud and isolated volcano than a smooth running river of clever theorems. Indeed, Brouwer refused to join the class of specialised academics, who swear allegiance to a particular topic. He felt free to invest his time and energy in a wide range of activities, running from mysticism, psycholinguistics, art, politics to long walks, swimming, solitary contemplation, to fighting injustice.

The scientific highlights, of course, are Brouwer's topological innovations and the creation of his revolutionary intuitionistic mathematics. Both are manifestations of his unparalleled power of reflection. His intuitionism clearly benefitted from, and was based on his mystic views.

In the following pages the life of this unusual scientist is sketched. The scientific highlights are his breakthrough in the young subject of topology that triggered the transition from the tradition of Cantor to modern topology, and the introduction and consolidation of constructive methods and philosophy under the name *intuitionism*. As a confirmed internationalist he got entangled in the interbellum struggle for the ending of the boycott of the German and Austrian scientists. And roughly at the same time he was drawn into the Formalism–Intuitionism conflict, know as the *Grundlagenstreit*, which found an untimely end in the so-called *War of the frogs and the mice*.

One should not get the impression that his life was one long string of conflicts, but is certainly true that his uncompromising opposition to injustice got him more than his share of problems.

The present biography is a revision of the earlier two volume biography published by the Oxford University Press. After these had, so to speak, passed their natural life span, the OUP gracefully agreed to allow me to publish the present version with Springer. The contents have here and there been updated, and some sections have been pruned.

I have in the Oxford Press edition expressed my gratitude to a large number of friends and colleagues and institutions, and I want on this place to say again how much the biography owes to them. I am indebted to Garth Dales who volunteered to proof read the first seven chapters.

Without the efficient and friendly support of Joerg Sixt and his staff my task would have been a heavy burden, they more than deserve my thanks.

In the mean time the Selected Correspondence of Brouwer (Brouwer 2011) (in an English translation) has appeared, so the reader will have access to a rich source of background information not available earlier.

Utrecht, the Netherlands
November 2012

Dirk van Dalen

# Contents

# Chapter 1
# Child and Student

It was into an optimistic country that Luitzen Egbertus Jan Brouwer was born—a country with a burgeoning science, literature, and social awareness. After having been one of the backwaters of Europe, the Netherlands was finding its way back into the mainstream of culture and commerce. Since the golden age of Huygens and Stevin, science had been more a playground for cautious imitators than for bold researchers. At the turn of the century all that was changing. Physics was flourishing—Lorentz, Kamerlingh Onnes, Van der Waals and Zeeman were all putting the Netherlands[1] in the forefront of modern physics, Van 't Hoff was doing so in chemistry, Hugo de Vries in biology, and Kapteyn in astronomy. The fact that Holland counted a number of its leading scientists among the early recipients of the Nobel prize may be sufficient evidence of the quality of the research in the sciences in the Netherlands.

Mathematics was slower to pick up the new élan. The best-known Dutch mathematician after Christian Huygens was Thomas Jan Stieltjes, who did not find recognition in Holland and who practised his mathematics in Toulouse. The mathematicians of the period were competent, but not on a par with their colleagues in Germany, France and Britain. Neither was their choice of subject very daring: the Dutch worked in the more settled parts of mathematics, from analytic-descriptive geometry to number theory and analysis. At the same time literature was freeing itself from the strait-jacket of nineteenth-century conventions, and a wave of new authors had already transformed the literary landscape. The explicit aim of the leading spirits was to 'push Holland high up into the stream of the nations'.[2] Almost inevitably the

[1]We will indulge in some 'abuse of language' by using 'The Netherlands' and 'Holland' as synonyms. This is definitely incorrect from the standpoint of geography. Holland is the collective name of the two provinces North- and South-Holland. There used to be a county Holland under the Count of Holland, but under the Dutch republic it became a province like any other. The concentration of trade and government, however, lent Holland so much prominence, that one often used 'Holland' instead of 'The United Republic'. In spite of the efforts of the later kings of The Netherlands, the habit of using 'Holland' as a pars pro toto has persisted. The phenomenon is not uncommon, think of 'England' and 'United Kingdom' for example. The reader is warned.

[2]*Nederland hoog op te stuwen in de vaart der volkeren.*

D. van Dalen, *L.E.J. Brouwer – Topologist, Intuitionist, Philosopher*,
DOI 10.1007/978-1-4471-4616-2_1, 

new literary trend in Holland was closely bound up with developments in the social movements. Many of the leading poets and authors were involved in the promotion of a better social climate. A number of them played a significant role in the socialist movement, which eventually led to the birth of the Social Democratic Party and, subsequently, the Communist Party.

The political scene at the turn of the century was mainly determined by the *Liberals*, who were the rightful heirs of the dominant movement of the nineteenth century, the *Protestants* (who were traditionally called the Christians), the *Catholics*, and the late-comers, the *Social Democrats*. Traditionally the Liberals made up the higher strata of society; the Protestants, Catholics and Socialists were instrumental in the emancipation of their respective sections of the population. History had equipped their respective parties with powerful charismatic leaders.

Generally speaking, Dutch society was experiencing a powerful upward thrust in its social structure. Progress had become the password of the day, and in this general movement there was a small but influential group that formed the backbone of the nation's new spirit: the schoolmasters. In the past they had been the favourite butt of pamphleteers and wits, for example, Multatuli, the reformist author, created the immortal schoolmaster *Pennewip*, a caricature of the fossils that used to educate the nation's youth. At the turn of the century, however, a whole new generation of schoolmasters had come to populate the schoolrooms of the country. The new teachers were, for the greater part, idealistic promoters of a better future, equipped with an unshakable belief in the beneficial influence of knowledge. The nation was blessed at the time with a fine body of teachers and an effective egalitarian educational system. A side effect of the strong admiration of schoolmasters for learning was that they often sent their own children to institutions of higher education; a surprising percentage of the scientific Dutch community actually came from schoolmasters' families.

Luitzen Egbertus Jan (*Bertus*, as he was called) Brouwer, was one of these schoolmaster's sons that was to put his stamp on Dutch society and culture; he was born on the 27th of February 1881 at Overschie, son of Hendrika[3] Poutsma (15 September 1852 Follega–3 May 1927 Utrecht) and Egbertus Luitzens Brouwer (17 April 1854 Bakkeveen–3 May 1947 Blaricum), schoolmaster at a primary school at Overschie. Hendrika was of Friesian stock; her earliest registered forebear Tammerus Gerhardi (1579–1644) was a minister of the Dutch Reformed Church at Joure and IJlst in Friesland. The Poutsma family tree is adorned with a good number of parsons and schoolmasters, and towards the end of the nineteenth century there is a definite tendency to rise to the higher strata of the teaching profession. Two of Brouwer's uncles became teachers at the Barlaeus gymnasium at Amsterdam, the finest school of the city. One of them, Hendrik Poutsma taught English literature and language, and wrote a classic textbook; he was awarded a honorary doctorate by the University of Amsterdam. The other uncle, Albertus Poutsma, taught Greek and Latin at the same school and eventually became its rector.

[3]The spelling of names is not uniform in the registers of older archives, e.g. 'Hendrika' is also spelt as 'Henderika', 'Hendrica', 'Henderica'.

The forebears of Bertus were all Friesian. His father was born at Bakkeveen in Friesland, and he in turn was the son of Luitzen Luitzens Brouwer (born 20 April 1813 in Duurswoude), who likewise was a schoolmaster (*onderwijzer der jeugd*). The latter was the son of another Luitzen Luitzens Brouwer (born in Duurswoude 1756 or 1757), who was a farmer and shepherd.

The Friesians were (and still are) known for their reliability, and they were welcome additions to many a profession in the western part of the country, the part that traditionally is known as Holland. The position of the Friesians in the Netherlands is somewhat comparable to that of the Scots in England. In view of the limited opportunities in Friesland, the more adventurous among them moved to Holland and settled there. For example, in the first half of this century the Amsterdam police force recruited a good number of its members from Friesland.

Bertus's parents were married on 8 April 1880. They immediately moved from Beetsterzwaag (Friesland) to Overschie, a small town which nowadays is part of Rotterdam, where their first child, Bertus, was born. The story goes that grandfather Brouwer came to see the baby.[4] He gravely looked at the child in the cradle, and spoke the memorable words: 'Let us hope that he can learn.'[5]

Bertus's parents were strict and honest people and none of the later extravagances of the sons seem to have visited them. Like all hardworking, sober Dutchmen they led a simple life in reasonable comfort, but without wasting money. Bertus's mother had indeed elevated saving to an art.

The most detailed information about the child and boy Bertus was provided by himself. At the age of sixteen he had to write a short 'auto-biography' as part of the initiation rites of the *Amsterdam Student Corporation* or *Fraternity*, in Dutch the *Amsterdamse Studenten Corps*.[6] The following lines are from this biographical sketch (September 1897):

> I was born on 27 February 1881 in Overschie. Here I lived for eight months, of which I naturally do not remember anything at all. I have never been back to that town, except for a few hours, when, four years ago, I went to Brussels, and stopped over in Rotterdam, skipping a couple of trains.[7] From there, one reaches Overschie in half an hour by horse tram along a road that curves through low, muddy, peaty meadows, criss-crossed with marshy drains and black bubbling ditches. The muddy cattle in those meadows feed themselves on the waste of the gin mills (which are centred in that area), the so-called

[4]Told by Louise Peijpers, the daughter from the first marriage of Bertus's wife. Although Louise was no witness to the earlier events of Brouwer's life, she had a perfect memory for the family folklore. A large part of the information on Brouwer's early years is based on oral communications of Louise Peijpers.

[5]*Als hij maar kan leren.*

[6]The main fraternity of the University of Amsterdam.

[7]The reader should note that this was rather exceptional; most Dutchmen hardly ever crossed the borders of their country, let alone at the age of 12! One should, however, bear in mind that Bertus was a precocious child, who was at the age of twelve already in the third grade of the high school (HBS). Foreign travel later became a second nature to Bertus (and his brothers).

'swill' which is put into tub-like containers standing here and there in the field. The village of Overschie consists, like so many villages, of two rows of houses built on each side of a wide road. It is distinguished by a dirty street, an ugly town hall, ugly private houses, a couple of dunghills, and a drawbridge in the middle.

Within a year of the birth of Bertus, the family moved to Medemblik, one of the old towns in North-Holland that had flourished on maritime trading in the old days of the Zuiderzee. The Brouwers lived in Medemblik for eleven years, during which time two more sons were born, Izaak Alexander (Lex) (23 January 1883) and Hendrikus Albertus (Aldert) (20 September 1886). Bertus's father taught in Medemblik at an elementary school, the so-called *Burgerschool*.[8]

The arrival of the second child, Isaac Alexander, known as Lex, seemed to have upset Bertus. According to the oral tradition of the Brouwer family, he blamed Lex for driving him from the comfortable cradle. Whatever may be true in this story, a fact is that as a boy, Bertus harboured a thinly veiled dislike of Lex. The latter came in for a fair portion of refined or not-so-refined pestering.

The relationship with the youngest brother, Aldert, was much more friendly, although they were known to fight occasionally. Even in their old age they annoyed and amused the members of the Royal Academy with their quarrels.

## 1.1 School Years

In Medemblik Bertus spent his early school years, and in 1890 he entered high school (the HBS) at the tender age of nine (which earned him posthumously an entry in the Dutch Guinness Book of Records). The student-biography also records his Medemblik years:

> My next abode did not rank much higher. I moved to Medemblik, where I lived for eleven years. In that town I learned to walk and speak, smashed a lot of things in the parental home, repeatedly fell down the stairs without—miracle, oh miracle—breaking my neck, and had the measles. [......]
>
> As I got older my amusements naturally changed, and the old ones gave way to looking at pictures, and soon also to the reading of stories from Mother

[8]The Dutch educational system knew one elementary school and a number of secondary ones. The elementary school (preceded by the *kleuterschool* (kindergarten)) was called the *lagere school*, age 6–12; the secondary general education schools were MULO (or ULO), literally *(Meer) Uitgebreide Lagere Onderwijs* ((further) extended elementary education), HBS, *Hogere Burgerschool* (higher public school) and gymnasium. The MULO took 3 or 4 years, the HBS 5 years (there was also a shorter, three year variant) and the gymnasium 6 years. The gymnasium was in a sense the direct descendent of the old grammar or Latin school, and the HBS was the product of the new ideas of the mid nineteenth century. The gymnasium was the training ground for future academics, whereas the HBS was intended to provide young citizens with the necessary skills for trade and industry. The MULO was a simplified version of the HBS.

Goose. I rarely went to the elementary school. Most things I learned at home, and I have few recollections of the school. Only, I can still see how the schoolmaster pulled one of my classmates by his hair through the schoolroom so that the boy passed out, and how then all of us got a day off. But all the sharper are the impressions that I still have from the time that followed; how I commuted with a season ticket by local railway from Medemblik to Hoorn and attended the high-school in the latter place.[9] I remember still very clearly that often, waking up, I saw that the train was about to depart in 10 or 5 minutes, and how I hurriedly slipped on my clothes, got my books and sandwiches, ran down the stairs, dived into the ally at the side of our house, then covered the road to the station, then running like a madman and finally, out of breath, flinging myself through the station into the train,[10] where I usually still had to lace up my shoes, and to finish tidying myself up—of which I had omitted various necessary parts in my haste.

Although we lived fairly close to the station, it was quite a job—especially in the severe winter of 1890–1891—to be on time at the station, and the hour of departure of our train, which was at first twenty past seven, was moved forwards at every new timetable, until it finally reached half past six. Fortunately, I was not the only one to bear this cross, but I had three fellow sufferers. The journey took one hour, and this hour was, if necessary, devoted to the learning of our lessons, but otherwise, out of boredom often, to forbidden actions. We posted ourselves on the platform of the train, which was strictly forbidden, or we fiddled with the equipment, for example, the gas lamps; we were even so childish as to bother our fellow passengers, for example, by pricking them through holes in the backs of the benches with a pair of compasses. Thus we were obviously in disrepute with the passengers and the railway employees, and the employees took revenge in any way where they saw the slightest opportunity. For example, if one of us forgot his season ticket, the guard gave no pardon and we had to pay the ordinary fare, and once the men pulled the following trick: in winter we usually stayed as long as possible around the stove in the waiting room, as it was ice-cold in the train, for we had ample time to board the train when we heard the whistle of the guard, since the train was always right in front of the waiting room. When our habit was noticed by the station master, the driver, and the guard, the train one morning rushed out as fast as it could without any announcement of its departure, and we were left behind. We did try to make a last minute jump on the running board, but the guards and the stoker prevented us from doing so. Of course, after this, the lingering in the waiting room was over. On the way to Hoorn we usually had little company to start with, but on approaching Hoorn the train slowly filled up with farmers, and at the end of the journey, especially on market days, we were packed like sardines. The proverb says 'the more the merrier', but we did

[9] Bertus was registered on the first of September 1890.

[10] The railway line Hoorn–Medemblik is nowadays operated by volunteers. In the Summer season one can make the trip in historic carriages drawn by an equally historic locomotive.

> not at all see our swelling company in that spirit. For all those farmers smoked like chimneys and at the same time they had a mortal fright of a draught, to the extent that they never allowed a window to be opened; and so we sat there amidst billowing smoke and in a terrible atmosphere. Thus we fostered a profound hatred for our fellow passengers, and so it is understandable that when we bothered them, there was a good measure of revenge involved. [......]
>
> We were the best of friends with the station master in Hoorn and all the staff of the station, and to that we owed the permission to push ourselves along the tracks on a trolley, or to make a ride on the freight train that left at a quarter past two to Purmerend while it was shunted; yes, even to act as a pointsman. Sometimes we played tag on shunting trains, jumped off rolling carriages, and up again, jumped from one carriage to another, climbed up on the roofs; in short, did all kinds of tricks which would have made our parents endure a thousand frights, could they have seen us there. Moreover, in winter we could skate without being bothered by other people on the smooth ice of a pond which belonged to the station and that could only be reached through the station. Usually, if it had snowed, we ourselves swept the rink before nine o'clock, on which we skated between school hours. Also at the station in Hoorn some mischief was practised, but here it never went so far as to spoil the good understanding between us and the staff of the station.

Bertus judiciously omitted to mention an incident that could have had a nasty ending: he once climbed onto the roof of the station in Hoorn, where he was seized with a spell of dizziness, so that he very nearly lost his balance.

Apart from Brouwer's own biographical sketch, little is known about his years in Medemblik. One would like to know about his childhood friendships, his development, his adventures in the quiet streets of the town and the expeditions into the countryside. Bertus's years in Medemblik are a closed book to us. There is just one minor but consequential detail that must be mentioned, as it plays a key role in Brouwer's later years.

In Medemblik there lived a family Pels with whom Bertus made friends. The daughter, Dina, was two years older than Bertus and she attended the same high school in Hoorn as Bertus. Dina entered the HBS in 1892, the same year that Bertus left for Haarlem, and she finished three classes. Subsequently she went to Amsterdam where she combined a job in a pharmacy with the training for the certificate of 'pharmacy assistant'. The training in those days was a matter of apprenticeship in one of the officially recognised pharmacies. Apparently Dina and Bertus met again when the latter enrolled in the university. As we will see later, the renewed relationship between the two had far-reaching consequences.

In 1892 the Brouwer family moved again, this time to Haarlem, the capital of the province of North-Holland. Brouwer senior obtained a position as headmaster of the MULO at Haarlem. This move was the last step upwards in his career. Lacking an academic education or the supplementary qualifications (the so-called *middelbare acten*) he could go no higher in the teaching profession. Haarlem had more to offer to Bertus than the sleepy town in the north.

As far as life in Medemblik itself was concerned, the older I got, the more I felt the unpleasantness of it. There were few boys for company, there was no surrounding countryside, there were no walks apart from the sea dike, sports were unknown, so that I had little recreation to go with my daily work. Thus, when we moved to Haarlem, I made a good switch as far as the town was concerned. In particular during the first year, there was a great deal for me to see; my lifestyle underwent a considerable change. The time reserved for learning was more and more cut back by other things. To begin with, walks took up a great deal of my time, because one can—if one is not overly prosaic—do quite a lot of walking in the countryside surrounding Haarlem before one starts taking it for granted. Already in *Den Hout*[11] it is possible to find ever new spots that catch the eye, even after having walked there for a hundred times. And then one can go to the dunes, where a hundred different hillscapes with ponds, villas, and copses unwind before one's eyes.

One can wander through the woods of Bloemendaal and Santpoort, and seek out the hollows in the dunes of Santpoort, and if one has had enough of all those sceneries of nature, one can start botanising, and find jewels of Dutch flora in the dunes. But apart from walks, also sports took up a great deal of my time, for soccer, swimming, cycling, and cricket soon found a keen player in me, and this gave me a lot of pleasure and a lot of excitement; for if one starts to practise a sport, it is easy to start racing, and many a soccer match or swimming contest deprived me of a night's sleep.

In the meantime I finished high-school (*HBS*); at the end of my high-school years, and the following year I learned Latin and Greek. Next I took an entrance examination for the sixth grade of the gymnasium, where I spent my last school year. This last year was not the least congenial year of my life; I was not overly pressed, I could devote a lot of time to sports, and I could get along perfectly with the boys of my class. There were eight of us, and now we have split up, four to Amsterdam, one to Leiden, two to Utrecht and one failed the final examination. And so a new period will now dawn for me.

I am a freshman (*groentje*), and I hope to become a student. Two thirds of the initiation has passed, and I have made many new acquaintances, talked a lot to them, and learned much from them. I have noticed how much I am lacking in general knowledge and moderation, and I have learned to respect men who, children of the same era but with more experience, could guide and advise me on the road which lies in front of me. Physically, I have, strictly speaking, not been bullied yet I have learned a great deal the hard way. This period is miserably exhausting for me, and I am glad that I have already had a long invigorating holiday, and that after one more week I can catch my breath again. One of the nuisances too, is all the work I get to do in this period: love letters, rhymes and proverbs in the various initiation journals, and not least, the autobiography of four pages without a margin.

---

[11] A wooded park in Haarlem.

> Fortunately no punishment has come on top of that, but I believe that if there had been, I would have dropped in my tracks. But let me put on a brave face; one more week and the barrier that separates me from student life has dropped.

In Haarlem the family lived in a new house on the Leidsevaart (the canal connecting Haarlem and Leiden); at that time the house was on the outskirts of the town. The HBS and the gymnasium were only a short distance from the house, so travelling belonged to the past.

Mrs. Brouwer took boarders, two schoolboys: Fer and Lau van der Zee, whose parents were in the Dutch East-Indies (the present Indonesia). No records survive of the relationship between the Brouwer boys and the boarders. The only remarkable fact to relate here, is that Brouwer later made use of the pseudonym 'Lau van der Zee' in some contributions to the student weekly *Propria Cures* and the magazine of the Delft students, see p. 62.

Bertus always had a sweet tooth, he loved sweets and treats; as the eldest boy he was usually given an extra helping of the custard, nonetheless he was so fond of this dessert that he sometimes bribed his brothers and the boarders into giving him their portions.

The musical education of the Brouwer boys was taken in hand by their mother; she came from a talented musical family,[12] and she gave piano lessons to her sons, who did not always meet their mother's standards. When the brothers played abominably, or when they had neglected to study at all, she would occasionally play the rod of the gaslight on the backs of her darlings. On one occasion the rod broke; Bertus had it repaired and presented it to her as a birthday-present.

Father Brouwer had the reputation of being a gifted pedagogue, but just the same he had difficulties in handling his own offspring. Bertus, in particular, did not get along with his father. At the time that they were living in Haarlem, Bertus now and then fancied spending a night in the dunes, something his father would not allow. Once, when the urge had become too strong, Bertus managed to lock his father into the cellar. When he was released, Bertus was already safely in the dunes. The relationship with his father remained uneasy his whole life. When Bertus was a well-established citizen he used to pretend an indisposition as soon as his father visited him, taking himself to bed and moaning as if in agony. The intimate friends, who were well aware of this play-acting, spoke jokingly of Brouwer's *vluchtbed* (a bed to flee to).[13]

For all their intelligence, the Brouwer boys were no softies; in their exploits they came well up to the Tom Sawyer-mark. Aldert was an excitable, impulsive boy, prone to accidents. Bertus himself was no stranger to daring actions; he was an inveterate climber in trees, buildings, etc. Even at the age of seventy-two, at a picnic

---

[12]The Poutsma family produced a number of scholars and artists, among whom was Geesje Poutsma, an older sister of Bertus's mother, who gave singing lessons and was a concert singer herself. See also Stuurman and Krijgsman (1995).

[13]Oral communication Mrs. F.J. Heyting-van Anrooy.

**Fig. 1.1** Family picture (1896). The boys from left to right: Lex, Aldert, Bertus. [Brouwer archive]

during a meeting in Canada, he upset his company by suddenly disappearing. To everybody's surprise he was discovered up a in tree.[14]

Bertus's school career was highly unusual; his exceptional intelligence and, no doubt, a certain amount of private coaching by his parents had enabled him to cut short his elementary school years and to get into high school three years in advance of the normal entrance age. The phenomenon of pupils skipping one or more grades was not unusual in the old educational system; Bertus's progress was, however, remarkable. The more so, since his report cards, right from the beginning, show him at the top of the class—and he stayed there during all his school years. In his first year at high school he ranked first in arithmetic, geometry, algebra, history, Dutch, French, English, and German; in natural history and geography number 2; only in art (drawing) he ranked as number 26 (out of 27 pupils). In the second grade of the HBS at Hoorn he ranked number 1 in all subjects except art. The mark that the Dutch educational system added for diligence (*vlijt*) is telling: whereas he scored in

[14]Oral communication J. Lambek.

Hoorn the top mark for almost all subjects, he was graded a 3 out of 5 for diligence in mathematics, but a 5 for the other subjects. This seems a fair indication that the mathematics curriculum at the school in Hoorn did not have much to offer to Bertus.

In Haarlem he entered in September 1892 into the third class of the HBS; when he moved up to the fourth class, he again was number 1 in the class. Even for a clever boy this was something of a tour de force, as he had taken in the meantime the entrance examination for the local gymnasium. The official record says that on 13 and 15 January 1894 he passed his written and oral examinations for admission to the first (sic!) class. The notice went on 'This candidate has to be a gymnasium student in order to be eligible for a grant of the St. Job's Foundation [see below]. He is admitted on the condition that he will promptly catch up in Latin and Greek.' And so Bertus learned his Latin and Greek while following classes at the HBS. In September 1894 he passed another entrance examination, this time for the third class of the gymnasium.

The school year 1894–95 was a busy year indeed, because Bertus compressed two high-school years into one year; at the end of the year he had mastered the total curriculum of the HBS and he took the final examinations (22, 23 and 24 July) with splendid marks (with exception of (again) arts, this time in the company of *cosmography*). Since he was not a regular student he had to take the examinations before a state committee (the so-called committee of experts (*deskundigen*)). The diploma was awarded on 9 August 1895. The following year he followed the lesson in class 4 of the gymnasium and at the end of the year he passed an entrance examination for class 6, so that he simply skipped a class. As a matter of fact he followed lessons in both parts, $\alpha$ and $\beta$—the literary and the science part—of the gymnasium (in the science part he was the only student!). Bertus did, as usual, very well, albeit that his father sternly admonished him with the words 'this must become a 4' in his report card, when Bertus had scored only $3\frac{1}{2}$ (out of 5) for German.

The gymnasium examination was conducted at the school itself by both the teachers and a committee of outside experts. The experts were as a rule university professors or lecturers, who spent part of their summer vacation travelling from gymnasium to gymnasium examining candidates. This examination spree was a traditional part of academic life; one was more-or-less expected to take part, and there was a modest fee. The system helped to maintain contact between the universities and gymnasiums, it was an implicit tool for control, and it enabled teachers to keep up their contacts with the professionals at the universities. Later in life Brouwer also regularly took part in this examination activity. There was a similar system with committees of experts for the HBS examinations. It is an interesting feature of the gymnasium examinations that they were primarily seen as entrance examinations for the universities and not as the crown on a high school career. This was literally and officially recognised in the *Nederlandsche Staats-Courant*, where we read in the issue of September 1897 that L.E.J. Brouwer had received a testimonial for admission to study in the faculties of theology, law, literature and philosophy, and also of Medicine, Mathematics and Physics.

Since a gymnasium diploma was the normal requirement for admission to the university, the decision of Bertus's parents to send the boy to the gymnasium was dictated by their wish to open for him the doors to an academic education. Another motive for the prolonged high-school education may have been the age of Bertus at the time he finished the HBS: at 14 years old, he would have been something of an oddity at the university. There is no doubt that Bertus's two extra years at the gymnasium were well spent. There are indications that he read and studied a lot in his spare time. Quite a number of more-or-less prominent Dutchmen had indeed been confronted by the fact that the HBS-diploma did not qualify them for university study; some outstanding Dutch scholars had nonetheless entered the University without a gymnasium diploma,[15] but this either required a special dispensation from the Minister of Education, or else one had to take an entrance examination. In 1917 the requirement of a gymnasium diploma for admittance to university was relaxed by law; the studies of medicine, mathematics and physics were opened to HBS graduates.

Judging from the marks that Brouwer earned at the HBS and the gymnasium, there was no specific field of study that was a priori excluded. Indeed, Brouwer had a deep love for languages and he cultivated his Latin and Greek during his whole career. His choice of the faculty of Mathematics and Physics could, however, hardly have been accidental, but the ultimate reasons for this felicitous decision remain somewhat vague. It can be said, however, that a posteriori his exploits fully justified the choice.

## 1.2 Student in Amsterdam

The sixteen-year-old Bertus enrolled at the University of Amsterdam, also called the Municipal University (*Gemeente Universiteit*). This university was an old institution in a new form. Traditionally the Netherlands knew only a couple of universities; the first university in the Low Countries was that of Leuven in the present Belgium, founded in 1425. After the Reformation, the University of Leiden was founded in 1575 at the instigation of William of Orange, as a reward for the tenacious resistance of the citizens of Leiden during the siege by the Spaniards. The universities of Groningen and Utrecht followed in 1614 and 1636. Higher education in Amsterdam was provided by the Atheneum Illustre, founded in 1632; its elevation to '*hogeschool*' was effectively blocked by the University of Leiden. In 1877 the Atheneum was transformed into the University of Amsterdam ('UVA' for short). There was, in Brouwer's days, one more university: the Free University at Amsterdam, founded in 1880 by the Dutch Calvinists. In addition there was the Polytechnic School at Delft, the former school of artillery of King William I[16] of the

[15]E.g. Korteweg, Van 't Hoff, Lorentz, Kamerling Onnes, Dubois, Zeeman, Zernike.

[16]This was the first Dutch King of the House of Orange, to whom the crown was offered after the fall of Napoleon. To outsiders, the history of The Netherlands may seem somewhat confusing. Until the French revolution the Princes of Orange were Governors of the Dutch Republic. Napoleon made his brother Louis Napoleon King of Holland; the Orange monarchy was introduced after Napoleon's fall. The famous William III was King of England, but Governor of The Netherlands.

**Fig. 1.2** Bertus. [Courtesy W.P. van Stigt]

Netherlands. It was elevated to *Technische Hogeschool* in 1903, and today it is the Technical University at Delft. The universities of Leiden, Groningen, and Utrecht were state universities, but the University of Amsterdam was a municipal university. As a consequence it was directly governed by the mayor, who was the chairman of the Board of Curators, and by the Council of the City of Amsterdam.

For a beginning student with aspirations in the sciences, the choice of the UVA was not an obvious one. The University of Leiden had gained a reputation in physics and astronomy with Kamerlingh Onnes and Lorentz as its star professors, who drew international attention. The young UVA had done well for itself by engaging Van der Waals, Van 't Hoff and Hugo de Vries. For mathematics there were few reasons to prefer one university to another; as pointed out before, Dutch mathematics was awaking from a long slumber and outstanding authorities were not easily found. So the choice of Amsterdam is difficult to explain. One reason might be its proximity to Haarlem, another its reputation as an exciting cultural-political centre in The Netherlands. Moreover, some relatives (the Poutsma uncles) lived and worked in Amsterdam, and this may or may not have been an extra argument in favour of the UVA. As a rule Dutch students were (and mostly still are) rather conservative with respect to their choice of university. Geographical arguments carried considerable weight, and mobility was markedly absent. Whereas students in neighbouring Germany usually changed universities before specialising, Dutch students tended to stick to their first choice.

Whatever the motivation may have been, on the 27th of September 1897 Brouwer registered as a student at the faculty of Mathematics and Sciences at the UVA. The young boy followed the example of most of his fellow students and joined the Amsterdam Student Corporation. The 'Corps' was, and still is, split into a number of clubs (*disputen*), debating or social clubs, sometimes of a specialised character. In Brouwer's days, and long after, membership of the fraternity was almost obligatory. Non-members, called 'nihilists', were considered to lack the essential ingredients that were believed to make a student something more than just a person who studies.

Bertus first signed up for the debating society NEWTON (23 October 1897). This was the society where the science students met and discussed, among other things, scientific topics. A month later he joined another fraternity (*dispuut*), PHILIDOR, the meeting place for chess players. A more important *dispuut*, however, may have been CLIO, a literary club, where Bertus met a number of fellow students who, in one way or another, were going to make their mark on Dutch society.

Although Bertus's intelligence was beyond dispute, one should not take it for granted that this by itself was enough to study at a university. The financial burden was far from negligible, and many a potential young scientist ended up as a shop assistant or bookkeeper. The income of a headmaster was scarcely sufficient to support a child at the university—let alone three! Bertus, however, was fortunate enough to obtain support from a private fund in Friesland, the *St. Jobsleen* at Leeuwarden.

This foundation is one of many private institutions that even today support worthy young students. Some of these funds offer grants to students of a particular geographic or religious background. The St. Jobsleen supported students of Friesian descent. The grant was awarded to Bertus for the first time in 1894, when he enrolled in the gymnasium in Haarlem, and he received DFl. 450 a year; when he entered the university the foundation doubled the grant. This was, by any standard, a generous amount, taking into account that a skilled labourer would consider himself well-paid with such wages. From this sum DFl. 150 was deducted to be paid out after the successful completion of the study.

The student fraternity brought Bertus into contact with a number of interesting fellow students, who helped him to extend his intellectual horizon. The most prominent among them was Carel Adama van Scheltema, the grandson of a clerical poet of the same name. Among the remaining members of Bertus's circle of friends, Jan van Lokhorst, Henri Wiessing, and Ru Mauve stand out for one reason or another. The most elusive among them was Jan van Lokhorst, a mathematics and physics student, who exercised a considerable influence over Brouwer. This somewhat unusual person dressed as an eccentric; on one occasion he sported a yellow suit with matching shoes. Jan van Lokhorst had introduced Brouwer, when still a gymnasium student, into the company of artists, in particular of Thorn Prikker, a well-known visual artist, who spent his later life in Germany, and Boutens, an influential poet. According to Wiessing:[17]

> Only the above mentioned Jan van Lokhorst, one of his contemporaries who later switched to Leiden and died young (1904), had any noticeable influence on Brouwer's attitude towards life. In the last years of his life this Van Lokhorst was already venerated and consulted by considerably older and already well-established authors and visual artists.

The death of Van Lokhorst is the subject of one of the many legends surrounding Brouwer. His stepdaughter, Louise, related that when Van Lokhorst's death was approaching, Brouwer felt inexplicably drawn to his friend, whom he found dying, in a small hut.

[17] *Bewegend Portret*, Wiessing (1960), p. 142.

**Fig. 1.3** A meeting of the fraternity CLIO (1897/98). Brouwer is top left. [Courtesy Letterkundig Museum, Den Haag]

Henri Wiessing, an enterprising young man of Roman Catholic origin, became a close friend of Scheltema and Brouwer. After his studies he became a journalist with far-left inclinations, and for some time he was the editor-in-chief of a progressive, left-wing weekly *De Amsterdammer*, known affectionately as *de Groene* because of the green colour on its front page; see p. 282. Brouwer sent from time to time contributions, in the form of a small article or a letter to the editor, to Wiessing's weekly, and he stood by him in a number of literary-political affairs. At the time of Brouwer's undergraduate years, Wiessing was infatuated with Adama van Scheltema.

Ru Mauve[18] was the son of the painter Anton Mauve; he studied medicine for some time in Amsterdam, but decided to prefer a simple, idealistic life. In 1898 he exchanged the world of study for that of a craftsman; he took up a job with the famous architect Berlage but soon changed his mind again and joined the much discussed commune of Frederik van Eeden (see p. 58). Having experienced the pleasures and miseries of life in a commune, he departed for Florence. After studying architecture he eventually enrolled in Delft, where he left without a diploma. Mauve remained a lifelong friend of Brouwer.

[18]See p. 59.

The poet, Carel Steven Adama van Scheltema, or ‘Scheltema’ as he was called for short,[19] was probably the most influential person in Brouwer’s early life; the two met in CLIO and NEWTON.

Scheltema was four years Brouwer’s senior, and already a man of the world. He was not as gifted intellectually as Brouwer; whereas the latter took the gymnasium in a gigantic stride, Scheltema had to struggle along. Among his fellow students Scheltema stood out by his striking personality. He had enrolled in the faculty of medicine in 1896, and although he soon discovered that he was not cut out to be a doctor, he duly took and passed the propaedeutic examination. Scheltema was a man of culture, blessed with a fair dose of charisma and authority. Already in 1897 he was elected to three important positions: member of the senate of the *Amsterdamse Student Corporation*, editor of the prominent student weekly *Propria Cures*, and chairman of the Student Drama Society. Indeed, his fervent wish was to become a professional actor. After his performance of the title role in Richard II on the occasion of the lustrum festivities in 1899, he joined a theatre company. His retiring disposition made him, however, ill-suited for an actor’s life, so he soon gave up the theatre.

After a short excursion into the world of art dealers, where he worked for the Van Gogh Art gallery, he once and for all gave up his quest for a regular occupation, and became a free-lance poet. He could afford to do so, mainly because his father, on his death in 1899, had left him enough to lead a modest but comfortable life.

Scheltema’s father had died of a tumour in the brain; and the experience was so traumatic that Scheltema was haunted for the rest of his life by the fear of a similar fate.

Scheltema had taken a keen interest in the young student Brouwer and, in fact, became the self-appointed mentor of Bertus. The friendship that ensued is reasonably well-documented by a collection of letters exchanged between 1898 and the death of the poet in 1924.[20]

During the first year of his study Bertus sampled some of the traditional activities of the university and the student societies. Wiessing described in his autobiography[21] his student friendships and provided some illuminating remarks on the young Brouwer, ‘a young and very tall Friesian from Haarlem, about whom—although only a student and no more than sixteen years old—rumours circulated concerning his mathematical knowledge’.

In contrast to Scheltema, and Wiessing himself, ‘this introverted ‘*éphèbe*’ remained ‘an obscure student”.[22] In the fraternity he not only shunned prominence, he avoided its members, in particular the prominent ones. He had declined to join the social club, which had invited him after the initiation period, and with the fra-

[19]The family name is ‘Adama van Scheltema’, a typically Friesian name. The suffix ‘a’ usually indicates Friesian descent. We shall stick to the abbreviated version ‘Scheltema’.

[20]Cf. van Dalen (1984).

[21]*Bewegend Portret*, Wiessing (1960).

[22]The traditional name for a student who seldom frequents the fraternity events.

ternity he had, after a short spell in the rowing club *Nereus*, hardly any ties, except through the study club NEWTON.

## 1.3 The Religious Credo

As a young student, Bertus joined the Remonstrant Church (*Remonstrantse Kerk*), a Protestant denomination that had its roots in a theological dispute in the seventeenth century. That particular dispute would have remained obscure, were it not for the circumstance that the religious rift among the Dutch Protestants (basically Calvinists) had its repercussions in social and political life. The Grand Pensionary Johan van Oldenbarnevelt (1547–1619), who favoured peace with Spain for the sake of economic expansion, had adopted the case of the 'Remonstrants', and, partly in reaction, Prince Maurits (son of William of Orange) had taken the side of the 'contra-Remonstrants'. The Remonstrants suffered defeat and have remained ever since a small but refined religious denomination.

Brouwer's choice to join the *Remonstrant Church* is somewhat puzzling, since his parents were members of the Dutch Reformed Church,[23] and presumably Bertus was baptised as such. There are no indications in his later life as to his religious affiliations, so there is no simple explanation of this step.

Before being solemnly accepted into the Church, Brouwer was, in accordance with the custom, requested to write a personal profession of faith. The text of this profession has survived; and it is highly significant in the light of Brouwer's later philosophical views. As a boy of 17 years old, he presented a coherent idealistic and even solipsistic view of his religious credo.

In the light of the following solipsistic reflections, it seems significant that already as a small child Bertus was occupied with the status of the ego. His mother used to tell that as a three-years old boy, Bertus asked the question 'What is I?' According to the experts most children start to discover the self at an early age, but hardly ever to such a degree that they formulate it explicitly.

His confirmation took place in March 1898, and Dr. B. Tideman was the Minister. The candidate for confirmation had to write down his private views on a list of questions or topics provided by the minister; as a rule the confirmands reproduced as well as possible what they had learned in the Catechism class. Not so Brouwer! The questions are unknown, but Brouwer's answers are interesting by themselves. A translation of the original text is reproduced below.[24]

> Point 14. What is the foundation of my faith in God? This is for me the main point of the profession of faith, and the only thing that may properly be called 'profession of faith of a *person*'. That I believe in God originates by no means in an intellectual consideration in the sense that I should conclude

[23]The one-time state religion.

[24]The order of the answers is Brouwer's.

from various phenomena that I observe around me the revelation of a 'higher power', but precisely in the utter powerlessness of my intellect.

For, to me the only truth is my own ego of this moment, surrounded by a wealth of representations in which the ego *believes*, and that makes it *live*.

A question whether these representations are 'true' makes no sense for my ego, only the representations exist and are real as such; a second reality corresponding to my representations, independent of my ego, is out of the question.

My *life* at the moment is my *conviction of my ego*, and *my belief in my representations*, and the belief in That, which is the origin of my ego and which gives me my representations, independent of me, is directly linked to that. Hence something that, like me, lives and that transcends me, and that is my *God*.

One should by no means read in the many words that I have used above, an intellectual *deduction* of the existence of God, for this belief in God is the bedrock, from which can be deduced, but that itself is not deduced. The belief in God is a *direct spontaneous emotion* in me.

Now I do think that this belief in God is of a somewhat different nature than the ordinary one, and this mainly because mine rests on a *Weltanschauung* that acknowledges only me and my God as living beings, of which I *know* myself, and sense my God, my master.

Furthermore, the representations which are given to me contain in themselves, for instance, also that there should be other egos, also with representations, but these are not real, they are parts of my representations, therefore they are *mine*. My representations are my life. Thus at this moment I live in the representation that I think of my life, and write a profession of faith, but *in* that life I do not find my God, my God is under[25] and outside my life, only the fact that I live, makes me sense my God, it is not *in the way* I live that I find my God.

This view includes my *immortality*, or rather, cannot consider mortality. For the concept of *time*, like *space*, belongs to my representations, whereas my ego is completely separate from those concepts. My relationship to my God is a dependent trust in him, who *makes* me *live*.

But the life that my God gives me to live can be thinking about things that I observe and the state of things that I see around me and having opinions about various matters, also so-called *religious matters*, but then these are representations given to me by my God, who is outside and above it; they cannot encompass my God, for they originate from him. Only the *sensing of my God* belongs to my proper religion.

However, here language is too awkward an instrument. The *sensing of God* and the *trust in God* is not a conscious thought, and hence a representation, for then it would again be situated outside God himself, but it is something

[25]under, i.e. is the *Urheber* of my life, [Brouwer's footnote].

that, as it transcends thought, cannot be thought, let alone written down; it is something that is tied to the unconscious ego, becomes conscious, that is receives representations, but is separated from those representations. Indeed, an image of it can come into being in a conscious thought, but only very vaguely.

This view of mine concerning the first point of the profession of faith renders the discussion of many other points superfluous.

[1–13] In the first place, a historic survey of religion can in my case contain nothing that guided me to my conviction, and hence has no place here.

[15] Objections to the acknowledgement of a divine rule in this sense, that I would not see how to reconcile with various things that I observe around me, do not exist for me. For my perceptions are part of my representations and none of these can by their nature be an objection, all of them are in *their* existence a proof of *God's* existence.

[16, 17] The characteristic feature of my conception of life in contrast to that of others has already been stated. I neither conceive life a burden I have to carry, nor a task to be fulfilled; no, my life is an accomplished fact, about which I cannot give an opinion. For to that end, I would have to view it objectively from the outside, but that I cannot do; for me, life is the great unique It to which I cannot assign properties, because nothing can be compared with it.

This view does not at all imply that my life should be a dull, blind, will-less letting go. The life that my God gives me to live can be rich in hope, anxiety and aspiration, full of passionate pursuit of ideals, and my own free will can be strong; all this, indeed, belongs to the representations that can be given to me.

[18, 19] I already mentioned at point 14 my unlimited trust in God, and my conviction of my immortality.

[20] Among the representations which my God gives me are those that make me at some moments feel intensively his existence, this is then followed by a strong self-confidence and a joyous courage to live. Each time when that awareness forcefully thrusts itself upon me, stirring my inner life, I may speak of *love for my God*. For me such moments of contact do not have the character of prayer, because my wishes and sorrows do not play a role, but, on the contrary, have totally disappeared for me.

So far I have been able to connect the points with my religious conviction. The remaining part has a totally different meaning for me.

To wit, my God has also given me the ambition to make my life, i.e. my representations, as beautiful as possible.

[21] from this it ensues that I am struck by the loathsomeness in the world that surrounds me, and is part of me, and which I will try to eliminate; also as regards the world of men. I can hardly call this *love of my neighbour*, for I detest most people; hardly anywhere do I recognise my own thoughts and spiritual life: the shadows of men around me are the ugliest part of my conceptual world. So, in theory, I will never sacrifice myself for another person; God has, however, given me feelings such as compassion, which sometimes force me to act in that direction.

Through the unconscious pursuit of beautifying my representations I have of course opinions on the being or not being useful of institutions in the human world, therefore I can also write about these points, even though I stray more and more from my religious conviction.

I approve of a church because, even though we do not hear in it our own conviction, it can direct our thoughts to fields where by our own action and thought happiness can be found. Ecclesiastical rule and dogmatism are of course phenomena of degeneration; I approve of religious forms for the simple crowd, to be subdued, in a reverential non-understanding, by a church that wants to dominate.

Once again, I approve of the church as the one that points out our task to us; to find in a religious conviction a staff to go through life. And this is the credo of my religious feelings and convictions, of which I have now given an account for the first time, and which I have ordered and sifted even though the unity and force have suffered by an arrangement in points that was not mine.

March 1898.
L. E. J. Brouwer.

This is not the place to give an extensive exegesis of the profession of faith, but let us just note a few interesting and important points.

In this document we can find Brouwer's views on his life, in his own formulation. The basic underlying idea is that life is just there; it is not within the competence of a person to put himself as a judge above it. Since the ego and life are almost synonymous, one cannot step outside life and view it from a higher position. He immediately goes on to reject a fatalistic view of life: 'the free will can be strong'. The views expounded by Brouwer are very similar to those of Indian philosophy and religion. At point [20] Brouwer describes the experience of feeling the existence of God. Here one recognises what traditionally have been called mystic experiences. The last section but one treats the relationship with world and the fellow human beings. It describes Brouwer's feelings for his fellow men in surprisingly frank terms. Apparently the intervention of God is required to give him feelings of compassion.

The final views on the church and its role are rather cynical, to say the least, but given Brouwer's basic view on the ego and its relationship to God, not without a point. The minister must have been surprised at such a confirmand, but fortunately the Remonstrant Church had a reputation for open-mindedness and tolerance.

Before leaving the topic of the credo, it is worthwhile to pause for a moment, and reflect on its status. The question one would like an answer to, is 'how original are the basic ideas?'

Some of the material has the flavour of Schopenhauer. Moreover, it reminds of Cusanus, when Brouwer points out the impotence of our intellect in the face of the problem of God and his existence. The analogy with *De docta ignorantia* suggests itself.

There is no definite answer to the above question. On the one hand Brouwer was highly original and unorthodox in his thinking; he had an unusually penetrating mind, as his later works shows. So it was not beyond him to develop a solipsistic view all by himself. On the other hand, he was a avid reader, and a superior school like the Haarlem gymnasium may very well have exposed the young Brouwer to ideas and traditions that could easily escape the untutored student. One should also keep in mind that Schopenhauer was very much en vogue around the turn of the century.

It would certainly not be beyond a clever boy like Bertus to assimilate the ideas of Schopenhauer. In the absence of convincing evidence, I would be inclined to give Brouwer credit for the originality of the credo.

From the above profession of faith one obtains a fairly accurate impression of the philosophical views of the young Brouwer. It appears that he had adopted a rigorous, Schopenhauer-like, view of the world, religion, and his fellow human beings.

The basic entity for Brouwer is his ego, and immediately after that there are the 'representations' (*voorstellingen*) of (or in) the ego. At this point Brouwer makes the radical choice for a strict idealism; there is absolutely no compromise with impressions from the outer world or representations of (or derived from) experience. The representations are inextricably bound up with the ego. Hence these impressions are autonomous in the sense that they cannot be checked against experimental or objective phenomena. The next step is not forced upon the ego, but is rather a matter of free choice, namely the recognition of God as that which is the source of the representations and of the ego. As Brouwer stresses, God is not deduced from the ego and its representations, but the belief in God is a spontaneous act of the ego. One could almost say that it *happens* to the ego.

At the points [16], [17] we can already note some of the characteristic points that we will meet again in Brouwer's booklet *Life, Art and Mysticism*, namely that one should accept the world as it is; it not something one can complain about, it is (in the later terminology) part and parcel of one's Karma.

## 1.4 Friendship: Adama van Scheltema

The years at the university were far from smooth for Bertus; although the actual study did not present any problems, he suffered from nervous attacks that were to plague him his whole life. Nonetheless he fervently pursued a great number of activities. In the summer of 1898 we find him in the infantry barracks in Haarlem.

Now here is a riddle. What, one would ask, is a boy of 17 doing in the army? He is too young and he is a student, so he has better things to do than to play soldier. Or to put it more positively: his first duty is towards Athena and not towards Mars. There is no definite answer to the question in the absence of data. The most likely solution to the 'army problem' seems a coherent strategy on Brouwer's side to get his army obligations out of the way before the beginning of his real career. He joined the army as a volunteer in 1898 with the rank of *aspirant vaandrig* (reserve officer to be). Combining the information from the National Archive (*Rijksarchief*) and Brouwer's correspondence, we conclude that he was enlisted in the fourth regiment infantry quartered in Haarlem. A letter of 14 August 1898 to Scheltema shows that army life was not as pleasant as Brouwer had probably hoped. Brouwer was no softy, he enjoyed rough sports, had no objections to outdoor life, and he had survived very well at school, although he was invariably much younger than his classmates (recall that for young children age differences are far more important than later in life). So it is possible that he had underestimated the hardships in a world that was probably alien to him. In the army things were not done by Brouwer's rules. Even his extraordinary intelligence would work against rather than for him.

He entered the army on 6 July 1898, and he obtained leave (*groot verlof*) on 21 September. On 27 August he was promoted to reserve corporal, and that was as far as he would go in the military world.

The decision to get done with the army as soon as possible lent an illogical feature to his army career. At the time the lottery system was still in operation, that is to say, it was determined by lot whether one was conscripted. For those who could afford it, there was the possibility to 'buy a replacement'—a person who had not been drafted, and who was willing to take over the military obligations of the conscript, usually this was some ignorant, underpaid yokel. As a result the army was not exactly pleasant company, to say nothing of its efficiency. Most eligible men would wait for the lottery result, and then contemplate how to handle the situation. Not so Brouwer; he reported on 11 December 1900 at the 'lottery board' in Haarlem, having already completed seven-and-a-half months military training. He was not so lucky to draw a blank, but according to the record, he was exempted from military service on the grounds of voluntary service,[26] that is to say the actual time served was to be deducted from the obligatory period. The record of the lottery board does not give much information. It listed his physical features: height 1.863 metres, oval face, blond hair and brows. The State Archive's records show that Brouwer was short-sighted at the right eye (0.5 dioptre) and that he was 1.848 meters tall when entering military service. For some reason people usually thought Brouwer to be taller than he actually was, estimates of 2 metre are no exception. The explanation is probably the fact that Brouwer was extremely slender, thus creating an illusion of great height. Of course, it is very well possible that in the following years he added a few centimetres to his length; after all, he was not yet nineteen when his measurements were taken.

The following letter makes it clear that even a short stretch of military service had been enough to fix Brouwer's view on army life for good.

> Your letter sounded to me like the far away familiar tolling of bells, waking me from a torpor into which I gradually, dulled, broken and kicked, had sunk, and filling me with nostalgia for all the endlessly beautiful things that I miss [...]. The military service is capable of first poisoning and then killing a soul within a few months.[27]

There was one particular event that year that must have varied the daily routine a little. As it happened, it was the year of the young Queen Wilhelmina's access to the

---

[26] *eigen dienst*.

[27] Brouwer to Scheltema, 14 August 1898.

throne. Her father had died when she was only ten years old, and her mother, Queen Regent Emma, had prepared her for her future duty as Queen of the Netherlands. On 6 September her inauguration took place in Amsterdam.[28] The route of the new Queen was lined with various parts of the army and volunteers from the student fraternities. One of the military men along the route was corporal Brouwer. Although he did not have fond memories of the army itself, the duty on the inauguration day filled him with some pride.[29]

To continue the story of Brouwer's military career, he was under arms off and on. Usually he spent part of the vacations in the army, to return to the university when the semester started again. In the State Archive there is a record of the time spent in actual service: from 15 December 1898 to 14 January 1899, from 23 March 1899 to 5 April 1899, from 15 June 1899 to 30 September 1899, from 7 July 1900 to 26 July 1900, and finally from 20 December 1902 to 9 January 1903. On 1 July 1903 he was transferred (in his absence) to the tenth regiment infantry. We shall see that his army periods were disastrous. They ruined his health and his nerves. It is not clear why and how Brouwer ended his military service. It is not unlikely that the authorities realised that the young man was not able to cope with life in the army.

The correspondence between Brouwer and Adama van Scheltema is an important, and almost the only, source, shedding light on the early years of Brouwer. It starts with a letter from Scheltema of 12 August 1898 and ends with a letter of Brouwer of 25 February 1924, almost two months before the tragic death of Scheltema.

The correspondence sets out on a slightly formal footing, the older student Scheltema addresses Brouwer as 'Dear chap' (*Beste Kerel*) and Brouwer writes *Waarde Scheltema*, a way of addressing that cautiously avoids formality on the one hand and familiarity on the other hand. The Dutch language had in the old days a refined spectrum of titles at its disposal, and '*waarde*' is the kind of opening that one used (and the more traditional still use) to address a colleague without offending him either by unsolicited familiarity or by haughty formality. Very soon, however, the tone of the letters changed into the informality of true friendship.

Bertus was fortunate to win the friendship and sympathy of the older and cultured student Carel, but Carel also gained a great deal from the company of his young friend. Carel, as the older and maturer person, guided Bertus into the world of art and into the more worldly aspects of life; while at the same time he benefited from the extraordinary philosophical and theoretical insights of Bertus. The friendship between Brouwer and Scheltema went largely unnoticed; of course their fellow fraternity members were aware of the close ties between the two, but their friendship, definitely, was not for public display. Although few who knew the young student Bertus would have foreseen it, he later acquired a large circle of acquaintances in all quarters of society. There was an extraordinary mixture of the cerebral scholar, jealous of his privacy, and the gregarious boyish man, with a hunger for company

[28]There is no coronation of the Kings and Queens of the Netherlands, but an inauguration. It takes place in the New Church on the Dam.

[29]Louise told that Brouwer used to recall the occasion with pleasure.

and talk. At no time in life was he deprived of friendships and relationships, but none of them equalled the friendship with Scheltema. This particular relationship had a tinge of the private and the sacred. As Brouwer himself expressed it at one occasion, his friendship with Carel Scheltema was a 'private friendship, one that was not entered into any collective'.[30] As in all real friendships, there were conflicts and disagreements, but on the whole Brouwer's relationship to Scheltema was one of the few true ideal friendships in his life.

Scheltema, commenting on the friendships of his student life, wrote in his diary:

> I can be brief about Bertus—initially reaching him a *paternal* and appreciating hand, I started feeling a warm affection for him as an equal—and subsequently in almost all respects as a superior. Indeed I still surpassed him in understanding of life, wisdom and experience and strength of soul and determination in visibility—but I had to rank his abstract capacities in the realm of knowledge and beauty higher than mine. And finally I had met a man whom I had to place above me, which could not have been the case with anybody else, unless very temporarily and self-denyingly. In the mean time a mutual human sympathy developed through these spiritual appreciations. It did not decrease my pride in the fact that I, as the only one out of those one hundred members of the fraternity, had seen and been sensitive to this man as a most extraordinary person—and even more, that I gradually proved to be the only one Bertus could have continual contact with, yea, to whom he attached himself as the only *friend*—and that was not because he did not see others, or could not find spiritual contact. I have not had more pure, more fundamental and more penetrating discussions with anybody—with nobody else I am spiritually as tranquil, and so close to beauty in vision or analysis.
>
> This is by far the greatest human being that I have met so far, and I don't think to meet a greater one—he is *in all respects* the paradigm of a man of exceptional genius—for being a *genius* he lacks the connection between his own mind and the world around him.
>
> The essential part of our relationship was my truly tireless endeavouring to make him come closer to the material world.—

This may suffice to indicate Scheltema's appreciation of Brouwer. He recognised at an early stage that the young student had all the characteristics of the great man-to-be, and he drew an intense intellectual and emotional satisfaction from the contacts with Bertus. This deep sympathy and appreciation enabled him to deal with the less pleasant aspects of Brouwer's personality. Brouwer for his part was deeply fond of his older friend, and at the death of Scheltema he summed up his feelings in the following words:[31]

---

[30] Wiessing (1960), p. 142.

[31] *Ter Herdenking van C.S. Adama van Scheltema* (In Memory of C.S. Adama van Scheltema). Note the reference to Homer, cf. Wiessing (1960), p. 106, apparently the members of the fraternity practised the 'handshake' that is referred to in the Iliad and the Odyssey by 'he grows him into the hand'.

> It was the reflection of your eyes, the inflowing grip of your hand, the warm engulfing of your voice, the peaty smell of your overcoat. It was the wild riches of your dream life, the confusing exuberance of your fantasy.
>
> But around you roamed the compelling force of determinateness, which you sensed and had to acknowledge, and you were determined and you wanted to understand, and to become a personality.
>
> You have understood much, and you have become a personality. And a part of your flourishing rhythm has become common property.

One may well guess that Scheltema served Brouwer as a guide and mentor in the most diverse matters, and that Bertus gladly accepted Scheltema's advice. For instance, when Brouwer had fulfilled his first period in the army he did not hesitate to turn to Scheltema: 'Tomorrow my cage will be thrown open, will you help me find a room?'.[32] When Bertus was depressed or ill, Carel tried to comfort him and to give the sort of advice that a student of medicine can give. Indeed, the student Brouwer had a poor health, and the letters to Scheltema contain a litany of complaints, disorders, nervous breakdowns, etc. For the greater part the poor health of Brouwer was due to the nervous tensions of the highly-strung, brilliant boy. The complaints follow each other in persistent succession until the fall of 1903, but from then onwards they occurred less often. The correspondence provides a depressing list of, mostly vague, complaints:

- Now self-control and a diet for the convalescing patient, and only a wrinkle of chastening and immunity will remain (20 September 1898).
- Because of some infirmity I have to stay in bed, for that reason I did not come to you; please drop in later when I am again allowed to see people (4 May 1901).
- I remember little from before September, I stand completely reborn, weak and free; not tied by desire, not by memory. But the disease has to wear off first. I have already mastered it with my will, and I will make it the mother of my working power (5 December 1901).
- I am here sometimes very poorly—in bed for a week at a stretch, and then 3 days sleeping badly (11 June 1902).
- ..., maybe I will get rid of the pain in the back, that tormented me terribly in Haarlem (23 August 1902).

It may come as a surprise that this same Brouwer, who suffered the penalties of a delicate nervous system with accompanying physical phenomena, would occasionally carry out downright silly experiments. At one time he wanted to find out how long it took before the ice in the water formed itself around a person's legs. So on a freezing day he took the test himself by getting into the river Amstel and waiting patiently. The story does not tell whether he waited long enough. Curiosity was always a powerful stimulus for Brouwer.

From the Brouwer–Scheltema correspondence it appears that Brouwer did the sensible thing: often, when his tormented body refused to serve him, he withdrew

[32]Brouwer to Scheltema, 20 September 1898, written in the barracks, one day before going on leave.

from the city and the university. The letters hint at intervals of recuperation, either on the Veluwe, the thinly populated area in Gelderland renowned for its healthy forests, its heath and its sands, or at the seaside, Den Helder, the national naval centre, or in 't Gooi, the enclave of utopists, health freaks and artists, east of Amsterdam. One wonders how he found time to study at all; as a matter of fact it took him three and a half years to pass the first examination, the *candidaats examination*, certainly no accomplishment to boast of for a student of his brilliance.[33]

On 16 December 1900 Brouwer took this first hurdle of the academic course. He was examined by a committee of his professors, and the diploma was signed by J.D. van der Waals and chemistry professor, H.W. Bakhuis-Roozeboom. In view of his excellent grades and matching performance at the examination, the committee granted him the *candidaats* degree *cum laude*. In the summer of the following year Brouwer widened his horizon by a longer trip to Italy. Following in the footsteps of countless predecessors he made this trip on foot.

After the *candidaats* examination, there was no marked improvement in health; but now Brouwer had decided to take action, to remain no longer passive under the cruel attacks on his body and mind. He had made up his mind to join the health fashion.

In 1902 he spent long periods in Blaricum in a pension that accommodated a loosely connected set of vegetarians, clean-living adepts, and the like. This was the locally celebrated Pension Luitjes, run by the couple Tjerk and Gerda Luitjes, which had already operated a vegetarian pension in Amsterdam before opening their establishment in Blaricum. Brouwer reported:

> Back from Blaricum, with the mixture of both sexes in vests with bare black feet with blue nails, the sunbathing of bare backs, and the gnawing of raw turnips and carrots; I am so miserable that I can not come tomorrow.[34]

Blaricum also appealed to his younger brother, Lex. In June 1902 Brouwer informed Scheltema that he was planning to share an apartment with Lex, but in September he reported that the plan was cancelled because his brother had decided to stay in

---

[33]The academic study in Holland basically consisted of two parts. The first was concluded by the '*candidaats*' examination and the second by the *doctoraal* examination. These examinations are roughly comparable to the BSc. and MSc. The titles connected to those examinations were *candidaat* and '*doctorandus*' (abbreviated as 'drs.' in titles). Some faculties had an extra examination at the end of the first year, the so-called propaedeutic examination—endearingly called 'propjes'. The time schedule was flexible. Three years for each examination was the average, but clever students could compress the period somewhat. A doctorandus had the legal right to present a dissertation to the faculty, and after a successful public defence he was awarded the title of 'doctor'. A 'candidaat' could enter the teaching profession, but the diploma was basically considered an intermediate stage on the way to the final doctoral diploma. In the well-stratified society before the Second World War each university diploma entitled its owner to a traditional civil title, a *candidaat* would (or rather could) be addressed as *Weledelgeboren Heer*, a doctorandus as *Weledelgeleerde Heer* and a doctor as *Weledelzeergeleerde Heer*. A professor was simply *Hooggeleerde Heer*. A female student/scholar had a similar title with '*Mevrouw*' or *Vrouwe* instead of *Heer*.

[34]Brouwer to Scheltema, 29 July 1902.

**Fig. 1.4** Aldert, Lex, and Bertus posing in the garden in Haarlem. [Brouwer archive]

Blaricum as a painter. Lex had given up his studies and had decided that the life of an artist was the right thing for him.[35]

Brouwer remained faithful to the practising of vegetarianism, albeit not in a fanatic way; he also became a lifelong adept of sun- or air bathing, fasting and the like. His poor health and nerves also drove him to visit spas. These institutions never reached the same popularity in Holland as they boasted in neighbouring countries. One short-lived spa, the *Bad Courant*, was in operation in Haarlem at the time the Brouwers moved from the town of Medemblik to Haarlem. So Bertus may have learned about the phenomenon in his hometown.

The disastrous, if brief, spells of military service greatly harmed Brouwer's health, both physically and mentally. The army proved an ordeal to the young man, his fellow soldiers pestered him and his superiors did not take him seriously. A letter of 7 January 1903, two days before his definite farewell to the draft, showed an exhausted Brouwer.

[35]Brouwer to Scheltema, 4 September 1902.

> Dear Carel,
>
> I only want to send you just a single note. There is of course no time to write a letter while in military service, moreover I am still half sick. After the stomach problem and the chest complaint with blood tinged sputum, and now a rash in my face. After performing my daily duties I drowse for a moment in a chair, stoically contemplate the course of my life, and say 'bah' and turn in. In order to have ideals in life a physically sound core is required, and I have lost that. I only ask for a small corner to wither away. Nonetheless I perform all the duties towards my body in a dull constancy; maybe it will work out alright, so that I re-enter the world with a shining look; or it is finished, and with that too I am in accord. I do not know myself which of the two I desire (The Cool Lakes of Death).[36] Should your ideals pine away, seek that state too. Addio.
>
> Bertus

Scheltema reacted resolutely. He wrote from Paris:

> Something positive has to be done. Huet is ancient history—you should go to Winkler or Werthem Salomonson.[37] It is a totally wrong idea that nothing more could be done to a body like yours, especially since the physical condition is the main thing. For example, your cold comes from anaemia, which can be improved by a regular use of iron pills.[38]

From that point onwards the complaints became less frequent; Brouwer remained however an easy target for attacks of a nervous nature, with their accompanying physical disorders. As a result of the physical experiences of his student years he had turned into a health freak, with a strict and even eccentric way of living.

While Brouwer suffered from a long list of complaints, his friend was well on his way to becoming a successful poet. Scheltema had a strong feeling of social responsibility. In as far as his poetry was concerned, this inspired him to write poems for the working classes, simple and understandable. Politically, it made him side with the young socialist movement.

The Social Democratic Workers (Labour) Party in Holland, SDAP,[39] was founded in 1894. It attracted a great number of intellectuals and artists, among others Henriette Roland Holst-Van der Schalk and Herman Gorter. The first was the grand lady of socialist literature in Holland. She had married into a family of artists. Herman Gorter was one of the first naturalistic poets in Holland; in the course of time he had acquired considerable authority in Dutch literary circles.

The students, in particular in Amsterdam, traditionally the red city of the Netherlands, were quick to turn to the new movement. A number of them joined the Socialist Reading Society (*Socialistisch Leesgezelschap*), where the texts of the founding

[36]Novel by Frederik van Eeden, *Van de koele meeren des doods*.

[37]Winkler was professor in diseases of the mind and psychiatry, Werthem Salomonson held an extraordinary chair in diseases of the mind.

[38]Scheltema to Brouwer, 17 January 1903.

[39]*Sociaal Democratische Arbeiders Partij*.

fathers of socialism were read and discussed, including the journals of the (in particular German) socialist movements.

A now almost forgotten, but at the time rather popular theoretical socialist, Joseph Dietzgen, was intensively studied by the students. Dietzgen was a socialist philosopher, born in Germany in 1828, who moved to the United States of America and died in 1888 in Chicago. By profession a tanner, he was a self-taught philosopher, author of *The essence of human brainwork. Excursions of a socialist into the domain of the theory of knowledge*.[40] Prominent Dutch Socialists–Communists studied and advertised his work, among them the astronomer A. Pannekoek and the above mentioned Henriette Roland Holst. Scheltema was particularly impressed by Dietzgen, and he was completely under the influence of his philosophy of socialism, somewhat to the chagrin of his friend Wiessing, who quickly developed an allergy for what he called 'the scientific Buddha of the despicable German social democratic bourgeoisie'. Wiessing characterised the particular brand of Dietzgen's socialist theory as 'the mysticism of disbelief'. Many of the students, including the members of CLIO, canvassed at the elections of 1901 for the SDAP, with its leader, Pieter Jelles Troelstra.

Scheltema's relationship to socialism was ambiguous, to say the least. He, an aesthete, far from the discomforts of the masses, felt absurdly uncomfortable in close contact with the less refined. Before his conversion to socialism, he wrote, for example, in 1897 from Visé in Belgium, to where he had retired to work, that the hotel was abominable, and that this was how it would be in a socialist state.[41] His opinion of his fellow men was at that time pronouncedly negative:

> [...] The friend above all; the sympathetic people, they constitute the world, and all those surrounding it are a shady bunch of scoundrels.

His eventual choice for socialism was that of the theoretician and the artist. The heyday of individualistic, impressionistic, naturalistic literature was past and Scheltema sensed a new perspective in the new ideals of social democracy; he formulated his conception for the first time in 1899 in *Propria Cures*, the Amsterdam student magazine, of which he was an editor. The literary output of Scheltema was twofold, he wrote poems and also books of a theoretical nature, including a travelogue of his Italian experiences. His poetry was written for a large audience, for workers, housewives, school children; it was extremely popular—at his death 65,000 volumes of his poems had been sold! Contrary to what one might think, these sales did not make Scheltema a rich man. He refused on principle to make money out of the working classes. Typically, his poetry volumes would cost only one guilder. Among the more theoretical works there was a study *Foundations of a new poetry* (1908), with its telling sub-title 'Essay of a social theory of art contra naturalism,

---

[40] *Das Wesen der menschlichen Kopfarbeit. Streifzüge eines Sozialisten in das Gebiet der Erkenntnis Theorie.*

[41] Bonger (1929).

anarchism, the movement of eighty[42] and its decadents'. The study was sharply critical of the exponents of the movement of 'eighty' and also of some of the leading spirits of the socialist movement. This book, which so explicitly singled out its opponents, was attacked violently. Brouwer wrote a penetrating review and sent it to Kohnstamm[43] for publication in the *Tijdschrift voor Wijsbegeerte* (Journal for Philosophy). For one reason or another the editors decided not to publish the review, to the utter indignation of Brouwer, who wrote to Scheltema[44]

> The jerks Meijer and Pen[45] have made notes in the margin on my very paper, like pencil scratches on statues of Donatello. I did not want it to survive the insult, and I burned it.

In general Brouwer was an attentive reader, and at times a severe critic, of Scheltema's work:

In 1905 Brouwer wrote a literary critique of the second of a series of three longer poems, *Londen*, *Amsterdam* and *Dusseldorp*. Like its successor above, it was rejected, but a copy survived in the archive of P.L. Tak, the editor of the *Kroniek*.[46] The review is interesting as Brouwer develops a sort of evolutionary schema in which socialist art *à la* Scheltema fits:

> The force of the new gospels is in their justified rejection of the one-sided positiveness of the old ones; but once they have made the old one perish, there remains only positiveness, which will cause in due time a reaction and they will be pushed back in turn.
>
> But the new will not force out the old if it does not first step into the forms of the old. Whoever presents something new in the forms of the old will make a career; but whoever makes people personally experience the new, will be reviled [...].
>
> A change of literary gospel is always preceded by an analogous change in the economy, where often the literary consequences of the old economy flourish on the illusion of the new one. The Dutch '80 of literature belongs to the '48 of the economy; before '80 the restoration in literary art was carried on by the liberals.[47] [...] And so the liberal literature is now carried on by the

---

[42] A literary movement in Holland that rebelled against the traditional school of the nineteenth century.

[43] Ph. Kohnstamm was a student of Van der Waals. He started his career in physics and switched eventually to pedagogy. He held chairs in Utrecht and Amsterdam.

[44] Brouwer to Scheltema 11 August 1908. There may actually have been more to the rejection of Brouwer's review than meets the eye, see p. 104.

[45] A disciple of the Leiden philosopher Bolland.

[46] A literary magazine. The review was posthumously published in Delvigne (1985).

[47] The 'Movement of 80' was a Dutch literary movement which introduced in the Netherlands the concepts and styles of the leading European artistic circles. It imported successful innovations such as naturalism, impressionism and the like. The movement of '80 dominated the Dutch literature far into the twentieth century. Brouwer compared this movement with the political-economical revolutionary ideas of 1848. He saw Scheltema as one of the first literary artists escaping the influence of '80.

> socialists Gorter and Roland Holst. Note that the old conceptless mood-poetry is now fed by the new economic gospel, and thus it provides this new gospel with old fashioned needs such as ideal-confirming art.

The implicit criticism of Gorter and Roland Holst, the figure-heads of socialistic art, was rather scathing—their poetry was seen by Brouwer (and by Scheltema!) as a belated exercise in bourgeois pre-socialist expressive art, and in this form supporting the new economic movement. Brouwer was inclined to see in the case of Gorter a strong visionary involvement with the new social gospel: 'strong and deep enough to break through the walls of bourgeois understanding; and so wild phantasms flower there on a ruin of intellect'; whereas in the case of Roland Holst 'the old logic, on the contrary, remained standing cleanly and squarely like a little house at the river Zaan'.[48]

Brouwer definitely had a point, but it was audacious, to say the least, to express it at the time. In contrast to the older generation of socialist poets, Brouwer presented Scheltema as the real socialist man of letters, who did understand the times and the modern world.

> The first realisation of socialism in literary art is Adama van Scheltema. Here no emotions are recognised, unless strongly felt as a necessary moment of interaction of his human nature with the living world around him and as a necessary parallel of his actions.
>
> In the meantime he is, more than is good for him, ahead of his own time.

This last sentence is an oblique comment on the fact that Scheltema, however popular as a poet of the common people, was not recognised by the literary world.

It should not come as a surprise that the review was rejected by the board of the *Kroniek*; the Gods at the Dutch Socialistic Olympus would not have been amused by irreverent remarks of a self-styled philosopher with mystical antecedents.

Objectively speaking, Scheltema will probably go down in history as one of the minor Dutch poets of our century, but that does in no way detract from his merits as a popular poet. He has added significantly to the appreciation of poetry, and art in general, in the working and middle classes. His theoretical work, however, has never attracted the attention it deserved.

Brouwer's opinion of Scheltema was, we guess, coloured by his friendship, but even more by extra-poetical arguments, such as his interest in the theoretical foundations of art, society and communication.

In view of Brouwer's relationship with Scheltema, it is hardly surprising that he was influenced by the latter's political views. In the early years Scheltema, like so many students, was impressed by Dietzgen, and some of Scheltema's enthusiasm for Dietzgen rubbed off on Brouwer.

Rather aloof, Brouwer followed his friends on their rounds of the workers circuit and even went so far as to join the Socialist Reading Society.

---

[48]The old heart of industrial North-Holland, reputed for its beautiful old Dutch houses, still to be seen in the local outdoor museum.

**Fig. 1.5** Carel Adama van Scheltema and Henri Wiessing. [Courtesy Letterkundig Museum]

The membership of that illustrious society, incidentally, played a trick on Brouwer. In 1901 on his walking tour to Italy, he was, in a mixture of eccentricity and practicality, dressed in a long cape, which he used when sleeping in the open air at night. One day in Italy he was arrested for vagrancy, and when the police found in his pockets the red membership card of the Socialist Reading Society, they were convinced of a good catch: a strange young man with a red card, printed in a cryptic nordic language, with the word 'socialistisch'. Brouwer was, however, released no sooner than he was arrested when a letter of the Italian mathematician Bianchi was found upon him, which contained the names of a number of famous scientists known to the local police magistrate. Brouwer used to narrate the story of his foot tour to Italy with justifiable pride.

One of the little adventures seems to come straight out of a classical adventure story: once, at the end of a day's hike, Brouwer had joined a small group of men who had found a suitable spot in the forest to spend the night. The company made a fire and prepared a meal. Brouwer was with proper hospitality invited to take part in the meal. He gladly accepted and joined in the general conversation at the fire. After a while he wrapped himself in his cloak and laid down to sleep. Listening to voices of his fellow travellers, he discovered that they were discussing how to rob the young foreigner. Waiting some time, he made his exit when everything was quiet. By the time he was missed, he was already far enough away to be safe.

The proper thing for a student at the beginning of the century was to join, if not the socialist party, at least the band of enthusiasts, who saw in socialism the road to a new and just society. Scheltema and Wiessing, each after their own inclination, became active supporters of the socialist cause, and for some time Brouwer joined them and frequented political meetings, albeit in a more contemplative spirit. He fostered for some time socialist sympathies. This is remarkable indeed, since the rigid materialistic views of, for example, Dietzgen contradicted everything he believed in.

The following quotation from his letter of 11 June 1902 to Scheltema gives some indication of his inclinations:

> *La société, c'est la blague*[49] and in this way it is actually rather funny—of course at the bottom there is the earnestness of life, and we are socialists, but that is only a dim basis for the grand totality that we try, according to our strength, to put somewhat in the right position.

But before two years had lapsed, he had cut his short liaison with socialism:

> My short-term socialist inclinations, dating back more than two years, have turned out not to be viable.[50]

Five years later his farewell to socialism was final. Korteweg, his Ph.D. adviser, was at that time apparently involved in the election campaign of the liberal party,[51] and Brouwer offered his assistance:

> Professor, I rarely think about politics, but my political sympathies are in the liberal, anti-democratic direction.[52] So if you can use the assistance of someone without experience in political work, then I will gladly help to support a liberal against a free- or social democrat.[53]

Later in life he became involved in local politics as an alderman for the Neutral Party in his hometown Blaricum.

The Brouwer–Scheltema correspondence shows us the melancholy process of the inevitable growing up of a young gifted, intelligent but impulsive young man; a

---

[49]Society is a joke.

[50]Brouwer to Scheltema, 18 January 1904.

[51]'liberal' has to be understood in the classical sense of the nineteenth century, *laissez faire*, limited government interference. It was neither conservative nor Marxist.

[52]There was a certain measure of anti-democratic feeling in the air at the time. The influential philosopher Bolland, for example, crusaded against democracy. It is possible that Brouwer also flirted with these ideas. Given the context of the letter, we may be sure that Brouwer referred to the Social Democrats.

[53]Brouwer to Korteweg 22 September 1909.

process with all its agony, yearning, despair and elation. One of the first letters gives us the picture of a confused, unsettled young man:

> I am too drowsy to experience my freedom, my mind is not capable of any activity, not yet delivered from the oppression of that most dark revelation of Adam's curse.[54]

Slowly, through the years, his physical state improved, and with it his mental state. Although there were numerous relapses, he slowly gathered the strength to face the world. After the utter resignation in his illness during the military service in 1903, the complaints had become less frequent, and a certain defiance of the problems of the world could be discerned. Having done with his military service, he returned to his studies:

> —after an absence of two years it required some determination, especially where any trace of love for the subject [mathematics] was lacking. [...] My work is done without illusions, but with a feeling of joy for the activity in itself.[55]

In a later letter, looking back, he lamented the loss of innocence and susceptibility:

> What a lack of tenderness, of childlike innocence, of abandonment in the words that I write down; I know it, I would be ashamed if I met myself as I was five years ago; but just as one cannot stop the growth of one's beard, so one cannot stop the growth of the philistine-tissue through one's soul. Then let me be great as a philistine! And follow my course alone, unfeelingly among the dead stones to the fair End. And so leave my trace on the thus melancholy world. That is, Ambition is born within me, perhaps! But in any case, one that knows to control itself, and to collect quietly material, until its time has come. **I will have to remain obscure for a few more years, then my grasp will be felt.**[56] Just because I feel the futility of all worldly things, no detraction or fear will disturb my course.[57]

That same year, on the sixteenth of June, he passed his final examination, the *doctoraal examination*, which earned him the title of *doctorandus*. Again, he passed with the highest honours, Cum Laude. He dryly informed Scheltema of this fact by a telegram containing just the word 'Cum'. In the years that lay between the first steps in science and the proud moment of being awarded the '*Cum Laude*', Brouwer had established himself as the equal and partly the master of his older friend. Their friendship had passed through all stages, reaching a stability that was not to be disturbed by the sometimes moody or even crude actions and remarks of Brouwer. The relationship between Brouwer and Scheltema had its touching moments, as well

[54] Brouwer to Scheltema, 20 September 1898.

[55] Brouwer to Scheltema 15 November 1903.

[56] My boldface.

[57] Brouwer to Scheltema, 18 January 1904.

as its small moments of unpleasantness. Their correspondence provides a revealing insight into the minds of the fast developing mathematical genius and the socialist poet. There are numerous small details that allow us glimpses of their habits, their reading, financial problems, ... For example, Scheltema strongly advised the 21-year-old Brouwer to read Flaubert's *Éducation sentimentale*[58] and a little later he asked him to return his copy of the *Contes* of Flaubert.[59] At another occasion Brouwer reported a visit to the theatre, and recalled the performance of Ibsen's *Wildente* (The Wild Duck). He wrote:

> Send me your new address. I hope that you will find solitary rest there; surround yourself with the books of your equals (*Ebenbürtige*) and congenials. I live with Pascal, Emerson, Madame Gimon and Montaigne. And do me a favour, if you do not yet know it, read the 'Journal of Marie Bachkirtoff'.[60] She has something of both of us, and she stands in between us.[61]

The fact that Brouwer had read the book of Marie Bashkirtseff is interesting. The author was a prodigy of Russian descent, who lived in Paris. When she died at the age of 23 she was greatly admired as a painter and author. Brouwer had recognised in her a kindred spirit—he was under no misconception where his own genius was concerned!

One should not get the impression that Brouwer's reading was restricted to the exalted regions of literature. Much later, writing from his bed he confessed:

> I don't feel very ill, but if I leave my bed for longer than one hour, my heart starts behaving in a funny way. So I stayed in all week and read all the volumes of the adventures of Arsène Lupin.[62]

In those days the stories of the gentleman-burglar Arsène Lupin were popular, but hardly the reading matter that the more conventional would expect in the hands of a learned scholar!

In 1903 Brouwer had gathered so much confidence in himself that he considered himself on a completely equal footing with his former mentor Scheltema. It was time to cut the umbilical cord, and to face the world on his own. He proposed to seal a formal union, as between two kings, to be concluded on Ascension Day. Apparently there had been some friction between the friends, and it is not wholly impossible that Scheltema saw, to his dismay, the ugly duckling starting to turn into a swan, with a private and virtually inaccessible inner world. Brouwer clearly realised that the mentor–novice relation was over.[63]

---

[58] Scheltema to Brouwer, 29 December 1902.

[59] Scheltema to Brouwer, 2 May 1903.

[60] Bashkirtseff (1888).

[61] Brouwer to Scheltema, 18 January 1904. 'Bachkirtoff' is Brouwer's spelling. Scheltema misspelt the name.

[62] Brouwer to Scheltema, 25 February 1916.

[63] Brouwer to Scheltema, 23 May 1903.

> *Il faut savoir séparer* (La Rochefoucauld) *C'est le privilège des grands esprits, de ne pouvoir se brouiller* (Voltaire).
>
> Carel, my rich poet, I have finished your book, but listen: In no realm are there two Kings, each has to live in his own country of subjects: that loneliness without their equals, that is what they are Kings for. But once a year they visit each other, and see their great contrasts, with nothing in common but the joint feeling of being royal, both being in the immediate grace of God; their associating can be nothing but showing each other the powers and glorification of their mutual domains.—And the only permanent grace, that, from the awareness of the kingship of each other, they can let flow to each other, consists of the rendering of obligatory courtesies and reporting outward appearances of their person and Kingdom.
>
> Carel, your realm is more summery than mine, and your people are more pacified—both our countries have been blessed by God with marvellous beauty.
>
> Well, after our discussion of Thursday night I believe that you are right, but the best idea for our having to live apart, is what I write down here.
>
> Let us meet every year on Ascension Day, and solemnly bathe in the cool sun of spring, and sup together, and exchange what the past year has brought us, and for the rest feel, invisible to each other, united in 'knowing the other to be a King'.
>
> So, brother, do you agree with this?
>
> Then, hail to You and Your Kingdom—Until 1904.
>
> Bertus

Two days later Scheltema answered; he accepted the separation as a tragic fate. The royal metaphor had immediately sunk in: the idea of Kings toiling in solitude for the happiness of the people, or to put it otherwise: of the best and the noblest working for the redemption of their fellow men, was certainly not a rarity. One did not have to read Nietzsche or the like, to see that the best of the nation were destined to play an important role as prophets and moral guides. It tells us a good deal about Scheltema that, although a socialist by conviction, he whole-heartedly embraced the idea of the responsible, benevolent king. In a sense this is not so surprising; the socialist in Scheltema was partly, perhaps largely, the product of his intellectual efforts, a fact that he was himself very well aware of. In fact, his socialism was a tour de force of a theoretical nature, rather than a spontaneous manifestation of the heart. In spite of his heart-warming poems, he remained a theoretician.

> —Sometimes I am desperately mad about my own desperate soberness! Then I would like to kick my own soul and for once have a touch of anarchistic looseness.[64]

Far from being a game of pretending, the matter of the two kingdoms was taken completely seriously by both correspondents; it expressed in a metaphorical way

[64] Scheltema to Wiessing, 21 July 1903.

their views on responsibility and society. As Brouwer put it, discussing the physical sorrows of the body:

> That is the misery of purification, [...], that holds the great soul back from soaring too high and that keeps it in the fear of God; of God, whom he has to serve in watching over God's children, asking nothing for itself—chosen are we—not for our pleasure in the world—we are the prophets, who, messengers between God and mankind, direct and inspire the development, the working, the growing, the flowering of it with the dewdrops that flow from our fingers—you walk earnestly and solemnly through your garden, and scatter them with firm and knowing gestures; I rush through my wilderness, and they roll, without my knowing it—indeed there are few who find them, but they are all the more valuable to them.[65]

A few weeks later he wrote again about the destiny of the exceptional person:

> —In the straight chain of the generations, where all the present is sacrificed to the future, nature sometimes allows itself to bear sidewards a barren beautiful flower, not connected with mankind that reproduces itself. Blessed is such a chosen one—mostly an eldest son, who in this way is sacrificed to Minerva—as long as he is aware of his consecration, and not worried about not being directed in the presence of all the strong, blindly directed ones in the chain. The flower supports nobody in the greater context, it has no other duty than to be beautiful, sufficient in itself: for the open side, bent away, is turned towards heaven, which can refresh itself in the beautiful appearance; but only God sees the tears within.[66]

This theme is stressed more than once by Brouwer. It is not farfetched to conjecture that the above view on the exceptional, chosen son was a reflection on his own perspective of life. In his case, but equally well in that of Scheltema, the view had a prophetic ring. The line of the generations did not continue through them. In November he returned to the topic:

> Ever working on, reading, and thinking—and in line with that thinking, harmonising one's life more and more, carried by resignation and faith in God—it is the bliss here on earth. My house is homely, striking and modestly comforting. And it is sacred to me—I could do no evil in it and have no evil thoughts—in it I am even friendly disposed to everybody. If a boring person visits me—in my house I do not find him boring—if I walk out-of-doors the next day, I do not understand how I could have suffered him.
>
> The ultimate harmonising of our life seems most difficult, slowest, most laborious with people of our sort. It seems that in the progress of the generations, of all parents the oldest child may not simply be sent along in the mainstream of the 'striving and pairing', but must be offered to the gods of consciousness,

[65]Brouwer to Scheltema, 9 August 1903.

[66]Brouwer to Scheltema, 26 August 1903.

> of the infertile consciousness in the worldly motion, as an opening sacrifice—as a sideways shooting flower, which has no further purpose for the growth of the trunk—in compensation those gods will forgo their title to the other children. So, let those consecrated sacrificial animals be aware of their role, and let them not be jealous of the rough rye bread of the herd.[67]

The two-kings episode is interesting for two reasons. On the one hand it illustrates the spiritual intimacy of the friends and their mutual appreciation, while on the other hand both accepted the inevitable divergence of their paths in the world. Scheltema was well aware of the intellectual and emotional potential of the younger man, while at the same time Brouwer realised that Scheltema—be it with a tremendous effort—had found his place in human society. When Scheltema pressed the young Bertus to accept the world:—'Try rather to see the reality than your own fantasies. The more you approach reality, the greater your chances of regeneration will be',[68] he well knew that utter isolation held destruction.

The idea of a select group of superior men, working for the salvation of mankind, was undeniably in the air: an intellectual elite as the new priests of the world. The same idea occurred somewhat later independently to Frederik van Eeden,[69] who was immensely pleased to find out that Brouwer had conceived for himself the idea of a cultural, spiritual elite.

The relationship between Brouwer and Scheltema was one of deep respect and spiritual understanding, and its tragedy was the irrevocable passing of the unity and similarity of spirits. It was Brouwer who repeatedly urged his friend to accept the inevitable divergence of the courses that their lives were destined to run. Both partners had to summon all their strength to accept the new relationship—kings in their separate kingdoms, aware of their duty and destiny.

Scheltema, who had just read Nietzsche's 'Birth of Tragedy', characterised their differences in personality as: 'You are *Dionysus*, I am *Apollinius*, and the society we live in is *Alexandrinius*.'[70]

This particular friendship, that lasted for slightly more than a quarter of a century, was of tremendous importance to Brouwer, and presumably to Scheltema. It was guarded as a private treasure; few knew about it, and eventually a few isolated remarks were left. Were it not for the lucky preservation of a substantial part of the correspondence, the relationship between Brouwer and Scheltema would have remained nothing but a poorly substantiated rumour.

---

[67] Brouwer to Scheltema, 15 November 1903.

[68] Scheltema to Brouwer, July 1901.

[69] See p. 243.

[70] Scheltema to Brouwer, 6 August 1907.

# Chapter 2
# Mathematics and Mysticism

## 2.1 Teachers and Study

Let us go back to the starting point of Brouwer's university life and the years of study. When the young boy enrolled in the University of Amsterdam, that university treasured a few great men in the sciences, the most outstanding amongst them being the physicist Johannes Diederik van der Waals and the biologist Hugo de Vries.[1] In mathematics there were no stars of the same order, but on the whole the students were in competent hands. Lectures in mathematics were given by Diederik Johannes Korteweg, A.J. van Pesch and in physics by Van der Waals and Sissingh. Korteweg, in a manner of speaking, had saved nineteenth century Dutch mathematics from an inglorious historical record. When the zoologist Hubrecht was presenting a survey of fifty years of exact sciences in the Netherlands at the occasion of the Inauguration of Queen Wilhelmina in 1898, only two lines of his 11 page essay were devoted to mathematics, and it was Korteweg's work that was referred to.[2]

Brouwer was doubtlessly influenced most by his mathematics professor, Korteweg (31 March 1848–10 May 1941), a man with a remarkable career, which was in a way characteristic of the first generation of scientists of the new era. The second mathematician in the faculty, Van Pesch, cut a rather poor figure compared to the impressive Korteweg. His lectures did not always measure up to the standards of his students. Wijdenes, one of Brouwer's contemporaries, told that when Van Pesch got into one of his muddles, Brouwer would get up, go to the blackboard, take over the chalk, and in his precise manner steer the lecture past the cliffs where Van Pesch had been stranded.[3] He did, however, not have the audacity to take liberties with Korteweg.

The mathematics training at the Dutch universities in the nineteenth century was in the hands of well-meaning professors, who followed at a safe distance the developments in the prominent centres. Apart from the well-known Stieltjes (who did

[1]The Nobel prize winner Van het Hoff had already left Amsterdam for Berlin.

[2]Ritter (1898), p. 70 ff.

[3]Communication of L. van den Brom.

D. van Dalen, *L.E.J. Brouwer – Topologist, Intuitionist, Philosopher*,
DOI 10.1007/978-1-4471-4616-2_2, © Springer-Verlag London 2013

**Fig. 2.1** Diederik Johannes Korteweg. [Brouwer archive]

not teach in Holland) there were no men of stature who could inspire the new generation of students. The rise of physics, and the sciences in general, in Holland, however, called for strong mathematics departments. But mathematics was still trying to catch up with the international developments. In the absence of outstanding pure mathematicians, who could have influenced the academic opinion, there was a general tendency to consider mathematics rather in her role of a handmaiden of the sciences, than that of the Queen of Science.

In this climate Korteweg started his studies at the Polytechnic School of Delft. As he was not particularly technically minded, he chose to break off his studies at the engineering school; not, however, without obtaining a certificate for teaching mathematics.

The situation in teacher training in Holland, before the wholesale reorganisations after World War II, requires some explanation. In Holland there were two roads to a teaching position at one of the nation's high schools (HBS) or gymnasiums: one could either obtain the normal degree of doctorandus[4] at one of the universities, or one could study individually a particular subject, ranging from the languages to the sciences, and get a special teaching diploma. The examinations for these subjects were conducted by a state committee, the diploma was called the *middelbare acte (MO-acte)* (secondary certificate), and the subject and level was indicated by a code in letters and numbers. The first secondary certificate for mathematics was the *MO-KI* acte and the second and higher one the *MO-KV* acte.

With this *KV* diploma in his pocket, Korteweg found a teaching position, and from 1869 until 1881 he taught at high schools in Tilburg and Breda, towns in the southern part of the Netherlands. In the meantime he prepared himself for the academic entrance examination, in order to study mathematics at a university. In quick succession he passed the entrance examination in 1876, the candidate's examination in 1877 at Utrecht and the doctoral examination in 1878 at Amsterdam, and without loss of time he defended in that same year his dissertation, *On the speed of propagation of waves in elastic tubes*; the doctorate was awarded 'cum laude'. At the age of thirty he was the first doctor of the young university of Amsterdam. His Ph.D. adviser was the physicist J.D. van der Waals. Three years later, in the same year

[4] Literally 'a person who should become a doctor'; a prerequisite for being admitted as a candidate for a doctorate.

that Brouwer his student-to-be was born, Korteweg was appointed a full professor in Amsterdam, holding the chair of mathematics, mechanics and astronomy from 1881 until 1913. From 1913 to 1918 he was an extraordinary professor.

Korteweg's mathematical production was impressive and wide-ranging, he published on such topics as theoretical mechanics, thermodynamics, the theory of voting, algebra, geometry, theory of oscillations, electricity, acoustics, kinetic theory of gases, hydrodynamics, astronomy, probability theory, actuarial science, philosophy, ... He was to a large degree the man who dealt with the mathematics behind the physical theories of his Ph.D. adviser, Van der Waals. Nowadays he is mainly known for the famous *Korteweg–de Vries equation* (1895), which he published together with his Ph.D. student, the mathematics teacher Gustav de Vries.[5] The equation describes the propagation of a solitary wave in a rectangular canal. The success of this equation should, however, not obscure his research on the folding of surfaces and on the Van der Waals surface.

As the chief editor of the collected works of Christiaan Huygens, (1911–1927) he combined his mathematical and historical interests; he solved the riddle of Huygens' sympathetic movement, concerning coupled oscillators.[6] Korteweg was a noble and generous man, who played a central role in the national institutions of learning—the Academy, the Mathematics Society, the Senate of the University of Amsterdam, and, of course, the faculty of Mathematics and Physics. We will meet his name again, when we reach Brouwer's dissertation.

The second person to exert a profound influence on Brouwer's career was Gerrit Mannoury (17 May 1867–30 January 1956), a man who had also come to mathematics via the detour of a teacher's career. Mannoury was the son of a captain of the merchant navy. He finished high school in Amsterdam in 1885 and obtained his teacher's diploma three months after the final high school examination. For comparison: the regular study for a teacher's diploma took 4 years! In 1886 he got an appointment at an elementary school in Amsterdam, and in 1888 he moved to a private educational institution at Noordwijk.[7] Three years later he obtained a position at a high school called the Public Business School (*Openbare Handelsschool*) at Amsterdam, this position was combined from 1893 until 1902 with an appointment as a private tutor of the son of Mrs. Henri Tindal (the widow of a newspaper tycoon). Between 1902 and 1905 the school society at Bloemendaal (*Bloemendaalse Schoolvereniging*) hired him; in 1905 he obtained a position as a teacher in Helmond. Finally, in 1910, he got a position at the high school in Vlissingen, where he taught bookkeeping, mathematics and economics. He also became the headmaster of the new evening school for business. On top of all that he worked from 1894 onward as an accountant.

While fully occupied as a teacher, he passed the numerous examinations that marred the life of many a school master (the so-called *acten*, discussed earlier). In

[5]The equation also occurred in the dissertation of De Vries. A recent book of Willink cites evidence that the role of De Vries was modest, to say the least; cf. Willink (1998).

[6]*sympatisch uurwerk*, Korteweg (1905).

[7]*Instituut Schreuders*.

**Fig. 2.2** Gerrit Mannoury. [Courtesy, J. Mannoury]

particular he obtained the diplomas for teaching mathematics in secondary schools, the *KI* and *KV* diplomas.

Even before passing these examinations he published original papers in mathematics. His paper *Lois cyclomatiques* (1898) introduced the new discipline of *topology* in Holland; it was followed by two more papers in the same area.[8]

The paper treats a generalised form of the Euler–Poincaré formula. Mannoury proved in this paper a theorem which Van Dantzig has called 'Mannoury's duality theorem'. In Hopf's words, 'The theorem expressed by the [indicated] formulas, which you correctly call 'Mannoury's duality theorem', belongs completely to the area of modern duality theorems, and the fact that Mannoury knew it in 1897 shows how far he was ahead of his time. It is a pity indeed, that he did not continue this work. He was very close to the duality theorems of Alexander'.[9]

At roughly the same time he familiarised himself with the new symbolic methods of Giuseppe Peano.[10] The latter had introduced a symbolic language for mathematics. From 1888 onwards, Peano had studied and advertised a formalism that is fairly close to our present-day logical notation. Although Frege preceded him by almost a decade, Peano's notation was a great improvement in terms of readability. Peano's best-known publication of the symbolic language was his *Formulaire de mathématiques* (1898), and he went so far as to publish his result on the solution of differential equations in the symbolic notation. Brouwer later somewhat scathingly remarked that Peano's paper was not read until someone translated it back into common language! Mannoury quickly saw the theoretical value of Peano's language, but he was sensible enough not to write his own papers in Peano's formalism. Thus, curiously enough, this schoolmaster without a formal mathematical training not only introduced topology in the Netherlands, but also symbolic logic.

He enrolled at the University of Amsterdam to study mathematics; unfortunately, in view of his daily teaching duties, he could not attend the lectures, so the study was far from simple.

[8]Mannoury (1898a, 1898b, 1900).

[9]van Dantzig (1957), p. 7.

[10]Peano (1895).

Korteweg, who recognised Mannoury's ability, gave him for some time private tutorials at home, on Sundays, and allowed him the use of his private library. Nonetheless, the limits of the combination of working and studying were reached before long; Mannoury gave up and never got a formal degree in mathematics. In view of his exceptional performance as a free-lance mathematician, Korteweg tried to further Mannoury's career. As a result he was appointed in 1903 *privaat docent* in the logical foundations of mathematics at the University of Amsterdam.[11]

The formal appointment to any of the positions of *privaatdocent*, *lector* (lecturer) or professor, required the appointee to give a public inaugural lecture. This was a formal occasion, attended by members of the senate in gown and cap. Mannoury presented his inaugural lecture with the title *The Significance of Mathematical Logic for Philosophy*, on 21 January 1903.

Brouwer has sketched the decisive role of Mannoury in his life, in the formal address, delivered at the occasion of the awarding of an honorary doctorate in 1946 to Mannoury:[12]

> As happens so often, I began my academic studies as it were with a leap in the dark. After two or three years, however full of admiration for my teachers, I still could see the figure of the mathematician only as a servant of natural science or as a collector of truths:—truths fascinating by their immovability, but horrifying by their lifelessness, like stones from barren mountains of disconsolate infinity. And as far as I could see there was room in the mathematical field for talent and devotion, but not for vocation and inspiration. Filled with impatient desire for insight into the essence of the branch of work of my choice, and wanting to decide whether to stay or go, I began to attend the meetings of the Amsterdam Mathematical Society. There I saw a man apparently not much older than myself, who after lectures of the most diverse character debated with unselfconscious mastery and well-nigh playful repartee, sometimes elucidating the subject concerned in such a special way of his own, that straight away I was captivated. I had the sensation that, for his mathematical thinking, this man had access to sources still concealed to me, or had a deeper consciousness of the significance of mathematical thought than the majority of mathematicians. At first I only met him casually, but I at least knew his tuneful name, which guided me to some papers he had recently published in the *Nieuw Archief voor Wiskunde*, entitled *Lois cyclomatiques*, *Sphères de seconde espèce* and *Surface-images*. They had the same easy and sparkling style which was characteristic of his speech, and, when I had succeeded, not without difficulty, in understanding them, an unknown mood of joyful satisfaction possessed me, gradually passing into the realisation that mathematics had acquired a new character for me. For the undertone of Mannoury's argument had not whispered: 'Behold, some new acquisitions for our museum of

[11]The position of *privaat docent*, similar to *Privatdozent* in Germany, brought the bearer of the title a nominal fee. Its main attraction was that it enabled one to keep a foothold in academic life, in the hope of a promotion.

[12]Brouwer (1946).

immovable truths', but something like this: 'Look what I have built for you out of the structural elements of our thinking.—These are the harmonies I desired to realise. Surely they merit that desire?—This is the scheme of construction which guided me.—Behold the harmonies, neither desired nor surmised, which after the completion surprised and delighted me.—Behold the visions which the completed edifice suggests to us, whose realisation may perhaps be attained by you or me one day.'

## 2.2 First Research, Four-Dimensional Geometry

Brouwer's relation to mathematics remained ambiguous for a number of years, only when the success of his work in topology blocked his retreat, did he definitely resign himself to a mathematician's life. As the laudation at Mannoury's honorary doctorate tells us, mathematics initially did not at all fulfil his expectations, mathematics as a clinical, sterile subject consisting of theorems and exercises did not in the least appeal to him.

At any rate, he decided to make the best of it; he had joined NEWTON, the club where students and their professors freely mixed. In 1899 he also became a member of the venerable *Wiskundig Genootschap*, the national mathematical society, founded in 1778, in the era of Progress and Enlightenment. Like all the venerable institutions of the Enlightenment the society had a motto: *Een onvermoeide arbeid komt alles te boven* (*Labor omnia vincit*, An untiring labour overcomes everything). The Mathematical Society, usually referred to as *WG*, traditionally met on the last Saturday of the month somewhere in Amsterdam. It was at these meetings that Brouwer fell under the spell of Mannoury, and it was there that he got to know the leading personalities in Dutch mathematics, if not personally then at least by sight. In spite of his earnest quest for the true living mathematics, he could not easily shake off his doubts. Nonetheless, he could not renounce his talent for mathematics. Even before his final examinations he had done original research in geometry. Wiessing reports that Brouwer gave a talk at a meeting of the science club NEWTON of the Student Corps:[13] 'only eighteen years old, he presented to the company at a meeting attended by Professor Korteweg, some theorems of four-dimensional geometry, found by him. Korteweg was completely confounded: "I don't know what to think of it", the professor said reflecting, "it is a great and ingenious discovery, or it is a mystification!"'

The content of Brouwer's talk unfortunately is unknown, but it is fairly certain that it contained the germs of three papers that were submitted to the Dutch Royal Academy by Korteweg:[14]

- *On a decomposition of a continuous motion about a fixed point O of* $S_4$ *into two continuous motions about O of* $S_3$*'s*,

[13] Wiessing (1960), p. 142.

[14] Communicated respectively at the meetings of 27 February 1904, 23 April 1904, 23 April 1904.

**Fig. 2.3** Brouwer (standing, left) at a dinner of a fraternity, probably NEWTON. The bald man on the right is Korteweg. [Brouwer archive]

- *On symmetric transformations of $S_4$ in connection with $S_r$ and $S_l$*,
- *Algebraic deduction of the decomposability of the continuous motion about a fixed point of $S_4$ into those of two $S_3$'s.*

The above papers, and a considerable number of his later papers, were published in the Proceedings of the Royal Dutch Academy of Sciences. Most of these papers were in fact published twice, one version in Dutch and one in English or German. There were actually two series of publications, the *Verslagen van de Koninklijke Nederlandse Akademie van Wetenschappen* and the *Proceedings of the Koninklijke Nederlandse Akademie van Wetenschappen*. Names have changed since then; when Brouwer started submitting his papers, the Academy was called the *Koninklijke Akademie van Wetenschappen te Amsterdam* (Royal Academy of Sciences at Amsterdam), during World War II the name was changed to *Nederlandse Akademie van Wetenschappen* and after the war a synthesis was arrived at: the *Koninklijke Nederlandse Akademie van Wetenschappen*. The institution itself was founded by the first King of Holland, Louis Napoleon, the brother of the Emperor Napoleon. At that time there were already a number of 'learned societies', such as the Dutch Society for the Sciences (*Hollandse Maatschappij van Wetenschappen*), the Utrecht Provincial Society (*Provinciaal Utrechts Genootschap*). These local academies were the fruits of the Enlightenment; they provided a platform for the scientific and commercial upper class of the Dutch Republic. When Louis Napoleon proposed to transform one of the existing societies into the national academy, he was met with protest and refusal, so he founded his own academy—which was considered an upstart by the older establishments. Eventually the Royal Academy superseded the local learned societies, the latter having remained to this day modest centres of the sciences and arts.

Before Brouwer was elected a member of the Academy, most of his papers were submitted by Korteweg, who, as a member, was entitled to present papers for publication in the Proceedings and the *Verslagen*.

The first of Brouwer's papers, mentioned above, was a treatise on rotations in four-dimensional Euclidean space. He showed by geometric means that a rotation in four-dimensional space can be obtained as the product of two rotations in three-dimensional space.

Freudenthal, in his discussion of the paper,[15] pointed out that the simplest way to treat the above transformations, is by means of quaternions. Brouwer first gave a geometrical proof and subsequently an algebraic one. Possibly Brouwer was not well versed in the geometrical applications of quaternions; whatever the reason may have been, he had opted for a laborious direct proof by geometric means.[16]

These very first mathematical publications made Brouwer, unwittingly, an actor in a priority controversy. It was his first experience of that kind, but, alas, not the last one. The topic of Brouwer's paper—a study of the orthogonal transformation group of a four-dimensional Euclidean space, had been the subject of investigations of a German mathematician, E. Jahnke,[17] a man of some weight, with a sizeable publication record.

Jahnke had spotted Brouwer's paper in the Proceedings almost immediately after its appearance; he wrote a rather condescending letter to Korteweg (15 March 1904) and magnanimously (and correctly!) assumed that Brouwer was unaware of Jahnke's publications on the subject, which, he said, already contained the results of Brouwer's paper. He acknowledged that Brouwer had obtained his results by new means, but he expressed his expectation that 'the author would use the opportunity to acknowledge in a short note in the same journal and if possible in the next issue my priority for the mentioned results'.

A letter from a man who had earned a reputation in applied mathematics, who was an editor of the *Archiv der Mathematik und Physik*, might have daunted a lesser spirit than Brouwer, but this young man was not to be silenced so easily. When

---

[15]CW II, p. 22.

[16]Brouwer's theorem in modern formulation reads $SO_4 \cong SU_2 \times SU_2/\pm(1,1)$. Another geometric proof is in Klein (1890), and a similar theorem can be found in Cartan (1914).

[17]At that moment *Oberlehrer* at the *Friedrich–Werderschen Oberreal Schule* and a *Privat Dozent* at the *Technische Hochschule Berlin*.

Korteweg duly informed Brouwer of the claims of Jahnke, Brouwer carefully studied Jahnke's papers and concluded that he had in no way invaded Jahnke's priority rights. He wrote a cool, polite, but unmistakably provocative letter:[18]

> From your letter, kindly transmitted to me by Professor Korteweg, and the enclosed papers, I see that my treatise interests you, and that earlier investigations of yours are connected with it.

Brouwer went on to explain in some detail to Jahnke the contents of Jahnke's and his own papers:

> The reading of your papers suggested the following remarks, which will certainly be plausible to you.[19]

And after spelling out the geometrical meaning of Jahnke's method (or rather, the lack of it) and of his own method,[20] he closed with:

> Thus I hope to have shown that our papers under discussion have nothing in common, but that your final result is a by-product of my principle.

Considering the provocation, Jahnke's reaction was rather mild; he demanded from Korteweg the publication in the Proceedings of a rejoinder from his hand. The latter had more faith in his own pupil's insight than in that of his German colleague; he promised Jahnke a note in the Proceedings, while at the same time asking Brouwer to write an exposition of the matter for the Proceedings. Jahnke's note[21] shows that he had still not grasped the geometrical meaning of the decomposition that Brouwer had obtained, but the affair ended quietly with Brouwer's algebraic derivation of the results.

Among Brouwer's papers there are some notes that comment on Jahnke's papers in a rather cutting way, he compared him to a man 'who has stumbled around without detecting anything but small traces, and who now sees that the thing itself has been found and that his traces have lost their value. Hence his hasty and anxious letter.' He went on:

> – *A* discovers that somewhere everything behaves just as in a magnetic field, and he even discovers that this field is remarkably simple. *B* finds the magnet and a very simple one at that; and says 'the matter is so and so'. Now *A* would not raise a priority claim against *B*, would he? At best one can say that Jahnke's researches suggested that there were two $R_3$'s. I have indicated those $R_3$'s (and not bothered with their properties, which Jahnke presents in full).

[18]Brouwer to Jahnke, 20 March 1904.

[19]The contents of the letter are incorporated in Brouwer (1904).

[20]See Freudenthal's commentary, CW II, p. 22 ff.

[21]Jahnke (1904).

– I would say: if a man finally deduces by means of boring observations, which rest on, and are combinations of, equally boring observations of a predecessor (Caspary)[22] *remarkably* simple results from all those complicated things, and finally somebody comes along and says that no complicated things are going on, but that something very simple is the matter, then he is ashamed and he withdraws himself. At least he does not raise a priority claim.

– When Newton found the law of attraction, and deduced the laws of Kepler from it, Kepler would not have wanted to diddle the credit from him.

In spite of all his self-assurance, the fledgling scientist showed the hesitance that every beginner has experienced: Brouwer asked Korteweg how to send out his preprints;[23] he did not know how to find the addresses of mathematicians he only knew by name. 'And furthermore, can I send a copy to people like Klein and Veronese? Or would that be presumptuous without an introduction?'

Brouwer was enrolled for almost 7 years, not exceptionally long in those days, but nonetheless too long for such a promising student. So, what kept Brouwer so long at the university? The blame cannot be put on the fraternities, for after his first year Brouwer scarcely frequented them. The cause was rather the military service; according to his own statement[24] his studies were interrupted for a good two years. We have already seen how Brouwer was drafted (cf. p. 20), and his dislike of the experience. Indeed, his physical and mental health was seriously put at risk. The military service put so much stress on him that at times he felt utterly desperate.

The actual time spent in the army did not exceed eight months, but each bout of military training apparently upset Brouwer so much that the recovery required time. The sensitive young man must have experienced military service as a kind of hell; it was not that he could not, or would not, cope with the physical hardships, as we have seen he had always enjoyed a good dose of rough soccer, and long marches did not tire him. It was rather the company that fed his distaste for the army. Even years later, during the First World War he recalled his national service with a shudder:

> My past service-time with the infantry is the darkest page of my life; from my equals I got little more than hatred, from my superiors little more than teasing and opposition; I repeatedly failed the examination for subaltern, and the consequence has been that after my military service for one and a half year I had a nervous disorder, and was not able to work; I recovered from it only very slowly.[25]

This letter, combined with the information of the Brouwer–Scheltema correspondence, confirms the picture of Brouwer as a man extremely susceptible to stress. The antique military establishment, not exactly known for its rationality or open-mindedness, clearly was not the environment to cherish an unorthodox—and proba-

[22]Caspary (1883).

[23]Brouwer to Korteweg, 14 May 1904.

[24]Brouwer to Scheltema, 15 November 1903.

[25]Draft of a letter to Lorentz 16 February 1918.

bly contrary, character like Brouwer. The result was a prolonged physical and mental breakdown. Only in November 1903 did Brouwer get into the rhythm again. He wrote to his friend Scheltema:

> Of course you have excused me for remaining silent for so long. I have been busy; returning to my subject after two years of absence required some dedication, in particular where any love for that subject was missing. By now I have gradually succeeded, and I row with long strokes towards my doctoral examination. My work is done without illusions, but with a feeling of cheerfulness on account of the activity itself.

Once he had resumed his study, the subject matter offered no problems. One guesses that his publications on four-dimensional geometry could not have failed to impress his examiners. The doctorandus-diploma was awarded on 16 June 1904, (cf. p. 33), and—as a mark of excellence, with the predicate 'cum laude'! The diploma was signed by A.J. van Pesch and D.J. Korteweg.

It did not take Brouwer long to make up his mind on the matter to his future activity. He soon informed Scheltema of his plans to start to work on a dissertation.[26] He planned to absent himself for some time, perhaps for a longer period, in order to

> recover the clear relations, in which I have to position myself *vis à vis* the various persons and institutions within my narrow social horizon, in order not to be distracted from the cultivation of my power and the development of my clairvoyance in the service of God.

Although the scales were by now definitely tipped in favour of mathematics, philosophy was still prominent in his mind:

> Next winter I will be either in Blaricum—where a cottage[27] is being built for me—working at a philosophical creed, that will be the prologue of my work—or in London, in the great British Library for my dissertation: 'The value of Mathematics' with the motto *Ουδεις ἀγεομετρὶκος εισιτω*.[28]
>
> I thank you for your well-meaning admonition to me at the gate of the paradise of freedom. Did I wish a kingdom on earth, then it would perhaps be good to wall in myself in mathematics, and to have me crowned like a pope in the Vatican, a prisoner on his throne. But I desire a kingship in better regions, where not the goal, but the motive of the heart is the primary thing.

## 2.3 Marriage

This exalted message was followed, only six days later, by a letter that contained a short but weighty message:

---

[26]Brouwer to Scheltema, 4 July 1904.

[27]See p. 59. This cottage was referred to as 'the hut'.

[28]Let nobody enter without the knowledge of geometry (Plato).

> Carel, my friend, in my life a thaw has set in. I have exchanged marriage vows with Mrs. R.B.F.E. de Holl. Greetings and hail to you, comrade
>
> Your Bertus

Whereas from our modern point of view there is no reason why a student who has only just finished his final examinations should not get married if he or she wishes to do so, in the old days marriage was not an affair to be rushed into. Prudence and tradition required a couple to save a substantial sum, and for the bride to collect a complete trousseau, before one could even contemplate matrimony. Middle class morality, in particular, was quite strict and specific in matters of engagement, marriage, children, etc. Handbooks of etiquette spelt out all the rules to a nicety; the number of sheets, teaspoons, ... was precisely indicated.

The fact that less than a year ago Brouwer still viewed himself as the eldest son, destined to remain without offspring, either shows that nature is stronger than theory, or that in Brouwer's view marriage did not necessarily entail procreation. It may be remarked here that the marriage remained childless, so the prophecy was fulfilled in spite of the marriage.

There are only hints in the correspondence of Brouwer and Scheltema concerning the other sex. In view of the long-standing tradition of fraternities to introduce students to all aspects of life, including those of the flesh, it is not unlikely that either Scheltema himself, or one of the other members of his fraternity companions, took Bertus' practical education in hand. A visit to one of the traditional establishments may not have been an obligatory part of fraternity life, but it was not actually frowned upon.

References to females are scant in the Brouwer–Scheltema correspondence. At the time of his depression in 1902, Brouwer wrote to his friend, after discussing Carel's recent volume of poetry, *Of sun and summer*, that 'there is a richer soul in your poetry, than in that of our modern poets'. These words were followed by an urgent and heartfelt counsel:

> ... Now, yet, a woman for you, Carel, you will not reach your destination before that; she will open up so much with her magic wand, I felt that again only just now thinking back to last year. I, too, feel homesick for the arms of a woman and a kiss. I would, even a month ago, never have thought that this could haunt me so much. Yes, it is really easily said that a man should live with his reason; a burden of thousands of years sits in mind and body, which compellingly shows him what to strive for, how he has to be.[29]

Scheltema calmly replied that women had no place in his life as an artist:

> ... you forget that I love my own art above all and that my life with it, and my fighting against it, is the life with, and the fight against, the powerful muse—the most cruel of all women. [...] And the muse does not tolerate any love but the love of friendship—and in that I am rich enough!

[29]Brouwer to Scheltema, 11 June 1902.

**Fig. 2.4** Lize, before her marriage to Brouwer, in the garden of the pharmacy. [Brouwer archive]

In the same letter he turned the table on Bertus and called on him to find a woman:

> —be it either to the act of pairing that gloriously relieves me like a bath—or the love of her, of whom you had often spoken, and who will certainly want you back. Maybe she will now be able to help you better than anybody else!—But in this case don't act without Huet or your present doctor...[30]

Scheltema apparently referred here to some earlier relation of Brouwer, which might be hinted at in the letter of 5 December 1901: 'In order to quell a growing tragedy, I have to get away from here.'

Brouwer's wife-to-be was Reinharda Bernardina Frederica Elizabeth de Holl (Lize) born on 5 August 1870, eleven years Brouwer's senior. She was the daughter of Eelbartha Johanna Jacoba de Holl-Sasse, widow of Jan de Holl. The latter was a medical doctor, who had his practice on the Overtoom in Amsterdam. He died young and left his widow with 7 children. He also left her the pharmacy which he had run together with the medical practice. Mrs. De Holl had decided to keep the pharmacy; according to the regulations she had to hire someone with the proper qualifications in order to guarantee the required expertise.

Lize had married Hendrik Frederik Peijpers, a former army doctor and, incidentally, a full cousin of hers, when she was still a young girl. Peijpers was sixteen years

[30]Scheltema to Brouwer, 12 June 1902.

her senior; he had first been working in the pharmacy for his aunt, the widow De Holl, with whom he was lodged. The marriage was far from happy, Peijpers did not want any children, and did not hesitate to carry out an abortion if and when Lize became pregnant. When she was once more pregnant, Lize managed to circumvent her husband's intervention, and the child, Anna Louise Elisabeth, was born on 26 March 1893. Soon after that Lize got a divorce and in the meantime she and Louise had moved in with her mother. It should be kept in mind that in those days a divorced woman with a child was in an extremely uncomfortable position. Her social status was far from enviable.

There are no elaborate accounts of Brouwer's courtship,[31] but the following story, as told by Louise, is undoubtedly authentic.

Bertus had renewed his friendship with a girl he knew from Medemblik, the earlier mentioned Dina Pels.[32] Dina, who was somewhat older than Bertus had found herself a place in the pharmacy of the widow De Holl, where she combined a full-time job with a training as pharmacy assistant.

The two made long walks and exchanged their experiences (a salient detail, reported by Louise, is that Bertus insisted on carrying Dina's purse, to return it safely at the end of the walk). Dina told about the proprietor of the pharmacy, the widow De Holl, and the routine at the shop. She also mentioned that a young divorced daughter with her child lived with Mrs. De Holl. Bertus' interest in the daughter and the pharmacy was soon aroused, and he devised a plan of action in the best romantic tradition: he climbed onto a roof in the neighbourhood of the pharmacy at the Overtoom,[33] in order to watch the object of his curiosity. The inspection must have led to a favourable conclusion, for a meeting was arranged, and on 10 July 1902 the two met. What Lize thought of this curious student is not known. It is reported that she had her doubts about the wisdom of marrying a man eleven years younger than herself. Brouwer, anyway, did not hesitate long. He marshalled all his charm and power of persuasion to win the heart and hand of the young divorcee, whom he often and fondly praised for her Memlinck[34] face. The campaign was successful. In spite of the negative advice of some of her friends, Lize accepted Brouwer's proposal. The two formed a striking couple, Bertus, over 1.87 meter tall, towered over Lize, who measured no more than one-meter-fifty. Two years later the marriage took place, two months after Brouwer's doctoral examination, on 31 August 1904—the birthday of Queen Wilhelmina (*koninginnedag*).

And so Dina Pels played a brief but decisive role in Brouwer's life. She later married a medical doctor in Alkmaar by the name of Formijne. She died at the age of fifty. Brouwer kept up the relationship with the Pels family—among the congratulations on the fortieth anniversary of his doctorate in 1947 there was a letter from

[31] Most of the information on the courtship and the marriage is from oral communications of Louise Peijpers. Confirmation of the information on Dina Pelswas provided by Mrs. J. Schout-de Waal.

[32] See p. 6.

[33] In other accounts of the same events the roof is replaced by a tree.

[34] Fifteenth century Flemish painter.

a member of the family. In 1953, on his tour of the United States Brouwer visited one of the Pels relatives, who had emigrated to the States. It may be stressed that Brouwer had a very strong loyalty to his friends and relatives; there are numerous testimonials to this fact. He always enjoyed a quick visit to those whom he had admitted in his personal circle, irrespective of status or gender. On the other hand he had no patience with potential bores; his motto was 'if you won't have anything to do with someone, pick a fight right away—it saves a lot of time'.

The marriage was a simple affair. Bride and groom took a streetcar to the city hall of Amsterdam, where the civil wedding took place. Both brothers Lex and Aldert had, on this special occasion, accompanied Bertus to the house of the bride; the three of them played leap-frog on the way.

At the dinner, which followed after the official part of the wedding, the uncle of the bride, Reverent De Holl, held a diatribe on the topic of the marriage of two students—one of them, moreover, the mother of a child!—who did not have a penny, and who had nonetheless commissioned the building of a house for themselves, unscrupulously borrowing money![35] The twelve year old Louise quietly snatched some dainty morsels from the table and stole away. In her memory the atmosphere at the dinner was stifling.

Scheltema, who in some way took part in the wedding, spoke bitterly of the occasion:

> I have not understood anything of your wedding and in particular of the embarrassing ceremony, and the, for me insulting, invitation of Poutsma[36] and the whole collection of people, one and all, that I found disgusting! Really Bertus, that day was a great sacrifice for me! I have made it without demurring because you insisted, and seemed to have your reasons...

After the marriage the young couple moved into the rooms over the pharmacy at the Overtoom, waiting for the completion of their Blaricum cottage.

The daily routines of the couple were not much changed by the marriage. Brouwer had started to work on his dissertation, and Lize was fully occupied as a pharmacy student. Mrs. de Holl had, as we said before, no licence to run the pharmacy. Therefore she had to hire a licensed pharmacist to run the professional part of the shop, although the management remained in her hands. Since the salary of such a *provisor* (as he or she was called) presented a serious drain on the finances of the pharmacy and the family—not to mention the space problems when a provisor happened to live in—Mrs. de Holl had considered the possibility of preparing Lize for the supervision of the pharmacy, with the intention of eventually letting her take it over. As a result Lize had enrolled as a student in the University of Amsterdam, and diligently studied pharmacy. Roughly a year later the pharmacy did indeed change hands and became the property of the young couple, cf. p. 194.

[35]Cf. the letter from Brouwer to Scheltema, 4 July 1904.

[36]Presumably this was one of the uncles, a teacher at the Barlaeus Gymnasium.

## 2.4 Bolland's Philosophy Course

Brouwer, clearly, was of two minds about his future scientific career; as we have seen above, he had not yet made a definite choice between mathematics and philosophy. If the text of his profession of faith had not been preserved, this would have come as a complete surprise, for there was at that time no other visible sign of interest in, or familiarity with, philosophy. The letter of the fourth of July to Scheltema underlines that, notwithstanding his early success in mathematics, Brouwer was totally serious about the role of philosophy in his work. This interest in philosophy explains his involvement in the following short episode, which was of such significance, that it was the immediate cause of his mystical-philosophical monograph *Life, Art and Mysticism*, cf. p. 64.

Philosophy in Holland, around the turn of the century was largely dominated by G.J.P.J. Bolland, a man with an spectacular career. He was basically a self-taught philosopher, a fast learner and an even faster user of new knowledge, a powerful protagonist in every sense of the word. After a colourful but tragic youth, he became a teacher in the Dutch Indies. Absorbing in a high tempo the philosophy of predominantly German thinkers, he soon developed into a formidable character in Dutch philosophy. His life looked like one long series of conflicts, most of which were of his own making. He acquired a certain notoriety by his extreme anti-Catholicism. In spite of his curious reputation, he was appointed to the chair of philosophy in Leiden, where he preached the philosophy of Hegel. Bolland was one of those legendary professors who tyrannised his audiences, his students, and probably his colleagues as well. He could, at his lectures, request specific members of the audience ('that person with the unborn face') to leave the hall because 'otherwise I cannot do my work'.[37] Nevertheless, he was a popular and inspiring lecturer, who did not hesitate to give his opinion on any subject.

His reputation as a forceful and interesting speaker made him much in demand for courses and talks. In view of Bolland's success at other universities, the Amsterdam students decided to invite the great man for a series of lectures. Brouwer joined the committee, and he soon was the major force behind the invitation.

On 15 January 1904 Brouwer wrote his first letter to Bolland. Even Brouwer must have felt some awe, for the letter is unusually timid. Bolland's part of the correspondence has not been preserved, so we can only try to interpolate his reactions. Apparently he consented to give the lectures, but not without his conditions. From Brouwer's third letter[38] we gather that only those participants were welcome who had bought Bolland's book 'Pure Reason'.[39] The letter shows that Brouwer was seriously worried about the size and the quality of the prospective audience:

> A great concentration of 'serious' listeners has not been secured. Many joined at first, who 'wanted to hear Bolland' for pleasure, but already the

[37]In this respect Bolland was not an exception. Cf. Wiessing (1960), p. 241.

[38]Brouwer to Bolland, 5 March 1904 (Boll. B 1904, 29. Leiden University Library).

[39]*Zuivere Rede*, Bolland (1904).

> condition of the purchase of the books—which was seen, in addition to the assumption of a considerable advance knowledge, a demand for activity, instead of pleasure—caused the number of prospective participants to dwindle to some twenty-five.

In spite of these discouraging messages, Bolland agreed to give his course. Brouwer went so far as to invite Scheltema to Bolland's lectures, although Bolland was a notorious anti-socialist. Brouwer told Scheltema that 'I so often heard the voice of Dietzgen in his words last night that I have to let you know'.[40] Scheltema resolutely declined the invitation: 'it must be most unpleasant to hear the Hegelian philosophy from this abominable man'.

The lectures were far from successful, as one can read in the student magazine *Propria Cures*. Whereas everywhere else Bolland was greeted by his audiences with hardly suppressed awe, the Amsterdam students shared no inclination to flatter Bolland's ego. When he made his entree he was met with smothered laughter and giggling. Bolland reacted predictably. He mercilessly attacked his audience—'you are nothing special, and I don't expect anything from you...'

*Propria Cures* reported the lectures in some detail; the issue of 15 October contained an introduction to the thoughts of the master in tones that echoed some of Brouwer's ideas (actually, he may have been the writer, but there is no certainty about that) including some mocking remarks about science, socialism and females, followed by a panegyric and a satire.

The lectures continued for some time, but the atmosphere in the lecture hall did not meet the speaker's expectations. For one thing, Bolland was used to be treated with reverence; he did not suffer from modesty, and he considered himself the most important philosopher in Holland, and so he expected his audience to treat him accordingly. His handling of disagreeable listeners (and it did not take much to be considered thus) was crude but effective. If the manners or behaviour of members of the audience were not to his liking, he could suddenly interrupt his lecture—'If those foetuses will not dispatch themselves, I cannot allow myself to continue', or words of a similar import.

He characterised his Amsterdam audience in various unflattering ways, and finally gave up the course, writing that he could only present his 'Collegium Philosophicum' to an audience of 'students' and not to one of 'spectators'. The *Propria Cures* copy of 16 December 1904 published Bolland's letter of cancellation, which (quite correctly) stated that the 'Collegium philosophicum' was offered to a circle of 'students', but not as a public performance for a fee, let alone for free. He definitely refused to lecture for an Amsterdam public that wanted to be amused and that did not even bother to procure the obligatory book, as they showed by their empty hands.

The issue of the 'Bolland Lectures' continued for some time to occupy the columns of *Propria Cures*; a number of comments and reader's letters were published that in turn defended or attacked the great man; there was even some excitement in Roman Catholic circles on the assumption that Bolland had abused the

[40] Brouwer to Scheltema, 8 October 1904.

catholic hearers or vice versa. Brouwer also took part in the discussion: he published a number of notes under the pseudonym *Lau van der Zee*.[41] In his contribution with the title 'Grounds of consolation' he concluded, after a flowery and convoluted commentary on Bolland, that 'Bolland's lectures have stopped at the point, where his path and the paths of his audience who were about to liberate themselves, parted... And thereafter he stayed away, just in time.'

Brouwer's relationship with Bolland is far from clear. The philosophical tradition in Holland eventually turned away from, and even against, Bolland. True, there was a hardy band of faithful followers, who devoted their undivided loyalty to the works and thoughts of Bolland, but gradually they were outnumbered and eventually forgotten.[42] Slowly the name 'Bolland' became a synonym for 'weird and unscientific', hence the present generation can no longer imagine the spell that Bolland cast over Holland. But anybody who takes the time to peruse the books of Bolland will find quite sensible thoughts (next to obscure passages). There is, for instance, a small monograph, *Intuition and Intellect*,[43] which contains quite sensible ideas, next to unfounded speculations. In particular, it showed that the mathematical layman, Bolland, was not as uninformed about mathematics as later commentators suggested.

Comparing Bolland's and Brouwer's writings, one can see that they share certain ideas, but whereas Bolland's text may be compared to the confused sounds of an orchestra that is tuning, Brouwer's philosophy is the crystal clear music of a transparent symphony. At some point, Brouwer studied Bolland's writings, together with P.C.E. Meerum-Terwogt, a contemporary of Brouwer, who had become a Bolland follower. He also visited the master personally.[44] There is no doubt that Bolland acted as a catalyst for the young Brouwer, but once the latter had become a philosopher in his own right, with his own programme, Bolland was no longer of any influence.

Nowadays there is little appreciation for Bolland's philosophical views, but during his life he exercised a considerable influence on his followers and adversaries. Almost all Dutch philosophical publications in the beginning of the century in one way or another paid a tribute to the recognised master and tyrant of philosophy.

This short episode is of some importance, as it shows that Brouwer was actively involved in matters of philosophy. He did not just want to dabble in philosophy, but wished to pay serious attention to the developments in that field. Already in his short observations in the student magazines one can discern his own private views on the basic issues of (in particular) moral philosophy. In the *Propria Cures* issue of 19 November 1904 Brouwer published under the name Lau van der Zee, a short

[41] One of the lodgers of the Brouwer family went by the name of Lau van der Zee, cf. p. 8. In the Brouwer archive there is a manuscript in Brouwer's handwriting, signed by Lau van der Zee and subsequently published in *Propria Cures*. So the identity of this 'Lau van der Zee' is well authenticated.

[42] The reader may find more about Bolland in a recent biography (Otterspeer 1995) (Dutch).

[43] Bolland (1897).

[44] Communicated by Mrs. N. Kapteyn-Meerum-Terwogt, the daughter of the above mentioned Meerum-Terwogt. No details of the conversation between Bolland and Brouwer are known.

note *On Morality (Excerpt)*. This note is almost a short preview of Brouwer's later lecture series *Life, Art and Mysticism*.

The main theme is the loss of the original innocence—'In passivity the world is a garden of marvel and joy and silence. There is no separation, no reality and one wants nothing.' The loss of this primordial equilibrium, in Brouwer's view, is caused by man's concentration on certain phenomena, the active directing of attention. 'The source of *All* is lost, one has been born.' Brouwer then goes on to indicate the modes of liberation, that is means to regain paradise. These modes are, strangely enough, distinct for man and woman. 'Moral' is to be found in the quest for the lost primordial state. It is remarkable to note that the female road to liberation is rather negative compared to the male one. This short note can, like the profession of faith, be seen as an overture to the mystical-moralistic book of 1905.

## 2.5 Among the Artists and Vegetarians

The year 1904 was an eventful one. Not only was it the year of the doctoral examination and the marriage, but also the year of Brouwer's settling in Blaricum, a small town (village) not too far from Amsterdam. The Brouwer–Scheltema correspondence contains a number of glowing references to Blaricum and its general surroundings. Brouwer evidently was infatuated with Blaricum, and in order to appreciate this phenomenon, we have to take a closer look at the town and the surrounding area.

Blaricum had been a desperately poor village populated by farmers and shepherds. Its soil was sandy and could not be expected to yield more than a scant crop.

At the end of the nineteenth century things started to change, the more affluent citizens of Amsterdam had discovered the charms of country life in an area where prices were still reasonable, and where the air was clear and healthy: the commuter had been born. At the same time *het Gooi*, a geographic unity, comprising Laren, Blaricum, Bussum, Huizen etc. attracted a number of artists. Somewhere in the eighteen-seventies the painter Jozef Israëls discovered the picturesque charms of Laren; in his wake more artists followed, Johannes Albert Neuhuys settled in Laren in 1883, and in 1886 he was joined by the painter Anton Mauve. The latter became famous for his paintings of the landscapes of Het Gooi, the flocks of sheep with their shepherd, the heath, and the small farmhouses and huts of the local population. He was so much identified with 't Gooi that one spoke of 'the land of Mauve'.

Gradually Laren and the neighbouring Blaricum became a well-known centre for painters; the list of resident painters of whom some had a more than local fame contains too many names to include them all. We must be content to mention a few: Jacob Kever, Frans Langeveld, Wally Moes, Jan Veth (who was an author as well), F. Hart Nibbrig, Arina Hugenholtz, Evert Pieters, F. Oldenvelt, Willem Dooijewaard, William Singer, Herman Heijenbrock. In the world of painting, Laren became known for its Laren School (*Larense School*). A special role in the history of Het Gooi was played by William Singer, the son of the American steel giant William

Singer from Pittsburgh. He had chosen to become an artist and to forego his rights as a successor to his father's steel industry. After some wandering through Europe, he alternatingly lived in Laren-Blaricum and in Norway. In Blaricum a magnificent house was built for him, which later became the town hall. After the Second World War, the widow of William Singer donated the funds for the founding of a memorial foundation and for the Singer Museum, which now attracts art lovers to Laren.

Art was vigorously promoted in Laren by the enterprising hotel keeper, Jan Hamdorff. No account of Laren would be complete without the mention of this enterprising individual, who governed the local art world as a benevolent autocrat, with a keen eye for the interest of his artists and of himself.

Much later Mondriaan and Van der Leck worked for some time in Laren. Laren and Blaricum not only attracted the adherents of the visual arts, but also considerable numbers of the Dutch literary society spent a part of their life in the idyllic villages and in the neighbouring towns. The poet Herman Gorter, the authors P.L. Tak, Frans Coenen, Victor van Vriesland, Carry van Bruggen, the couple Henriette and Richard Roland Holst lived in Het Gooi, and last but not least the famous, but somewhat controversial author, psychiatrist, philosopher, philanthropist Frederik van Eeden. We will meet the latter again in connection with the so-called 'significs'.

At roughly the same time there was an invasion of a totally different kind: the advance of the communes and of the health fanatics. A number of these communities, usually called *colonies*,[45] founded on idealistic, mostly socialistic and/or religious bases, have made history; they have influenced life in het Gooi to no small degree, although nowadays they are considered just a curiosity in the local history of the region.

The best-known colony, was *Walden*, founded by the above mentioned Frederik van Eeden (1860–1932). Van Eeden was well-known, and not only in the Netherlands; he had studied medicine and through his own efforts he had become the first psychiatrist in Holland. He was a sensitive man, the author of a number of books, dramas, and poems, with a keen social conscience, rejecting, however, Marxism as an acceptable basis.[46] His colony was named after Walden Pond in Concord, Massachusetts, where Thoreau carried out his famous experiment. Thoreau's book had fired Van Eeden's socio-romantic imagination. In 1898 he bought some land in Bussum from an ex-patient, and started to master the practical and theoretical problems of running a colony.

Following Henry Thoreau, he had a cottage built for himself, followed by a number of cottages for members of the colony. In 1899 Walden opened its doors. It attracted a mixed group of people, consisting of a number of disciples and patients of Van Eeden and some farmers. Walden suffered from the usual defects; idealism and love of mankind are no substitutes for organisation and leadership. Van Eeden, who was often absent, was not cut out to be the practical leader of a group of colonists. The enterprise was a financial disaster, in particular for Van Eeden. In 1904 the colony, as an official institution, ceased to exist.

[45]*kolonies*. Cf. Boersen (1987), Heyting (1994).

[46]There is a two-volume biography of Van Eeden (in Dutch) by Fontijn, cf. Fontijn (1990, 1996).

The project had, however, caught the imagination of the Dutch people. Walden became something like a catch word; socialists and communists condemned the idea on principle, and ethical idealists treasured its memory—much as the true romantic adores ruins.

The second movement was centred around another charismatic personality—Professor Jacobus van Rees. Van Rees was the son of a social-liberal historian, Professor Otto van Rees, and the father of the painter Otto van Rees. Inspired by Tolstoi, he became a religious-anarchist, active in the fight against military service, the killing of animals, alcohol and tobacco. In 1899 he helped to found the colony of the International Brotherhood in Blaricum. The colony was an agriculture enterprise, which functioned for a brief period. A lack of expertise, discipline, and the barren soil eventually finished off the colony.

The original inhabitants of Blaricum and Laren were not altogether pleased with the presence of what they called 'reds' and 'grass eaters' (*plantenvreters*). The colonists dressed and behaved in objectionable ways, and they were, in the eyes of the hard-working indigenous population, lazy and ignorant. Nonetheless, there was a good deal of tolerance, after all, the communes brought the shopkeepers business. Violence only erupted during the big railway strike in 1903, when it was rumoured that the colony people would stop the steam train (*Gooische Stoomtram*), the connection of het Gooi with Amsterdam.

The adherence to the fundamental Christian-anarchist principles of the 'International Brotherhood' gradually eroded, and even more so after a number of Friesian socialist farmers had joined; they clearly wanted to combine socialism with successful farming—'they needed livestock for dung and they had bought along guns to shoot the rabbits that ate the cauliflower'. Eventually the Brotherhood was dissolved in 1911.

Already before that time, Brouwer had bought a strip of land from Professor Van Rees, situated along the Torenlaan in Blaricum, and it was at this spot that Brouwer, while still a Ph.D. student, had a hut built. It was designed by his friend Rudolf Mauve, the son of the painter Anton (see p. 57). At the end of October 1904 the wooden cottage was ready to receive its occupants; it was a charming construction of a modest size, basically one room plus a kitchen, and a bedroom upstairs. It had a thatched roof and was situated in a wooded lot. The location exactly answered Brouwer's dreams; a secluded spot in the middle of a romantic landscape. The privacy was later increased by an enclosure of rush-mats.

In the seclusion of his private domain Brouwer enjoyed the pleasures and rites of a healthy life. He practised a number of traditional health activities, such as vigorous exercises to improve the circulation of blood and oxygen, open-air baths, mostly in the nude, sleeping—weather permitting—in the open air, swathed in wet sheets.

The eating habits and the food were subjected to a strict regime. In this respect the couple were well-matched: Bertus and Lize both practised vegetarianism. Lize was, as all sources confirm, well-informed about vegetarian diets, traditional herbal cures, and the like. The marriage of Brouwer certainly could not have been more felicitous with respect to his lifestyle. Lize was known in the village for her knowledge of herbs and as a pharmacist she used to prepare bottles of herb cures; even in old age,

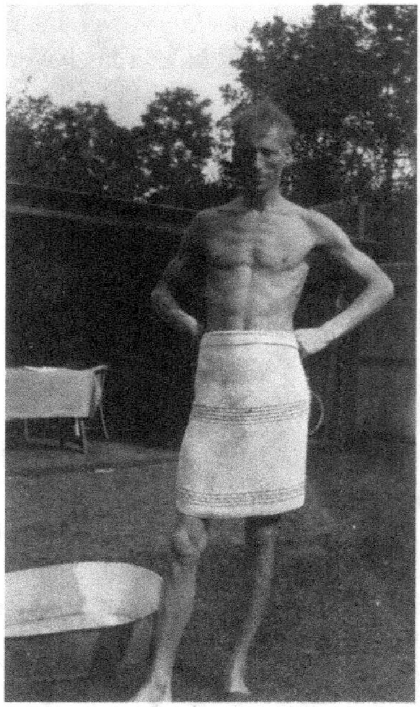

**Fig. 2.5** The healthy life in Blaricum. Brouwer in his garden. [Brouwer archive]

she could be seen stirring a huge cauldron with a brew of all kinds of herbs. Under her guidance the pharmacy also dispensed homeopathic medicines.

The more irreverent could not resist the temptation to associate the earnest, tawny old lady, going about her business of preparing potions of herb drinks, with the dark images of the old fairy tales.

Bertus had acquired his knowledge of the vegetarian kitchen in a German health clinic of Doctor Just, in the town Jungborn in the Harz. He had been a long-time sufferer of complaints of the nose; his father saw this as the cause of his son's long drawn out studies, and he ordered the boy to have his nose treated, but after it was flushed, things became even worse. The doctor proceeded to prescribe him seven goose eggs, daily, and one and a half pounds of steak, in order to shore up his general condition. The result was that Bertus felt more sick than before. Finally he went to Just's health clinic, where he was introduced to the secrets of diets, vegetarianism, open air baths, exercises, etc.[47] He adhered to the vegetarian diet the rest of his life, albeit for pragmatic reasons. He was not dogmatic enough to resist the temptation of an occasional bite of meat or chicken.

Among Brouwer's papers there are some notes that illustrate the eating habits of the Brouwers. An undated list, probably from the early years in Blaricum, gives detailed instruction for the daily diet; there is an enumeration of wild herbs for each month, for example, 'April: scurry grass; lady's smock; wild sorrel; stinging-nettle; dandelion; plantain; lamb's lettuce; onions and the like; carrots, lemons, ... September: acorns; beech-nuts, cabbage-lettuce, cucumbers; endive; onions c.s.; French

[47]Oral communication Louise Peijpers.

**Fig. 2.6** Bertus and Lize in their Blaricum garden. [Brouwer archive]

beans; lemons....' The procurement of food products was a matter of serious consideration. The choice of rice, for instance, was not left to chance: 'brown rice (Van Sillevolt, rice-huskers, Rotterdam)'. The note gives general rules, based on a list of very detailed instructions for the choice of fruits and nuts, arranged per day for the various seasons. In combination with Brouwer's adherence to vegetarianism, the items give some insight into his daily routine. The self-chosen Spartan lifestyle is illustrated by the following rules:

> – In case of momentary fainting one takes according to one's need a juicy fruit, or milk, or milk and bread.
> – No more departures [of the rules] allowed as a guest.
> – No more cleansing baths.
> – Never eat by artificial light.
> – Always: Once per week swimming in the open.
> – Once a week play football or practice another intense physical exercise (preferably with danger and fights).
> – To bed only after fasting for 3 hours.
> – In case of fever immediately the fruit diet.
> – Sleep at least from dusk to midnight.
> – Rise as early as possible.

The reader may perhaps wonder if Brouwer tackled the food matter in dead earnest. It should be borne in mind that his student years were one prolonged misery of medical problems. He had, clearly, decided to fight the physical weakness of his body by a systematic regime. And it may be said that the method proved successful. Although he had his breakdowns and illnesses from time to time, he boasted a wiry, lithe body without any trace of fat.

The whole atmosphere of Blaricum and Laren and the congenial housing in the hut must be viewed as the ever changing and yet permanent background of

Brouwer's life. No offers from famous universities were able to uproot him. Blaricum was his irreplaceable home. After his fame had spread, he did not even have to leave the village; the established and the newcomers came to knock on the door of his hut.

## 2.6 The Delft Lectures

During the years 1904 and 1905 Brouwer suddenly displayed an interest in the cultural side of student life in Delft. He published a number of short notes in the Delft student weekly, and ended by giving a series of lectures.

His first note was a short comment on Frederik van Eeden's book 'The Joyous World'.[48] In this note Brouwer opposes the view that an improvement of the economic circumstances will result in a morally and ethically better world, 'A bad father beats his child; to improve the father by anaesthetising the child, and thus taking away the pain, proves hopeless. Therefore it would be necessary to raise the ethical level, then, slowly, the Joyous World will grow. And for that reason the book of Van Eeden is the deathblow for any Marxist.'

The note on Van Eeden's 'Joyous World' was soon followed by an ecstatic exhortation to attend Bolland's lectures, which were to be held in Delft.[49]

As we have seen, Brouwer was involved in the organisation of Bolland's lectures in Amsterdam. Although Bolland's performance was not exactly successful in the nation's capital, Brouwer saw no reason not to promote him in Delft, where an active student association had a tradition for organising cultural events.

Brouwer send an exalted letter, under the pseudonym of Lau van der Zee, to the student weekly:

> ... Then a shining star will appear, the joyful sign of hope, then you will rise to higher regions, to God's glory, although this will not be attained in this life.
>
> Bolland sees that star clearer than you will learn to see it; the veils disappear before his eyes. And thus he can lead you on the difficult shining way to God's throne.

The note was (correctly so, one might say) judged incomprehensible, so that in a next issue an article appeared in which the author calmly outlined arguments for the role of philosophy in Delft, adding that 'We cannot forego the occasion to show the author of this exhortation, Lau Van der Zee, our appreciation for his laudable efforts, although we consider his way of operating most dubious.'

It is certainly surprising that Brouwer showed such an interest in the promoting of philosophy in Delft. There are two partial explanations: in October 1904 he was still making propaganda for Bolland in Amsterdam, and he may have decided to give

[48] *Studenten-weekblad* 6 October 1904.

[49] *Studenten-weekblad* 17 November 1904.

his fellow organisers in Delft a helping hand, and furthermore he had connections in Delft. In fact, his brother Aldert had registered as a student in Delft in 1903.

There is a third possibility: Brouwer was preparing a philosophical exposition ('a philosophical creed that will be the prologue of my work', cf. p. 49) as a part of his dissertation. It is not unlikely that it did not take him long to realise that the material was not exactly suited for a faculty of mathematics and physics. As we will see (p. 92) even a modest philosophical motivation did not survive the axe of his adviser. So he may well have considered Delft as a suitable platform for his philosophical message. Any of these reasons or a combination, may have sufficed for his involvement. It is not known how Brouwer got himself invited to give a series of lectures in Delft. It seems plausible that his brother introduced him in the local student association *Free Study* (*Vrije Studie*), and that Brouwer sufficiently impressed the governing body to get himself booked for a series of lectures, the first of which was held on 29 March 1905.

The lectures were well attended and apparently quite successful. The organisers were somewhat sceptical about the text that was going to appear in print—'The oral lecture was necessary indeed for a proper understanding, I think that in print much will seem incomprehensible.'[50]

Brouwer, although no doubt seriously trying to get his message across, could hardly suppress a quiet amusement at the behaviour of the audience (which, by the way, listened patiently for 3 hours(!) on end). He reported to his friend Scheltema that:

> The lecture will be printed at the request of the Delft public. When I send you the booklet, you will read it, won't you, and you will not lay it aside unopened in fear or disgust? Why then didn't you mix with the public as a solitary, darkly watching enemy among all those others, who were either stupidly frightened or admiring, or did not understand, or got angry. If you had seen how a couple of girls, in the second break, cried that they could not bear it any longer and demanded to be taken home, your nostrils would have flared, you would have snorted of hatred.[51]

The promised book itself was written and produced at an incredible speed, the publisher already advertising it in the weekly of 8th of June:

*BROUWER. Leven, Kunst en Mystiek.*

with the chapters

I The sad World
II Introspection
III The fall caused by the intellect
IV The reconciliation
V The language

---

[50] *Studenten-weekblad*, 6 April 1905.

[51] Brouwer to Scheltema, 7 April 1905.

VI Immanent Truth
VII Transcendental Truth
VIII The Liberated life
IX Economics

The language of the treatise is partly borrowed from the majestic language of the bible, and partly it uses the expressive emotional language of the literature of the turn of the century, albeit with the personal Brouwerian flavour. If the reader feels that Brouwer's language is unusual and convoluted, he is certainly right. Brouwer had no mercy on his readers, even his Dutch is hard to read, and the extraordinary length of his sentences, with many subordinate clauses, was notorious.

There are a number of main themes, with which the book is concerned, all of which, however, can be retraced to the central theme.

The main point of *Life, Art and Mysticism* is the truly mystical doctrine that man's ultimate goal and challenge is total introspection—a turning into oneself (*zelfinkeer*). All the remaining points and chapters are elaborations of that particular task. Before Brouwer turned to this central issue, he first presented in a 'pedagogical' first chapter the disastrous influence of man on the world and nature in general. It is called 'THE SAD WORLD', as a wordplay on Van Eeden's 'The Joyous World'. The violations which Brouwer describes, will nowadays generally be recognised as such, but at the time of writing little understanding could be expected. On the contrary, the things that Brouwer condemned, would be greeted by most as miracles of progress. Practically speaking, in Brouwer's young days the crimes against man and nature had only just begun. The wholesale poisoning of complete areas and seas, the destruction of the natural balance in the system of our rivers, just to mention a few topics, has reached nowadays a dimension that one could not have imagined in 1904. Hence, one may well assume that messages, as contained in 'THE SAD WORLD'—if they would find readers at all!—would be brushed aside as totally irrealistic, and as scare mongering. One cannot do better than read Brouwer's text, of which parts are reproduced here. A translation, unfortunately, does not do justice to Brouwer's prose, which is extremely solemn.

> The Netherlands came into existence and was preserved by the deposit of silt of the rivers; a balance between the dunes, the delta, the tides and the discharge of the water was established—a balance in which temporary floodings of parts of the delta were incorporated. And in that land a strong human race could live and endure. Meanwhile, people were not content, they built dikes along the rivers to regulate or prevent the flooding, changed the river courses at will in order to improve the drainage or the shipping routes, and in the meantime cut down the forests. Small wonder then, that thus the subtle balance of the Netherlands was undermined, that the Zuiderzee was eaten away, that the dunes were slowly but inexorably washed away. And that nowadays ever harder labour is required to protect the country from total destruction. And does it not seem curious to observe, how this self-inflicted labour is not only accepted in resignation, but that it is even lent a lofty cachet of a task imposed in the name of God or Inexorability?

> The people originally lived separated, and each tried to preserve for himself his balance in the supporting environment of nature, amidst sinful seductions; thàt filled their lives, no interest in each other, no worry about the morrow. Hence, also, no work and no grief; no hatred, no fear; also no pleasure. Meanwhile, one was not content; one sought power over each other, and certainty about the future. Thus the equilibrium was destroyed, ever more sore labour for the suppressed, ever more infernal conspiracies for the rulers, and all are the suppressed and the rulers at the same time; and the old instinct of separation lingers on as pale envy and jealousy. [...]
>
> It is part of the balance of eternal and omnipresent life, that everyone will be called from this life on earth, when his time has come; and until that time [he is] physically and spiritually ill, as befits his evil mood of thrift, thirst for power, vanity and fear; once more, one is not content with this, one tinkers with the body by means of medicines and prescribed ways of life, and with the souls by hypnosis and suggestion, thus disturbing the purgatory of the lusts, and destroying the balance between psychological responsibility and physical constitution; the body is degenerated from the morale to such an extent, that one can indeed no longer be held responsible for one's crimes, for one's actions in this world. Although medical science boasts in recent times of the prolonging of the (incidentally, far too short) span of human life, what is the value of it? It is as sad to leave this life after one's time, as before one's time—and death? 'Nature never destroys without returning something better for it'. [......]
>
> The life of mankind as a whole, is an arrogant eating away of its nests all over the perfect earth, a meddling with her mothering vegetation, gnawing, spoiling, sterilising her rich creative powers, until it has gnawed away all life, and the human cancer withers away over the barren earth.
>
> They call the folly in their heads, which accompanies thàt, and which turns them insane: 'Understanding the world.'

'THE SAD WORLD' sketches in dramatic tones the degeneracy of man, who has exchanged his natural stability for the sinful state of never ending subjugation of nature and his fellow creatures. The chapter serves as a contrast to the next one, which sketches the possibilities and virtues of introspection. By turning one's attention away from the world, to the inner world of the self:

> ...the passions become silent; you feel yourself pass away from the old exterior world, from time and space and all other manifold things. And the eyes of a joyous silence, which are no longer tied, open up.

This chapter contains descriptions of the mystic experience, that seem to put it beyond doubt that Brouwer was no stranger to the experience. It paints in the words of a visionary the victory of the introspecting Self over the sad World. This chapter and later ones contain a number of quotations of Meister Eckehart and Jakob Böhme; this shows that Brouwer was well acquainted with the old European mystics.

The inner world, that the Self can obtain access to, is a boundless, chaotic mixture of fantasy worlds.

> And in that merging sea of colours, without separation, without permanence and yet without movement, that chaos without disorder, you know a Direction, which you follow spontaneously, and which you could just as well not follow. You recognise your 'Free Will', in so far as it was free to withdraw itself from the world, in which there was causality, and then remains free, and yet only then has a really determined Direction, which it reversibly follows in freedom. [...]
>
> The phenomena follow each other in time, bound by causality, because you yourself want, shrouded in clouds, the phenomena in that regularity.

The passages in 'INTROSPECTION' do not yet have the preciseness and conciseness of the later explanations of 'move in time' and other notions, but in a poetic way the nucleus of Brouwer's foundational credo is expressed here.

In Brouwer's opinion the sorry state of the world, including man himself, is caused by the interference of the intellect. Chapter III, THE FALL THROUGH THE INTELLECT, deals with the phenomenon of man's effectiveness in matters of domination of the world.

> Intellect renders men in the Life of Desire, the diabolic service of the connection goal–means between fantasies. While in the hold of the desire of one thing, the intellect hands them the pursuit of another thing as a means to that end; thus for the shifting of the riverbed: the making of a dam; giving vent to one's jealousy on another: setting his house on fire; ...

Here Brouwer formulates for the first time his *end to means principle*, which was going to play an important role in his overall philosophical considerations.

Whereas, the intended domination that is implicit in this 'leap from end to means', is already in itself objectionable to the introspective person, Brouwer points out a serious inherent shortcoming of the end-to-means transition. The transition, he says, is always slightly 'off key', so that repeated use of it, eventually leads to effects that were not desired.

> The act, which seeks the means, now, always somewhat overshoots the target; the means has a direction, which makes an angle, albeit a small one, with that of the target; it thus works, except in the direction of the target also in other dimensions; an effect, that, if the attention were not isolated from it, could perhaps be experienced as very harmful; but more: the attention gradually loses sight completely of the end and henceforth only sees the means. And in the sad world, where together with the Intellect, Drilling and Imitation are born from Fear and Desire, and nobody any longer surveys the whole human bustle. Many come to know that, what originally was a means, only as an end in itself; they pursue, let us say, an end of second order; with which perhaps again a means will be discovered, and that again makes a slight angle with its corresponding end. If the alluring leap from end to means is thus repeated several times, then it can easily happen that eventually a direction is pursued,

> that apart from its deviations in other dimensions, makes moreover an obtuse angle with the very first direction, and so counteracts it.

The chapter provides a whole catalogue of disastrous consequences of this practice, some of them are now generally recognised, whilst some—even today—would be considered unrealistic exaggerations. For example, Brouwer's views on nature are forerunners of the present ecological tenets:

> Does not industry originally deliver her products with the end to create in nature an environment of maximally favourable conditions for human life? In that connection, it was neglected that those products were themselves manufactured drawing on nature, in which, for this end, interventions were made in a disturbing manner. The balance of the conditions of human life was violated to a greater detriment, than the industrial products could ever benefit us. All the required wooden material, for instance, has led to the disappearance or degeneration of so much forest, that in the temperate zones hardly any crop for human consumption grows spontaneously. And more: we started to view the generating of industrial products as an independent goal. And in the pursuit of that goal, created as a means, a new industry of instruments that facilitate the old industry: a further blow to the old balance. In addition we recklessly started to collect the raw materials in remote countries, giving rise to trade and shipping, with all their physical and moral horrors, and the mutual suppression of nations.

In principle the same lines could have been written by modern environmentalists and reformers, and they were certainly in the minds of nineteenth century utopists and reformers. So far, Brouwer's indictment of the intellectual human imperialism differed from the moral programs of his predecessors and contemporaries mainly by the fact that his philosophical-ethical principles were far deeper and more radical. His rejection of the human pursuit of domination of the world, nature and fellow humans, was total and well-argued. He did, however, not stop at castigating society for the crude exploitation of economic, social and political powers, but went on to draw the ultimate conclusion. Namely, that even in the domain of the intellect and the mind, man was perpetrating the iniquities of the destruction of the natural balance. He extended the theme outlined above by adding science to the list of culprits perpetrating the abomination of iterated jumps from ends to means. Science is introduced as a means to further industry, and in turn becomes an independent subject; it then is followed by the 'foundations' of the science under consideration, which in turn is followed by 'epistemology'—'but the embarrassment ever increases, until all heads are reeling'. As for the scientists who take part in this regression, Brouwer comes to the merciless conclusion that

> Some of them give up quietly in the end; having thought, for example, for a long time about the intangible link between the intuiting of consciousness, which evolves with life out of that Anschauungs-world, and the Anschauungs-world, which itself only exists by and in the forms of the intuiting consciousness—an embarrassment, stemming from one's own sin of

> establishing an Anschauungs-world—then they put the 'I' which was self-created just like, and simultaneous with, the Anschauungs world in the opening, and say: 'Yes, there should, of course, remain something incomprehensible, because it is 'I' who must understand.

In a sharp indictment Brouwer accused an immense catalogue of established practices—science, the industry of stimulants and of pleasure, the misuse of art and religion, the medical industry and profession ('The medical industry was in the right hands with barbers and quacks, ...').

Clearly, the conflict between the life of the mystic in self-contemplation and the ambitious world of improvement and domination, presented Brouwer with a real and significant problem. A sincere person like him could not just expose the undermining of worldly life; he had at the very least to consider solutions. The chapter, 'RECONCILIATION', offers such a solution in a remarkably mature way; it avoids the pitfall of action for transforming the sad world into a better one—'each attempt to eliminate the non-balance only causes a shift of the non-balance'.

The solution which Brouwer offered, consisted of a reconciliation with the straying world, a resigned life in which pain, labour, desire and fear belong to one's fate. One should not frivolously add to the burden of one's karma, but one should neither wish to be better than one is—'that would be a voluntary following of evil desire'. Nor should one wish to improve the world beyond what it is—'that would be evil lust for power'.

These considerations may offer an answer to the vexing question, how one can live in a world like ours, without an abject betrayal of all that is good and sacred. The ideal of detachment, as preached by mystics and Buddhists alike, points a way out of the horrible dilemma 'collaborate or resist'. Brouwer, doubtlessly, must have considered the problem of reconciling the contemplative life of a mystic and that of the academic scientist. He was far too sincere just to ignore this fundamental problem. We should at the same time keep in mind that on many occasions nature was stronger than principles. We shall see Brouwer rush off to rescue the innocent and to fight noble battles for justice. Only the unimaginative live by the book!

It should not come as a surprise, given Brouwer's views on inner life, communication and language were secondary notions. There is a special chapter 'LANGUAGE' to which many of his later philosophical insights can be traced. The basic claim is that there is no communication between souls.

> No two persons will experience exactly the same feeling, and even in the most restricted sciences, logic and mathematics, which can properly speaking not be separated, no two [persons] will think the same thing in the case of the basic notions from which logic and mathematics are built. Yet here the will is parallel in the two, for both there is the same forcing of the attention by a small insignificant area in the head. [...]
>
> But the use of language becomes ridiculous, where one deals with the finer gradations of will, without living in that will; just as when so-called philosophers or meta-physicians discuss among themselves morality, God, consciousness and free will; people who [...] share no finer movements of the soul, ...

It seems tempting to conclude that Brouwer must have advocated an abstinence from communication; there is not much supporting evidence, however. He practised great care in his scientific communications, even to such an extent that they became difficult to read. Moreover, in daily life he was an inveterate conversationalist. In view of the earlier remarks on 'Reconciliation', there is nothing paradoxical in this. On the contrary—a person who is aware of the weaknesses of communication will probably take extra care in his use of language. Later developments will shed more light on this topic.

The Chaps. VI and VII deal with IMMANENT TRUTH and TRANSCENDENT TRUTH; the first truth 'points in the world at the consummated Karma of the world', the second points, in the world, at the personal life: 'Immanent truth clarifies, transcendental truth makes devout.'[52]

The chapter on Immanent Truth has acquired a measure of notoriety because of its view on women. It is a theme that in the underground folklore belongs to the *chronique scandaleuse* of an otherwise respectable science. The chronicles of intuitionism have always passed over *Life, Art and Mysticism* in an embarrassed silence; the mystic views of the founder of mathematical intuitionism were thought to be a liability that might very well detract from the objective virtues of intuitionism. The resulting picture of intuitionism showed a somewhat flat pragmatic practice, which—whatever one may think of the mathematical subject—did not do justice to its historical roots. At one point there was, however, justified cause for reticence—the topic of the female. Brouwer's conception of the role of women in the world is rather dated; this may surprise those who think of Brouwer as the revolutionary innovator—but, whereas this characterisation is certainly apt in relation to his topological work, he may be considered a conservative in a number of philosophical matters.

The fact that Brouwer's intuitionism was considered new and revolutionary, can be simply explained by the observation that people had neglected their inheritance of idealistic philosophy, so that after a spell of the fashionable formalism (in mathematics) and neo-positivism (everywhere else), Brouwer's doctrines seemed to the less informed the newest thing, instead of a return to nineteenth century idealism. In fact, Brouwer was basically a conservative; *Life, Art and Mysticism* was definitely a protest action against the prevailing optimism of Progress.

His views on women can be classified as equally conservative, although one must understand that his views are part of the total mystic view that is being propounded.

The chapter *Immanent Truth* deals with the aspects of the resigned life in the world with all its conflicting desires and interests (as opposed to Transcendent Truth, which deals with life, disengaged from the influences of the world) and treats among other things the influences of various art forms and the burdening of the karma

[52]"If Truth points in the world to the personal life, free from the ties of fear and desire, where the bliss and wisdom and the quiet rejoicing of the timing in upon oneself flourish on modesty, poverty and quiet fulfilment of duty in this life on earth, which is one's own accomplished karma, then it is Transcendent Truth."

by 'avarice, ambition, and ... the illusion of woman'. "Immanent truth enlightens, transcendent truth makes devout."

Immanent truth breaks through even in science, which has alienated what is perceived in the outer world from the self. Science builds outside life a mathematical-logical substrate, a chimaera, and in life it builds a veritable tower of Babel. But nowadays, says Brouwer truth which is breaking through, returns the centre of gravity from what is perceived to the observer. 'Copernicus brought the rotation of the heavenly bodies to the earth: it will yet be placed man's own body'. 'All this', he adds, 'is of course no use: it leaves the world as dim as before; it is no 'Turning into itself', no turning to free truth, but the appearance of truth in the garb of folly'.

There is nothing new in the repudiation of the female by mystics and hermits. The history of the church provides numerous examples of saints who had a keen eye for the dangers of female company. In fact Christianity is not the only religion to take (or at least, to have taken) a rather defensive view towards women. So, in itself, the claim that the female burdens the karma of the man is quite in line with the tradition of ages. Brouwer, however, went beyond this observation (which, of course, carries an unmistakable danger sign: beware!) by classifying woman as a creature of a separate level. Her true function is to ward off disturbing influences from the man's karma, although she is, paradoxically, the greatest temptation to him. The role of the woman is a serving one:

> Humble she will be, humbly she will wish to take all ignoble work from his hands, all work other than the pure indulgence of the faculties of his body, in which he walks the earth; without the wink of an eye she will give her life, to save his balance.
>
> Serene will be her eye, tenaciously and patiently she lives on, and does what serves the beloved. Her body will be unwrinkled, motionless, without passion to seduce, not conscious that it seduces and yet so unbearably seductive in its taunting composure, that no man can endure it.
>
> The Venus of Milo shows in a clear, pure way that karma of the woman, of the quiet, desireless, unconscious, and yet so infernally seductive woman.

According to Brouwer, a woman can also burden her Karma, for example, with 'male activities'. Noble institutions will degenerate when women intrude, so the work of man, when taken over by women will necessarily be degraded. By the usurpation of parts of the prerogatives of man, the woman acts against her karma. In a long catalogue he lists all the shortcomings of the woman. If she is independent, she looses her femininity and burdens her karma, and if she is truly female, she is a shadow of her beloved and is guilty of naivety:

> In worldly matters and worldly convictions she will naively follow the beloved, and defend views, unthinkingly copied, as objectively indisputable axioms against all objections from third parties; in disputes with such a woman the ridiculousness of language as a means of reaching an agreement clearly appears in the form of the notorious 'feminine logic'.

Whereas the philosophical and mathematical conclusions that Brouwer would eventually be drawing from his mystic insight and convictions were original and even revolutionary (perhaps in a counter-revolutionary way) his views on women, it seems, were rather modelled after the prevailing views of the nineteenth century. Many of the cliché's that turn up again and again in the treatises on the weaker sex,[53] appear in Brouwer's essay; women are temptresses, endangering male purity—pale, without expressive lines. Even Ophelia, the preoccupation of the Victorian period, is called as a witness.

The passages on women are certainly not the strongest in Brouwer's book, but at least he was not guilty of building a scientific argument, which was not uncommon in this particular area.

His message was mainly one of a moral nature, a warning from an ascetic mystic to the world. To the young man with the exalted ideals of the introverted seeker, women personified the dangers to man's karma. No doubt his personal emotional life and history had influenced his philosophic outlook. As a result, woman was not only the spectre of the fall from the ardently sought inner peace, but she was furnished with all the paraphernalia of the temptress and of the weak. Woe to him that succumbs to the distracting charm of the female:

> But truth in art shows the distinct lines: man should avoid, ignore woman; but the woman should live in the man, holding herself insignificant, powerless and worthless, and sacrificing everything to the beloved. A real woman is pale, supple, without expressive lines, with dull, dreamy eyes: she has no muscular strength, and cringes from nothing. And a man who turns to a woman, has lost his life.

The choice of images and of words suggests that Brouwer was acquainted with the nineteenth-century literature on the role of the woman. Many of the themes and descriptions have a familiar ring to the connoisseur of the Victorian era. It is tempting to conjecture that his stern views on the role of the weaker sex may have been the result of female attacks on the bastion of Bertus himself. Given the lack of facts, this has to remain what it is: a conjecture.

Many have wondered about Brouwer's theoretical aversion to women and his rather progressive daily outlook; in the twenties he was one of the first mathematicians (maybe the first) in Holland to engage a female assistant, and he admired his female colleagues in the academic world. He was on good terms with the renowned Emmy Noether and with Olga Taussky, and he certainly did not avoid female company, neither in the context of science, nor in his private life. Presumably he conceived the sermons on the distracting female in the framework of the avowed goal of the mystic: the unconditional introspection, whereas—as we have seen above—the actual life in the world required the sincere mystic to suffer the ways of the world, under pain of loss of karma through pride.

There is also a brief mention of science and truth:

> Furthermore immanent truth breaks through in science as well. It has separated the observed things from the ego, and placed it in a Anschauungs-world,

[53]Cf. Dijkstra (1986).

> which is thought to be independent of the ego, and which has lost the connection with the Self, which alone feeds and directs. Thus it builds outside life a mathematical-logical substrate, a chimaera, and within life, a tower of Babel with its confusion of tongues.

Here the adoption of an independent outer world appears as an attribute of immanent truth.

*Life, Art and Mysticism* contains a number of further remarks on society and its organisation. Brouwer's conception of work as something noble and lofty clashed forcefully with more progressive ideas. We have already mentioned the inroads of women into the labour market and its negative consequences:

> The gradual usurpation of certain forms of work by women will go inexorably hand-in-hand with a degeneration of that work into an ignoble state.

The graduate student, who only a few years ago attended socialist and communist meetings in the company of Scheltema and Wiessing, apparently had realised that certain features of socialism were incompatible with the world of the mystic. It should be fairly evident that socialism, with its preoccupation with the world and its socio-economic features, has no call for a mystic credo, and conversely the mystic would consider the materialistic socialist as one of those unfortunate, unavoidable parts of his Karma. Brouwer viewed the consequences of socialism with some mild horror:

> Until quite recently the state, and public life, were viewed as something honourable, even metaphysical; and a position in society was considered a noble task; [...] But the socialistic movements have in the last century washed away that aspect of 'honourable', [...].
>
> When, as the endpoint of the socialistic degeneration, the state will have turned into a well-oiled automaton, well—then the administration will perhaps be completely left to women.

The final chapter, 'ECONOMICS', must be seen as a logical conclusion of the preceding chapter. The proper attitude in life towards the theoretical contemplation of life and society, of the human who has taken to heart the call to introspection and detachment, is one of renunciation.

> There is one more thing, which the free life should be careful not to become tainted with, as long as its ties with society last: economics.

For, inherent to economics is, according to Brouwer, the idea that 'foolishness and injustice' are essential—otherwise economics would be superfluous. Thus the intellectual study of the ways and laws of misfortune and injustice will not attract the free man. Desire of property evidently turns the attention outward;

> ..., for he who views something as desirable or deplorable, views it as something outside himself, as part of a world which exists independently and persistently, as part of a fixed inalienable possession, that one can cultivate, take

> care of, clean, raise, as one can with one's flowers or chickens. Exerting influence outside oneself, be it for improving the world or for one's own power, is: blinding, vanity, thirst for power.
>
> The Free rather view their fellow beings as delusions which solicit compassion, disturb the path of life, which are to be borne like guilt, for their freedom does not suffer them around. And the Free cautiously slip past them.

Society is, in this perspective, an artificial web of power and domination, complete with its moral justification in abstract terms of 'suppression', 'justice', 'rights', ... Here, Brouwer appears in the cloak of the radical anarchist, denouncing the social theories that legitimise the powers that be. The theorists are vigorously criticised:

> They are talking about 'Human Rights' as if man brought rights into his life, and more than miserable duties, as a punishment for being born.
>
> They are talking about 'labour', its necessity, and the happiness it brings. As if the labour of mankind were something else, but a blind convulsion of fear for what is no evil, and of desire for what brings misery.
>
> The economists and leaders of the people also love to talk about a 'future state of the deliberately co-operating people'; this would be possible for people without fear and desire, but those would not work, and a world of such persons would not exist.

Thus this final chapter once more confirms the basic tenets of the earlier ones: meddling with the organisation of the life of human beings is doomed from the start. The final words sum up the lesson:

> he, who knows not to possess anything, not to be able to possess anything, not to attain stability, and who resigns in resignation, who sacrifices everything, who gives everything, who no longer knows anything, wants anything, wants to know anything, who lets everything take its course and who neglects everything, to him will be given everything and to him is opened the world of freedom, of painless contemplation, of—nothing.

The Delft lectures and the subsequent book went largely unnoticed; there was a review in one of the daily newspapers, a rather devastating one, and there it stopped. Brouwer sent complimentary copies to his friends and colleagues. Two reactions have been preserved; his friend Scheltema disagreed on principle:[54]

> I received your booklet shortly after my letter and I have started to read it. So far I did not read with the aversion I expected, and I wish many a Philistine this literature as a refresher—but you know how 'heartily' I disagree with you, how, for example, the conscious social democrat immediately rejects the premise you start from (that is that the original animal-like human life is the happiest imaginable one, and at least should be worth pursuing most) on the

[54]Scheltema to Brouwer, 16 May 1905.

> contrary, stresses that the happiness of the human society will only *begin* after the *conclusion* of the era of barbary in which we still live ...

A surprisingly mild reaction considering Brouwer's volleys at socialism!

Korteweg, Brouwer's Ph.D. adviser, was more critical:[55]

> That I am very much interested in you and hence appreciate the sending of your slim volume, therein you are certainly not mistaken.
>
> Whether I shall read it? I thumbed through it, but it is not the reading that I wish or that is good for me. It is true that there are abysses very close to us, but I do not like to walk on the brink of them. It makes me dizzy and less able for what I need to do. Whether it is good for you, I don't know. So much is certain, that I would rather see you walk along other paths, even though I find it sometimes also hard to follow you there, where you cut so deeply through fundamental matters.

*Life, Art and Mysticism* earned its author a reputation of eccentricity, although rather by an obscure oral tradition than by direct acquaintance with the text.

The question that immediately comes to mind is, was Brouwer serious about it all? Considering the available evidence, the answer is probably 'yes'. He may have exaggerated here and there to provoke the audience, but by and large the mysticism was genuine. In addition to his pronounced views on life, the complacency of progress and the progressives was probably a valuable source of stimulation.

## 2.7 Family Life in Blaricum

To Louise, who was only 12 years old at the time of *Life, Art and Mysticism*, the whole matter seemed mysterious. She was attending in Amsterdam the school of a certain *Master Gerhard*, which was conducted on socialist principles. The fee for the school was forty cents a week. With this education, based on the principles of clean, healthy, idealistic socialism at the turn of the century, she must have been puzzled, to put it mildly, by her stepfather's rather unusual, gloomy views.

Like any child, she would go exploring in the house when she was left by herself. She recalled that she was drawn by a magnetic power to the cupboard in which the manuscript of *Life, Art and Mysticism* was stored. She cautiously climbed on a chair and read bits and pieces of it, closing the cupboard and jumping down as soon as she heard a key in the lock. Mysterious as the contents were to her, she understood enough to guess that this book should not get into the hands of Master Gerhard, and so she told Brouwer that she knew about the manuscript, asking him to promise that he would not show it to her teacher. Eventually Master Gerhard got wind of the book, as a result poor Louise was expelled from the school on the grounds that her parents did not conform to the socialistic ideals.

[55] Korteweg to Brouwer, 13 May 1905.

**Fig. 2.7** The hut in Blaricum. [Photo Dokie van Dalen]

Since she was no longer at school after this incident, Louise was made responsible for the housekeeping in Blaricum—something she hated. The hut had little comfort, just one bedroom upstairs. Louise had to sleep on a wicker chair downstairs. Brouwer worked downstairs or in the garden, often reclining in his characteristic pose in the same long wicker chair.

Lize stayed in Amsterdam during the week, managing the pharmacy. Brouwer stayed home to work and went over to Amsterdam for his teaching, visits to the library, occasional meetings with fellow mathematicians, and for his regular concerts in the Concertgebouw. When it was more convenient to stay in Amsterdam, he joined Lize in the apartment over the pharmacy.

Once a week the mathematician Hendrik de Vries came to visit. He usually brought his violin, which he played well, and Brouwer accompanied him at the piano. Louise's role was to play the conductor. The lessons of mother Brouwer had born fruit. Bertus became an ardent amateur pianist. He loved Beethoven, and according to Lize, his favourites at the time were the piano scores of the Beethoven Symphonies.

When Brouwer got it into his head to order a day of fasting, Louise was not allowed any food at all, not even yesterday's leftovers. She soon discovered, however, that Brouwer dodged the fast by nibbling nuts from the drawer of his desk. From then on she took her preparations, and bought enough biscuits[56] to survive the day.

As part of her education, Brouwer decided to teach Louise how to cook. The first lesson was how to prepare rice; it ran as follows: 'rice is a product from the Indies;

[56]frou-frou, a special kind of wafer.

it is carried by a ship and there negroes sleep in it. Therefore I want you to wash the rice ten times before you boil it' (Brouwer knew very well how lazy Louise was). Needless to say Louise did not follow the instructions. There are many little stories and events that show the uneasy relation between Brouwer and Louise. To mention one example, Brouwer told Louise to eat by herself in the kitchen when Lize came over for dinner.

Louise also had to walk from time to time to her school in Amsterdam. And sometimes it happened that Brouwer chose to teach her French; sitting at a terrace in Amsterdam, he would quite sternly take her through the drill, not hesitating to slap her if her attention relaxed.

The relationship between Brouwer and Louise was problematic from the beginning. Brouwer was easily irritated by Louise and considered her lazy, stupid and stubborn. She certainly was far from industrious, but Brouwer's treatment of her did little or nothing to improve her attitude. As long as Brouwer lived, they had fierce clashes about everything. In the beginning Brouwer had the advantage of his age, but gradually Louise began to free herself from the pressure of her stepfather. And eventually the two battled on equal terms. Whereas Brouwer had made it a principle to avoid unpleasant and obnoxious people, he could not avoid Louise. For Lize the situation was extremely painful, because Louise was her only child, and she did not want to sacrifice the happiness of Louise to that of her husband, but neither did she want to hurt Bertus. The surviving correspondence between Lize and Louise shows how difficult it was for Lize to steer a safe course between daughter and husband. Louise told how, when she was still living at home, Brouwer sometimes introduced her to his visitors with the words: 'and this is my silly daughter'. The sad thing was that it was not intended as a joke. In return Louise thought that her stepfather was himself as mad as a hatter; in her opinion all the fuss at the house in Blaricum only served to keep him quiet. If no prophet is recognised in his home country, then certainly he is not recognised at all in his own home! Only very late in life, long after the death of her stepfather, she changed her views on him.[57]

Even in her old age Louise's memory was wonderfully clear, she provided many facts that at first sight seemed curious. A complicating factor was that she was very confused about religious matters. The details she provided on daily life and the personal history of Brouwer and the crowd around him have, however, mostly been born out later by independent confirmation. By extrapolation I have come to view her information as generally reliable.

[57]When in 1981 a conference was dedicated to the centenary of Brouwer's birth, a short biographical article appeared in the Dutch weekly *Vrij Nederland*, van Dalen (1981). Reading this biography, Louise suddenly realised that Brouwer was not a fool after all. As she regularly communicated with the spirits of the departed, she noted that Brouwer's spirit had found rest after this public recognition.

# Chapter 3
# The Dissertation

## 3.1 Preparations and Hesitations

> Mathematics, rigorously treated from this point of view, and deducing theorems exclusively by means of introspective construction, is called intuitionistic mathematics.
>
> Brouwer (1948)

While the excursion into philosophy and mysticism was going on, Brouwer quietly continued his mathematical research. He read much and tried to get acquainted with modern developments. Although his principal teacher, Diederik Johannes Korteweg, was a man of quality and of good taste, his own work was almost exclusively in the domain of applied (or at least, applicable) mathematics.

There is no doubt that Korteweg offered a solid mathematical education, but on the whole, the choice of topics was rather conservative. The marvellous and exciting innovations from Paris and Göttingen were not taught or discussed. Students were definitely not brought in touch with the newest developments in pure mathematics.

We have seen that Brouwer had unusually strong philosophical views, which were, if anything, far from progressive or fashionable. Although he had come to terms with the world in the sense that a crusade for the spiritual liberation of mankind would only result in a burdening of his karma, and that thence a 'live and let live'-policy was the proper choice, he could not just shrug off the preaching habits he had acquired during the *Life, Art and Mysticism* period (which, by the way, partly coincides with the research for the dissertation).

There are a number of documents relating to the preparation of the dissertation, in the first place a series of notebooks, in which Brouwer jotted down his ideas and comments, and in the second place a synopsis he made for a first version of the dissertation. In the synopsis a selection of the material in the notebooks is sorted into chapters.

The notebooks contain a mixture of all sorts of topics. It is interesting to see that when he started out, the mystical–philosophical considerations were still uppermost in his mind. Gradually the non-mathematical remarks give way to purely mathematical topics, to re-appear suddenly somewhere in full force. In particular at the

D. van Dalen, *L.E.J. Brouwer – Topologist, Intuitionist, Philosopher*,
DOI 10.1007/978-1-4471-4616-2_3, © Springer-Verlag London 2013

end the philosophical content increases again. The notebooks are not dated, so it is difficult to make specific guesses about timing and progress.

Here and there in the notebooks one can find schedules, plans, outlines for reading, research and writing. They are instructive in the sense that they shed some light on Brouwer's reading. They also illustrate his self-discipline. The following schedule, written on the inside of the cover of the first notebook, may serve as an example:

1 week Russell
1 week non-Euclidean geometry (Klein, Lie)
1 week Dedekind and Cantor
1 week Poincaré and *Revue de Métaphysique*

---

Elaborate notes on Sundays

Like so many passages in the notebooks, this schedule is struck out, probably because the work had been done. Did Brouwer stick to the schedule? It is not impossible. Brouwer is known, in particular in his early years, as a hard worker, with explosive bursts of activity; nonetheless, four weeks for this amount of reading and the corresponding compiling of notes seems a tall order.

On page 1, in the margin, another list of things-to-do for the dissertation is given:

1. Foundations, Russell; Couturat, algèbre de la logique; D. Hilbert in Heidelberg Congress; Hilbert, found. of geometry, Teubner.[1]
2. non-Eucl., Klein, Lie
3. Numbers, Cantor, Dedekind.
4. Metaph., Kant, Poincaré, Couturat
5. Projective views on vector- and potential field
6. Work out notes
7. Hankel complex numbers
8. Hertz. Cf. Poincaré.

Only Hankel and Hertz have not been treated in the notes, the rest is discussed fairly extensively.

It is clear from the notes that Brouwer was still occupied with his mystic-philosophical views. The form these views take in the context of mathematics are not all that different from what we have seen in the preceding chapter.

As we have seen, 'the highest attainable', state in life is the return to the deepest home of the self, the ultimate introspection (p. 64). The exodus from the original state of the soul[2] is what Brouwer calls 'sin'. The manifestations of sin, in this sense, are manifold. To mention a few: seeking of power over nature, the domination of fellow beings, the exploitation of one's talents. Since mathematics was, in

---

[1]Brouwer cites in his dissertation the *Festschrift* version of the *Grundlagen der Geometrie*, but he privately used the French translation in *L'Enseignement Mathématique*.

[2]One would be inclined to say 'mind', but in Brouwer's later work 'mind' is given specific meaning, cf. Brouwer (1933a), van Stigt (1990); hence it is better to avoid the term here.

Brouwer's eyes, party to the domination of nature, it was one of the tools of sin—unless practised for more lofty purposes, that is the free unfolding of the self in playful development.

The notes are pervaded with a reticence to take part in a sinful practice, or rather, to collaborate (no matter how pure his personal motives were) in a worldly design that he could not approve of.

We shall reproduce a number of Brouwer's moral comments here:

> Whoever practises mathematics with a conscience (even though that is often out of fear, or undeserved respect for others and for theories suggested by others) finds few things, although these few may perhaps make a brilliant impression. He can produce much more if the place of the conscience is taken over by the mentality of the businessman who is just 'marketing'.
>
> The activity of modern mathematics is forbidden, much as all excesses [...], and publishing, much as the marketing of meat extract.
>
> Let the motivation behind mathematics be the craving for the good, not passion or brains.
>
> Most investigations of mathematics (also Lobachewski's and Riemann's geometry) is the marketing of "rather pretty" goods. And just as little this will bring blessing as free competition and advertising in the liberal world of merchandise.
>
> The role of foundational research must be: *given* the temptations of the devil, who is the world and its categories, to appreciate the true value of the world, and to relate it constantly to God.
>
> Not worrying about the 'foundations', and just doing mathematics is the same as: not worrying about economy and economic morals, and just doing business and earning money and making a career.
>
> In both one can be very clever, and yet a zero.
>
> One should refuse to do mathematics, but since this point has been reached, one should refuse to do the next step, that is mathematical logic.
>
> My own, spontaneously observed, life has no fixed laws, but it is a miraculous play of chance. Just as one of the elements of this world of chance, there floats the dark cloud of the mathematics-practising rabble which acts in this world in such a way that it can only react in laws (that is something from the underworld).

The above mentioned notion of 'sin' occurs frequently in Brouwer's early writings, in particular in his notes and letters. Later in life, when he became, whether he liked it or not, part of the academic establishment, the moral overtones disappear, to reappear suddenly in his great post-Second World War address, *Consciousness, Philosophy and Mathematics*. The keywords in his early description of sin are 'centralisation', and 'externalising',[3] both of which indicate the passage from free, undirected contemplation in the self to the concentration on certain aspects, and to the

[3] *centraliseren* and *veruiterlijken*.

positioning of the contemplated (or experienced) concepts in an independent outer world. In other words, sin is a form of treason to the self (or ego) and not something that concerns church or society.

The contents of the notebooks and the subsequent resumé make it clear that Brouwer had not shed his convictions of the year before, when he presented his views on the world and on mankind in *Life, Art and Mysticism.* Indeed, the moral aspects are far from neglected in the notes.

Sprinkled through the notebooks are remarks on the topics that he valued above all. He must have felt the paradoxical position he was forced to adopt: he had to write a dissertation on a topic in mathematics, and thus to become an accessory to the sinful power game of science, while at the same time preserving his loyalty to the detached life of the mystic.

The dissertation episode is in a way the watershed in Brouwer's relationship with the world. He did not give up his mystic ideals or his inner convictions, but he realised that he had to establish a pact between himself (his ego) and the world. This is not uncommon in students who reach the end of the blessed period of uncompromised innocence, but in Brouwer's case it was a really painful transition. The Brouwer–Scheltema correspondence betrays the growing pains and the sadness of the loss of independence. In a letter Brouwer refers to 'the coarse mansion of society', where they would have to 'light its chandeliers, and grace its door-posts'.[4]

Not so long ago, Brouwer had still been making his often biting comments on the topics that occupied his thoughts. There are undated notes, which possibly belong to the same period as *Life, Art and Mysticism*, which illustrate the difficult process of transformation from defiant student to respectable scholar. They are the aphorisms of the young mystic, untainted by collaboration with the corrupting powers of the world. A few examples may give the reader an impression of the audacious views of the young man before his surrender to convention:

> One could see as the goal of one's life: Abolition and delivery from all mathematics.
>
> Improve the world, or make it beautiful? Aren't we humans the earth's bacteria of decay, which help and speed up the process of decomposition for which the time had come?
>
> Most people become socialists in order to join a suitable milieu and make relations in it.
>
> The vulgar ones do not know that language–truth and understanding–truth do not exist; thus their passion concentrates itself on 'wanting to twist the truth'.
>
> Understanding between 2 people is gradual; but mathematical understanding is something like 'yes' or 'no' just like sleeping is something like 'yes' or 'no'.
>
> That mathematics and its applications are sinful follows from the intuition of time, which is immediately felt as sinful.

[4]Brouwer to Scheltema, 8 July 1907.

The rich and poor are the complementing elements of a world fallen in sin. He who knows to back out of it will be neither rich nor poor.

The feeling of dignity and being honoured is analogous to the erroneous feeling of ease, of power, of being armed, and other feelings, which tempt one to fall asleep.

The degeneration of the practical mathematical activity can be its desire to see mathematical systems (theoreticians) and the desire to realise mathematical systems (tinkerers, agitators and intriguers). The normal organising talent has not yet to be considered a degeneration.

Sorrow is the inability to find in one's head the mathematical system that gives rest. Only he who wants rest therefore knows sorrow.

Mathematics justifies itself, needs no deeper grounds than moral mysticism.

The space of animals and trees is not Euclidean, does not even have a group of motions.

Isolated remarks of this sort also occur in the notes for the dissertation. In a way the genesis of the dissertation is the story of temptation, as experienced by all hermits and saints. No matter how much Brouwer fought the evil influence of the world, the fascinations of mathematics proved stronger than his Spartan views. Brouwer's views were permeated with the awareness of the sin of externalisation with its intended domination of the world and nature. It is no wonder that trade and business figure prominently on the list of sins; even the Jews, the traditional exponents of small and big business, are mentioned in passing: 'How the Jews dominate the farmers with the help of mathematics, and how the farmers do the same to the animals. Mathematics as a part of the technique of culture, put on the market.'[5] It is as if we witness Brouwer's losing battle against the temptress, Mathematics. His notes contain fierce attacks at the sins of externalisation, but they make the rather sad impression of rear guard actions. The final, printed version of the dissertation is completely neutral. The anger of the young Brouwer had found an outlet in the fierce indictments of his contemporaries Russell, Hilbert, Couturat and others. The parting shots at the sinful world are a curious mixture of sadness and indignation.

The notebooks show that Brouwer had an extensive reading program; that he went through a great many publications, with the following central topics:

- Axiomatic Geometry and Foundations of Geometry
- Non-Euclidean Geometry
- Foundations of Mathematics, Logic and Set Theory
- Potential Theory.

It is interesting to read how Brouwer found by himself a way through the undergrowth of new developments and contradicting positions of the mathematics of the turn of the century. From the dissertation and the notes, one can more or less piece together his sources, and see by whom he was influenced.

[5]From the summary of Brouwer's notes for the dissertation.

The influence of Hilbert is clearest in the treatment of the Lie group section of the dissertation and in the discussion of formalism and axiomatics. At the time Brouwer wrote his dissertation, Hilbert's *Festschrift* (*The Foundations of Geometry*) had not yet dominated geometry to the extent that the memory of Hilbert's predecessors had become obliterated. This can be seen in Brouwer's notes and the dissertation; indeed Brouwer quotes carefully the axiomatic researches of Klein, Pasch, Pieri, Schur, Vahlen, and Veronese. Of course, Lie's work takes a central place in the dissertation, but that is no surprise, as Lie groups are treated explicitly. The influence of Hilbert's *Grundlagen der Geometrie* has been so massive that later generations were under the impression that the axiomatic treatment of geometry was Hilbert's personal achievement. At the beginning of the century this optic distortion was not yet common. Hilbert's *Grundlagen der Geometrie* clearly derived a large part of its popularity more from its forceful presentation, which fitted the mood of the times, rather than from its novelties.[6]

The alternative approach to geometry was picked up by Brouwer directly from the works of Lie and Klein.

In the foundational part, Brouwer mostly dealt with Russell's *An Essay on the Foundations of Geometry*, Hilbert's Heidelberg address *On the Foundations of Logic and Mathematics* and a number of papers of Poincaré. He also read the papers of Bernstein, Cantor and Zermelo.

It is easy to see that in the case of the foundations of mathematics, Brouwer stuck to his own philosophical–foundational views, and that he measured, so to speak, the authors in the field by his own yardstick.

For that reason, it is somewhat disappointing that Brouwer did not join the discussions of the day. For example, with respect to Zermelo's axiom of choice, he only remarks that the axiom cannot be right because, if we ask a person to make choices from a collection of (non-empty) sets, it is unlikely, that if the person is presented the same set twice, he will make exactly the same choice. Hence Brouwer objects to the functional character of 'choice'.

The potential theory did not find a place in the dissertation; it was published in a series of papers in the proceedings of the Academy.[7]

The most direct information concerning the dissertation is provided by the correspondence between Brouwer and his Ph.D. adviser, Korteweg. The choice of supervisor was not difficult; the only alternative to Korteweg was the mathematics professor Van Pesch, a scholar who practised mathematics at a safe distance from the frontline of contemporary research.

Korteweg, on his part, had no hesitation to accept the young man as his Ph.D. student; already before Brouwer had finished his regular studies at the university, three papers of his on transformations in four-dimensional space (cf. p. 46) had been communicated by Korteweg to the Academy, so that the latter was eminently aware of the quality of the young man.

[6]Cf. Freudenthal (1957).

[7]Brouwer (1906a, 1906b, 1906c).

The preparations for the dissertations were carried out in parallel with the research on vector distributions and potential theory. The last two topics were safely out of the way when in September 1906 Brouwer started the actual writing of his dissertation in earnest. The last paper of the series was communicated to the Academy on 29 September 1906.

The series consisted of two papers on potential theory, *The force field of the non-Euclidean spaces with negative curvature* and *The force field of the non-Euclidean spaces with positive curvature*, and one paper on higher dimensional vector-fields, *Polydimensional vector-distributions*.[8] In the latter Brouwer proved, among other things a generalisation of Stokes' theorem to arbitrary finite-dimensional Euclidean spaces. In 1919 Brouwer returned to the subject, pointing out that the proof of his 1906 paper also established the non-metric form of the higher-dimensional Stokes' theorem. In the original paper no references were given, but in Brouwer (1919k), Poincaré is mentioned as having enunciated the theorem already in 1899, 'without a proof however'. In a footnote Brouwer referred to earlier publications of Poincaré in which the simultaneous vanishing of both sides of the Stokes equation is stated.[9]

The paper *The force field of the non-Euclidean spaces with negative curvature*, contained a novelty that has escaped the general attention: on page 5 Brouwer introduced the parallel displacement (without giving it a name) years before the notion officially entered into the literature. Brouwer must have attached some value to this invention, since he indicated the definition by a line in the margin in all his reprints.[10] Struik told me that he stumbled on Brouwer's definition more or less by accident, when checking some literature.[11] He informed Schouten, who was surprised that the notion had been introduced (in a special case) before he and Levi-Civita had formulated it (independently). It was Struik's impression that Brouwer was not aware that his formulation of the notion of parallel displacement preceded that of Levi-Civita and Schouten before somebody called his attention to the footnote in Schouten's *Ricci-Kalkül* mentioning Brouwer's priority (at least for the case of constant curvature).

The potential theory papers were communicated by Korteweg to the Academy between 26 May and 29 September 1906. This productivity shows that Brouwer, in spite of the pressure of time, managed to think of other things than the coming 'promotion'.

## 3.2 Under Korteweg's Supervision

Korteweg's role as a Ph.D. adviser was, I guess, restricted to competent and critical comments. The topics were Brouwer's own choice, and there was little overlap

[8] Brouwer (1906a, 1906b, 1906c).

[9] Cf. Poincaré (1899) (*Méchanique céleste III*), Poincaré (1887, 1895).

[10] Cf. CW II, pp. 58, 69, 71, 78, 83, 86.

[11] D.J. Struik to Van Dalen, 16 May 1992.

between the contents of the thesis and the work of Korteweg. Korteweg must have been somewhat cautious with this remarkable student. He agreed with the choice of the topics, provided 'enough mathematics was left in'. In a letter[12] Brouwer reminded Korteweg of his permission, defending himself against the, real or imagined, scepticism:

> You know well, that when I selected my topic two years ago, it was not a matter of inability to tackle a more 'common' one, but only because I felt an urge towards this subject: it originated spontaneously in me.

Fortunately for history, a fair portion of the correspondence between Brouwer and Korteweg has been preserved, so that we can reasonably well reconstruct the last stages of the writing of the dissertation.

The first letter that mentions the 'Promotion',[13] that is the official defence of the dissertation and the awarding of the doctorate, is the one of 11 January 1906. It deals with the grant of the St. Jobs Foundation. Since Brouwer was already twenty-five years of age, he was reaching the limits stipulated in the regulations. He argued, however, that one of the rules of the Foundation allowed an escape, so that a promotion at the age of 26 years would not disqualify him for another year of support. The Foundation was fortunately wise and liberal enough to grant him an extra period until 12 November 1906; Korteweg's strong recommendation had swayed the opinion of the regents of the Board.

Korteweg was putting a lot of trust in the capacities of Brouwer, although his student was exceptionally brilliant and intelligent, it was a daring gamble. With less than a year to go, and not a syllable on paper, Korteweg was promising more than the average Ph.D. adviser would be willing to guarantee. To make things worse, Brouwer was tied up with his research papers until the last minute. As a matter of fact, Brouwer was still in the weeks before the summer vacation of 1906 making the final improvements of his above mentioned paper, *Polydimensional Vectordistributions*, in which he wanted to correct some mistakes in the preceding Dutch version. He had already received the proofs, and the desk editor claimed that changes were against the rules. Eventually Korteweg's help had to be enlisted in an effort to get the corrections done; he really had to put his foot down to protect his student. In his words, 'One does not force a man to go round with his tie in disorder, even if he gets company—because he tied it the wrong way in the morning!' Moreover, he said, Brouwer was already greatly upset, and it could easily set him fretting, 'It is difficult for a person of a conscientious disposition if he has to leave mistakes in his work, and we must be somewhat careful with Brouwer. The last few years he has been alright, but not so that it doesn't worry me any more.'

---

[12] Brouwer to Korteweg, 5 November 1906.

[13] The formal ceremony of the defence of the dissertation and the awarding of the doctor's degree is called the *promotie* in Holland. The tradition of the public defence of the dissertation before the faculty has, with minor changes, been preserved until to-day.

The desk editor grudgingly gave in, 'I don't want to cause Brouwer to be ill—and so it has to be done'.

Finally, on 7 September, Brouwer[14] reported that he had stopped reading new material, and that he had started to order his notes into chapters.

The draft listed 8 chapters:[15]

- Axiomatic Foundations
- Examples of Axiomatic Uniqueness Proofs
- Genetic in the Mind, and Empirical Foundations
- Mathematics and Liberation of the Mind, and Philosophical Valuation of the various Exact Sciences
- Mathematics and Society
- The Construction itself (the Building proper[16])
- So-called Philosophical Foundations
- Criticism of Set Theory

By that time it was just plainly impossible to finish the dissertation before 12 November; presumably Korteweg again exerted his influence to protect Brouwer's income, for in the letter of 18 October Brouwer thanks Korteweg for approaching the St. Jobs Foundation.

Korteweg, indeed, had convinced the regents that in this particular case an exception should be made. The person was clearly more than ordinarily gifted, and financial problems would probably put him under severe stress. So, he proposed to end Brouwer's grant on 11 November, and to pay the deducted DFl 1.500,–[17] after the completion of the Promotion. This proposal was almost immediately accepted by the Board.

On 16 October Brouwer submitted a tentative list of chapters to Korteweg—it differed rather from the above draft:

(1) The construction of mathematics.
(2) Its origin in connection with experience.
(3) Its philosophical meaning
(4) Its founding on axioms

---

[14]Letter of 7 September 1906 to Korteweg.

[15]– *Axiomatische Grondslagen – Voorbeelden van Axiomatische Uniciteits-bewijzen*
– *Genetisch in de Geest en Empirische Grondslagen*
– *Wiskunde en Geestesvrijmaking en Philosophische Waardering der verschillende Exacte Wetenschappen. (Verband tussen Levens-Logica en Wiskunde)*
– *Wiskunde en Samenleving*
– *De Opbouw zelf*
– *Zogenaamde Philosophische Grondslagen*
– *Aanmerkingen op de Mengenlehre*

[16]Brouwer used the terms *'opbouwen', 'bouwen', 'wiskundig gebouw'*. in his Dutch publications; the literal translation is 'building', 'mathematical building'; we will stick to the more usual (but also less colourful) 'construction'.

[17]The result of the yearly deductions of DFl 150,– since 1897, cf. p. 13.

(5) Its value for society
(6) Its value for the individual.[18]

Within a month this list was further reduced to three chapters which constituted the final version of the dissertation:

I The Construction of Mathematics
II Mathematics and Experience
III Mathematics and Logic

The whole episode of the preparation of the dissertation is typical of a dignified and noble academic past, a time when letters could be exchanged the same day, when professors had time to read drafts of dissertations at short notice, when publishers produced hand-composed proofs within days. Considering that in September 1906 Brouwer had not put any part of the dissertation worth mentioning on paper, and that in October he was still arranging the material, when he had already told the publisher to start printing, it is a small miracle that he eventually finished in February 1907. In the middle of October he had finished Chap. I, and he more or less counted on Korteweg's kind co-operation, so that he could send it to the printer in a week's time.

The second chapter was, however, an altogether different matter; it dealt with the place of mathematics in the outer world, and the first version must have been something of a shock to Korteweg. He had read, or at least glanced through, the philosophical opus *Life, Art and Mysticism*, but probably had hoped and expected that Brouwer would keep his philosophical activity and his scientific activity strictly separated. In Brouwer's case the philosophy was, however, the basic ingredient that made the mathematics work. Here Brouwer and Korteweg had interpreted their agreement in different ways. Brouwer had understood Korteweg's condition 'as long as there is enough mathematics left' as a lower bound on the mathematical content, without stringent restriction on the philosophical part. Korteweg on the other hand wanted mathematics, and possibly philosophy, but only if it was of the traditional sort—say Kant, Poincaré, Frege, Russell. Brouwer expressed his view in a letter:

> [You] suspected that [the topic] would strongly drive me into philosophy, which it indeed did. To the extent that I even, at times, completely lost sight of mathematics. But what I have brought you now, exclusively treats *how mathematics roots in life*, and how, therefore, the points of departure of the theory ought to be; and all special subjects of the dissertation derive their meaning in relation to this fundamental thesis.[19]

[18](1) *De opbouw van de wiskunde.*
(2) *Haar wording in verband met de ervaring.*
(3) *Haar philosophische betekenis.*
(4) *Haar grondvesting op axioma's.*
(5) *Haar waarde voor de samenleving.*
(6) *Haar waarde voor het individu.*

[19]Brouwer to Korteweg, 5 November 1906.

Brouwer had no intention to lead a double life, mathematics during working hours, and philosophy as a gentlemanly occupation for the leisure hours. On the contrary. In Chap. I he had already put the *ur-intuition* on the stage and in Chap. II he intended to present an eloquent description of the role of science, that is mathematics, in the domination of the environment by man. Indeed, the introduction to Chap. II, as originally conceived by Brouwer, was a clear and coherent elaboration of the moral issues in *Life, Art and Mysticism*. 'The struggle for the domination of nature and fellow humans', according to him, 'is strikingly different from the brutal assimilation and destruction practised by other creatures. The secret of the success of man, lies in his potential for objectivisation of the world.'

> The primeval phenomenon is simply the intuition of time, in which the iteration of 'thing-in-time, and one more thing' is possible, but in which (and this is a phenomenon outside mathematics) a sensation can resolve into constituent qualities, such that a single moment of life is lived as a sequence of qualitatively distinct things. One can, however, restrict oneself to the simple observation of those sequences as such, independent of the emotional content, that is from the various gradations of frightfulness and desirability of that which is observed in the outer world. (Restriction of the attention to intellectual contemplation). Then the strategy of the goal-oriented activity of humans, is to substitute the means for the goal (the preceding one for the succeeding one in the intellectually conceived sequences) when the instinct deems the chances of the means better in the struggle. [......] However, in general the strategy, consisting of the observation of causal sequences,[20] and in connection with this, the shift from goal to means, is successful, and gives mankind its power. Indeed, if the capacity were not effective, it would not be there, just as a lion would not have paws if they were not effective.
>
> One succeeds in discovering regularity in a restricted domain of phenomena, independent of other moments of life, and latent under intellectual observation. In this way one succeeds to find a weak spot in nature, to render in this way an enemy helpless in some essential part of life. In order to maintain the certainty of an observed regularity as long as possible, one often tries to *isolate* systems, that is to keep away that what is observed as interfering with the regularity; thus man creates much more regularity in nature, than originally occurred spontaneously in it. He desires that regularity because it makes him stronger in the struggle for existence, because it enables him to predict, and to take his measures. (Rejected parts)[21]

[20] In his later writings, for example Brouwer (1929a, 1933a), Brouwer introduces the term 'causal sequence', which denotes the 'equivalence class' of sequences under the identification by the subject. One may safely conclude from the use of 'sequence' here, that Brouwer had this sort of sequence in mind. The individual, isolated sequence has to be supplemented with a notion of identification in order to receive a certain stability. We will anticipate the terminological practice of Brouwer's mature intuitionism in so far that we will freely use 'causal sequence' from now on.

[21] van Stigt (1979, 1990), van Dalen (2001b).

In the above lines Brouwer describes his key concept, the 'leap from end to means'.[22] It involves by itself a measure of externalisation: observed sequences are detached from the observing ego, they are lumped together in 'similar' or 'identical' sequences, which are known as *causal sequences*. These causal sequences are essential in the shift from goal to means. Because causal sequences are no longer accidental, one-time, observation (or sensation) sequences, but rather more-or-less stable sequences which we recognise on the basis of earlier observations, we may, when we wish to attain an event $B$, and we have learned that in a particular causal sequence another event $A$ invariably precedes $B$, try to realise the means $A$, and trust that the causal sequence will take us to $B$.

The first, and partly rejected, version of Chap. II was, like *Life, Art and Mysticism*, based on strong moral views. Mathematics, in the wider sense intended by Brouwer, is a prime tool for the subjugation of nature—and fellow humans—and as such it is despicable: 'He who dominates, is already cursed, and they are cursed qualities that help to dominate', and 'Since the assimilation of the environment removes it [mankind] ever further from the natural situation that originally supported mankind, each conquered and adapted environment becomes ultimately intolerable for mankind.' (Rejected parts)

The topic 'language' was—understandably—also treated; this time Brouwer's views were somewhat milder than in *Life, Art and Mysticism*. Language is presented as an imperfect tool of communication, which could never guarantee the evocation of the same sensations in different individuals, but which could at least direct the mathematical actions of the listeners in the direction desired by the speaker. The mystical writer, however,

> will carefully seek to avoid everything that smacks of mathematics or logic; otherwise weak minds might easily be led to mathematical believing or mathematical acting, outside the domain where either society, or their personal struggle for life requires it, and thus come to all sorts of foolishness. (Rejected parts)

Chapter II had to render an account of the pleasant, but often-thought unreasonable, success of mathematics in handling the world, say physics and the other natural sciences. Brouwer did not duck this responsibility. He showed how his philosophy provided a natural explanation of the sciences, and the place of mathematics in them.

In his man-based system, physics was no exception in the general framework; the physical phenomena were to be handled as causal sequences of a specific kind.

> Now, should it not be surprising, that one really succeeds, not only to observe causal sequences that reappear again and again, but that so many groups of phenomena, that affect the naive senses in a totally different way, can be brought under some general viewpoints, which correspond with simply con-

[22]Brouwer uses the Dutch *sprong* ('jump' or 'leap'), probably to indicate that the phenomenon is not a gradual continuous shift, but a discontinuous transition. In later publications Brouwer calls this the *mathematische Handlung* (mathematical act) Brouwer (1929a), and *cunning act* Brouwer (1949c).

structible mathematical systems? That would be a miracle indeed, but let us keep in mind, that the physicist only deals with the projections of the phenomena on his measuring instruments, which are all manufactured following a similar technique from fairly similar *rigid bodies*, and that it is not surprising that the phenomena are forced to share a similarity with either similar 'laws', or no laws. The laws of astronomy, for example, are no more than the laws of our measuring instruments, when they are used to follow the course of the heavenly bodies. (Rejected parts)

The Ph.D. adviser (*promotor* in Dutch) Korteweg, had the challenging task of guiding his student in the proper academic directions. Teacher and student had regular discussions, and (fortunately for us) a frequent exchange of letters. After Brouwer's letter of 16 October, they must have had a fairly heated discussion, for on 18 October Korteweg writes that he spent his time re-reading the first five pages of the manuscript. He admitted that after the oral explanation '...everything appeared in a different light, and even seems very well and clearly formulated', and he regretted his hasty judgement.

—'I am sorry that, when we talked I still was of another opinion, so that I felt it necessary to tell you a few things which must discourage you, although that was not my intention. So now temper my regret of that action by working calmly but steadily at your second part.'

Korteweg and Brouwer did not quite see eye-to-eye on the foundational topics of the thesis. They exchanged a good many letters and probably had even more discussion sessions.

On Sunday(!) 4 November there was another discussion between student and teacher at the home of Korteweg. Apparently Korteweg had again expressed his doubts as to the admissibility of a number of topics, for the next day Brouwer wrote a long letter[23] defending his approach.

In the emotions of the moment, Brouwer even forgot the civil opening, and plunged at once into the discussion. The first sentence contains the barely veiled suggestion that his adviser might be out of touch with the important developments in mathematics: he sent Korteweg the issue of the *Göttinger Nachrichten* that contained the famous problems which Hilbert had presented at the International Conference of Mathematicians in Paris in 1900.[24] These problems, which became known as *Hilbert's problems*, were intended as a challenge to, and homework for, the mathematicians of the twentieth century. And indeed, they have occupied a host of mathematicians; the problems often opened up new and important avenues in mathematics.

Brouwer, in his letter, claimed the (partial) solution of three of the problems; we shall see to what extent he did solve any of them, but it is quite clear that he did not wish to become a doctor on the strength of some routine research. He immediately went for the highest honours!

[23]Brouwer to Korteweg, 5 November 1906.

[24]Hilbert (1900), Browder (1976), Gray (2000).

The problems dealt with are the first, second and fifth. Problem 1, the *Continuum Problem* went back to Cantor (who at one time thought he had a solution); it is usually presented as the *continuum hypothesis*: Cantor had conjectured that *infinite subsets of the continuum are either countable or have the same power as the continuum*. Problem 1 asked to prove or disprove the Continuum Hypothesis.

Problem 2 posed the question of the *consistency of arithmetic*, that is it asked for a proof that no contradictions can be derived from the usual axioms of arithmetic (as formulated, for example, by Peano). And, finally, problem 5 asked to rid the theory of Lie groups of the traditional differentiability conditions.

Somewhat defiantly, Brouwer wrote, 'I send you this book, because I thought to perceive that you had some doubt whether the topics in my dissertation were, after all, worth the effort.'

Apparently, Korteweg had demonstrated a certain lack of confidence in the choice of topics. The letter of Brouwer eloquently defended the contents of the thesis. It is most likely that Korteweg was worried about the philosophical passages; he might, with good reason, feel some misgivings about the reception of this work in the faculty. The reference to Hilbert's problems was clearly designed to put him at ease. Brouwer was, however, not the person to give up his philosophical views without a struggle, and so the greater part of the letter is devoted to a defence of the foundational parts. In reply to Korteweg's questioning of the place of Kant in a mathematical dissertation, he pointed out that Russell's *Essay on the Foundations of Geometry* and Couturat's *Les Principes des Mathématiques* were largely devoted to discussions of the Kantian principles, 'and Poincaré points out that the present foundational controversies are a continuation of the old mathematical–philosophical struggle between Kant and Leibniz'. 'Why avoid the name of Kant', he said, 'when his ideas are discussed in mathematical texts, for example, of Russell and Couturat?'

It was, however, not just the pure philosophical content that bothered Korteweg; even the views of his student on the topic of physics—usually deemed on safe middle ground—were disturbing.

Korteweg must have commented on that particular point too, for Brouwer explicitly defended his views by referring to Poincaré, who had expressed similar opinions. Poincaré had in *Science et Hypothèse* declared that the only meaning of 'the earth rotates' is that 'in order to classify some phenomena in a simple way, it is most convenient to assume that the earth rotates'. Brouwer wholeheartedly supported Poincaré's view:

> And I think that something like that is far from absurd; on the contrary, that it convinces anyone at first reading. The system of the heavenly bodies is nothing but a mathematical system, freely built by us, of which people are proud, because it is only in this way effective in controlling the phenomena.[25]

The philosophical credentials of Brouwer had been questioned the day before by Korteweg, who had maintained that he was not certain at all that his student had

[25]Brouwer to Korteweg, 5 November 1906.

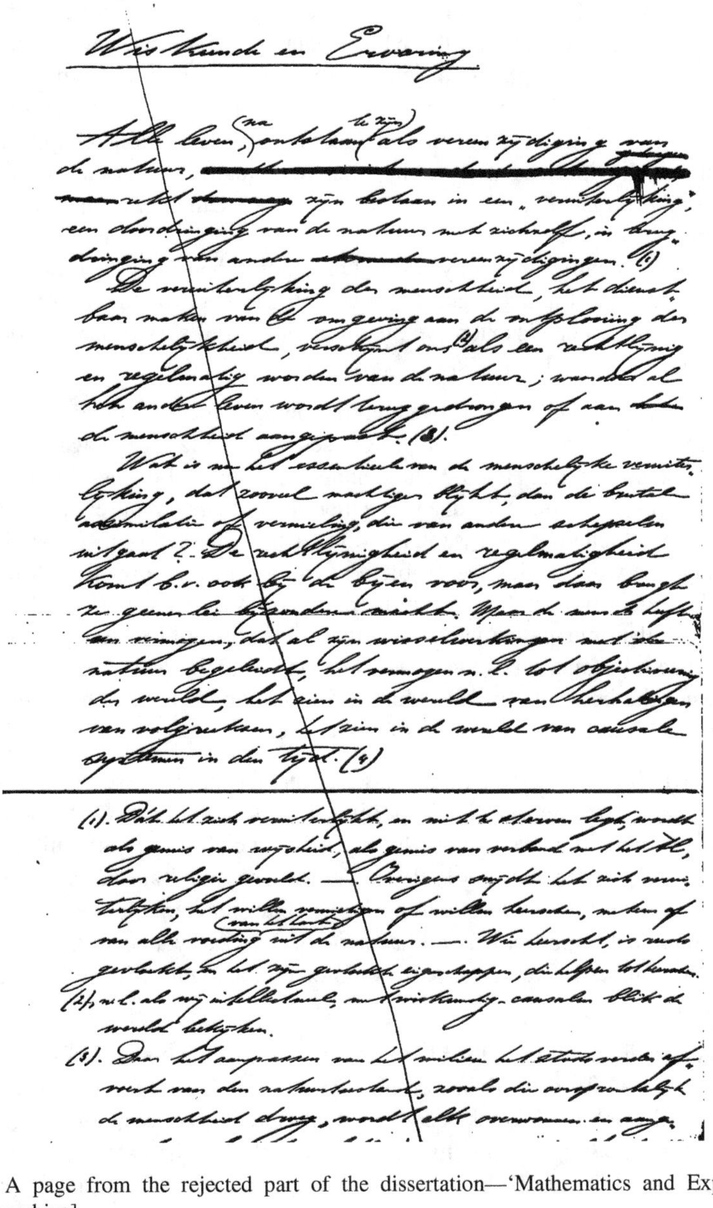


**Fig. 3.1** A page from the rejected part of the dissertation—'Mathematics and Experience' [Brouwer archive]

studied Kant thoroughly enough to be entitled to an opinion. Brouwer felt that he could not let this pass: 'I can, of course, not give you that certainty, but I can anyway *tell* you that I have read all of the *Critique of Pure Reason* and studied many parts (including those that are relevant to my dissertation) repeatedly and seriously.'

Korteweg, wisely, took his time; he replied to Brouwer that same day[26] that he would answer after he had read the whole manuscript, adding that 'I did not for a moment doubt the thoroughness of your mathematical preliminary studies.'

Brouwer could not, however, bring himself to wait for Korteweg's comments; he dispatched another extensive defence on 7 November. 'The second chapter', he said, 'essentially dealt with two topics: (a) how the mathematical experience accompanies the essentially-human actions; (b) to what extent the mathematics of experience can be a priori.'

The material dealing with physics and astronomy had only been inserted as an example of (a), and, if it had to go, Brouwer would rather drop the rest of Chap. II as well. He also motivated his extensive treatment of Russell's *Essay on the Foundation of Geometry*:

> he is the only one who writes about the philosophical foundation of mathematics, and (at least most of the time) uses an *exact* language against which one can be up in arms. Hegel, Schopenhauer, Lotze and Fechner don't do that (I just drop a few names) if I discussed them I would get on purely philosophical terrain, which you don't want, nor do I wish so. Russell is, as far as I know, the only one who tackles the problem of the a priori, equipped with mathematics. And, criticising him, I could remain on *my own ground*, which, in my eyes, is all the time mathematical, and in this criticising I found opportunity to emphasise my standpoint in various directions. And that was for me the main point, not the book of Russell, which I find important for its character only, but which I find for the rest an absolute failure.

On the whole Brouwer was in his dissertation critical of Russell in a polite, academic way, as the reader can check for himself. But in the résumé, which Brouwer made before the actual writing of the text, Brouwer found it difficult to control himself. He repeatedly speaks of 'Russell's nonsense' and at one place he cannot stand it any longer: 'To put something higher than mathematics, you must feel a non-mathematical intuition; Russell has none of that, and yet he starts to bullshit.'

On 11 November Korteweg gave his opinion on the manuscript so far—in particular Chap. II.

> After receiving your letter I have again considered whether I could accept it as it is now. But really Brouwer, this won't do.[27] A kind of pessimistic and mystic philosophy of life has been woven into it that is no longer mathematics, and has also nothing to do with the foundations of mathematics. It may here and there have coalesced in your mind with mathematics, but that is wholly subjective. One can in that respect totally differ with you, and yet completely share your views on the foundations of mathematics. I am convinced that every supervisor, young or old, sharing or not sharing your philosophy of life,

[26] Korteweg to Brouwer, 5 November 1906. Note that in those days one could dispatch a letter and receive the reply the same day!

[27] *Maar waarlijk, Brouwer het gaat niet*, 11 November 1906.

> would object to its incorporation in a mathematical dissertation. In my opinion your dissertation can only gain by removing it. It now gives it a character of bizarreness which can only harm it.

Korteweg had clearly been rather piqued by Brouwer's hints as to his expertise. His retort was gentle but firm:

> You inform me of all sorts of matters which could not possibly be unknown to me, as a regular reviewer[28] of the *Revue de Métaphysique et de Morale*, as if they were things that I would not know. You thought to have understood that you were not allowed to use the name of Kant, even where it concerned opinions of Kant on mathematics, and you thought that I found the view 'that astronomy is nothing but a convenient summary of causal sequences in the reading of our measuring instruments' absurd. No, not *that* view; I admit that one can present the matter in that way, although in my opinion the general law of gravity has little to do, indeed, with our measuring instruments which led to its discovery, than that these make measurement *überhaupt* possible; but that the similarity of the laws which are valid in very different parts of physics would find its origin in the similarity of the used instruments; it was that claim that appeared absurd to me.

Although Korteweg insisted that the passages dealing with Brouwer's philosophy of life should be struck out, he did appreciate the comments on Kant, and he saw no objections to the more traditional treatment of the philosophical issues in mathematics. In spite of his firm stand concerning the more mystical parts, the letter was rather conciliatory. In particular, he wrote, his questions did not in any way imply superficiality on Brouwer's part; they were exactly what they should be: requests for elucidation.

Brouwer's reply, two days later, is interesting because it contained an elaboration of his views on the role of mathematics in physics. Little of this discussion has reached a wider audience, thus contributing to the popular misconception that Brouwer's theory could not, and did not, provide a foundational basis for applied mathematics.

The following passages from the letter of 6 November 1906 could be read independently of their philosophical (that is Brouwerian) background; one could even imagine a positivist writing it. It is good, however, to keep in mind that it is an integral part of Brouwer's idealist philosophy.

> You think that [...] the general law of attraction has preciously little to do with the instruments, which have led to its discovery; but are laws anything but summaries of phenomena by induction, means for the control of phenomena, and existing only in the human mind? In itself the law of attraction exists only with respect to the Euclidean space, and that only exists through an efficient, but arbitrary extension of the domain of motion here on earth. Without

[28]For the *Revues semestrielles*.

> solid bodies on earth, the law of attraction would not exist, and the relation between both is laid down by the astronomical measuring tools. The law of attraction exists in relation to the astronomical phenomena, as molecules in relation to the state equation; both appear to summarise efficiently a group of phenomena, and to be effective as a tool for prediction; only the law of attraction wins in simplicity from the molecular theory. But once again: the law of attraction is a hypothesis; the distance from the earth to the sun is just as much a hypothesis.

With respect to the issue of 'similar instruments yield similar laws', Brouwer pointed out that:

> There is no difference between the electromagnetic field of a Daniell-element, projected on our measuring instruments, and that of a Leclancher-element; but if we consider it without bias, we must expect that there must be as much difference between the fields, as between copper sulphate and ammonium chloride; but on our counting- and measuring instinct, operating by means of certain instruments, they both act in the same way; it turns out that a similar mathematical system can be applied to both, but it is only the lack of suitable instruments, which so far has prevented us from finding other mathematical systems which can be applied to the one field and not to the other. In each phase of development of physics, the measuring instruments—which 'have been found suitable', form a restricted totality, relative to the totality of measuring instruments which 'could be found suitable to govern all kinds of other, as yet unknown, phenomena'. Parallel to this, we have that 'the 'mathematical systems which have already been applied to nature' form a restricted totality relative to the totality of mathematics, that 'would be applicable to nature, if physics had been sufficiently extended'.—And where now every restricted group of mathematical systems has its invariants, it is to be expected that every restricted group of physical phenomena has its invariants, exactly on the grounds of those restrictions, in the form of laws or principles which are valid for all phenomena of that group.

The letter continues to elaborate in a convincing way the role of mathematics, in which, after all, the physical laws are formulated. Brouwer argues that physical measurements yield outcomes in suitable rigid mathematical groups (for example the group of rotations in the plane, when measuring torsion), and hence that

> Each physically measurable quantity is eventually reduced to a measurement in a rigid group, and it is *the laws of those measurements which are looked for in all sorts of different circumstances.*

Some people have questioned the role of Korteweg in the process of Brouwer's development; Brouwer's best-known student, Heyting, was for example completely unaware of Korteweg's influence on Brouwer; he was, on the contrary convinced that Korteweg had played no role whatsoever in Brouwer's scientific development. In the sense of modifying Brouwer's philosophical views or even determining the topics of the thesis, this is certainly correct, but in the sense of guiding the young man, Korteweg doubtlessly played an important role—as the cited correspondence bears out.[29] By asking the right questions, he forced Brouwer to refine his arguments and formulations. Moreover, a dictatorial Ph.D. adviser would have put off Brouwer. It is unthinkable that one could have forced him to forsake his personal ideals in mathematics.

In spite of his personal convictions, Brouwer could see Korteweg's point that the highly subjective philosophical sections would not enhance his credibility among his fellow scientists, and so he dropped them—but he kept the manuscript among his papers.

He realised of course, that without the basic motivation Chap. II had lost some of its force:

> After their sudden appearance on the foreground as a replacement of their late leader, it was not immediately possible, to dress them all, so that they together could by themselves save the performance.[30]

It is true that the reading of the dissertation without the benefit of the 'rejected parts' leaves one with the uncomfortable impression that Brouwer's arguments were a bit ad hoc.

## 3.3 On the Role of Logic

Although the agreed date of the defence of the dissertation came unpleasantly close, adviser and student still continued their exchange of arguments and questions. The final topic to be discussed in their letters was that of logic. In his letters of 18 and 23 January, Brouwer answered questions and objections of Korteweg. Korteweg must have concluded from Brouwer's philosophical discourse that logic was seen as a part of, or possibly an application of, psychology—a not uncommon misconception at the end of the nineteenth century. Frege and Husserl had already demolished the bastion of psychologism, but it seems fairly certain that Brouwer was unaware of those particular writings.

In his answer to Korteweg[31] he dispelled any doubts on this point: 'From your characterisation of theoretical logic as a part of psychology, I understood that I had expressed myself rather vaguely, for it was my express intention to show theoretical logic, although it is a science, has under no circumstance significance for psychology.'

---

[29]The correspondence was after Korteweg's death deposited in the Amsterdam University Library by Mannoury, but this, apparently, was not known to Heyting. That the Korteweg files were unearthed was the result of a hint of Bastiaan Willink, a nephew of Korteweg, who was aware of the existence of the Korteweg archive.

[30]Brouwer to Korteweg, 11 January 1907.

[31]Brouwer to Korteweg, 18 January 1907.

In the next letter[32] he elaborated his views on logic. In complete opposition to the then prevailing beliefs, he argued forcefully that mathematical reasoning is *not* logical reasoning. Mathematical reasoning, in his view, consisted of (mental) constructions, whereas logical reasoning took place in the realm of language. Thus, he stated in the dissertation that 'Mathematics is independent of Logic' (Diss. p. 125) and 'Logic depends on Mathematics' (Diss. p. 127). On the whole his view of the role of logic was rather dim. He considered it as a secondary activity dealing with linguistic figures that could on principle not be expected to take the place of mathematical constructions, and in those cases where one could easily supply the missing constructions, as in the syllogism 'all men are mortal, Socrates is a man, hence Socrates is mortal', the result is not very informative.

It would be a mistake however to think that Brouwer dismissed logic altogether. This becomes clear on the very first page of the third chapter. The first instance where the traditional concatenation of syllogisms becomes non-trivial is where the so-called *hypothetical judgement* of logic seems to be called for. Let us say that two statements $A$ and $B$ describe structures **A** and **B** to be constructed. In the case of 'if $A$ then $B$' one seems to assume the construction of **A** in order to construct **B**. From Brouwer's constructive point of view that would be highly undesirable. It may be the case the we know little about constructions for **A** and **B**, while we have no problem accepting that any construction for **A** will yield a construction for **B**. All we need, so to speak is a construction that will turn any construction for **A** into a construction for **B**. Here is a simple example: if a point lies outside a circle, then there is a tangent through this point to the circle. If we find, or are given a construction that establishes that the point is outside the circle, then we know how to carry out the construction of a tangent. Or, if 0=1, then 1=2. Here we even know that there is no construction for 0=1, but we know very well how to show that 1=2 if someone were to give a construction for 0=1. We see here the germ of the 'proof = construction' notion of Brouwer. Unfortunately Brouwer's formulation is not wholly unambiguous, so that the consequences of this idea are not totally clear. The more precise formulations had to wait for Kolmogorov and Heyting.[33] It is quite clear that here the proof-construction interpretation for the implication is intended.[34]

Brouwer's view of the use of logic in mathematics was downright pessimistic: 'I show in the beginning of the chapter that mathematical arguments are *not* logical arguments, that it only makes use of the connectives of logical argumentation out of poverty, and that it will keep the accompaniment of the language of logical argumentation alive, long after the human intellect will have outgrown the logical arguments.' In spite of this harsh indictment, he apparently did reconsider the role of logic after he finished the dissertation, for a year later, in the 'Unreliability'-paper, he presented a milder view.

Above all, his objections were directed against the shift of attention from mathematical arguments (that is mental constructions) to the study of regularities in the

[32]Brouwer to Korteweg, 23 January 1907.

[33]See van Dalen (2004), van Atten (2008).

[34]Note that the construction interpretation for the remaining connectives is far less problematic.

accompanying *language* of mathematics. Even when the system of mathematical–logical language is shown to be free of contradictions, its subject matter is 'language', and there is no immediate connection with mathematics proper, for example arithmetic (Diss. p. 132).

A more argued case against logic was presented in the letter of 23 January to Korteweg:

> One knows nowadays very well, that if one derives by logical reasoning something concerning the outer world that was not so immediately a priori clear, it is for this very reason totally unreliable; for one does not believe any longer the underlying postulate that the world must be a finite, albeit very large, number of atoms, and that each word should represent a (hence also finite) group, or group of groups, of atoms. In other words, one knows very well that the world is not a logical system, and that it does not lend itself to logical reasoning; one knows very well that each debate is really humbug, and that it is only to be settled for mathematical problems, but then *not through logical reasoning* (although this may seem to be so in an inadequate language; how misleading this appearance is, may be seen in the case of axiomatic foundations and transfinite numbers) *but through mathematical arguments.*
>
> Theoretical logic does not teach anything in the present world, and one knows this, at least sensible people do. It now only serves lawyers and demagogues; not to instruct others, but to fool them. And that this can be done is because the vulgar unconsciously argue as follows: that language with logical figures exists, and so it will presumably be useful—and they meekly will let themselves be deceived by it; just as I heard some people defend their gin drinking with the words: 'why else is there gin?'. Whoever has illusions of improving the world can just as well crusade against the language of logical reasoning, as against alcohol and no more is it a 'strange company' that does not drink alcohol, than it is a 'strange company' that does not argue logically. Although I believe that perhaps no abuse is entrenched more than that which is blended with the most popular parts of language.

It is instructive to compare the above paragraph to Brouwer's mystic writings. The tone is more tempered (in spite of the reference to gin drinking), and there is an argument, not just indignant rejection.

Brouwer, in his letter, calmly and clearly defends the priority of mathematical, that is constructive, argument over logical proof. He uses an example of Euclidean geometry: an isosceles triangle has an acute angle. The logical interpretation is that the set of isosceles triangles is a subset of the set of acute angled triangles (Brouwer goes so far as to reduce the problem to subsets of $\mathbb{R}_6$), but the mathematical interpretation is that one concludes from the construction of an isosceles triangle that it has an acute angle.[35] In spite of his unmistakable constructive intentions, Brouwer

[35] Here again the proof-construction interpretation of the implication may be conjectured.

made at this point an appeal to reductio ad absurdum (if the angle be right or obtuse, the construction fails). In fact the treatment of logic in a constructive context suffered from too much haste.

The part about the demagogue reveals a glimpse of the emotions underlying Brouwer's scientific activity. Realising that the contents of the letter were rather unconventional, Brouwer apologetically added: 'Perhaps I have after all expressed my intentions more clearly in this somewhat wildly written letter, than in the subdued text. But, maybe, the text will appear in another light after this letter. That would please me very much.'

The explanation above of the shortcomings of logic, in particular the consequences of the non-finiteness of the universe, recurs in later writings, including the dissertation itself; the finiteness assumption underlying the belief in logic is mentioned on page 130 of the dissertation.

In spite of the 'down with logic' rhetoric, Brouwer saw a positive role for logic. He envisioned a constructive revision of the subject. The dissertation contains the germ of the 'proof = construction' idea.

The dissertation was finished in the nick of time; in spite of some printing problems, Brouwer managed to meet the faculty deadline. It reflected Brouwer's ambitions by dealing with a multitude of topics. The first chapter contained the 'real' mathematical material. Basically it contained the partial solution to Hilbert's problem no. 5, the treatment of Lie groups without the differentiability conditions, and problem no. 1—the continuum problem. Brouwer showed that 'independent of the differentiability, [...] there is only one construction for the one, resp. two-parameter continuous, uniform groups' (Diss. p. 35), while he left the three parameter case as an open problem.

After an investigation of the *Helmholtz–Lie Raumproblem*, Brouwer turned to the continuum problem (see p. 90). The first chapter closed with a brief treatment of non-Archimedean uniform groups on the continuum, and the non-Archimedean geometries.

The topological part of the dissertation was, so to speak, Brouwer's ticket to the mathematical community. An improved version was submitted to the *Mathematische Annalen*, and it was the topic of one of Brouwer's talks at the international congress in Rome in 1908. His work on Lie groups immediately drew the attention of the experts (see p. 127).

There is one particular feature of Chap. I which we have not mentioned so far, and which even in 1907 must have puzzled the readers. At the end of the nineteenth century it had become an accepted insight that the real numbers should be introduced as a derivative of simpler number systems—the rationals, the integers and ultimately the natural numbers. This was the result of the efforts of Weierstrass, Cantor, Dedekind and others, the so-called arithmetisation of analysis.

Brouwer did not follow this new trend; he firmly declared that the continuum and the natural numbers were both given as parts of the *ur-intuition of mathematics*:

> the substratum of all perception of change, which is divested of all quality, a unity of continuous and discrete, a possibility of the thinking together of

several units, connected by a 'between', which never exhausts itself by the interpolation of new units. (Diss. p. 8)

Thus the continuum is immediately given by intuition, and not reduced to the discrete. Over this intuitive continuum a scale can be constructed of a denumerable set of discrete points (for example the dual (or binary) scale). Points of the continuum may be approximated by approximation sequences—'which can, however, never be regarded as finished, thus have to be considered as partly unknown'. Brouwer recognised the possibility that the scale need not automatically be everywhere dense (that is to say, the intuitive continuum need not be Archimedean), but he constructed, by 'brute force', segments in which there were no point of the scale. Or to put it positively, two points are viewed as distinct,[36] when their dual approximations differ after a finite number of digits. The continuum will turn up again, when we discuss the continuum problem.

## 3.4 Mathematics and the World

Chapter II, which had suffered severely under Korteweg's axe, expounds Brouwer's views on the relation of mathematics to experience. In spite of Brouwer's rather solipsistic views in the Profession of Faith, the formulations in this chapter are fairly neutral, that is an outer world is introduced without comments. One may assume that at this point Brouwer did not want to be drawn into ill-timed philosophical discussions. It should be observed that a neutral terminology by no means excluded an idealist—nor a realist—reading.

The causal sequences take an important place in this chapter, and Brouwer argues that here mathematics plays its role in the struggle for life.

The basic scheme of Brouwer's grand design was to handle causal sequences in such a way that regularities occur, and that predictions can be made.

Here the individual makes clever use of the end-to-means jump (which is no longer diabolical in the thesis). After remarking that, there is no guarantee that in a causal sequence the end will *always* be obtained after realising the means (the rule may no longer apply) Brouwer holds forth that:

> in general the tactics, consisting of the consideration of the causal sequences and on the regression from the end to the means, where it is easier to intervene in the means, appears to be efficient, and provides mankind with its power. One succeeds discovering regularity in a restricted domain of phenomena, independent of other phenomena, which can thus remain completely latent in intellectual consideration.
>
> In order to maintain, as long as possible, the certainty of the regularity, one tries to *isolate* systems in the process, that is to keep away that what has been seen to disturb the regularity. Thus man makes much more regularity in

[36] In later writings Brouwer would use 'apart' for this positive notion.

> nature than originally occurred spontaneously in it. He *wishes* that regularity, because it strengthens him in the struggle for existence, because it enables him to predict, and to take his measures. [...]
>
> The intellectual consideration of the world widens its scope because one builds from the ur-intuition of the intellect abstract mathematics, independent of direct applicability. And thus one has a stock of virtual causal sequences at hand, which are only waiting for an occasion to be projected on reality. (Diss. p. 82)

So, the external world (one may think of the physical world) is viewed as a huge configuration of causal sequences, and the individual will try to optimise his environment (the 'struggle for life'). It is therefore of importance to make the end-to-means jump more and more perfect. For this purpose, the individual tries to eliminate events and phenomena that might jeopardise the transition from means to the end. For instance, one drops a ball in vacuum, or one observes the sky from an isolated mountain top. The mathematical treatment will have the best chance of succeeding in those cases were isolation can be realised. Of course, there are innumerable cases where no regularities can be called forth.

Brouwer's next observation is that things may get so complicated that the steps to be taken to obtain isolation cannot immediately be seen, or it might not be clear what the suitable means is for a given end. It is here that pure mathematics intervenes. The individual starts to generate virtual causal sequences that are not directly observed (or not the direct result of sensations). These sequences may be easier to handle, and the mathematical treatment will then yield results that can be applied in the case of concrete, actual causal sequences. In this way the individual develops a stock of mathematical theories that can be called on at any moment. Euclidean geometry is an example of such a successful mathematical theory (that is the stock of man-made imaginary causal sequences); it suffices for a large number of applications.

The most useful instances of this kind of mathematical manipulation are to be found in those cases where 'a large number of causal sequences are subsumed under one viewpoint by means of a mathematical system, called a *law*, built with the help of mathematical induction. The difference between two sequences falling under it then only depends on the difference in values of the parameters occurring in the law.' (Diss. p. 84)

We see that causal sequences and the end-to-means jump thus become the cornerstone of Brouwer's universe and its properties. In the course of time he refined the arguments, but he remained faithful to the original idea.[37]

The first part of Chap. II deals with this notion of causal sequences, laws and predictions (for example the mechanistic explanation of nature). The last part is concerned with philosophical matters, mainly in the context of Russell's *Essay on the Foundations of Geometry*. The discussion of Russell's *Foundations* is preceded by Brouwer's own views on *objectivity* and the notion of a priori.

On Brouwer's view 'objectivity' of quantities or laws simply means the 'invariance under the simplest or the most common interpretation'. For example:

[37] Brouwer (1929a, 1930, 1933a, 1949c).

> one usually calls mass objective, and thinks of its indestructibility; we have seen, however, that masses are nothing but coefficients which simplify through their introduction, the mathematical image of nature, and which remain invariant under mathematical transformations, which represent natural phenomena. Were one to find, however, natural phenomena which can most easily be represented by assuming the masses to be variable, then one could only keep calling them objective, on the grounds of their invariance under a *very important group of phenomena* in the image of nature. (Diss. p. 95)

So, objectivity gets a specific and well-determined meaning in Brouwer's view. Both physical time and space are, under this interpretation, highly objective. The matter is different for the notion of a priori; if one understands by a priori 'existence independent of experience', then on Brouwer's view, all of mathematics is a priori, Euclidean as well as non-Euclidean. If one understands by it 'necessary condition for the possibility of science', then since

> scientific experience finds its origin in the applications of intuitive mathematics to the real world, and since there is, except for experimental science, no other science than just those properties of that intuitive mathematics; we may call nothing a priori but that one thing, which is common to all mathematics, and which, on the other hand, is sufficient, to build all mathematics on—the intuition of many-one-ness, the ur-intuition of mathematics. And since this coincides with the awareness of time as change per se, we may say: *The only a priori element in science is time.* (Diss. p. 98)

Thus Brouwer sets himself to 'rectify and update' Kant's views. On pages 114 and 115 he spells out the inadequacies in Kant's arguments for the *a priority* of 3-dimensional Euclidean space.

He not only rejects Kant's assumptions (we cannot get external experiences except in the empirical space; empirical space is the 3-dimensional Euclidean space), but also the argument leading to Kant's conclusion.

We note that Brouwer added the observation that 'Properly speaking the building of intuitive mathematics *per se* is an action (*een daad*) and not a theory (*een theorie*) ....' The reader will recognise here Hermann Weyl's later dictum: 'mathematics is more an action than a theory' (*mehr ein Tun als eine Lehre*), cf. p. 313.

The discussion of Russell's *Essay* consists of a series of criticisms on methodological and mathematical issues (for example Russell's claim that a variable constant of space curvature is unthinkable).

## 3.5 Observations on Set Theory and Formalism

Chapter III, MATHEMATICS AND LOGIC, is largely a criticism of the developments of the last decennium. As we have seen, in Brouwer's view mathematics was the construction of 'mathematical buildings', and as such it was a language-independent activity. Logic, which was not (yet) exclusively identified with formal mathematical logic, dealt with descriptions of mathematical states of affairs, and according to the majority of mathematicians, mathematics could not do without logic in order to establish its theorems. Brouwer turned this around, and boldly asserted that *Mathematics is independent of Logic, and Logic is dependent on Mathematics* (Diss. p. 127).

Brouwer's criticism was directed against four recent developments in the foundations of mathematics:

1. The founding of mathematics on axioms.
2. The theory of transfinite numbers of Cantor.
3. The logistic of Peano–Russell.
4. The logical foundations of mathematics according to Hilbert.

The axiomatic method was criticised on fundamental grounds, that is to say, the properties and the study of the axiomatic systems, with their notion of consequence, cannot replace the intuitive mathematics of (mental) constructions. Furthermore, the basic desideratum of the axiomatic method, which says that one can always find a *mathematical* interpretation for a consistent logical system, is not substantiated. Brouwer was, however, cautious enough not to go too far. He did not state that there were cases where such a mathematical interpretation ('building' in his terminology) could not be provided; he merely stated that the existence

> was not established by the axiomaticians, not even for the case that the given conditions include that the thing that is looked for is a mathematically constructible system; for example, it has nowhere been proved that, if a finite number has to satisfy a system of conditions, of which it can be shown that they are not contradictory, that such a number then also exists (Diss. p. 142).

Brouwer added in a footnote that therefore it was not certain at all that 'every particular mathematical problem must necessarily admit an exact solution, be it that the answer to the given question can be given, or that the impossibility of the solution, and thus the necessity of the failure of all attempts, is established'. This conviction, *Hilbert's dogma*, was put forward by Hilbert in his lecture *Mathematische Probleme*, with the remark that every mathematician is at heart convinced of its correctness. Brouwer summarily declared it 'unfounded' in his Thesis no. 21.[38] In the light of his doubt of Hilbert's dogma, it is rather surprising that a few pages earlier Brouwer had acknowledged the *principle of the excluded middle* (PEM) as vacuous or trivial. This principle, also going by the names *principium tertii exclusi*, principle of the excluded third, and *principium tertium non datur*, was one of the basic principles of Aristotle's logic, together with the *principium contradictionis*. The

[38] A dissertation was in Holland always supplemented with a list of theses, which were presented without proof, and which had to demonstrate the proficiency of the candidate in a wide area. The theses also functioned as a concession to those members of the faculty who had to take part in the examination, but could not spare the time to read the dissertation.

latter asserts that one cannot at the same time have $A$ and *not* $A$ (i.e. $\neg(A \wedge \neg A)$),[39] and the former states that one has either $A$ or $\neg A$, and there is no third possibility (i.e. $A \vee \neg A$). Brouwer had accepted the principium contradictionis as evident.

In his view, the words of a mathematical argument of a logician are nothing but an accompaniment of the wordless building process. When the logician concludes a contradiction, the building mathematician observes that the building process is blocked. The latter does not need the principium contradictionis for this observation (Diss. p. 142). The principle of the excluded middle, he claimed, was correct but uninformative:

> 'A function is either differentiable, or not differentiable' tells us nothing, it expresses the same thing as the following: If a function is not differentiable, it is not differentiable (Diss. p. 131).

This statement will surprise the modern reader; how could Brouwer accept PEM and reject Hilbert's dogma?

Although a conclusive answer will be hard to give, a possible solution to this riddle may be conjectured. C. Bellaar-Spruyt was teaching philosophy in Amsterdam when Brouwer took his courses. Bellaar-Spruyt did not publish much, but he influenced Dutch philosophy by his firm defence of Kant, and his equally firm attacks on the then predominant materialistic philosophy, embodied by the Utrecht philosopher Opzoomer.

There is no conclusive evidence that Brouwer took the courses of Bellaar-Spruyt, but it is plausible that he did. In the Brouwer Archive there is a picture of Bellaar-Spruyt, which Brouwer kept for all those years, and—more relevant, Brouwer's own philosophy has evident Kantian features. A philosophically interested, curious student would certainly attend the courses of his local philosophy professor, in particular if he had the reputation of an independent thinker, like Bellaar-Spruyt.

The logic course in Amsterdam was also in the hands of Bellaar-Spruyt, and his lectures have been published. Here one finds a hint towards the solution of the above mentioned problem. PEM is explained by Bellaar-Spruyt by means of an example: 'If you deny that Alexander was a great man, well, then you have to acknowledge that he was not a great man. Both opposite judgements, A. was a great man and A. was not a great man, cannot both be false'.[40] Replace 'Alexander is a great man' by 'the function is differentiable', and you get Brouwer's example. It is plausible, and even likely, that Brouwer learned this particular form of PEM from Bellaar-Spruyt. Needless to say that this reading does not pass the criteria of the constructivist.

There may have been more influence of Bellaar-Spruyt on Brouwer than just this case of PEM. Some Brouwerian themes already occur in Bellaar-Spruyt's lectures. Here are some examples: (i) logic is different from psychology. Brouwer would wholeheartedly agree, but where Bellaar-Spruyt sees logic in its normative role, Brouwer only views logic as a means of recording the original mathematical constructive activity. (ii) In the section on 'The pernicious influence of language on

---

[39]From now on we will follow the conventions of logic, and write $\neg A$ for *not* $A$.

[40]Bellaar-Spruyt (1903), p. 18.

the clarity of our thoughts', Bellaar-Spruyt discusses a phenomenon that very much looks like a case of 'the leap from end to means': 'It is indeed a familiar psychological law that if a state of mind $A$ is often followed by $B$, the arising of $A$ leads to the arising of $B$.'[41]

Although it is still a big step from a simple observation to the systematic use in the context of causal sequences, one may guess that the young Brouwer amply received impressions in the lectures of Bellaar-Spruyt.

Summing up, one is tempted to see Brouwer's view of PEM as a thoughtless adoption of the formulation that was given in the course. It could, of course, also be that he hesitated to destroy one of the pillars of Aristotelian logic—after all, a more than two-thousand-year old heirloom!

Among his private notes there is some reference to PEM, that seems to indicate that he already mistrusted the principle. To jump ahead of our story, soon after he finished his dissertation, he hit on the correct reading of PEM:

> this requires that each assumption is either correct or incorrect, mathematically formulated: that for each assumed incorporation of systems in some way into each other, either the termination [success], or the encounter of an impossibility can be constructed. The problem of the validity of the principium tertii exclusi is thus equivalent to the problem of the *possibility of unsolvable mathematical problems*.[42]

In general, he argued, 'the principium tertii exclusi is, as yet, not reliable in infinite systems'. The paper also contained the first instance of his celebrated 'unsolved problems', that were at the basis of his later counterexamples, now known as *Brouwerian counterexamples:*

- Is there in the decimal expansion of $\pi$ a decimal that in the long run occurs more often than others?
- Are there in the decimal expansion of $\pi$ infinitely many pairs of consecutive decimals which are equal?

In both cases we (still) have no clue as to the truth of the statements, hence PEM cannot be considered to hold for them.

The above reflections were embodied in a paper, *The unreliability of the logical principles*, published in a semi-obscure Dutch philosophical magazine[43] (the same journal that had rejected Brouwer's review of *The foundations of a new poetry* of Scheltema[44]). As a consequence it was read by few colleagues, even in Holland. The paper was received with a certain mistrust; after long deliberation the editorial board consented to publish it. Its hesitation may have had something to do with the fact that Brouwer was a self-made philosopher, and with the friction between

[41] Bellaar-Spruyt (1903), p. 62.

[42] Brouwer (1908b), p. 5.

[43] *Tijdschrift voor Wijsbegeerte* (Journal for Philosophy) (Brouwer 1908b).

[44] Cf. p. 29.

Brouwer and the philosophical community, resulting from an attempt of Brouwer and Mannoury, two years before, to found a philosophical journal.

Maas & van Suchtelen, the publisher of Brouwer's dissertation, had approached Mannoury with a proposal to establish the first philosophical journal in Holland. Mannoury, after some consideration, answered that he was prepared to set up such a journal provided Brouwer would join him. Subsequently, the two explored the Dutch philosophical landscape in order to find out if the project was viable. They soon met determined opposition from the side of the professional philosophers, who probably did not want to see the Dutch journal in the hands of philosophical amateurs, so the various philosophical societies suddenly felt the urge to start a journal themselves. When Maas & van Suchtelen was confronted with the situation, it decided to make the best of it by joining sides with the professionals. Kohnstamm, who was the spokesman for them, made the condition that Brouwer and Mannoury would not be asked to join the editorial board.

The editor in charge of the paper, the physicist Kohnstamm, made it quite clear that his firm stand in the meeting of the board had saved the day. He also used the opportunity to say that, in his opinion, Brouwer had completely missed the point of the PEM, 'I cannot see that your observations on unsolvable mathematical problems violates the PEM. It seems to me that the matter is of the same sort as the question whether a square circle should be considered round or angular.[45] This would not be the last manifestation of incomprehension that Brouwer was confronted with in his long struggle for a renewed mathematics.

Whereas Brouwer's objections to logic and its uses in mathematics were of a general philosophical nature, his criticism of Hilbert's formalism had quite specific technical points. As the role of Hilbert and his formalism will become prominent in the nineteen-twenties, it is worthwhile to go into Brouwer's arguments in the dissertation. The criticism in the dissertation was directed at Hilbert's Heidelberg address,[46] *On the Foundations of Logic and Arithmetic*, which had appeared in 1905. Hilbert had set himself the task of completing what he had started with his *Foundations of Geometry* (1899), that is to show the consistency of number theory—on which the consistency of geometry rested. The Heidelberg talk could be considered as a try-out of a new method for consistency proofs. The systems treated were only small fragments of logic and arithmetic, but it was the method that counted. Hilbert had indicated means to show that in the formal systems under consideration one could not derive the formula expressing a contradiction. The reader who is not familiar with this type of argument may consider the following problem: is it possible that by using the usual rules of elementary algebra, one ends up with a formula with more left than right brackets? The answer is 'no', and one easily proves this by means of complete induction. So here one proves something about objects of the language of mathematics (that is formulas, strings of symbols), instead of the usual objects of mathematics, such as numbers, functions, triangles, ... Hilbert's aim was

[45] Kohnstamm to Brouwer, 3 January 1908.

[46] Hilbert (1905).

to use this technique to show that in a fully formalised part of mathematics (say arithmetic) one could not derive a contradiction (say '$0 = 1$'). The subject of 'deriving' was in this context to be taken in its formal sense, as, for instance, expressed in Hilbert's Paris address of 1900:

> When we are engaged in investigating the foundations of a science, we must set up a system of axioms which contains an exact and complete description of the relations existing between the elementary ideas of that science. The axioms so set up are, at the same time, the definitions of those elementary ideas; and no statement within the realm of the science, whose foundations we are testing, is held correct unless, it can be derived from those axioms by means of a finite number of logical steps.[47]

Brouwer viewed Hilbert's scheme to free the foundations of mathematics from intuition as misguided. He argued that if one proves by means of our intuition that in the formal-logical description of this intuition one cannot derive a contradiction, then one has not made any progress:

> Who will prove a mathematical theorem, by deriving it once more on the strength of the theorem itself, and will then say 'now the assumption is justified as well'? (Diss. p. 172).

Popularly speaking, we can paraphrase Brouwer's objections as: you need intuitive mathematics to justify the (consistency of) its formalisation (in particular, you need induction to justify induction) and consistency of the formalisation does not justify the underlying intuitive arguments (constructions). This insight was lacking in Hilbert's paper. Some thirty years later, Gödel drove home the message by means of technical logical arguments.

In addition to the criticism of Hilbert's meta-mathematics, Brouwer gave a meticulous analysis of the various levels involved in the proof theoretical treatment of mathematics, (Diss. pp. 173–175). He distinguished 8 levels in Hilbert's exposition of 1904, a classification that in precision out did both Hilbert–Bernays' and Tarski's later language–meta-language distinction.

He gave an 'enumeration of the phases which were mixed up in the logical treatment of mathematics':

1. The pure construction of intuitive mathematical systems, which, if applied, are externalised by viewing the world mathematically.
2. The language parallel of mathematics: mathematical speaking or writing.
3. The mathematical consideration of the language: logical language constructions, built according to principles from ordinary logic or its extension to the logic of relations, are observed, but the elements of those language buildings are linguistic accompaniments of mathematical constructions or relations.
4. No longer thinking of a meaning of the above mentioned logical figures; and the copying of the construction of those figures by a new mathematical system of

[47] Hilbert (1900).

the *second* order, for the time being without a language which accompanies the constructions; it is the system of the logisticians, which easily becomes susceptible to the figure of contradiction at the least free generalising extension, unless Hilbert's precautions are taken against it. And these precautions make up the real content of Hilbert's treatise.[48]

5. The language of logistics, that is the words which accompany and motivate the logistic building activity; Peano, indeed, takes, as much as possible, care to tie the accompanying thoughts to symbolic signs. Nonetheless the system remains decomposable in the proper construction, and the principles according to which the construction develops itself. Even though those principles are symbolically formulated, such formulations must be considered as heterogeneous with respect to the further formulas, to which the first ones are applied—not as formulations, but as intuitive acts of which the adjoined formulations are only accompaniments.

   Hilbert needs those intuitive language acts, thus also the accompanying language, more than Peano, because he wants to prove the non-contradiction of his logistic system *in itself*, something Peano does not care about.

   The verbal content of Hilbert's treatise up to page 184, V. belongs to the fifth phase.
6. The mathematical consideration of that language. The explicit performance of this step is something essential with Hilbert, in contrast to Peano and Russell; he notes, looking back at his own words, logical figures, which develop according to logical and arithmetical principles, including among other things, the theorem of complete induction. The elements of these logical figures, such as the words *mehrere, zwei, Fortsetzung, an Stelle von, beliebig*, etc., are a linguistic accompaniment of construction acts in the above mentioned mathematical system of the second order.
7. No longer thinking of a meaning of the elements of above mentioned logical figures; and the copying of the construction of those figures by a new mathematical system of the *third* order, for the time being without accompanying language.

   Hilbert carries out the transition from 6. to 7. in his mind loc. cit. page 184 and 185 under V, first paragraph.
8. The language accompaniment of the mathematical system of the third order which motivates and shows the non-contradiction of it.

   This phase is, in the *words* of the above mentioned paragraph loc. cit. page 184, 185, the last one found with Hilbert.

One could go on even further, but the mathematical systems of even higher order would all be roughly copies of each other; it thus makes no sense to pursue the matter further.

For that matter, the previous phases, from the third one on, are not of mathematical interest either. Mathematics only belongs in the first one; it cannot remain free of the second one in practice, but that phase remains a non-mathematical unconscious act, be it guided and supported by *applied*

[48]Hilbert (1905).

*mathematics* or not, but never gaining priority with respect to the intuitive mathematics. (Diss. p. 173 ff.)

The promised solution of Hilbert's problem no. 2, the consistency of arithmetic, is not given the prominence that one would expect. This is perhaps not surprising, for, from Brouwer's point of view, the effective construction of the mathematical system automatically protected it from contradictions. He pointed out that the early axiomaticians (including Hilbert) were quite satisfied with the procedure: '...the advancing of mathematical systems as existence proofs for the logical systems, implies that one still saw that the mathematical system required no further existence proof than its intuitive construction' (Diss. p. 137).

Brouwer's views become visible, so to speak, in his discussions of the axiomatic method and of Hilbert's formalism; he probably considered the proper consistencymethodology (that is model construction) so natural and obvious, that he did not want to spell out the underlying details.

It is virtually impossible to do justice to all parts of Brouwer's dissertation, so for the moment we have to be content with a short final glance at Brouwer's reaction to set theory. Set theory was the hot topic of the turn of the century, it aroused the curiosity of the mathematical community, and although it had a long incubation period before it was accepted as a useful and self-evident part of daily practice (and mathematical training) it occupied some of the finest minds of the period. The founding father was Georg Cantor, professor in Halle. Cantor had, after some experimenting with sets that played a role in 'real mathematics', that is sets that occur in number theory, algebra or analysis (including the new discipline, topology), opened up an inexhaustible universe of sets, sets of sets, etc. His fundamental theorem on power sets held that there are essentially more subsets of a set, than it has elements. For finite sets this was not particularly shocking, a direct calculation shows that $\{0, 1\}$ has 4 subsets, namely the empty set, $\{0\}$, $\{1\}$, and $\{0, 1\}$, and, in general $\{0, 1, 2, \ldots, n-1\}$ has $2^n$ subsets. For infinite sets there is a problem: it is no use to count them. So instead, Cantor introduced the concept 'as large as', or 'equivalent' (*gleichmächtig*). Two sets are equivalent if they can be brought into 1 *to* 1-correspondence.[49] This is a reasonable generalisation of 'as large as' for finite sets: $\{0, 2, 7\}$ is as large as $\{\pi, 12, \sqrt{2}\}$ and there are obvious 1–1 correspondences. Cantor's famous theorem tells us that the power set $\mathcal{P}(A)$ (that is the set of all subsets of $A$) is always larger than $A$. To be precise: $A$ is equivalent to a subset of $\mathcal{P}(A)$, but not to $\mathcal{P}(A)$ itself. As a result there is no such thing as a 'largest set' in Cantor's universe. The set of natural numbers, $\mathbb{N}$, is infinite, and it is smaller than $\mathcal{P}(\mathbb{N})$, which in turn is smaller than $\mathcal{P}(\mathcal{P}(\mathbb{N}))$, etc. Indeed, $\mathbb{N}$ is 'the smallest infinite set', we call sets equivalent to $\mathbb{N}$ *denumerable*. Elementary set theory tells us that the continuum is equivalent to $\mathcal{P}(\mathbb{N})$.

Furthermore, Cantor generalised the notion of counting beyond the existing practice, he introduced *transfinite numbers*. The first transfinite number is $\omega$ (that is, $\omega$ comes after all the natural numbers) and the first few transfinite numbers (ordinal

[49] That is, if there is a bijection (1–1 mapping) from one onto the other.

numbers) look like $\omega, \omega+1, \omega+2, \dots, \omega+\omega\ (=\omega\cdot 2), \dots, \omega\cdot 3, \dots, \omega\cdot n, \dots$ These ordinal numbers are all denumerable. Cantor took the bold step to consider the set of all denumerable ordinal numbers (traditionally called *the second number class*); this set is itself another ordinal number: the first non-denumerable ordinal $\omega_1$. Cantor showed that an unbounded number of transfinite numbers exist in his universe.

At the time Brouwer's dissertation was written, this material was all fairly new; lots of mathematicians considered Cantor's creations as if they were inmates of some asylum rather than useful, let alone necessary, citizens of the mathematical kingdom. An influential mathematician like Poincaré was publicly sceptical of set theory; all the same he was excellently informed about the new developments, and he gave a beautiful talk on set theory in Göttingen.[50]

Set theory was rocked around the turn of the century by the so-called paradoxes. Cantor, Burali-Forti, Zermelo and Russell had shown that uncritical handling of certain set-forming operations led to contradictions. The 'set of all sets', the 'set of all ordinal numbers', and the like, spelled disaster. When the dust had settled somewhat, it was Zermelo who got the debate on set theory under way again by publishing his miraculous proof of the well-ordering theorem,[51] which states that each set can be well-ordered, that is put into an *ordering like an ordinal*. Technically speaking, a set is well-ordered by an ordering relation $<$ if each non-empty subset has a first element. Well-ordered sets are exactly in the form that corresponds to the transfinite 'counting' of sets. Cantor had posed the question if all sets could be well-ordered; the more conservative mathematicians were inclined to doubt this; they usually pointed out that there was not even the faintest clue how to well-order the continuum. Brouwer, in his dissertation, refutes the well-ordering theorem by pointing out that in the case of the continuum most of the elements are unknown, and hence cannot be ordered individually—'So this matter also turns out to be illusory.' (Diss. p. 153).

Zermelo's proof was based on a new principle, the *axiom of choice*, which postulated, that given a set of non-empty sets, there was always a (choice-) function that picked an element from each of these sets. The matter caused quite a stir in mathematics; objections of various kinds were raised by eminent scholars, Peano, Borel, Poincaré, Schoenflies, ....

Zermelo's second great achievement was the formulation of an axiom system for set theory. It turned out to be so adequate that, up to minor improvements, it is still the main basis for modern mathematics.

With respect to set theory, Brouwer maintained that the larger part of the creations of Cantor was beyond the realm of the mentally (and thus mathematically) constructible. Examples of (according to Brouwer) meaningless word play are the *second number class* and the higher *power sets*. In Brouwer's opinion, mathematical objects had to be constructed; one can thus easily imagine that considerable parts

[50]Poincaré (1910).

[51]Zermelo (1904).

of Cantor's universe had to be jettisoned. Brouwer had no problem accepting the countable ordinals, but he balked at the second number class as a whole (indeed his supplementary thesis no. XIII boldly asserts 'The second number class of Cantor does not exist'). Cantor introduced the second number class as the totality (*Inbegriff*) of all countable ordinals, which according to Brouwer 'cannot be thought, that is to say, cannot be built mathematically'. Brouwer argued that constructions based on the 'and so on' are legitimate only if the 'and so on' refers to an iteration of the same thing at most $\omega$ times. In the case of Cantor's second number class there evidently is no $\omega$-iteration, nor, says Brouwer, is there a repetition of similar things (Diss. p. 146). 'Thus Cantor is here no longer on firm mathematical ground.'

Similarly, Brouwer rejects the unrestricted exponentiation of sets.[52] Brouwer's argument is simple: we cannot imagine (think), for example, $2^{2^{\mathbb{N}}}$. It is worth noting that he does not mention the power sets $\mathcal{P}(A)$.[53] It would be a matter of wishful reading to conclude that he had already realised that power sets and exponents are, constructively speaking, different things. It was probably the influence of Cantor, but it is equally well possible that Brouwer just did not care for 'the collection of all subsets'. It would not fall under his generation principles of sets anyway.

A conceptual analysis of the mathematical construction process led Brouwer to the insight that:

> we can create in mathematics nothing but finite sequences, and further, on the ground of the clearly conceived 'and so on', the order type $\omega$, but only consisting of *equal elements*,[54] so that we can never imagine the *arbitrary* infinite binary fractions as finished, hence as individualised, because the denumerably infinite number of numerals behind the (decimal) point cannot be viewed as a denumerable number of *equal* things, and finally the intuitive continuum (by means of which we subsequently have constructed the ordinary continuum, the *measurable continuum*). (Diss. p. 143).

The explanation of 'and so on' may seem somewhat puzzling. On a literal reading a denumerable sequence like '$1, 2, 3$, and so on' could only be viewed as '$1, 2, 3, 3, 3, 3, \dots$' But it is the *operation of successor* that is repeated, so one does get '$1, 2, 3, 4, 5, \dots$' As a consequence the 'and so on' here requires an operation or a law. Hence one cannot expect to get arbitrary sequences (think of the choice sequences which Brouwer introduced later).

In spite of his critical view of Cantor's set theory, Brouwer was no bigoted foe of the topic as such. As his topological papers were to show, he was fully aware of the usefulness of sets and their machinery. In the dissertation he did not preach abolition of set theory, but rather a thorough overhaul of the subject.

---

[52] $A^B$ is the set of all mappings from $B$ into $A$.

[53] Cantor himself systematically discussed subsets in terms of their characteristic functions (*Belegungen*).

[54] Diss. p. 143. [Brouwer's note]: Where one says: 'and so on', one means the arbitrary repetition of the same thing or operation, even though that thing or operation may be defined in a complex way.

We have seen that he claimed in a letter to Korteweg to have solved the continuum problem. His proposed solution was based on an analysis of the possible set-creating operations on the domain of point sets.

As the continuum problem was supposed to be a central issue in understanding the nature of the continuum, it is clear that definition or construction is of the utmost importance. In classical mathematics (say, Cantorian set theory) there are various precise, and equivalent, definitions, so that a treatment of the problem can at least be entertained on the basis of an exact specification. In Brouwer's mathematics, this is problematic indeed. Since the continuum was given outright by an act of the individual intuition, there seemed to be little hope for the treatment of the problem.

The intuitive continuum was, so to speak, an amorphous mass out of which individual points could be picked, but totalities of individualised elements could only be created in denumerable quantities (Diss. p. 62). In his words: 'The continuum as a whole was intuitively given to us; a construction of it, an act which would create as individualised, 'all' points of it by means of mathematical intuition is unthinkable, and impossible.'

Brouwer's view of the continuum shows a suggestive similarity to the mystic experience of the initial chaotic state of the individual. In this state, also, there are no sharp bounds, everything is flowing and amorphous. From that point of view it is indeed implausible, if not impossible, that all mathematical structures are made up of individual, sharply distinguished, elements. So the continuum, being traditionally the flowing, continuous medium, must from Brouwer's point of view, have been the structure par excellence created by a human faculty analogous to the original mystic state.

On the basis of the ur-intuition (see p. 98), Brouwer distinguished three set-generating principles for subsets of the continuum:

1. Combinations of finite sets, sets of the order $\omega$, and sets of the order $\eta$ (that is finite sets or sets that are similar to the natural numbers or to the rationals).
2. The 'completion to a continuum' of a set of order type $\eta$, which is dense in itself.
3. The complement of a dense in itself subset of type $\eta$, with respect to the continuum.

These three construction principles, according to Brouwer, yielded only countable sets and sets equivalent to the continuum, and hence the continuum problem was answered in the affirmative (Diss. p. 67). Actually, Brouwer was more cautious then the above suggests; he wrote 'seems to be solved'. In Chap. III, Brouwer returned to the continuum problem (Diss. p. 149), see below. This time to show that the problem was not well-posed. For he argued that neither 'the set of points of the continuum', nor 'the set of all numbers of the second number class' exist as mathematical objects, so their equivalence is unthinkable. On the other hand one can turn the whole problem into a logical problem: introduce the relevant sets as hypothetical objects and show that the continuum hypothesis is non-contradictory.

The reflections on the possible point sets, and their generation principles had led Brouwer to consider another kind of 'pseudo' sets, or cardinalities. Evidently there are totalities that cannot be exhausted by successive selection of points, the continuum is an obvious example. The new notion was based on the idea that, although one could not exhaust such totalities, one could, with luck, always find, after an exhaust-attempt, another point in a systematic way. In fact Cantor's diagonal method yields such a systematic method for the continuum.

Brouwer coined a new term for this phenomenon:

> We mean by a *denumerably unfinished set* one, of which only a denumerable group can be indicated in a well-defined way, but for which then, according to some previously defined mathematical process, new elements may be derived from such a denumerable group, which belong to the set in question. (Diss. p. 148).[55]

The notion of 'denumerably unfinished' makes the impression of an interesting digression, almost an afterthought in the context of the continuum problem. It plays, however, no role in Brouwer's actual mathematics. The notion occurs a couple of times in later papers, to disappear, once the choice sequences have made their entry. After that it is mentioned in passing, but without significant or helpful comments. The idea behind this innovation was to apply the predicate 'denumerably unfinished' to 'collections' that could not be handled as sets according to the above mentioned construction principles. Since the denumerably unfinished sets were not really sets, Brouwer decided, as a compromise, to treat them as a 'manner of speaking': 'We can, however, introduce these words as arbitrary expressions for a known intention.'[56] As examples of denumerably unfinished sets he listed: the totality of all ordinals, the totality of all definable points of the continuum, the totality of all possible mathematical systems. The comparison of denumerably unfinished sets, according to Brouwer, was simple: every two of them were equivalent. On the other hand, he admitted in a footnote, (Diss. p. 149), that one could just as well consider denumerable and denumerably unfinished sets as being equivalent, as each denumerably unfinished set may be mapped onto the ordinal $\omega^2$. This seemingly paradoxical statement was clarified by pointing out the difference between 1–1 mappings between two enumerable sets and those between a denumerable and a denumerably unfinished set, the latter are themselves 'unfinished'.

On the basis of the above considerations, Brouwer recognised the following cardinalities:

1. the various finite ones
2. the denumerable infinite
3. the denumerably unfinished infinite
4. the continuous.

In this more refined context, Brouwer reconsidered the Continuum Problem. In the tradition of set theory there were basically two (equivalent) formulations of the continuum problem. One did not mention ordinals and cardinals: *each infinite subset of the continuum is either denumerable, or equivalent to the continuum.* The

---

[55]Observe the similarity with Dekker's productive sets, cf. Rogers (1967).

[56]This is similar to the notion of 'class' in Zermelo–Fraenkel set theory; a class is not a set, but it is a convenient fiction.

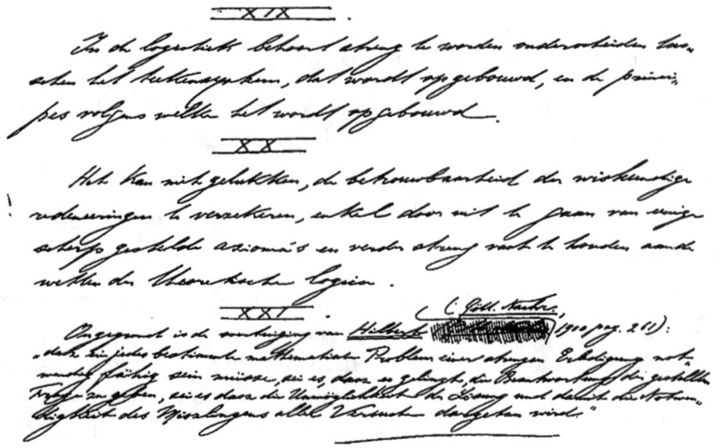

**Fig. 3.2** The last three of the obligatory theses of the dissertation. XXI rejects Hilbert's dogma. [Brouwer archive]

other one made use of cardinals and ordinals: $2^{\aleph_0} = \aleph_1$, in words *the continuum is equivalent to the second number class*. The second formulation was problematic for Brouwer. As neither the second number class, nor the 'set of all points of the continuum' exist in the proper sense as sets of individual points, the problem had no meaning in this form. The problem could, of course, be turned into an exercise in logic by abstracting from the meaning and trying to show the consistency of the continuum hypothesis (Diss. p. 150). That is to say, since the set of *all* points of the continuum does not make sense, one can consider the set of all definable numbers of the continuum. This is a denumerably unfinished set and hence equivalent to the equally denumerably unfinished second number class. Hence, under this interpretation, the equivalence turns out to be trivially true.

In 1908 Brouwer attended the International Mathematical Congress in Rome, where he gave two talks, one of them with the title 'The possible Cardinalities'.[57] In this talk Brouwer expressed himself more cautiously. The phrase 'continuum problem' had disappeared. But the above classification is by and large upheld. He concluded that there is only one infinite cardinality—that of the denumerable sets, but he mentioned the *denumerably unfinished* cardinality, which is a *method* rather than a set, and the *continuous* one, which is something finished, but only as a *matrix* and not as a set.[58]

Finally, Brouwer upbraided in his dissertation Poincaré (with whom he agreed on most issues, and whom he greatly admired) for 'not adopting the intuitive construction as the sole foundation of his critique'. Brouwer refers here to the famous (or notorious) quote 'in mathematics the word exist can have only one meaning; it means freedom of contradiction'. (Diss. p. 177).

[57] *Die Moegliche Mächtigkeiten* (Brouwer 1908a).

[58] Brouwer (1908a), p. 571.

The dissertation ended with a summary, which opened with a bold declaration, vaguely reminiscent of Cantor's slogan *The essence of mathematics is in its freedom*:

> Mathematics is a free creation, independent of experience; it is developed from a single *a priori* ur-intuition, which can be called both *constancy in change* and *unity in multitude*.

The summary deals only with the foundational issues of the dissertation. It recapitulates three fundamental points:

> (i) The application of mathematics, in the form of projecting mathematical systems, is also a free act of the human being.
>
> (ii) Mathematical definitions and properties are not to be studied mathematically. They are to serve as support of our memory, and in communication.
>
> (iii) A logical construction of mathematics, independent of mathematical intuition, is impossible.

The main body of the dissertation was followed by the obligatory theses; they were partly concise formulations of fundamental ideas of the dissertation, and partly claims about topics in mathematics, not connected with the themes of the dissertation. In all there were 21 of them. The first draft of the Theses contained 39 items. The number was, most likely under the guidance of Korteweg, reduced to the abovementioned 21.

There is a draft of a letter from Brouwer to Professor J. de Vries (Utrecht University), in which Brouwer gave a short resumé of the contents of the dissertation, together with some elucidations. In this letter Brouwer comments on Cantor's set theory. The following quote is interesting, but puzzling:

> With respect to the question whether actual infinite sets exist, and if 'yes', which ones?
>
> Here I neither subscribe to the opinion of Poincaré, who rejects the actual infinite out of hand, nor do I accept all the transfinite sets of Cantor, but I acknowledge denumerably infinite sets, and with a restriction, the continuous cardinality, and finally, with another restriction, a new cardinality, which I call denumerably infinite unfinished. I expose however, all the higher cardinalities of Cantor as a logical chimera. At the same time I try to strip transfinite set theory of its parasite parts, such as transfinite exponentiation, the theorem of Bernstein with its applications, and more; all of which result from the false logical foundations of set theory. In this connection I can formulate:
>
> 1. Actual infinite sets can be created mathematically, even though in the practical applications of mathematics in the world only finite sets occur.

This view on the actual infinite is a plausible consequence of Brouwer's concept of 'ur-intuition', for the continuum as a whole is immediately given to the individual, and it undeniably is infinite. The denumerable sets can, however not be considered as actual (that is 'finished') infinite.

This particular statement on 'actual infinite' did in fact occur explicitly in the original list of theses as no. 38. Although this particular thesis was dropped, there is an explicit acknowledgement of the existence of actual infinite sets in the dissertation (p. 176); unfortunately Brouwer did not elaborate the matter.

## 3.6 The Public Defence

The grand apotheosis took place on 19 February 1907. On that day Brouwer defended his dissertation in public before a delegation of professors from the faculty. His dissertation had, as tradition and the rules decreed, been printed and distributed. The publishers, Maas & Van Suchtelen (Amsterdam–Leipzig), and their printers, the steam printing shop Robijns & Co (Nijmegen), had done a marvellous job. The font and the choice of paper would have been the delight of any bibliophile.

In the old days the formalities of the 'promotion' took place in the auditorium (*Aula*) of the University at the Oudemanhuispoort, a solemn hall, reminiscent of a church hall, where the doctor-to-be occupied a small lectern in front of the imposing rector's lectern in the shadow of a statue of Pallas Athene. The actual examination and defence took three quarters of an hour; the candidate had to defend his dissertation including the theses against the objections of the faculty members (professors only).

The candidate was accompanied by two helpers, *paranymphs*, dressed in tails, like the candidate. The paranymphs were placed on both sides of the lectern, where they stood like statues during the ceremony. The professors filed in, preceded by the beadle with his staff decorated with medals, the soft clinking spreading an almost devotional atmosphere. Once the professors were seated, the examination was opened, according to the local Amsterdam tradition, with the so-called *opposition from the floor*: persons from the public were allowed to attack the dissertation and the theses. This sounds more liberal than it actually was (and still is). These opponents from the floor were picked in advance by the candidate and the adviser, and the approval of the rector had to be obtained before any such opposition could take place. As a consequence, this part of the 'opposition' was often a set up. The candidate could easily arrange the questions beforehand with the opposition, and most of the time this indeed happened. In Brouwer's case, the opposition from the floor was carried out by Mannoury and Barrau (a fellow student, who was a bit older than Brouwer). Both had informed Brouwer of their questions[59] and Brouwer had, as appears from some sheets in the Brouwer archive, meticulously prepared his defence. Mannoury's objections were directed against the use of irreducible concepts in mathematics (Diss. p. 180) and of infinite (albeit potentially infinite) sets, in particular the use of 'and so on'. Barrau questioned the notion of the continuum as a matrix, and wanted to see the discrete as the basic notion in mathematics. It may be assumed that Brouwer got the better of these opponents from the floor.

The promotor also used the occasion to give vent to his doubts. In his opinion Brouwer's wholesale desertion of Kant's a priority of space was ill-motivated. The

[59]Brouwer to Korteweg, 16 February 1907, Mannoury to Brouwer 13 February 1907.

intuition of time was, according to him, not rich enough to allow geometry of all dimensions. Thus he proposed to allow for an intuition of displacement (transformation) in addition to the ur-intuition of time. He observed that although the candidate had no use for an extra intuition in addition to the intuition of time, it by no means followed that therefore it did not *exist*. This extra intuition of space had, according to Korteweg, to be locally Euclidean. Brouwer's reply has not been recorded, but presumably he would not have had much difficulty in warding off this blow too.

And so at the age of 25 he acquired the degree of doctor in mathematics and physics, with the predicate Cum Laude. He was assisted by his paranymphs Carel Adama van Scheltema and Ru Mauve, two friends of the old days.

When the examination was over, Korteweg could show his delight at the success of his pupil, addressing the 'young doctor' with words of warm praise. Given Brouwer's preference for the deep foundations, he was not surprised that his student came to know

> ... those most recent investigations of men like Cantor, Hilbert, Poincaré and Peano, where they occupy themselves with investigations in those subterranean parts of this building, where for many of those who are used to work on the higher stories in clearer light, there seems to be a twilight that seems almost impenetrable. And, certainly, those researches would not have been valued so highly, if they had not been conducted by men, the guidance of whom one had learned to trust by their mathematical work of a different nature.

Korteweg was perfectly aware of the gigantic task that Brouwer had taken upon his shoulders. He knew that his student had written his dissertation as a mathematical credo. He expressed in his speech the hope that Brouwer might be in the position to elaborate his ideas. In spite of his recognition of the foundational programme of his Ph.D. student, Korteweg could not help wondering why Brouwer had not simply continued the research that just preceded the dissertation, but he accepted and admired Brouwer's choice. Nonetheless he urged Brouwer to devote some attention to the work at the higher floors, for which the talent was certainly not lacking. Prophetic words indeed! As we will see later, Brouwer took the implicit message to heart.

There is no mention of the traditional celebration after the promotion; usually the 'young doctor' invites his Ph.D. adviser, the paranymphs, and a selection of his friends and relations for a dinner and possibly a party after the 'promotion'. It is likely that Brouwer observed the tradition, but one cannot be certain. Scheltema, one of his paranymphs, fled immediately after the occasion from the city to the solitude of the Veluwe. Like Bertus, he was not made for boisterous parties. A week later the two friends automatically resumed their traditional exchange of birthday wishes (Scheltema's birthday was actually the day before Brouwer's). Scheltema had observed his friend well enough to see the changes in his life. He wrote, 'may the next year bring you closer to reality than you were—you are already somewhat closer than before'. He added, 'and, please don't give up on your plan to visit those professors abroad'.

The family was probably impressed by the learned member of the Brouwer–Poutsma clan. One of Brouwer's aunts presented him at the occasion of his doctorate with a slim volume of Eastern mysticism, *Le Livre de la Voie et de la ligne-droite de Lao-Tse*, by Alexandre Ular.[60]

The reception of Brouwer's dissertation in Holland is hard to judge; it is a reasonable conjecture that it was rather mentioned and quoted than understood. Brouwer was way ahead of his fellow countrymen; even Mannoury, who was without doubt the most competent reader, misunderstood Brouwer on certain crucial issues. In a couple of reviews, one for the mathematics journal and two for a wider readership, he discussed the basic issues of the dissertation. The latter, more philosophical in nature, ended with a downright repudiation of Brouwer's claims of the reliability of (his!) mathematics:

> No, Brouwer, logicians do not ensure the reliability of the 'mathematical properties', but neither will you through your continuity intuition, for the simple reason *that it does not exist*. Mathematics is a human artefact, a human conception, in which there is no truth but that which relates to human language, intention and society. Your book is an action of thought-courage and a consequence of an acquired higher insight, but that thought-courage and that insight, ... they are 'unfinished'. Dissociate yourself (indeed *completely*) from all conventions and agreements, from the language and all word-constructs, and I am certain that you will arrive at the acknowledgement (which is the only true foundation of mathematics): there is no immutable truth and no immutable measure for truth—there is no absolute unity, no absolute space and no absolute time, there *is* no mathematics.[61]

These were hard words to swallow for Brouwer, even (or maybe, in particular) coming from a friend. Nonetheless he replied in a mellow mood:[62]

> Is there no mathematics? And what about your objective criticism which is therefore so mathematical? Designed in accordance with norms and hence understandable? And in particular: why was there the age-long uncomfortable feeling with respect to the axiom of parallels, and not with respect to the fundamental property of arithmetic?[63] Because the first is mathematised reality, and the latter mathematised mathematics, free from reality, which here means 'not experienced by me'.

The above skirmish did in no way detract from the friendship between Brouwer and Mannoury. They remained the best of friends for the rest of their lives.

---

[60]Ular (1902), The book of the way and the straight line of Lao-Tse.

[61]Mannoury (1907).

[62]Brouwer to Mannoury, 1 August 1907.

[63]That is 'all countings of a finite set yield the same number'.

# Chapter 4
# Cantor–Schoenflies Topology

## 4.1 The Geometry of Continuous Change

The evolution of geometry may, if one wishes to do so, be viewed as an everlasting struggle to free the subject from the shackles of convention, to allow the study of more and more general figures and properties. Where originally straight lines and circles were studied, soon conics were introduced and gradually more general curves entered the domain of geometry: the spiral, the conchoid, the cycloid, the sine-curve, curves with no tangents at all, and so on. This expansion of the domain of geometry went hand in hand with a more liberal attitude with respect to the notion of 'the same' (or 'similar'): are only congruent triangles the same, or transformations which are 1–1[1] and continuous both ways; the technical term for these mappings is *homeomorphism* or *topological mapping*.

The geometry on a rubber balloon is a fine example of topology; one may stretch, distort, inflate the surface as long as one does not cut or tear it. It is clear that notions like 'straight line' or 'angle' lose their meaning in this example. There are, however, quite a number of geometric properties that remain invariant; for example, the number of intersections does not change, and an $O$ cannot be turned into an $F$.

Let us illustrate how one works in topology by looking at Euler's theorem (which, by the way, can be traced to Descartes). We will look at a simple form.

Consider a convex polyhedron (that is, a surface that is built up from flat pieces, with intersection lines and points, without dents or holes; think for example of a cube). There happens to be a fixed relation between the number of faces ($F$), edges ($E$) and vertices ($V$): $V - E + F = 2$.

We immediately see that if the polyhedron is stretched, bent, dented, etc., this relation between $F$, $E$ and $V$ remains the same. This insight is used to give a simple proof of Euler's theorem.

Let us demonstrate the method on a simple polyhedron, the cube, Fig. 4.1.

[1] Such a transformation $f$ can be reversed. To be precise: there is another transformation $g$, so that $f$ followed by $g$, and $g$ followed by $f$ take points to themselves.

D. van Dalen, *L.E.J. Brouwer – Topologist, Intuitionist, Philosopher*,
DOI 10.1007/978-1-4471-4616-2_4, 

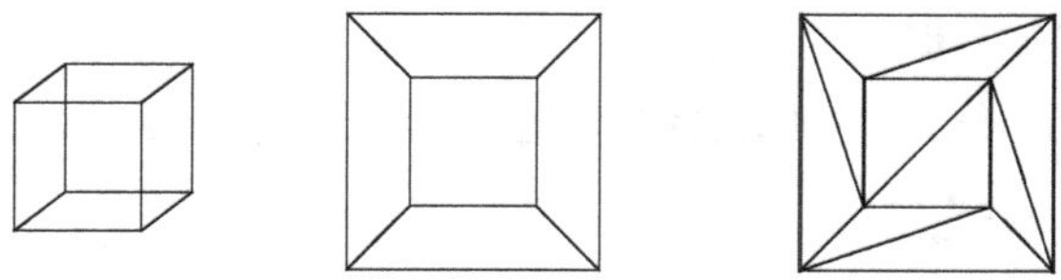


**Fig. 4.1** Euler's formula for the *cube*

We cut away the front face and stretch the remaining part of the cube until it lies in a plane. With a bit of extra stretching we can get the edges straight again. The figure now looks like the middle figure of Fig. 4.1.

Instead of proving the above relation, we now have to show $V - E + F = 1$, because one face has disappeared.

The next trick is to split the faces into triangles; see the right-hand part of Fig. 4.1. Note that we each time introduce one extra edge and one extra face, so the outcome $V - E + F$ for the triangulated polygon is the same as the one for the original one.

Now imagine that we build up the triangulated polygon starting with one of the triangles and adding fitting triangles until we have recovered the polygon.

We note that, given a polygon, the introduction of a single triangle introduces one extra edge and one extra face, or one extra face, one extra vertex, and two extra edges, or two vertices, three edges, and one face.[2] In each case the new additions cancel each other out in the Euler formula. Hence the Euler relation holds for the triangulated polygon, and thus also for the original one. So we are done.

The reader will have noted that the procedure uses essentially a number of tricks that Euclid would have frowned upon, that is angles and lengths are completely ignored, and straight lines were bent and stretched ad libitum, etc. This is one of the lessons of topology; forget about Euclidean notions, but deform your figures continuously (and 1–1).

After Euler the subject of topology remained dormant for almost a century; in 1833 Gauss observed 'Of the Geometria situs, conjectured by Leibniz and into which only a few geometers, Euler and Vandermonde, were allowed a glimpse, we know after one and a half century little more than nothing'.

The name '*topology*' was introduced by Listing in his *Vorstudien zur Topologie* (1847–48); the alternative name *analysis situs*[3] survived for more than half a century, and was used by prominent topologists such as Poincaré and Schoenflies, but gradually it has become a historical curiosity.

After Listing the interest in topology gradually increased; it was noted that many geometrical phenomena were not really dependent on the standard notions of traditional geometry. Geometry itself was relaxing its attitudes towards its proper content. The insight that there was more than Euclidean geometry, and the systematic use of transformations in projective geometry, affine geometry, and so on, culminated in Felix Klein's above mentioned *Erlangen Program*, which characterised the

[2] Alexander Soifer pointed out an omission in the first formulation.

[3] Actually a misnomer, cf. Freudenthal (1954).

**Fig. 4.2** A Jordan curve

various geometries as the objects and notions invariant under specific transformation groups.

This program opened the way for a more systematic approach to topology, in all its forms. The subject had at that time also received vigorous impulses from the combination of analysis and geometry, in particular from Riemann's work on the foundations of analysis, differential geometry and Riemann surfaces—originally designed for the study of the global behaviour of analytic functions. Riemann's geometric investigations were also the point of departure for the particular line of research in the foundations of geometry that nowadays is listed under the heading of *Riemann–Helmholtz Raumproblem*, a project of determining the Euclidean and non-Euclidean geometries by means of their transformation groups. This particular direction in topology and geometry was followed by Sophus Lie, Killing, Elie Cartan and others; the theory of Lie groups is the mathematical offspring of the Riemann–Helmholtz project.

In France the subject of topology was taken up by Camille Jordan, whose name is most of all connected with a particular kind of curve in the plane; this so-called 'Jordan curve' is obtained by applying a topological transformation to the unit circle. In popular terms, one considers the unit circle as a rubber band, which is placed in the plane without cutting the band and without introducing self-crossings.

Jordan stated the following basic theorem: a Jordan curve divides the plane in two components, that is there are two parts such that one cannot traverse from one part to the other via a (continuous) path without crossing the Jordan curve. The theorem will appear completely trivial to the outsider; however, when prodded, he will probably not even know how to begin to prove it. Jordan's own proof (1887) did not quite pass the test of correctness, and eventually Oswald Veblen gave the first rigorous proof in 1906.

The key word in topology is 'continuous'. The notion of continuity had been scrutinised in the nineteenth century by a number of mathematicians. After a quiet existence as a rather obvious concept, something like 'a curve is continuous if one can draw it without lifting one's pencil from the paper', it became a central notion in analysis and subsequently in geometry. Precise definitions were furnished (along the lines of the $\varepsilon$-$\delta$-tradition) and most of the basic properties of continuous functions on the real line, or the plane, were established. A well-known example is the intermediate value theorem of calculus: 'if a continuous function $f$ is negative at 0 and positive at 1, then there is a point $a$ between 0 and 1 such that $f(a) = 0$'. The theorem may not seem terribly exciting at first sight, but to appreciate its significance one must have a clear insight into (i) the nature of the real line, (ii) the nature of continuity.

At the time that Brouwer turned his attention to topology the grand master in the field was Henri Poincaré, a man with wide-ranging interests. He had come to topology through his investigations of differential equations, the qualitative behaviour of functions defined by such equations required radically new tools—which Poincaré found in topology. The spaces that were investigated by Poincaré, Lie, Élie Cartan and others were still closely related to Euclidean spaces of arbitrary dimensions. They were obtained by judiciously pasting together pieces of Euclidean spaces after suitable treatment (such as bending or twisting); the resulting spaces were called manifolds (or varieties, *Mannigfaltigkeiten* in German).

A man like Poincaré, who commanded almost the whole of mathematics, brought a considerable part of the mathematical arsenal into play when studying topics in topology; for example, he usually supposed surfaces, etc. to be differentiable (or more). A new generation, with Brouwer in the forefront, would set topology free from those strong extraneous conditions; But not only would the study of Euclidean spaces be liberated from too stringent conditions, the spaces themselves would also be subjected to drastic generalisation.

In the eighteen-seventies topology had begotten a curious child: set theory. Georg Cantor had introduced certain transfinite arguments in his study of Fourier series, and after studying sets of real numbers and sets of points of Euclidean spaces, he started to consider arbitrary sets and functions. In 1908 Zermelo provided the final touch: an axiom system for set theory. Set theory, in its turn, enriched topology with an abstract setting in which neighbourhoods could be postulated, so that the traditional definition of 'continuous' could be mimicked (Fréchet). This so-called 'set-theoretic topology' allowed a considerable generality that was very convenient in diverse fields of application.

Brouwer entered the stage when most notions were still rather rudimentary and not universally adopted; indeed, his topological investigations played an important role in setting better standards for the discipline. He was self-taught, and thus it is not surprising that his familiarity with the subject of topology showed some gaps; he was, however, a fast learner. Although he had marked preferences for certain parts of topology—for example the topology of planes, spheres, and manifolds in general—he knew how to handle topology in the broadest sense. We will consider here his contributions in the early period, before the First World War.

## 4.2 Lie Groups

Brouwer's first encounter with topology was in an area that was already well understood at the end of the nineteenth century: the theory of the Lie groups, named after Sophus Lie, the Norwegian mathematician who uncovered their properties. Lie groups consist of transformations, depending continuously on a finite number of parameters, of finite-dimensional manifolds (the reader may think of Euclidean-like spaces).

'*Group*' is the technical term for some set with a binary operation (usually called 'product' after the multiplication of numbers) such that:

(i) there is an identity element, that is an element $e$ such that $a \cdot e = e \cdot a = a$ for all elements $a$;
(ii) the product of two elements again belongs to the group;
(iii) the product is associative, that is $a \cdot (b \cdot c) = (a \cdot b) \cdot c$;
(iv) every element $a$ has an inverse $b$, that is $a \cdot b = b \cdot a = e$ ($b$ is usually denoted by $a^{-1}$).

There is no lack of examples of groups: the integers, for instance, with addition and subtraction form a group. Addition is associative: $m + (n + p) = (m + n) + p$, the identity element is 0, that is $0 + n = n + 0 = n$ and finally $-n$ is the inverse of $n$, that is $n + (-n) = 0$.

Similarly, the positive real numbers, with multiplication and inverse, form a group: $a \cdot (b \cdot c) = (a \cdot b) \cdot c$, $a \cdot 1 = 1 \cdot a = a$ and $a \cdot a^{-1} = a^{-1} \cdot a = 1$. Along the same lines, vectors in a fixed Euclidean space form a group under vector addition.

Groups play an important role in almost all areas of mathematics, in particular in geometry, where they occur naturally as transformations. For example, the congruences of elementary geometry (those transformations of the plane that do not change length) form a group. The main idea of composition and inverse in the case of transformations is that 'what one can do in two steps, one can also do in one step' and 'each step can be reversed'; it explains, for example, why congruence is an equivalence relation: if triangle $\Delta$ is congruent with triangle $\Delta'$, and $\Delta'$ with $\Delta''$, then $\Delta$ is congruent with $\Delta''$; if $\Delta$ is congruent with $\Delta'$, then $\Delta'$ is congruent with $\Delta$, and $\Delta$ is congruent with itself.

In the original approach to the theory of Lie groups, not only the algebraic features of the groups played a role (that is those concerning the operations of the group by themselves) but also properties that belong to the area of analysis (or calculus), such as continuity and differentiability. Lie groups, in the style of the nineteenth century, were transformation groups of locally Euclidean manifolds, depending on a finite number of parameters; Lie had required that the parameters of the product of two transformations were twice continuous differentiable functions of the parameters of the factors. He proved that under those conditions, these functions were even analytic. An easy way to express this is: the operations are analytic (similarly, 'continuous', 'differentiable' etc.).

This differentiability condition gradually came to be viewed as rather artificial, and when David Hilbert presented in 1900 his list of mathematical problems to the International Conference of Mathematicians, the elimination of differentiability conditions from the theory of Lie groups was on the fifth place of the list.

In modern terminology, a Lie group is an analytic manifold with a group structure so that the operation $xy^{-1}$ is analytical. Hilbert's problem number 5 asked to replace the latter condition by '$xy^{-1}$ is continuous'.[4]

One guesses that Brouwer, when looking for suitable topics for his dissertation, looked through Hilbert's list and decided that this fifth problem suited his taste and capacities. There is among Brouwer's early papers none that leads up to this particular problem, so the choice must have been more or less spontaneous. As we have seen, Brouwer felt that his Ph.D. adviser, Korteweg, did not fully appreciate the contents of the dissertation, and, trying to convince him, he simply referred to Hilbert's problems, claiming that he had solved three of them, including the fifth one. Indeed, he treated the one-parameter case in the dissertation and derived the differentiability, rather than postulating it. As a result the basic results of Lie applied also to this instance of a more general setting.

A year after obtaining his doctor's degree, Brouwer attended the International Conference for Mathematicians in Rome, where he gave two talks, one on Hilbert's fifth problem and one on the possible cardinal numbers in constructive mathematics.[5] In the first talk, *The theory of finite continuous groups, independent of the axioms of Lie*,[6] he carried out the elimination of the differentiability (with the help of an auxiliary argument, spelled out in *About difference quotients and differential quotients*[7]) and gave a geometric-topological classification of the groups. The uniform differentiability notion, introduced by Brouwer, surpassed from a methodological point of view the traditional 'continuously differentiability'; it did not find its way into practice.[8] It is nowadays adopted in constructive analysis.

In view of the importance of the subject and the prior involvement of Hilbert, it is not so surprising that in the same year a substantial paper of Brouwer, *The theory of finite continuous groups, independent of the axioms of Lie, I*, Brouwer (1909c) (which extended the dissertation and the Rome talk) was published in the *Mathematische Annalen*. The second part appeared a year later, also in the *Mathematische Annalen*, and dealt with the two-parameter case. At the end of the paper Brouwer remarked that he now had all the tools for enumerating the groups of two-dimensional manifolds, and promised to do so in a next communication. This third part, however, never appeared, although he mentioned in a letter of 9 April 1924 to Urysohn that

> Due to numerous distractions, the manuscript of my third communication on *The theory of finite continuous groups* (*Die Theorie der endlichen kontinuierlichen Gruppen*) unfortunately still waits in my drawer for the finishing touches, which indeed I hope to be able to make in the near future.

---

[4]The problem was eventually solved in 1952 in the affirmative by A.M. Gleason, D. Montgomery and L. Zippin. As a consequence one may now define a Lie group as a topological group over a locally Euclidean manifold.

[5]Brouwer (1908a, 1909b).

[6]*Die Theorie der endlichen continuierlichen Gruppen unabhängig von den Axiomen von Lie.*

[7]Brouwer (1908c).

[8]Freudenthal, CW II, p. 101.

Just looking at the dates of publication, one might get the erroneous idea that the sequel to Part I was a matter of routine. As a matter of fact, Brouwer suffered a serious setback—which eventually proved a boon. In a letter of 14 May 1909 to Hilbert he reported that a couple of months ago ('last winter'), when he was about to send Part II to the *Mathematische Annalen*, he had discovered that Schoenflies' investigations on Analysis Situs,

> on which I had based my work in the most extensive way, cannot be vindicated in every part, and that also my group theoretic results became questionable.

And thus Brouwer decided to occupy himself thoroughly with Schoenflies' topology. The result was the celebrated paper *Zur Analysis Situs*,[9] which 'rejected or modified some parts of Schoenflies' theory, and completely rebuilt other parts'. We will return to this paper on page 140.

After this substantial excursion into the realm of topology of the plane, he wrote to Hilbert on 26 July 1909 that the manuscript of Part II would follow in a month's time.

It is quite certain that the two papers did not exhaust Brouwer's results on Lie groups. According to Freudenthal,[10] he was aware of the difficulties in the three-dimensional case, such as presented by Antoine's set[11] and the horned sphere of Alexander.[12]

In the summer of 1909 Brouwer finally met Hilbert in the flesh. He had missed him at the Rome Conference, where (according to the list of participants) Hilbert was absent, but now Hilbert spent part of the summer vacation in the sea-side resort Scheveningen, now a part of The Hague. The meeting, a precious gift to a young mathematician, made a lasting impression on Brouwer. He wrote in elated terms to his friend Scheltema:[13]

> This summer the first mathematician of the world was in Scheveningen; I was already in contact with him through my work, but now I have repeatedly made walks with him, and talked as a young apostle with a prophet. He was only 46 years old,[14] but with a young soul and body; he swam vigorously and climbed walls and barbed wired gates with pleasure. It was a beautiful new ray of light through my life.

The famous mathematician and his young admirer must have found enough to talk about. Brouwer's ideas about Lie groups and topology were of course close to Hilbert's interest. Later in life Brouwer often referred to this first meeting. He used

[9] On Analysis Situs (Brouwer 1910e).

[10] CW II, p. 117.

[11] Antoine (1921).

[12] Alexander (1924).

[13] Brouwer to Scheltema 9 November 1909.

[14] 45 years, actually.

to point out that foundational matters also came up in their conversation; in particular he explained his various mathematics, language and logic levels to Hilbert. It is rather plausible that he told Hilbert his objections to the Heidelberg talk.[15] After all, why else mention the levels? Olga Taussky reported that Hilbert visited Brouwer in his Blaricum cottage.[16] As far as we know, the 1909 trip to Scheveningen was Hilbert's only visit to Holland, so in 1909 the great man of mathematics sat down in Brouwer's cottage or under the trees of Blaricum to discuss mathematics with his enthusiastic admirer—and future challenger.

As Brouwer had mentioned in his dissertation, Hilbert had already in 1903[17] determined the Lie group of the Euclidean plane independent of differentiability conditions; Hilbert's arguments were briefly summed up in Brouwer's dissertation. When he reworked his results for the *Mathematische Annalen*, he probably returned to Hilbert's paper, and checked it in detail. In a cordial letter of 28 October 1909[18] he explained to Hilbert that the 1903 paper required a number of corrections and addenda, and went on to spell out the details. After his excursion into Cantor–Schoenflies topology, the topological details required for the shoring up of Hilbert's paper presented no problem. It is not known if and how Hilbert reacted to this flood of good advice; no answer to this letter has been found. Hilbert revised the paper, which was republished as Appendix IV to the *Foundations of Geometry* (from the third edition (1909) onwards). The revised version avoids the difficulties pointed out by Brouwer, but whether in consequence of Brouwer's letter or independently, cannot be ascertained. Hilbert, apparently did not further discuss the matter with Brouwer, for as late as 16 June 1913 the latter wrote a letter to Hilbert:

> I recently read that a fourth edition of your Foundations of Geometry is to appear. Have the remarks concerning Appendix IV [that is the paper from the *Annalen* 56], which I sent to you in the fall of 1909, been taken into account? I would in any case be glad to assist you in putting the relevant sections right if you should wish so, and if the imprimatur has not yet been given.

He mentioned the subject again in his note on the history of dimension theory in 1928, at the time when the relation with Hilbert was already beyond repair: in a comment on the proper dating of mathematical contributions (which was at issue in that paper) he pointed out that it was beyond dispute that one should put the date of the set-theoretic foundation of geometry concurrent with "Hilbert's original paper of 1906, and not with the reprint of this note, in which an evident oversight in the underlying axioms was corrected on the basis of indications in a letter of mine of October 1909". The fact that Brouwer kept the draft among his papers seems to indicate that he attached some importance to this piece of topology and geometry; the circumstance that it concerned one of the greatest mathematicians of the day will have added to its value.

---

[15] See Brouwer (1928c).

[16] Taussky to Van Dalen, 1991.

[17] Hilbert (1902, 1909) and later editions.

[18] Brouwer to Hilbert, 28 October 1909, reproduced in CW II, p. 102 ff.

There was a long and friendly correspondence between Hilbert and Brouwer. The latter admired the great statesman of mathematics as no other. The above letter of 28 October 1909 illustrates Brouwer's reverence for the elder and wiser man, whom he described in 1909 to his friend Scheltema in glowing words.[19] Only the letters of Brouwer in Hilbert's archive, and some drafts, have survived. They all testify to an unreserved and cordial relationship between Brouwer and Hilbert. In view of the later conflicts between Hilbert and Brouwer, there have been conjectures as to earlier irritations and frictions. No traces can be found in the pre-*Grundlagenstreit* correspondence.[20]

Brouwer sent reprints of his papers to Hilbert with warm dedications. Some of these can still be found in the archive of the Mathematics department of the Nagoya University. Here are a few examples: the first Annalen paper on Lie groups bears the inscription 'in warmest admiration from the author' (*in wärmster Verehrung vom Verfasser*); invariance of dimension paper—'To professor Geheimrat Hilbert in warmest admiration from the author' (*Herrn G.R. Hilbert in herzlichster Verehrung vom Verfasser*); the second paper on Lie groups—'To his dear and admired professor Geheimrat Hilbert, L.E.J. Brouwer' (*Seinem lieben und verehrten Herrn G.H. Hilbert, L.E.J. Brouwer*). The dedications at the papers are certainly no empty phrases. Brouwer indeed deeply admired Hilbert; he was totally sincere in his love (which is the right expression, in spite of its melodramatic reputation). The later rejection of Brouwer by Hilbert must have been all the more painful.

## 4.3 Publishing in the *Mathematische Annalen*

Brouwer's first paper on finite continuous groups, *Die Theorie der endlichen kontinuierlichen Gruppen, unabhängig von den Axiomen von Lie, I*[21] published in the *Mathematische Annalen*, got a rather patronising review in the review journal *Jahrbuch über die Fortschritte der Mathematik*,[22] and this led to a curious correspondence between the author and the reviewer, Friedrich Engel, the close associate of Sophus Lie, and co-author of the three volumes *Theorie der Transformationsgruppen*. The surviving correspondence has been reproduced by Freudenthal in the *Collected Works II*. The letters show a striking conceptual rift between the new and the old generation. Engel, as one of the founding fathers of the subject, was so strongly entrenched in the traditional analytic approach that Brouwer's general

[19] Brouwer to Scheltema, 9 November 1909.

[20] The only hint that could throw doubt on the good relationship is a letter of Bernays to Freudenthal, relating to the above mentioned letter of Brouwer to Hilbert. In this letter Bernays more or less expressed his surprise that Hilbert and Brouwer were still good friends at that time. There are, in my opinion, no historical facts that would support this. But it is, of course, possible that Hilbert expressed in private conversation criticism of Brouwer.

[21] Brouwer (1909c).

[22] The volume for 1909 appeared in 1912.

set-theoretic approach was in effect unreadable for him.[23] The review opened with Engel criticising Brouwer's definitions as being somewhat vague and too restrictive, moreover, Engel thought that there were too many loose ends. 'But', he remarked, 'it would be unjust to demand that such a difficult problem would really be completely solved at a first go'.

A review of this sort was just the thing to provoke Brouwer. From the formulation it can be guessed that Engel did not fully grasp Brouwer's definition, and that he was in some confusion as to the local behaviour of the transformation in the neighbourhood of the identity and its global group theoretical behaviour. Brouwer reacted immediately with a cold and slightly patronising letter, asking Engel for chapter and verse:[24]

> Concerning the first point,[25] I would be grateful if you would be so good as to tell me which extended version of the problem you have in mind; for, I did not succeed in getting a clear idea from your hints.

The reader should keep in mind that in the time between the publication of the paper and the review, Brouwer had turned from an unknown beginner into one of the foremost experts in topology. The review was somewhat condescending in tone, and Brouwer must have felt that he could not leave it at that. At the time of this correspondence Brouwer was in a somewhat ambiguous position: his mathematical talent was well-recognised in the world, but in Amsterdam he was still at the mercy of the board of the university and the city council, as just another extraordinary professor. Nonetheless, I guess, he would have reacted as he did under any circumstances. Even as a beginner he would have taken up the gauntlet and accepted the consequences.

In reply to Brouwer's letter, Engel frankly admitted that he could not keep up with Brouwer's set-theoretical formulations, but he insisted that the clarity of formulation left much to be desired:

> Maybe they are clear to the set theoreticians incarnate (*eingefleischte Mengentheoretiker*), but I must say 'the expressions of the system sound obscure to uncircumcised ears'.

Engel's letter[26] illustrates the fact that Brouwer and Engel were worlds apart; their terminology and way of thinking, the first one topologically and the second one analytically, is so different that a common view was virtually excluded.

---

[23] In Freudenthal's words: 'Engel, [...] who could not grasp a group except in its analytic setting, and Brouwer, who had shaken off the algorithmic yoke and from his conceptional viewpoint could not comprehend his correspondent's difficulties. Manifolds and one-to-one mappings as substrate and action of Lie groups instead of Cartesian space and many-valued mappings was indeed a great step forward, though for older contemporaries of Brouwer it was too much' (CW II, p. 142).

[24] Brouwer to Engel, 21 January 1912.

[25] That is the excessive restrictions.

[26] Engel to Brouwer, 28 January 1912, CW II, p. 144.

The first two letters in the correspondence make a somewhat tense impression, Brouwer barely hid his annoyance at being lectured by an elder statesman, for a—in his eyes—perfect paper; the style of the letter is polite, bordering on the cynical. Whereas the letter may indeed display deep reverence, it could equally well be construed as mischievous. The closing sentence, 'I would highly appreciate to reach an agreement concerning the above, with a group theoretician of your authority', might have been written 'tongue in cheek', but it was probably sincere. Engel, in turn, seemed to sway between appreciation and mild sarcasm, writing that he had not intended any disapproval, but rather had, in the words of Lessing, evaluated the product of the Master, 'doubting in admiration, and admiring in doubt'. At the time of the exchange of letters Brouwer's fame had already reached the corners of the mathematical world, so that Engel must evidently have been aware of the merits of his correspondent, therefore it is quite likely that Engel was prepared to accept Brouwer's mathematical authority, but not without questioning. He closed his letter with the remark that he could not quite see the gain of investigations like Brouwer's in relation to the ingenuity and toil that were invested, expressing the hope that 'you would now also occupy yourself in group theory, for there is much to be done'. As mentioned above, the main point of confusion in Engel's understanding was the relation between local and global. The following letters returned to this matter. The tone of the letters, however, had greatly improved, but Brouwer still insisted that, even though Engel's objections were mainly the product of a difference in mathematical culture, readers of the review would get the impression that the paper contains actual gaps.[27] He felt that he was entitled to a rehabilitation and therefore he kindly asked Engel to correct possible misconceptions in the review of the second part of the paper.[28] The letter also contained a birds-eye view of the basic topological notions, including an elegant definition of connectedness.

Engel, indeed, set out on his journey to Canossa—without actually reaching it; the review of the second paper contained a somewhat reluctant retraction of the earlier statements, followed by a complaint that 'In general, I cannot dissemble that I am beset with a mild horror in the face of such colossal generality of the investigation, and the large number and variety of required inferences'. The second review only confirms the impression of the first one, namely that the master of the traditional school of Lie groups was separated by a wide gulf from the newcomer with his abstract topological methods.

It seems not unlikely that the wish , both of them expressed at the end of this correspondence—to talk things over in person—was fulfilled at one time or another. Whether this indeed happened and how profitable it was, we shall probably never know.

There is one more paper in this series dealing with Lie groups, this was the text of an address to the Dutch Physical and Medical Society.[29] It provided an elegant

---

[27]Brouwer to Engel, 6 March 1912.

[28]*Die Theorie der endlichen kontinuierlichen Gruppen, unabhängig von den Axiomen von Lie, II.*, Brouwer (1910b).

[29]*Het Nederlandsch Natuur- en Geneeskundig Congres.*

and short characterisation of the various geometries of Riemann's program.[30] The paper shows the hand of the master in mixing geometrical, topological and group-theoretical arguments.

The technique of eliminating differentiability conditions was applied once more in the paper *On a theory of measure*,[31] in which Brouwer sharpened the results of G. Combebiac on a certain functional relation.

## 4.4 Fixed Points on Spheres and the Translation Theorem

The investigations on Lie groups took place at the same time as work on continuous mappings on spheres, work on vector distributions, and the above mentioned research on the topology of the plane.

On 27 February 1909, Korteweg presented to the Academy of Sciences the first paper by Brouwer of a series 'Continuous one–one transformations of surfaces in themselves'.[32] The question that Brouwer asked himself was 'whether this [that is the one–one continuous mapping of the sphere into itself] is possible without at least one point remaining in its place'?

The method used in this paper is wholly based on the available techniques of plane topology; it involves a detailed study of the behaviour of a family of concentric circles and their images. The best-known results of the paper are:[33]

> A continuous one–one transformation in itself with invariant indicatrix [that is orientation preserving] of a singly [that is simply] connected, two-sided, closed surface possesses at least one invariant point.

and its companion

> A continuous one–one transformation in itself of a singly [simply] connected, closed surface leaves at least one point invariant.

As one can easily see, a two-sided, simply connected closed surface need not have a fixed point under a one–one continuous orientation inverting mapping[34] (think of a mapping that takes each point of the sphere to its antipode). Brouwer's

[30] Characterisation of the Euclidean and non-Euclidean motion groups in $R_n$, CW II, Brouwer (1909d) (originally in Dutch).

[31] Brouwer (1911f).

[32] Brouwer (1909e).

[33] Brouwer (1909e), pp. 10, 11.

[34] Popularly speaking one fixes the orientation on a surface by indicating a direction on a little circle, the orientation at other places is determined by shifting this circle all over the surface. If this is possible, that is if one never can get via different routes two circles circulating in opposite directions around one point, the surface is called orientable, and it has an orientation. The sphere is an example of such a surface. The Möbius strip is an example of a one-sided non-orientable surface (take a rectangular strip of paper, twist it and glue the ends together).

papers on the topology of surfaces are connected to his research on Lie groups; the fixed point theorem for the sphere is, for example, used in Brouwer (1910b) (p. 193).

The paper on surface mappings was the first in a series of eight papers in the Proceedings of the Academy, the last one appearing in 1920. The pre-war papers are of the same nature, they use elementary techniques; the post-war papers use the new tools that Brouwer had developed himself before the war.[35]

The second paper in the series deals with arbitrary two-sided surfaces; it ends with Brouwer's first, and (as he soon realised) incorrect, version of the *translation theorem*: a 1–1 continuous mapping of the plane without fixed points is a continuous image of a translation.

The translation theorem was, so to speak, the natural supplement to the stream of fixed point theorems; it spelled out what one could expect in the absence of a fixed point for transformations of the plane (where, of course, no fixed points are guaranteed, think of an ordinary shift). At the time, the translation theorem was something of a dashing exploit, fit to be told in small groups of professional mathematicians, huddled in the bar after a strenuous day at a the conference. The theorem has attracted the attention of topologists ever since; there are a number of new and clever proofs around nowadays, see for example Franks (1992).

The papers following this communication were all concerned with continuous bijections of surfaces. Unfortunately, the early ones suffered from one and the same defect: Brouwer had relied on Schoenflies' monograph *The development of the theory of point sets II*,[36] often affectionately referred to by the insiders as the *Bericht*. Schoenflies' book was, together with the Encyclopaedia paper of Dehn and Heegaard, the latest word on topology; in particular Schoenflies dealt with all the items that were relevant to the foundations of analysis, such as the basic notions of convergence, open, close, dense, perfect, ..., including the notion of 'curve'. As it happened, Brouwer's work made generous use of properties of curves. Much to Brouwer's disappointment, a good deal of the treatment on curves by Schoenflies showed serious defects. As late as 1912 Brouwer somewhat crossly remarked[37] that the survey, *On one–one, continuous transformation of surfaces into themselves*,[38] which he had presented at Hilbert's request to the *Mathematische Annalen*, was based on unsubstantiated claims of Schoenflies:

> In carrying out the ideas sketched in the second communication[39] on the subject, I found out that in some points in the course of the demonstrations indicated there, the Schoenflies theory of domain boundaries, criticised by

[35] When the publisher, at the end of the series, mixed up the numbering, the numbers 7 and 8 were erroneously published as 6 and 7. Brouwer patiently corrected the numbering by hand in the off-prints.

[36] *Die Entwicklung der Lehre von den Punktmannigfaltigkeiten II*, 1908.

[37] Brouwer (1912h), p. 360.

[38] *Über eineindeutige, stetige Transformationen von Flächen in sich*, Brouwer (1910g).

[39] Brouwer (1909g).

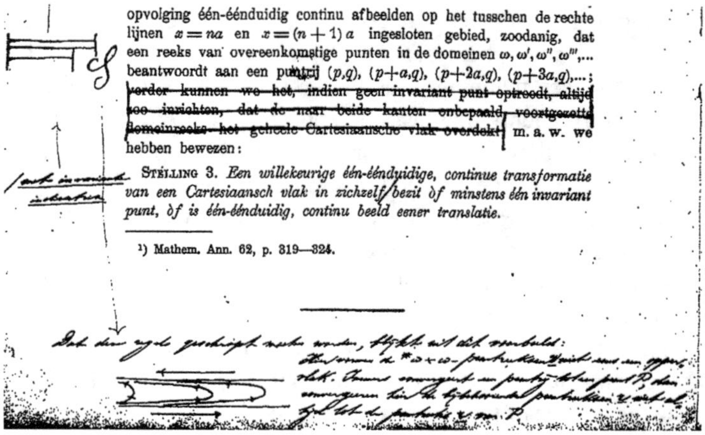

opvolging één-éénduidig continu afbeelden op het tusschen de rechte lijnen $x = na$ en $x = (n+1)a$ ingesloten gebied, zoodanig, dat een reeks van overeenkomstige punten in de domeinen $\omega, \omega', \omega'', \omega''', \ldots$ beantwoordt aan een puntrij $(p,q)$, $(p+a,q)$, $(p+2a,q)$, $(p+3a,q), \ldots$; ~~verder kunnen we het, indien geen invariant punt optreedt, altijd zoo inrichten, dat de naar beide kanten onbepaald voortgezette domeinreeks het geheele Cartesiaansche vlak overdekt~~ m. a. w. we hebben bewezen:

STELLING 3. *Een willekeurige één-éénduidige, continue transformatie van een Cartesiaansch vlak in zichzelf bezit òf minstens één invariant punt, òf is één-éénduidig, continu beeld eener translatie.*

[1]) Mathem. Ann. 62, p. 319—324.

**Fig. 4.3** The handwritten corrections at the end of Brouwer's paper (in Dutch) *Continuous one-one transformations of surfaces in themselves*. [Brouwer archive]

> me,[40] still plays a role, so that the Theorems 1 and 2 formulated on p. 295 and likewise the 'general translation theorem' based on them, and enunciated without proof in *Mathematische Annalen* 69, p. 178 and p. 179, cannot be considered as proved,[41] and a question of the highest importance is still to be decided here. The 'plane translation theorem' stated at the end of the second communication and similarly in *Mathematische Annalen* 69, p. 179 and p. 180, has meanwhile been proved rigorously by another method.[42]

The correct proof of the translation theorem in the *Mathematische Annalen*, is a complicated piece of plane topology in the traditional style. It is remarkable that Brouwer formulated the theorem in rather sweeping terms, so that a reader could (without reading the proof) easily be misled. In 1919 he published a short note[43] with the proper formulation, and a counter example to the superficial reading. This, by the way, was already inserted in handwriting by Brouwer in his reprints of the second communication (Dutch version). In all likelihood those corrections were inserted soon after publication.[44]

The proper formulation of the translation theorem requires more topology than our exposition allows. Let it suffice to say that a continuous 1–1 mapping of the plane without fixed points splits the plane into a number of strip-like parts disjoint

[40]Cf. Brouwer (1910e).

[41](Brouwer's note) Already the property of p. 288 that the transformation domain constructed in the way indicated there, determines at most two residual domains, vanishes for some domains, incompatible with the Schoenflies theory.

[42]Brouwer (1912b).

[43]Brouwer (1919n).

[44]Cf. CW II, p. 220, and Freudenthal's comments on p. 219.

from their images, so-called 'translation fields'. Freudenthal, in his editorial comments in Brouwer (1976) (p. 218), wrote that 'In the twenties and thirties Brouwer stated orally his belief that the original proof could be salvaged. It would be worthwhile to try it because its idea, in particular the construction of a transformation field at one stroke, is more attractive than that of Brouwer's final proof'.

A reader of Brouwer's papers will be struck by the conscientious manner in which corrections and emendations were provided. Brouwer, apparently, was striving for perfection in his papers, even after they had appeared. A number of private copies of reprints carry handwritten corrections or additions, that often appeared as supplements to later papers, or as individual corrections, even when only trivial slips or misprints were concerned.

## 4.5 Vector Fields on Surfaces

One month after the communication of the first paper on the continuous 1–1 mappings of the sphere, Korteweg presented another paper of his student: *On continuous vector distributions on surfaces*.[45] The paper was the first of another series of papers, all published in the proceedings of the Academy. It dealt with vector distributions on spheres in a generalised setting, that is to say, on homeomorphic images of spheres. Brouwer's approach was based on Peano's existence theorem for differential equations;[46] he investigated singularities of vector fields by means of differential equations. There is a close connection between this series of papers and the fundamental papers of Poincaré on the qualitative theory of differential equations (*Mémoire sur les courbes définies par une équation differentielle I, II, III*, 1881, 1882, 1885).

Poincaré had introduced a whole new aspect to the theory of differential equations. Whereas traditionally one studied solutions locally, that is in the neighbourhood of a point, he initiated the study of global behaviour:

> Finding out the properties of differential equations is a task of major importance. A first step has been made by studying the proposed function in the neighbourhood of one of the points of the plane. The time has come to go further and study the functions in the whole plane [...]
>
> The study of a function thus includes two parts: (1) The qualitative one, as it may be called, where the curve is studied geometrically, (2) the quantitative one, where values of the function are calculated numerically.[47]

It appears that Poincaré's papers were somewhat slow to be absorbed into the daily routine of mathematics; Brouwer, anyway, had not read them when he embarked on the investigations of vector fields. But he had heard the presentation of

[45] Brouwer (1909f).

[46] Peano (1890). Brouwer also cited the simplification of Arzela (1896).

[47] Poincaré (1881). Translation by Freudenthal. CW II, p. 282.

**Fig. 4.4** Jaques Hadamard. [Courtesy M. Loi]

Poincaré's talk at the Rome conference. During the conference Poincaré fell ill, and someone else had to read his address. Poincaré had planned to give at this conference his views on 'The future of Mathematics', with 'Differential equations' as one of its topics. Poincaré had warmly advocated the 'qualitative discussion of curves defined by a differential equation'; in view of Brouwer's enthusiastic report of Poincaré's talk,[48] it seems plausible to look here for the motivation of the choice of a topic of Brouwer's subsequent research.

Although Brouwer was an admirer of Poincaré, and had read quite a number of his publications (in particular the more philosophical ones), he had somehow missed the topological œuvre of Poincaré, to the detriment of both Brouwer and topology, one might say. There are two letters in the correspondence between Hadamard and Brouwer which show that Brouwer was not aware of Poincaré's work in the area. Hadamard, thanking Brouwer for two reprints—the above paper on vector fields and possibly one of the papers on mappings of the sphere—suggested an easy solution to the fixed point theorem on the sphere by applying the main result of the paper on vector fields. In the same letter he drew Brouwer's attention to Poincaré's *Mémoire* of 1881. Brouwer's reply of December 24, 1909 contained a refutation of Hadamard's suggestion and a grateful acknowledgement of Hadamard's reference to Poincaré's work. In the same letter he announced another approach to a proof of the fixed point theorem on spheres, 'Reading the memoirs of M. Poincaré cited by you, I got another idea'.[49]

Brouwer's investigations were, however, no duplications of Poincaré's earlier work; his viewpoint was a truly topological one, whereas Poincaré assumed some algebraic and analytic conditions. The use of Peano's existence theorem is telling, since it requires in the differential equation $dy/dx = f(x, y)$ only the continuity of $f$.

The results of Brouwer were of a more general nature than Poincaré's; in particular he did not assume any uniqueness conditions (which are implicit in Poincaré's

[48]Cf. p. 199.

[49]Cf. Johnson (1981), p. 154. The letter from Hadamard was undated, but it seems likely that it was answered promptly by Brouwer.

paper), since Peano's theorem did not require any. The price to pay was, however, a degree of complexity that robbed him of potential readers. In the long run Poincaré's results became the standard reference for the qualitative theory of differential equations. It is not wholly impossible that the place of publication, that is the Proceedings of the Amsterdam Academy, was less than fortunate. It did not have a circulation among mathematicians comparable to that of ordinary journals; as a rule periodicals of the sort of academy proceedings, which (at least in those days) published papers from a variety of disciplines in the sciences, were stored in the general sections of central libraries, and not in the mathematics sections. Also, for whatever reason, no comprehensive presentation of the material of the vector distribution series was submitted to the *Annalen*, or any other journal.

Freudenthal, in his comments in CW II (p. 282), points out that where Poincaré used uniqueness conditions and assumed algebraicity and simple singularities, 'Brouwer on the contrary chose the utmost generality and even in specialising he admitted more general types than Poincaré had done. Thanks to a simpler design, Poincaré's work has been influential in the long run, but it is strange that Brouwer's contributions have been entirely overlooked'.

The main results of Brouwer's first paper on vector fields were as follows

**Theorem 1** *A vector direction varying continuously on a simply connected, two-sided, closed surface must be indeterminate in at least one point*

And from this follows directly:

**Theorem 2** *A vector distribution anywhere univalent and continuous on a singly* [i.e. simply] *connected, two-sided, closed surface must be zero or infinite in at least one point.*

If we project the complex plane stereographically on the sphere, a complex function becomes a vector distribution on the sphere. So we can also interpret the above result as follows:[50]

**Theorem 3** *A univalent, continuous function of a complex variable being nowhere zero or infinite and without singular points cannot exist.*

Freudenthal observed that 'theorem 2 is usually ascribed to H. Poincaré, 1881, though in fact Poincaré asserted and proved it under quite restrictive conditions.[51] Brouwer generalised it to higher dimensions and proved it by more adequate methods.'[52] Brouwer's first theorem became popular as the 'hairy ball theorem': one cannot comb a hairy tennis ball without getting a crown.

---

[50] Brouwer (1909f), p. 856.

[51] CW II, p. 282.

[52] Brouwer (1911c), p. 112.

If Poincaré's basic papers were slow to attract attention, Brouwer's work on vector fields remained buried in the pages of a poorly read journal—to the disadvantage of the mathematical community.

The first paper was followed by two more with the same title;[53] they contain Brouwer's closer analysis of the nature of finite sets of singular points. The second communication contains the structure theorems for the singular points and for the behaviour of the field in the neighbourhood of the singular points. Apart from the classification of singular points, the paper contains a few novelties such as the introduction of the winding number by purely topological means, and its use in the study of the structure of zero points in relation to the behaviour of the field in its neighbourhood,[54] as illustrated by the next theorem

**Theorem 5** *The total angle which, by a circuit of a simply closed curve enveloping one point zero, the vector describes in the sense of that circuit, is equal to* $\pi(2+n_1-n_2)$*, where* $n_1$ *represents the number of elliptic sectors,* $n_2$ *the number of hyperbolic ones, which appear when a vicinity of the point zero is covered with tangent curves not crossing each other.*

Furthermore the paper contains the first appearance of homotopic changes of vector fields.[55] The third communication continues the study of finitely-many singular points under special conditions, for example, in the absence of simple closed tangent curves (Brouwer introduced in this paper the picturesque and suggestive term 'irrigation fields').

Following a suggestion of Hadamard,[56] Brouwer applied the winding number to the (finite) set of singular points, obtaining generalisations (in an exclusively topological setting) of Poincaré's results from the *Mémoires* quoted above.

Brouwer apparently, and not unreasonably, felt that there should be a connection between the fixed point theorem on the sphere[57] and the existence of singular points of vector distributions. The obvious trick that comes to mind first, that is the construction a vector field out of a continuous one–one mapping via the great circle connecting them, was suggested in Hadamard's letter[58] and rejected by Brouwer; he spelled out his arguments in the third paper on vector distributions,[59] and gave a correct procedure to get the fixed point theorem on the sphere in the orientation preserving case from the theorems on singular points.[60]

---

[53] Brouwer (1910c, 1910f).

[54] Brouwer (1910c).

[55] Cf. Freudenthal, CW II, p. 302.

[56] Cf. Johnson (1981), p. 153.

[57] Brouwer (1909e).

[58] Hadamard to Brouwer 24 December 1909.

[59] Brouwer (1910f), CW II, p. 314.

[60] Section 3 of Brouwer (1910f). See also Hadamard's acknowledgement in his appendix to the second edition of Tannery's *Introduction à la théorie des fonctions II* (dealing with topological applications of the Kronecker index), Hadamard (1910).

Looking at those papers through modern eyes, one cannot fail to be impressed by the approach (and the stamina) of Brouwer; he undertook the investigation of the properties of singular points with his bare hands.

All the methods involved are available to any student with a basic knowledge of differential equations and a sniff of general topology. The advance with respect to older research, for example Poincaré's, consists of a consistent exploitation of the topological viewpoint. One will find in Brouwer's papers certain arguments that have gone out of fashion, for example the use of the second number class and infinitesimal circles and the like, but on the whole the papers are very readable.

Five years later Brouwer published in the journal of the Dutch Mathematical Society an application of the above theory to the projective group: 'On the orthogonal trajectories of the orbits of a one-parameter plane projective group.'[61] At that moment his first topological period was over. His attention was already fixed on foundational matters.

## 4.6 Analysis Situs and Schoenflies

The name of Schoenflies has already come up earlier in these pages in connection with the development of topology around the turn of the century. It is time to have a closer look at him and his work.

Arthur Moritz Schoenflies was born in Landsberg an der Warthe (now Gorzow, Poland) on 17 April 1853. He studied with Kummer in Berlin and after his Ph.D. and *Habilitation* became an extra-ordinary professor in Göttingen (1892) for applied mathematics. In 1899 he was appointed full professor in Königsberg (Kaliningrad), and in 1911 he moved to Frankfurth am Main, where he had a chair until 1922. He died there on 27 May 1928.

Schoenflies was a geometer at heart. He had started his career in mathematics with a number of investigations of a fairly traditional kind, but in the wake of Jordan he entered into a more modern era with his study of groups of motions (*Bewegungsgruppen*). In this area, he carried out an extensive classification. This line of research was close to another of his activities: the study of crystal-structure by group theoretic means. The latter was laid down in a comprehensive book *Crystal Systems and Crystal Structure*.[62]

The bulk of Schoenflies' mathematical production is however concerned with the set theory of Cantor. He published extensively on the subject and was at the turn of the century recognised as the outstanding expert in the field. The DMV, the German Mathematics Society, asked him to prepare a survey (*Bericht*) of set theory (including point-set theory, that is topology). Part I, the development of the theory of point sets, appeared in 1900 and Part II in 1908. For some time both Schoenflies'

---

[61] Brouwer (1915).

[62] *Krystall Systeme und Krystallstruktur* (1891); a new edition was published in 1984!

monographs were the only comprehensive texts on set theory in the wider sense—Hausdorff's book appeared in 1914.[63] It is true that many textbooks on analysis or function theory contained sections explaining the basic notions of set theory, but these usually stuck to the immediately applicable parts of the subject.

Both in set theory and topology, Schoenflies had produced a considerable œuvre. He was, however, in the dubious position of a man on the borderline of two cultures; he belonged to the first generation of set theorists, and in ideas and method he was very close to Cantor. He lacked, however, the penetrating insights required to push the subject beyond its initial phase. More imaginative mathematicians, above all Zermelo, created the framework and concepts that made set theory the subject we know today. Schoenflies, the active champion of the new disciplines, was allowed to see the promised land, but he was not the one to lead mathematics into it. Where Zermelo showed set theory the way, Brouwer gave topology its ticket for the twentieth century. Already in 1905, in the animated discussions following Zermelo's proof of the well-ordering theorem, Schoenflies had been taken to task by Zermelo: 'his paper, however, contains further errors and misunderstandings that cannot be ignored here'[64]—and, again, in 1910 he was the victim of a vigorous dressing down, this time by Brouwer.

The new methods and concepts introduced by the latter changed the subject almost beyond recognition. One could well say that, after the introduction of the new Brouwerian methods, Schoenflies' approach became something of a curiosity.

Nonetheless, Schoenflies' role in topology should not be underestimated; his survey monographs are examples of a painstaking sorting and collecting of facts and conjectures in a new area, where the wheat had not yet been separated from the chaff. Looking back, we can say that Schoenflies took, in preparing the surveys of set theory and topology, the role of midwife upon himself. More so, one would be inclined to say, in the case of topology, than in the case of set theory.

From 1899 onwards Schoenflies had written a series of papers with the intention of providing a systematic treatment of the basic concepts that were beyond Cantor's and Jordan's work. In particular his *Contributions to the theory of point sets I, II, III*[65] are clear, easily readable expositions. The great survey of 1908, *The development of the theory of point sets, II* collected all hitherto published material in one volume.

Schoenflies formulated his goal in the first of the above contributions:

> As one of the most general problems of the theory of point sets, we can point to the task of formulating and establishing set-theoretically the fundamental theorems of analysis situs, and setting forth the relationships which exist between the set–theoretic–geometric and the analytic modes of expressing these concepts and theorems. The paradoxical results, as they occur, for

---

[63]There were other texts, for example, Young and Young (1906) but they had only a modest influence.

[64]Zermelo (1908), Schoenflies (1905).

[65]Schoenflies (1903, 1904, 1906).

> example in the one–one mapping of continua and in the Peano curve, have completely destroyed the naive ideas of analysis situs. All the more, we must demand that set theory provide a substitute and define the basic geometrical concepts in a way that returns to them their natural content, characteristic of analysis situs. Even if the much maligned intuition, is no source of proof, it still seems to me that it is a goal of research to reconcile the content of geometrical definitions with the content of intuition—at least in the domain of analysis situs.

As to the proper subject of research, Schoenflies was quite clear. Following Klein's Erlangen Program he stated, in justifying his new definition of the notion 'connected', that Analysis situs can be considered as the science which studies notions invariant under univalent and continuous mappings (obviously he had topological mappings in mind).

The *Beiträge*-papers deal with the topology of the plane, and Schoenflies treats the major topics of the day, for example properties of plane curves, polygons, the structure of perfect sets, and Jordan curves.

It is not known exactly when Brouwer started to study Schoenflies' topology, but the topics of both topological mappings of surfaces and of vector fields presupposed a lot of Schoenflies' material. The earliest mention of Schoenflies is in Brouwer's 1908 Rome talk,[66] where he indeed prided himself on being the first person to do something useful with topology:

> Only the groups of the two-dimensional manifold can be determined with the help of the results so far of Schoenflies, and I believe that thereby for the first time an application of these results is given.

Clearly Brouwer was not slow to absorb Schoenflies' papers; in his dissertation he referred to the first volume of the *Bericht*, and he had read and digested the 1908 survey of Schoenflies in the year of its appearance; and so it is fair to say that he knew the latest in topology (but, as we have seen, by no means all the classical papers).

At the meeting of the Dutch Mathematical Society of 30 October 1908, Brouwer gave a talk with the title *On plane curves and plane domains*.[67] Although the manuscript with the same title in the Brouwer archive is undated, there is little doubt that it is the text of the talk, because Brouwer refers to the Schoenflies proof of the invariance of the Jordan curve 'earlier this year'. Brouwer presented an exposition of the various notions involved in plane topology, such as Peano curve, closed curve, Jordan curve, domain, approximating polygon, and accessibility. He pointed out the mistakes in Schoenflies' recent proof and gave a proof of the Jordan–Schoenflies theorem.

The text shows that Brouwer had learned his plane topology, and also that in ease and insight he was second to none.

---

[66] Brouwer (1909b).

[67] *Over vlakke krommen en vlakke gebieden.*

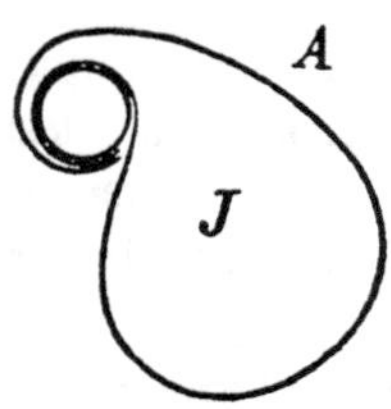


Fig. 1.

**Fig. 4.5** A domain without a closed boundary curve, Brouwer (1910e)

His investigations of topological maps of spheres and vector fields had brought home to him the questionable state of Schoenflies' foundation of topology, so he set himself the task of getting the topic back on its feet before proceeding to the topics that were really attracting him. In a letter to Hilbert, he gave the motivation for his investigation of the basic notions of topology:[68]

> To regain clarity, it was first of all necessary to check thoroughly the theory of Schoenflies involved here, and to ascertain precisely on which of its results one can build in full faith. This was the beginning of the enclosed paper, which has serious consequences for various parts of the theory of Schoenflies, and more or less recreates some parts.
>
> It seemed fitting to me that, if possible, it should be published at the same place, where Mr. Schoenflies had originally published his paper, therefore I send it to the editorial board of the *Mathematische Annalen* for publication, and at the same time I will send a copy to Mr. Schoenflies.

The paper that Brouwer sent to Hilbert, was his pioneering *Zur Analysis Situs*,[69] which changed the landscape of elementary topology. It consists of two parts: the first one contains counterexamples to a number of Schoenflies' statements, and the second one redevelops some parts of the topology involved. Apart from its scientific value, it has a certain curiosity value, for it is the only paper in the *Mathematische Annalen* with coloured illustrations! These illustrations were, by the way, a source of worries to the printers, the editor and to the author; they were discussed repeatedly.

Although Brouwer was clearly annoyed at having relied on doubtful material, he went out of his way not to affront Schoenflies—after all he still was a newcomer without a position in the academic world. After stating in the paper that 'several of his [that is, Schoenflies'] results were false, and others, 'indeed correct, but insufficiently supported', he cautiously added in a footnote: 'I stress specifically that this paper does not try to diminish in any way the high value of the researches of Schoenflies.'

Some of the paper's counterexamples to claims or theorems of Schoenflies are still relevant, but the evolution of the basic concepts in topology has made some others less interesting. Nonetheless, one should keep in mind that Brouwer's counterex-

[68] Brouwer to Hilbert, 14 May 1909.

[69] Brouwer (1910e).

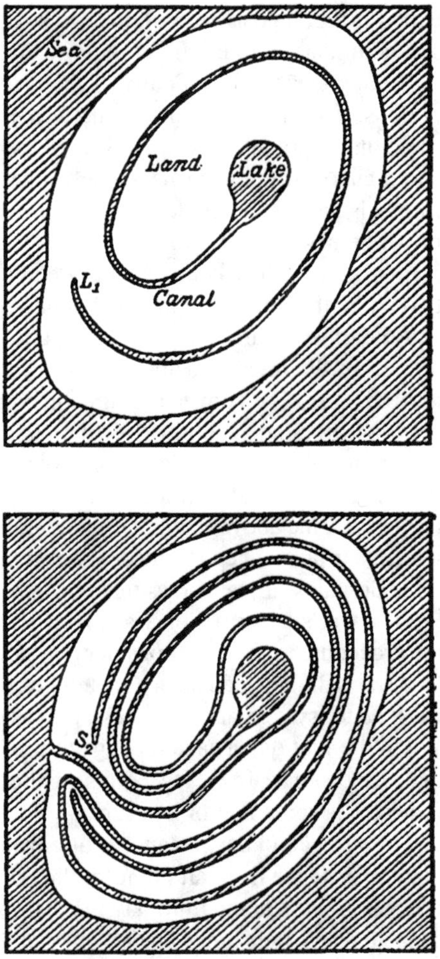


**Fig. 4.6** Wada's island and canals. See Brouwer, Collected Works 2, p. 367

amples themselves played an important role in the cleaning-up of the foundations of topology.

For instance, in the *Bericht* Schoenflies stated that 'One easily recognises the correctness of the following theorem: VI. The boundary of a domain (*Gebietsgrenze*) is a set which is closed [...]'.[70] Brouwer's counter example is simple enough to please, even today a beginning student: The domain, $J$, 'is obtained if one continues a simple curve arc to both sides, and wraps both extensions, without letting them meet, like a spiral around one and the same circle'.

The best-known construction of the paper is that of a curve which splits a square into three domains. That particular example gave rise to an intensive study of so-

[70] A domain is a connected open set. Schoenflies (1908), p. 112.

called indecomposable continua[71] (i.e. continua that cannot be cut into two proper sub-continua), a tradition in topology that reached well into the forties.

To appreciate the difficulty of the problem, it will be helpful to introduce a few notions. The simplest kind of curve that we know is a line segment, and since we are doing topology, it seems plausible to call the result of bending and stretching still a curve. If the result is not self-intersecting, we speak of a simple open curve. In precise terms: a simple open curve is the homeomorphic (or topological) image of a line segment. A closed curve is the homeomorphic image of a circle. Now Jordan's theorem tells us that a closed curve (or a Jordan curve) splits the plane in two parts, the interior (which is bounded in size) and the exterior. Both the interior and the exterior are connected, and one cannot go from a point of the exterior to a point of the interior without crossing the curve, which is the common boundary of both parts. The converse of this theorem, established by Schoenflies tells us that if a closed bounded subset $C$ of the plane determines two open domains, such that each point of $C$ is accessible from both domains, it is a Jordan curve. So it looks as if closed curves and a splitting in two parts are hand in glove. Schoenflies used the property of the above set, $C$, as the definition of a closed curve. One can imagine the general surprise when Brouwer showed that there are closed common boundaries to *three* domains. It just shows that our normal geometric intuition has a lot to learn!

Brouwer's construction was given an amusing form by Wada,[72] who visualised the procedure as follows: consider in a saltwater sea an island with a freshwater lake. One starts digging canals in a systematic way such that alternatingly one extends the fresh water system and the salt water system:

On the first day a canal is dug starting at the lake, and not meeting the sea, such that the distance from the points of the shore to the fresh water supply is $< \frac{1}{2}k$ (where $k$ is some unit of measure).

On the second day, a canal is dug from the sea, not mixing with fresh water, so that the distance from all points at the border of the lake or the first canal to the salt water is $< \frac{1}{4}k$. On the third day the digging is resumed at the end of the fresh water canal and an extension is made such that the distance from the shore and the borders of the salt water canal to the extended first canal is $< \frac{1}{8}k$, etc.

This process is continued countably many times; the result is (eventually) a curve which separates the fresh and the salt water. One can also start with two lakes, say with blue and red water). The result then is one curve that separates three 'domains' of water.

The final part of Brouwer's paper contained a critical examination of the main topological notions and their properties as treated in the last chapter; notions such as *domain, accessibility, curve*, etc., *connectedness*, *Jordan curve*, etc., and indications how to repair the lacunae.

The paper closed with a brief catalogue of false statements followed by two claimed, but not satisfactorily proved facts, including the elusive invariance of the closed curve.

---

[71] A continuum is a compact connected set with at least two points.

[72] Yoneyama (1917).

The last mentioned problem was settled by Brouwer himself in a later paper.[73] He had, as a matter of fact, already mentioned this open problem in his inaugural address of 1909. The 'Analysis Situs'-paper did not only mark the end of a period in point-set topology (aptly called *Cantor–Schoenflies topology* by Freudenthal), but it was also a turning point in Brouwer's career. It drew attention to a promising, ambitious mathematician, who could use some support in his attempts to find a position in the Dutch academic world.

Brouwer was, deservedly so, proud of the above-mentioned exploits in set-theoretical topology (in the plane). An extra bonus was the attention paid by the incumbent grand master of topology, Schoenflies. It definitely had the pleasant effect of strengthening Brouwer's position on the job market (see p. 205). Schoenflies accepted the collapse of his Magnum Opus with grace, albeit reluctantly. He insisted to publish his comments in the same volume of the *Annalen*, following Brouwer's paper.[74]

The correspondence about the gaps in Schoenflies' *Bericht*, of which only Schoenflies' letters have been preserved, shows a confused and hurt Schoenflies.

It is painfully clear that he had not seen the pitfalls of his own speciality, and even after Brouwer's observations, he remained convinced that only marginal corrections would do.

In his letters he tries to save as much as possible, and to minimise the impact of Brouwer's criticism. The first letter, dated 27 May 1909, from Schoenflies, is still rather optimistic about the extent of the damage. After thanking Brouwer for the copy of the manuscript submitted to the *Annalen*, he gave vent to his disappointment, 'My delight that you have carried the study of my *Bericht* to such depths is, alas, not without a bitter flavour.'

He admitted that a number of geometrical shapes of curves, under his definition as 'boundary of domain', had escaped him. He then went on to fill the remainder of the letter with corrections and 'refutations' of Brouwer's criticisms.

But this was just the beginning; letter after letter was exchanged—without much effect. Schoenflies had great difficulty in following Brouwer's arguments. Apparently Brouwer soon lost his patience, for we find in the letters of Schoenflies signs of exasperation: 'Would it then really be impossible for me to convince you in writing?' (14 August 1909); 'It pains me, so to speak, that you could think, that I mean by the outer boundary of a domain, what you mean by it. I have given you no cause for a thing like that. Moreover I had believed, on the basis of all my papers in this area, to be protected from that!' (12 September 1909).

After this letter Brouwer must have lost control of his patience altogether, for Schoenflies opened his next letter with the words 'To my strong disappointment, I must start to say that I must decidedly insist to be spared letters of the kind of your last letter. You have neither reason nor occasion for that' (13 December 1909).

By now the first couple of revisions of the proofs had already been handled by the *Annalen*; it became rather embarrassing to go on much longer. Nonetheless,

---

[73]Brouwer (1912i).

[74]*Bemerkung zu dem vorstehenden Aufsatz des Herrn L.E.J. Brouwer.*

Brouwer must have protested once more, as can be inferred from Schoenflies' reply (19 December 1909): 'If you think that you "in a sense have the right to demand the deletion of the words indicated by you", and furthermore, that I had acted against our agreement, I must defend myself most emphatically. I can in no way acknowledge this.'

The exchange stopped essentially here, and so Schoenflies' reaction to Brouwer's paper was in final form roughly seven months after Brouwer submitted his paper.

Although Brouwer had cleared the troubled waters of Schoenflies' topological work, the two reluctant companions were not through yet. Destiny would bring them together once more.[75]

After this 'animated' correspondence, the harmony between Schoenflies and Brouwer was restored through Hilbert's mediation. In the end the two authors agreed to send out combined reprints (see also p. 206).

Brouwer continued to work in this area, although it had to give way to more challenging problems when Brouwer discovered the new topology, to be described in the next chapter. Among other things, there is a lovely, elegant proof of the Jordan Curve theorem;[76] Brouwer's proof was adopted by Hausdorff in his *Grundzüge der Mengenlehre*. In 1923 Erhard Schmidt also presented an elementary proof, which could compete in elegance with Brouwer's proof. Schmidt's proof is completely elementary and uses only the tools of point-set topology. This paper also contains the theorem that subsequently became known as the 'Phragmen–Brouwer theorem', which states that the frontier of a connected component of the complement of a compact, connected set is connected. The terminology is poorly chosen; Phragmén's only paper in topology contains just the theorem that a closed point set with no connected subset, does not split the plane.[77]

The modern reader would prove the Phragmén–Brouwer theorem, like much of Brouwer's topological work, using homology.[78]

The revised version of the proof of the Jordan Curve theorem was sent to Hilbert on the fifteenth of October 1909, with a covering letter that neatly summed up Brouwer's research aims of the moment:

> *Sehr geehrter Herr Geheimrat,*
>
> Enclosed, I send you once more my proof of the curve theorem, and I hope very much that you find the presentation satisfactory.
>
> Allow me to point out to you that we have no certainty about the measure of validity of the theorem *that the one–one continuous image of a closed curve again is a closed curve*,[79] and that there seems to be here a really difficult problem. (By a 'closed curve' we mean a point-set, which determines two

---

[75] Cf. p. 226.

[76] Brouwer (1910a).

[77] Phragmén (1885), cf. Freudenthal, CW II, p. 383.

[78] Cf. Dieudonné (1989), p. 207.

[79] Proved in Brouwer (1912e, 1912i).

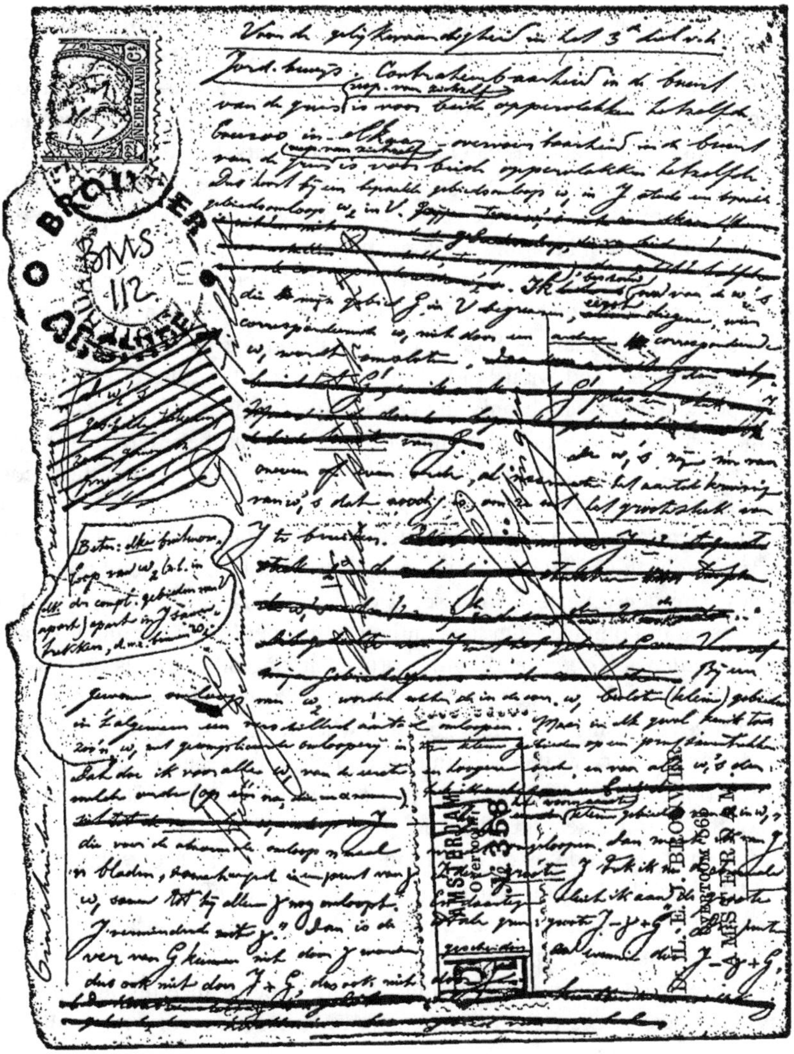

**Fig. 4.7** Topological notes on an envelope. [Brouwer archive]

domains in the plane, and which is identical with the boundary of both of these domains.)

There neither is a proof, that the one–one continuous image of a sphere determines two and only two domains in space.[80] Nobody, anyway, doubts the correctness of this claim.

---

[80]Proved in Brouwer (1911e).

Recently I found that, if one asks for those continuous transformations of the sphere into itself, which are univalent in one direction and *at most* two-fold in the other direction, such a transformation is always a one–one continuous image of the function $x' = x^2$. This gives me hope, that the whole Riemann theory of algebraic functions can be founded on Analysis Situs.[81]

This might perhaps even be the starting point for the theory of analytic functions, if one asks for such correspondences, which are built up from domainwise one–one and continuous correspondences.

Singular points, then, are those for which no neighbourhood exists in which the correspondence is one–one and continuous. Anyway, here a mass of problems belonging to Analysis Situs arises: what can the nature of the sets of singular points be, and what kind of singularity can there be in each point; which point sets, in which only the continuity is known, can one admit, so that the non-singularity of the points of these sets may be inferred from the continuity; and in the first place, whether one can correlate each of these correspondences to an analytic function.

Thus I return once more to the question you brought up in our discussion,[82] namely the behaviour of an analytic function in the neighbourhood of a nowhere dense perfect point set $\mu$ of exceptional points. Some of that I find treated in Pompéiu: 'On the continuity of functions with complex variables', *Annales de Toulouse* 1905. He proves among other things that, when the function is uniformly continuous, and the set $\mu$ has measure zero, continuity suffices to ensure analyticity, but that when the measure of $\mu$ is not zero, the function can very well be continuous in those points, but not analytic.

He also introduces a new notion, the 'length' of a set, and derives on the basis of this notion a second criterion.

Allow me to remind you of your promise, to have reprints of yours (on the Dirichlet principle, integral equations and the Obituary of Minkowski) and of Mr. Zermelo (Foundations of set theory) and Koebe (uniformisation) sent to me? I express in advance many thanks.

My wife sends you warm greetings. In the hope of meeting again, with many greetings for *die liebe Frau Geheimrat*, whose recovery we would be very pleased to learn.

Yours truly
L.E.J. Brouwer.

Indeed, the coming years would show that Brouwer was as good as his word—topology would never be the same after the eventful years preceding World War I.

Brouwer investigated a number of topics in the Cantor–Schoenflies tradition; among others the Cantor–Bendixson theorem (which was called 'Cantor's fundamental theorem' at the time). He generalised the theorem in various directions, in

---

[81] Brouwer (1919d, 1919e, 1919j, 1919m).

[82] Scheveningen, September 1909, cf. p. 125.

particular he proved a generalisation by replacing the role of points in Cantor's theorem by closed components of a set.[83] In the 1911 paper Brouwer formulates and proves his *reduction theorem*, which he used to provide simple proofs of earlier theorems (including the theorem of Janiszewski and Mazurkiewics on irreducible continua). The topology papers contain many little gems that earned the author a well-deserved reputation as an original and ingenious topologist, slowly Brouwer's innovations have been absorbed into more general treatments; and the modern topologist would hardly be aware of their parentage. To mention just one case: in Brouwer (1910d) the first example of a topological group which was not a Lie group is given. The example is an early (perhaps the first) limit construction (in the categorical sense).

There is a hitherto unknown sequel to Brouwer's activity in the more traditional topology:

In 1912, when Brouwer was already fully occupied by his revolutionary new insights in topology, he returned to the more concrete problems of the past. In that same year, the sudden and unexpected death of Poincaré shocked the mathematical world. After an operation, which was by itself successful, the great mathematician had died from an embolism, only fifty-nine years old. Shortly before his death his last paper appeared in the Rendicotti; on 9 December he had sent it to the editor with the comment that—contrary to his custom—he submitted an unfinished paper on a problem which he considered of the highest importance. This paper contained a statement, which has become known as 'Poincaré's last theorem':

> A measure-preserving homeomorphism of a circular ring in the plane unto itself, so that the inner circle is rotated clockwise and the outer circle anti-clockwise, has fixed points.[84]

Poincaré's theorem (or conjecture at this stage) would, of course, appeal to Brouwer, who had proved a number of fixed-point theorems on surfaces in the past years. Brouwer immediately set to work at the problem and convinced himself of the solution; there are a number of notes in his archive in which he attacked the proof. There are even a few sheets of a paper which he intended to present to the Academy. In a letter of 31 September 1912, however, he wrote to Korteweg that 'It turns out that the solution of the Poincaré problem, before it is written in a matured form, it will take many more weeks. Before that time I would rather keep it under me; please don't discuss it with anybody. Only after I have finished the complete version, I will present an outline to the Academy.' There is no record of any publication of Brouwer on the topic. The many obligations that Brouwer had accumulated started to take their toll. In December 1913 he complained in a letter to Korteweg that he lived under a constant pressure on account of all kinds of obligations to foreign mathematicians, he had not been able to carry on his research which 'is resting already for one and a half years.' Brouwer had mastered the baffling problems

[83]Brouwer (1910d, 1911a).

[84]Poincaré (1912).

of topology, but not the art of saying 'no'. In his capacity as the leading topologist, his advice and judgement were all too often sought and given. The Poincaré problem was solved and the solution published by Birkhoff in 1913. Brouwer reacted by sending Birkhoff a postcard—"Many thanks for your magnificent proof of Poincaré's geometric theorem. You remark with Poincaré that the existence of an invariant point implies immediately the existence of a second invariant point. I do not see how this can be deduced from the property of the preservation of areas."

# Chapter 5
# The New Topology

## 5.1 Invariance of dimension

Although mathematics had been an exciting human enterprise at all times, the nineteenth century was particularly rich with excitement, often bordering on shock.[1] On numerous occasions old beliefs were shattered by the improved critical spirit of the century. The already heavily tried mathematical intuition underwent another traumatic experience in 1878 when a short paper appeared in which a professor at Halle proved an incredible result: there is a one–one mapping from the side of a square to the square itself (or from $\mathbb{R}$ to $\mathbb{R}^2$).

The author, Georg Cantor, had more surprises in store for the mathematical world, but this result, hidden under the non-descript title 'A contribution to set theory',[2] was a straightforward attack on common geometric intuition: 'one dimension is essentially different from two', or 'two independent variables cannot be reduced to one'. Given the ingrained intuitions and habits of traditional geometry there was enough reason for surprise at this new result.[3]

The result of Cantor seems to show that dimension is not the immutable rock it was always supposed to be. No mathematician prior to Cantor would have expected to fill up the square with a line, or to unwind the square as a line, like one unpicks a sweater. So was this the end of dimension as mathematics knew it? Dedekind tempered Cantor's excitement by pointing out that the bijection that Cantor had exhibited was highly discontinuous, and that the moral of Cantor's theorem should not be overestimated. Dimension would be in serious jeopardy if Cantor's mapping were continuous—if, roughly speaking, one could fill a square with a pencil without taking it from the paper and without passing through a point twice.

[1] This chapter makes essential use of Freudenthal's comments in Volume II of the Collected Works and of the paper *The Problem of the Invariance of Dimension in the Growth of Modern Topology I, II* of Dale Johnson.

[2] *Ein Beitrag zur Mannigfaltigkeitslehre.*

[3] The history of this particular theorem is published by Emmy Noether and Jean Cavaillès, Noether and Cavaillès (1937).

D. van Dalen, *L.E.J. Brouwer – Topologist, Intuitionist, Philosopher*,
DOI 10.1007/978-1-4471-4616-2_5, © Springer-Verlag London 2013

The mathematical community was rather inclined to keep dimension on its throne, and so it was generally conjectured that there could not be a bijection from the line to the plane, continuous in both directions.

Even stronger, Cantor's contemporaries would have been inclined to bet that one could not even fill up the square with a continuous curve that would be allowed to intersect itself (that is the pencil might pass through the same point more than once). The general feeling of mathematicians was that our ordinary lines, surfaces, solids, etc., are continuous in nature, and hence a comparison of their dimensions should take this property into account. So far this was an intuitive conviction; one of the tasks of the new discipline 'topology' was to make these ideas precise. The ultimate outcome was that Cantor's '*Gleichmächtig*' ('equivalent', 'of the same cardinality') was too general and arbitrary to reflect dimension faithfully.

Almost immediately after Cantor's paper appeared, efforts to save the invariance of dimension (and hence dimension itself) were made. Among the mathematicians who attacked the problem were Lüroth, Thomae, Jürgens, Netto and Cantor himself.

For an extensive discussion of the history of dimension the reader is referred to D.M. Johnson's magnificent papers of 1979 and 1981; for the present it suffices to record the achievements of Lüroth, who proved that one cannot give a topological mapping of a one, two or three-dimensional space onto a higher dimensional one (1878, 1899). None of the other attempts succeeded. The proof for the three dimensional case was, however, so complicated, that everything seemed to point to the inevitable conclusion that either the general problem was so hard that it ranked among the great unsolved problems, or that it required a totally new approach.

The majority of practising mathematicians, however, was not disturbed by the state of affairs, and they were assured by no less a person than Schoenflies, in his chapter in the *Encyklopädie der Mathematischen Wissenschaften I*, 1898, that the problem had been settled by Netto and Cantor: a one–one correspondence between distinct $R_n$ and $R_m$ is never continuous. One wonders if Hilbert ignored the problem of the invariance of dimension on the authority of Schoenflies, or whether he dismissed it as 'just another problem'; whatever his reasons were, the problem does not occur in his famous list of 1900. This may have been a matter of expediency, but it was certainly not the consequence of a lack of importance, as Brouwer's remark in his Rome talk shows (see below).

Another aspect of the dimension enigma was discovered by Giuseppe Peano, the man who had created a symbolic language for mathematics, and who had proved the best existence theorem so far for differential equations. He had exhibited in a three-page note a continuous mapping from the side of a square onto the square itself, 'On a curve which fills a plane area' (1890). In plain words, one can fill in the whole square with a pencil without taking it from the paper.[4] Peano's example was followed by more, so-called, *space-filling curves*.[5] All these examples added to the urgent feeling of inadequacy of plain geometric intuition.

[4]There is a fairly simple geometric representation of the above mapping, Johnson (1979), p. 171 and Young and Young (1906), pp. 165, 291.

[5]Among others Hilbert and Moore, cf. Young and Young (1906).

Brouwer was aware of the significance of the problem of the invariance of dimension in 1908 at the latest, when he gave his address at the Rome conference; after defining Lie groups, he added in a footnote:[6]

> Whether $p$ [that is the number of parameters] is an invariant for each group, is an open problem as long as the 'non-applicability' of two spaces with distinct dimension numbers, is unproven.

There is a number of sources for Brouwer's research plans in this period, among them is the inaugural lecture *The nature of geometry*,[7] which Brouwer gave on 12 October 1909 at the Amsterdam University at the occasion of his appointment as a *privaat docent*. It contained his views on geometry, including topology. Brouwer's idea of geometry was closer to the Riemann tradition than to the Euclid–Hilbert tradition. The influence of Klein's Erlangen Program is evident:

> Geometry is concerned with the properties of spaces of one or more dimensions. In particular, it investigates and classifies sets, transformations and transformation groups in these spaces.[8]

He succinctly summarised the role of transformation groups.

> Finally, it can often be shown that figures and operations with which we became acquainted in the smaller group, can be completely defined by properties invariant for the larger group; they can then be *more generally characterised*, though perhaps it may be useful for special questions afterwards to consider the smaller group. An example is supplied by the potential functions in two dimensions, which were at first characterised in Euclidean geometry. After a point at infinity had been added, and the conformal group had been introduced, it could be shown that such a function is completely defined by its invariants for the conformal group, and that the Euclidean group is inessential for it.[9]

In plain words, a larger group may solve some of the riddles that the small group leaves unexplained. The phenomenon was well-known through the Cayley–Klein approach to projective and non-Euclidean geometry.

It was in this vein that Brouwer studied geometry, and it also was his approach to group theory. Whereas the topic, properly speaking, is a part of algebra, Brouwer looked at groups with the eyes of a geometer.

The above considerations are general in nature, but they reflect Brouwer's own research activity. Of late he had been investigating groups larger than the traditionally considered groups, namely groups of topological transformations, that is of

---

[6]Brouwer (1909b), p. 297.

[7]*Het wezen der meetkunde*. The translation of 'wezen' by 'nature' is somewhat flat. *Wezen* expresses something more, like 'essence'.

[8]Brouwer (1909a), CW I, p. 116.

[9]Brouwer (1909a), CW I. p. 117.

those one–one transformations that are continuous (preserve limits, in the intuitive approach) in both directions. As we have seen he strove for methods of logical simplicity, shunning extraneous concepts;[10] in particular he stripped away (or tried to do so) differentiability conditions whenever they were not essential, thus carrying out Hilbert's wishes, as expressed in the fifth mathematical problem of 1900, to new areas.

The reader, who is used to modern topology, will have to keep in mind that the terminology took a long time to establish itself. A lot of notions have gone through a long evolution, causing a good deal of confusion on the way. One such notion is 'connected'; as we will see later, Brouwer got into trouble over the interpretation of this notion. Likewise, he used 'one–one continuous' for what we now call a 'homeomorphism' or 'topological mapping'.[11]

The 1909 inaugural lecture is a most helpful document: it gives us a glimpse of Brouwer's private research program for topology. While presenting a survey of 'modern geometry', it lists a number of key problems:

- 'investigation of the general character of a system of several one–one continuous transformations [of a set] into itself'.
- 'a classification for analysis situs of the sets of points in a space from the viewpoint of analysis situs'.
- 'An immediately related problem is, to what extent spaces of different dimension are distinct for our group [of topological mappings]. Most probably this is always the case, but it seems extremely hard to prove, and probably will remain an unsolved problem for a long time to come.'
- 'in spite of many efforts no satisfactory proof has as yet been given for the seemingly very slight extension of Jordan's theorem, that the one-to-one continuous image of a closed curve is again a closed curve.'—'No more are we certain that a closed Jordan surface, that is a one-to-one continuous image of a spherical surface, splits the three-dimensional Cartesian space into two domains.'

A number of the above topics would soon be successfully attacked by Brouwer, although he apparently was at that time far from certain to succeed where others had failed. In particular he was rather aloof as regards the invariance of dimension. Nonetheless, within half a year the situation changed dramatically. Not only did Brouwer solve the problem of the invariance of dimension, but he enriched topology with (at least) two tools that enabled him to attack a whole range of problems.

The birth of the new topology has been recounted in Freudenthal's *The cradle of Modern Topology, according to Brouwer's Inedita*.[12] The first recorded evidence of the new ideas was found by Freudenthal in an exercise book of Brouwer, that had a label *Potentiaaltheorie en Vectoranalyse*, when he was editing the topological part

---

[10]Not to be confused with *Methodenreinheit* (purity of methods) of older generations. Brouwer was quite prepared to use whatever means were available.

[11]The first term was introduced by Poincaré (1885), the second one by Brouwer, Brouwer (1919f, 1919g).

[12]Freudenthal (1975), cf. CW II, p. 422 ff.

of the Collected Works. The editing had already progressed to the final stage, when this notebook came into his hands; its title did not spell any surprising revelations, but when he opened it, two drafts of letters fell out. One was the draft of a letter to Hilbert, dated 1 January 1910, and the other of a letter to Hadamard, dated 4 January 1910. The letter was written during the Christmas holidays,[13] spent with his brother Aldert in Paris. It shows that Brouwer had already developed two of his new methods that appeared in print only in 1911: the *degree of a mapping* and *simplicial approximation*. The letter to Hilbert ran as follows:

Paris, New Year's morning 1910.

Dear *Herr Geheimrat*,

Best wishes for you and your dear wife for the New Year, for your health and for your scientific activity.

I am staying here for the Christmas holidays with my brother, the geologist; unfortunately my wife could not accompany me. In the middle of January my classes start again and I shall return.

The harmony with Mr. Schoenflies has been re-established, and mainly through your intervention. I am enclosing the last two letters to him, which I have answered by saying that I am satisfied with his last version, and that I consider the matter closed.[14]

May I make a few comments on the univalent (not necessarily one–one) continuous mapping of a sphere $\kappa$ onto a sphere $\lambda$? If one subjects it to the condition that it should be continuous both ways, then it is a one–one continuous image of a rational function of a complex variable [...] By this condition of the two-way continuity I mean that a closed Jordan curve around a point, $L$, of $\lambda$, which converges to $L$, corresponds, for each point, $K$ of $\kappa$, of which $L$ is the image, to a closed Jordan curve around $K$ that must converge to $K$.

Now, if we have two univalent mappings of a sphere (or a more general closed surface) $K$, satisfying these conditions, onto a sphere, $L$, and a sphere, $M$, then the question is raised what extra conditions should be satisfied so that the correspondence between $L$ and $M$ is a complex algebraic one in the sense of Analysis Situs.

If I return again to the general univalent and continuous correspondence between two spheres, then a finite number $n$ can be indicated as its degree such that all correspondences of the same degree, and only these, can be transformed continuously into each other. In particular all correspondences of degree $n$ can be continuously transformed into rational functions of a complex variable of degree $n$.

In order to define this degree, we introduce homogeneous co-ordinates $x, y, z$ on $\kappa$ and homogeneous co-ordinates $\xi, \eta, \zeta$ on $\lambda$, and consider first the univalent mapping, which is determined for each domain by a correspondence

$$\xi : \eta : \zeta = f_1(x, y, z) : f_2(x, y, z) : f_3(x, y, z),$$

[13] CW II, p. 421, see also Brouwer to Scheltema 3 December 1909.

[14] Concerning the 'Analysis Situs'-paper, cf. p. 144.

where $f_1, f_2, f_3$ are polynomials.

If we fix a positive orientation on both spheres, and choose in each point of $\kappa$ this positive orientation, then each generic point of $\lambda$ occurs $p$ times with positive orientation and $q$ times with negative orientation. One can show that $p - q$ is a constant for all generic points, which we call the degree of the mapping.

If the correspondence between $x, y, z$ and $\xi, \eta, \zeta$ is not determined by polynomials, then one can approximate them by such polynomials, and one easily shows that those approximating correspondences have a constant degree, which we therefore also assign to the limiting correspondence. This degree is always a finite, positive or negative number. In particular the one–one continuous mapping of the sphere into itself has degree $+1$ if the orientation is not changed, $-1$ otherwise.

Now you know my theorem that every *one–one* continuous transformation of the sphere into itself which does *not* change the orientation, always has at least one fixed point. This theorem can be extended in such a way, that every *univalent* continuous mapping of the sphere into itself which does not have degree $-1$, always has at least one fixed point.

And I succeeded in extending the theorem to $n$-dimensional spheres. Then it reads: every univalent continuous transformation of the $n$-dimensional sphere into itself has at least one fixed point. The transformations of degree $+1$ form an exception for odd $n$, and those of degree $-1$ for even $n$. On the one hand for odd $n$ and inverted orientation, on the other hand for even $n$ and unchanged orientation, one–one transformations therefore necessarily have a fixed point Even more general is the result for univalent continuous mappings of the solid sphere into itself, for these have a fixed point anyway.

Once again best wishes for both of you,
as ever yours faithfully,
L.E.J. Brouwer

This letter documents a spectacular breakthrough of Brouwer, soon after his exchange of letters with Hadamard. The work on vector fields, that preceded the discovery of the degree of a mapping, had brought Brouwer close to earlier work of Poincaré, and Hadamard had pointed out to him the relevance of Poincaré's investigations, including that of the indicatrix (introduced by Leopold Kronecker, 1869, and probably known to Gauss as early as 1840). The bold idea of stripping away the analytical notions, had, roughly speaking, occurred in the short interval between his letter to Hadamard of 24 December and 1 January. There is a letter with a similar content to Hadamard (4 January 1910). Freudenthal, in his comments in the Collected Works II (p. 422, ff.) points out that the exercise book and the letters to Hilbert and Hadamard provide invaluable information on the birth of the mapping degree, and on the role of Hadamard.

What has now also become clearer is Hadamard's role as midwife. Before I had never understood how Hadamard figured in this story and why Brouwer

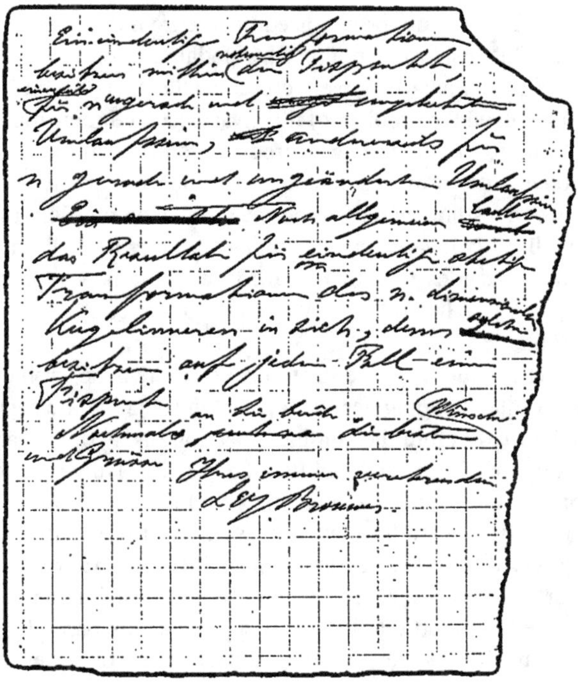

**Fig. 5.1** Page of a draft of letter to Hilbert, 1 January 1910. [Brouwer archive]

regarded him so highly in this context. Hadamard's work in this part of topology, Hadamard 1910, looks rather old-fashioned. Its strong dependence on analytic tools would have hampered rather than stimulated true topology. It was Brouwer's achievement to have shaken off the yoke of analysis from topology. In the genesis and maturation of his ideas, however, his intercourse with Hadamard must have meant more to Brouwer than can be expressed by mere citations and quotations.[15]

Hadamard's influence is acknowledged by Brouwer in his third paper on vector distributions. Hadamard, for on his part, in his Appendix *On some applications of the Kronecker index*,[16] gives the credit for the method used in this paper to Brouwer. As Freudenthal put it: 'According to [the draft for the letter *Brouwer to Hadamard* 4.I.1910.] Y17 he learned the deeper reason for it [the connection between fixed points on the sphere and singularities of vector distributions] in the correspondence with Hadamard, which does not mean that Hadamard revealed it to him.'

In a letter to Kneser of 21.XII.1925 Brouwer returned to the exchange of ideas between him and Hadamard, 'The content of both my paper "über Abb. v. M." and the book of Tannery was discussed around Christmas 1909 in Paris. On that occasion I have, referring to a paper that was at the time partially in print, but waiting for a

[15]Freudenthal in CW II, p. 425.

[16]Hadamard (1910).

final formulation, pronounced the theorem that was for the first time proved by you .... This instance explains why Hadamard referred to 'Brouwer's theorem'.

Freudenthal's discovery corrected the plausible but false impression that Brouwer first solved the problem of the invariance of dimension, and subsequently discovered the notion of 'mapping degree', possibly distilled from the invariance proof. The above letter provides convincing evidence that the mapping degree came first—the letter does not even mention the dimension problem as such! So, why did Brouwer publish his Invariance of Dimension-paper first? Whatever his reasons were, it was not a bad idea to start the overture to the new topology with a resounding trumpet blast. What better way is there to attract attention than to solve a famous open problem?

The solution of the invariance of dimension-problem, as a matter of fact, made use of the mapping degree, without introducing the concept by name. The paper was submitted before the summer vacation of 1910, probably before July, and it seems likely that Brouwer submitted it directly to Hilbert, who then passed the manuscript on to Blumenthal, the managing editor of the *Annalen* to process it for publication. A word of warning may, however, be in order. Brouwer's research was of a rich variety during the hectic years of 1909 and 1910. It is not certain when Brouwer started to investigate the invariance of dimension, but in a letter of 18 March 1910 he wrote to Hilbert that he was preparing a new paper for the *Mathematische Annalen* in which he had partly solved the invariance of dimension, in so far as he had shown that spaces of even and odd dimension are not homeomorphic. The paper was never published, and no drafts have survived, so we do not know what Brouwer's approach was in this particular proof. In all likelihood he found the complete solution before he had finished the manuscript of the partial one.

The manuscript of the invariance of dimension paper was sent to the *Mathematische Annalen* in June 1910, and the paper appeared in the issue that was dated 14 February 1911.

Although the material of the 'invariance of dimension', was rather out of the way for his Dutch fellow mathematicians, Brouwer presented a complete proof in a beautiful didactic form, at the October meeting of 1910 of the Dutch Mathematics Society. The manuscript, which has been preserved, shows us a clear exposition that could nowadays easily be used in an undergraduate course.[17] The reactions to Brouwer's talk are not known, but it seems plausible that it conveniently raised the status of the young *privaat docent*, whose promotion to lecturer had just been turned down by the local authorities.[18]

Between the submission of the dimension-paper and its publication, something happened that added a tragic and dramatic note to the history of the invariance of dimension.[19]

---

[17]Freudenthal (1979).

[18]Cf. p. 213.

[19]Freudenthal (CW II, p. 435 ff.) has given a thorough historic and mathematical analysis of the invariance of dimension episode. The reader is referred to Freudenthal's comments for more technical details.

Blumenthal had made a trip to Paris in the summer vacation of 1910, where he met Henri Lebesgue. The latter was a prestigious mathematician, who, among other things, created modern measure theory; the Lebesgue measure and integral have been household words in mathematics ever since their introduction. Lebesgue contributed to many areas in mathematics, in particular also to the young discipline of topology, which, as a matter of fact, touched on measure theory in quite a number of points. Blumenthal, aware of the importance of the topic, told Lebesgue about Brouwer's exploit. In a letter of 27 October 1910 Blumenthal informed Hilbert of his meeting with Lebesgue:

> We made a very nice trip to Paris during the summer holiday. Unfortunately, however, I did not see any mathematicians, they were all on holiday. That is to say, I have made acquaintance with Lebesgue, who happened to be in Paris. He is a very interesting man and told me that he possessed already for some time (*seit langer Zeit*)—not just one, but several proofs of the theorem of the invariance of the dimension number, which Brouwer has now proved in the *Annalen*. He has sent me one of those proofs for the *Annalen*, which looks very amusing. I have not examined it closely for the correctness of the underlying idea, for one can depend in the matter of details on such a shrewd man. If you want to check the paper in detail, it is at your disposal.

We do not if Hilbert occupied himself with the matter, but when in February 1911 Brouwer's spectacular paper appeared, it was immediately followed by an extract from a letter of Lebesgue to Blumenthal, *On the non-applicability of two domains belonging respectively to spaces of $n$ and $n + p$ dimensions.*[20] The tone of the message is somewhat patronising. There is a ring of quiet amusement and superior insight in this note to the mathematical world, which had in all the years failed to find a proof:

> Recently, when you talked to me about the proof of the impossibility of establishing a one—one continuous correspondence between the points of two spaces of $n$ and $n + p$ dimensions, a proof by Mr. Brouwer that the *Mathematische Annalen* will publish, I have indicated to you the principle of several proofs of the same theorem. I will exhibit the simplest of these proofs; I will not occupy myself here with drawing from the argument other consequences than the theorem in question itself.

The principle which Lebesgue referred to was the marvellous *paving principle*[21] which reads:

> If each point of a domain $D$ of $n$ dimensions belongs to at least one of the closed sets $E_1, E_2, \ldots, E_p$, finite in number, and if these are sufficiently small, then there are points belonging to at least $n + 1$ of these sets.

[20] *Sur la non-applicabilité de deux domaines appartenant respectivement à des espaces à $n$ et $n+p$ dimensions (Extrait d'une lettre à M. O. Blumenthal)*, Lebesgue (1911a).

[21] *Pflaster Satz.*

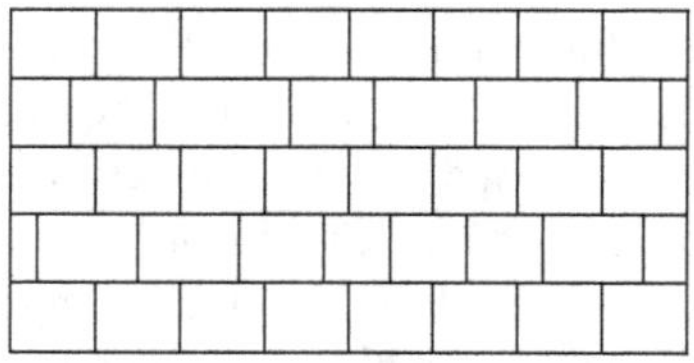

**Fig. 5.2** Paving principle

**Fig. 5.3** Henri Lebesgue. [Courtesy l'Enseignement Mathématique]

By 'domain', Lebesgue means closed, bounded, connected set. The principle is illustrated in the plane by the following example, Fig. 5.2, of the pattern of a pavement (which is the worst possible case: minimal overlap).

One immediately sees that there are points belonging to three tiles (we consider tiles with boundaries). The invariance of dimension is indeed an immediate corollary to the paving principle. Unfortunately, the paving principle itself is by no means trivial, at least not the way Lebesgue thought it was.

The records do not show whether Brouwer was informed of this letter before the publication, but it is highly unlikely. Brouwer would immediately have reacted! So we may assume that he was confronted with Lebesgue's letter when he opened his copy of the *Mathematische Annalen*—a nasty shock indeed! Imagine having settled an outstanding fundamental open problem of the day, only to be ridiculed by the 'simplest possible solution'. Brouwer indeed was furious; however, he quickly realised that Lebesgue's proof contained a serious gap. No matter how brilliant the paving principle was, as far as Lebesgue was concerned it had the status of a conjecture. Nonetheless, Brouwer was upset; apparently few mathematicians could fathom the methods involved, and hence there was a serious possibility that—also in view of Lebesgue's barely hidden boasting—the credit would go to Lebesgue with, maybe, Brouwer as a fair second.

The reader has to bear in mind that the intricacies of topology were lost on the average mathematician. At best they would think of the geometrical figures and objects that were familiar from traditional geometry, but these did not display the pathologies that tricked even the best in the field. Now Brouwer's paper, which nowadays most readers would find clear and easy going, possessed a precision and a Spartan mode of argument that were unfamiliar to the mathematicians of the nineteenth century, who were accustomed to the pleasant narrative style of, say, Felix Klein. So a reader, presented with the papers of Brouwer and Lebesgue, would be tempted to agree with the Lebesgue paper, which was written in a beautiful convincing style, with a whiff of 'this is all very simple for those who would care to follow the indications given here', and skip Brouwer's paper as 'a difficult proof of the same thing'. Blumenthal was not the only one to fall for the temptation: in the *Fortschritte* the reviewer of Lebesgue's note—the same Engel who struggled with Brouwer's Lie group paper—virtuously cited the paving principle and accepted the paper on Lebesgue's authority. It is probably fair to say that the material was beyond him and most of his contemporaries, possibly even beyond Lebesgue. Even though the word 'priority' hardly crops up in the correspondence about the dimension problem—name not the rope in the house of the hanged man—it would not be human to claim that the following conflict was just about truth and correctness, and not about priority. More dispassionate scholars may point out that priority is not everything and that in the end history will see that justice is done. But unfortunately that is not always the case. There are enough examples where history has stuck the wrong label on theorems, methods, proofs, etc. Usually this concerns an item of local importance, and seldom a major result; it is unthinkable that, say, the proof of Fermat's last theorem would be attributed to the wrong person. In the period we are describing here, the dimension problem had a somewhat comparable status, it was one of the big challenges of modern mathematics. The fact that nowadays the theorem is part of a first course in algebraic topology is irrelevant, in Brouwer's days there were no tools available, and one had to go at it with bare hands. In the case of the dimension problem Brouwer had every right to be upset; it would already have been bad enough to find out that a well-known mathematician had a couple of solutions in his drawer, but it would really hurt if the priority would go to the wrong person on the basis of an unproven conjecture!

A key figure in the controversy that followed the publication of Brouwer's and Lebesgue's papers, was Blumenthal, a generally respected mathematician, who was active in complex function theory and applied mathematics.

He was born in Frankfurt a. M. in 1876, and studied mathematics in Göttingen with Klein, Sommerfeld and Hilbert. Blumenthal was the first Ph.D. student of Hilbert (1898), and he wrote his *Habilitationsschrift* in 1901.

He also studied for some time in Paris with Emile Borel and Camille Jordan. Since 1905 he had a chair in Aachen, not far from the Dutch border and close to Maastricht. Blumenthal and Brouwer became close friends and the latter used to visit Aachen regularly. Blumenthal was the managing editor of the *Mathematische Annalen*, the daily matters of the journal were completely in his hands, although he routinely informed Hilbert of the editorial matters. In effect he was a key figure of the *Mathematische Annalen* for the impressive period of 32 years. His involvement in the Brouwer–Lebesgue controversy was a natural consequence of his position. Either Brouwer had sent his manuscript right to Blumenthal, or Hilbert had passed it on. One would not blame Blumenthal for talking to Lebesgue about Brouwer's invariance proof, after all an editor may discuss papers with third parties. If Blumenthal can be blamed at all, then it is for not sending Lebesgue's note to a referee.

That, however, was a bit difficult, because, apart from Brouwer, there was virtually no competent person to judge this kind of topology. His 'Analysis Situs'-paper had made that much abundantly clear. Anyway, it would have only have been fair to inform Brouwer about the Lebesgue note, since Lebesgue was also informed about Brouwer's paper.

Brouwer reacted immediately, he analysed Lebesgue's note and saw that the proof did not work. In early March, he submitted a short note to that effect to Hilbert, *Remarks on the invariance proof of Mr. Lebesgue*. The copy of this note bears 'Accepted Hilbert' in Hilbert's hand. Blumenthal, referring to this note in a letter of 14 March 1911 to Hilbert, could not match Hilbert's scientific composure. He was uncomfortably caught in the middle of a controversy that he could not quite fathom:

> I find the Brouwer–Lebesgue affair highly unpleasant, and in fact I am, on the whole, on the side of Lebesgue. That is to say: Lebesgue says explicitly that he assumes certain theorems; those have to do with certain linear equations and inequalities, and they will presumably be provable, in other words, the difficulty does not seem to be in that part, and the whole arrangement of the proof of Lebesgue is in my opinion, taking everything into consideration, a feasible and beautiful route to get to the dimension proof. Whoever reads Brouwer's note, does not altogether have this impression; the note has in my opinion an unfriendly and unpleasant ring.

Blumenthal advocated the withdrawal of the note, and he begged Hilbert to step in, should that be necessary. The matter was further complicated because

> On top of all that I, for one, like Lebesgue (according to an earlier communication) am not able to understand Brouwer's proof.

As a matter of fact, Brouwer's note was neither unfriendly nor impolite, at most somewhat condescending—possibly matching Lebesgue's tone. It only stated that Lebesgue's proof was defective, indicating the gap. Brouwer did not provide a counterexample, but confined himself to the remark that 'in any case considerable further elaborations are required'. Furthermore, Brouwer pointed out that Lebesgue's reference to Baire's work was off the mark, as 'the unproved theorems to which the problem is reduced,[22] are deeper than the problem itself'.

Lebesgue's reaction to Brouwer's note was brief, and showed that he did not yet fully grasp the difficulties:

> If I understand the remark of Mr. Brouwer correctly, it comes down to: I have announced that I will provide the facts that I qualified as evident, that does not replace a proof of these facts.
>
> I do agree with Mr. Brouwer on this point, I only add that I have not completely written out my proof, because I promised some time ago a paper on this topic to the secretary of the *Société Mathématique de France*.

[22]By Baire.

> I do recognise that my formulation is very poor, since Mr. Brouwer has been able to believe that I had not seen the necessity of proving everything, and that till to-day it seems useful to him to point out this necessity to other readers.

Lebesgue's letter had convinced Blumenthal that Brouwer had nothing to fear. Blumenthal explained in his letter of 25 March[23] that:

> ... whatever Lebesgue may stress, nobody doubts or disputes your priority for this fundamental proof [...] Lebesgue is, in his own opinion and that of the world, not your rival but your follower (*Nachfolger*).

As to Lebesgue's position, he wrote

> ... on the one hand Lebesgue's letter shows that he has a clear idea of the proofs of the tentatively accepted theorems, on the other hand he writes himself literally: Writing down the complete proof does not take long, and I am about to do so, but truly, it seems impossible to me to make my results look as ready-to-eat chunks (*bribe à bribe*) and I think that your readers, more generous than Mr. Brouwer, will give me credit until the appearance of my definitive memoir.

Blumenthal added a little sermon:

> From this communication it seems to follow, that it would not be right to publish your note before you have convinced yourself, that not just Lebesgue's note in the *Annalen*, but really *his whole line of proof* is defective. I am convinced that Lebesgue will put the manuscript that he had prepared, at your disposition for inspection. If necessary, I would be prepared to mediate in this respect. I would indeed like to draw your attention to the fact—and here I come to the heart of my impressions—that *your note has been formulated in a very rude form*, and that everybody must necessarily interpret it in such a way, that you consider the gaps stressed by you as irreparable, that is to say that you consider Lebesgue's proof as *false*; for false and incomplete is the same thing in this case. In my opinion, you can only accuse a man of Lebesgue's position of this, if you are completely certain of your ground.

In the same letter Blumenthal announced that Lebesgue had withdrawn his reply to Brouwer's note, and written a new one. This document is not extant, so we cannot judge the contents, but it is not unthinkable that Lebesgue started to realise that the difficulties were not just minor formalities.

Brouwer immediately answered that, in view of Lebesgue's promise to provide a complete proof, he was happy to withdraw his note. In a letter of 31 March to Hilbert,[24] Brouwer expressed his gratification, he could now drop the matter. This was the more pleasing, he said, since letters of Lebesgue and Blumenthal showed

---

[23] Cf. Johnson (1981), p. 191.

[24] CW II, p. 440.

**Fig. 5.4** Blumenthal and Brouwer. [Brouwer archive]

that readers could misinterpret his action as a priority complaint, 'which it was not intended to be'. The statement may seem strange, but Lebesgue had satisfied Brouwer that there was no priority claim involved. Reconstructing the whole discussion, one is left with the impression that Brouwer was angered by the flamboyance, bordering on arrogance, of the older colleague, and by the apparent injustice of Lebesgue getting away with claiming a theorem without so much as a proof.

The appreciation of the mathematical world was, of course, another matter. Even though Lebesgue had made Brouwer's priority clear, there were others who did not have a clear grasp of the situation. For example, L. Zoretti, in a review of Schoenflies' survey of point set theory[25] stated that 'Chapter *V* contains an interesting study of the correspondence between two domains of $n$ and $n+p$ dimensions, and the invariance of the notion of dimension. Very recent investigations of Messrs. Baire, Lebesgue and Brouwer have made a decisive step in the matter ...'.[26] Although the statement is so unspecific that it could hardly be wrong, it conveys to a reader the impression that somehow the mentioned persons had contributed in roughly the same proportion. As it happened, Brouwer had in the *Annales d'École Supérieure* pointed out a number of mistakes in a topological paper of Zoretti, which may have coloured the judgement of the latter, but it is equally well possible that Zorretti simply had failed to understand the matter. There was a similar misjudgement on the part of Emile Picard, when he presented in 1911 a report to the *Académie des Sciences* on the works of R. Baire:

[25] Schoenflies (1908).

[26] Zoretti (1911).

> In 1907, in two notes entitled *On the non-applicability of two continua of $n$ and $n+p$ dimensions*,[27] Mr R. Baire has given a method to study the matter alluded to in the title. Some lacunas remained in the proofs which the author proposed to clear up. He was earlier than Brouwer and Lebesgue, who by different roads arrived at the theorem which was clearly anticipated by Baire, to wit that one cannot establish a bijective correspondence between a continuum of $n$ dimensions and a continuum of $n+p$ dimensions.

One wonders to what extent Picard, an outstanding expert in the theory of functions, and familiar with mathematics at large, obeyed the academic tradition and polished up his report.

The *Fortschritte* soon realised that its reviewer had been a bit careless about the Lebesgue note; in its 1914 volume a brief correction was published (by Blaschke):

> In the *Bericht* on p. 419 on the paper of Lebesgue it should be noted that the proof of Lebesgue is not correct. There is a gap, which the author so far has not filled. More can be found in a paper of L.E.J. Brouwer in *J. für Math.* 142, 151.[28]

Brouwer was under some apprehension, as he himself had not yet found a proof of the paving principle, but his private notes show that he was almost certain that Lebesgue misjudged the intricacies of this part of topology, and that he would not be able to give a proof. Lebesgue, as a matter of fact, sent Brouwer some more details, but declined to publish them in the *Mathematische Annalen*.

Lebesgue, who had evidently quite sound and effective intuitions on the matter of dimension, did not stop to provide a proof of the paving principle, but went on to present an alternative proof of the invariance of dimension to the *Académie des Sciences* (23 March 1911). The argument here is completely different, and it introduces the concept of 'linking varieties'. Again the underlying ideas are marvellous, but the note does not give satisfactory proofs or even proof sketches. Apparently he failed to grasp the value of the gem he had found.[29]

Brouwer, upon reading the above paper *On the invariance of the number of dimensions of a space and on the theorem of Mr. Jordan for closed varieties*,[30] quickly spotted its weak points, and prepared another critical note for the *Mathematische Annalen*.[31] In the accompanying letter of 9 May to Blumenthal, Brouwer motivated the submission of the new note to the *Mathematische Annalen* by referring to Lebesgue's curious behaviour. Not only had Lebesgue failed to observe common courtesy (again!) in the matter, he had also disassociated himself from an earlier

---

[27] *Sur la non-applicabilité de deux continus à $n$ et $n+p$ dimensions.*

[28] The dimension paper, cf. p. 174.

[29] Alexander (1922) and Alexandroff and Hopf (1935), Chap. XI elaborated the ideas involved.

[30] *Sur l'invariance du nombre de dimensions d'un espace et sur le théorème de M.Jordan relatif aux variété fermées*, Lebesgue (1911b).

[31] Johnson (1981), pp. 198, 199.

promise to provide a complete version of the *Annalen* note, at the same time disowning all earlier proof attempts as hasty. These earlier attempts, said Brouwer 'are teeming with incorrect arguments, and are irreparably false'. And also the second *Comptes Rendus* note contained an 'incurably' false proof, to which Lebesgue stubbornly clung in spite of Brouwer's remonstrations. The long and short of it all was that Brouwer saw it as his unpleasant duty to publish the note on Lebesgue's invariance paper, but he said that he would withdraw the note if the editorial board would consider the note not to be in the general interest.

This note 'Comments on the invariance proofs of Mr. Lebesgue' (9 May 1911) and its successor (written in French for the benefit of Lebesgue) of 11 June 1911 were not published. The last one contained not only a critique of Lebesgue's papers, but also an account of the discussion so far. Brouwer complained that after he had withdrawn his note to the *Mathematische Annalen*, Lebesgue had not fulfilled his promise to provide a correct proof of the paving principle, but only sent inconclusive material:

> ...I studied his arguments repeatedly, but they remained obscure to me, and for the rest contained nothing that could not at a first glance be perceived by anyone.

Repeated questions were not honoured, so Brouwer concluded that it was impossible to continue the correspondence.

Blumenthal, to whom, most likely, those notes were addressed, was by that time convinced of Brouwer's viewpoint. He asked Brouwer[32] for an elaboration of his critique (in French) in order to confront Lebesgue.[33] There is a draft of Brouwer's answer; he complied with Blumenthal's request, but he insisted on providing the details for Blumenthal's private use only.

> For I have already so extensively and so often explained everything in writing, that I cannot tell him anything new.
>
> Exactly for this reason, the following explanation of his attitude has forced itself more and more upon me: that he directly after my first letter has recognised his lapsus, but that he was too vain to admit it, and that his further behaviour was determined by the hope of finding, perhaps later, a proof of the assumed statement, and by the necessity to gain time.

In the letter Brouwer suggested that Blumenthal should urgently press Lebesgue to provide a proof. This stratagem would force Lebesgue to furnish a proof, with the risk of committing new errors, or to plead failure. The letter ended with a request not to tell Lebesgue that Brouwer in the meantime had obtained a proof of the paving principle, because the latter considered it Lebesgue's obligation to clean up his own arguments first. A similar sentiment is expressed in the letter of 5 November 1911 to Baire, in which Brouwer told Baire that 'I have myself found a proof of the lemma

[32]Blumenthal to Brouwer, 16.6.1911.

[33]Johnson (1981), p. 203.

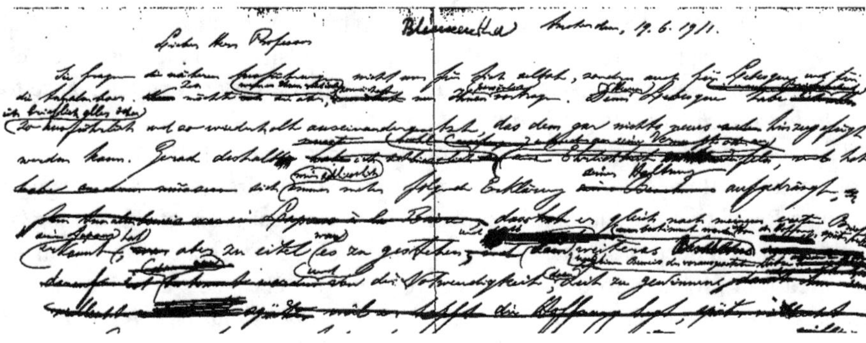

**Fig. 5.5** Draft of letter to Blumenthal, 19 June 1911. [Brouwer archive]

of Lebesgue a couple of days after its publication, but I believe that I should not publish it and give Mr. Lebesgue the opportunity to carry out his task himself.'

In the *Nachlass* of Brouwer there is a number of small scraps of paper, scribbled with Brouwer's fine detailed handwriting. Some of them are concerned with the Lebesgue affair; and the item Y4, dated 1912 by Freudenthal,[34] contains a proof of the paving principle.

In a letter of 2 July to Blumenthal, Brouwer wrote down (in French) the crucial step in the inductive argument for the paving principle. The note evidently was intended for publication, but it never appeared; finally its proof was published in the paper *On the natural notion of dimension*.[35]

On 8 July Brouwer drafted a letter to Blumenthal, in which he pointed out 'another gap in Lebesgue's so-called third proof' in the *Comptes Rendus* note.[36] Whether this letter was actually mailed is hard to say, since there is no Blumenthal Archive, but the fact that Brouwer kept the draft among his papers suggests strongly that it was mailed. Blumenthal, in his answer of 14 July, admitted that his mediation attempts had failed, and that he therefore saw no objections to publishing Brouwer's Note on Lebesgue's proofs. He remained, however, cautious to the extent that he advised Brouwer to adopt the title 'On my and Lebesgue's papers on the invariance of the number of dimensions'. He declined however to publish Brouwer's proof of the paving principle—'I cannot do that before I have given Lebesgue a generous opportunity to speak for himself. For I intend to request from him his *Mémoire étendu*.' Of course, he went on, you are free to publish your note elsewhere, for example, 'under supervision of the great Van der Waals in the *Verslagen*'.

Brouwer had some reason to feel that he was let down in order to save the face of his adversary. He drew his conclusions and published the proof and significant new material in the *Journal für die reine und angewandte Mathematik*[37]—a fateful

[34]CW II, p. 440.

[35]*Über den natürlichen Dimensionsbegriff*, Brouwer (1913a).

[36]Lebesgue (1911b).

[37]Brouwer (1913a).

choice as we shall see later. Brouwer had made up his mind on the choice of journal in November, as appears from a letter of Hellinger, the assistant of the editor Hensel. This 'desertion' was quite a step for Brouwer, who had a very strong allegiance to the *Mathematische Annalen* and Hilbert!

The correspondence did, however, continue; in a letter of 19 August 1911, Brouwer told Blumenthal that he had reconsidered the first part of Lebesgue's *Comptes Rendus* note—and that he thought the theoretical basis of the linking of manifolds was a deep problem. His re-assessment of the concept of 'linking manifold' led him probably to the more positive view of Lebesgue's contributions in *On the invariance of the number of dimensions of a space and on the theorem of Mr. Jordan for closed varieties*.[38] The second part of the same letter contained the sketch of an elegant alternative proof of the paving principle that made no use of the mapping degree; it was never published.

The name of a second prominent French mathematician has come up in the above story, that of René Baire. Baire had made a name for himself in the theory of real functions, where he had systematically investigated classes of discontinuous functions, now known as the *Baire classes*. Darboux had already initiated the study of discontinuous functions, but Baire had gone on to develop a systematic, abstract viewpoint. His work, together with that of Émile Borel, is the hard core of the so-called descriptive set theory.

In his note in the *Mathematische Annalen* Lebesgue had referred to notes of Baire in the *Comptes Rendus*,[39] suggesting that Baire's methods would yield a straightforward proof.

> Without doubt, Mr. Baire has not developed his proof; but it seems to me that, if one takes into account the hints given by Mr. Baire, there is nothing left to solve but difficulties of detail that are not very serious.[40]

Freudenthal pointed out that Baire and Lebesgue were not on speaking terms,[41] so Lebesgue's reference to Baire could hence be seen as an attempt to belittle Brouwer and tease Baire at the same time.

Brouwer had, on seeing Lebesgue's note, sent reprints of his *Mathematische Annalen* paper to Baire, probably inquiring after Baire's alleged proof. Baire's first letter dates from 28 October; it contains no mathematics, but polite congratulations on the progress made in analysis situs. He declared that because of other commitments and a prolonged illness, he had not been able to pursue his research.

In his reply of 30 October, Brouwer diplomatically approached Baire concerning his proof of the invariance of dimension:

---

[38]Sur l'invariance du nombre de dimensions d'un espace et sur le théorème de M. Jordan relatif aux variété fermées. Lebesgue (1911b).

[39]Baire (1907a).

[40]Lebesgue (1911a), p. 168.

[41]CW II, p. 439.

> Concerning your publication of 1907, I suppose that you yourself do not agree with the lines that Mr. Lebesgue devoted to you at the end of his false proof [...]
>
> The important theorems that you formulated in 1907 seem to me of a much more subtle character than the invariance, which is to me the most fundamental, but also the most crude, property of the analysis situs of $n$ dimensions.

He went on to say that he was unfortunately obliged to correct Lebesgue's statement, for it would, if correct, strip his own 1911 paper of all importance; this disagreeable task was not to be construed as a criticism of Baire. There is a short exchange of letters between Brouwer and Baire, but although Baire clearly resented being used by Lebesgue—their friendship of the days of the *École Normale* had gone stale when Lebesgue had gone in for career making[42]—he stuck to the view that, although he had not done so, one could without serious difficulties, but maybe at the cost of some lengthy exposition, prove the required statements on the basis of his 1907 *Comptes Rendus* notes. 'On the other hand', he added, 'these statements don't form such a complete set as your statements 1, 2, 3 of p. 314'.[43]

As time went by, Brouwer lost sight of Lebesgue and the matter of the proof; and only when in 1923 Urysohn called his attention to Lebesgue's 1921 paper, Brouwer felt obliged to point out that this proof was basically his own 1913 proof, but complicated by Lebesgue's presentation.

The Brouwer–Lebesgue controversy more or less bogged down in a stalemate; Lebesgue clearly was not willing to concede the point, but he was equally unable to meet Brouwer's challenge. He eventually published a proof of the paving principle in 1921. In his *Notice sur les Travaux Scientifiques* of 1922 Lebesgue went so far as to state that

> On the occasion of the publication of the work of Mr. Brouwer, I indicated[44] the principle of an argument that also allows us to establish that theorem; I have recently[45] developed the proof.

The whole affair leaves a sad impression; on the one hand the older and established mathematician treated the topic and the newcomer as another occasion to demonstrate his superior intelligence, thus losing sight of common civility towards newcomers, but—and this is a more serious and purely mathematical shortcoming—also completely misjudging the complexity of the problem. On the other hand, the younger man could not distinguish between irritating vanity and intentional foul play. On top of that, Brouwer's somewhat inflated sense of justice demanded an unconditional surrender; the loss of face of his opponent did not worry him. There is a striking contrast between Brouwer's genuine mystic ideal of detachment from the world, of introspection and non-interference with fellow human beings, and the

[42]Baire to Brouwer, 5 December 1911.

[43]The $n$-dimensional Jordan theorem, Brouwer (1911e).

[44]Cf. Lebesgue (1911a).

[45]Cf. Lebesgue (1921).

burning wrath that always was close under the surface, and that easily erupted at the least provocation. Brouwer bears a striking similarity to the character from Shaw's 'Androcles and the Lion', Ferrovius, with the strength of an bear and the temper of a mad bull, 'who has made such wonderful conversions in the northern cities'. Like Ferrovius, he could not resist the temptation of a battle against injustice and evil men—'When I hear a trumpet or a drum or the clash of steel or the hum of the catapult as the great stone flies, fire runs through my veins: I feel the blood surge up hot behind my eyes: I must charge; I must strike.'

This fighting spirit remained with Brouwer all his life—however, for all the agony it caused his adversaries, he certainly got his share in the suffering.

It is likely that, Lebesgue did not want to rob Brouwer of his priority, but he would not have minded changing the perspective of the dimension problem to his advantage by belittling Brouwer's efforts. One cannot escape the impression that the mathematical authorities of the day did little to see that justice was done.

Blumenthal, as the managing editor of the *Mathematische Annalen*, did not display the decisiveness that was required of him in his function. His wish to keep on the good side of a recognised authority like Lebesgue may have obscured his judgement. It is also remarkable that in a spectacular dispute such as the one above, none of the older statesmen of the *Mathematische Annalen* stepped in. One might suppose that Hilbert would have seen the importance of Brouwer's methods, if not the theorem—also in view of the prior correspondence. But, alas, the whole matter was left to Blumenthal, who did not carry enough weight to address the problem. The episode also illustrates how little the mathematicians of the day understood the intricacies of topology. It is no exaggeration to say that only in the twenties a measure of understanding of Brouwer's methods started to spread. The influential book of Felix Hausdorff, *Grundzüge der Mengenlehre*[46] of 1914, that introduced large numbers of readers to the beauty of set theory and topology, remarked (p. 461) that 'The brevity of Brouwer's papers, which often forces the reader to fill in many details by himself, is most regrettable, in the absence of other impeccable and extensive expositions.' Brouwer had little patience with those complaints; he even felt rather offended at the suggestion (for example by Hausdorff) that his topological work should be redone. The full impact of Brouwer's topological innovations was not felt before the twenties. Although the delay may have been partly due to the First World War, the main reason was that it was generally viewed as difficult and inaccessible. Heinz Hopf reported that in 1917 Erhard Schmidt discussed Brouwer's topological methods in a course at the University of Breslau,[47] and when Schmidt moved to Berlin in 1920, topology became a regular part of the curriculum in Berlin. In America J.W. Alexander had grasped the new methods and started, from 1915 onward, adding his own contribution in the early twenties.

In spite of all the negative features of the conflict we described, there is at least one positive side effect: Lebesgue's obstinacy prompted Brouwer into feverish action; some of the papers of this period are reactions to Lebesgue's claims. In July

[46]Basics of Set Theory.

[47]Hopf (1966).

**Fig. 5.6** Winding number

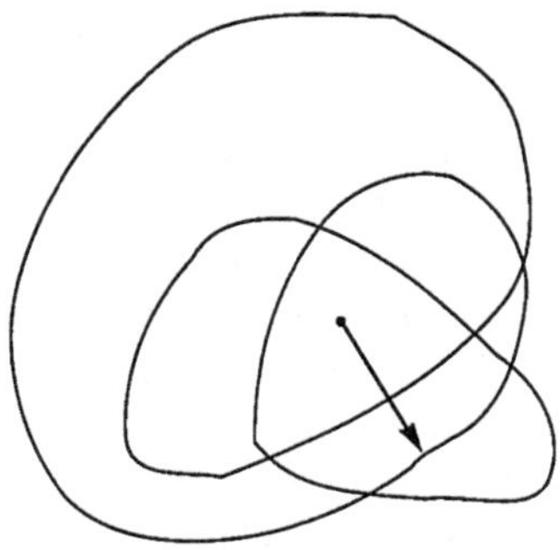

1910 Brouwer submitted his fundamental paper *On the mapping of manifolds*[48] to the *Mathematische Annalen*. It was the detailed elaboration of the facts mentioned in the letter of 1 January 1911 to Hilbert. The importance of this paper can hardly be overestimated; it contains virtually all the tools of the new topology. It was, so to speak, a short but exhaustive course in the topology of the future. Among the many concepts introduced here, we find that of *simplex*,[49] open and closed manifolds (based on simplexes), ($n$-dimensional-) *indicatrix*, *simplicial approximation*,[50] *mapping degree, homotopy of mappings* (under the name *continuous deformation*), *singularity index* (under the name *degree of the simplex*). The basic theorems, for example the preservation of the mapping degree under homotopy, are all proved by direct geometric methods. The converse of this invariance theorem was established a year later in *Continuous one–one transformations of surfaces into themselves, no. 5*; this paper, in contrast to the earlier ones in the series employed the new tools and generalises earlier results; it thus takes a special place in the long series.

## 5.2 The Fixed Point Theorem and Other Surprises

The famous *fixed point theorem* of the ($n$-dimensional) ball is the quiet apotheosis of the 'mapping of manifolds'-paper. The theorem that made Brouwer's name a household word in mathematics and other disciplines was presented in one line, without any comments. Did Brouwer think that such a spectacular result, like good wine, did not need any recommendation, or was he showing off in modesty? We shall never know.

Of all Brouwer's topological achievements, the fixed point theorem has appealed most to the imagination of non-specialists. If mathematicians had coats of arms,

[48] *Über Abbildung von Mannigfaltigkeiten*, Brouwer (1911c).

[49] Poincaré had already defined the notion in Poincaré (1899). Brouwer does not quote Poincaré, so presumably he was not aware of the paper.

[50] Note that in the letter to Hilbert of 1 January 1910, Brouwer still uses polynomial approximations.

Brouwer's should have contained a fixed point.[51] A number of popularisations have been put before the public. One of these illustrates the surprising omnipotence of topology: consider a cup of coffee which is stirred slowly in an arbitrary manner, then when the coffee has again assumed a state of rest at least one 'coffee point (particle)' has returned to its old position. The stirring instructions are necessary to prevent discontinuities; one does not want to have drops flying around.

The mapping degree was the prime instrument in Brouwer's topological investigations. It is surprising, and hard to explain why Brouwer, who had all the equipment he needed, never made use of, or developed, algebraic topology. In particular one wonders, why he never used homology theory, and all its marvellous conveniences. 'Instead', Dieudonné remarks,[52] 'he used his discovery to define rigorously the concept of *degree* of a continuous map, and then proceed, mostly by fantastically complicated constructions, relying exclusively on that notion, to prove the famous *Brouwer theorems*'. The trained topologist will, of course, see the implicit use of algebraic methods, but Brouwer did not make the important step to expand his insights into a systematic machinery for the algebraisation of topology. He was and remained a geometer.

Part of the power of Brouwer's methods resided in his *simplicial approximations*; 'it may be said that Poincaré defined the *object* of that discipline,[53] but that it is Brouwer who invented *methods* by which theorems about these objects could be *proved*, something that Poincaré had been unable to do'.[54]

Although Brouwer's mapping degree was a new phenomenon in topology, it was foreshadowed on certain respects by earlier notions. There was, for example, the 'winding number' of a path around a point: walk around the point and count at the end how many revolutions you have made. The walk may be totally disorganised, going backwards and forwards in an unpredictable way.

Note that if one looks from the point $a$ and follows the moving point, the backward and forward neatly cancel out each other, so that the complicated figure above really comes to one revolution. The notion is captured by the following integral in complex analysis:

$$\frac{1}{2\pi \mathrm{i}} \oint_C \frac{dz}{z-a}$$

which measures the angle swept by the vector following the movement of the point on the curve.

The idea of the mapping degree can be simulated by wrapping a rubber band around a cylinder. With a little bit of imagination, one can think of the band doubling backwards and forwards. What one gets in this case is a continuous mapping of the circle onto itself (looking in the direction of the cylinder). The mapping degree

[51] The reader need not worry, nobility has not been created in the Netherlands for more then a century.

[52] Dieudonné (1989), p. 161.

[53] Simplicial topology.

[54] Dieudonné (1989), p. 168.

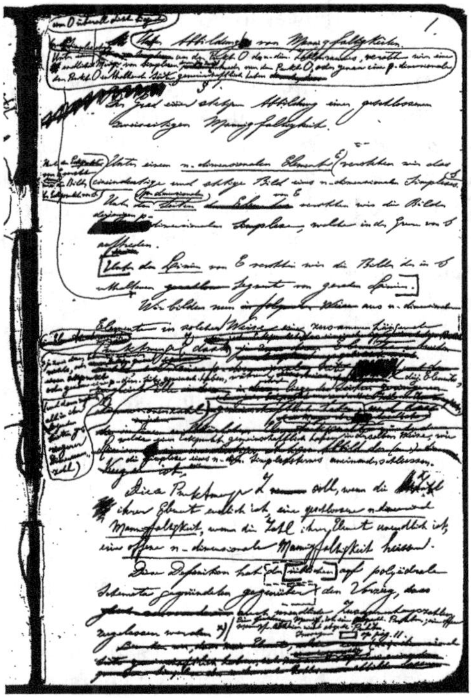

**Fig. 5.7** Bouwer's notebook with a first version of 'Mappings of manifolds'. [Brouwer archive]

counts the number of times a point is covered, cancelling the 'backwards' against the 'forward'.

For two-dimensional pictures like the image of a circle, one has a reasonable geometric intuition, but for higher dimensions, it gets harder. Moreover, the phenomenon of the Peano curves had frightened people off. Mathematicians had come to distrust their geometric intuitions. Nonetheless, Brouwer succeeded in defining the correct notion; or, to be more precise, he proved that a higher-dimensional analogue of the winding number is a constant for a manifold (for example a surface) and then considered its properties in all kinds of situations. Using the mapping degree, he proved a wealth of new facts.[55]

A problem presented by Brouwer's topological research is that it is not immediately obvious what the underlying motivation is. Brouwer's published papers are no help in that respect. It looks as if he solves hard problems just because they present a challenge. This may partly have been responsible for the unfounded impression that Brouwer only practised topology to gain status and a secure job. There are a couple of clues, however. We have seen that his research into vector distributions may have been triggered under Poincaré's influence (cf. p. 134); another clue is provided by

[55]After the ascent of homology theory, a much simpler definition of the degree of a mapping became available: let $f$ be a continuous mapping from $M$ to $M'$, where $M$ and $M'$ are compact, connected, oriented (pseudo-) manifolds, then for $f^* : H_n(M; Z) \rightarrow H_n(M'; Z)$, we have $f^*([cM]) = c[M']$ where $c$ is the mapping degree.

a letter of Brouwer to Poincaré, related to the 'invariance of domain'. In this letter, undated, but probably written in the autumn fall of 1911 (and only preserved in draft), Brouwer wrote 'My proof of the invariance of the $n$-dimensional domain was inspired last year by the reading of your 'Method of Continuity' of volume 4 of the *Acta Mathematica*'. The letter shows that Brouwer had familiarised himself with the geometric–analytic work of the grand master of European mathematics, and that in the middle of his general topological work he was looking for concrete problems areas. The 'Method of Continuity' is a tool from the domain of uniformisation and automorphic functions (see below). It was devised to solve the basic problems in the field, but until Brouwer stepped in, progress was blocked by topological difficulties. Poincaré and Klein were actively looking for means to make the method work. They were joined at the beginning of the century by Paul Koebe. We note in passing that Koebe was one of the main speakers at the Rome conference; hence it is very well possible that Brouwer had heard his talk on automorphic functions, and that he had as a consequence looked into the problems of uniformisation. The method of Continuity was also mentioned in the draft of the letter to Baire (5 November 1911). We will return to this particular topic of uniformisation later in this chapter.

In the paper *Proof of the invariance of domain*,[56] Brouwer considered the problem 'does a topological mapping map a domain in an $n$-dimensional space onto a domain?'

This paper contains the first public reaction to Lebesgue's dimension papers; in a footnote he characterises Lebesgue's *Annalen* paper as 'insufficient' and the *Comptes Rendus* note as 'qua content identical with mine:—the deviations complicate the line of thought'.[57]

As Freudenthal pointed out, this is a strange formulation, 'Lebesgue's proof is not correct, and his method differs quite a bit from Brouwer's.[58] But Brouwer read Lebesgue's proof through the spectacles of his own knowledge, sympathetically filling in the gaps with ideas of his own, and then stated that it was his own proof, though needlessly complicated.'

The letter of 5 November 1911 to Baire discussed the invariance of domain, as claimed by Baire. Brouwer, however, could not see that Baire had solved the difficulties involved.

> As far as I can see, the indications, given by you in the *Comptes Rendus*, leave untouched the principal difficulty. For a long time I had searched for a proof; for $n = 3$ it is easy, for arbitrary $n$, I succeeded only last summer, by means of an argument that I afterwards have found again in the note of Lebesgue (*C.R.* 27 March 1911) where, by the way, it is in an almost unreadable form—and inexact—if one takes it literally.

The *invariance of domain* comes to the following: the image of a domain (a connected open set) under a topological mapping is again a domain. The theorem was

[56] *Beweis der Invarianz des n-dimensionalen Gebiets*. Brouwer (1911d).

[57] Brouwer (1911b).

[58] CW II, p. 443.

proved by Brouwer in his *Mathematische Annalen* paper in 1911 by means of the *No Separation Theorem*, which in the simple case of the plane tells that an arc of a Jordan curve (think of an arc of a circle) does not split the plane in two parts. The $n$-dimensional case is completely analogous.

Schoenflies had proved as early as 1899 the invariance of domain theorem for the plane, and Baire and Hadamard had shown that the theorem for arbitrary dimension was an immediate corollary of the $n$-dimensional Jordan theorem (which was still open at that moment). As we have seen Lebesgue had suggested that, up to some technical details, the work of Baire basically yielded the invariance of dimension.[59]

Brouwer had already discussed the matter with Baire in the above-mentioned correspondence, and in the paper *Proof of the invariance of the n-dimensional domain* he added in a footnote that:

> In so far as the developments of Mr. Baire, cited by Mr. Lebesgue, are concerned, the theorems to which the problem is reduced there [that is in the paper in the *Mathematische Annalen*] are deeper than the invariance of dimension.

Indeed, a fairly straightforward argument derives the invariance of dimension from the invariance of domain.

In the meantime Brouwer had realised that the theorem could also be derived from the second part of Lebesgue's *Comptes Rendus* note (cf. p. 164) and Baire's earlier work. However, in a two-page note *On the invariance of the n-dimensional domain*,[60] he gave a short direct proof, based on the degree of a mapping.

On 24 February 1912 Korteweg communicated to the Dutch Royal Academy, *KNAW*, a paper of Brouwer that gave an exposition of his version of 'linking manifolds', which he had discovered independently.[61] This paper *On looping coefficients*[62] is, like all of Brouwer's topological work, geometric in nature. The proper way of dealing with this topic would be to use homological arguments, but those were only much later applied by Alexander. As it is, Brouwer's exposition is somewhat tortuous.

Already in a letter of 15.10.1909 to Hilbert, accompanying the manuscript of the Jordan curve theorem,[63] Brouwer had remarked to Hilbert, that the seemingly evident theorem that 'the one–one continuous mapping of a closed curve is again a closed curve' was open, and that it seemed to present a 'genuinely hard problem'. To appreciate the problem one has to realise that the 'closed curve' was defined in the Schoenflies tradition. It was simply a bounded subset of the plane determining two domains of which it was the common boundary. Clearly, under this definition the problem is not trivial. It was settled by Brouwer in his paper *Proof of the invariance*

[59] Baire (1907a, 1907b).

[60] *Zur Invarianz des n-dimensionalen Gebiets*, Brouwer (1912c).

[61] See the letter to Baire of 5.11.1911, Johnson (1981), p. 218.

[62] Brouwer (1912j).

[63] Brouwer (1910a).

*of the closed curve* in 1912 in the *Mathematische Annalen*, and summarised in the *Comptes Rendus*.[64] The technique of the proof is close to homotopy (no mapping degree, this time); Brouwer used in his paper the notion of *Zyklosis* (which probably goes back to Listing, 1847). The theorem of the paper is somewhat more general: the number of domains determined by a bounded closed connected planar point set is invariant.

The notion of mapping degree was also employed by Brouwer in the sequel to his series of papers *Continuous one–one transformation of surfaces in themselves*,[65] the fifth communication. The paper introduces the notion of *transformation class*, which we would now call homotopy class. This notion is used consistently in the statements of the paper. The content of the paper is that, for mappings of spheres into themselves, the mapping degree determines the homotopy class.

The last paper in the series, dealing with dimension, was published by Brouwer in 1913 in the *Journal für die reine und angewandte Mathematik* (see p. 165), a journal with a more restricted circulation than the *Mathematische Annalen*. It bore the title *On the natural notion of dimension*,[66] and it contained a fully fledged abstract definition of dimension. It will be discussed later in the context of the post-war developments of topology.

Brouwer briefly returned to topology after the First World War, but his main achievements belong in the period discussed here. One may well wonder why Brouwer, after his imaginative breakthrough failed to exploit the homological and homotopical features that were hidden in his work—and that had been envisaged by Poincaré! Indeed, in 1922 Alexander published a paper (the first draft was already finished in 1916) in which he used homological methods to prove his duality theorem, and which extended some of the Brouwer theorems.[67] Brouwer's reluctance to adopt the tools of homology theory may have been rather the result of a personal disposition than an inability to do so. After all, a man of his calibre could not very well miss the things that were staring him in the face.

On the whole, Brouwer was oriented towards the continuous and the geometrical. Even in his foundational work he eschewed the finitary, and attacked—broadly speaking—the problems of the continuum and Baire space (or Cantor space) rather than those of arithmetic or of a finitary nature. There are no publications in pure algebra among Brouwer's papers. It is clear that he knew his algebra, but he mainly applied it in geometrical context, that is to *support* the geometry, not to *supplant* it. Another explanation of his neglect of the new algebraic machinery may be Brouwer's inclination to find satisfaction almost exclusively in revolutionary innovations. His œuvre contains few, if any, routine, run-of-the-mill papers. After introducing some new idea, concept or tool, he experimented enough to be satisfied that it was worthwhile, but he was not the man to exploit his new ideas (let alone those of others) in

---

[64] *Sur l'invariance de la courbe fermée, Beweis der Invarianz der geschlossenen Kurve*, Brouwer (1912e, 1912i).

[65] Brouwer (1912h).

[66] *Über den natürlichen Dimensionsbegriff.* Brouwer (1913a).

[67] Alexander (1922).

an endless series of papers. That routine of mathematics just did not appeal to him. The grinding out of mathematical theorems was not his idea of mathematics. His concept of mathematics was rather that of the gentleman practitioner, or even better, the scientific artist, who would, in bursts of creativity, deliver high-quality products; but, as often as not, idle around.[68] His attitude is illustrated by a remark in his paper on the history of dimension theory:[69]

> I have restricted myself to the laying of the foundations of the theory of dimension, and refrained from further dimension-theoretic developments, on the one hand because with the proof of the justification theorem[70] the intended purpose had been reached, on the other hand because an intuitionistic realisation of the subsequent considerations (in the first place those which can be grouped around the 'sum theorem' and the 'decomposition theorem') was, in contrast to the justification theorem, not plausible.

It seems plausible to conclude that Brouwer was ready to leave the field to others after the basic facts had been established.

## 5.3 The Karlsruhe Meeting and the Continuity Method

The next episode in Brouwer's career has to do with the so-called *automorphic functions* and *uniformisation*. The topic is a traditional part of complex function theory and its subject matter is of the greatest importance for the theoretical understanding of analytic functions.

The uniformisation problem[71] can be illustrated by a simple example: 'does $f(z) = z^2$ have an inverse function?' In the traditional high school mathematics the answer is trivial: 'yes, take $g(z) = \sqrt{z}$'. Everybody will have experienced some qualms: why $\sqrt{z}$ and not $-\sqrt{z}$ ? This is, in a nutshell, the problem. So we could also have asked 'are there solutions of $w = z^2$ for $z$, how many are there, and what is their connection?' The solution to this problem goes back to Bernhard Riemann. The basic idea is to take the two-solutions idea seriously, and grant them each a limited autonomy. Take two copies of the domain, that is the complex plane, and give one to $\sqrt{z}$ and the other one to $-\sqrt{z}$, or in polar coordinates: $f_1(z) = \sqrt{r}e^{(1/2)\mathrm{i}\varphi}$ and $f_2(z) = -\sqrt{r}e^{(1/2)\mathrm{i}\varphi}$, where $z = r \cdot e^{\mathrm{i}\varphi}$.

Each solution seems fine, until we discover that $f$ gives to the same point two values, depending whether we consider it as $z = e^{\mathrm{i}\varphi}$ or $z = e^{\mathrm{i}\varphi+2\pi\mathrm{i}}$. Geometrically speaking, if we travel around the origin on a circle, we do not get the same output for $f_1$ after one full revolution; the same holds for $f_2$. The idea now is to combine both

---

[68] Cf. p. 190, Wiessing (1960), p. 143 ff.

[69] Brouwer (1928e).

[70] That is, Brouwer showed that $\mathbb{R}_n$ has dimension $n$.

[71] For a precise and general definition of uniformisation the reader is referred to the literature, for example Nevanlinna (1953), Beardon (1984).

functions (and both domains): make one revolution on a circle in the first copy of the $z$-plane and use $f_1$, then switch to the second plane, make another revolution and use $f_2$. Exactly where we switch is not so important, say at the negative imaginary axis. We now have a decent function defined for all $z$ which acts as the inverse of $z^2$, and which also treats the two possibilities, $f_1$ and $f_2$, on an equal footing. The $z$-plane has been replaced by two sheets which are glued on along a line.

The geometric surface one gets by this cutting and gluing is called a *Riemann surface*, after its inventor. The process of extracting one decent function as the inverse of $w = z^2$ is called *uniformisation*. Some of the greatest minds of the late nineteenth and early twentieth century devoted much of their energy and ingenuity to solving the general problem: how to uniformise general analytic functions of the form $f(z, w)$.

The algebraic functions are the traditional examples; for example how to get $w$ as a function of $z$ satisfying $w^3 + z^3 - wz = 0$.

The theory of uniformisation became the meeting point of all the technical know-how of nineteenth-century mathematics: Riemann surfaces, group theory, topology, function theory, algebra, potential theory, .... The two masters of the subject were Klein and Poincaré, who in playful competition tried to outwit each other.

The history of the theory of automorphic functions is covered by a veil of romance, drama and heroism, in particular through Klein's vivid description in his history *Development of Mathematics in the 19th Century*, Vol. II.[72] Both Klein and Poincaré were looking for the final solution of the fundamental problem of automorphic functions from 1880 onwards, like a couple of King Arthur's knights searching for this Holy Grail—and bent on getting there first. In rapid succession Poincaré and Klein overtook each other, publishing while the ink was still wet. The stress took a heavy toll on Klein; in 1882 his health collapsed, and Poincaré had the field to himself.

A great many other mathematicians took part in the quest for the Holy Grail of uniformisation: Schottky, Fuchs, Koebe, Fricke, Bieberbach, to mention just the more prominent ones. Of those, Koebe and Fricke made it the main task of their lives to break the deadlock that the subject reached after the long series of successes of Klein and Poincaré. The efforts of those two masters of the *fin de siècle* were directed to the theory of *automorphic functions*. There is a close connection between automorphic functions and Riemann surfaces. It was expected that through the theory of automorphic functions one could solve the basic problems about Riemann surfaces; for Riemann surfaces were the key to the uniformisation problem.

Automorphic functions are complex functions which are invariant under a group of transformations of the complex plane. An example is the exponential function $e^z$; it is invariant under a vertical translation over the length $2\pi$. Indeed $e^{z+2\pi i} = e^z$ for all $z$.

Now Riemann surfaces (at least the compact ones) can be viewed as spheres with handles, and the nineteenth century way to 'tame' them, was to cut them up,

[72]Klein (1927).

**Fig. 5.8** Felix Klein. [Courtesy Niedersächsische Staats- und Universitätsbibliothek Göttingen]

in such a way that they could be flattened as a piece of the plane. For example, a torus (which can be topologically transformed into a sphere with one handle) can be cut twice, so that it becomes a flat piece (think of the inner tube of a bicycle tyre). The correspondence between automorphic functions and Riemann surfaces was embodied in the fact that an automorphic function maps *fundamental domains* onto a cut-up Riemann surface. For a fundamental domain, one may think of the smallest part of the domain that is repeated (like a tiling) by the transformations. Think of $e^z$: if one knows the values for all $z$ between the real axis and the line $y = 2\pi i$ (its fundamental domain) then one knows all values of $e^z$.

The question that remained to be solved at the end of the century was: 'does one get all Riemann surfaces in this way?' The process of passing from automorphic functions (and their groups) to Riemann surfaces was a continuous one, and Klein and Poincaré had hoped that exploiting this continuity would yield the desired answer. This route to a solution was called the *continuity method*; Klein tried to establish the result by the technique of counting parameters—some kind of dimension argument—one may think of the kind of use of dimension arguments in linear algebra. This counting method had however become highly suspect after Cantor's dimension lowering arguments; see p. 149.

In 1907 Poincaré and Koebe independently solved the uniformisation problem along other lines. But such was the power of tradition that the automorphic function approach was still pursued after the hunting season was closed. Klein and Koebe remained active in attempts to save the continuity method. There is correspondence which shows that as late as 1910 the topic was still discussed. Whereas Poincaré had moved on to other areas after the successful wrapping up of the uniformisation episode, Koebe remained active in the area. He was considered to be the outstanding expert in the field, and he certainly did not disagree on that point.

But even the indefatigable Koebe seemed to have had enough of the topic by around 1910, and to be looking for new challenges; Hermann Weyl reported in a

**Fig. 5.9** Paul Koebe. [Courtesy Niedersächsische Staats- und Universitätsbibliothek Göttingen]

letter to his Dutch friend Pieter Mulder 'Now Koebe has done with uniformisation, and he is looking for another topic.'[73]

Paul Koebe, the prince of the automorphic functions, was an ambitious function theorist, one year younger than Brouwer, born in Luckenwalde on 15 February 1882. He studied mainly in Berlin, where he wrote his dissertation under H. A. Schwarz. Subsequently he moved to Göttingen, where he was successively a *Privat Dozent* and later an extra-ordinary professor. In 1911 he was appointed full professor in Leipzig, where he stayed until his death on 6 August 1954.[74]

Koebe's mathematical œuvre consisted almost entirely of complex function theory, with emphasis on uniformisation and automorphic functions. Between 1905 and 1941 he wrote 68 papers, among which there is a number of lengthy ones (eighty pages is no exception). When Brouwer was still one of the nameless academic crowd, Koebe had already earned fame by his solution of the uniformisation problem, which put him in a bracket with Klein and Poincaré! During his life he got his share of recognition; a paper of his was awarded a prize by the King of Sweden, he was awarded the Academic prize of the Royal Prussian Academy of Sciences (1914), and the Ackermann–Teubner prize (1922). Koebe was well aware of his importance and apparently did not try to hide it, which made him a natural target for some irreverent gossip.

Brouwer's sudden involvement in the defunct theory of automorphic functions at first seems rather surprising. Of course, Brouwer knew his complex function theory, and he had demonstrated his ability to handle the problems of analysis; yet, there were other things to consider for a budding topologist. The first hint of an interest in the problem area of uniformisation may be found in his letter of 15 October 1909 to Hilbert (cf. p. 140), in which he asked for some reprints, including a couple of Koebe. Apparently Hilbert had discussed a number of topics with Brouwer during his vacation in 1909 in Scheveningen (cf. p. 127). It is not unlikely that Hilbert brought up the solution to the uniformisation problem, which at that time was barely two years old. If so, he could very well have called Brouwer's attention

[73] Weyl to Mulder, 29 July 1910.

[74] Cf. Kühnau (1981).

to the continuity method. There are no documents that confirm Brouwer's interest in automorphic functions, but he must have considered the problem.

Light is shed on the activity of Brouwer by a letter from Brouwer to Poincaré, written after the Karlsruhe meeting (see below), most likely in the early days of December 1911. Not only did Brouwer state that his invariance of domain theorem was inspired by Poincaré's *Méthode de Continuité* in the *Acta Mathematica*—he also wrote that he had recognized that the continuity method was a consequence of the invariance of domain, but after reading that Poincaré considered his 'exposition of the continuity method as perfectly rigorous and complete', he had started to fear that he had poorly understood Poincaré's memoirs. And so, he said, he had restricted himself to an oral communication at the Karlsruhe meeting of the German Mathematical Society, instead of adding it as an application to the invariance of domain-paper.

Thus Brouwer saw the solution late 1910 (or early 1911), but did not quite trust his judgement, and left the matter until the Karlsruhe meeting. Although he knew his complex function theory as well as the next man, he was no expert in automorphic functions. During a visit to Aachen he discussed the matter with Blumenthal, as appears from the last lines of the letter of 19 June 1911 that deals with Lebesgue's defective proof of the paving principle:

> ...and finally, I would like to learn from you which theorem of Analysis Situs that you mentioned to me in Aachen, as strictly necessary for the Continuity Proof of the existence of polymorphic functions on Riemann surfaces. From your latest letter I seem to conclude that it is not the Jordan theorem.

Blumenthal answered the question two months later, replying that he was no longer up to date in the theory of automorphic functions, but that he would be surprised if just the Jordan theorem, or even its converse, would do the trick.[75]

By that time Brouwer must already have solved the problem, for he was at the last minute added as an invited speaker to a symposium on Automorphic Functions at the annual meeting of the German Mathematical Society, DMV, from 27 to 29 September 1911. The first announcement listed Klein, Fricke and Koebe as speakers; Brouwer featured as an invited speaker in the second announcement. It is likely that he informed the organisers (in particular Klein) of his justification of the continuity method, and was immediately included in the list of speakers.

Brouwer's contribution to the automorphic function-session consisted of a redemption of Poincaré's original uniformisation proof, by means of the filling of two gaps. The first one is of a technical topological nature, and the second one required the invariance of domain, which Brouwer had already proved, but not yet published, although Klein, of course, knew about it.[76]

---

[75]Blumenthal to Brouwer, 26 August 1911.

[76]The Invariance of domain-paper was sent to the *Mathematische Annalen* on 14 June 1911, published November 1911.

Brouwer's talk was followed by a discussion, in which Koebe made a number of remarks. The reader who checks the printed report of the meeting[77] will be struck by the lack of coherence between Koebe's claim—namely that he had been able to prove the continuity property for *all* of Klein's fundamental theorems using Klein's general method—and Brouwer's rejoinder. The explanation is given in a letter from Brouwer to Hilbert of 24 February 1912:

> *Lieber Herr Geheimrat!*
>
> I am asking you for help and protection in a very unpleasant matter. [......] On 2 January I sent Koebe a copy of my letter, which I sent to Fricke in December and which has been submitted to the *Göttinger Gesellschaft der Wissenschaften*, and received roughly a week later the enclosed card. This card was followed on 14 February, not by the promised manuscript, but by the enclosed letter (together with my reply), in which I have underlined the statements to which my rejoinder refers.
>
> Koebe can, however, not really mean the indicated claim, just as little as anybody who has heard my Karlsruhe talk. Therefore I can only perceive in Koebe's statement his objective to lend the matter in his next note the appearance, as if my letter to Fricke contained certain ideas, which I should have learnt from discussions with Koebe, whereas the real state of affairs in Karlsruhe was that I presented a complete continuity proof for the 'Grenzkreisfall', but that Koebe only contributed a vague inkling, that perhaps something could be done with his *Verzerrungssatz*[78] in the continuity method.[79] He said in the session of 27 September at the end of my talk the following: 'Since by my *Verzerrungssatz* nothing can happen under continuous modification of the modules, in my line of thought the efforts of Mr. Brouwer in the case of the difficulties of the invariance of domain and the absence of singularities of the module manifold are superfluous.' Whereupon I (emphatically) retorted: 'The *Verzerrungssatz* can only extend the result that Poincaré reached for the case of the *Grenzkreis* (boundary circle) and thus at the same time extend my continuity proof to the most general fundamental theorem; my contributions are, just as before, required in their full extent.' Thereupon Koebe pronounced the following nonsensical words: 'What Brouwer has done, I do with Poincaré series', and then Klein closed the discussion.

The report in the *Jahresberichte* contained a totally different version of Koebe's remarks. No mention was made of the *Verzerrungssatz*, but in Koebe's contribution to the discussion the claim was made that already before the meeting Koebe had carried out the continuity proof for *all* the fundamental theorems of Klein.

---

[77] JDMV 1912.

[78] Cf. Behnke and Sommer (1955).

[79] *..., Koebe aber nur eine gewisse Ahnung, dass sich etwas mit seinem Verzerrungssatz in der Kontinuitäts Methode lasse, mitbrachte.*

Brouwer's reaction, according to the report of the discussion, consisted in the remark that the topological difficulties of the continuity proof came down to the absence of singularities in the module manifold, and the invariance of domain, which were solved in Brouwer's talk for *all* of Klein's fundamental theorems. He added that these were neither mastered, nor avoided, by Koebe's claim, but that a combination of Brouwer's method and that of Koebe was sufficient to cover the general case of Klein's fundamental theorem. The formulation suggests that both Koebe and Brouwer edited their parts in the discussion long after the meeting.

It is all the more surprising therefore that Koebe in the *Fortschritte* of 1912 (published in 1915), reviewing the report in the *Jahresberichte* used the opportunity to modify his discussion contribution in the direction of Brouwer's letter to Hilbert: 'Koebe states that he can carry out the continuity proof for *all* fundamental theorems by means of the *Verzerrungssatz*'; Brouwer's remark is not mentioned at all. Koebe was having a field day anyway in this particular volume of the *Fortschritte*: he reviewed four of his own papers plus a paper of Plemeljs on uniformisation. His own papers got lengthy extensive reports (at least in comparison to the traditional length of the reviews in the *Fortschritte*), and even Plemeljs' review was used for unabashed self-promotion.[80]

Brouwer described, in his letter to Hilbert, his experiences during the Karlsruhe meeting:

> Only after longer private discussions, in which also Bieberbach, Bernstein and Rosenthal participated, Koebe learned after the talks, from 27 till 29 September, from me which partial result (by the way, formulated by Klein in the *Mathematische Annalen* 21, and at that time called the Weierstrass theorem) can be obtained via his 'Verzerrungssatz' and which part, to be settled by my contribution, still remains necessary. And, as the above mentioned gentlemen know precisely, in these discussions I have mentioned <u>all</u> the details of my present note.
>
> However, already at that time several warning voices said to me: 'All that, you have explained now to Koebe, you will only with the greatest effort be able to claim as your property, as soon as he has understood you', and indeed Koebe displayed some symptoms that seemed to bear out those voices. So when I was at home again I wanted to refrain from any publication on that particular topic, which is anyway rather far from my interest and with which I only casually concerned myself at the request of Klein, in order to avoid an unpleasant fight with Koebe.
>
> Only after Blumenthal prodded me, and after I had, moreover, heard that Klein would like to see a publication from my hand, the note of the thirteenth of January[81] came forth.

[80] He quoted the author: 'The simplest and most natural of all these proofs [of the uniformisation theorem] is the first one given by Koebe.' Modesty was not one of Koebe's defects.

[81] Brouwer (1912d).

A great number of letters were exchanged in the Brouwer–Koebe conflict, but not all of them have been preserved; Brouwer wrote to Poincaré, and Koebe turned to Klein and Hilbert, and Brouwer and Koebe exchanged a number of letters.[82] As Brouwer explained in a letter to Hilbert on 31 May 1912, he had reluctantly intervened in a matter that he had rather wished to avoid:

> For a better preparation of our coming discussion, two letters are enclosed, that explain to you the somewhat obvious question, why I had anything to do with Koebe at all. As a matter of fact my Karlsruhe contribution to the continuity proof consisted of two parts, of which the first one (the invariance of domain) had already been submitted for publication by itself in July,[83] whereas with respect to the second part (the extension of the group set to the set of automorphic functions with *m* poles), according to the enclosed letter of Bernstein (...) Koebe claimed priority. Since this second part did not seem very deep to me anyway, I of course hesitated to publish it, although Blumenthal pressed me to do so. Finally, in the beginning of November, I sent the manuscript of my talk to Fricke, with the question if he thought the contents were new, and deserved to be published. The information on Koebe (marked in pencil) that it contained, complicated the situation to the extent that, when shortly thereafter, both Blumenthal and Fricke and also Klein (namely indirectly through Fricke) requested me to publish, I could not possibly do so, without contacting Koebe in order to get more certainty and clarity concerning his accomplishments, for otherwise I would be in danger that Koebe would accuse my publication of triviality and myself of plagiarism. In the correspondence with Koebe I then obtained only evasive answers to my specific question. The only thing I could get out of him was a mutual agreement to edit our notes on the continuity proof in [full] co-operation.[84] How he later broke his word and dragged out the matter, is known to you.

When Brouwer realised that his excursion into the kingdom of automorphic functions was going to be more than a quick trip, he wanted to make certain that he was not committing an error, so he wrote Poincaré,[85] asking him about details of his version of the continuity method. Although he was not personally acquainted with Poincaré, we may safely assume that the latter was aware of his work; Korteweg had written at some time in 1911 a letter to Poincaré, asking for a letter of recommendation for his student Brouwer. Since Korteweg's draft (which is the only surviving document from that correspondence) is undated, Poincaré may or may not have received the extra information on Brouwer before Brouwer wrote to him. It seems likely that Brouwer's solution to the dimension problem, coupled with the Lebesgue conflict, had not escaped Poincaré's attention.

---

[82] Of which only a few have survived.

[83] Brouwer (1912f).

[84] I.e. the notes for the *Göttinger Gesellschaft* submitted by Klein and Hilbert on 13 January 1912.

[85] Undated letter, mentioned above.

Of the correspondence between Poincaré and Brouwer only a few letters have survived. One might wonder why Brouwer did not simply write to Klein, with whom he had already corresponded; one answer may be that Brouwer wanted to ask some specific questions relating to Poincaré's work, another might be that he trusted Poincaré's judgement in this matter more than Klein's.

The answer was dispatched on 10 December, and contained a brief indication of Poincaré's solution.[86]

On 2 January 1912, Brouwer had indeed sent Koebe his manuscript of the *Letter to Fricke*. The latter played for time and did not send his manuscript in return, but instead wrote on 2 February 1912 a letter to Brouwer in which he listed his objections to certain passages of Brouwer's manuscript, even suggesting a formulation of a footnote that would do more justice to Koebe.

The manuscript of the 'Letter to Fricke' was already in the hands of Klein, who submitted it on 13 January to the *Nachrichten der Göttinger Gesellschaft*. There is a draft of a letter from Brouwer to Klein, which shows that Klein had already gone over an earlier version (the real letter that Brouwer wrote to Fricke, most likely), and suggested some changes,[87] Brouwer must have met Klein's wishes.

It is not a coincidence that Hilbert presented at that same meeting a note of Koebe on the same topic 'The foundation of the continuity method in the area of conformal mapping and uniformisations'. The two notes were the ones that both parties agreed 'to edit in full consultation'. Brouwer had kept his part of the agreement, as we saw above, but Koebe simply ignored his obligation. Brouwer, not surprisingly (and probably quite correctly) concluded that he who refuses to send a promised proof, does not have one.

The 'Letter to Fricke' contained a summary of the theorems required for the continuity proof of 'the general fundamental theorem of Klein'. There are six theorems listed, the last two of which were singled out by Brouwer for further attention. One of them (no. 6) was literally the 'invariance of domain theorem', and the other one dealt with a technical aspect of Riemann surfaces, and Brouwer indicated how the latter could be avoided by a slight modification of the continuity proof. As to the remaining theorems, he remarked that Poincaré had already proved them in the case of the *Grenzkreis*, and that

> for the most general case only Theorems 3 and 4 still resisted an exhaustive proof; in the meantime this gap too will be closed in forthcoming papers of Mr. Koebe.

Koebe objected to this particular passage; in the above mentioned letter of 12 February 1912 he suggested the following formulation to Brouwer:

> for the most general case, in particular for the Theorems 3 and 4 and A,[88] the exact justification is still lacking, which however, Mr. Koebe, as a consequence of his tentative communication in the *Gött. Nachr.* (see in particular

[86] Cf. Alexandrov (1972), Zorin (1972).

[87] Cf. Freudenthal's commentary; CW II, p. 581 ff.

[88] One of the objections of Koebe to Brouwer's note.

also the newest communication 'Foundation of the continuity method in the area of conformal mappings and uniformisation' (1912)), has succeeded in solving completely, as he will shortly exhibit extensively in the *Mathematische Annalen*. The proofs, given by Mr. Koebe, also cover the only case considered by Poincaré, that of the *Grenzkreisuniformisierung*, and point—through the liberation from the idea of limit polygons and closed continua, introduced by Poincaré and borrowed from Klein–Fricke[89]—to the decisive *life-giving*[90] progress, which at the same time means a return to Klein's old viewpoint of non-closed continua, which was vehemently attacked by Poincaré! Koebe's continuity method, by the way, yields another remarkable fundamental innovative distinction with respect to Klein, as Koebe in fact does not use Theorem 4, although, as Mr. Koebe told me, this theorem can indeed, with the help of the *Auswahlkonvergenzsatz*, be proved in the framework of Koebe's methods of proof.

N.B. [This] can best be incorporated as a footnote, since it is not part of the letter.

Brouwer, quite sensibly, refused to comply; he sent a sharp reply to Koebe,[91] in which he answered some of Koebe's objections, for example Koebe's claim that Brouwer's method covered only the case of closed manifolds:

> Fortunately I am still in the possession of the abbreviated text of my Karlsruhe talk, which I enclose, so that you can no longer maintain that I used in Karlsruhe in the talk or in discussions the 'closed' manifolds of Poincaré!
>
> That you could make such a statement, by the way, only proves that modern set theory must be absolutely unfamiliar to you. For the developments of Poincaré that work with the so-called 'closed' manifolds are pure nonsense and can only be excused by the fact that at the time of their conception there was not yet any set theory.

The letter closed with

> 'Why don't you send me a copy of your manuscript, as I did, and as you promised me?'

Ten days later Brouwer sent Hilbert the above cited letter, enclosing Koebe's letter. Brouwer complained that the promised manuscript did not come forth. The letter was accompanied by a copy of Brouwer's refutation of Koebe's points 'on which the rejoinder bears, which I have marked in blue (everything else is nonsense).'

Alas, Koebe again failed to send his manuscript; instead he wrote Brouwer (6 March 1912) that he was looking forward to the published version of the Karlsruhe-talk, but that he protested against the 'letter to Fricke', for: 'the arbitrator-like exposition given there does not give you permission to put the accomplishments

---

[89] Fricke and Klein (1897, 1912)

[90] *Lebenspendend.*

[91] 14 January 1912, see CW II, p. 585.

of Poincaré and me in an unworthy and false light'. The promised exchange of manuscripts was not even mentioned! Thus Brouwer gave up hope of ever getting Koebe's manuscript. On 9 March he wrote to Hilbert:

> After sending you my latest letter, I got the enclosed card from Koebe. It neither brings the retraction of his false claims, demanded by me, nor the promised proofs of his note that he owes me. I now must give up hope of his return to common sense, and I therefore ask you to have my note for the *Göttinger Nachrichten* printed. All the same, it is important to me to answer here, for your information, Koebe's objections against my note. [...] Here Koebe is moving around in a circulus vitiosus; for on the one hand he demands me to praise his as yet unpublished work extensively, on the other hand he tries to prevent me from learning its contents.

After pointing out that Koebe was not aware of Fricke's *Würfelsatz* and its consequence for Brouwer's proof, and a fact about singularities, he closed the letter with:

> That the forthcoming note of Koebe does not contain falsehoods or insinuations is after all even more in Koebe's interest than mine; for, in a possible refutation, I could probably not avoid disgracing him irreparably.

Brouwer, nonetheless, was not unreasonable; he inserted on 20 May 1910 a small change in the first proof of the 'Letter to Fricke':

> ..., in the meantime Mr. Koebe has told me that in forthcoming papers he has completely filled the gap*)
>
> ---
>
> *) There is already a tentative remark on theorem 3 in a note, submitted to the *Göttinger Nachrichten*

As Brouwer had written to Hilbert,[92] he did not object to quoting Koebe in this particular detail, because

> I have myself verified in all detail that Theorem 4 can be derived completely and generally from the *Verzerrungssatz*.

When this 'Letter to Fricke' appeared in the *Nachrichten*, Brouwer was confronted with an unpleasant surprise: the contested passage, quoted above, was changed into

> ... in the meantime Mr. Koebe has succeeded in filling this gap too.[2)]
>
> ---
>
> 2) Cf. his tentative note of 13 January 1912 in the *Gött. Nachr.* 'The foundation of the continuity method in the area of conformal mapping and uniformisation' and 'On the foundation of the continuity method' in the *Leipziger Berichte*.

---

[92] Brouwer to Hilbert, 9 March 1912.

The change is not spectacular, but it slightly strengthens the statement, and it suggests that Brouwer had seen, and agreed with the paper quoted. It does not, however, give credit, let alone priority, to Koebe for his Karlsruhe claims. Freudenthal described the episode in the *Collected Works*, vol. II (p. 575):

> Brouwer's note of 1 May 1913 makes an allusion to another incident. Brouwer (1912d) had appeared with a change in the main text and in a footnote. The change implies a more positive acknowledgement of some claims of Koebe, though not of those related to the continuity proof. This *unauthorised* change was signalled by Brouwer.
>
> Oral tradition tells a cloak-and-dagger story about this footnote: On some dark afternoon in March an unidentified person wearing a large hat, a turned up collar, and blue glasses called at *Dieterick'sche Univ.-Buchdruckerei W. Fr. Kaestner* in Göttingen, the printing office of *Göttinger Nachrichten*, and asked for the printer's proof of the next issue. He got it, and after a while he gave it back and left. The identity of this person has never been determined, nor is it known whether he made any change in Brouwer's reading proof, which of course disappeared after printing. I do not know how much of this story is true. To a trustworthy friend of mine who years later asked him about this incident, Koebe explained it as a trick somebody played on him. Though the revised edition of the footnote gives information which at that time was not publicly available, the hypothesis that it was a practical joke cannot at all be excluded in the Göttingen ambiance. Koebe was a picturesque character whose honesty and frankness forbade him to disguise his greatness as a mathematician; in order to escape embarrassing admiration he travelled incognito, and he often said that in his birthplace Luckenwalde the street boys called after him *Da geht der grosse Funktionentheoretiker.*[93]

Freudenthal told that when he was a student in Berlin, Bieberbach once heard that he came from Luckenwalde—'So you are one of the Luckenwalde street boys who run after Koebe to call "there goes the famous function theorist", are you?' (which earned Koebe the nickname of *der grösste Luckenwalder Funktionentheoretiker*[94]). Koebe considered himself the rightful objective of unlimited admiration. He preferred not to register in hotels as 'Koebe', he travelled incognito because he could not stand waiters and housemaids asking him whether he was a relative of the famous function theorist,[95] and one of his celebrated sayings was *Europa spricht davon—Koebe versendet Separate.*[96]

Van der Waerden added another characteristic anecdote about Koebe: one time Landau gave a party for his Göttingen colleagues and Koebe was also present. Landau, who enjoyed jokes, at one point asked all his visitors to write on a piece of

[93] There goes the great function theorist.

[94] The greatest function theorist of Luckenwalde.

[95] Freudenthal (1984).

[96] All of Europe talks about it: Koebe is mailing reprints.

paper the name of the person who thought himself the most important mathematician. All notes but one read: 'Paul Koebe' and one read: 'Paul Koebe, and justly so' (*Paul Koebe und mit Recht*); Koebe did not sacrifice truth to modesty.[97]

It should perhaps not come as a surprise that Koebe, as a professor, had pronounced views on the relative importance of the various topics in the curriculum. For example, he told his students that they did not have to take Van der Waerden's classes. In Van der Waerden's eyes, this made him not just vain, but also rude.

> That such a character should be the target of a practical joke, in a place where practical jokes were not unusual, is not a farfetched hypothesis. In a letter to Klein of 25 May 1914, Koebe complained about gossip spread about him in Göttingen, related to the sticker in Brouwer (1912g). It seems incredible that for nearly two years Koebe had been unaware of the whole affair, but this does not prove hypocrisy. Gossip, if unjustified, goes a long way before it finally reaches the ears of the ones incriminated. From Brouwer's correspondence with Klein one can infer that Brouwer discovered the unauthorised change in Brouwer (1912d) in the last week of June 1912. Koebe persevered in withholding the manuscripts of his publications even after they had been referred to in Brouwer (1912d) without Brouwer's knowledge. It seems that even Klein did not succeed in changing Koebe's mind (Letter to Klein of 1 July 1912).[98]

The violation of his final proof sheets was a traumatic experience; Brouwer never again trusted the sanctity of the printer's shop again. It was not unusual for him to demand that his proofs be locked into a safe and the keys given to him.

Whoever pulled the trick on Brouwer, and by implication Koebe, must have been well informed. He changed exactly the part that Koebe had objected to. Koebe must somehow have been left out of the gossip circuit; in a letter of 25 May 1914, he wrote to Hilbert that he had been informed that Brouwer had put a sticker in his reprints, which contained a tacit improper reference to him.[99] 'As a result', he complained, 'all kinds of insinuations are circulating in Göttingen'.

We are running ahead of our history, however.

At the end of March Koebe wrote a soothing letter to Hilbert,[100] saying that he was about to send his proof sheets to Brouwer (but apparently he did not do so) together with Brouwer's own copy of his Karlsruhe lectures—but he had not been able to read that note more in detail because of lack of time. Anyway, he assured Hilbert, 'There is not the least reason to worry about a threatening priority conflict'.

Koebe's main tool was the *Verzerrungssatz*, and it enabled him to get round Poincaré's closure trick, but not in all cases. By the time he came to write his note for the report of the Karlsruhe meeting, he must have known that his initial belief,

---

[97] Van de Waerden to Van Dalen, 25 February 1992.

[98] Freudenthal in CW II, p. 575.

[99] See CW II, p. 571. The sticker summed up Brouwer's grievances mentioned above.

[100] Koebe to Hilbert, 29 February 1912.

that the *Verzerrungssatz* could replace the invariance of domain, was unfounded. The aggressive tone of his letter of 12 February 1912 to Brouwer gradually made way for a more objective one, and in Koebe's paper 'On the theory of conformal mappings and uniformisation'[101] Brouwer's view is done full justice.

When the report of the Karlsruhe meeting appeared, Klein committed a beautiful understatement in his introduction, by stating that the talks were followed by a 'stimulating discussion'. Brouwer was less than satisfied with the outcome, as may be clear from his sticker action. He was in particular incensed by the fact that Koebe had used the time between the receipt of Brouwer's manuscript and the final printing of the issue of the Jahresberichte to rewrite the history of the continuity method.

The brief but violent incursion of Brouwer into the domain of uniformisation and automorphic functions ended formally with a paper *Über die Singularitäten-freiheit* der *Modulmannigfaltigkeit*,[102] in which he attacked the problem of the singularities in a direct way.

The commotion over the Karlsruhe talks soon died down. Fricke, in the Foreword to Volume II of the 'Lectures on the theory of automorphic functions' (1912), lavishly praised Koebe (who indeed gave the impetus to the automorphic function saga, after the Klein–Poincaré episode) and the 'through his set theoretic papers renown[103] L.E.J. Brouwer'. The report on Brouwer's achievements was nonetheless somewhat inaccurate. Brouwer had already in the autumn of 1910 informed Fricke and/or Klein of the progress involved in his invariance-of-dimension theorem, but the final result, the invariance of domain, was in the 'Lectures on automorphic functions, II' still announced as an open problem.

Brouwer did not forget the incident, and when he returned to the subject in 1918,[104] he inserted a footnote with a number of historical comments (partially coinciding with the sticker in the reprints of the Karlsruhe report): 'The quotation on page 604 of it [that is the letter to Fricke], Brouwer (1912d), that refers to future publications of P. Koebe (who was at New Year 1912 in the possession of a copy of my letter to R. Fricke) has been inserted after the completion of the proofs by a person unknown to me,[105] without my knowledge or compliance; the footnotes in question became known to me only after their publication ...'.

In 1922 the Brouwer–Koebe conflict had a late revival connected with Klein's collected works.[106] Vermeil, an assistant of Klein, was preparing the third volume, which also dealt with automorphic functions. He asked Brouwer's permission to reproduce his part of the Karlsruhe proceedings. Brouwer answered, somewhat crossly, that the report of the genesis of the continuity proof of the general fundamental theorem of Klein was grossly misrepresented, and so permission could only

[101] Koebe (1914).

[102] On the absence of singularities of the module manifold, Brouwer (1912f).

[103] *rühmlichst bekannt.*

[104] Brouwer (1919i).

[105] *von einer mir unbekannten Hand.*

[106] Klein (1923).

be granted if Brouwer's own rectifications were incorporated. He added that Vermeil could, when in doubt, ask Bieberbach and Bernstein about the matter; both were present at the meeting. Bieberbach, said Brouwer, had made his view clear by referring to the Brouwer–Koebe continuity proof, whereas Koebe tended to advocate the terminology 'Koebe-proof' (or if it must be, 'Koebe–Brouwer proof').

Klein, in his comments in the *Gesammelte Mathematische Abhandlungen III*, on the continuity method, did not want to commit himself to either of the contestants (after more than 10 years!), and so when he came to the crucial point (p. 734) he appealed to the alphabetic convention, and wrote 'Brouwer–Koebe'. Klein's summing-up did Brouwer full justice, as it did Koebe. But one may guess that the style of Koebe was closer to Klein's heart, which hearkened back to the golden time of the eighteen-eighties. Even Koebe, first in the letter to Klein of 30 March 1912, and later in his publications, gave Brouwer his due, albeit as an outsider who had the good luck to provide an essential tool. Koebe's wording of Brouwer's role in the theory of automorphic functions could easily be interpreted by a lesser man than Klein as if Brouwer more or less unwittingly had provided a key tool for the theory of automorphic functions, but that it took a man like Koebe to realise the significance of the 'invariance of domain theorem' for the continuity method. In his survey paper *The essence of the continuity method* of 1936[107] he referred to Brouwer in a way that did little to clarify the history of the continuity method:

> This, also in the following, for the final general foundation of the continuity method, important theorem [invariance of domain], was proved in fact for the first time by Brouwer.

The reader who consults Klein's History of the mathematics of the nineteenth century will find to his surprise that only Koebe is credited for saving the continuity method.

In *Zum Kontinuitätsbeweise des Fundamentaltheorems*,[108] however, Klein gave full credit to Brouwer. He even adopted Brouwer's presentation of the solution of the continuity method.

Brouwer's name has almost disappeared from the theory of uniformisation, and even the textbooks that mention his name do not make clear what his contribution was. It certainly is the case that nowadays uniformisation is carried out by more advanced methods, so the modest place of Brouwer in this respect is not so surprising after all. Even Koebe's name is no longer guaranteed a place in the treatises on the topic.

This chapter has told the story of the birth of the topology of the twentieth century. Brouwer's role in the process has undoubtedly been of the greatest importance; he gave topology new tools and showed how to use them; doing so he had a view of the promised land, but he did not enter it himself. With his knowledge of topology and algebra he could easily have developed algebraic topology, and thus reaped the

[107] *Wesen der Kontinuitätsmethode*, 'after lectures held at the meetings of the German Mathematical Society in September 1913 in Vienna and September 1935 in Stuttgart', Koebe (1936).

[108] On the continuity proof of the fundamental theorem, Klein (1923).

harvest that was within his reach. That he did not do so is a surprise that is hard to explain. One possible explanation is that he was a geometer by nature; he liked to 'see' his mathematics rather than to embed it in a calculus. This is undoubtedly a valid point; nonetheless one can imagine that, if Brouwer had been familiar with Poincaré's ideas about homology, he might have been carried away by the elegance and power of these tools. Who knows what power a beautiful theory may have over a man? On the other hand Brouwer did not have the common urge to exhaust an area of research, a career like Koebe's, consisting of a life of toil, polishing results, extending the theory a bit, trying an alternative approach, ..., was not his idea of mathematics. There was too much of the artist and the free spirit in him to be tied to a rigorous, ambitious research program. Finally, we must not forget that the foundations of mathematics were beckoning—there work was to be done as well.

There is a penetrating sketch of Brouwer's views in the book of Wiessing,[109] who loosely interviewed his friend during one of Brouwer's rambles on the heath. It clearly shows Brouwer's aversion to the treadmill of the mathematics industry.

> W: What kind of a figure are you in the world? What is your place in mathematics of the present and of the past? [...]
>
> B: I could only give you an impression in very global terms, but I am willing to try. You must imagine [...] that in the course of the centuries an increasing number of more-or-less mutually independent branches of mathematical thought have emerged, some of which, that up to now have found no material applications, may perhaps be used in connection with certain physical phenomena. [...] Such a future subservience to the physical sciences is in my opinion fairly unlikely for the branch of mathematical thought that I try to get accepted alongside the existing branches, and which distinguishes itself from those, among other things, by not treating a [particular] subject and by not recognising axioms and postulates. [...] Basically my mathematical thinking is non-sensory internal architecture. You may compare these forms of thought to music or poetry. My first inklings of the possibility of such a mathematics emerged, I think, from discussions with my teachers at the time of my HBS[110] and gymnasium study. But only in my dissertation of 1907 I have started to give these thoughts a definite formulation. Since then this mathematics, nowadays called 'intuitionistic', has developed with interruptions. The recognition in professional circles of this work of mine came only in a rather slow tempo, with many ups and downs. It has by no means found general acceptation! Many view it, even now, as charletanism. There are also people who say that it may be correct, but that it is totally uninteresting and not even new. If I had not, now and then, written about 'ordinary mathematics', I don't think a place at a Dutch University could have been found for me.
>
> W: But you have had, already in our early years, offers of professorships from abroad, I remember that all too well.

[109] Wiessing (1960), p. 143 ff.

[110] High school, see p. 4.

B: Yes, that is right, but in thought and activity I feel far too much a Dutchman and in particular a Friesian Dutchman. And I would rather live here between Dutch friends to enjoy, and Dutch enemies to see through, than far away among strangers!

W: What kind of mathematics do you teach now after all as a professor?

B: Mainly ordinary mathematics, which I transpose here and there, depending on the degree in which the topic and the receptivity of the audience allows this, into my own system of thought. Just now and then, when a particularly gifted and interested group of students presents an occasion, a course consisting exclusively of my own work evolves, and this course may extend over several years. On such an occasion I try to educate students to whom I could eventually trust the further extension and dissemination of my theory.

Several former and present members of my audience (among them some foreigners, who have come to live here in Blaricum for that purpose) give me hope in that respect. And yes, if this will be fulfilled, that will give me great pleasure! Because I indeed really love mathematics for something other than mathematics. That is for the clear light it sheds sometimes on the general problems of life. And ultimately it is in the first place the problems of life that make my natural flow of thought find its way.

W: Could you formulate the special character of your mathematics in a way that is a bit clearer for laymen?

B: Well, let me try once more like this, although it will again be clumsy: consciousness gains access to free creation—which is my mathematics—as soon as it knows itself autonomous and immortal, ignoring objective knowledge and common sense. A condition, in my opinion, for all creation of truth and beauty.

W: I can imagine that you were occasionally called a charlatan.

B: Me too. All the more, as I said, since I don't like mathematics and it basically bores me.

W: [...] What would you, if it comes to that, rather have done than practise mathematics?

B: That is hard to say. Let me say: to have no subject and to let my thoughts roam freely. Every attachment to a subject brings, as you will agree, that your realm of thought suffers a certain mutilation. And it is obvious that then one can only have pleasure in such a profession, if one is, as I sometimes observe with some people, supported and driven by ambition or conviction. But that has never been the case with me. Anyhow, life demands that you choose a profession. Well, then I think that science is for a man like me, who is by nature solitary, not such a bad sanctuary. One is less dependent on the public, and one can more easily preserve one's solitude, than if one takes up literature or the visual arts, not to mention music. For no matter how much pleasure and satisfaction art by itself may give to a person, society, I think, demands more violating concessions from artists than science does. But I may have reached this conclusion also because I have no talent for practising art.

Even in his intuitionistic enterprise of the twenties, he could not bring himself to grind out the routine material that was required to give the program a good start in life.

The conflicts we have seen in this chapter had a double role in Brouwer's scientific life. The positive influence was that they forced him to stick to the subject, and to use his ingenuity for the purpose of widening and deepening his ideas. The influence of Lebesgue has been (although Brouwer would have been loath to admit it) beneficial. Without the constant challenge of Lebesgue's bluffing, Brouwer could have left the field much earlier.

The same cannot be said for Koebe's role; Brouwer would not have engaged in a competition with Koebe, he had little taste for the kind of function theory that Koebe practised. As he said himself, he considered his contribution to the theory of uniformisation a pleasing piece of fall-out, but no more. From the correspondence it appears that Brouwer would gladly have left the matter at his Karlsruhe talk. It was the incredible childishness of Koebe to claim the whole area of uniformisation and automorphic functions as his personal domain that triggered Brouwer's fighting spirit. The moral misbehaviour of the king of automorphic function theory angered him more than anything else.

These conflicts took a heavy toll; Brouwer was ever prone to the return of his nervous breakdowns. Moreover, they tended to give Brouwer the reputation of a difficult man. He certainly did not belong to the meek of this world, but his conflicts were always the result of some instance of injustice, be it towards himself or towards others.

The first and intense topological period ended as suddenly as it began. After the dimension paper of 1913 Brouwer returned to his first love, the foundations of mathematics. Even here there are external as well as internal factors that influenced the course of his activities. If World War I had not isolated him from the international mathematical community, he might possibly have remained active in the field under the pressure of fellow topologists, but when his ties to Göttingen were temporarily cut, no significant impulses influenced him.

# Chapter 6
# Making a Career

> ...and that his brilliant reasoning power would rise to the level of intuition, until those who were unacquainted with his methods would look askance at him as on a man whose knowledge was not that of other mortals.
>
> A. Conan Doyle. *The red-headed league*

## 6.1 Financial Worries

Any graduate of a Dutch University, until long after World War II, was confronted with the pressing question how to reconcile earning a living and practising one's subject. For some disciplines this was less problematic, for example medicine and law, but in general a university degree did not guarantee a future in one's own subject. A master's (*doctoraal*) degree in most of the sciences and the languages could at best earn the recipient a position at one of the high schools or gymnasiums.

Brilliant physicists or chemists might be lucky and get a university position, but even they often had to serve a spell as a high school teacher. Of course, there were industrial positions, but they were scarce. Even the famous Van der Waals and Lorentz had, before their academic career, to teach at high schools. Mathematics offered even less hope for a scientific career; there were no industrial jobs to speak of (apart from an incidental position with an insurance company) and the only way that led to an academic career ran via the teaching profession—a route followed by most academics. The successful *doctorandus* (cf. p. 40) taught for a living, while studying and publishing in the spare hours, working for a Ph.D. degree, and hoping and waiting for some old professor to retire or die. The bright side of this practice was that the standard of teaching at high schools and gymnasiums was generally high; many a future professor taught Euclid or Homer at a high school or a gymnasium—and was fondly remembered for it.

Doctorandus Brouwer was not looking forward to such a career, for one thing, he was temperamentally ill-suited for the drill and the discipline of school life. But above all, he had drunk from the fountain of mathematical knowledge. He was in no doubt about his capacities, and he had, as we have seen with *Life, Art and Mysticism*, an almost fanatical drive for preaching morals and mathematics.

D. van Dalen, *L.E.J. Brouwer – Topologist, Intuitionist, Philosopher*,
DOI 10.1007/978-1-4471-4616-2_6, © Springer-Verlag London 2013

The grant that had enabled him to study in reasonable comfort, did not extend beyond the doctorate; on 14 June 1907. The St. Jobs foundation transferred a final sum of 224.10 guilders, and closed the books on Brouwer. This did not mean that Brouwer was entirely without financial means: on 1 September 1905 he had bought his mother-in-law's pharmacy, and so, one would think, his wife and he could expect a reasonable livelihood. Appearances were, as so often, misleading, as we shall see.

The pharmacy had a long history; the house itself dated back to about 1600. It was originally a well-sized tavern, called 'The Angler', formerly *The Barn Dance*.[1] At that time it belonged to the town of Nieuwer Amstel (the predecessor of the present Amstelveen). From the last part of the eighteenth century onwards it was occupied mainly by surgeons. The father of Lize, Jan de Holl, set himself up in 1873 at the Overtoom as a family doctor with a surgery and pharmacy. The large building was not only a home for a rapidly expanding family, but it also housed those assistants and servants who were interns. In 1880 Jan de Holl died, 46 years old, leaving behind a widow and seven children (cf. p. 51). The widow, having no training or qualifications in the medical profession, had to close the medical practice, but she continued the pharmacy with the help of a so-called 'provisor', that is a professional pharmacist who carried the responsibility for the pharmaceutical side of the shop. This was required by law, and it guaranteed the professional expertise. It was, however, a heavy financial burden on the widow De Holl-Sasse. Some of these provisors lodged with Mrs. De Holl, a circumstance that was grudgingly accepted by the family De Holl. The pharmacy had to provide a livelihood for Mrs. De Holl and whatever children were still dependent on her. Lize and her daughter Louise had, after the divorce from doctor Peijpers, moved in with Mrs. de Holl. Louise was effectively raised by her grandmother while Lize studied pharmacy. When Brouwer bought the pharmacy, he had to accept a substantial financial responsibility towards his mother-in-law and her dependents.

The acquisition of the pharmacy by Brouwer was therefore certainly not the solution to his financial problems, on the contrary! The salary of the provisor still had to be paid, and Brouwer had accepted the contractual obligation of an annuity of 1100 guilders for Mrs. de Holl-Sasse. His friend Carel Adama van Scheltema, among others, had to stand surety. The transaction confronted Brouwer with problems that his academic training had not prepared him for—at the last minute Brouwer discovered that the transaction carried a notary fee of 800 guilders, a considerable amount, which was due within a week of the transaction.[2] It was again Scheltema, who had to rescue his younger friend, he lent the amount to Bertus on security of his house in Blaricum.

> I am terribly sorry that I have to scrounge like this: a friend that you hardly ever see, and then to get all sorts of horrible chores from him, must be painful and disturbing for you with your soul that is halfway that of a grocer. But your

[1] *De Hengelaar*, formerly *De Boerendans*. We gratefully acknowledge the generous help of Mr. and Mrs. J.A.L. van Lakwijk-Najoan, who provided valuable information on the pharmacy of Brouwer.

[2] Brouwer to Scheltema, 23 August 1905.

free poet's soul is indeed above such a thing, and 'generous' means 'royal'.[3] I just wanted to say that I know that such sacrifices bother you, and that I value your sacrifices. But you know that if the occasion arises you can count on me as firmly as you would on yourself.

Well, bye now. The money must be there, and at once.

Now, just do this calmly for me—with such a receipt with a pledge you truly put nothing at risk—and then go on and live your regular cautious artist life, that I, in fact, fear to disturb.

Bertus.

Scheltema complied with a heavy heart; he could, in all fairness, not see, why he should be a security to guarantee Brouwer's mother-in-law a lifelong annuity of 1100 guilders. We note in passing, that Mrs. de Holl lived until a ripe old age, so that, in as far as annuities are a kind of a gamble, Brouwer certainly was not a winner. His relationship with his mother-in-law was always strained, to say the least. The matter of the periodic payments definitely did not endear her to him. She occasionally suffered from mental breakdowns, and spent some time in mental institutions. On the whole she was a good and loving mother and grandmother, but the relationship with Brouwer remained uneasy. Keeping in mind that Brouwer did not even get along with his own father, it would be too much to expect him to develop the obligatory filial love for his mother-in-law.

Although one should not speak of poverty, life was definitely not without its financial problems for the young couple. Matters improved when Lize, after successively passing her doctoral exam in September 1907 and her special pharmacists exam in December 1907, established herself in the pharmacy at the Overtoom, and ran the shop as a fully licensed pharmacist without the help of an expensive provisor. Lize's doctoral diploma caused no stir in the sober household; Brouwer reacted no more enthusiastically than if buying a new pair of shoes; Louise gave the day a festive touch by buying a small bottle of wine and by cooking some floury potatoes which her mother liked so much.[4]

In spite of the resulting improvement of the financial basis of the pharmacy, the pecuniary worries persisted. One should keep in mind that in the first part of the century pharmacies were in general not the goldmines they became later. Moreover, even at the best of times, Lize was a poor hand at management. The correspondence between Brouwer and Scheltema, time and again, mentions late repayments, new loans, etc. Scheltema himself lived modestly but well on the money his father had left him, but after an American investment of his had plummeted (1907) he had to avoid extravagances.

By the time Brouwer's dissertation approached its final stages, he had already lost the unencumbered freedom of the student; he had published respectable mathematics, and he started to view the scientific community with different eyes. The

[3]The pun is lost in translation: '*royaal*' stands for 'generous', 'handsome' and 'royal'. Recall that Scheltema and Brouwer considered themselves kings of spiritual realms, cf. p. 34.

[4]Oral communication Mrs. Peijpers.

**Fig. 6.1** Passport photographs of Bertus and Lize. [Brouwer archive]

transition from student to breadwinner did not leave him cold; society started to exact its dues:

> Life is a magic garden. With wondrous, softly shining flowers, but among the flowers the gnomes are walking, and I am so afraid of them. They stand on their head and the worst is that they call out to me that I must also stand on my head; once in a while I try to do so and die with shame; but then sometimes the gnomes cry that I do it very well, and that I am indeed a real gnome too. But on no account will I fall for that.[5]

Indeed, Brouwer was getting his share of recognition in Dutch mathematics; his brush with Jahnke had done him no harm, on the contrary—Korteweg's opinion of his student was confirmed, and the mathematicians P.H. Schoute (the author of a successful book on higher-dimensional geometry, professor in Groningen) and W. Kapteyn (professor in Utrecht, the brother of the famous astronomer) started to notice Brouwer's achievements.

After finishing his dissertation, Brouwer published a kind of philosophical appendix *The unreliability of the logical principles* (cf. p. 104). This paper contained the revolutionary rejection of the general validity of the principle of the excluded third—it remained virtually unknown until it was translated in the collected works in 1975. While continuing his study of modern mathematics, Brouwer prepared two papers for the International Mathematics Congress in Rome; both were offshoots of his dissertation, one on Lie groups and one on Foundations.

Most of the information on the period 1907–1908 is to be found in the correspondence with Scheltema and with Korteweg; the letters exchanged between Brouwer and Scheltema tell a tale of friends growing up into the world, and of dreaded but unavoidable separation. The friends were no strangers to the frailties of mind and body—almost verging on hypochondria:

> That what we have in common also starts to whither, if the soil, our common youth, dries up. Must we now resign ourselves to that, and each for him-

[5] Brouwer to Scheltema, 7 September 1906.

self let things go in the coarse mansion of society, and light its chandeliers, and grace its door-posts?[6]

Even Scheltema, the artist, was caught up in the general malaise of defeated youth and past:

> I have given up the struggle for the 'superman'—I could reach him now and then, but not lastingly [...]. You wrote in a recent letter that we drift apart—perhaps, but there is such a deep difference in principle between us:—the other day I read Nietzsche's 'Birth of Tragedy' and then I thought of you and I felt things clearer; read it some day if you do not know it. You are *Dionysus* and I am *Apollinius*, and the world we live in is *Alexandrinius*.[7]

The characterisation may not have been so far off the mark, a good deal of Dionysus could probably be found in Brouwer!

Just before Lize's final exams, the Brouwers made in August 1907 a walking tour through Belgium. Brouwer was a lifelong devotee of long walks. The tour to Italy has already been mentioned, but also in Holland he used to ramble through the landscape, sometimes accompanied by his youngest brother Aldert. There are reports that they made long hikes together through North- and South-Holland.[8]

Physical exercises were always a favourite pastime of Brouwer; he would gladly undertake some exhausting extravaganza. In the winter of 1908, for example, he skated, in a fierce wind and on poor quality ice, all the way from Amsterdam to Rotterdam and back.[9] The exertion proved too much; three days of high fever followed, and he asked for a notary to make his last will. The fever, however, disappeared as suddenly as it had come.

When Brouwer was not abroad, he spent his days in the hut in Blaricum, or in Amsterdam in the apartment over the pharmacy, where he often occupied himself with the administration of the pharmacy. Lize regularly stayed at the Overtoom, where she supervised the pharmacy. There is a well-known (and well confirmed) story that at times, when Lize was in Amsterdam and Bertus in Blaricum, Lize would take a pan with food to the point of departure of the *Gooische tram*, where it was put next to the driver's seat. At the other end Bertus would then pick it up at the stop almost in front of his house. The Gooische tram was the normal means of transport between 't Gooi and Amsterdam and Brouwer was a regular passenger.

This tram had acquired a certain notoriety as a result of the not infrequent accidents; it received the nickname of *Gooische moordenaar* (murderer of 't Gooi). The fearful potential of the tram was brought home to Brouwer, when Lily van der Spil, the fiancée of his brother Aldert, was run over by it, and lost both legs.[10]

---

[6] Brouwer to Scheltema, 8 July 1907.

[7] Scheltema to Brouwer 6 August 1907.

[8] Oral communication C. MacGillavry.

[9] Altogether a distance of more than 130 km.

[10] Brouwer to Scheltema, 2 June 1909.

Lily had studied *bouwkunde* (architecture) in Delft, where she met Aldert. The two married on 2 June 1909; they had three children, and the eldest of them was named after Bertus.[11]

In the flurry of all the new mathematics that he absorbed, Brouwer worked wherever he could, but he definitely preferred the peaceful surroundings of Blaricum; he loved to bask in the sun in his garden with a wide brimmed straw hat, working on an old-fashioned draft board and, depending on the weather, in a state of partial undress.

It is surprising to read in a letter of Scheltema, that this man with the sharpest mind in the country, did not escape the lures of the irrational. The poet, who was in fact a far more sober-minded person than his friend the mathematician, scolded Bertus when Lize wrote to him that Bertus was upset on account of a fortune-teller's prediction: 'shame on you for that silly nonsense'.[12] Even in the time of his budding productivity, Brouwer found time to keep up the correspondence with his old friend. The friends went on to deplore the steady deterioration of their original innocent state and to provide each other with words of solace and advice. Brouwer, in passing, mentioned his successes in mathematics, and Scheltema kept Brouwer informed about his literary achievements. There is one, somewhat enigmatic, theme that keeps cropping up in their exchange of letters: Scheltema repeatedly reminded Brouwer of some promise, described in vague terms. Before leaving for Italy, where he intended to spend a prolonged visit, he urged Bertus: 'If at all possible, fulfil your promise before I leave again.'[13] And, indeed, in May of the next year, Brouwer reports that the promised object is ready, and that he had prepared one for himself as well. In his reply the next day from Florence (sic) Scheltema dwelled on the promised object and spoke of a talisman, that embodied a 'possibility' that can give rest, 'How many abysses of the soul go by, that alarm less with a talisman' In a later letter he talked of 'Charon's pennies' and 'non plus ultra's'. One gets the strong impression that Scheltema, who feared and abhorred the tortures and indignities of a pitiless disease (his father had died of a brain tumour—this had left an indelible impression on Carel!) had asked Brouwer for two suicide pills from the pharmacy, one for him and one for his wife.

After some pressing reminders, Brouwer prepared the coveted objects, 'The promised object is ready for you. I have at the same time made one for myself too.'[14] And on 7 July 1909 he wrote Scheltema that 'The talisman and taliswoman (forgive me the lugubrious joke)—good for half a century—are quietly waiting. Perhaps that certainty will already put you at ease!'

---

[11]Like his father, he studied geology. He ended his brilliant career as the Chairman of the Board of the Royal Shell.

[12]Scheltema to Brouwer, 11 October 1908.

[13]Scheltema to Brouwer, 4 September 1908.

[14]Brouwer to Scheltema, 12 May 1909.

## 6.2 First International Contacts

In April 1908 Brouwer travelled to Rome to attend the international conference of mathematicians; there is a short report of his experiences and impressions in a letter to Korteweg.[15] The letter is well worth reading, as it paints the emotional impression of a young man who suddenly finds himself in the presence of the great ones.

> I am very glad that I made this journey. In the first place because it made me feel more healthy and vigorous; but even more because of the admiration and respect that I felt for mathematics under the impression of this Congress, seeing and hearing the heroes of abstraction, and by the aura of 500 honest thinkers acting on me. Poincaré was a revelation to me, and also Darboux and Picard made strong impressions. In general I recognised in the impressive heads practically all those, for whose work I had developed the highest regard. But furthermore, the sight of the persons generously provides hints as to the choice of authors that I will read later on. For example, I will never turn to Mittag-Leffler for instruction after seeing his superficial pompous face; but certainly to Darboux, of whose work I know as yet nothing.
>
> In general it seemed to me that the French are mostly the leaders, mostly command the central parts; Hadamard and Borel as well as the three mentioned above. They, rather than the Germans, seem to me to possess an instinct for what is truly important, a kind of aristocracy in the choice of their topics. To reach a point from where you can, like Poincaré, give an address on 'The future of mathematics' (*L'avenir des mathématiques*),[16] of which everybody feels the reliability as a guidance in his work, seems to me the highest ideal a mathematician can strive for. My respect for the Italians and Americans has not increased, and I am convinced more than before, that they cultivate unimportant parts, without much feeling for the direction in which the main body is moving.
>
> I did not get much pleasure out of my talks; the one on continuous groups was the last one of the morning, and the five preceding speakers were so boring that after each talk a part of the audience disappeared, and there were only thirteen left at my talk. Nonetheless there was one, who, as Versluys[17] said, followed [the talk] attentively from beginning to end, and whose manner of applauding—again according to Versluys—clearly showed that he had fully understood the purport. But when, after a few brief words with the chairman, I left the podium, the man had disappeared, and I have never seen his face, or found out who he was. The other talk on cardinalities had a much larger audience, but due to lack of time I was allowed to talk for only 10 minutes, so that

[15]Brouwer to Korteweg, May 1908.

[16]As a matter of fact, Poincaré fell ill during the conference, and someone else had to read his lecture. It is likely that Brouwer had nonetheless seen, or perhaps even met, Poincaré earlier during the conference.

[17]Professor in Delft, the only other registered Dutch mathematician.

it was so vigorously condensed that it was not done justice. But on account of the mass of other impressions, I did not get much out of my own talks.[18]

On the way back Brouwer stopped in Milan, presumably to see the city and the art collection. He had an extensive knowledge of Italian art and architecture, having already visited Italy before, and later in life he was a regular visitor. He loved to talk about his Italian impressions, and even in his old age he retained very sharp memories of Italy; he could give instructions of a surprising precision to his friends: 'when you come to the cathedral square in (say) Lucca, turn right at the little barbershop at the corner, enter the small alley opposite the post office and then...'

When Scheltema spent a year in Italy, Brouwer advised him:

> Listen Carel, I feel obliged to give you a gentle warning concerning Rome. It could be that you will not fare like me, but I was so fed up after a stay of 3 weeks in Rome, that I must advise you to make your arrangements for a stay of a year with great reservation.
>
> I always had a hard time leaving Florence; there I would leap for joy, my whole life long; in Rome I would get on the train with a sigh of relieve; it is a suffocating, an oppressive, an evil place on our beautiful earth [...].
>
> Don't leave Florence before visiting Siena and Lucca. In the first there are a few admirable Sodoma's, in Lucca the most beautiful Breughel in the world. And in particular, the landscape around Siena with its cool red angular mountains strikes one as miraculous, after the singing blue and white of Florence.[19]

And half a year later he wrote:

> Have you already been in Paestum? That really is a place where the belief in the reality of beauty created by man can be absorbed. I would rather live in Naples than in Rome.[20]

Scheltema and his wife spent the year 1909 in Italy, and a charming book of impressions was published in 1914.

## 6.3 Climbing the Ladder

Back in Amsterdam, Brouwer resumed his freelance activities; he tried to get a foothold in the few mathematical enterprises that were open to a nameless beginner. He started, for example, to do review work for the editors of the *Revues*

[18]Looking at the list of the participants of the Congress, Castelnuovo (1909), one gets the impression that Brouwer could have met almost everybody that was going to be of interest for his later career. Emile Borel, Felix Bernstein, Blumenthal, Carathéodory, Cartan, De Donder, Dehn, Hahn, Hardy, Levi-Civita, Koebe, Hadamard, Hilbert, Poincaré, Zermelo. Even Jahnke was there. But even though the congress had not reached the size of the modern mega-meetings, it is hard to conclude from the list of participants who met whom.

[19]Brouwer to Scheltema, 3 December 1909.

[20]Brouwer to Scheltema, 11 June 1910.

*Semestrielles*, a Dutch review journal, and he hinted to Korteweg that he would be prepared to take over some of the regular review duties,[21] should Schoute or Kluyver wish to stop ('although, of course, I do not ask for it').

At the same time he put out his feelers in Groningen, where P.H. Schoute, who was soon to retire, held a chair. When he expressed his awe for the high status of a professor to Korteweg, the latter reassured him: the scientific virtues and vision were the prime criteria, but he warned his student that routine duties would certainly conflict with his present studies:

> There is only one aspect of your consideration that I think I have to react to. I concerns the 'loftiness' of the position of a professor. In my opinion the position of a professor is neither higher nor lower than any other one. As with any other position, one has to look for the man who will best carry out the duties. That here the scientific virtues and the scientific insight come in the first place is clear. But not *only* those. If we had one big university instead of four, it would be different. Then one could afford the luxury of professorships, such that each scientifically prominent man—for there are not so many of those—could find a place.[22]

Korteweg pointed out that the position of professor brought many teaching duties and other obligations that would probably conflict with Brouwer's research activities. But he made it clear that he would respect Brouwer's decision, no matter how it would turn out. Korteweg had a reason to be concerned: Brouwer was fully emerged in his topological research. In 1909 he published his first paper on Lie groups in the *Mathematische Annalen*, and at the same time his series of papers on transformations on surfaces and on vector distributions started to appear in the proceedings of the Academy. At this moment in life Brouwer needed all the time he could find to fulfil the promises of his topological genius.

As it happened, Brouwer was spared the difficult choice, since the Groningen University could only offer him a one-year teaching job until the chair could be filled by a more senior man.[23] He turned the offer down, and thanked God that he had the courage to avoid scientific suicide and to cut short his career. As he remarked to Scheltema;[24]

> It was not at all that attractive: when the term was over, some professor from Delft would be appointed in Groningen, and I could have moved to Delft, where a professor is something like a supervisor at drawing classes.

With the example of the physics department in mind (where Zeeman's succession of Van der Waals had been successfully negotiated), Brouwer sounded Korteweg on the possibilities of a position as a lecturer or an extraordinary professor in

[21] Brouwer to Korteweg, 1908, undated.

[22] Korteweg to Brouwer, 8 November 1908.

[23] In 1909 Frederik Schuh was appointed. He was also a student of Korteweg, six years older than Brouwer.

[24] Brouwer to Scheltema, 1 March 1909.

Amsterdam. The answer was disappointing; no extension of the mathematical staff (consisting of Korteweg and Hk. de Vries[25] was considered in the near future.[26]

In view of Brouwer's unusual qualities, Korteweg had no problem to get, in April 1909, the young Brouwer admitted as a *privaat docent*. Even this modest step in his career caused Brouwer a great deal of anxiety, when the formalities had to be completed in June.

In a letter to Korteweg of 8 June he ventilated his many doubts. He was afraid of squandering his energy on matters of secondary importance, whereas he had decided to dedicate himself fully to mathematics, in order to attain the best he could offer. Korteweg must have tried to arrange a position for Brouwer that allowed him time to concentrate on his research and to teach a few advanced courses of his choice. But that was not what Brouwer had in mind. He considered *privaat docenten* in extra-curricular subjects to be social climbers, and he hoped, he said, that Korteweg knew him well enough by now to understand that social status did not interest him. He was willing to be a *privaat docent*, but only if he was of any use to Korteweg or the faculty:

> As far as my personal wishes are concerned, I hope, at least as I am still immature as a mathematician, as I am now, to remain excused from teaching; and even later if I am not needed. But in any case, I will never be able to teach classes, where I am at the mercy of the *bon plaisir* of the audience. I have seen that successively with Van der Waals, Jr, Mannoury and Kohnstamm, and I have felt more and more that I am not capable of something like that.[27]

Brouwer viewed his appointment as *privaat docent* with some mistrust; it was another step on the road to his assimilation into the system, but, as he wrote to Scheltema;[28]

> I finally succumbed to the pressure to become a privaat docent here in Amsterdam, in some subjects that have my sympathy. If it suits me, and I suit them, then it will probably result in a lecturer's position or an extraordinary professorship with a very restricted duty, and without official tasks. Well, that is, I believe, in the long run fairly acceptable to me.
>
> Moreover, when they held out their hand to me after I turned down Groningen, I could not very well turn my back on them any longer. Yet, it made me melancholic, when I read my appointment to *privaat docent* in the newspaper; and I almost started to cry for feeling sorry for myself. It is, after all, a 'Sic transit', but my youth passes, and I work more and with more pleasure than before.

---

[25] Another student of Korteweg. He had been a professor in Delft for one year before he was appointed in Amsterdam (10 October 1906).

[26] Korteweg to Brouwer, 16 February 1909.

[27] Brouwer to Korteweg, 8 June 1909.

[28] Brouwer to Scheltema, 12 May 1909.

And so, at the age of 28, Brouwer made his first step up the academic ladder. Compared to newcomers at German universities he was already fairly old, and although he had displayed sufficient mathematical talent and revolutionary ideas in his dissertation, no great feats had put his name on the mathematical map. Compared to, for example, Koebe, who became famous at the age of 25 for solving the uniformisation problem, and Von Neumann, who was to revolutionise set theory at the age of 22, Brouwer was a slow starter.

At the time of his appointment to *privaat docent*, he was entering the field of topology, and gradually mathematics became his true love. He could not muster the same feelings for society:

> I did not live for long in Amsterdam; being a respectable man oppresses me, and after my classes I flee to 't Gooi and gloriously pace the heath as a tramp. That works out wonderfully, as long as I don't have the misfortune to run into a student here in my vagabond outfit. In class I always look spick and span. I am getting ever fonder of my subject, but I detest society more and more.

The relationship between Korteweg and Brouwer was apparently of the sort that allowed for the inclusion of extra-mathematical activities; when Korteweg, who was mildly involved in politics, discussed the coming parliamentary elections, Brouwer—surprisingly—offered his support in campaigning for the Liberal Party.[29] Whether he actually took part in any campaigning is unknown. Anyway, it left its traces in the correspondence in the form of a political positioning in the Dutch party system.

Brouwer expressed his gratitude to Korteweg by presenting him with the collection of all the papers that Korteweg had communicated so far to the Academy, bound in a small volume,[30] with the inscription

> To prof. D.J. Korteweg in grateful memory of the submission to the Academy of these essays of his student.
>
> L.E.J. Brouwer September 1909

The period following the dissertation seems to belie Brouwer's fervently claimed principles. There is no calm resignation to fate; even though he claimed in his letter of 12 May 1909 to Scheltema that a minor academic position of limited visibility would satisfy him. No sooner had the position of *privaat docent* become his, then he started to worry, complain and look for a better position. Even in the case of a highly strung, well principled young man like Brouwer, the flesh turned out to be weak—even a practising mystic apparently wants his comfort and recognition!

Job-hunting kept Brouwer's mind continually occupied, and in July he had found another career perspective: the Teyler Museum in Haarlem was looking for a candidate to fill the position of director of Teyler's physical cabinet and editor of the

[29]Cf. p. 32, Brouwer to Korteweg, 22 June 1909.

[30]Kept in the University Library at Amsterdam.

*Archives du Musée Teyler.* Brouwer judged this to be an excellent place for a person like himself with no urge to teach and with an extensive research program. He expected that a director could devote almost all his time to research.[31]

One can imagine his interest in the position—having lived in Haarlem, he knew the oldest museum in Holland well, and it was, as it is to-day, a delightful institution. It was the first science museum built as such, a monument from the days of the Enlightenment, founded in 1778 by a rich merchant couple, Teyler Van der Hulst. It has a marvellous collection of scientific instruments, for example the electrical machine of Van Marum (1784) and an substantial art collection, containing paintings, prints and drawings. An important part came from the collection of the 17th century Queen Christina of Sweden.

The position of director would certainly have allowed ample time for personal research. However, in spite of a modest measure of lobbying, the application was not successful.

## 6.4 The Shortcomings of Schoenflies' *Bericht*

Mathematically speaking, all was well with Brouwer; he was rapidly finding his way into the impenetrable fortress of topology. He was conducting investigations in the topology of surfaces, vector distributions and in general topology. The research in general topology was the direct consequence of his work on Lie groups; as we have seen (p. 140), Schoenflies' monograph on point-set topology was far from reliable, so that Brouwer had to redo the material for himself.[32] The result was the famous 'Analysis Situs'-paper. Brouwer had submitted it to the *Mathematische Annalen*, and this led to an exchange of letters with David Hilbert, the leading editor. Brouwer was undeniably flattered by Hilbert's attention.

> I received the enclosed postcard from Hilbert, from which I conclude with somewhat mixed feelings of satisfaction, that my latest Academy communications[33] have drawn his attention.
>
> Somewhat mixed, as I say, because I value that publication much less than my other work: anybody whose thoughts would wander in that direction would have found the results. The value of a mathematical composition lies, like that of every work of art, in its penetration, and not in some surprising and popularly comprehensible result, no more than that the value of a Dutch painting lies in the little windmill.[34]

---

[31] Brouwer's idea was not as eccentric as one might think: Lorentz filled the position of director from 1912–1920 after he had given up his full professorship at Leiden.

[32] For the mathematical aspects of the *Bericht* episode, see Sect. 4.6.

[33] Brouwer (1909e, 1909f) the first paper of the series on continuous maps of surfaces into themselves and the first one of the series on vector distributions.

[34] Brouwer to Korteweg, 22 May 1909.

The 'Analysis Situs'-paper, in the meantime, held the promise of some tangible recognition. Brouwer had sent a copy of the manuscript to Schoenflies, who returned an elaborate letter—in Brouwer's words: 'probably because I pressed him hard, ...'. The paper made a clean sweep of set-theoretical topology, as presented by Schoenflies in a prestigious monograph, *The Development of the theory of point sets, II*. 'Nonetheless', wrote Brouwer to Korteweg, 'I am glad to have finally a bite, and to receive more than a kind postcard at my request'.[35]

Brouwer had indeed scored; Schoenflies was one of the leading set theorists, in spirit the successor of Cantor. He had mainly carried on the Cantor tradition in point-set topology. Schoenflies was an acknowledged expositor; he was the author of a large number of papers on point-set topology and abstract set theory. His work culminated in two influential monographs on set theory, which had the ill-fortune to become obsolete soon after their publication. Schoenflies was a close friend of Hilbert, which partly explains the role of the latter in the relationship between Brouwer and Schoenflies.

Schoenflies, clearly, was taken unawares by Brouwer's manuscript. His letter to Hilbert, written almost directly after the receipt of Brouwer's paper,[36] shows that he suspected that after all these years of preparations the control over the field of topology was slipping from his fingers. He was in the position of a general who at his victory banquet is informed by a common soldier that the battle is not won but lost:

> The paper of Brouwer has made me half glad, half sad. Glad—because I see that the younger generation has started to study the *Bericht*; sad, because among the abundance of geometrical forms, a possibility has escaped me. For this is the essential content of Brouwer's paper.

He sadly acknowledged the shortcomings of the *Bericht*. In a moving passage, he reminded Hilbert of the adverse circumstances under which both *Berichte*[37] were produced.

> You have by now also experienced that one can reach a point where the mind starts to fail. Alas, I have not had the good sense to pause in such a case, but I devoted all my energy to reach the goal.

It appears that Schoenflies had corrected the proofs of the last *Bericht* during a summer-trip, missing some of the finer points in the atmosphere of vacationing.[38] The letter closed with an expression of annoyance at the formulation of the opening of Brouwer's paper; in Schoenflies' opinion it rather suggested that the author considered the *Bericht* as largely incorrect. Schoenflies went on to say that this impression was, however, contradicted by the accompanying letter of Brouwer (which he enclosed).

---

[35]Brouwer to Korteweg, 18 June 1909.

[36]Schoenflies to Hilbert, 22 May 1909.

[37]of 1900 and 1908.

[38]Brouwer to Hilbert, 24 June 1909.

**Fig. 6.2** Arthur Schoenflies. [Courtesy Niedersächsische Staats- und Universitätsbibliothek Göttingen]

Notwithstanding Brouwer's lethal criticism, he still hoped and expected that minor revisions would suffice. In that optimistic spirit he was already considering the revision of the *Bericht*. He clearly underestimated the damage discovered by Brouwer.

Whereas Schoenflies' letter to Hilbert had the character of a lament to a trusted old friend, the first reaction to Brouwer was (understandably) more guarded. Taking his time to think the matter over, Schoenflies had concluded that the situation was not hopeless. He suggested some minor adaptations, but he could not make sense of the figures of the manuscript, partly because the letter did not match the text, and also he did not see the point of Brouwer's criticism of his notion of 'curve' and 'accessibility'. Brouwer's suggestion to publish his criticism together with a reaction of Schoenflies, was gladly accepted.[39]

The latter had so far recovered that in the same letter he undertook a defence of his earlier results against 'admonitions which seem to me partly unjustified, partly exaggerated'. It clearly distressed him to be taken to task by a much younger man, who did not bother to hide his disapproval, and who had repeatedly used the term *ungenügend* (inadequate) in connection with Schoenflies' topology. In the years before the war there has been an extensive correspondence between Schoenflies and Brouwer. The first group of letters dealt with the Analysis Situs paper, and the second group concerned the new editor of the *Bericht*. Unfortunately the correspondence is far from complete; most of Brouwer's letters are missing.

In a prolonged exchange of letters, Brouwer tried to convince Schoenflies of the shortcomings of his proofs, and the latter continually grasped at straws to defend the remnants of his *Bericht*. He did not realise that something was basically wrong, but attributed the gaps in the proofs to the nature of the *Bericht*, which he saw as

[39] Schoenflies to Brouwer, 17 June 1909.

a quick, convenient and, above all, readable introduction of the reader to the problems and results of topology. In particular he assumed that the attentive reader could provide further details when necessary. Here, evidently, was a clash between two scholars with different notions of exactness. Schoenflies belonged to the older tradition in which the foundations of topology were taken for granted, and where geometrical intuition often took the place of mathematical argument. Indeed, Brouwer's (counter-) examples showed that the sophistication of topology required more than the conviction of 'first impressions'.

Reading the correspondence, one gets the impression of a growing exasperation of the older man. When all of his arguments bounced off Brouwer's granite wall of exactness, he must have felt that the younger man was not susceptible to reason—'Should it then really be impossible to convince you in writing',[40] and 'I can now only wait and see if you will stick to your arguments concerning Chap. IV, Sect. 12. Nonetheless, I hope to have finally convinced you that here the error is on your side.'

But Brouwer did not give up; he kept trying to set Schoenflies right. Schoenflies felt that his dignity was at issue, rather than his topological expertise. Brouwer may have been under the impression that Schoenflies tried to save his face by making light of the inadequacies of his work. Here he misjudged Schoenflies, who protested: 'Anyway, I don't think anyone could think that I try to belittle your paper or to diminish my mistakes. In my opinion the opposite is the case.'

Schoenflies had turned to Hilbert for protection from this young Turk, and Brouwer had likewise appealed to the undisputed Master of Mathematics, who as editor-in-chief of the *Mathematische Annalen* had a vested interest in the matter. He informed Hilbert that Schoenflies did not fully realise the defects of his presentation, and that, should he stick to his views in the forthcoming paper, some extra space would be required to enlarge upon the details of the defective parts. Brouwer strongly stressed the importance of the topics, so the corrections in his opinion were really necessary. He wrote to Hilbert:

> I am awfully sorry that Mr. Schoenflies feels insulted by something that rather resulted from my appreciation, but I don't want [to run the risk] that someone could reproach me with the slightest semblance of justification, that I publish scientific trivialities, especially in the *Annalen*; nor do I want to allow that the supplement of Mr. Schoenflies contains even the slightest innuendo in this sense, and I kindly beg you to make this explicitly clear to Mr. Schoenflies, and to allow him extra space only on the ground of the objective content of my remarks.[41]

Schoenflies' excuse that he did his proof-reading in a hurry during his vacation, was interpreted by Brouwer as an implicit complaint that Brouwer had resorted to nit-picking—the idea being that between gentlemen one does not mention the smaller mistakes.

---

[40]Concerning decomposition theorems, Schoenflies to Brouwer, 14 August 1909.

[41]Brouwer to Hilbert, 24 June 1909.

> With respect to isolated inaccuracies, which do not impede the general flow of the theory, Mr. Schoenflies, if I understand him correctly, calls attention to the fact that he did the proof-reading during a summer trip, that this explains the minor slips, and that it is a petty thing of me to locate them. I thought, however, that even these little corrections were worthwhile, in view of the great importance of the subject.

and he continued:

> In order to rule out that Mr. Schoenflies could feel the least bit hurt, I just would like to insert at the end of my introduction the following note (warmly meant, by the way): 'I explicitly stress that this paper does not seek to diminish in any way the great value of Schoenflies' discoveries, but rather that it tries to emphasise the value more clearly.'

In the paper itself this note was modified to:

> 'I expressly stress that this paper does not seek to diminish in any way the great value of Schoenflies' discoveries. It is just that the considerable consequences have given me reason for this criticism, which by the way, does not essentially concern the largest part, namely the theory of simple curves.'[42]

In the bargaining that followed, Brouwer softened parts of his argument and further sugared the introduction, in the hope that this would pacify Schoenflies.

The affair went on for the better part of a year, and eventually Hilbert got fed up with Schoenflies' defence tactics; at least, in December Schoenflies complained about the rather unfriendly tone of Hilbert's letter. And in reaction to Hilbert's exclamation, that the readers had already lost interest in the matter, he retorted that they had not even seen the papers, and that he would, when necessary, revise his attitude with respect to the *Annalen*. It appears that Hilbert felt no compunction in letting scientific interest prevail over old friendship. By the end of December, both parties had reached an agreement, and the *Annalen* could publish the papers.

The paper sent shock waves through the mathematical community. Hardly anybody had guessed that the road of topology contained so many pitfalls. For Brouwer, the Analysis Situs affair had after all a pleasing consequence; the paper contributed substantially to his status in the mathematical community.

Brouwer's exploits had already drawn the attention of the reigning king of mathematics, David Hilbert. Brouwer's work on Lie groups and his subsequent research in topology had satisfied Hilbert that he was dealing with a clever young man; the talks in the dunes at Scheveningen (cf. p. 125) doubtlessly had reinforced this impression. To be noted by the Göttingen group of mathematicians was already a feat by itself, but Brouwer's luck did not stop there. He found himself a niche in the community, if not in the hearts, of the Göttingen mathematicians in a remarkably short period. The correspondence with Hilbert soon led to plans to visit Göttingen in person. Already in 1910 he made plans for a visit in the summer vacation, but

[42] Brouwer (1910e).

**Fig. 6.3** David Hilbert. [Courtesy Niedersächsische Staats- und Universitätsbibliothek Göttingen]

the trip was cancelled because of family matters. A year later he made his entree in Göttingen; he wrote to Hilbert that 'I will be in the Harz for a couple of weeks and on the way I'll stay for a few days in Göttingen. I am very much looking forward to get to know the people and the situation over there'. The next year Brouwer again visited Göttingen, this time accompanied by Lize: 'Next Sunday I will come to Göttingen via Löhne, Hameln, Elze.' Again he combined his visit with a stay in the Harz mountains; the *Brocken* was his destination. In this short period Brouwer managed to make many friends for life, Felix Bernstein and Hermann Weyl, to mention two of them. He became one of the 'extra-territorial' members of the Göttingen group. Brouwer immediately made himself popular with the company; he was an expert conversationalist with an inexhaustible fund of stories and a sharp insight into most branches of mathematics. Göttingen benefitted by acquiring the support of the leading topologist, but Brouwer paid the price of a considerable increase in workload, consisting mostly of refereeing jobs and general advice.

## 6.5 Privaat Docent

With the first term of the academic year 1909–10, Brouwer's duties as a *privaat docent* had started. Traditionally, *privaat docenten* (like lecturers and professors) gave an inaugural lecture in order to present their professional views on their subject. This was a formal address delivered in the auditorium (aula) of the university, in the presence of colleagues, friends and future students. Brouwer gave his lecture on 12 October 1909 with the title 'On the nature of geometry'[43] a beautiful programmatic exposition of the latest in geometry, including the space-time geometry under the Lorentz transformations, complete with moving systems and clocks. Curiously enough, there is no reference to Einstein, and only Lorentz with his 'relativity

[43] *Over het Wezen van de Meetkunde*, cf. p. 151.

postulate' and transformations is mentioned. Brouwer drew from this geometrical-physical theory the conclusion that the a priority of space and time, and hence of Euclidean space and time, had become untenable. Later, in the collection *Mathematics, Truth and Reality*,[44] Brouwer added 'This lecture, were it given at present, would undoubtedly bring up in its first part the 'general theory of relativity', introduced after 1909, which would not have influenced the epistemological conclusions'.

The last part was devoted to Analysis Situs (topology), and contained a list of prominent problems, see p. 152. We note in passing that the better part of those problems was to be solved by Brouwer within a few years!

The writing of the inaugural lecture had to be squeezed in between many other activities. In a letter of September 1909, Brouwer complained to Korteweg that he was in the middle of the quarterly administration of the pharmacy and that the correspondence with Schoenflies had entered a new phase—at that moment only two weeks were left for composing his text.

The inaugural lecture ends with a credo that strikingly reflects Brouwer's personal taste, and that had a prophetic ring. After discussing the foundation of geometry in the light of Analysis Situs he stated that:

> Thus, one does not have to ban co-ordinates and formulas entirely from other theories if one succeeds in basing them on analysis situs, but the formula-free, the 'geometric' treatment will be the point of departure, the analytic one becomes a dispensable aid.
>
> It is the possibility and desirability of this priority of the geometric treatment, also in parts of mathematics where it does not yet exist, that I primarily wanted to point out in the preceding pages.[45]

Two days later his classes began: projective geometry and analytic geometry. These courses remained his regular contribution to the mathematics curriculum during the first years; a short (and no doubt polished) description of Brouwer's teaching can be found in the student's almanac, which provided among other things reports on the courses of the year before. The almanac of 1911 describes the contents of Brouwer's courses in projective geometry (for first- and second-year students in physics) and the 'suppletion' course in analytic geometry. The projective geometry course contained a fair amount of material: for example an axiomatic treatment, the 'numerical-model', the introduction of cross ratios and the derivation of the Euclidean and non-Euclidean metrics. The reporter remarked that:

> The modest size of the audience was an indication that the course may have been too much for freshmen, perhaps also because the high-speed teaching of Dr. Brouwer made it difficult to acquire a good insight into this topic, that was completely new to us. But those who have taken the trouble to work through it till the end, will certainly agree with our great admiration for this course of Dr. Brouwer.

---

[44] *Wiskunde, Waarheid en Werkelijkheid*, 1919.

[45] Brouwer (1909a), p. 23.

The report of the next year ends on the same note:

> It seems to us that the audience would have benefitted more from Dr. Brouwer's lectures, if those would have been not so fast.

The gist of these reports is confirmed by most of Brouwer's students; his courses were models of elegance and precision, but patience with the less gifted student did not belong to Brouwer's virtues. A former student[46] reported a comical event in the late thirties. The audience was left far behind by Brouwer during a certain course, so the students took courage and sent a delegation to Brouwer with the request to reduce his tempo of lecturing. After listening politely to the delegation, Brouwer answered: 'Alright, I will talk as slowly as the gentlemen think.'

In October 1909, Brouwer had a mild difference of opinion with Korteweg; he had found a considerable simplification in a derivation in a paper of Van Uven, the mathematics professor at Wageningen, and asked Korteweg to present a short note to the Academy. Korteweg resolutely refused: 'An alternative derivation of a formula does not belong there.' In spite of Brouwer's insistence, Korteweg stuck to his guns; the note was not published.

As we have seen in the preceding chapter, Brouwer was fully immersed in his research in topology; he had already obtained his results on fixed-points on the sphere, and he had opened the cleaning-up operation in elementary (plane) topology. And now on the instigation of Hadamard he had started to read the *Mémoires* of Poincaré, and this was bearing fruit in the form of simpler proofs of some of his earlier results:

> the result appears here as a surprise, whereas in the original proof I gradually constructed the transformation, thus forcing myself step-by-step to admit the invariant point.[47]

In spite of the growing recognition of Brouwer among mathematicians, there was no comparable academic recognition at home. He made himself useful in Amsterdam by his teaching, and by taking care of the collection of mathematical books and journals in the University Library. In those days there was no mathematical institute and no independent mathematics library; the central building of the university in the Oudemanhuispoort contained lecture rooms and 'faculty rooms', the latter serving mostly as a meeting place for faculty members and as a cloakroom for those who had to lecture. Those professors who had no laboratories or clinics dropped in to give their courses and went home again to conduct their research. Brouwer was no exception to this pattern; the pharmacy was his pied-a-terre in Amsterdam, but his heart drew him to Blaricum.[48]

The academic routine dragged on, and Brouwer, who was totally absorbed by his topological research, grudgingly fulfilled his tasks.

---

[46] F. Kuiper.

[47] Brouwer to Korteweg, 24 December 1909. The letter was accompanied by a copy of Brouwer's letter to Hadamard.

[48] See p. 203.

Korteweg, in the meantime, tried to improve the prospects of his brilliant student. He saw that the average teaching load in Amsterdam surpassed the level that was generally thought acceptable. By appointing Brouwer, one could kill two birds with one stone, the teaching load would be reduced, and the mathematical genius of the young man would be rescued from real and imaginary dangers.

In order to underpin his arguments for an extension of the number of mathematicians at the University of Amsterdam, Korteweg conducted a small investigation into the general situation in mathematics in Holland. The results are interesting, because they shed light on academic mathematics in the Netherlands at that time.

We reproduce here Korteweg's list of the mathematics departments:

*University of Amsterdam*[49]
Professors: J. D. Korteweg and Hk. de Vries
Teaching duties: mathematics, mechanics (that is mathematical physics, mechanics), astronomy.

*University at Leiden*
Professors: J. C. Kluyver and P. Zeeman[50]
Teaching duties: mathematics, mechanics.

*University at Utrecht*
Professors: J. de Vries and W. Kapteyn
Teaching duties: mathematics.

*University at Groningen*
Professors: P. H. Schoute and F. Schuh
Teaching duties: mathematics.

The load varied from 8 hours a week (Leiden) to 10 hours a week (Amsterdam).

There were more mathematics professors in the Netherlands than the ones just mentioned, to be specific, in Delft and Wageningen, but Korteweg did not want to compare those (technological and agricultural) institutes with the universities.

The number of graduates in mathematics, physics and astronomy in Amsterdam was also steadily rising:

| | Candidaats degrees | Doctoraal degrees |
|---|---|---|
| 1880–1884 | 4 | 2 |
| 1885–1889 | 5 | 4 |
| 1890–1894 | 1 | 4 |
| 1895–1899 | 11 | 3 |
| 1900–1904 | 10 | 5 |
| 1904–1905 | 9 | 9 |

[49]The Amsterdam University was a municipal university. The other ones were state universities.

[50]Not the famous physicist.

After ample consultation, Korteweg and De Vries addressed the City Council of Amsterdam with a request for an additional lecturer. The arguments were not unreasonable, and they could be (and are) repeated at any time and any place: too much teaching, supervision of Ph.D. theses and research. Moreover, Korteweg argued, this exceptional man, Brouwer, 'who can be considered as the equal of the best mathematician of our time', should be attached to our university.[51]

The authorities, however, did not share Korteweg's views; even though Korteweg had pointed out that the university could have Brouwer for one thousand, say fifteen hundred, guilders, they did not see the justification of an extra lecturer.

The rejection hurt Brouwer; he had just been passed over in Delft, where 'a younger and lesser man' was appointed.

Already in June he had written to Scheltema that he was fed up with the job, and that he wanted to quit after the summer vacation. He had expected an extraordinary chair, but since nothing of the sort was considered, he felt no obligation to go on with teaching for a marginal fee. The old dream of a quiet hermit's life was returning:[52]

> ...do you remember Watt's principle? It says that the vapour-pressure in a container is determined by the lowest temperature occurring in it. Likewise, the intensity of our sensing and thinking is determined by the most impure of the activities that we are involved in, no matter how little. He who teaches two courses a week, withers like every professor, even though he is for the rest surrounded by the fragrance of flowers, and shone upon by the sun.
>
> Thus I now ponder the question, how I can decently, without quarreling, get rid of my job. For the rest I am turning more and more into a home-lover, and I hardly leave any more my beloved terrain, where I dig, weed, prune and philosophise. Yesterday and the day before I have tarred, in the blazing sun, my faithful in a scraggy reddish brown.
>
> Maybe this is caused by the advent of old age, but again and again all my desires subside, and make place for cheerful self-satisfaction, and the world that interests me contracts more and more. Last year I was still enthusiastic to break the bread with the star of the mathematicians,[53] but now I find that idiotic too, and I no longer admire my colleagues.

In spite of his aversion, Brouwer did not give up his position, marginal as it was. With varying hope he muddled through the coming years, no doubt keeping his morale up by his research activities and the international recognition that started to come his way.

[51] Letter to the city council, 6 October 1910.

[52] Brouwer to Scheltema, 11 June 1910.

[53] David Hilbert, cf. p. 125.

## 6.6 Korteweg's Campaign for Brouwer

In 1910 Korteweg undertook another action for his student. If, so he reasoned, the City Fathers do not see the importance of furthering the career of the gifted Brouwer, and, worse, if they do not realise the importance of securing the services of such an exceptional scholar, then they may be helped in getting the right perspective by an act of recognition from an unsuspected side. And so he started a campaign to get Brouwer elected to the Academy of Sciences; some skilful lobbying got him the support of his colleagues in the Academy and of the Dutch mathematicians. Moreover, he approached some of the leading international mathematicians, Hilbert and Poincaré, pointing out to them that, although the attempt to find a position for Brouwer had failed, the situation was not hopeless, and that a membership of the Academy was a powerful recommendation that could not so easily be ignored. Both Hilbert and Poincaré reacted positively, Hilbert's answer is preserved in the form of a draft written on Korteweg's letter.[54] Apparently Poincaré was not fully aware of Brouwer's latest feats, but he was sufficiently impressed to write a flattering recommendation.[55] Hilbert on the other hand had seen almost all of Brouwer's work in topology, and he had a sharp eye for outstanding mathematics. His letter is so much to the point that it would be a pity not to reproduce it here:

> I am very pleased that you will try to get Mr. Brouwer a chair at the University. I wish that you may succeed not only in the interest of Brouwer, but also of science. For, I consider Brouwer a scientist of unusual talent, of the richest and most extensive knowledge and of rare ingenuity. The area in which Brouwer has been particularly active is that of the theory of point sets, a theory, that, as you know, interacts in an essential way with almost all disciplines of mathematics, and therefore its development is one of the most important tasks. Moreover, it is Brouwer's characteristic not to contend himself with easy successes that are offered by researches of a general nature, but he attacks each time a special, difficult, deep problem, and he leaves off only when he succeeds in obtaining the fully satisfactory solution. I am thinking here in particular of his solution to the problem of the finite continuous groups and his marvellous proof of the Jordan curve theorem.[56]
>
> That your small country repeatedly produces in the most varied areas outstanding scientists, is a pleasing phenomenon that may fill you with pride. But this phenomenon imposes on the government and the leading authorities the duty of special efforts. I hope therefore that your efforts on behalf of Brouwer will be successful.

The letter beautifully illustrates the confidence of the old style academic in dealing with the authorities: one simply has to tell them their duties to science!

---

[54] Korteweg to Hilbert, 6 February 1911.

[55] The letter of Poincaré has not been found.

[56] Brouwer (1910a).

But even Hilbert's authority did not have the magical power that one would nowadays assume. The matter was further complicated by the circumstance that the Academy already had a fair share of mathematical members, and Brouwer's appointment would either increase the number of members, or it would be at the cost of some other discipline.

In the meeting of 25 March 1911 Brouwer was proposed for membership by a most impressive array of scholars: Korteweg, W. Kapteyn, Cardinaal, Hendrik de Vries, Schoute, Jan de Vries, Kluyver, Lorentz, H.G. van de Sande Bakhuyzen, Zeeman, Van der Waals, J.C. Kapteyn, Hugo de Vries, Winkler en Hubrecht.

At the next special meeting, 28 April 1911, the voting took an unexpected course. In the first round the largest number of votes was cast for Brouwer (21), but still less than required to get elected. And so a next round of voting took place. This time another candidate, Kuenen, got the required statutory 24 votes and was elected. In the final round Brouwer was only two votes short of the required number, and thus only Kuenen collected enough votes to join the Academy.

Korteweg reported (not quite truthfully) in October to Klein that only one vote had failed to get Brouwer into the Academy.

The parallel struggle for recognition by the curators of the university was not easily decided either; for one thing, it involved money! In September 1911 Brouwer, irritated by the earlier rebuff from the City Council, informed Korteweg that he would gladly continue his lectures on analytic geometry, since that was a personal favour to Korteweg, but that he refused to lecture on projective geometry, since that should be a favour to the authorities.[57] Apparently, Korteweg succeeded in calming down his hot-headed student—the course on projective geometry was given after all.

However prominent Brouwer had become through his majestic topological papers, in the privacy of his home, he was again beset by thoughts of the futility of it all. In a letter of 7 November 1911, he bared his soul to the only friend who had seen glimpses of the real Brouwer, the by now famous poet Carel Adama van Scheltema:

> ...Although I am nowadays fairly fertile, and have gradually gathered some international fame and envy, you must not get an overly serious impression of my work. For I have, as ever, the intimate certainty that mathematical talent is of the same sort as an abnormal growth of the nail of the big toe.
>
> Yet, at congresses I play for the popes of science the role of an enthusiastic ensign, but, when I paint *in flammender Begeisterung* (with burning enthusiasm) conversations rich with thoughts, with the prospects that inspire my work, my apparently absorbed gaze quenches itself with the monomania of their features, and sees in some of them inconsolable, imprisoned heroes, in others poisoning kobolds, and in the latter the unprecedented hangmen of the first. And while I am physically permeated with the sensation of being in hell, my eyes beam with sadistic lust of sympathy.

[57] Brouwer to Korteweg, 10 September 1911.

> My productivity therefore will never bring forth a grand creation, because it is exclusively fertilised by a mocking dissection of what exists.[58]
>
> None of my colleagues will, however, fathom this, although some of them eventually get ill at ease in my presence; they go round and speak evil.

The above passage provides a revealing glimpse of Brouwer's insight into his own motives, but it also shows the traditional tendency of the high-minded to flog themselves mercilessly, to confess evil thoughts. Whereas Brouwer certainly was not then or later in life, a dull-but-respectable citizen, he was not as contrary as he wanted Scheltema to believe. Indeed, he could never bring himself to take science deadly seriously, but he definitely had come to love his subject, and he had rather strict ideals concerning it. Even his bleak views on the scientific community are artistically exaggerated. No doubt, his rather unusual behaviour, choice of words, of topics, would surprise, or even shock, people, and make him an easy target for gossip or worse, but his relations with his colleagues were not disastrous; with some of them he had lifelong relationships of friendship, and in general he recognised honesty and ability.

Brouwer's mildly cynical view of his learned colleagues, as described in the above letter, sometimes manifested itself when his natural inclination for mocking ran out of control under the influence of blatant pompousness, often combined with insignificance. One such instance was recorded by an eyewitness: at a conference, during the lecture of some pretentious mathematician, Brouwer had taken the precaution to stand at the back of the hall, so that he could unobstrusively leave if things got unbearable. When, however, the speaker ventilated a particular shocking enormity, Brouwer could no longer control himself and managed to fall flat on his back, as if struck by thunder.[59]

If anything, it was his honesty and his high standards that, combined with his emotional character, lent him a reputation of 'difficult'. We have already seen his conflicts with highly-respected members of the mathematical community; many more are to follow.

In Dutch mathematical circles Brouwer was by now fully recognised. He had presented two talks in 1908 and 1910 at meetings of the Dutch Mathematical Society,[60] and in 1911 he was the speaker at the November meeting with his talk *The theorem of Jordan for n dimensions*.[61] If his position in Dutch mathematics can be measured by the role he played in the professional organisation, the *Wiskundig Genootschap*, we note that he was appointed as a member of the Committee for prize

[58] A surprisingly modest position for a man who has just solved one of the famous problems of Cantor!

[59] Oral communication E. Hölder.

[60] In the October meeting of 1908 he gave a talk *On plane curves and plane domains* (see p. 139), and in the October meeting of 1910 he gave another talk, this time *The invariance of the number of dimensions*.

[61] 25 November 1911, published in Brouwer (1911e).

essays, together with Ornstein, the theoretical physicist from Utrecht, at the meeting of 26 April 1913. Shortly afterwards, at the April meeting of 1914, Brouwer was elected President of the Mathematical Society for a two-year term.

Korteweg, in spite of the earlier rebuffs, refused to accept the verdict of the city council of Amsterdam as final. He set out to organise another campaign for Brouwer's appointment, this time he not only argued the practical need for more teaching support, but also the exceptional qualities of the candidate. As in the case of Academy campaign, he collected some weighty references. This time he solicited recommendations from Hilbert, Poincaré, Emil Borel and Felix Klein. Borel answered somewhat guardedly, 'The papers that he has published so far are interesting and sometimes deep, they allow one to hope that he will one day arrive at important discoveries. . . .'[62] Borel's judgement is certainly surprising: Brouwer had already changed topology beyond recognition—what else could one want? Perhaps Borel was aware of Brouwer's foundational ideas (after all, Borel was at the Rome conference, and he could have attended Brouwer's lecture, or read the paper) and expected spectacular contributions. Borel also wanted to know if the position that was envisaged for Brouwer was a newly-created one, or if there was a competition involved; the wording of the recommendation, he said, would depend on the nature of the position. Korteweg patiently explained that he was thinking of a new position. The final recommendations have apparently not been preserved.

In the meantime the lecturer Van Laar,[63] a contemporary of Brouwer, had resigned and Korteweg saw an opportunity to propose Brouwer as his successor. In the faculty meeting of 24 April, Korteweg vigorously defended the view of the mathematicians that a replacement for Van Laar was necessary, the teaching for chemistry and biology students required a separate course and an extra teacher. He proposed Brouwer 'who in a few years time had been able to become one of the leading mathematicians of our time'. The faculty, no doubt conveniently impressed by the letters of recommendation of Hilbert, Klein, Poincaré and Borel, unanimously supported Brouwer for a position of extraordinary professor. The switch from 'lecturer' to 'extraordinary professor' rather suited Brouwer; he was not looking forward to the treadmill of teaching mathematics for applications.

Not only had Korteweg to embark on a long march along the various dignitaries of the City Council; he also had to convince the faculty that mathematics required and deserved reinforcement. The operation 'a chair for Brouwer' was carried out parallel to a new initiative for Brouwer's membership of the Academy.

This time Brouwer' candidacy was proposed at the meeting of the physics section of the Academy on 30 March 1912, by Korteweg, Schoute, W. Kapteyn, Jan de Vries, Kluyver, Cardinaal en Hendrik de Vries. At the special meeting of 26 April

[62] Borel to Korteweg, 2 January 1912.

[63] The faculty was divided on Van Laar. Some found him useful, but others considered him to suffer from delusions of grandeur, for example, he insisted that an honorary doctorate was due to him. There was no support to keep him on. In fact, he was a man with a fine reputation in thermodynamics. He indeed got an honorary degree in Groningen in 1914. For more information see van Emmerik (1991).

1912, Brouwer was this time elected in the first round. He again got the largest number of votes, this time 37, hence he was elected straight away. The meeting went through 7 rounds of voting before the last member passed the prescribed threshold of 24 votes.

The Minister of Internal Affairs informed the Academy on 15 May that Her Majesty the Queen had approved the appointments of the gentlemen de Sitter, L.E.J. Brouwer, Boeke, J.C. Schoute, van Hemert en Wertheim Salomonson.

In early 1912 things started to move in the University; the curators, in particular their president, the Mayor of Amsterdam, began to see the request of the mathematicians with more sympathetic eyes. Korteweg and De Vries had visited the Mayor,[64] and convinced him that Brouwer was 'extremely competent' and that the connection between the University of Amsterdam and Brouwer should be given a firmer basis. In May the faculty was asked to submit a formal request for Brouwer's extraordinary chair. From then on everything went smoothly. In July 1912 the appointment was made official. Of course the decision had been favourably influenced by the consideration that the final proposal only mentioned an extraordinary chair, with a yearly salary somewhere between 2000 and 2500 Dutch guilders, which—as Korteweg pointed out—was less than the salary of a high school teacher (HBS).

After some more deliberations, Brouwer was indeed appointed. He expressed his joy in a letter to Klein,[65] thanking him for writing a supporting testimonial.

At the time of his appointment Brouwer had already obtained more fame and recognition in mathematics then all his Dutch colleagues together. Nonetheless the title of 'professor' acted as a strong booster of his self-confidence.

On 14 October 1912 he gave his inaugural lecture[66] in the auditorium of the University. This time the topic was the foundations of mathematics. The title *Intuitionism and Formalism*[67] is significant for Brouwer's views on the topic; according to him the prime contenders for the foundations of mathematics were the formalists and the intuitionists—no mention of logicism, in particular Russell, is made. The lecture is interesting as a review of the situation in the foundations, but apart from some refinements of certain views, it is basically a reformulation of the dissertation.[68] It owed its influence to an English translation, published by Arnold Dresden in the *Bulletin of the American Mathematical Society*. This lecture made the name *intuitionism* a household word for a particular mathematical–philosophical tradition that goes back to (at least) Kant. In his inaugural address Brouwer introduces intuitionism as a largely French philosophy of mathematics, which suffered from the defects of Kantian principles. His own brand of intuitionism is introduced with the words:

---

[64]Cf. Korteweg to the Mayor, 24 March 1912. The letter was accompanied by the recommendations of Hilbert, Klein, Poincaré and Borel.

[65]Brouwer to Klein, 21 June 1912.

[66]Published in 1912, reprinted in Brouwer (1919b) and translated in Brouwer (1913b).

[67]The terms 'formalist', 'intuitionist' were already in 1893 introduced by Felix Klein in his 'Evanstone lectures', Klein (1893). Klein's meaning does however not coincide with Brouwer's.

[68]Cf. van Dalen (1999b).

**Fig. 6.4** Brouwer at the occasion of his inaugural address in the main building of the university at the *Oudemanhuispoort*. [Brouwer archive]

> However weak the position of intuitionism seemed to be after this period of mathematical development, it has recovered by abandoning Kant's a priority of space but adhering the more resolutely to the a priority of time. This neo-intuitionism considers...[69]

In fact he wisely stuck to the name 'intuitionism' for his views, which from now on represented mainstream constructive mathematics. The term 'neo-intuitionism' had in fact been introduced earlier in a review of a book of Mannoury.[70]

The lecture can be seen as the apotheosis of Brouwer's first foundational program, but it also shows glimpses of the new intuitionism. There is relatively little new material; it consisted mainly of a survey of the various foundational views and of an exposition of the intuitionistic criticism. The discerning reader will however note some tension in the text. Brouwer appears to be moving from the first stage of his intuitionism to the more mature second stage. At some places he is still using the terminology of his first program, whereas at other places he has already moved

---

[69]Here the urintuition is spelled out again.

[70]Brouwer (1910h).

on to a more liberal and richer intuitionism. We shall come back to the inaugural address on p. 233.

One interesting point worth mentioning is that Brouwer briefly compared the real numbers of the intuitionist and the formalist; the first, he said, recognise only decimal expansions given by laws, whereas the formalists allow expansions 'determined by elementary series of freely selected digits'.[71] Although one might conclude that this was a wholesale condemnation of arbitrary sequences (and the admissibility of 'choice') one should keep in mind that Brouwer is talking about the continuum. Here an undetermined sequence would not yield a well-determined, individualised real number. This would not rule out arbitrary sequences as legitimate objects. We will see that he later reconsidered this view, cf. p. 236. Already in the 1914 review of Schoenflies' revised *Bericht*, he accepted arbitrary choices as legitimate objects for intuitionists.[72]

Considering the number of projects (with accompanying conflicts) Brouwer was involved in, one is inclined to think that job hunting would have been the least of his problems. But one should keep in mind that his mathematical activities, including an almost incessant travelling, heavily burdened his financial reserves: furthermore the annuity of Brouwer's mother-in-law (the substantial sum of 1100 guilders) fell heavily on the budget of the Brouwer couple. It may be remarked in passing that Brouwer made himself a perfect nuisance in the matter of the annual payments. He was often late and had to be reminded of his obligations by the family of Mrs. de Holl. The pharmacy was far from a gold mine, and the fee of a *privaat docent* was not sufficient for the lifestyle of a leading, travelling and letter-writing mathematician; the extra income of a professorship, was most welcome. In fact, his brother Aldert, who worked as a geologist in the Dutch Indies, financially supported Bertus. To put not too fine a point on it, Bertus was slowly accumulating a debt!

One should also not underrate the psychological effects of the lack of recognition at home: his contemporaries (who were not in the same mathematical league!) were getting chairs, and he was left out. There are no specific complaints about persons, but there is no doubt that Brouwer thought himself passed over by the establishment. This was partly due to his personal preference for the geographical location. Not only was he strongly attached to Amsterdam and to Het Gooi as a perfect environment, but he was also subjected to a boundary condition that nowadays is generally recognised, but that was rather uncommon in his days: his wife had a job. Not only just a job, but one that could not so easily be transferred to another town—one has to keep in mind that the pharmacy had a tradition, and that a pharmacy, say in Groningen, would not have the same emotional value.

So when Korteweg in April 1913 warmly recommended Brouwer for a mathematics chair in Groningen, which had became vacant upon the sudden death of P.H. Schoute, Brouwer was faced with a difficult decision.

Korteweg had reached years ago the conclusion that his student should, if at all possible, be kept in Amsterdam; moreover, he was ready to act on his own if the

[71] Brouwer (1913b), p. 92.

[72] Brouwer (1914), p. 79.

authorities did not share that view. Since the negotiations with the curators did not yield palpable results, he generously proposed that he and Brouwer should switch chairs. At the time Korteweg was 65 years old,[73] and he had found the recognition that was due to his solid mathematical production. He felt that at this point it was of tantamount importance to put Brouwer's career beyond risk. The thought may seem strange from our present viewpoint, sooner or later a university, be it in Holland or abroad, would make him a fair offer. But one has to keep in mind that there was always the possibility that Brouwer could bid farewell to mathematics and academia at any time; the mystic in him was still powerful enough to make him give up the worldly affairs of science and mathematics.

Korteweg presented his plans to the curators of the university, who invited him and De Vries to their meeting of 28 June for further explanations. The curators expressed their concern for Korteweg's pension rights, and when that was settled to everyone's satisfaction, they wondered if the faculty would after all, even with Brouwer and De Vries as full professors and Korteweg as an extraordinary professor, require an extra mathematics professor. Korteweg confessed that he had not even consulted the faculty on his present proposal, but that the faculty would doubtlessly react quickly to a request of the curators for support. He made it clear that under the present circumstances his proposal was motivated by the offer of a Groningen chair to Brouwer. The curators were apparently convinced of the importance of keeping Brouwer in Amsterdam, and so they sent the proposal with their official backing to the city council.

Lize, in a letter to Brouwer's brother Aldert and his wife Lily in the Dutch Indies, soberly summed up the situation:[74]

> Last week Friday Professor Schoute in Groningen died. Bertus can be appointed there as an ordinarius;[75] of course we cannot accept it, but it may lead to a rise in salary or an appointment to ordinarius.

The choice between Amsterdam and Groningen worried the Brouwers a good deal; the weighing of all factors kept them busy. Lize had made up her mind.

> I am not in favour [of Groningen]. It would cause great changes, Blaricum would no longer be useful and everything should be sold. The pharmacy would also be over. Groningen is in the north, but Bertus would again wish to live in the country in Anlo or Gieten. The main advantage would be that we should at last have a regular household. This way of life is not easy. Bertus finds the travelling back and forth rather a welcome change. Yet I have an idea that a regular household will not suit Bertus. *Enfin*, for the time being we can only wait.

Brouwer was thus confronted with a difficult dilemma: go to Groningen and get a substantial raise, or stay in Amsterdam. While conducting high school examinations

[73]The retiring age was 70.

[74]Lize to Aldert and Lily, 4 May 1913.

[75]i.e. a full professor.

in The Hague (recall that university professors spent part of their summer vacation touring the country, conducting oral examinations at gymnasiums), he asked Hilbert for advice.[76] Should he accept the offer from Groningen, which left him totally free to teach and conduct his research, but

> where I will find in the petty provincial town fewer sympathetic colleagues than in *Amsterdam*, i.e. in a lively big city, which has always been intimately connected with my life, where I have my cosy home in Blaricum, and where the dunes are close by.

Amsterdam had in the meantime adopted Korteweg's scenario: Brouwer was offered Korteweg's chair. With the chair went, however, the obligation to teach 'mechanics'; Brouwer feared that his research would suffer under the demands of this particular piece of applied mathematics. From the letter to Hilbert, one gets the impression that Brouwer had already made up his mind, he was horrified of an exile in (what was usually considered) the northern wilderness—'The small-town contacts with the conventional pressure must be terrible in Groningen, and there is no nature around at all', he wrote to Hilbert. But above all, his heart was in Amsterdam and Het Gooi.

A month later the matter was determined by 'force'; the University of Groningen issued an ultimatum, and Brouwer opted for Amsterdam. Korteweg had deftly manoeuvred Brouwer's chair through the faculty; on 26 June he told the faculty that the University of Groningen demanded Brouwer's decision before next Saturday. The board of the University did not object to the arrangement, so he told. And thus, the faculty unanimously accepted the proposal. Now Brouwer's toil of years of research was rewarded by a full professorship at the University of Amsterdam, a young university, but one which had already gathered a considerable amount of prestige, and which had added now a precious pearl to its crown. Brouwer thus became a full professor with all the prestige it carried in those days; in addition it brought him a salary of 4000 guilders per annum (+ 500 guilders lecturing fees).

While the Universities of Groningen and Amsterdam were considering to offer positions to Brouwer, there was another university interested in the same young scholar. The University of Göttingen was looking for a successor to the eminent mathematician Felix Klein, who was about to retire. The mathematics department wished to fill the vacancy with a mathematician with a geometric approach to mathematics, and not just in the narrow sense, but in intimate connection with all areas of mathematics. The faculty listed three candidates, who had mastered the present prominent problems, which emerged from the connection of set theory and function theory: 'What are the most general ideas one has to form about the continuum and the structures which are contained in it?' The three candidates were Carathéodory, followed at a distance by Hermann Weyl and Egbertus Brouwer (ex aequo). The choice of the faculty clearly shows that Hilbert, as the most influential man, was sufficiently impressed with Brouwer, to think him good enough for Göttingen. The

[76]Brouwer to Hilbert, 16 June 1913.

available material strongly suggests that Brouwer was not aware of the Göttingen proposal (else Korteweg would have used it as another strong argument to convince his curators).

One should not get the impression that Brouwer was after the comfortable life of a research professor; he certainly had his problems combining a flourishing scientific practice with the aspects of teaching, but on the whole he, then as well as later, took his teaching duties quite seriously. Apart from the routine calculus courses for the science faculty, he gave courses on topics closer to his research interests. In 1913 he taught a course on the theory of functions, and he introduced a novelty—undoubtedly inspired by the German example: namely a seminar in pure mathematics for the older students. The seminar consisted of talks by students on recent advances in mathematics, and it was obligatory for the *doctoraal* exams.

Furthermore, Brouwer was—against his wishes—sucked into the committee-circuit; he had to attend faculty meetings, official exams, academy meetings, etc. On top of that he was rather occupied with writing reviews for the *Revues Semestrielles*, a Dutch review journal. Brouwer and his colleague Hendrik de Vries drew some comfort from a certain degree of innocuous tomfoolery that they practised at exams and meetings. De Vries excelled in providing comments under his breath that heavily taxed Brouwer's facial muscles.

Brouwer's teaching consisted at that time of a number of topics. A considerable part of his time was taken up by the standard course Differential- and Integral Calculus for chemistry students. As Lize put it,

> ...it has to be taught fairly superficially, this is very dull for Bertus. He requires the students to make exercises, and is of course very cross if this is done poorly. Last week he had a young lady, who in this respect gave little reason for being content; he made her work at the blackboard for half an hour, so that she almost fainted. As a contrast he called somebody to the blackboard, whom he knew would solve the problem in 5 minutes.[77]

The sudden elevation from beginner to expert, and from *privaat docent* to professor, seemed to have taken Brouwer by surprise. As so often happens, the newcomer was loaded with tasks that soon accumulated to more than he could handle. A number of these tasks were undertaken in the expectation of a better position, but some were indeed almost forced upon him. The signs of the pressure on Brouwer soon began to tell; on New Year's Eve he more or less desperately wrote to Korteweg, reporting on the prize essays of the *Wiskundig Genootschap* (Dutch Mathematical Society), that

> With the continuous pressure, under which I live as a consequence of all kinds of obligations with respect to foreigners, for the fulfilment of which time is lacking me, and as a consequence of the standstill of my mathematical researches (to which in my opinion, the task of a professor and member of the

[77]Lize to Aldert and Lily, 4 May 1913.

Academy forbids me to resign), I am forced to refrain from all unnecessary work.[78]

The situation, however, did not improve, and Brouwer kept complaining. And so it is hardly surprising that Korteweg deemed it necessary to set his gifted student right.[79]

Amice, De Vries already told me how much Göttingen[80] occupies your time and I quite understand that at this moment you do not want to take on a lecture.[81] Anyway, my request was only a consequence of my endeavour to raise the level of these talks as much as possible, and I half and half expected that you would excuse yourself this time.

Less did I expect your outburst. That your professorship suits you so little distresses me very much. I view this, however, as a subjective phenomenon, indeed connected with your great talents, like everything in a particular human being is more or less connected, but I don't consider it inseparable from such talents.

In my opinion our physicists, who are members of international Academies, and for a prolonged period had no fewer official encumbrances than you (Van der Waals, Lorentz, who took over [Kamerlingh-] Onnes' course for the medical students) prove that. It is thus difficult to accept that six courses a week, partly elementary in nature, a few examinations a month (with an ample four months of nearly undisturbed vacation (sic!)) would prevent anybody from expecting scientific work, also of the highest level.

If this is the case with you, then there is for you really nothing but to accept a German chair, and that occasion will not fail to come up, although I expect that also there, 'hampering' influences will arise, if you are so sensitive to them. It is another question if you could not, if you come, with me, to the conclusion that the difficulty is to be found *in yourself*, do something or other to lessen the conflict.

For example, prepare your courses in the vacation, so that you are always well ahead, and that you have to prepare yourself each time only briefly. That would take away much of the nervousness and agitation that otherwise is inherent to the teaching of new material for the first time.

The reaction of Brouwer to this fatherly advice is not known, but a month later Korteweg wrote another letter in the same vain. The direct cause was Brouwer's wish to withdraw from the prize-essay committee of the Mathematical Society. His success in the world of mathematics had its consequences: the demands from academic and mathematical organisations started to absorb all his time and energy. And

[78]Brouwer to Korteweg, 31 December 1913.

[79]Korteweg to Brouwer, 4 June 1914.

[80]Brouwer's research contacts were mainly with Göttingen; moreover, he was involved in refereeing for the *Annalen* even before he became a member of the editorial board.

[81]At a meeting of the Math. Soc.

in an act of desperation he had blamed the stagnation of his mathematical research on the professorship. Korteweg put things into a more realistic perspective, when he wrote[82]

> Whereas under different circumstances I would be most pleased with your honourable appointment to editor of the first among the mathematical journals of the world,[83] this is now not whole-heartedly, and for more than one reason.
>
> *In the first place* I view the work with which you are swamped by the Göttingen people as a very serious and *lasting* hindrance for the continuation of your own work, and yet you will be in the long run, also in Germany, be judged *by that.*
>
> *In the second place* I foresee that you will absent yourself more and more from Dutch mathematical life, whereas one expects, in my opinion justly so, the opposite attitude of a Dutch professor [...]
>
> *In the third place* I fear that you will look for the cause of your diminished fertility there where it is not, or only for a small part: in your professorship, and this will thus appear more and more as a pure nuisance. [...]
>
> I must concede one thing, in order not to become unfair with respect to you. Namely this, that elementary courses seem to present you with great difficulties because they arouse your impatience, and seem to make you temporarily unfit for other work.

Korteweg's diagnosis was perfectly to the point, and, in view of later developments, prophetic. Indeed, the lure from Göttingen was an unparalleled temptation,—to be elected to the mathematical elite! The price was dear: Brouwer spent a great deal of time refereeing papers for the *Annalen.* And, necessary and honourable as it might be, it was not sensible to exchange research for supervision.

Korteweg, who had given up his chair for his student, had good reasons to be bitter, but his main worry seemed to be the fate of this student. He pointed out that, contrary to Brouwer's complaints, the Netherlands and its universities were on the whole a congenial environment for the young scientist, and that Brouwer had indeed profited from the liberal possibilities.

The reaction of Brouwer has not been preserved, but we may assume that he accepted the consequences of his position; he did not give up his chair, nor did he severe his fresh links with the *Annalen.* The outbreak of the war solved the problem of the claims of Göttingen by brute force: the demands of the *Annalen* were temporarily reduced to more moderate proportions.

## 6.7 Schoenflies Again

So far Brouwer had led the life, not only of a fundamental researcher, but also of a free-lance philosopher and a moralist. Lately another burden, that had nothing to do

[82] Korteweg to Brouwer 13 July 1914.

[83] Brouwer was appointed editor of the *Mathematische Annalen*, cf. Brouwer to Klein 10 July 1914.

with his administrative or teaching duties, had been added to already overcrowded schedule: the co-operation with Schoenflies.

The comprehensive survey of Schoenflies, commissioned by the German Mathematical Society, had presented a state of the art in set theory and topology. The two volumes appeared in 1900 and 1908. The volume on topology had attracted Brouwer's attention; he had studied it closely and used it as the background material for his own research. As we have seen, the monograph was not proof against Brouwer's scrutiny, which had revealed a number of serious lapses. The 'Analysis Situs'-paper had resulted from this study.

In view of the importance of the topic, and also of the so-recently discovered mistakes in the topological part of the *Bericht*, Schoenflies started to prepare a new edition.

Brouwer, as the recognised expert on topology, had been receiving from various quarters the request to write a book on set theory, since the existing books and encyclopaedia contributions were too superficial and undependable. When Brouwer was visiting Göttingen in the summer of 1911, he was confronted with new attempts to get him to write such a book; he suggested that a satisfactory solution could be reached with little loss of time 'if he were given the opportunity to supervise the book of Schoenflies during the preparation, and when necessary to improve and supplement it'.[84]

At the time Brouwer was fully engaged in research, and the writing of a book would seriously interrupt it. When that same summer Brouwer visited Fricke in the Harz, the latter offered to mediate between Brouwer and Schoenflies. This, at first sight, fortunate solution proved disastrous; Schoenflies wanted to restrict Brouwer's role to the correction of mistakes, and Brouwer, naturally, wanted to improve the book in depth and to update it.

The ensuing bickering upset both parties, but Brouwer suffered most; his quick and emotional mind could not cope with the slow pace of the older man, who fought a war of entrenchment against Brouwer's innovations, and who often failed to grasp Brouwer's corrections and improvements. He would, for example, ask Brouwer to prepare part of a chapter, only to tell him later that the text should be incorporated in another chapter; Brouwer reacted,

> That is not how Schoenflies can make use of my time, which has its value, just as much as his time. I am neither his assistant, nor I am helping him out of personal friendship, but because I find it important that a good book on set theory should be produced, and because I am at the moment the most competent person in this area.[85]

Again and again, Schoenflies would try to escape the iron control of Brouwer. He apparently did not come up to the standard of exactness that was set by Brouwer in topology, and by Zermelo in set theory. Even Hilbert's influence was not sufficient to

[84]Brouwer to Hilbert, 16 April 1913.

[85]Ibid.

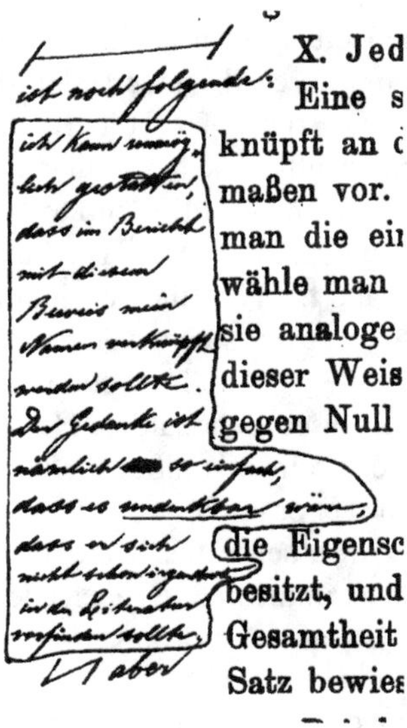


**Fig. 6.5** A correction of Brouwer in the proofs of Schoenflies' *Bericht*. [Brouwer archive]

keep Schoenflies on the right track. He regularly abbreviated, reformulated, or even cancelled Brouwer's corrections, 'He starts to abridge industriously much of what, after an endless correspondence, finally was formulated correctly. He must be *very* overworked, for he makes mistakes, for which any student should be ashamed.'[86]

Schoenflies kept rehashing Brouwer's contributions, and relapsing into his old mistakes. In reaction, Brouwer would repeatedly appeal in desperation to Hilbert, begging him to bring Schoenflies to reason. Schoenflies, also turned to Hilbert with the urgent plea to rescue him from the hands of this merciless Dutchman, just as he done at the occasion of the 'Analysis Situs'-paper. Unfortunately the correspondence concerning the revised version of Schoenflies' book is deplorably incomplete, but enough is left to understand the mutual irritation of both men.

Schoenflies enjoyed a reputation as an excellent expositor, and he understandably reserved the right of the final wording of the manuscript. Brouwer's censorship was far from complimentary for a man of his age and status; in fact he repeatedly tried to rebel. Once he sent in despair Hilbert a proof sheet, with a sentence underlined in red, that, according to Brouwer, contained a logical mistake, challenging Hilbert to find it—'If you find it, I will gladly give you a present.'[87] More than four letters were exchanged between Brouwer and Schoenflies to discuss this particular sentence!

The whole affair was taxing Brouwer's patience heavily, and some of the letters are therefore rather harshly formulated; at one occasion Schoenflies had, for example, inserted a statement to the effect that a certain theorem could be proved using Brouwer's methods, whereas Brouwer had given a counterexample to the theorem. Utterly astounded Brouwer demanded to know if he was to be quoted as the in-

[86]Brouwer to Hilbert, 16 June 1913.

[87]Schoenflies to Hilbert, 8 July 1913.

tellectual father of the proof of a falsity. Given Brouwer's emotional sensitiveness, one can easily understand Mrs. Brouwer, when she added a post script to a letter from Brouwer to Hilbert; 'If my husband does not become insane on account of Schoenflies, he has to thank you for it.'[88]

The new edition of the *Entwickelung der Mengenlehre und ihrer Anwendungen* appeared in 1913.[89] The book must have been a sad disappointment to the experts in the field; all the developments of the last few years were ignored. Almost none of Brouwer's achievements in the 'new topology', or even in Cantor–Schoenflies topology, were incorporated. It was as if the shock of the Analysis Situs paper had paralysed Schoenflies. He left out all the material concerning curves, dimension, etc. Brouwer's influence apparently did not go far enough to add material, only to correct or to omit.

In a sense the book was already obsolete by the time it appeared. It was almost immediately superseded by Hausdorff's *Grundzüge der Mengenlehre* (1914). Also, as far as Brouwer's wishes were concerned, the book was a disappointment—no account of his recent advances in topology was given, and the demand for a coherent exposition remained. His stratagem, to incorporate the newer material intoSchoenflies' book had utterly failed. Although Brouwer is mentioned here and there, there are hardly any recognisable traces of his mastery of set theory and topology.

In the introduction Schoenflies profusely thanked Brouwer for his 'unselfish and abundant' support, pointing out that not only he, but 'the collective mathematical world' was in debt to Brouwer for the exact and useful form of the book.

Brouwer had lost his last struggle with Schoenflies, when he tried to force Schoenflies to insert a statement to the effect that the co-operation of Brouwer did not mean that he had given up his constructive convictions in favour of the set-theoretic ones; Schoenflies flatly refused to do so. It seems plausible that the suppression of Brouwer's intensive involvement in the preparation of the volume was the result of a compromise.

Some of the objections of Brouwer can, however, be found in his review of 1914, in which he not only objects to some parts that become meaningless when seen from an intuitionistic viewpoint, but also pin-points some weak spots in the 'classical' treatment.

With this item the first act of Brouwer's mathematical career ends; a number of topics will reappear in the sequel.

There is no doubt that the co-operation with Schoenflies heavily taxed Brouwer's resources; the material in itself was not a serious obstacle, but the endless bickering, pointing out mistakes, giving new proofs, only to find that Schoenflies did not

[88] '*Wenn mein Mann durch Schoenflies nicht verrückt wird, verdankt er es Ihnen. Ihre Lize Br.* Mrs. Brouwer to Hilbert, 11 September 1913.

[89] Although the title page says 'published jointly with Hans Hahn', Schoenflies is the only author. Hahn acted as a critic and adviser (albeit not as prominently as Brouwer), he was to write a second volume on real functions which never appeared. In 1921 Hahn published his own monograph *Theorie der reelle Funktionen I*, Hahn (1921).

grasp the point and preferred his own defective arguments—it all wore out Brouwer. A more equanimous man would have borne his cross quietly, but Brouwer—with his fierce convictions of right and wrong—was easily tempted to see insults, and hence to repay in force. Conflicts in general affected him strongly, to the extent of physical afflictions. So it is not unthinkable that Schoenflies' hard headedness had worn out Brouwer's resilience to the extent of the drying up of the fountain of topological productivity. The complaints of Lize, who knew her husband well enough, are telling. She complained in the above-quoted letter to Aldert and Lily, that the Schoenflies project had cost Bertus the priority for Poincaré's last theorem.

The ultimate recognition, which generously redressed all the hardships of the past years, had come in July 1914, when he was co-opted by the *Mathematische Annalen* as an editor. In a letter of 10 July to Felix Klein, Brouwer expressed his appreciation for this 'high honour'. The appointment was indeed the crown on his scientific work; the journal carried a prestige that surpassed that of scientific societies and academies. Under Klein's regime the *Annalen* had become the leading journal in mathematics, and the editorial board was more than just a collection of editors: in Carathéodory's words it had been moulded into

> a kind of Academy, .... That was in my opinion the main reason why the *Annalen* could claim to be the first mathematics journal in the world.[90]

As we have learned from Korteweg's letters, Brouwer was already working for the *Annalen* before his elevation to the rank of editor. Brouwer told Schoenflies in 1921 that he was an informal editor (*Mitarbeiter*) from 1911 to 1914.[91]

At the age of 33 Brouwer had now reached the highest position in the republic of mathematics, ten years after his first papers, and in spite of his attacks on the prevailing views on the foundation of mathematics.

[90] Carathéodory to Courant, 19 December 1928.

[91] Brouwer to Schoenflies, 17 January 1921.

# Chapter 7
# The War Years

The Great War that brought tragedy and hardship to the warring nations, passed by the borders of the Netherlands, a country that took its neutrality for granted. The Dutch mobilised on the first of August 1914, all political parties supported the government in its energetic policy of neutrality. The average citizen had only vague ideas concerning the reality of war, after all, since Napoleon's defeat, the nation had only known the short skirmish of the Belgian uprising, and the far and exotic military exploits in the colonies. The European conflict was generally considered to be 'none of our business', and sympathy for the Entente and the Central Powers was fairly evenly distributed.

Brouwer, who had visited in July cathedrals in France with his stepdaughter Louise, had returned home before the outbreak of hostilities, firmly resolved to stay at home. But soon his curiosity got the better of him, and in early August he travelled to the war scene, where he reportedly saw quite a bit during a stay of a week.[1]

The immediate consequence of the war for Brouwer was a disruption of his international contacts. In the last few years he had come to consider Göttingen as his second scientific home; ties of admiration and friendship bound him to Hilbert, Klein, Blumenthal and others. And, although it was not impossible for a subject of a neutral nation to travel in the territory of the belligerent nations, it was far from simple to obtain the required visa and permits. So, in effect, his contacts were limited to correspondence.

Whether the isolation was the main cause, or perhaps the lure of the foundations, the war years saw a return to the problems that were raised in his dissertation, and that had been left unsettled. War or no war, the University went about its business, and Brouwer had to carry his teaching load like his fellow professors. Apart from the standard courses (which rather bored him—but which he always took seriously), he taught a number of courses on subjects of his own choice. In 1912, whilst still a *privaat docent*, he had started a course on algebraic functions, and after his appointment as an extraordinary professor he introduced courses on Theory of Functions

[1] Brouwer to Scheltema, 15 August 1914.

D. van Dalen, *L.E.J. Brouwer – Topologist, Intuitionist, Philosopher*,
DOI 10.1007/978-1-4471-4616-2_7, © Springer-Verlag London 2013

(*Functiereekening*), which were actually courses on point-set theory. The University catalogue announced the 1912 course as *Functiereekening*, but Brouwer listed the course in his own notes as *Point sets*. In addition he gave in his first professorial year another course: *General Set Theory*.

From 1913 onwards he taught also the courses on Mechanics, inherited from Korteweg. Through the years he offered courses in *Higher Mechanics* (in two parts, I and II) and the *Theory of Oscillations*. Furthermore he regularly taught courses on geometry, described in the catalogue as follows:

> *Projective Geometry I* (1913–14)—Projective coordinates, Projective and reciprocal transformations, invariants, Conics and sheaves of conics, higher algebraic curves, multiplicity of intersections, connection between equations in point and line coordinates, multiplicity of main points of a tangent, Plücker relations, $n$-th degree dependence of point systems, genus of algebraic curves, theory of unicursal curves.
> *Projective Geometry II* (1914–15)—Continuous groups, non-Euclidean Geometry, Foundations of Geometry.
> *Set Theory* (1915–16)—Finite and infinite cardinal and ordinal numbers, the fundamental operations with well-ordered ordinals of the first, second and third domain, general theory of well-ordered sets, intuitionistic generation of point sets, solution and internal decomposition of point sets.

In particular the kind of set theory that had been initiated by people like Baire and Emile Borel (what we now would call the theory of real functions, or descriptive set theory) was close to Brouwer's personal interests. This course was at first intuitionistic in the spirit of the dissertation, only after 1915–16, when Brouwer knew how to handle his new intuitionism, based on choice sequences, it became fully intuitionistic in the modern sense.

There is not much material that helps us to form an idea of the courses during the first years of Professor Brouwer; the first notable piece of evidence is a set of handwritten course notes of his 1912–13 lectures on set theory. It sheds light on the evolution that Brouwer's foundational views underwent in the years of introspection and search.

## 7.1 Sets and Sequences—Law or Choice?

In order to appreciate the solitary quest of Brouwer for the truly constructive mathematics one has to go back a little in history and view the early approaches to constructiveness as advocated mainly by Borel.

Virtually all mathematicians will recognise the numbers $1, 2, 3, 4, \ldots$ as effectively given. The German mathematician Leopold Kronecker even went so far as to say that 'the natural numbers are made by God, and everything else is the work of man', which was the poetic expression of his belief that the only legitimate mathematical objects (on the side of number systems, geometry was not included) were those that could be reduced to natural numbers. We know lots of those objects: integers (as pairs of natural numbers, for example, $(5, 2)$ stands for 3 and $(2, 5)$ for $-3$), rational numbers (as pairs $(n, m)$ with $m \neq 0$, for example, $(5, 2)$ stands for $\frac{5}{2}$).

Irrational numbers can also be given by means of natural numbers, for example, if they are roots of polynomial equations, but in general they do not have such simple representation. Therefore, Kronecker banned the irrationals that were not decently representable from mathematics, a feat that did not contribute to his popularity among his contemporaries. Fortunately he was one of the giants of algebra and number theory, so his authority did not depend on his foundational views. Kronecker was an outspoken man, and he did not hesitate to attack well-respected and well-established colleagues. His attacks on Cantor and Weierstrass were notorious. He was the hornet of the end of the nineteenth century, and he did not hesitate to call the work of certain contemporaries totally devoid of meaning. In the eyes of most mathematicians Kronecker was an eccentric, albeit brilliant, man, but his rejection of that what was generally accepted as progress gave him the reputation of a reactionary and kill-joy. Kronecker stressed over and over that only mathematical objects that could be admitted were those constructible in finitely many steps. It was not unusual to contrast in Kronecker's days the 'algebraic' and 'logical' methods, the latter were abstract, such as Cantor or Dedekind used, the first were finite and effective.

After Kronecker, the French mathematicians, led by Poincaré and Borel, started to cultivate certain constructive arguments. Poincaré, as we have seen, was close to Brouwer in certain respects. The French constructivists stressed in particular the 'definability in finitely many words'. Their view, which was generally shared by mathematicians of a constructive leaning, was that 'infinite objects', i.e. objects that could not be built in finitely many steps from natural numbers, should be presented by a finite definition, so that there was a finite guarantee (building instruction) for their existence.[2] Remarks to that effect can be found in the literature. Otto Hölder, for example, in his book *The mathematical Method* explicitly demanded, that if one wants to give a set with infinitely many points, one has to prescribe a law.[3] Thus, for instance, the irrational number $\sqrt{2}$ is effectively determined because there is a specific procedure to calculate all decimals. A real number whose decimals were successively determined by flipping a coin (e.g. 0.0011101010111110...) was thus not be admitted.

Brouwer seemed to share this view still in 1912, when he gave his inaugural lecture 'Intuitionism and Formalism':

> Let us consider the concept: 'real number between 0 and 1'. For the formalist this concept is equivalent to 'elementary series of digits after the decimal point'; for the intuitionist it means 'law for the construction of an elementary

[2]An analysis of the ideas of the French constructivists can be found in the excellent and instructive survey (Bockstaele 1949).

[3]*Die Mathematische Methode*, Hölder (1924), p. 98. He refers to an earlier statement of the same tenor in 1892.

> series of digits after the decimal point, built up by means of a finite number of operations'. And when the formalist creates the 'set of all real numbers between 0 and 1', these words are without meaning for the intuitionist, even whether one thinks of the real numbers of the formalist, determined by elementary series of freely selected digits, or of the real numbers of the intuitionist, determined by finite laws of construction.[4]

Before we conclude from this quotation that Brouwer rejected sequences of 'freely selected' numbers, let us recall that the above refers to the individual points of the continuum and not to the so-called unknown points. These points must be given by a law in order to turn them into distinguishable, individual points; a freely selected decimal expansion would (viewed in that stage of Brouwer's programme) not yield a specific well-determined point. Indeed, two pages later Brouwer asserts that the intuitionist may admit, on the basis of the intuition of the linear continuum *elementary series of free selections* as elements of construction.

The real numbers of the inaugural lecture are given by decimal expansions. Such an approach is more restrictive than the treatment in his dissertation. It is not implausible that Brouwer chose the 'decimal' approach on didactic grounds. After all, the lecture was for an audience of non-specialists. It took Brouwer until 1921 before he publicly renounced the adequacy of decimal expansions for representing the full continuum.

It should be pointed out that the terminology of *Intuitionism and Formalism* is that of 1912, that is to say, before Brouwer had started his own modern brand of constructivism. So, 'intuitionist' is a fairly loose term, referring at this time both to the French school and to his own 'improved intuitionism'. It is not simple to compare Brouwer's first intuitionism with the earlier and contemporary forms of constructivism or intuitionism. Borel, for example, insisted on the 'finite definability' condition for mathematical objects, but this notion hardly plays a role in Brouwer's writings. With Brouwer, 'constructiveness' and 'algorithm' are more or less immediate—one recognises a law when one sees one. That is to say, Brouwer's algorithms are basically the result of his natural number construction (the two-ity intuition) but he did not bother to investigate the possible algorithms. Brouwer agreed with Poincaré on the topic of mathematical induction, but he vigorously disagreed with Poincaré on the identification of 'existence' and 'non-contradiction'. With respect to the continuum Brouwer differed from his fellow constructivists, in that he recognised a special 'continuum-intuition'. Something like that can be found in Borel's work, who recognises a 'geometric continuum' as immediately given, Borel (1928), p. 16.

There is one momentous distinction that separates Brouwer from his predecessors: he saw that Aristotelian logic and constructivism do not go together. No matter how much inventiveness was invested by the French school, it would never produce a coherent constructive mathematics while sticking to the principle of the excluded middle, proof by contradiction and similar sins.

---

[4]Brouwer (1913b), p. 91.

Before World War I, all constructivists, including Brouwer, agreed on the point of individual objects: they had to be given by a law. Similarly, Brouwer accepted only those sets that could be obtained in specific ways, as indicated in his dissertation (Cf. p. 111). From this point of view it is natural that Brouwer rejected, along with the majority of his fellow constructivists, sequences generated by choices, be it that he mentioned them in disguise, that is as the unknown numbers. There was another objection to choice sequences, to wit that they were associated with subjectivism. So the case for choice sequences seemed a lost one. Nonetheless they did appear in the nineteenth century in the literature, mainly as pedagogical examples that illustrated objections against free choices. The earliest examples are to be found in the book on function theory of Paul Du Bois-Reymond[5] in the discussion between the *Idealist* and *Empiricist*: the first explicitly considers decimals determined by throwing a die. As far as we can see, Du Bois-Reymond's choice sequences were ignored; they did not generate any discussions. Not only were they ignored, they were in fact forgotten; I was greatly surprised when I happened to hit on a 'lawless number' in Du Bois-Reymond's book!

The topic of 'choice' became popular after Zermelo's epochal proof of the well-ordering theorem based on the axiom of choice;[6] this paper triggered a heated debate on the legitimacy of the notion of choice in mathematics, a discussion that reverberated long after set theory had returned to 'business as usual'. Borel, in particular, discussed the status of series determined by choice; in his eyes an uncountable number of choices made no sense at all, but countable choices were allowed if there is a procedure that guarantees that each choice indeed will be made after a finite amount of time.[7] In 1912 he even considered choices made consecutively by a large number of persons; he did so, however, mainly to illustrate probabilistic and pedagogical issues.

On the whole, one may conclude that constructivists were not inclined to accept choice elements, and classical mathematicians, say set theorists, were not interested in them.

Brouwer was worried all the same by this state of affairs, and for a quite specific reason: at best the continuum of the semi-intuitionists[8] is denumerably unfinished, and hence must have measure 0. This, obviously, was totally unacceptable for him. As late as 1952 Brouwer advanced this as the prime argument against the semi-intuitionists. Furthermore, the foundational scene was haunted for years and years by the paradox of Richard.[9] This paradox, which used only the most simple tools, highlighted the inherent dangers of the notion of definability. Consider real numbers

---

[5]*Die allgemeine Functionentheorie*, 1882.

[6]Zermelo (1904).

[7]Borel (1908b), Troelstra (1982).

[8]We will from now on call the French intuitionists, Borel, Lebesgue, Baire, Poincaré, Hadamard, and their associates *semi-intuitionists* in accordance with present usage. Brouwer used in his later publications the name *old-intuitionists* for his predecessors.

[9]Richard (1905).

the sequence of decimals of which are defined by finitely many words; these definitions can be put into a linear sequence $D_0, D_1, D_2, D_3, \ldots$ (roughly in the same way as the words in a dictionary are put into the alphabetical order) then we can define a new number as follows: the $n$-th decimal of $r$ is a 0 if the number defined by $D_n$ has an $n$-th decimal which is not 0, and it is 1 if this decimal is 0. Here, evidently, a real number is defined, but it is distinct from all the numbers which were defined by $D_0, D_1, D_2, \ldots$. Hence the list cannot contain all definitions: Contradiction!

This paradox seriously worried the mathematical world: Poincaré, Russell, and others, went to great lengths to avoid the phenomenon. Only much later, when the notion of definability was better understood, did the fear of Richard's paradox subside.

Poincaré banned Richard-like phenomena from mathematics by restricting its methods to so-called predicative ones. The peculiarity of the above defined number is that it is a definable number which uses in its definition all definable numbers, hence also itself. Poincaré saw such notions, which he called *impredicative*, as the bane of mathematics. Russell, too, tried to safeguard mathematics by imposing his *vicious circle principle*: 'no object shall be defined in terms of a collection that contains it'. This particular issue 'predicative vs. impredicative' is completely absent from Brouwer's writings, but he was no doubt familiar with it, since Richard's paradox was treated in the courses of his friend and teacher Mannoury. Although Richard's paradox may not have been uppermost in his mind, it is not improbable that it added to his dissatisfaction with the state of intuitionism.

Slowly, Brouwer's attitude towards infinite sequences started to change; in his review of Schoenflies's *Bericht* on set theory[10] he explicitly mentioned sequences of choice as legitimate objects.

The course on set theory in 1912–13 was still fairly conventional; however strong Brouwer's philosophical convictions were, he had not yet discovered how to do justice to them in mathematics. Brouwer's private notes were kept in a black exercise book; he conscientiously wrote down the course material on the right-hand page and recorded emendations, corrections, remarks on the left-hand page. He always used a fine steel pen, and produced an incredibly fine but still legible writing, often meticulously crossed out in the pattern of a fine grid. Extra lines were inserted in an even finer handwriting, and sometimes pencilled remarks (unfortunately less readable) were added. Notebooks such as this one present us with a kind of 'history' of the development of his thoughts. This particular notebook starts at the first lecture on 5 November 1912 (oh golden days of leisurely teaching!). Its first line defines the cardinal number of a set, closely following the pattern set by Cantor. On the whole the choice of material is fairly traditional; cardinals, ordinals, orderings, basic point set topology, and measure theory are treated. The constructive view point is observed, but rather as a limitation of traditional methods than as a new approach to mathematics. Repeatedly Brouwer corrected himself later by inserting criticism of parts of the course, e.g. after showing that finite sets yield the same number under

[10] Brouwer (1914), p. 79.

all countings (the fundamental property of finite sets), he went on to show that for infinite sets 'exactly the opposite holds'. But afterwards he noted on facing page 'the weak point, into which we will not enter ... is the tacit assumption that each set is finite or infinite', and on the same page he spotted another weak point, namely why is a set with a denumerable set of points removed again a set? This remark, by the way, shows that in 1911 Brouwer still stuck to the notion that only sets generated in a specific way, as indicated in the dissertation, make sense. The well-ordered sets are not explicitly defined in the notes (presumably Brouwer used the notes partly as a summary of the lectures, so that a few notes would often be enough to keep track of the course), but knowing Brouwer's partiality to Schoenflies's book, we may assume that he used the classical standard notion; in his remarks after the lecture he questioned this notion. One remark runs 'According to my Schoenflies-criticism there can, by the way, be no other ordinals than countable ordinals' and a few lines later he suggested: 'Maybe this is correct after all [that is the argument that all countable ordinals belong to the second number class]; for we can demand for well-orderings that they are brought about by the two generation principles; but then the theorem is self-evident, a tautology; that is perhaps the cleanest view (*zuiverste opvatting*).' Here one can see the genesis of Brouwer's theory of ordinals.

Brouwer did not use the Cantor–Bernstein theorem, which according to him 'must be regarded as an open problem, ... but the proof of this theorem is considered inconclusive by many mathematicians', hence he defined 'as large as' by '*A* can be injected into *B* and *B* into *A*'.

The real breakthrough came in 1916. In the academic year 1915–1916 Brouwer taught an advanced course, the theory of point sets (listed in the university catalogue as 'Set Theory'), which started in January. According to the catalogue it was continued the next year, but Brouwer's private notes suggest that he started the course all over again. The university catalogue gave the following description of the 1916–17 course:

> Point sets and internal sets. Deductions and derivations of point sets. Inducible properties of point sets. Covered and measurable point sets. Measurable properties which do not define point sets. Connection between measurable properties and internal limiting sets.

Brouwer's handwritten notes are clearly those of the 1915–1916 course, but with emendations added in the margins and on the backs of the pages. It is of course possible that Brouwer made these notes for purely private purposes, but it seems far more likely that he gave an improved course in 1916–1917 and started again from scratch, but this time using new insights.

The 1915–1916 course was basically an improved version of the 1912–1913 course. However, during the course, and most likely before the new course, Brouwer must have struck by a revolutionary insight that changed the whole perspective of the course.

The private course notes of Brouwer witness the revelation that shaped modern intuitionism: the original text was written in the characteristic hand with a steel pen, but, added in pencil in the margin of the very first page, there was the casual formulation of a revolutionary breakthrough, taking place in a quiet class room in Amsterdam.

After announcing that he was going to do mathematics from an intuitionistic point of view—admitting that he had himself often applied the *principium tertii exclusi* in his work, 'which then probably has not yielded correct, but only non-contradictory results'—the notes in the margin go on:

> First say something about sets defined *mathematically* and by *comprehension*. A mathematical thing is either an element of a previously constructed fundamental sequence $F$ (governed by induction, like the sequence $\rho$) or a fundamental sequence $f$ (which is not finished and not governed by induction) of arbitrarily chosen elements from $F$ or a finite set. With such a sequence one can work very well, if one always has *at each phase* for the *finite thing d or the fundamental sequence r* that is derived from it, to work with a suitable initial segment of $f$ ($r$ is then in general *also* never finished) [...] A set is now a law by means of which a $d$ or an $r$ is derived from $f$; this $r$ can then, for example, contain also relation symbols (e.g. ordering ones), so that the law can, for example, lead to well-ordered sets or other ordered sets, or to functions (one can indeed not get the set *of* ordered sets or the set of well-ordered sets). In addition one can accept *pseudo-sets* defined by comprehension, better called *species*, and one can call one species of a higher cardinality than another one, or two species equivalent.[11]

The definition reappears in full prolixity in the first *Begründungs*-paper of 1918 (cf. p. 304). It is helpful to compare both formulations to see that finally the choice-feature had gained a firm foothold.

The significance of this concept of set and species is that Brouwer had found another way to deal with the genesis of sets. Already in the dissertation particular methods for creating new sets were given, but here is a new method which makes use of the notion of arbitrary sequence.[12] The basic distinction between sets (later called spreads) and species is introduced in the 1916–17 notes. The mere sanctioning of choice sequences is by itself not yet remarkable, for Du Bois-Reymond and Borel had already considered sequences of choices, before Brouwer. In a crucial respect Brouwer goes beyond those authors; he indicated how one can use choice sequences in actual mathematics, showing that it was not a mere curiosity without any uses.

Brouwer had in the above note formulated the key to the success of the choice sequences: if you do not know how a sequence is going to be continued, use only the first values that are given so far. This already demonstrates a piece of common sense that goes a long way in everyday mathematics: suppose that a real number

[11] $\rho$ is the sequence of natural numbers; by 'governed by induction' Brouwer means 'given by a law'.

[12] Brouwer was not quite consistent in his use of 'fundamental sequence'. Usually (and certainly in the early foundational papers) this was a lawlike sequence, but from time-to-time Brouwer also speaks of fundamental sequences of arbitrarily chosen elements. One has to be careful in reading his papers, usually the meaning is clear from the context, but not always!

| | 0 | 1 | 2 | 3 | 4 | 5 | 6 | 7 | 8 | 9 | 10 | 11 | 12 |
|---|---|---|---|---|---|---|---|---|---|---|---|---|---|
| $f_0$ | **1/2** | 47 | 18 | 0 | 1 | 0 | 3 | 3 | 7 | 10 | · | · | · |
| $f_1$ | 0 | **12/13** | 7 | 2 | 4 | 1 | 0 | 89 | 4 | 7 | · | · | · |
| $f_2$ | 0 | 5 | **6/7** | 3 | 7 | 1 | 1 | 1 | 0 | 7 | · | · | · |
| $f_3$ | 5 | 8 | 3 | **19/20** | 2 | 5 | 4 | 7 | 2 | 1 | · | · | · |
| $f_4$ | 3 | 9 | 45 | 2 | **3/4** | 7 | 1 | 0 | 4 | 6 | · | · | · |
| $f_5$ | 0 | 0 | 45 | 23 | 76 | **1/2** | 0 | 76 | 31 | 95 | · | · | · |
| $f_6$ | 64 | 0 | 1 | 1 | 4 | 63 | **3/4** | 6 | 9 | 1 | · | · | · |
| $f_7$ | 0 | 6 | 9 | 4 | 45 | 5 | 1 | **7/8** | 9 | 0 | · | · | · |
| $f_8$ | · | · | · | · | · | · | · | · | · | · | · | · | · |
| $f_9$ | · | · | · | · | · | · | · | · | · | · | · | · | · |

**Fig. 7.1** Cantor's diagonal method

is given by a converging sequence of rational numbers $a_0, a_1, a_2, \dots$ (a so-called Cauchy sequence), where the $a_i$ are chosen arbitrarily, that is to say the choices are not restricted as long as one takes care that the sequence converges. Here is an example: first choose a natural number $n$, and after that choose from $n-1, n, n+1$, say $a_1$; next choose from $a_1 - \frac{1}{2}$, $a_1$, $a_1 + \frac{1}{2}$, say $a_2$; next choose from $a_2 - \frac{1}{4}$, $a_2$, $a_2 + \frac{1}{4}$, etc.

Now we want to multiply the resulting number by 2. This is most easily accomplished by multiplying all the approximating rationals $a_i$ by 2. The moral is that for the $n$-th approximation of the resulting number, one only needs the $n$-th approximation of the original.

So far, so good: it allows one to use choice sequences in practice, but somehow it is a cautious approach: if you can reduce an operation to an operation on initial segments then you are home. But can it always be done? This basic question was answered by Brouwer only a few pages later:

> The impossibility of mapping all elements of $f_1$ to distinct elements of $r$ follows from the fact that the choice of the elements of $\rho$ would have to take place at a certain point of the (forever unfinished) choice sequence, and in this way all continuations of such a *finite* choice-branch determining the element of $\rho$ will obtain the *same image in* $\rho$.[13]

This remark is the essence of Brouwer's Second Proof of the non-denumerability of the set of all natural number sequences, which has found a modest place in the margin of the note book. Thus Brouwer gave a beautiful and bold argument that replaced the famous diagonal argument of George Cantor. The latter argument is one of the most powerful inventions of the nineteenth century, and it is surprisingly simple.

Consider the set of all infinite sequences $f$ of natural numbers, that is to say, each $f$ is a sequence of natural numbers. Assume that this set is countable, that is, assume that it can be written as $f_0, f_1, f_2, f_3, \dots$. Then one can put the sequences and their values in a sort of infinite chess board, see Fig. 7.1.

[13] $f_1$ and $\rho$ stand for the sets of all choice sequences of natural numbers and the set of the natural numbers.

Now consider the values in the diagonal: $f_0(0), f_1(1), f_2(2), f_3(3), \ldots$; we change these values as follows: add 1 to each of them, thus we obtain a sequence $d$ with $d(i) = f_i(i) + 1$. This sequence does not occur in the list $f_0, f_1, f_2, f_3, \ldots$, for, if $f_m = d$ for some $m$, then $f_m(m) = d(m) = f_m(m) + 1$, which is impossible! The result can be summed up as 'the set of all infinite number sequences is not denumerable'.[14]

Now Brouwer had in the 1912–1913 and 1915–1916 courses faithfully reproduced Cantor's diagonal argument,[15] but after the recognition of choice sequences as sound mathematical objects, he realised that the nature of choice sequences forbade an enumeration of all choice sequences for a totally different reason. Indeed, this would require the assignment of natural numbers to choice sequences, such that distinct choice sequences would correspond to distinct natural numbers. Now let 273 be assigned to the choice sequence $f$. Then the assignment is effected after, say, $f(0), f(1), f(2), \ldots, f(2001)$ have been given. But then any two sequences that start with $f(0), f(1), f(2), f(3), \ldots, f(2001)$ are assigned the same natural number. Hence one can never satisfy the requirement that distinct choice sequences correspond to distinct natural numbers under the assignment. In simple mathematical terms, Brouwer's argument says 'All functions from $\mathbb{N}^{\mathbb{N}}$ to $\mathbb{N}$ are continuous, and hence there is no bijection from $\mathbb{N}^{\mathbb{N}}$ to $\mathbb{N}$.'

The above 'fact' that Brouwer invoked in the quotation above has become known under the name *Brouwer's continuity principle.*[16] The insight that the mathematical universe contains not only infinite sequences given by laws, but also arbitrarily chosen sequences, was a significant step towards a viable constructive mathematics. The 'poor but honest' doctrine of Kronecker and the semi-intuitionists was very upright and respectable, but it could never without some extra foundational insight progress beyond a cautious fragment of traditional mathematics.[17] One can imagine that such a practice did not appeal to an imaginative person like Brouwer; but the main objection to a life in the shadow of classical mathematics was that it did not make use in a proper way of reflection on the power of human thinking.

Between the discovery of the virtues of choice sequences and their appearance in print some time lapsed. The new approach was however, albeit in a rather cryptic way, already announced in a short note with corrections and addenda to the dissertation.[18]

[14]Cantor (1892).

[15]Brouwer had seen that the diagonal method is perfectly acceptable to a constructivist; this is implicit in the inaugural lecture of 1912; it is explicit in the set-theory lectures, which were constructive (in the spirit of the dissertation); Brouwer provided the non-constructive arguments with the *caveat* 'this is not constructive'.

[16]'Functions from Baire space to the natural numbers are continuous'. In traditional mathematics this is, of course, not the case.

[17]Note, however, that Bishop managed to regain a good deal of mathematics in the Kronecker–Borel tradition. The price to be paid was a general strengthening of assumptions.

[18]Brouwer (1917a).

> However, it has lately become clear to me as I hope to explain in a paper that will appear before long, that the limits of set theory can be extended.

Brouwer's 1916–1917 course on set theory can indeed be considered as the watershed between the, fairly negative, intuitionism of the dissertation and intuitionism as we know it today.

## 7.2 The International Academy for Philosophy

At the same time that Brouwer was discovering a new basis for intuitionism, he was drawn into a philosophical enterprise that has gone down in history as 'Significs'. It brought together a remarkable and mixed company. Significs did not come out of the blue, but it had its roots in the last years of the nineteenth century. The main actors in this ambitious and idealistic project were Frederik van Eeden—a well-known Dutch author, Gerrit Mannoury—the self-made mathematician and philosopher, Jacob Israël de Haan[19]—a man of letters and of law, Henri Borel[20]—a sinologist and journalist, and Brouwer. Each of these persons led a colourful life, and the significs group still stands out as the most imaginative, and indeed the only, collective philosophical scheme in the Dutch tradition. The intellectual drive behind Significs was to be provided by Mannoury, but the pioneer of this particular brand of philosophy of mind and language was Frederik van Eeden.

Van Eeden was born in Haarlem in 1860. He came from a well-to-do family; his father had taken over the family's bulb trade.[21] Frederik studied medicine in Amsterdam, where he was a rather successful student; he became Rector of the Student Corps, the highest honour in the world of students.

Ever since his school years, Van Eeden had felt literary ambitions, and when still a student, he wrote a masterpiece, which was the main source of his literary fame: *De kleine Johannes*[22] (1887) was translated into numerous languages; it appeared in English under the title 'The Quest'. Dutch literature at that period was in a phase of transition from the lofty tradition of nineteenth century Dutch letters to a more impressionist style. The old guard had founded an authoritative literary journal *De Gids* (The Guide), and the new generation expressed its sentiments in a counter journal, *De Nieuwe Gids*, perhaps not the height of originality but sufficiently adequate as a symbol of protest. Van Eeden's talents did not escape the young poets and authors, and so it was more-or-less natural that he was invited to join the select group of founders of *De Nieuwe Gids*.

---

[19] Among friends he was called 'Jaap' or 'Joop'.

[20] The Dutch sinologist and author. This Henri Borel will frequently be mentioned in the history of Significs. The reader should not confuse him with the French mathematician.

[21] There is a two volume biography of Van Eeden, Fontijn (1990, 1996).

[22] Little John.

As a medical doctor, Van Eeden was equipped with a good dose of curiosity and initiative; indeed, he, together with a colleague, opened in Amsterdam the first practice for psychiatry (1887). He fostered a keen interest in almost all phenomena that had to do with the manifestations of the mind; he was equally interested in hypnosis, spiritualism and occultism. As if his medical practice and his literary activities were not yet enough, he undertook an experiment of practical social reform by founding an utopian colony, and engaged in philosophical studies. His early occupation with philosophy produced a neat monograph *Rhetorical foundation of understanding*,[23] a treatise containing Van Eeden's view on ratio, communication, mind, mathematics, mysticism, etc. Although Van Eeden was not a trained philosopher, he was sufficiently well-informed to understand the developments of his day. In particular his views on mysticism and on mathematics are on the surface vaguely similar to Brouwer's views; he fully realised that one did not exclude the other.

Van Eeden was strongly influenced by Victoria Lady Welby, a sometime Maid of Honour to Queen Victoria (who was also her godmother).[24] When Van Eeden participated in 1896 at a conference for experimental psychology in London, as a delegate for the *Société d'Hypnologie et de Psychologie*, he met Lady Welby, who had presented a paper *The use of the 'Inner' and 'Outer' in Psychology: Does the Metaphor Help or Hinder?* She invited Van Eeden to Denton Manor, where they could discuss topics of mutual interest. Lady Welby's research centred around the notion of 'meaning', the functioning of signs in ordinary communication. Gradually a friendship developed between Van Eeden and Lady Welby that made him stand out among her numerous correspondents, such as Bertrand Russell, Ferdinand Tönnies and C.S. Peirce; their friendship was also Van Eeden's ticket to the higher strata of English society. This did not, however, dull his social conscience. Far from it! He gradually conceived the plan of founding a commune, or, in the usage of the day, a *colony*. Under the influence of, among others, the leading expert of Dutch flora and geology, Jac. P. Thijsse, he had come to a strong appreciation of a way of life in balance with nature.

After reading Thoreau's 'Walden' his mind was made up; he was going to found a colony in which he expected to interest people of a like mind. In 1898 Van Eeden bought some land in Bussum, a small town in Het Gooi,[25] not far from Laren and Blaricum, and soon a few cottages were erected: the colony 'Walden' was a fact. The colony attracted a number of people, for example, Nico van Suchtelen, the author of a number of novels and later director of the *Wereldbibliotheek*, an idealistic publishing company, and also Brouwer's friend, Rudolf Mauve, the architect who designed Brouwer's cottage. The colony became the best-known institution of its sort in Holland, no doubt through the fame of its founder. It was, in contrast to the colony of the International Brotherhood in Blaricum, not based on religious

[23] *Redekunstige grondslag van verstandhouding*, published in *Studies* 1–3, 1897.

[24] Victoria Alexandrina, the Hon. Lady Welby, wife of Sir William Welby-Gregory, 4th Baronet (1837–1912). I am indebted to Adrian Mathias for the information on Lady Welby.

[25] Cf. p. 57.

principles, but a purely social-humanitarian enterprise. Poor management and the unavoidable bickering, aggravated by financial problems, soon put an end to the undertaking. Van Eeden indeed lost a sizeable part of his mother's fortune on his pet project. Although the experiment could not be seen to support the colony-idea, the name 'Walden' has ever since been cherished under a romantic veil. Van Eeden could not bring himself to settle down for the management of his social experiment; various other causes claimed his attention, and so the idealistic experiment foundered, not least because of negligent leadership.

In spite of the failure of the Walden colony, Van Eeden saw a special role for himself in a world in spiritual confusion, and slowly the idea of a ruling cultural elite pressed itself onto him. In 1910 his path crossed that of some like-minded individuals, Poul C. Bjerre, a Swedish psycho-therapist and author, and Erich Gutkind,[26] the author of a remarkable book, *Sideric birth—Seraphic wanderings from the death of the world to the baptism of the deed.*[27] After reading a book of van Eeden, Erich Gutkind had sent his own book, in which 'The essence of man was set free from all naturalist determinedness and constituted above all nature and physics in the transcendent essence',[28] to Van Eeden. Van Eeden was enthralled by this work, in which he recognised a spirit in tune with his own. After meeting the author in person (at first Van Eeden ascribed the 'Sideric birth' to Buber), he wrote enthusiastically to Lady Welby

> I am here now in daily conversation with Volker, the author of *Sideric Birth*. There is your man, the man the world is looking for, or rather waiting for, without looking or knowing.[29]

Van Eeden and Gutkind set themselves assemble a small band of noble minds, they talked to Gustav Landauer, an anti-marxist anarchist, author and philosopher, and Martin Buber, the Jewish religious philosopher. They prepared a 'Call', an appeal to all the best minds in the world, to come to an understanding. 'We shall dub it: A call to all kingly men, for the transcendent conquest of the world, by deeds of royal love ... Our motto will not be 'proletarians of all nations unite!', but *noblest* of all nations unite![30] It was Van Eeden's firm conviction that significs, the study of signs, communication and meaning, could and should play a key role in the council of the 'Men of a Royal Mind' (*Koninklijken van Geest*), that he and his new companions envisaged. Van Eeden had not realised however that Gutkind, Buber and Landauer did not share his views on language at all. On the contrary, Gutkind stressed the central place of 'the breaking down of the fetters of language', appealing to Mauthner's 'Critique of Language'.[31]

---

[26]Pseudonym *Volker*.

[27]*Siderische Geburt—Seraphische Wanderungen vom Tode der Welt zu Taufe der Tat*, Volker (1910).

[28]Gutkind (1930).

[29]Van Eeden to Lady Welby, 7 October 1910.

[30]Van Eeden to Lady Welby, 16 November 1910.

[31]Mauthner (1906).

The promised call was published by Gutkind and Van Eeden in 1911, *Conquest of the world through heroic love*.[32] Each author had written his own part, and, as to be expected, Gutkind reproduced the ideas of his 'Sideric Birth', while Van Eeden stressed the importance of significs. In spite of the lack of positive reactions, Van Eeden pressed on, and in 1914 he again travelled to Berlin in order to further the realisation of a circle of noble minds. In consultation with Gutkind, Rang, Buber, Walther Rathenau and Max Scheler, the founding of the so-called *Forte Kreis* was scheduled for the autumn of that year. The meeting place was to be *Forte dei Marmi*, a small place north of Pisa, and the list of (tentative) participants was impressive: Barlach, Bjerre, Henri Borel, Buber, Däubler, Dehmel, Van Eeden, Gutkind, Landauer, Mombert, Florens Rang, Rathenau, Romain Rolland, Stehr, Susman.

However, the war superseded all plans, and soon the high-minded feelings gave way to more nationalistic ones. Thus the Circle came to its end even before it was founded. Rang and Gutkind fell prey to the virus of militant nationalism and the citizens of the belligerent nations readily forsook their supra-national ideals. The hostilities effectively prevented any further action in the planned direction, but Van Eeden, not willing to betray his ideals, did not quite give up hope. He directed his attention for the time being towards his native country. And so we find suddenly in Van Eeden's diary an entry (22 October 1915):

> Beautiful weather. Yesterday I dined with Borel at Professor Brouwer's, the mathematician. A sympathetic man, with the charming manners of a genius. It is curious that so often mathematical genius goes hand-in-hand with freedom of judgement and noble character. A refined, sharp, spiritualised head, clean-shaven, with young wrinkles. He is only 34 years old. Dressed in a white linen suit. In a conversation he at times, as immersed in thought, sits down on the floor. He had already conceived the ideal of the *Forte-Kreis* for himself. His wife is a fine type, with a high forehead, and brown eyes, like a model of the primitives, Van der Weyden or Metsijs.

Van Eeden was immediately taken with Brouwer; he had recognised a kindred spirit who might be won for significs and for a possible substitute for the Forte-Kreis. It is interesting that Van Eeden's impression of Lize Brouwer conforms to that of Brouwer, who spoke fondly of her Memlinck face.

Brouwer, on his part, appreciated Van Eeden and reciprocated his friendship. The relationship between the two men lasted for the better part of the rest of Van Eeden's life, but although the author surpassed Brouwer's friend Scheltema in fame and influence, this friendship was not to reach the same intensity and purity. The circumstances were, of course, totally different. Brouwer had become a famous man, albeit not in the eyes of the average Dutchman. He had reached a stage in life where recognition was no longer as urgent as it had been for the beginner of the period after his dissertation. After all, what could a man with an international reputation, who was accepted as an equal by men like Hilbert, Klein and Poincaré, gain from

[32] *Welt-Eroberung durch Helden-Liebe*, van Eeden and Gutkind (1911).

a connection with a Dutch author? The answer is probably that the popular fame of a scientist can never match that of an artist (of whatever kind)—at least during his lifetime. There was a certain mystique and notoriety surrounding the charismatic Van Eeden, that, added to a certain similarity in philosophical and social-cultural outlook, made Brouwer feel flattered by the attention of the great man. Van Eeden, definitely, was a man recognised by his fellow countrymen as an outstanding character—although the opinions ranged from 'superior author and playwright' to 'social amateur' and 'poseur'.

The Great War was seen in Holland from the detached viewpoint of a neutral spectator; the Dutch were somehow convinced that neutrality was, since the Napoleonic wars, their rightful rôle in history, and the horrors of war reached them only through the reports and misery of the bands of refugees from Belgium. For Brouwer the war was a challenge of a moral and ethical nature. His first brush with the facts and fiction of the events of 1914 was in connection with the so-called *Declaration of the 93*. After the first military actions of the Germans on the western front, a number of accusations of violations of the code of war started to circulate in the allied press. For quite a number of otherwise quite reasonable German scholars, this imputation seemed a wilful act of slander, designed to denigrate a nation with recognised cultural standards. In reaction to the allied reports, a number of scientists published a pamphlet, 'Appeal to the World of Culture' (4 October 1914).[33]

The 'Appeal' protested against accusations that laid the blame for the war at Germany's doorstep, and that painted the Germans as wilful destroyers of the Belgian cultural heritage (for example, the sack of Leuven[34]). In a six fold *Es ist nicht wahr* (it is not true) the 93 scholars vehemently denied the charges. The list of the signatories contained—a cynical coincidence!—one of the former candidates of the Forte-Kreis! Interestingly, the list contained only one mathematician: Felix Klein had become the victim of temporarily enhanced national feelings and misleading information, cf. p. 328.

The recent events even found a modest echo in the scholarly confines of the Dutch Royal Academy; the minutes of the meeting of November 1914 reveal that Brouwer, under the heading of 'other business', called the attention of the meeting to 'the phenomenon that scientists of renown of various nationalities had published manifestos in their quality of scientist, creating the impression that the leaders of science had reached, *on the basis of their scientific reflections*, the conclusion that a military and fiscal separation between the nations is necessary to protect the higher interests of culture'. He asked if the Academy could not prod the International Association of Academies into making a statement on the issue. After some discussion the chairman acknowledged the importance of the issue, but 'answered Mr. Brouwer that perhaps individual persons, rather than an official body, such as the Academy, could do something to repair this impression with the public'.[35] The manifesto of

[33] *Aufruf an die Kulturwelt*, Kellermann (1915).

[34] August 1914.

[35] *Verslag van de gewone vergadering der wis- en natuurkundige afdeling XXIII (1e gedeelte)* p. 828.

the 93 was to haunt the German scientists for years to come; it tainted them with the image of servants of military imperialism. A good many of them soon regretted signing this declaration without access to reliable information.

Of course, the manifesto of the 93 was not the only, and not even the first, excursion of scientists into the uncertain domain of politics. To Brouwer, actions of this patriotic sort were anathema; he had, in line with his views of *Life, Art and Mysticism*, developed a strong disapproval for demagogic arguments and notions, such as 'fatherland' (*vaderland*). He considered the moral-emotional content of such terms as totally misleading and harmful, a cloak under which interests of certain groups were pursued. In short, he had become a thorough internationalist. His views were spelt out in a review under the title *Anti-nationalistic Literature*[36] of two books, *In the light of the flame of war*[37] by Frederik van Eeden, and *Patriotism, Philanthropy and Education*[38] by Carry van Bruggen. Brouwer upbraided both authors for being spineless, because they still considered nationalistic manifestations as a product of an autonomous life-tendency of the individual. According to Van Eeden these were a desire for self-denial and absorption in a larger community, and according to Van Bruggen they constituted a quest for support of one's own shaky pride and self-confidence:

> The authors thus maintain, rather nebulously, the insight that the state is the institute of regulated parasitism; in the first place of the community of citizens on the spontaneous intellect-less life within the confines of the state, in the second place of the prosperous surface of that community on its needy core, and that the durability of this institution rests solely on the fact that anyone who shares in its management, gets a stake in its continuation.

Brouwer rejected the solutions of Van Eeden and Van Bruggen, to wit the cultivation of the common intuitive sense of justice and an anti-national drill at the elementary schools, and pointed out that 'the perpetuation of the fatherlands by the interested parties can only seriously be impeded after eradicating the word-of-power 'fatherland' itself'. Brouwer claimed that the abolishing of 'words-of-power' is the most effective way of fighting social injustice.

> The suppression of one's fellow man by means of private property and trade would be impossible without words-of-power, such as *entrepreneur, interest rate, profit*. And the morphinising of the conscience required for the perpetuation of social injustice would be impossible without words-of-power like *happiness, religion, art, civilisation, genius, duty, reliability*. These *means of defence of injustice*, among which the word *fatherland* is one of the most powerful, can only be destroyed by ending the present *anarchy in the formation of words*. And thus there emerges the primary social task, imposed on us by our conscience: the founding of an *institute for language-reflection* that

[36]Published in *de Nieuwe Amsterdammer* of 3 February 1917.

[37]*Bij 't licht van de oorlogsvlam.*

[38]*Vaderlandsliefde, menschenliefde en opvoeding.*

> will first have to brand the *words-of-power of injustice* in extensive analytical manifestos, and that subsequently will have to occupy itself with *the forming of words for the realisation of the common intuitive sense of justice*, as soon as the eugenics for this sense of justice will have created a possibility for development.

From this review Brouwer emerges as stern a judge of the arrangements of human society as befits the erstwhile author of *Life, Art and Mysticism*. Comparing, however, the above exhortations with the message of *Life, Art and Mysticism*, one observes a major distinction. The new element is the role of language, and even a language institute is advocated—this foreshadows the International Institute for Philosophy. The negative role of words-of-power must rather be viewed through the eyes of the mystic than those of a socialist, pacifist etc. The objectionable terms, in Brouwer's perspective, are nothing but tools of the established powers in the struggle for domination of mankind, a domination which is in direct conflict with the non-interference principles of the mystic. Brouwer's proposal is simple enough: remove the terms, and then the oppressor will be robbed of his efficient means of indoctrination. There is, of course, no promise that the immoral effects will disappear, but at least their perpetuation will be made harder. Brouwer was quite consistent in this particular view; after the Second World War the same advice is repeated.

The earlier tendency of Brouwer to blame the evils of human society on its unlimited growth is repeated forcefully in the above review. He judged Van Eeden's appeal to 'common intuitive sense of justice' a sad illusion:

> So long as the health of men languishes so sadly, that repulsion is the dominating intuitive feeling they instil in each other, and, as a consequence, everybody's instinct for self-preservation compels him to find a place where he does not have to get closer to his disgusting fellow citizens than he chooses—for which only a parasitic constitution offers chances. And there is only a prospect of an improvement of human health after an end has been put to the anarchy of *procreation* and the population density has been reduced to such an extent, that the atmosphere of free breathing of single individuals gets a bearable scope.

This passage may elicit a smile from the modern reader, after all Brouwer lived in a country with fewer than 6 million inhabitants, against the estimated 16 million of today. Generally speaking, life in his days in Holland was, with respect to privacy and ecology, a paradise compared with the present situation. So what was there to complain of, one would be inclined to ask? Probably this is an instance where an exceptional person with prophetic gifts is able to read the signs of the times better than politicians and administrators.

The above is quite characteristic of Brouwer's views on the inadequacies of mankind with respect to the world and life on it.

**Fig. 7.2** Lize and Bertus in travel outfit. [Brouwer archive]

## 7.3 Family Life

In the early years of the war Brouwer's little household underwent some change: a new house mate joined the family.

His stepdaughter Louise had been a source of anxiety to Lize and Bertus; she was not an easy child to get along with, and the intellectual prospects of her education interested her preciously little. Not being cut out for an education in a gymnasium or HBS,[39] it was decided that she should enrol in a so-called domestic science school.[40] This particular school on the Zandpad in Amsterdam had a splendid reputation; it attracted girls from various social strata, including daughters from the better families. The girls were taught all they should know about running a household. Various careers were open to the graduates of this school; Lize and Bertus had decided that Louise should select her courses so that she could become a teacher in laundry techniques.

Louise joined the school as an resident student. Little is known about her progress, but eventually she left the school without a certificate. Brouwer, who did not like to leave things to chance, had decided that it would be wise to keep an eye on Louise's friends. He therefore asked for the list of students. He studied this list

---

[39]Cf. p. 4.

[40]Huishoudschool.

carefully, and found two suitable girls on the list; one of them was the youngest daughter of a notary. Brouwer let Louise know that she could bring this girl home to Blaricum. The young lady, Cor (or Corrie) Jongejan, was more or less the opposite in most respects of Louise. She was fun-loving and flippant; she preferred the tennis court over classes, and at school she was cheeky. In contrast to Louise, she had finished highschool (HBS) before entering the present school. Like Louise, she did not make much progress in the domestic sciences, much to the chagrin of her parents, who despaired of Cor's future. Under these circumstances, Brouwer's interest in the girl appeared to them something of a godsend. It occurred to them that perhaps this stern professor could knock some order and respect into the pretty head of their daughter. Brouwer was approached by them with the request to take the supervision of Cor in hand and somewhere in 1914 Cor joined the Brouwer household as a kind of general secretary-factotum.

Louise could not stand Cor, and the new arrangement was not to her liking. Most sources confirm that Louise's opinion was mostly dictated by (an understandable) jealousy. Cor was an engaging person, who could make pleasant conversation and who easily made friends. The somewhat sullen, sombre Louise had to content herself with a modest place in Cor's shadow.

Cor was about twenty-one years old when she became Brouwer's assistant. She performed all kinds of secretarial duties; for example, she copied the manuscripts of Brouwer, at first by hand and later on a typewriter. In addition she copied (most of) the manuscripts that Brouwer handled for the *Mathematische Annalen*. In the beginning she was rather awed by the impressive scientist, but eventually her jolly nature overcame her timidity, and she started to treat her employer on a less formal footing. In fact Cor became an intimate friend of Brouwer; she often accompanied him on his travels, and Lize came to accept her as a normal part of family life.

The family of Cor was less pleased, and the ties with Cor were cut. Nonetheless, when her father was diagnosed with a terminal illness in 1918, Cor went back to him and helped to nurse him in his last months of his life.

Cor inherited the tidy sum of 60.000 guilders from her father. Back in Blaricum she handed the money over to Brouwer, with the words 'I don't know what to do with it'. Brouwer invested the money in various ways for her, for example, after the war he bought a house for her in the Harz (Harzburg), which served as a *pied à terre* for visits to the area, which was close to Göttingen and to Brouwer's usual health clinic.[41]

The avowed international traveller Brouwer found the limitations, brought by the war, hard to bear. He desperately longed for an occasion to cross the borders again. Although travel abroad was not easy, he did manage to make a short trip to Germany in the spring of 1915. For a person like Brouwer, who loved travelling, and who felt

[41] The house is still there; an old man who had lived in Harzburg his whole life, vividly remembered Professor Brouwer *mit seiner Kusine*. He recalled that Brouwer offered his father to invest the family capital of only 18 Marks into Dutch guilders, which was strictly forbidden by the currency laws in this period of galloping inflation after World War I, just to help them out. The father refused the offer.

at home in the most diverse countries, being cooped up in Holland was one of the worst aspects of the war. In March he wrote Schoenflies a postcard, asking what the travelling conditions were; he had read in the Dutch newspapers 'that travelling in Germany for non-Germans was not quite safe, and that one runs the risk of suddenly being locked up for several days, for a closer inspection of one's identity, even if one has a passport'. Schoenflies, with whom he had excellent relations once the editing of the Set Theory *Bericht* was out of the way, must have reassured him, for he visited Schoenflies in Frankfurt am Main during the Easter vacation.[42]

## 7.4 An Offer from Leiden

In the meantime a vacancy had come up at the University of Leiden and H.A. Lorentz, the leading Dutch physicist, tried to win Brouwer for Leiden. He personally advocated Brouwer's candidacy for the mathematics chair:

> nothing would please me more than that you could decide to change Amsterdam for Leiden, ... we all think it of the greatest importance for the flourishing of the faculty. In particular Kluyver, de Sitter, Ehrenfest and I would appreciate collaborating with you. You may be assured that you will be received warmly and with open arms.[43]

The two knew each other well enough, both being members of the Royal Academy, of which Lorentz was President. When Brouwer was still an unknown *privaat docent* Lorentz had reacted to the inaugural lecture of Brouwer in 1909, and this led to an exchange of views on the theory of relativity.

The offer from Leiden had its attraction for Brouwer. The Leiden faculty was apparently convinced that the addition of this prestigious mathematician to its staff was worth a few sacrifices. It told Brouwer[44] that they would not object if Brouwer were

> to put into practice the view that the discharging of the duty of a professor rather consists of the pursuit of his own scientific work and of being available for students who work on their own, who seek supervision and advice, than of regular courses in stereotyped theories, which have already for years found clear expositions in books.
>
> An oral clarification of the offer was summarised by my colleague Ehrenfest with the words 'Thus materially nothing will be required of you other than your presence.'[45]

[42]Brouwer to Schoenflies, 28 March 1915, 10 June 1915.

[43]Lorentz to Brouwer, 11 June 1915.

[44]As reported in a letter from Brouwer to Zeeman, 19 June 1915.

[45]*Also materiell wird von Ihnen nichts weiteres verlangt als dass Sie da sind.*

This generous offer contrasted sharply with the practice in Amsterdam, where, according to Brouwer, the teaching load was not taken so lightly. He complained that 'since the acceptance of my duties I have been handicapped in my research activities in a discouraging manner'. It seems unlikely that Brouwer took the offer seriously, but it set him thinking about his own situation in Amsterdam.

Without taking Ehrenfest's statement too literally, he would at least like to obtain some freedom from the board of the Amsterdam University to organise his activities after his own insights on the basis of the Leiden model. That is to say, no undergraduate teaching, and mostly supervision and consultation where advanced students were concerned. The Leiden offer gave him some bargaining power, which he used to ask the Board of the Amsterdam University for two specific commitments:

1. To raise the strength of the teaching staff in mathematics to a level comparable to that of the state universities, at the retirement of Korteweg in three years' time (N.B. Amsterdam had 2 full chairs and 1 extra-ordinary chair, but the teaching duties included mathematical astronomy and theoretical mechanics).
2. On account of his weak health, Brouwer could not live in a city, but according to the rules he had to live in Amsterdam. Thus he felt obliged to keep a town house and a house in the country. He therefore requested permission to take up his residence in the country.

Brouwer had already taken the initiative to discuss the matter with the municipal authorities, and on 18 June he had an audience with the Mayor of Amsterdam. The Mayor was, he wrote to Zeeman,[46] favourably disposed, but he could not possibly go so far as to promise an extra lecturer immediately.

The Mayor was as good as his word, Brouwer's case was discussed in the meeting of the curators of 25 June. No objection was raised to Brouwer's request to be released of the obligation of living in Amsterdam. The remaining requests were another matter: the curators flatly refused to discuss the distribution of chairs after Korteweg's retirement. The board also devoted their attention to the mode of teaching proposed by Brouwer. In the absence of facts it was hard to form an opinion, they agreed, but nonetheless they questioned the idea of a professor not teaching regular courses, but making himself available to students at certain office hours. One easily recognises here an echo of Ehrenfest's ideas. The curators were not particularly happy with such a radical innovation, 'For very gifted and diligent students this may be possible, but for the great majority this certainly is not sufficient'. The curators decided that they could not approve of a manner of teaching where the professor would only be available for consultation, but they did see that some consideration was necessary to enable Brouwer to continue his researches. Finally they had no problem granting the mathematicians an extra 500 guilders a year for a seminar library.

Thus in the end Brouwer obtained permission to live in Blaricum, but in most other matters he probably had to content himself with promises of a sympathetic

[46]Brouwer to Zeeman, 19 June 1915.

**Fig. 7.3** Zeeman, Einstein, and Ehrenfest. [Courtesy Nationaal Archief Noord-Holland]

reception of later proposals. The fact that, after all, he preferred Amsterdam with its teaching load over the generous offer from Leiden, strongly suggests that the combination 'Amsterdam–pharmacy–'t Gooi' outweighed almost any other situation.

The faculty was more than pleased that Brouwer had turned down the offer from Leiden down; the chairman, at the meeting of 6 February 1916, expressed his pleasure that Brouwer 'had withstood the siren call of Leiden, and thus was saved for Amsterdam'.

The recognition that Brouwer had already received in the form of an editorial post at the *Mathematische Annalen* was further confirmed when Blaschke asked him in November 1915 to write a book on the new developments in topology for the publisher Teubner.[47] The invitation suited Brouwer's plans, and he answered that he would gladly write a book on Analysis Situs.[48] However, no drafts or manuscripts of this book exist, and it is doubtful if he ever started to work on it. For at the same time he was rebuilding mathematics from a constructive viewpoint, and soon most of his active research time was taken up with his new intuitionism.

[47] Blaschke to Brouwer, 4 November 1915.

[48] Brouwer to Blaschke, draft 19 November 1915, Cf. CW II, p. 410.

## 7.5 Van Eeden and the International Academy

When Van Eeden and Borel met Brouwer in September 1915, the two were already involved in a project that envisaged nothing less than a philosophical academy. The war had made it quite clear that philosophical considerations had given way to the nationalistic ones, and that the Union of the Noble of Spirit had proved illusory.

Independent of Van Eeden a small group had proposed a meeting in the summer of 1915 that should combine people of all denominations to express religious feelings across the borders of the various creeds. Borel and Van Eeden made contact with the leading personalities, Countess Van Randwijk-de Jonge and a broker in tea, J.D. Reiman. The initiative was taken to found an Academy of a philosophical nature, and a committee was installed, consisting of Van Asbeck, Van den Berg van Eysinga, Blok, Brouwer, Van Eeden and Reiman. Brouwer probably met Van Eeden at the first meeting of this committee (21 September 1915). Since Van Eeden soon saw in Brouwer a supporter of the Forte-Kreis idea, Borel and Van Eeden started to visit him in order to get better acquainted. Brouwer was made chairman of the committee, and Reiman secretary.[49] Van Eeden reported in his diary:

> Brouwer presided [24 October 1915]. He was sometimes exceedingly naive in his plans. But it was good all the same. This is already an elaboration of the Circle.

The committee chose *Oud Leusden* as the site for summer schools, with room for cabins for the students. Oud Leusden is a village just south of the town of Amersfoort, and, to run ahead of our story, it still is the location of the 'International School of Philosophy'. The founding of a philosophical institute put all those involved in a state of pleasant excitement; meetings were held, and plans and rules were proposed. The project, however, did not work out the way it should have. In January 1916 a conflict arose; Brouwer's way of chairing the committee had met with incomprehension and disapproval. His proposal of an ascetic lifestyle for the school added considerably to the confusion. The relation with the philosophical establishment in Holland was also a moot point. The Van Eeden group wished to exclude the Bollandists from the school, something that antagonised the company at large. Bolland, the philosophical giant from Leyden, had refused to take part in the activities, since he considered a committee with a poet (Van Eeden) in its midst, 'strange company'.

The conflict erupted at the meeting of 10 January 1916; Brouwer clashed with Reiman, who matched Brouwer in authoritarian behaviour. Brouwer wanted to get the action on its way by consolidating a core of the five active members—Henri Borel, Bloemers, Van Eeden, Reiman and himself. The remaining members would then be encouraged to join the group if they accepted its principles. Reiman did

[49]The history of the founding of the philosophical institutions, mentioned here, can be found in van Everdingen (1976), Schmitz (1990a), Heijerman and van der Hoeven (1986), Fontijn (1996). Furthermore, Van Eeden's diary is an invaluable source of facts and of running commentary, van Eeden (1971).

not agree to this procedure, and Brouwer was forced by the meeting to give up the chair; thereupon Borel, Brouwer and Van Eeden left the meeting. The rift could not be bridged, a reconciliation attempt on 27 February failed, the parties clashed again, and each went its way.

Reiman and his group went on to found on 13 February 1916 the 'International School of Philosophy' at Amersfoort. The school had a building erected at the Doodenweg in Leusden; the architect was K.P.C. de Bazel, who was one of the leading architects of the period in Holland. In June 1917 the chairman Reiman opened the main building. The same building is still standing today, and the school still functions more-or-less as intended. The International School of Philosophy has been the scene of some distinguished meetings, and the list of speakers boasts a impressive cross-section of the philosophical community. The following list is a small selection: Adler, Bernays, Bloch, Buber, Cassirer, Church, Dürckheim, Heidegger, Koyré, Landgrebe, Naess, Quine, Reichenbach, Schweitzer, Tagore, Tarski, Tillich. The loss of the initiative to Reiman irked Van Eeden, he noted in his diary that 'Reiman is a dwarf compared to the giant Brouwer'[50] and after the final clash he described Reiman with the words:

> The man is the dupe of his all too great ideas. He cannot bear them. He feels himself a chosen one, a tool of God, and hence becomes a fanatic and upsets everything.

Brouwer, with his over-sensitive nature, suffered intensely—in Van Eeden's words:

> Now it became clear to me how extremely sensitive he is. The person of Reiman makes him sick. It gives him visions, and he sometimes shouts 'Reiman!' to get rid of him.

The group around Van Eeden went its own way; Van Eeden tried to revive the Forte-Kreis formula in the form of an International Academy for Philosophy. It was the small but select group that had left the fold: Borel, Brouwer, Mannoury and Van Eeden. The organisation of flourishing societies was not their strong point, and they rather formed an exquisite band of individualists. We have already met Mannoury; he was in a sense, in an extra-curricular way, the mentor of Brouwer. His philosophical interests made him stand out in Holland, but in addition he was a prominent leftist and writer–pamphleteer. Together with Brouwer, he represented the philosophical expertise of the group. Henri Borel was a man of many trades, albeit mainly literary; he was a (free-lance) sinologist and earned a living as a journalist and author. His literary fame rested on two early novels *Het Jongetje* (The little boy) and *Het zusje* (The little sister), (1899, 1900), and numerous books and brochures on China, Asia and its religion, art and philosophy. Borel was an expert gossip; he freely mingled with the leading men of letters and the artists of the day, there is many a volume of correspondence of Borel with one of his artistic or intellectual relations. His contacts with Van Eeden went back to 1889. He lived at Walden for some time in 1905. After meeting Brouwer in 1915, he grew very much attached to him.

[50] van Eeden (1971), Diary 11 January 1916.

The composition of the group guaranteed more-or-less that significs was to be one of the, if not *the*, central topics of the International Academy.

The next steps soon followed; on 6 March Van Eeden, Bloemers and Brouwer discussed the principles of the Academy, and at the end of the month a circular letter was sent out to prospective members, explaining the reason for parting from the Amersfoort-group, and announcing their plan to found an Academy of a different nature. This academy should consist of a limited number of members, to be elected by co-option, and a board of trustees 'made up of local, that is Dutch, persons who are well-known and sympathise with the plan'. The circular ended with a call for members of this board. The result must have disappointed the initiators: only Mannoury sent in a positive reply; that did, however, apparently not stop them.

The acquaintance with Van Eeden had considerably widened Brouwer's circle in the society of 't Gooi. He was, of course, already part of the local scene. In the small towns with their artists and intellectuals, he was recognised as a special character, the tall, lean professor in mathematics—a very clever person no doubt, but hard to follow. After joining the circle of friends of Van Eeden, Brouwer's contacts multiplied. He met the main characters of the Walden episode and also a number of members of the fringe. Life was full of little events, tea's, birthday parties, friendly visits, discussions and the like. There is an enormous difference in Brouwer's lifestyle before and after the First World War; one might guess that the international isolation combined with the somewhat Bohemian atmosphere of Het Gooi gave Brouwer a taste of a rather un-Dutch 'dolce vita'.

The School of Philosophy might be a serious matter, but it stopped Brouwer nor Van Eeden from enjoying the pleasures of the micro-society of Het Gooi. Van Eeden's fame made him a natural centre of a small court of admirers, and the curious Brouwer, with his appetite for personal relationships (which seldom touched his inner soul), enthusiastically took part in the meetings. Although he was on Van Eeden's turf, he saw no reason to share Van Eeden's preferences or dislikes.

Van Eeden was no stranger to conflicts,[51] and at that time he was in the final stages of a conflict with his secretary, Holdert. Brouwer, not influenced by Van Eeden's aversion for Holdert, developed friendly relations with Holdert and his wife Gerda.

The famous Van Eeden attracted a rich variety of visitors and followers, ranging from devote admirers to curious loafers. He was in particular surrounded by artists and quasi-intellectuals with a penchant for 'deep thoughts'. Brouwer watched the circle around Van Eeden with quiet amusement, and from time to time took part in the exchange of wit and profundities, often sitting with crossed legs on the floor or reclining in a picturesque pose. Some members of Van Eeden's entourage became regular visitors or friends of Brouwer, for example, the aforementioned Holdert and Borel.

One could regularly find at Van Eeden's place a couple by the name of Langhout, which lived in 't Gooi not far from Van Eeden. At one point Langhout apparently

[51]Cf. Fontijn (1990).

had tried to impress the company with 'deep' conversation; he had defended the view that artists should suffer in poverty in order to create good works of art. Van Eeden recorded in his diary that

> Langhout did this, hyper-idealistically, with the exaggeration of a superficial person. And Jaap [de Haan] could not bear to hear this from somebody whom he had heard described as a sluggard by Brouwer himself.

Tine, Mrs Langhout-Vermey, was eventually to become a close friend of the Brouwer family. Brouwer already knew Tine before her marriage; he enjoyed her conversation and her company. He thought she made a colossal mistake marrying Langhout, whose only virtue was that he was well-to-do; indeed, on the day of the marriage, Brouwer went all the way to the Town Hall to talk her out of it.

In fact, Brouwer and Tine became more than just 'good friends'. She occasionally accompanied him on foreign trips, thus taking the place of Cor Jongejan. Lize was perfectly aware of Bertus' attachments, but far from being jealous, she was often relieved that 'Bertus was taken care of'.

Brouwer and Van Eeden were for a time fairly close; they met regularly and discussed all and sundry. Van Eeden, for example, visited Brouwer at the Amsterdam Academy on 28 February 1916 to discuss a couple of matters. He recorded in his diary that Brouwer told him about Lorentz' talk:

> … who now has found the connection between gravitation and electromagnetic phenomena. An event as important as the laws of Keppler … I talked with him until half past six. And again and again, tears come to my eyes, by the feeling of gratefulness for his understanding, for the space he gives me …[52]

Brouwer, curious as ever, also accompanied Van Eeden to spiritistic seances. The latter suffered from guilt feelings after leaving his first wife. The subsequent death of his son Paul drove him in desperation to frequent the sessions of a spiritistic clairvoyant. Van Eeden was strongly influenced by the clairvoyant, a young girl. She played an important role in Van Eeden's later conversion to the Catholic Church. In this exalted company Brouwer was probably rather out of place; his curiosity and the complete absence of the required awe was not appreciated. When he attended on 30 August 1918, a day of light, as the seances were called, he was told to hold his tongue, because 'the days of light were no common seances'!

At the same time Van Eeden's diary tells us that remorse did not work lasting changes in a man like himself; it appears that Brouwer and Van Eeden shared a lively interest in the other sex, for example, they attended the concerts of the pianist Henriette Roll together, and were both under her spell. Van Eeden, it seems, did not appreciate the presence of a younger, and virile, competitor for the lady's attention.

The already disturbed relationship between Borel and Van Eeden did not help to set Van Eeden's mind at ease. He must have complained in a letter to Lize about

[52]Ibid. 28 February 1916.

the influence of Borel on Brouwer, for in a letter of 18 December 1917 she tried to reassure Van Eeden. 'Bertus is on a cycle tour', she wrote, 'so I have to use the occasion to take away your distress'. According to her everything was part of a malicious plan of Borel; Bertus had broken off his relations with Henriette, and any talk of a new amorous relation with an unnamed lady (probably one of Van Eeden's female admirers) was totally unfounded. 'Bertus always tells me everything, and I am not aware of anything.' This little episode illustrates the tensions that existed even among 'men of a royal mind'!

In the meantime the relationship between Van Eeden and Borel had deteriorated; each went out of his way to point out the inferiority of the other. There is a passage in Van Eeden's diary in which he reflected on the proper attitude with respect to the fair sex, apparently inspired by Borel's behaviour:

> In connection with the case Borel [...] I will in future demand two things from those who wish to count themselves among my friends: (1.) that they avoid needless talk, keep themselves under control and mind what they say; (2.) that they maintain their self-control in sexual matters. These two things are demands of discipline and dignity. He who does not meet these demands and does not make all efforts to do so, not only harms himself but also his friends. His place is not in the company of the Good and Royal. Women acquire power over the man who is above them through their sexuality. If the man tolerates this power, he sinks and loses his sense of self-respect [...]. For the masses it doesn't do harm, there the values are the same. But in the company of the good, the woman should also be good. There are women whose only power lies in their sexuality. We should avoid those. What is innocent among common people, is dangerous among the Good and Royal. Among them the greatest discretion and chastity must be the rule. [...] and this is not only meant for Borel, but also for Brouwer.[53]

Clearly, Van Eeden was worried about Brouwer, whom he deeply loved and admired. Brouwer could joke and flirt like the next man, albeit with refinement and style, and Van Eeden apparently was inclined to view this as demeaning. Van Eeden's concern about sexual matters did not just spring up overnight. He enjoyed female company and was an easy prey of his amorous urges; being often surrounded by female admirers, he did not lack occasion to practise his romantic inclinations. Van Eeden left his first wife after more than twenty years of marriage. The stern demands on others may very well have reflected the realisation of his own frailty.

At this time Van Eeden started to read Brouwer's Delft Lectures, *Life, Art and Mysticism*; since the book was not likely to be on the counter at just any book shop, one may well presume that the author had drawn Van Eeden's attention to it, or—more likely—given him a copy. Van Eeden was thunderstruck by the lectures, which were rather close to his own thoughts and writings. He immediately set himself to write a belated review of the book, which appeared as a series of essays, which were

[53] Diary Van Eeden, 11 August 1916.

printed in the weekly, *De Nieuwe Amsterdammer*,[54] under the title *A Potent Brew*.[55] In Van Eeden's words one can still hear the resonance of the shock that must have been felt by other readers of Brouwer's book as well:

> These one hundred pages of Dutch prose are indeed the most powerful, but also the most terrible, in my opinion, that have been published in this century. They are beautiful and deep and full of truth. But they are fiercely revolutionary, completely hostile to our whole society. They conflict directly with the order, the religion and law of the people. In this manner they are similar to many prophetic words—and it would be an almost ridiculous inconsistency, an unforgivable inanity, of mankind that poisoned Socrates, stoned prophets, crucified Jesus and burned Bruno, to let this formidable doom-brewer walk free without hanging him or at least interning him behind barbed wire. But look—the man is a Professor at the University of Amsterdam and a member of the Royal Academy ...

Van Eeden's life was a perpetual series of social and philosophical actions and positions, on all occasions presented and defended with a strong emotional force. His discovery of Brouwer and his mystic-philosophical views came at a time when he was under the spell of his ideas of an intellectual elite; *Life, Art and Mysticism* fitted his views of the moment perfectly, and Brouwer's conversation must have contributed considerably to Van Eeden's admiration for ideas which were so close to his own. The review series reflect Van Eeden's admiration for his new friend, the only point where he refuses to follow Brouwer is in the latter's stern judgements on females.

It was the second time that fate brought Brouwer into contact with a poet–author; compared to the strict privacy of his friendship with Adama van Scheltema, the relation with Van Eeden had another dimension. It could not possibly have escaped Brouwer that Van Eeden was intellectually no match for him, but it was no small matter to be a close associate of the most controversial poet and social reformer of the period. It did, however, not compare to the intimacy of the friendship of his youth. All the same, the relation with Van Eeden lasted and played an important role in this phase of his life. The two went swimming, exchanged visits, went on a cycle tour—and founded an academy. Indeed the thirty-six-year-old mathematician and his friend, who was twenty-one years his senior, made a long cycle tour all the way to Limburg, some two hundred and fifty miles in all. Van Eeden was rather depressed and the ride with a younger and more energetic friend heavily taxed him; there is a charming description in his diary, where he honestly noted that 'I enjoyed it when Brouwer had a flat tyre. So little pleasure did the trip give me.'

The preparations for the 'Academy' were set in this atmosphere of informality and friendship, with its small gossip and jealousies. The serious rift between Borel and Van Eeden did not break up the company, but was characteristic of the tension between the participants, who all saw themselves as superior minds.

---

[54]Cf. p. 14.

[55]*Een Machtig Brouwsel*; 'brouwer' is Dutch for 'brewer'.

**Fig. 7.4** The cyclist Brouwer. [Brouwer archive]

In view of the modest size of the founding group, new members were more than welcome. The first addition was the theoretical physicist Ornstein, professor at Utrecht, introduced in 1917 by Brouwer. Ornstein attended a few meetings, but soon gave up significs.

The statutes of the Academy, together with an invitation to join, were sent at the end of December 1916 to a number of prominent foreigners; the list of invitees contained among others the names of the candidates for the ill-fated *Forte-Kreis*, Walther Rathenau, Martin Buber, Gustav Landauer, Erich Gutkind, Ernst Norlind, and some more names mainly suggested by Brouwer. The mathematicians Schoenflies, Peano, Birkhoff, and Mittag-Leffler subscribed to the ideas of the Institute. In addition, Rathenau and the authors George Davis Heron and Allen Upward reacted positively. This rather meagre response did not put off the initiators, but did not spur them into action either. This situation changed when Brouwer got Mannoury to join the group. It is fair to say that Mannoury took the organisation in hand and saw to the realisation of the high-minded intentions.

In his book *De Hollandse Significa*,[56] Walther H. Schmitz has analysed Brouwer's conceptions that led to the original 'Academy'-approach. He has shown

[56] Schmitz (1990b).

that Brouwer's proposal was based on an earlier treatise of the philosopher–sociologist Ferdinand Tönnies, 'Philosophical Terminology',[57] in which an international academy of scientists is advocated. This academy of Tönnies was to be assigned a very special task: the wholesale upkeep and creation of terminology.[58]

Brouwer's idea clearly had a good deal in common with Tönnies's Academy. The founding declaration of the academy that was envisaged by Van Eeden bears the stamp of Brouwer. Content and formulation of the plan are reminiscent of *Life, Art and Mysticism.* And, like its predecessor, the Forte-Kreis, the Academy was to be based on merit and expertise.

The documents relating to the enterprise give ample information on the goal and the means of the academy, which eventually was baptised *International Institute for Philosophy*.[59] Its goal was 'the renewal of the valuation of the elements of life of the individual and society'.

The task of the academy was summed up as follows:[60]

1. To create words of spiritual value for the languages of the Western nations, and thus to give a place to spiritual values in the mutual understanding of the people of the West (therefore a 'declaration of spiritual values of human life').
2. To indicate in the present legal order and in the commercial production developing under its protection, the elements that most strongly suppress or stun the spiritual tendencies. And to propose restrictions which are desirable on that account, on the sphere of influence of legal order and technology.
3. To mark in the principal languages the words that suggest spiritual values for notions that ultimately root in the pursuit of personal security and comfort. And, for that reason, to purify and make precise the goals of democracy in the direction of a world-state with exclusive administrative competence.

It seems plausible to see Brouwer's hand in the above points; the influence of Van Eeden and Borel was probably restricted to minor details.[61]

Brouwer's interest in the academy project can be seen as a continuation of the line of thought of *Art, Life and Mysticism.* The mystic, apparently, was willing to suspend pure introspection for the sake of improving his fellow humans. In spite of his declared distrust of the power of language and communication, the part played by Brouwer in the signific enterprise shows that, far from renouncing the use of language, he had made up his mind to 'make the best of it'!

The group had set itself a quite ambitious goal; it wanted to avoid the mistakes of earlier philosophers who had attempted to carry out similar tasks single-handed, thus restricting the desired reform of language and thought to the author and a handful of readers. The social impact of these projects had therefore been negligible. The

[57] Tönnies (1899a, 1899b, 1900, 1906).

[58] Cf. Schmitz (1985), Tönnies (1906).

[59] *Internationaal Instituut voor Wijsbegeerte.*

[60] Manifesto of March–April 1916, cf. Schmitz (1990a).

[61] Cf. Schmitz (1990a), p. 220 ff.

present group saw its common activity as superior in that respect, since 'the group of independent thinkers would take on the same task *jointly*', thereby the 'insights *formed in mutual understanding* will have obtained automatically a linguistic accompaniment, *suitable to find a place in the mutual understanding of the masses*'. One could say that the members considered their exchange of ideas and formulations as a useful experimental testing ground for the use of their linguistic proposals in society at large.

On the last day of 1917 Brouwer and Mannoury crowned the months of discussion and planning at the notary's office by officially registering the *International Institute for Philosophy at Amsterdam*, on behalf of Bloemers, Borel, Van Eeden, De Haan, and Ornstein. Thus the Institute was born; it was a small but select band. Unfortunately the international character did not extend beyond its name; the war may have been to blame for that. There were a number of sympathisers, but the actual deliberations took place in Holland, and so we may take it that the international aspect was for the time being more theory than actual practice.

The International Institute for Philosophy was, in legal terms, a foundation (*stichting*). It had set itself a quite specific goal: the above mentioned *renewal of the valuation of the elements of life of the individual and society.* The means for realising this goal were:

> 1. The founding and maintaining of an International Academy for practical Philosophy and Sociology.
> 2. The founding and maintaining of a school for the dissemination of the notions and notion-relations[62] formed by the Academy.

The prospectus of the institute was sent to several thinkers, among whom were Martin Buber, Gustav Landauer, Erich Gutkind, Ernst Norlind and Walther Rathenau. Most of them were unaware of the history of the manifesto, in particular they could not be aware of the role that Brouwer's ideas played. Those who were familiar with Brouwer's earlier philosophical publications would have grasped the underlying intentions, but even Mannoury did not understand Brouwer's aims. He wrote much later[63] 'What Brouwer meant exactly by those 'words of spiritual value' of point 1, and how he planned to put those words in the service of the spiritual reformation of the world [...] never became quite clear to me.' Thus it is not surprising that the other invitees were uncertain about the enterprise. Both Buber and Gutkind reacted critically. Buber, who was already acquainted with Henri Borel, directed his comments to the latter. Borel had enthusiastically described the personality of his new friend, 'He is a professor of mathematics, and it is indeed this mathematics that has made him a mystic.'[64] When Buber sent his comments on the manifesto, Borel passed the letter on to Brouwer.[65] The objections of Buber were directed against

[62] begrippen en begripsverhoudingen.

[63] Mannoury *Nu en morgen. Signifische varia.* (unpublished) 1939. Cf. Schmitz (1990a), p. 226.

[64] Borel to Buber, 7 December 1916.

[65] Buber to Brouwer, 1918, Gutkind (1919).

Brouwer's idea of 'word-creation'. He questioned the possibility of the creation of words by a collective:

> The creation of a word is for me one of the most mysterious processes of spiritual life, indeed I admit, that in my perception there is no difference in essence between what I call here creation of a word, and that what one may call the emerging of the Logos. The genesis of a word is a mystery, that takes place in the kindled, opened soul of the world-dreaming, world-thinking, world-discovering man. Only such a word, generated in the spirit, can be generated in man. Therefore it can, in my opinion, not be the task of a community to make it. It rather seems to me that a corporation like the one planned by your friends, should and must only allow itself the cleansing the word as a goal. The abuse of the great old words is to be fought, not the use of new ones learnt.[66]

Buber's idea of 'word' is separated by oceans from Brouwer's; for Buber it is a mythical object, directly connected with the knowing subject and the designations. For Brouwer words are utterances intended to influence fellow men:[67]

> The word of the Occident does indeed in various cases have in addition to its material, a spiritual value, but the latter is always subjugated to the first, and where the first one has acquired a more certain and permanent orienting influence on the activity of society, in the sense that it induces isolated individuals, in their pursuit of physical security and material comfort, to hinder each other as little as possible and wherever possible to support each other; the latter lacks any influence on the legal state of affairs (except possibly when it is used itself for the devious realisation of injustice); therefore its influences are weak, fleeting and local. Words which have exclusively a spiritual value and which are suitable for the in- and exhaling of the *Weltgeist* and for the orientation on the observance of Tao are non-existent in Occidental languages; should they exist, then their influence would be paralysed by the mutual physical hatred, rooted in mutual distrust of the purity of their birth, of people who live too closely together, and which hampers the pursuit of material comfort by the single individual only mildly, but which considerably hampers the in- and exhalation of the *Weltgeist*. The introduction of the first word of exclusively spiritual value into general human understanding will, as a phenomenon, be inseparably connected with the understanding of the intolerability of this physical hatred, and thus will be the immediate cause of a legal regulation of human procreation. Only then will a possibility for this introduction be provided, if the 'Mystery of the genesis' of the word concerned, has not been taking place in the single individual, but in the mutual under-

[66] Buber to Borel, 17 March 1917.

[67] Brouwer to Buber, 4 February 1918.

standing of a society of clear feeling and sharp-thinking people, who are, by the way, not too close in the material sense.

Yours truly,
Prof. Dr. L. E. J. Brouwer

Clearly, Brouwer and Buber missed each other's point; this is certainly surprising, as Brouwer was a patent mystic, but he drew the line quite a bit earlier than Buber. Whereas Brouwer considered language as a rather external phenomenon that only fleetingly touched the inner ego of the individual, Buber was inclined to include language in the spiritual domain; thus an understanding, or even a compromise, was unlikely.

Gutkind's objections were of another kind; he doubted if European languages could simply incorporate or assimilate words of a high spiritual value. 'In any case there should be available [...] a higher form of society; the simple research society does not qualify.'[68] Brouwer replied:

> The word cannot wait for the higher form of society, because the higher form of society waits for the word. For on the one hand the timely creation of new words, fitting the new values, participates to a great extent in each liberation process, in each revolution; on the other hand, it is the lack of language-critical reflection, which is the cause that the self-revolutionising of society still proceeds discontinuous and with devastating crises, and which yields with respect to the necessities a continuous, enormous backlog, which keeps mankind in lasting tension.

It may be remarked here that Brouwer was not one of those scientists who preach the overriding importance of the word, and then go on expressing themselves with a minimum of regard for precision. His personal use of language was of a legendary refinement. Mrs. Vuysje, the daughter of Mannoury, described Brouwer's conversation as an almost artistic process: Brouwer could in the middle of a sentence pause to savour a particular expression, and to replace it after consideration by a more suitable one. She remembered him, standing in the garden stretching out his hand, tasting a particular word and contemplating fitting equivalents. The result was not a faltering flow of words; Brouwer had mastered the technique of refinement to such a degree that he managed to produce gentle, flowing sentences of incredible length without discernible interruptions. Of course, he could also lose his way in one of those excessively intricate sentences which he loved so dearly. Max Euwe recalled with admiration the beautiful, long sentences of Brouwer. Until the end of his life Brouwer had this precious gift of formulation.

The Institute from its inaugural meeting on 12 September 1917 onward met monthly at Walden where Van Eeden lived, at Mannoury's, or at Brouwer's, be it in the pharmacy in Amsterdam or his house in Blaricum. The attempt to gather an international group had utterly failed the whole organisation was in the hands of the

[68]Gutkind (1919).

small group that had been involved from almost the beginning of the project. The driving force from now on was Mannoury, who eventually came to dominate significs; he had become an extraordinary professor in Amsterdam in 1917, probably with the active help of Brouwer and apparently against the will of Korteweg. The minutes of the faculty meetings and the available correspondence suggest that a rift had appeared between Brouwer and his former promotor.

After long and earnest preparations and discussions, the International Institute for Philosophy finally got under way in 1918, observing all the necessary formalities to a nicety. Van Eeden, the most influential member at the national level, published a number of essays on significs in the weekly *De Groene*, and gave a talk on *Intuitive Significs* at the University of Amsterdam (13 March 1918). Brouwer announced this talk in *Propria Cures* with a brief introduction to the subject; the message of it was by-and-large identical with his comments on Buber:

> The means of production and the legal system of contemporary society accept human individuals as mutually completely separate centres of fear and desire, fighting each other or co-operating in the fight against third parties. Therefore nobody can be happy in the *activity of this unholy society*. And those who search for happiness or holiness elsewhere find in the words of contemporary languages, which are after all nothing but the command signals of the social labour rules, no impulses for energetic thoughts; at best they find in their sound and rhythm sources of moods, poor in consequences. If, now and then, in restricted groups better means of production and legal systems were created, which allowed for a holier way of life of the individuals, then these could not be lasting, because their maintenance and enforcement must rely on the *common language* as a means of communication, so that these reformers were forced to composes the signals of their new society out of that of the rejected society, and thus remained subordinate to the suggestions of the latter. *Intuitive significs* is concerned with the creation of *new words*, which form a code of elementary means of communication for the systematic activity of a new and holier society.[69]

In August 1918 Brouwer, Van Eeden, De Haan and Mannoury appointed Paul Carus, Eugen Ehrlich, Gustav Landauer, Fritz Mauthner, Giuseppe Peano and Rabindranath Tagore as members of the associated Academy. The first practical activity of the Institute was the compilation of dictionaries for the legal and socio-economic aspects of society; the topic turns up regularly in various unofficial documents, and it was discussed at numerous meetings, but concrete advances were not achieved. After the First World War new initiatives were taken.

---

[69] *Propria Cures* 9 September 1918.

## 7.6 Faculty Politics

In Amsterdam Brouwer had also become involved in faculty politics. In 1916 a vacancy came up in the chemistry department and the chemistry professor in charge displayed some tactics that rather put off Brouwer. The man went so far as to declare at one point that candidate X had better rights to the position than candidate Y, but that he nonetheless preferred Y as 'it was in *his* interest to promote Y, who would assist *him*'.

At roughly the same time Brouwer had again approached the curators of the university with his desiderata for the increase of the number of professors at the mathematics department. Since Korteweg was to retire in 1918, plans for his succession were being made, but Brouwer had independently presented his wishes to the board. Brouwer's plans were apparently overly ambitious; at least the curators, at their January meeting in 1917, observed that Brouwer had admitted in conversation that he had realised that he was asking too much. Nonetheless Brouwer made it clear that he would appreciate it if the university were to allow the mathematicians to appoint a lecturer. After ample discussions with De Vries and after a (more perfunctory) consultation of Korteweg, Brouwer suggested the appointment of Mannoury. Brouwer had already scored a minor victory in 1916: the university had granted him an assistant (for 600 guilders a year) beginning January 1917.[70] But Brouwer wanted more, and he had a clear goal: a superior mathematics department. Should the University of Amsterdam prove reluctant, he could easily move somewhere else. It was no secret that the state universities aimed each at two chairs in mathematics, and the minister seemed favourably disposed to grant Utrecht two chairs; Leiden would then soon follow. The curators of Amsterdam University seriously considered the possibility that 'Brouwer and Mannoury would then go to Utrecht, and Amsterdam would not be able to get such first-class mathematicians as replacement'. This might eventually lead to the departure of Prof. Hendrik de Vries. A proposal to urge the city council to prevent the departure of Brouwer, and to grant the immediate appointment of a lecturer, who could be promoted to full professor after Korteweg's retirement, was, however, not accepted, turned down, as the curators were of the opinion that nothing should be done to upset Korteweg; they well realised the obligation of the university to this father of Amsterdam mathematics. So it was decided that the matter would be arranged in agreement with Brouwer's wishes after Korteweg's retirement. At the January meeting the chairman informed the curators that the City Council had no objections to Brouwer's desiderata.

The lecturer position being out of the question, the faculty sent a warm recommendation for Mannoury's appointment as an extra-ordinary professor to the curators. The faculty lavishly praised Mannoury's geometrical work, 'of which one does not know what to admire more, the penetration of his intuition, or the powerful deductive ability'. In particular his 'surprising simplifications in the proof of Dedekind of the fundamental theorem of formalist arithmetic' was mentioned with approval. The faculty even went so far as to refer to Mannoury's verbal contributions

[70] Brouwer to Zeeman, 8 May 1916.

to science: 'In particular it has been noted at the meetings of the Dutch Mathematical Society, that Mannoury, if he is present, always takes part in a masterly manner in the discussion, and subjects of the most diverse character are equally perfectly commanded.' The faculty did not mention that Mannoury, in spite of his recognised ability had so far published only seven papers in mathematics, one book and an inaugural lecture as a 'privaat docent',[71] and that after 1901 he had not published a single research paper in mathematics.

The style and content of the above recommendation betray Brouwer's hand.[72] The reader will recognise Brouwer's formulation, cf. p. 43, of the laudation for Mannoury at the occasion of his honorary doctorate.

There was a second candidate: J. Wolff, Brouwer's fellow student, who passed his final examinations a year after Brouwer.

The proposal did not meet with serious difficulties, and on 24 April 1917 Mannoury was appointed. On 10 October he gave his inaugural lecture, *On the social significance of the mathematical form of thinking*,[73] in the presence of the Brouwers, Joop de Haan and his wife, and Frederik van Eeden.

Brouwer had in the past years seen enough of faculty politics to advocate more stringent arrangements. On 3 April 1917 the faculty, following a proposal of Brouwer, took two important decisions. It split the faculty into sections: (a) mathematics, physics, astronomy; (b) chemistry, ...; (c) biology, .... Furthermore a committee for the expansion of the faculty was installed, with Brouwer and Zeeman as members of section (a). Brouwer devoted a considerable amount of time to the reorganisation, but as it dealt mostly with the expansion of the chemistry department, we shall not go into the activities of the committee.

In the final years of the war the relationship between Brouwer and Korteweg seemed to have deteriorated, and both the negotiation for the Defence Committee (see below) and the procedure for Mannoury's appointment led to suspicions on Brouwer's part that Korteweg had turned against him. In how far this was the case cannot be ascertained. It is not unthinkable that Korteweg did not share Brouwer's enthusiasm for Mannoury. Although it was not hard to appreciate Mannoury's originality and sparkling conversation, it was almost exclusively directed towards politics and significs, and it could not have escaped a sharp observer like Korteweg that Mannoury's attention was no longer directed at mathematics. The cause of the estrangement cannot easily be pinpointed. In 1915 there was no cloud on the horizon; Brouwer and Korteweg had lively discussions on mathematical topics. Brouwer was at that time studying Korteweg's paper 'On the stability of periodic plane orbits',[74] and he wondered if Korteweg had missed a case. Korteweg truthfully answered that he had difficulty in recalling his considerations of 29 years ago, but he immediately returned to the problem of 1886 and started thinking of solutions to Brouwer's

---

[71] On the significance of mathematical logic for philosophy, 1903.

[72] Faculty to Curators, 9 January 1917.

[73] *Over de Sociale Betekenis van de Wiskundige Denkvorm.*

[74] *Ueber Stabilität periodischer ebener Bahnen, Wiener Berichte* 1886. Brouwer to Korteweg 15 October 1915.

questions. A day later Brouwer confirmed Korteweg's idea, writing 'I just found that your conjecture is correct, and that the possibility of a system of perturbed orbits, which I had in mind, is ruled out'.[75]

A little later Brouwer and Korteweg discussed a problem of triple tangents to algebraic curves. In rapid succession letters were exchanged,[76] and in the last one Brouwer wrote:

> I have been able to fix now the matter of the general presence of double and the general absence of triple tangents, even in a fairly simple way, so that it surprises me that I have not hit on the same idea on any of the three occasions upon which I 'discussed it in class'.

During this exchange some insensitive remarks must have been made by Brouwer, for on 11 December Korteweg referred to Brouwer's card of 2 December: 'A small remark about your card of 2 December. That we should henceforth never talk about scientific topics seems to me a rough remedy. It would, as I feel it, cause a distorted relation between us. The word 'talk' is not intended by me in contrast to a 'serious discussion', but to 'writing'. Even if I talk about a scientific topic, I am really 'serious'. Even though what I say it is not always correct, I strive for its correctness.'[77]

So much is certain that Korteweg did not bow to the whims of Brouwer and did not hesitate to correct him in the faculty meeting, if the occasion arose. In 1918, at the time that the new chairs in mathematics were arranged, the relation had already deteriorated to an intolerable extent. Brouwer complained in a letter to Mannoury about Korteweg's behaviour, 'K. has since two or three years no longer any cooperation from me, since I saw that almost all his efforts serve to cripple Dutch mathematics, efforts which he will doubtlessly continue to his dying day.' The correspondence suggests that Mannoury's appointment raised questions that could not just be brushed away.

One year later Mannoury became the successor of Korteweg as professor in Analytic and Descriptive Geometry, Mechanics and Philosophy of Mathematics. His course on philosophy was regularly devoted to significs, and so the topic gained a modest but secure foothold in Holland.

Mannoury was a kind and understanding teacher, revered by his students; he taught and published in a very personal style. There is a large number of publications from his hand of a philosophical–social nature; his inaugural address, with its title '*On the social significance of the mathematical form of thinking*' set the tone for his activities in the university and in society. After a long career he bade his farewell to the university with a valedictory lecture, entitled *The beauty of mathematics as a signific problem*.[78] After his retirement in 1937 he devoted himself to the writing

[75] Korteweg to Brouwer, 17 October 1915, 18 October 1915, Brouwer to Korteweg 19 October 1915.

[76] Brouwer to Korteweg, 29 November 1915, 1 December 1915, 2 December 1915.

[77] The reference to Brouwer's card is erroneous. There must have been a verbal exchange.

[78] Mannoury (1937).

of a comprehensive text on significs;[79] he remained unusually active. In spite of all good intentions, the signific movement was to remain an almost exclusively Dutch affair, and after Mannoury's death it gradually withered away.

Although Brouwer had already more than enough duties to keep him occupied, he felt that he had to keep an eye on the quality of the staff at the mathematics departments in Holland. And so he was naturally interested in the appointments of mathematics professors elsewhere. Strictly speaking, an appointment was a local matter, and external advice was usually purely a matter of courtesy or prudence and left to the discretion of the local faculty. Brouwer, however, considered the choice of mathematics professors, no matter at which university, of national interest. As a consequence he was always willing to give advice, or, as some would say, to meddle.

When in 1916 the mathematics chair at Utrecht became vacant through the death of Kapteyn, the physicist Ornstein wanted to appoint Brouwer. Apparently Brouwer was vaguely interested; he was willing to consider the matter if a third mathematics chair would be forthcoming. He was clearly aiming at a mathematical centre in Holland that could stand comparison with existing international centres. The faculty committee nominated two respectable, but somewhat unexciting candidates: Van Uven (who had a chair at the agricultural institute (*hogeschool*) and Rutgers. Ornstein opposed this choice, and put down his own list: Valiron, Denjoy, Rosenthal and F. Bernstein.

Brouwer, who was not a member of the committee, suggested a list of Dutchmen: Mannoury, Wijthoff, Droste, Rutgers, Boomstra and Bockwinkel. After ample discussions *en petit comité*, it was decided that a short list, consisting of 1. Fréchet, 2. Bernstein, would be proposed to the faculty, and, should the faculty turn this down, a second short list: 1. Droste, 2. Boomstra, as a kind of reserve. In March 1917 Brouwer again interfered; he now recommended to the committee the following candidates: 1. Denjoy, 2. Valiron, 3. Bernstein. Brouwer's number one was a young French function theorist with a wide interest. His œuvre contained papers on topology, set theory, function theory, measure theory, .... In topological matters he was close to Brouwer's interests. His fame was based in particular on his process of 'totalisation' and for the Denjoy integral.

The faculty adopted Brouwer's suggestion, and Denjoy was approached. He made two conditions: a salary of 6000 guilders and an annually renewable contract. The faculty was far from happy, because they feared, rightly so, that Denjoy did not want to give up his French position, and would only use Utrecht as a temporary base. However, the faculty appreciated Denjoy's reputation as a competent analyst of the French school, and the appointment was duly made on 19 July 1917. He indeed combined a French and a Dutch academic position, which must have made life pretty uncomfortable for him. During the war he regularly made the trip between Holland and France by sea, going via England. The dangers of this routine were far from imaginary, and on one occasion his boat to England was torpedoed; the faculty recorded in its minutes with some relief that Denjoy had safely reached Harwich.

[79]Mannoury (1947, 1948).

Brouwer and Denjoy worked in perfect harmony; they exchanged papers and manuscripts and Brouwer presented Denjoy's work to the Royal Academy, and nothing seemed to hint at the eventual violent disruption of their friendly relations.

In the period following his topological research Brouwer had, voluntarily or not, become involved in a number of activities outside the academic sphere. We have seen some of this above; some of Brouwer's actions may have been the unconscious fall-out of his recent contacts with artistic and political circles, but underlying all his projects and plans there was his strong conviction of right and wrong, his aversion to injustice, and a deep passion for the underdog. His rationality saved him from the fate of Cervantes' hero, but nonetheless he became the champion of a number of causes where a more sober and calculating man would have remained aloof. The period of the First World War was perhaps richer than usual of challenges to a man with the moral and social conscience of Brouwer. The following episode is another example of his involvement in politics.

## 7.7 The Flemish Cause

The direct cause for his activity in this case is somewhat unclear; Brouwer may have had some information from Belgian refugees, but it is more likely that his attention to the position of Flemish higher education was called forth by third parties. The correspondence with Schoenflies seems to indicate that certain German circles were trying to obtain support for German interference in civil matters in the occupied part of Belgium. On 1 July 1916, Brouwer wrote that he wanted to talk to some Belgians and to collect some data before he could answer Schoenflies' question. In his next letter, Brouwer said that he had talked to numerous Flemings, and that he had diligently studied the Belgian Constitution and the Law of Higher Education. The problem for which Schoenflies, or somebody else in Frankfurt, had tried to enlist Brouwer's assistance, was that of the position of higher education in Flanders.

In 1914 there was still no university for the inhabitants of Flanders with Flemish as an official language. There had been harmonious efforts by the Flemish community to redress this wrong, but the German invasion had brought the cooperation to an untimely end. Some Flemings suspended all actions and remained neutral, and others used the German intervention to get a Flemish University realised.

After reading the relevant legal documents Brouwer soon concluded that 'to my vivid disappointment I have *not* been able to convince myself that the German occupying authorities have the right to make the Gent University Flemish against the wishes of the Belgian Government'.[80] And thus he claimed the 'in dubiis abstine' for himself as a politically neutral person. It is not clear what exactly was asked from Brouwer; it could be that efforts were made to find professors for the Flemish University at Gent. He added, strictly confidentially, that he could possibly find Dutch colleagues, who were '*also politically*' and in particular in the matter of the

---

[80]Brouwer to Schoenflies 5 August 1916.

Belgian matter, completely on the German side, and who would probably be willing to make the desired efforts.' The letter was followed by another one[81] in which he explained that the The Hague Convention did not allow the German occupation authorities to change the legal status, as long as the public order did not force them to do so, and thus turning the University of Gent Flemish (*Flamandisierung*) could not be justified. He refused therefore to co-operate in this matter, 'no matter how unsympathetic the Belgian Government is to me, in that it has refused the Flemings this right [of their own university] for 80 years'.

Although he did not wish to interfere with the internal affairs of the Belgian state, Brouwer did publish an open letter to the Belgian Government in *De Groene Amsterdammer*, which contained a summary of the injustice done to the Flemish community. He concluded that the inclination of many of the Flemish to feel a quiet sympathy for the, unjustified, German efforts to turn the University at Gent into a Flemish university, was excusable in the face of the suspicion that the Belgian government 'would after the war, forgetting that four-fifth of the Belgian defensive army consisted of Flemish men, violate the Flemish rights as before'. Therefore he invited the Belgian Government to declare that it would recognise Flemish rights after the war by adopting an earlier proposal along those lines.[82] He closed the letter with the rather provoking words

> In this way it [the Belgian government] would provide to everyone's satisfaction the proof that it has the moral courage to refuse unconditionally, not only to hand over the total Belgian population to German imperialism, but also half of the Belgian population to French imperialism,

In the same year Brouwer considered an action against the German government on the issue of the use of Belgian forced labour. From the correspondence it appears that he could not get convincing corroboration; some rumours said that Belgian unemployed workers went voluntarily to Germany because of the high wages. Not only in the above cases, but in general, Brouwer was extremely careful to investigate the matter at hand before undertaking any action, and so in the latter case he probably did not want to proceed without sufficient grounds.

In 1917 Brouwer organised another protest action, this time in Holland against the treatment of a Dutchman, J.C. Schröder, who had been sentenced to three months in prison for making his political views public. Brouwer asked his colleagues at the Academy to sign an address against the judicial action, which 'can endanger in this way our complete constitutional freedom of the press, one of the main guarantees of our personal freedom'.

---

[81] Brouwer to Schoenflies 27 August 1916.

[82] De Nieuwe Amsterdammer 19 June 1916.

## 7.8 Air Photography and National Defence

In the middle of the war a totally unexpected thing happened: Brouwer showed national feelings. For a mystic with avowed internationalist views this certainly is surprising. The review article (p. 246) with its scathing remarks on the word 'fatherland' would have almost ruled out a sudden rallying to the defence of the nation. Nonetheless, there is a curious episode in which Brouwer had a brief infatuation with defence matters.

When Brouwer travelled in Germany at the occasion of his visit to Schoenflies, cf. p. 250 (which, as far as we know, was his only wartime visit to Germany), he learned that the Germans, and presumably the other belligerents, employed numerous young mathematicians for the transformation and measurements of air reconnaissance photographs. Furthermore, he heard that many Academy members were enlisted as scientific advisers of the general staff.

Wishing to kill two birds with one stone, that is to say, to stay out of the regular army and to serve his country, he had decided to start his own investigations in photogrammetry. He was completely serious about the importance of a systematic use of mathematical methods as a means to extract exact information from air reconnaissance photographs, which usually were taken under varying angles, and showed all kinds of distortions. As early as September 1915 he sent a memorandum to the minister of defence, drawing his attention to the importance of photogrammetry and offering his services.[83] The letter was answered by the chief of staff, General Snijders.[84] The general said that he was much indebted to Brouwer for his suggestions, but the army had at its disposal excellent topographic maps (1 : 25.000), and that it was very well possible to indicate the precise position of an object, observed from the air, on such a map. Here the matter ended as far as the army was concerned.

That Brouwer was not just day-dreaming may be inferred from the fact that he seriously studied the available literature, and prepared a number of papers, which were published in aviation journals.[85] After the rejection of his proposals by the army, there was little Brouwer could do. Another chance to promote his idea came, however, in 1917, when Professor Jaeger, a physicist from Groningen, agreed with Brouwer that it would be profitable for the nation if proper use were made of the intellectual capacities of its academic specialists. He wrote to Brouwer, asking him to get Zeeman interested, so that a small group of academicians, including Brouwer, could consult Lorentz on the matter. Apparently there was considerable discontent in academic circles with the firm intentions of army headquarters to ignore the possible impact of science on warfare and military management. As Jaeger expressed it: 'The military high command has had for three years the possibility of improving the army by more efficient use of the intellect that has been enlisted. The result was totally nil, because there is no initiative at all [...]. Thus, nothing can be expected from

[83] Brouwer to Minister of Defence, 18 September 1915.

[84] Snijders to Brouwer, 12 October 1915.

[85] Brouwer (1916a, 1916b, 1916c, 1917b, 1919l).

**Fig. 7.5** H.A. Lorentz. [Courtesy Nationaal Archief Noord-Holland]

the common sense and initiative of the military command, a fortiori not in times of panic. And thus it has to come from us.'

Formally, the Academy had every right to approach the government, as the rules of the Academy specifically allowed it to advise the government on matters of national importance. Zeeman supported the idea, and sounded Lorentz on it. When Lorentz also agreed, action was taken.

It may be remarked in passing that Lorentz wanted to go further and proposed that all professors would be excused from military (or para-military) service,[86] in order to be available for special duties as advisers. An attempt to get a dispensation for members of the Academy foundered, however. In December 1917, the Minister made it known that he saw no reason to put Academy members on a par with, members of the judiciary, civil servants of the social service, aldermen, and the like.[87]

Soon after the October meeting of the Academy a meeting with the responsible minister, Lelie, was arranged. The Academy was represented at this meeting by Brouwer, Jaeger and Schoute (the meteorologist). The government reacted positively on the proposals, and what would seem more natural than that the temporary committee, in particular the originators, Jaeger and Brouwer, would joining the executive committee? The authorities decided otherwise, however, and neither Brouwer nor Jaeger were included.

Brouwer was despondent; he was back at the dark depressions that ruined his student years. He started to draft letters to Lorentz, whom he held for a large part responsible for the unfortunate outcome. He was also convinced that his old teacher, Korteweg, opposed his plans; he may have had a point there.

Turning things over in his mind, Brouwer got more and more excited. The horror of another period in the army arose in his mind's eye, and he frankly declared that

[86]In 1915 the so-called *landstorm* was created, an organisation comparable to the National Guard. It was intended for volunteers who had completed their first military training, but who were not called up in the mobilisation.

[87]Brouwer to Zeeman, 17 December 1917.

this was more than he could stand.[88] He sat down and summed up all his grievances in a first draft, three days later he finished the letter itself, which was much more composed in nature; the draft reveals some of his strong emotions. He felt insulted because 'his unwritten rights as an initiator' were violated, without even a warning or an apology. In particular as the president of the Academy, Lorentz had the duty to guard the rights of the members. At this moment it seemed that Holland would remain neutral, and that the risk of being drafted had vanished. That was, in Brouwer's opinion, no excuse for the army to drag its feet in the matter of photogrammetry. For if at any moment Holland should get involved in an armed conflict, it was obviously too late to start build up a photogrammetric service.

The extent to which Brouwer's mental stability was threatened may be illustrated by a draft of the following letter from Mrs. Brouwer to Lorentz (possibly not sent, there is no copy in the Lorentz' archive).

> Dear Professor Lorentz, I am worried about my husband; he is so hurt that he is passed over as a member of the Commission for Scientific Research for National Defence, that I am at a loss what to do with him. Since he has heard this, he cannot do anything, and not so much because of the matter itself, but rather because of the hostility of his elderly colleague and teacher, whom he surmises is behind it. As you know he initially took on this new study in order to get, in case of a war, the opportunity to offer the country his services in a more scientific manner than otherwise his rank of corporal would require. Gradually he has pursued the study with an ever increasing interest, and prepared the instalment of the above mentioned committee. When the funds for the practical elaboration of his plans were made available, he was very glad, and quietly awaited, without suspecting any harm, the outcome. He thought to be able to get to work soon, and was full of plans for the future laboratory. And now this. Each day he wants to write to you, but gets so enraged each time, that he cannot find the right words. What will be the end? He will not resign himself to it, which seems even to me impossible. How everything turned out that way, can be guessed, but not ascertained. But an explanation will have to be given, why the *auctor* of the whole conception has been removed from the arena. How often do I deplore, that the hospitable and safe Leiden has beckoned him in vain, as a quiet hiding place for unrest and conflict. It would have been the fulfilling of my dearest wishes in the interest of my dear husband. And yet it were not the personal reasons or matters of health, also mentioned in your presence, but commitments of an indebtedness to those who had him appointed in Amsterdam, and who allowed him everything that the other side offered in advantages, [...].

This letter confirms the nature of Brouwer's depression. He felt deprived and insulted. Lize deplored in the letter the obligations that had effectively prevented Brouwer from accepting the chair in 'the hospitable and safe Leiden'. The letter

[88]Cf. p. 48. The passage occurs in the draft but not in the final letter (Brouwer to Lorentz 16 January 1918).

also hints at 'remaining years of struggle with another colleague', this is most likely a reference to Korteweg, who could be quite stern with Brouwer, and who deplored what he considered to be the *prima donna* airs of his student. Brouwer for his part grew dissatisfied with the limited scientific exchange with his old teacher. A degree of disillusionment on both sides cannot be denied, and Brouwer may very well have shown a lack of filial respect for the older man. But to what extent Korteweg's opposition was imaginary cannot be ascertained.

On 26 January Lorentz talked to Brouwer about the choice of the members of the committee and it appeared that Brouwer was prepared to accept the inevitable. Lorentz must have been struck by the importance Brouwer attached to the membership, so that he must have pulled some strings. On 15 February Lorentz wrote to Brouwer, informing him about the discussions in the planned committee, and asked if he was willing to join the project as an extraordinary member of a subcommittee, the one 'for the terrain photographs made by aircrafts'. Lorentz confessed that he felt rather out of place in the committee itself, 'surrounded by experimenters and technicians', but that the minister had insisted that the committee should contain some members of the academy.[89] Brouwer accepted Lorentz' offer, and so he found himself in due time an extraordinary member of no less than two subcommittees.

The Royal Decree of 20 February 1918 appointed Lorentz (chairman), Zeeman (secretary), S. Hoogewerff, C.A. Pekelharing and F.A.F.C. Went to the board of the 'Scientific Committee of advice and research in the interest of public welfare and defence' (*Wetenschappelijke Commissie van advies en onderzoek in het belang van Volkswelvaart en Weerbaarheid*).

Brouwer figured on the membership list of the various subcommittees of the Scientific Committee as an extraordinary member of the subcommittees for X-rays, and of the one for Photogrammetry. Lorentz himself was the chairman of the latter one.

On 2 May 1918 Lorentz wrote to Zeeman that he had discussed the photogrammetry problems with the other members of the subcommittee, Brouwer and Schoute (See above). The three had agreed that it was desirable to add a military expert to the subcommittee, the name of Major H. Walaardt Sacré, commander of the military aviation section at Soesterberg,[90] was mentioned.

Furthermore the members of the subcommittee thought it a good idea to have also a physicist on the subcommittee.[91] They suggested F. Zernike, who had already, as an assistant of Kapteyn, occupied himself with measurements of photograms. It fell to Brouwer to sound out Zernike; this apparently was unproblematic, since Zernike was duly added to the subcommittee.

Brouwer, Schoute and Zernike did not lose much time. On 20 June, Brouwer reported to Lorentz, that they had visited Soesterberg. They had found that the pho-

[89] It is a fact that Lorentz could not say 'no' to requests from the government. Another instance is his involvement in the 'Zuiderzee project', which took up much of his valuable time.

[90] A military airfield of long standing. After the second world war an American squadron was based at Soesterberg.

[91] As already suggested in Brouwer to Lorentz, 19 February 1918.

tographic service mostly took vertical shots, but the staff was readily convinced that in a time of war one had often to work with photographs taken from various angles, and that it was desirable to be able to draw conclusions from these photographs too. It was agreed that Lieutenant Meltzer would make a number of photographs of the city of Utrecht from various angles. It would then fall to Brouwer to transfer the information to the maps. The plans were, however, thwarted because there was no gasoline available for the flight!

Again in July the three visited Soesterberg, and this time inspected the cameras and the equipment. As a fair number of the cameras came from aircraft that had been interned on Dutch soil, they were far from uniform, differing in focus, means of orientation, etc. The previous arrangement was repeated, but before any results became available, the sad news arrived that Meltzer had died.[92]

In order to secure the investigations, Brouwer proposed to Lorentz to add two physicists from Zeeman's laboratory, Van der Harst and Bosch, as reserve officers to the Soesterberg staff.[93]

Brouwer's own investigations led him to inquire with our national aeroplane builder, Anthony Fokker, into the state of photogrammetry. Fokker sent copies of papers on the importance of air photography for surveys and on the Cranz–Hugershoff method. He told Brouwer that 'all experiences which were gained in Germany during the war, in particular also a number of excellent experts, who belong to the top in the area, are at my disposal'.[94] In the same letter he offered his cartographic services for future projects in air photography, in particular in the colonies.

The more the work advanced, the less place there was for a gifted amateur, such as Brouwer. He slowly withdrew from the area, but not without at least taking part in some of the photography runs. This early experience with actual aviation remained one of his treasured memories.[95] In 1920 he became a member of a committee that had to report to the ministry of war on 'air photography' and 'air cartography'. On 13 July a letter was sent to the minister which, Brouwer remarked, would probably be the final burial of the subcommittee.

During the last year of the war Brouwer had gathered enough insight into the intuitionism new-style, to start the preparation of a series of expositions on the subject. The first one appeared in 1918 and the following issues in 1919 and 1923; they were all published as individual pamphlets in the *Verhandelingen* (Acts) of the Royal Academy, two of them under the title *Begründung der Mengenlehre unabhängig vom logischen Satz vom ausgeschlossenen Dritten* (The founding of set theory independent of the logical proposition of the excluded third) and one on function theory with a similar title. The choice of his title was curious, and it invited misunderstanding. An outsider could easily get the impression that this was a logical project with a weakened axiomatic basis. As we have seen, Brouwer's

[92]Brouwer to Lorentz, 8 September 1918.

[93]First having obtained Zeeman's advice. Brouwer to Zeeman, 12 July 1918.

[94]Fokker to Brouwer, 24 March 1919.

[95]Oral communication C. Emmer, Brouwer's friend and family doctor.

constructive mathematics incorporated genuine *mathematical* principles (involving continuity and transfinite phenomena), so the choice of name could hardly be called felicitous. Maybe Brouwer had not yet made up his mind as to a proper name for his project, but it is also possible that he considered the name 'intuitionism' a bit grand for a one-man business. The term 'intuitionism' occurred frequently in his early foundational papers, but there it still refers to a vague kind of constructivism *à la* the French school, a restriction of classical practice, rather than an autonomous, legitimate mathematics. Only in 1920 did Brouwer start to refer in writing to his mathematics as 'intuitionism', cf. Brouwer (1919h).

One has to keep in mind that Brouwer was not after the founding of a school of constructivism in a small corner of the building of mathematics, but indeed after a complete revision of all of mathematics. Hence a label, no matter how tasteful, would always reduce his efforts to just another local specialisation. The emergence of the term 'intuitionistic mathematics' may therefore have been a by-product of the foundational conflict of the twenties.

The war years, although they kept Brouwer in Holland, did not bring him much spare time. His contacts with the group around Van Eeden, his teaching, research, his professional contacts in the University and in the Academy, kept him busy enough. Furthermore he cultivated a large circle of, if not friends, at least close acquaintances—another time consuming matter!

His contacts with his closest friend, Scheltema, remained rare. On one occasion the Scheltema family called in Brouwer's help to knock some common sense into the head of Carel's brother Frits, who had developed a virulent pro-German attitude. If and how Brouwer reacted has not been recorded.

The contacts between the two were mainly kept up through correspondence, Brouwer sadly sighed, 'I would be so happy to talk to you sometime, but Blaricum and Bergen are so far apart, and my spare days get so scarce.' Indeed, meetings were becoming a thing of the past. Even the birthdays were no longer occasions for visits. Brouwer reported in a letter of 25 February 1916, that he could not visit Scheltema on his birthday; he was not actually ill but

> if I stay out of bed for more than one hour, my heart starts to behave in a funny way, so that I stayed in bed for a whole week and read all the volumes of the adventures of Arsène Lupin.[96]

This minor detail shows that Brouwer, in spite of his acquired status, could still enjoy the pleasures of the adventure novels in the shade of recognised literature.

Scheltema had by this time become the pre-eminent socialist poet, with an immense readership, his poems were to be found in the homes of the working class, but not there alone. They were set to music and sung in the homes all over the Netherlands and by choirs of all parts of the population. In the years 1917–1918 he wrote a number of theatre plays, *The Nude Model*, *Great days in Knollenbroek*[97] and *The*

---

[96]The gentleman-burglar Arsène Lupin, was the hero of detective novels of M. Leblanc.

[97]A fictitious village.

*gilded Pretzel.*[98] The Brouwers got free tickets for the premiere of the *Great Days in Knollenbroek* and they intensely enjoyed the farce. Brouwer complimented his friend, 'Altogether I am grateful to you for your play, and I believe that you have shown yourself up to this genre.' A year later the Brouwers attended the performance of *The Guilded Pretzel.*[99]

The war had both positive and negative effects on Brouwer. It had given him a chance to concentrate in isolation on the tough problems of a revision of mathematics in the sense of his programme, but it had kept him from his natural audience: the Göttingen mathematicians. It also had introduced him into the circle of artists and intellectuals around Van Eeden. Mathematicians will probably deplore this as a waste of time, but it satisfied a need of Brouwer to take part in social-cultural life. After all he had strong convictions on a rich variety of issues. His moral position might have become more balanced, but it certainly had not evaporated.

So at the end of the war Brouwer had re-thought his position, both in the esoteric domain of mathematics and in the world of his fellow human beings.

[98] *Het Naakt Model, Grote dagen in Knollenbroek, De vergulde Krakeling.*

[99] Brouwer to Scheltema, 27 September 1917 and 8 September 1918.

# Chapter 8
# Mathematics After the War

The end of the First World War was greeted by Brouwer with relief. He immediately set out to re-establish his international contacts. During the war years he had lost contact with his foreign colleagues, and whether induced by his isolation, or as a consequence of his changed interests, he had almost totally dropped his research in topology. There was also a marked drop in international correspondence; among the remaining letters (which, of course, may totally misrepresent the actual number) there are only a few to and from Blaschke, Birkhoff, Carathéodory, Mittag-Leffler, Scholz, Kerékjártó, Buber, Study, Denjoy and Schoenflies. The correspondence with the last mentioned had miraculously sprung from their rather acid exchanges at the time of the Analysis Situs paper and the new edition of the *Bericht*. The two men had become good friends, and although Brouwer did not take Schoenflies quite seriously in his topological capacity, he had come to appreciate his older correspondent as a good friend and colleague. During the Easter holidays of 1915 Brouwer even managed to overcome the bureaucratic obstacles to visit Schoenflies in Frankfurt a.M.

We should not be surprised that Brouwer was keen to revive his old contacts. The Göttingen mathematicians were, however, one step ahead of him; he was elected to the *Gesellschaft der Wissenschaften zu Göttingen*, and in a letter of 28 August 1919 he warmly thanked Hilbert, whom he guessed to be the instigator of this honour. This was his first official recognition by a learned society outside the borders of his native Holland. As a matter of fact, Brouwer had already been put forward to the *Göttingen Gesellschaft* in 1917 by Hilbert, Landau and Klein. Brouwer was recommended for his topological work only:

> To Brouwer are due, in the first place, important contributions to set theory, the general solution of the problem of the invariance of dimension, and the generalisation of the Jordan curve theorem to space and $n$ dimensions; he has moreover applied his set theoretic methods with success to existence theorems from the theory of automorphic functions.

The proposal was for some reason withdrawn and submitted the next year, when it was crowned with success.

D. van Dalen, *L.E.J. Brouwer – Topologist, Intuitionist, Philosopher*,
DOI 10.1007/978-1-4471-4616-2_8, 

Brouwer was keenly interested in the developments in Europe after the war. After the armistice between the belligerents was signed (11 November 1918), he wrote identical letters to Klein and Hilbert, expressing his hopes for the future:

> May the healthy heart of your fatherland overcome the present crisis; and may the German lands soon prosper to unknown bloom in a world of justice.[1]
>
> This is wished you by
> your Brouwer.

As a matter of fact, Brouwer was seriously worried by the turn things were taking. Like most Dutchmen, he had abhorred the war, and having many friends in the camps of both the Central Powers and the Entente, he was keenly aware of the misery on both sides. At the end of the war one had not to be a trained politician to recognise the mood in the world; the victims of the German attack had made up their mind to exact war damages from the losers, and to make certain that Germany's role as a middle European power would be over.

Already in May of 1918 Brouwer and Van Eeden had approached the American consul in the Hague with a plea for a just peace. Brouwer had at the occasion proposed a conference of the scholars of the belligerent nations. Two months later another meeting was organised with a representative of the American legation.[2] Needless to say, their good intentions did not accomplish much. The event illustrates that Brouwer was willing to put his ideals into practice, irrespective of their chance of success. However unlikely it was that the two men were to succeed where even President Wilson failed to temper the feelings of revenge of the victors, one should keep in mind that the couple, Van Eeden–Brouwer, had no doubts about their duty. The spirit of the foundered Forte Kreis was still very much alive in their minds, and they took their obligations as spiritual leaders in a torn world extremely seriously.

The return to mathematics did not prove easy. After the hectic years of the great topological breakthrough, Brouwer had, partly because of the interruption of his international contacts and partly because his mind looked for other challenges, left the mainstream of the new mathematics that he had helped to create himself. We have seen his involvement in philosophical projects, and his return to the foundations of mathematics. Also, his academic duties and his sociopolitical conscience had to a considerable extent claimed his attention. So when he could return to the beckoning haven of topology, a multitude of other interests and obligations stood in the way. Politics, albeit not along any party-lines, ever tempted Brouwer to actions and views for a better world. In addition to that kind of activity, Brouwer had developed an interest in academic affairs. It could not escape anybody that he was by far the strongest man in the mathematical wing of the faculty. Korteweg used to have a certain natural authority over him, but with advancing years Brouwer started to

[1] Brouwer to Klein, and Hilbert, 25 November 1918. '*Möge das gesunde Herz Ihres Vaterlandes die heutige Krise überwinden, und mögen die deutsche Landen alsbald zu ungekanntes Blüte gedeihen in einer Welt der Gerechtigkeit! Das wünscht Ihnen Ihr Brouwer.*'

[2] Diary of Van Eeden, 22 May 1918, 24 July 1918.

resent the ideas of his old teacher, cf. p. 266. It is not difficult to see that Brouwer gradually became more and more interested, and bogged down, in academic organisational matters, appointment committees, and the like.

Although the war had not touched the Netherlands, the country did not escape its by-products; a general scarcity had made itself felt, food was rationed and the inevitable war-profiteering reared its head here and there. When the guns fell silent at all fronts, a wave of revolution and social unrest swept through Europe, scourging in particular Germany; it even reached Holland.

The most prominent socialist politician at that time was Pieter Jelles Troelstra. He was the leader of the *Sociaal Democratische Arbeiders Partij*. In this position he had to steer a middle course between the radical proletarians and the fashionable intellectual leftists. The latter were not quite taking Troelstra seriously; from the pinnacles of theoretical purity, they critically followed his exploits. Leading personalities like Henriette Roland Holst called him disparagingly, 'Troeleman'.[3]

On the same day that the armistice was signed in Compiègne, 11 November 1918, Troelstra called on the workers to take over the state. But neither the party nor the workers were prepared to take the required drastic actions. The Dutch revolution misfired, and Troelstra admitted his miscalculation two days later in the parliamentary debates. The more cautious and conservative part of the nation had shown its distaste for revolutionary adventures by a mass demonstration in support of the Royal Family on 18 November. As one can imagine, there was an understandable fear among the progressive intellectuals that Troelstra's rather rash imitation of the historical events elsewhere in Europe would result in a freezing of social progress in Holland. Among the reactions, we note an open letter to the Dutch citizenship, in *De Groene Amsterdammer* of 30 November, bearing the signatures of a number of prominent citizens, including Brouwer's, warning that the events of November were not isolated phenomena, and that it was an illusion to hope and expect that modest concessions would rid society of a discontent proletariat. The letter ended with an eloquent appeal to the press to provide a counterbalance to the conservative press:

> I think that we still are at the eve of great events. It will partially depend on the position and insight of the bourgeoisie, along which roads the coming changes will be guided. And in that process, the information given by the press will be of great influence. Let the inclination of a part of that press, to make an effort to keep the material interests of the bourgeoisie intact as long as possible with fake concessions or with force, be somewhat compensated by statements from the other side. Imponderabilia carry weight too.

This letter called attention to a number of problems that were apt to be overlooked in the euphoria after the failed revolution. It was published in the weekly of Brouwer's friend from his student days, Henri Wiessing. Wiessing had already become a prominent man in Holland; his political preferences were clearly on the far left. As a student, he had been active in the social democratic scene, and later he,

[3] A non-existent word, sounding like a pet name for a toddler, Wiessing (1960), p. 201.

like so many of his generation, became a communist, although *not* a party member. After a short career as a correspondent for the *Algemeen Handelsblad*,[4] Wiessing became, at the age of twenty-nine, editor-in-chief of *De Amsterdammer*.

After a conflict of a political nature, Wiessing left *De Amsterdammer* and founded his own weekly, *De Nieuwe Amsterdammer*, popularly known as *De Groene (Amsterdammer)*.[5] Wiessing succeeded in attracting writers of good standing, for example Van Eeden. Brouwer too wrote from time to time in *De Groene*; his choice of this particular magazine was probably more a consequence of his friendship with its chief editor than of its political colour. It would go too far to see in Brouwer's choice of *De Groene* a commitment to a particular political view, or even to the views of Wiessing; it was probably a matter of personal loyalty and partly a matter of convenience. As today, it was not always easy to find an outlet for one's ideas (in particular when they tended to be eccentric and radical, as Brouwer's views often were). When *De Groene* ended its short life in 1920 in a cloud of financial problems, Brouwer wrote a summary obituary in the last issue:[6]

> You have dared to wish to cultivate a magazine that would serve ideals instead of interests, principles instead of party leaders. Such a thing conflicts with the laws of the dynamics of society.

Brouwer had already published a review in *De Groene Amsterdammer* in 1915; it is rather curious that Brouwer had done so, since the subject was a mediocre pamphlet with the impressive title *Geometry and Mysticism*. It was written by a mathematics teacher, H.A. Naber. In spite of the trifling occasion, Brouwer could not let an opportunity go by to ventilate his insights.

> As the making and observing of mathematical forms in the *Anschauungs*-world is a preparation for, and a consequence of, the *intellectual* self-preservation of man, and since theoretical mathematics can only be defined as the activity of the *intellect in isolation*, and since, furthermore, mystical vision only begins after the intellect has gone to sleep, neither practical nor theoretical geometry can have anything to do with mysticism.
>
> Therefore, already the above title betrays in the author the lack of the first principles of epistemological insights that have in the last twenty years become widely accepted among the practitioners of mathematics and physics, in particular thanks to Poincaré, and in which our time, no matter how coarsened otherwise, possesses a wisdom that all historical periods of civilisation have lacked.
>
> The text of the book is for the greater part no better than the title makes us fear; for example, the following three facts are put forward in one of the three

[4]For a long time there were two quality newspapers that stood out in the Dutch newspaper world, the *Algemeen Handelsblad*, based at Amsterdam, and the *Nieuwe Rotterdammer Courant*. The two merged in the nineteen seventies.

[5]De '*Mosgroene*'.

[6]The *Groene* was later properly resurrected; it is still flourishing today in its own modest way.

> lectures: that a nice two-pronged fork can be drawn with the help of a pentagram; that in the old days by the word 'cross' sometimes a fork was meant; that the Pythagoreans revered the letter Y. Conclusion, hold your breath: the Holy Grail is really a divining rod.
>
> In addition to its repugnant title and text, the book contains, however, a collection of very beautiful illustrations; details of the cathedral of Amiens and other buildings; furthermore striking form rhythms from the animal and vegetable kingdom. The geometric analyses of some of the pictures made by the author are not uninteresting, and they suggest the conviction that what causes the aesthetic sensation in architecture is the impression that a large wealth of forms is governed by a few surprisingly simple geometric relations.

The last paragraph of the review is interesting as it hints at Brouwer's aesthetic views:

> . . .so the emotion of beauty is a purely intellectual sensation, and it is only the physical disturbance in the central nervous system caused by the above that awakens in our diseased bodies the frozen consciousness of God; a condition that can indeed also be produced by other physical means, such as isolation, fasting and the adopting of various Eastern attitudes.

It may be pointed out here that, for the average mathematician or academic, Brouwer's mystical convictions were all but invisible. The most strikingly philosophical passages had been removed from the dissertation, and the only philosophical paper, so far, had appeared in a rather obscure Dutch philosophical journal. The early book, *Life, Art and Mysticism*, had a rather limited circulation, and no more than a handful of Dutchmen were aware of its existence. Brouwer only started to publish his philosophical convictions in the late twenties. In the company of the significists, Brouwer freely discussed intuitionistic and philosophical matters, but little of it reached the general public.

## 8.1 How to Appoint Professors

In July 1918 Brouwer stirred up the otherwise quiet pond of the Dutch universities by launching a revolutionary proposal; in the *De Nieuwe Amsterdammer* of 27 July he published a letter to the editor which discussed the procedure for appointing professors. The Dutch universities and institutions of higher education had only at the end of the nineteenth century joined the company of prominent scientific institutions in neighbouring states. Around the turn of the century, the corps of academic teachers contained a number of excellent scientists, but at the same time there was a group of rather invisible scholars that may have excelled in teaching a somewhat cautious curriculum. This rather inhomogeneous company obtained its new members via co-option. In practice, if a vacancy occurred, the faculty got together, discussed the possible candidates, and after ample consideration proposed a candidate to the Board of Curators. There was no obligation to consult outside experts, neither was there a generally accepted check-list of required qualities. In fact this system survived in its basic form until the nineteen seventies. One did not apply for a chair; one was asked. As we have seen in the case of the appointment of Denjoy, Brouwer did not hesitate to interfere in matters of other faculties if he considered it necessary (and if he could make himself heard). He was evidently dissatisfied with the general procedures for the appointment of professors, in particular the stress on teaching. And so he formulated his views in the 'Groene':

> THE ARRANGEMENT OF APPOINTMENTS OF PROFESSORS
>
> Since in each country there ought to be institutes for scientific education, and that thus the principal character of the universities should not be abandoned in favour of their secondary task of vocational training, at least the full professors of the universities should exclusively be appointed on the grounds of their scientific merits. All other personal properties are, for the proper fulfilment of their position, of infinitely less, and moreover of a significance, which can only be established *a posteriori*.
>
> Now, how can guarantees be obtained that a comprehensive view of the scientific importance of the appointee has indeed been the decisive factor? By motivating the recommendation with an analysis of the scientific work of all eligible persons, carried out by qualified committee, under the obligation of publication in a scientific journal that is read by the international colleagues, to whom the committee thus becomes answerable. [...]

The editor in chief, Brouwer's friend Wiessing, had not taken the risk of promoting a discussion that would fall flat, so he had sent letters to all the members of the Academy, requesting their views on the appointment issue.[7] He put it to them that 'Where it is for fellow countrymen of an importance which hardly could be overrated, that the appointment of the scientific leaders in the country is made by means of a system which gives the greatest possible certainty that the most competent will be appointed, it seemed to us the responsibility of our weekly not to be content with the publication of professor Brouwer, but to ask the opinion of his confrères of the Royal Academy.'

The next issue brought two positive reactions. One of them, Ariëns Kappers,[8] gave his unqualified support to Brouwer's proposal, pointing out that 'students not only learn to practice a branch of science, but chiefly to further this branch of science, that they learn to work at the frontier of the field, and that their teacher encourages them and leads them in this respect'.

The other writer observed that the existing old-boy network was a serious obstacle for the proper functioning of the universities, and that thus Brouwer's proposal was a painful infringement on the academic etiquette.

*De Groene* of 25 January 1919 contained a selection of the replies to Brouwer's challenging proposal. Only five out of more than forty respondents were supporting Brouwer; the objections mainly concerned the public features of the proposed

[7] Wiessing to members of the KNAW, 21 October 1918.

[8] A former fellow member of CLIO, cf. p. 13.

proceedings. The reactions ranged from 'good idea, but it won't work' to 'nonsense'. Some of the respondents used the occasion to point an accusing finger at the City Council of Amsterdam, which indeed handled appointments of professors in a manner that differed from the procedures of the state universities. There had been occasions where the Council had not hesitated to ignore the proposals made by the professionals, be it by changing the order of recommendation, or by an appointment that had little to do with the advice of faculties or trustees.

A particularly negative reaction to Brouwer's proposals read:

> The preparation of professor's appointments, as desired by Prof. Brouwer, will, in my opinion, result in a big fuss. With that item we are already well-stocked.

Zeeman observed that under Brouwer's proposal gifted young candidates with few publications but with a bright perspective would not stand a chance. It is in this context worth noting that Brouwer himself was in such a position at the time that Korteweg tried to secure a position as a lecturer for him.

Brouwer replied in *De Groene* of 1 February, refuting the objections one by one. He referred, in reaction to the claim that 'In our country there is hardly a case known that a scientist of more than standard importance has not been appreciated', to the embarrassing history of Stieltjes:

> T.J. Stieltjes, undoubtedly the first Dutch mathematician of the nineteenth century and perhaps of all time, was here regularly passed over at the occasion of professor vacancies and had as a consequence to work under extraordinarily difficult circumstances, until France took pity on him and offered him a chair at Toulouse, where he died young. Had there been a system of appointments, as proposed by me, Stieltjes would have been saved for his fatherland and been spared many difficulties; not improbably he would have lived longer, and accomplished more. This gain, colleague..., would not have been bought too dearly with a little more fuss.

Brouwer also observed in his reaction that the 'Groningen system' of asking the advice of sister faculties in the case of vacancies was put forward as an argument that the old system worked perfectly well. He did not see the weight of this remark, as few faculties followed the Groningen example. Moreover, he added, 'I know a faculty in which a member who brought up the desirability of the Groningen system had little success, and got the remark "What is the use, everybody only recommends his own students".'

And here the matter rested. Nothing was done and professor's appointments remained largely the concern of small groups in the local faculties.

## 8.2 The Return to Topology

Although Brouwer had spent the war years mainly investigating foundational matters, and had more or less given up topology, topology had not given up Brouwer, and, the war being over, it was natural that mathematicians should turn to him for criticism and advice.

His own postwar activities in topology were largely 'unfinished business' from the first topological period. Apart from a number of corrections or additions to earlier papers, Brouwer published an elementary proof that a function, continuous on a closed subset of $\mathbb{R}^n$, can be extended to a continuous function on the whole space. This proof had the characteristic features of Brouwer's approach; whereas De la Vallée Poussin and Harald Bohr had proved the same result by means of series or integrals, Brouwer's proof contained just the bare essentials. The theorem had, however, already been proved by roughly the same methods by Tietze in 1914; it is now known as the Tietze extension theorem. Brouwer had missed the paper, which, by a curious coincidence, was published in the same journal that published Brouwer's last pre-war topology paper! Brouwer acknowledged Tietze's priority in a note a year later.[9]

Another paper that appeared in the *Mathematische Annalen* dealt with the topological aspects of the Lebesgue measure; it showed that measurable sets in the plane lacked topological invariance.[10] By means of an intricate construction Brouwer gave examples of closed, nowhere-dense subsets of the unit square that could be topologically transformed into sets with any measure between 0 and 1. Similar results had already been established for the real line, for example, by Bohl and Carathéodory, and by Brouwer himself. He had communicated this fact to Blumenthal in a letter in 1913. The measure theoretic style and technique is that of the Schoenflies *Bericht* from 1913, which is not surprising as Brouwer was in a certain way the godfather of that book.

Brouwer's main efforts in the postwar period were directed towards the topology of surfaces. The first postwar paper in the series 'On one-one continuous transformation of surfaces in themselves' (sixth communication),[11] presented to the Academy on 30 November 1918, seemed the continuation of the old series, but it was in fact wholly different in style and method. The paper was presented as an elaboration of a footnote in the letter to Fricke of 1912.[12] It offered a purely topological proof, and an extension, of an analytical theorem of A. Hurwitz[13] on birational transformations of Riemann surfaces of genus 1. The letter to Fricke belongs to the Koebe affair. In the above paper Brouwer used the opportunity to set the record straight on the matter—in particular the incident of the faked note, cf. p. 186.

The topology of 1918 was no longer the same discipline it had been before the war. The germs of the early years had developed into an independent subject, the significance of which for mathematics at large was fully realised. When mathematics resumed its normal life after the war, there were a good number of specialists in

[9]Brouwer (1918b, 1919c), Tietze (1914).

[10]Brouwer (1918c).

[11]Brouwer (1919i).

[12]Brouwer (1912d).

[13]Brouwer (1919i), Hurwitz (1892).

the subject; without pretending completeness, we mention M. Dehn, P. Heegaard, H. Weyl, F. Hausdorff, A. Schoenflies, M. Fréchet, W.J. Alexander, H. Tietze, G.D. Birkhoff, W. Dyck, A. Denjoy, and O. Veblen.

More and more mathematicians realised the importance of the subject, and incorporated topological methods and terminology in their papers and books. At the beginning of the twenties there were already some influential books and monographs dealing with the young discipline: A. Schoenflies, *Bericht über die Mengenlehre I* (1900, 1913), *II* (1908), M. Dehn and P. Heegaard, *Analysis Situs, Enz. d. math. Wiss. III* (1907), F. Hausdorff, *Gründzüge der Mengenlehre* (1914), H. Weyl, *Die Idee der Riemannschen Flächen* (1913), O. Veblen, *Analysis Situs* (1916), and W.H. Young and G.C. Young, *The theory of Sets of Points* (1906). The landscape of topology after the First World War became, thanks to the seminal work of pioneers such as Poincaré and Brouwer, a bustling scene. Brouwer's new methods were catching on, although it took some time before they were fully understood. A new generation of young topologists joined the field, and among the first to turn to Brouwer was a young Hungarian, Bela von Kerékjártó, who had already given a talk in Göttingen. Kerékjártó had submitted a paper to the *Mathematische Annalen*, and it was handled by Brouwer as the editor mainly responsible for topology. The topic probably stirred some of the old enthusiasm in Brouwer—he published in the years 1919–1921 some 14 papers on the topology of surfaces. The first one, in the *Mathematische Annalen*, immediately followed Kerékjártó's paper.[14] It opened with the words:

> The results of the preceding paper of Mr. v. Kerékjártó were known to me for several years; I told Mr. Bernstein in 1911, among other things, the following two theorems:
>
> 1. Every periodic one-one and continuous transformation of the sphere with invariant indicatrix is topologically equivalent to a Euclidean rotation.
>
> 2. Every involutary [i.e. periodic] one-one and continuous transformation of the sphere with inverting indicatrix is topologically equivalent to a Euclidean reflection in the centre, or in a plane through the centre.

Brouwer pointed out that his old proof was not simpler than Kerékjártó's, but that it was easier to generalise. He listed a number of generalisations in the paper, which combines the techniques of Riemann surfaces and combinatorial topology in a manner of quiet routine, but it does not introduce new unexpected insights. The topic was pursued further in a paper presented to the Academy in Amsterdam.[15] In a subsequent series of papers Brouwer carried out the classification (enumeration) of continuous or topological mappings of various surfaces, for example the torus and the projective plane.

Brouwer was sufficiently taken with Kerékjártó to try to find him a position in Holland. In 1921 he asked Ehrenfest if there was a possibility to get him a temporary job in Leiden; the answer, regrettably, was negative.

[14] Brouwer (1919m), Hurwitz (1892).

[15] Submitted, 29 March 1919. Brouwer (1919j).

That Brouwer was not blind to the mathematical shortcomings of the young man appears from the covering letter of a recommendation for Kerékjártó, written to F. Riesz.[16] He pointed out that Kerékjártó was '(both in human and scientific respect) restless and rash', and that he was not certain that he would ever 'conquer these traits sufficiently to become a dependable researcher and a useful teacher'. Being the responsible editor for most topological contributions to the *Annalen*, Brouwer had seen ample evidence of Kerékjártó's performance as a researcher: 'Of all his submissions so far, only a small part was, after thorough revision, fit for publication: the greater part had to be rejected without any hope of redemption.'

Brouwer's interest in topology got a second impulse from the young Danish mathematician Jakob Nielsen, who had learned his topology from Dehn in Kiel. After serving in the war in Belgium and Turkey (in the German army, he became a Danish subject only after the war) Nielsen visited Göttingen in the summer term of 1919. There he submitted a paper to Klein for the *Mathematische Annalen*, and Klein duly asked Brouwer to have a look at it and sent him the manuscript.[17]

Brouwer answered a month later that the contents were valuable, but that the paper had to be straightened out both in presentation and precision. Nielsen soon contacted Brouwer, and a correspondence ensued.

Apparently Brouwer did a thorough job analysing the paper, for Nielsen replied that, already in 1912, he was well aware of the weaknesses in the dissertation, but that under the pressure of the time schedule he could not afford to elaborate the matter.[18] Brouwer was truly impressed by Nielsen, and he did not hesitate to express his appreciation.

The short episode of the Nielsen paper is instructive because it illustrates how Brouwer handled papers for the *Mathematische Annalen*. On the twenty-first of October he reported to Klein that he had started to check the paper, adding that he was certain that correct and direct proofs would emerge, provided he could be sure that the author would not be allowed to discuss the printing of the paper simultaneously with the managing editor. This seems to suggest that editors and authors were sometimes confronted with conflicting instructions; Klein, it appears, was sometimes easier to convince than his editorial staff. One can easily imagine the possibilities for conflicts!

Brouwer added that

> Only because Carathéodory strictly maintained this principle with respect to Kerékjártó, I have been able to get something good out of the young Hungarian and only because Blumenthal was too lenient with Juel, a lot of confused nonsense could be published.[19]

Brouwer did not leave it there. He even went so far as to say that he would welcome a guarantee that a paper, sent to him to handle, could only be accepted by

[16]Brouwer to Riesz, 22 November 1921.

[17]Klein to Brouwer, 12 September 1919.

[18]Nielsen to Brouwer, 18 October 1919.

[19]Brouwer to Klein, 21 October 1919.

him—especially in this case where the author had contacts with Göttingen. Klein's answer (if there was any) is not known. For Nielsen, Brouwer's editorial activity did work out all right; on 9 November the paper was accepted.[20] The paper dealt with fixed points of topological transformations of the torus. Nielsen's work on fixed points matured in 1927, he established a lower bound for the number fixed points of continuous maps for certain surfaces. His work has been generalised to wide classes of topological spaces, and the minimal number of fixed points has become known as the *Nielsen number*.

Nielsen's work had re-awakened Brouwer's interest in the topic. He wrote a paper, immediately following Nielsen's paper in the *Mathematische Annalen*,[21] extending the results to continuous mappings of the torus and the Klein bottle. Both authors showed that the minimal number of fixed points is determined by the homotopy class.

The last paper of the postwar series of Brouwer's papers on the topology of surfaces dealt with the characterisation of continuous mappings of finitely connected surfaces.[22] In March Brouwer presented another paper on one-one continuous maps to the Royal Academy; this paper also dealt with the minimal number of fixed points.[23]

Although Brouwer had gradually turned his mind almost exclusively towards intuitionism, topology was to re-enter his life dramatically in 1923, when suddenly a young topological genius, Paul Urysohn, emerged and continued the research where Brouwer had left off. This story will have a prominent place in later chapters.

With the return of mathematics to 'business as usual' journals too were active in picking up the threads. Brouwer in his role of editor of the *Mathematische Annalen* got his share of new papers, among which there was one of a clever young Dutchman, Jan Schouten. Schouten had become an expert on tensor calculus, and he had submitted a paper which applied his techniques to the theory of relativity. After consulting the Bonn mathematician Study, Brouwer rejected the paper. In a letter to Felix Klein, the chief editor of the *Annalen*, he wrote, 'in the first place because the author does not understand the art of presentation, and in the second place (what is more important) because his achievements consist, briefly said, of the veiling of the results, found earlier by inventive authors, in a new (but thick and opaque) robe'.[24] Klein, who probably had more affinity with this brand of differential geometry, found it difficult to see Brouwer's point of view. He turned to Hermann Weyl for a second opinion. Weyl confessed that he was not quite without prejudice in the matter,

> Brouwer is a person I love with all my heart. I have now visited him in his home in Holland, and the simple, beautiful, pure life in which I took part there

[20]Nielsen (1920).

[21]Nielsen's paper was submitted on 15 January 1920 and Brouwer's on 20 January 1920.

[22]Brouwer (1921b).

[23]Brouwer (1920).

[24]Brouwer to Klein, 19 September 1919.

for a few days, completely and totally confirmed the impression I had made of him. He is certainly not domineering; the violence of his behaviour towards Koebe as well, at present, towards Schouten, I believe, is mainly based on the circumstance that his being revolts against the impurity that it discerns here. [...] I don't know enough about the facts to play the judge here; but I think that Brouwer's reproach is not wholly without grounds. [......]

His big paper on the *Analysis zur Relativitätstheorie* contains, I believe, nothing new. (I must admit that I could not work my way through the symbolism of Schouten.) The way in which he mentions there the notion of parallel shift (mentioning Levi-Civita only casually, whose paper of 1917 he claims to have read only after finishing the manuscript, although he gets the Rendicotti sent from Palermo, and this paper is made use of in other papers which he quotes) does not quite make a reasonable impression.

The rejection was followed by a vitriolic exchange of letters between Brouwer and Schouten and then the matter petered out. Finally, in 1929, a reconciliation was reached through the mediation of Van der Waerden and Weitzenböck.

## 8.3 The Offers from Göttingen and Berlin

The above mentioned visit of Weyl to Brouwer was not a mere coincidence. Brouwer was playing a subtle game; the post-war activities had brought certain changes at the German mathematics department. In Göttingen a vacancy had arisen as a consequence of Hecke's move to Hamburg, and in October 1919 the faculty had offered Brouwer a chair.[25] Without exaggeration this could be called the apex of Brouwer's career: at the age of 39 he was offered a tenure position at the world's leading mathematics department. The offer, by the way, once more underlined the undisturbed harmony between Brouwer and Hilbert. Where Klein might have had his doubts as to the desirability of the gifted but obnoxious hothead, Hilbert was not plagued by any such doubts. We may safely assume that he was only vaguely aware of Brouwer's foundational program. Apart from the translation of his Inaugural address and the Rome paper, only a few foundational remarks of Brouwer had appeared in English or German.[26] To Hilbert he was a topologist with a side interest in foundations. Brouwer's reaction to this offer of a chair in Göttingen has not been preserved; this is rather surprising, after all, the event certainly must be considered the ultimate recognition of his status. The absence of documents could possibly be explained by the fires that have destroyed part of Brouwer's private archive.

There were more parties interested in the Dutch topologist; it never rains but it pours: the philosophical faculty of the University at Berlin had also decided to

[25] Akten der Phil. Fak. Az. II PH/36–d, Dozenten Generalia, 30 October 1919.

[26] German: Brouwer (1908a, 1914, 1918a), English: Brouwer (1913b), Dutch: Brouwer (1907, 1908b, 1918a).

make Brouwer an offer; on 19 December it proposed to the Minister of Science, Art and Education to fill the vacancy, left by Carathéodory's departure,[27] by Brouwer, Herglotz or Weyl (in this order). Brouwer was recommended warmly for his work in topology,

> His papers deal mainly with Analysis Situs, for which he has provided the firm foundation, long sought in vain. Among the big results of his research we only mention the proof of the invariance of dimension and the decomposition of space by means of closed surfaces, the theorem of invariance of domain, the theorem of the existence of fixed points of continuous deformations of the ball, and the far-reaching group theoretic researches on continuous transformations. These theorems, of which Brouwer had already made the most fertile applications for the proof of the basic uniformisation theorems in function theory, promise also undreamed-of consequences for the modern theory of physical and astronomical differential equations. Brouwer is equalled in the originality of his methods by none of the mathematicians of the younger generation. Brouwer has a most extensive and deep knowledge at his disposal, not only of pure, but also of applied mathematics. He speaks German fluently, and has given excellent talks in the German language for academic societies. His political persuasion is on the whole sympathetic towards Germany (*deutschfreundlich*).[28]

Given the choice between Berlin and Göttingen, the former may have appealed more to Brouwer than the latter, and indeed for the same reason that Amsterdam appealed to him. The cultural atmosphere of Berlin probably suited him better than that of the (after all) provincial town of Göttingen. Berlin and Göttingen both had a reputation for mathematical research and for top mathematicians, but Göttingen had Hilbert, and Hilbert was the universally accepted icon of mathematics. Hilbert's influence was due to a number of factors; his mathematics was superior not only in the sense of quality, but also of elegance, choice of topics and innovative character. As an organiser and teacher, he was second to no one, and a worthy master in the tradition set by Klein. Mathematically speaking, Göttingen would thus have been the natural place for anybody offered the choice. Brouwer's apparent but undocumented preference for Berlin, therefore, probably had something to do with non-mathematical attractions.

Hilbert summed up the matter quite aptly in a letter to Weyl, in which he declared himself unable to see why Weyl should choose Berlin:

---

[27] Carathódory left Göttingen in 1918 for Berlin. In 1920 he accepted a call from Athens, and subsequently went to Smyrna in order to organise the founding of a mathematics department in the new university.

[28] Bestand Philosophische Fakultät, Humboldt Universität, Nr.1468, Bl.313. The recommendation was signed by Carathéodory, Erhardt Schmidt, Planck, H.A. Schwarz, Nernst, Cohn. The qualification 'deutschfreundlich' was taken quite seriously after the war. The German nation felt itself misunderstood and surrounded by hostile nations.

> Which I can quite well understand in the case of Brouwer and Landau—Brouwer only wanted temporarily to stay in Berlin, and getting to know Berlin, just as the nimbus of getting an offer in the capital, were his motives...[29]

As it happened, Brouwer was visiting Berlin in late December 1919 in order to re-establish, on behalf of the newly established International Academy for Philosophy, the contacts with the people from the Forte Kreis. He was wished *bon voyage* by his friend Henri Borel with the words: 'Please give Gutkind and Rang, a gem of a man, a firm handshake, and a respectful kiss from me to Mrs. Gutkind.'[30]

Brouwer's friends and colleagues were seriously worried that the honour of a chair at one of the great Universities of Germany (and of the world), would be more than Brouwer could resist. Jacob Israël de Haan wrote to Van Eeden in January 1920, 'I hear that Brouwer is going to Göttingen, quite a blow.' De Haan was misinformed about the place, but not about the threat that was posed to Dutch mathematics, and, what was closer to his interests, to significs. Lorentz, too, showed his concern about Holland's possible loss of Brouwer. He wrote Brouwer that he hoped that Brouwer would stay in Holland.[31]

Brouwer was indeed using this trip to Berlin with the intention to engage in some bargaining. He was offered enough incentive to start a new life in Berlin, but he would be perfectly willing to remain in Amsterdam, provided the City Fathers could make him a similar offer. He did visit the Gutkinds, as intended. We will return to that part of the visit in a later section.

In a flurry of negotiations Brouwer had to hurry back to Amsterdam to attend the faculty meeting on the twenty-first. Five days later Brouwer had a personal interview with the Mayor of Amsterdam, Mr. J.W.C. Tellegen. The results must have been satisfactory, for Brouwer felt that he had a reasonable prospect of obtaining a suitable compensation from the City of Amsterdam in exchange for turning down the Berlin offer. He left for Berlin almost immediately, much to the chagrin of his friend Frederik van Eeden, who wrote in his diary 'I heard that Brouwer was in the country and left again for Berlin. He had wounded himself with some glass, when he kicked in a window in his house, because it was closed. And that without telling me. He loves me after all in another way than I love him.'[32]

In Berlin considerable pressure was put on Brouwer to get him to join the faculty. Von Mises, writing from Dresden about mathematical business, expressed his hope that Brouwer would accept, or maybe would already have accepted, the Berlin

[29] Hilbert to Weyl, 16 May 1920. ETH Hs.91: 606.

[30] Borel to Brouwer, 20 December 1919.

[31] Lorentz to Brouwer, 25 January 1920.

[32] Van Eeden's diary, 27 January 1920. Brouwer was indeed short tempered where such obstructions as closed doors or the like were concerned. Brouwer used to drop in occasionally to see a journalist of the *Handelsblad*, living in Blaricum, with whom he was on good terms. Once, when he did not find the man in, Brouwer got so angry that he kicked in a window (oral communication, Mr. Crèvecoeur).

offer.[33] It was generally seen as a matter of considerable prestige to get Brouwer attached to the university.

The Board of Curators of the University met on 2 February and discussed the threatening loss of Brouwer for Amsterdam. As the Mayor was by City rule the President of the Board of Curators, he could give the information to his fellow curators 'straight from the horse's mouth'. He listed the conditions that Brouwer had formulated, and that were submitted by the faculty: an extra credit of 10,000 guilders for books and journals for the mathematical seminary, two lecturers and a raise of Brouwer's salary to the legal maximum. The Curators decided to ask the City Council for one lecturer and to consult the faculty on the other two items.

Brouwer, knowing the date of the meeting of the Curators, lost no time in inquiring after the results of the meeting. He wrote from Berlin to the Mayor[34] that he had been received in such a friendly way and with such offers and concessions, that, should he have to disappoint the University of Berlin, he would insist on breaking the news in person, rather than to resort to a written statement. The letter also contained a business proposal with respect to the request for a mathematics library. Would it not be possible, he asked, to convince the Board of the City to provide him with the money he had asked for immediately, so that he could get the books and journals for a much better price than when ordered from Holland? Even though the gigantic sum of 10,000 guilders would buy the university a nice collection, he could buy two or three times the quantity of books on the spot by direct negotiations.[35] He planned to order them in person in Leipzig (where the publisher Teubner had its home base). We have already seen that Brouwer judged an independent mathematics library of the greatest importance, cf. p. 251. He wished to follow the Göttingen example of a library where mathematics staff and students could have direct access to books and journals, without the involvement of an extra organisation that had to get the books out of the storage rooms, to register the books, etc. In short, he wanted a collection that was optimally geared to the needs of the researcher and the student. In his letter to the Mayor he called this collection the '*handbibliotheek*' (reference library).

The Mayor answered a week later, he was happy to inform Brouwer that things had gone well for him.[36] The City Council had shown no objections to grant Brouwer's wishes, provided Brouwer would stay in Amsterdam. There were, however, a few minor details to be settled. In the first place the Mayor could not just put the promised 10,000 guilders at Brouwer's disposal. At the very least Brouwer would have to confirm his commitment to stay in Amsterdam. In the second place the transfer of such sums to a foreign country required special caution. These matters could possibly be settled later. As to the lecturers' positions, the Mayor had

[33] Von Mises to Brouwer, 28 January 1920.

[34] Brouwer to Tellegen, 4 February 1920.

[35] Although the galloping inflation was still two years away, the German *Reichsmark* was not as strong as it used to be, and certainly weaker than the guilder.

[36] Mayor to Brouwer, 12 February 1920.

hinted at the meeting of the City Council that some money might be saved by offering the lecturers' positions in combination with a part-time teaching position at one of the municipal gymnasiums or high schools.

A week later Brouwer was back in Holland; he had definitely chosen to stay in Amsterdam. The German colleagues still had high hopes of winning over Brouwer, as a letter of Schoenflies showed. He had already asked Brouwer if it was true that he had accepted the chair in Berlin, but Brouwer answered him that he would stay in Amsterdam—he had received a very attractive offer.[37] The rumours about Brouwer's Berlin chair circulated nonetheless through Germany. The *Jahresberichte* reported that Brouwer had accepted the chair in its 1919 volume: *Professor Dr. L.E.J. Brouwer an der Universität Amsterdam wurde zum o. Professor der Mathematik an der Universität Berlin ernannt* (JBDMV, 1919. p. 61 Italics). Van Eeden could set his mind at ease; the signific enterprise would not founder on account of Brouwer's departure. He wrote in his diary that Brouwer had dropped in to see him on the nineteenth, and that on his return home Brouwer had found two laundry baskets full of mail. Brouwer settled with relief in his beloved Het Gooi. Van Eeden visited him on the twenty-fourth at home and joined the Brouwers three days later for Brouwer's birthday party. Brouwer was approaching forty and had reached the top position in Dutch mathematics, the maximum salary, and the largest department in the country. His demands had been fulfilled, and he could now build himself a department comparable to that of Göttingen.

The official aspect of Brouwer's agreement was efficiently taken care of. The Curators were told by De Vries that two lecturers were really necessary, and so the faculty obtained the promise of the appointment of two extra lecturers. This eventually proved, as we will see below, the cause of some embarrassment.

The lecturer positions remained vacant for the time being. Brouwer had, as we will see, made new plans. The City Council held its part of the bargain and voted Brouwer the maximum salary, 10,000 guilders. The official decision was conveyed to Brouwer on 27 April.

## 8.4 The Academy—How Denjoy Was Elected

In 1919 new members of the KNAW were to be elected, and some lobbying had already been going on. Already in January, Hk. de Vries had asked Korteweg to support the candidacy of his former Ph.D. student F. Schuh, and to write the required letter of recommendation. De Vries was doing so on behalf of a number of fellow Academy members, including Brouwer. Frederik Schuh was six years Brouwer's senior. He had defended a dissertation on enumerative geometry one year before Brouwer's doctorate. After being accepted as a 'privaat docent' in Groningen, his career took him first to Delft (1907–1909), then to Groningen (1909–1916) and finally back to Delft (1916–1945). In 1906 he had been second on the list for a chair

[37] Schoenflies to Brouwer, 16 February 1920, Brouwer to Schoenflies, 22 February 1920.

in mathematics in Amsterdam, when the vacancy left by Van Pesch was filled by Hk. de Vries. Schuh was a mathematician with a traditional expertise in geometry, analysis, number theory, statistics, etc. His publications dealt with almost all topics under the sun, but they did not earn him a niche in the Pantheon of mathematics. In Holland he was well-known for his efforts to raise the standards of secondary mathematical education. Whole generations of mathematics teachers were raised on books of Schuh. In addition he was a good expositor who also exercised his talents in the more mundane papers and magazines. When still a third year student, the University of Amsterdam had crowned his prize essay in mathematics. By 1920, Schuh had published a considerable number of papers on various subjects, for a large part in the area of applied mathematics, but also on the theory of algebraic curves. Korteweg did point out that 'Not all the papers on the list are of course of the same value. A smaller number has been written for mainly pedagogical reasons. Yet, also in these, the astuteness of the author, who has carried out a useful task for our future mathematicians with these papers, is to be praised.'[38]

In 1919, apparently, the general opinion was that Schuh would eventually make his mark (and, to be fair, compared to some of the incumbent members of the academy, he was quite an acceptable candidate) and so the lobbying for Schuh started. The candidacy of Schuh was supported by Kluyver, Kapteyn, Jan de Vries, Brouwer and Hk. de Vries. Some time after the start of this campaign, Brouwer must have felt that the Academy deserved a better candidate, and he proposed Denjoy. The recommendation of Denjoy sketches in flowing words his merits in 'almost all areas of function theory and set theory', but in particular his invention of the notion of 'totalisation', which closed the chapter on the inversion of differentiability. This document was signed by Kluyver, Cardinaal, Hk. de Vries, Jan de Vries, Korteweg, Kapteyn, Lorentz and Brouwer, so it had a solid support. In the May meeting of the section for the physical sciences, neither Denjoy nor Schuh got enough votes to get elected. Six rounds of voting did not yield a positive outcome for either of them, Schuh being far ahead of Denjoy all the time.

In 1919 a leading foreign mathematician got elected: Brouwer had, with the support of Hk. de Vries and J. de Vries, put forward David Hilbert's candidacy. The recommendation took only half a page. For a man like Hilbert this was more than enough! Brouwer simply stated that

> David Hilbert, professor at the University of Göttingen, who has added many mathematical theories, of which some did not exist at all before, others had been in a chaotic situation during longer periods, as monuments of crystalline simplicity to the spiritual property of humanity. These theories are in particular concerned with invariants, algebraic number fields, integral equations, the principle of Dirichlet, and the axiomatising of geometry, arithmetic, and physics.
>
> It is not surprising that the Bolyai prize, which is awarded every five years to the scholar whose total oeuvre can be considered to have exerted the greatest influence on the general development of mathematics, and which was for

[38]In his second recommendation of 1920.

the first time awarded in 1905 to Poincaré, has been presented for the second time in 1910 to Hilbert.

Of course, Hilbert was elected without any hesitation. In May 1919 he was appointed as a foreign member of the KNAW.

The Secretary of the Academy, Van Bemmelen, worried by the number of candidates for the ordinary membership, and the possible unpleasant discussions, had proposed a voluntary reduction of the list for the next year. The proposal was adopted, and in consequence Brouwer dropped the Denjoy proposal in favour of Schuh when Korteweg called a meeting of the mathematical members and advocated a reduction in line with Van Bemmelen's proposal.

When, however, some members of the Academy revoked the agreement, Brouwer no longer felt obliged to support Schuh, and he again pushed the candidacy of Denjoy. Another private reason for reconsidering the matter was the information that Denjoy's announced departure for Strasbourg had become uncertain and, if realised at all, would take place much later. Brouwer's *volte à face* probably irked some of his fellow academicians, for on 27 March 1920 he distributed a printed pamphlet among the members of the section mathematics and physics, which contained the above arguments. It closed with the lofty statement that the withdrawal of the Denjoy candidacy in favour of Schuh was, in his eyes, a disgraceful injustice, which his conscience allowed him to support only if, and as long as, a considerable effect could be expected for the cause of preventing the greater injustice of unreasonable elections of academy members. As this justification had disappeared, he no longer felt free to suppress Denjoy's name on the list, as he was a man 'whose genius and international fame could not be compared to Schuh's many and all-round merits'.

The candidacy of Schuh was this time supported by Korteweg, J. de Vries, Kapteyn, Cardinaal, Hk. de Vries, Brouwer and Kluyver, and the candidacy of Denjoy by the same members plus Lorentz. Brouwer again wrote the letter of recommendation, in which he praised Denjoy as 'one of the most gifted younger representatives of the mathematical school of Paris', who had 'created, as an application of Cantor's transfinite induction process, the operation of *totalisation*, of which one can say that it, with its capacity for inverting, without exception, any differentiation, has ended a period of development of the theory of real functions'. Korteweg was the spokesman for the Schuh appointment; he vigorously supported him, but to no avail.

Brouwer's eloquent arguments, and (one would like to think) Denjoy's superiority won the day. The correspondence contains a cordial letter from Denjoy in which he expressed his gratitude for the appointment and for Brouwer's efforts, which he conjectured were behind the operation.[39]

This short episode may serve to illustrate the influence of Brouwer among his colleagues—it is no minor feat to succeed in getting a candidate elected against a gentleman's agreement of the majority! As to the comparison of Schuh and Denjoy,

[39]Denjoy to Brouwer, 29 April 1920.

**Fig. 8.1** Hermann Weyl. [Courtesy Niedersächsische Staats- und Universitätsbibliothek Göttingen]

one is now inclined to side with Brouwer, but at the time the choice was not all that clear to the members concerned. Whatever the justification was, the outcome was painful for Schuh, who never was admitted to the Dutch Olympus of scholars.

## 8.5 Negotiations with Hermann Weyl

After having turned down the offers of Göttingen and Berlin in exchange for far-reaching promises of the City of Amsterdam, Brouwer started to contemplate the use he should make of the present opportunity to build a strong mathematics department. His explicit goal was to found a mathematical institute in Amsterdam that could compete with Göttingen, and there is no doubt that the concessions made by the City Council were concrete signs of his successful bargaining. Later in life Brouwer used to refer to 'the City Council's promise of a second Göttingen'. In this form no evidence has been found in the archives, but there is little doubt that the promises made in 1920 embody what Brouwer used to call 'his Göttingen'. It is more than likely that in the discussions with the Mayor, a man of consequence in his twin role of President of the Board of Curators and of Mayor of Amsterdam, terms of this sort were used. His first attempt to raise the level of the Amsterdam mathematics department was to offer a chair to Hermann Weyl.

Brouwer and Weyl met no later than the summer of 1912, during Brouwer's visit to Göttingen. There is a postcard from Brouwer to his friend Piet Mulder (5 June 1912), signed by Lize, Weyl, Weitzenböckand Bernstein, which documents the contact of the two. This card also solves a minor riddle in Brouwer's files. The archive contains a little notebook with notes on the problem of the schoolgirls. This problem was formulated in 1850 by Kirkman. It runs as follows: *Fifteen young ladies in a school walk out three abreast for seven days in succession: it is required to arrange them daily, so that no two shall walk twice abreast.* The problem was, of course, generalised to arbitrary numbers of 'girls and days'. The required lists of arrangements are called Kirkman systems.

Apparently Brouwer was interested in the problem in his student years. At least, there is a postcard from P.H. Schoute of 1 December 1903 in which the literature on the problem is listed. Mulder had been working on the problem for his Ph.D. thesis,[40] and this must have reawakened Brouwer's interest. Brouwer wrote to Mulder that 'For a couple of days I am taking part in the local routine, which is quite pleasant for an outsider. I have explained the 15 schoolgirls to the company here. See you soon.' This is the only instance where Brouwer actually tried his hand on a combinatorial topic. Since there are no recorded results, we may assume that he did not pursue the matter. Most mathematicians have their taste and preferences and Brouwer was no exception, he was too much of a geometer to get seriously involved in other areas.

Weyl was thoroughly familiar with Brouwer's topological work, there is an explicit reference to Brouwer's work in the 'Idea of the Riemann Surfaces' (1913).[41] In the introduction he generously acknowledged his debt to Brouwer's innovations in topology: 'Three events decisively influenced the composition of my book: the fundamental papers on topology of L.E.J. Brouwer, dating back to 1909, ...'[42]

Also, at the time of the uniformisation conflict, Felix Klein had asked him to write a critical evaluation of the Brouwer–Koebe controversy,[43] so he was thoroughly familiar with Brouwer's mathematics. In July of that same year Weyl reported in the Göttingen Mathematics Society on Brouwer's work on the invariance of dimension and of domain. Weyl was four years younger than Brouwer, he had studied in Munich and Göttingen, and he was highly regarded by his teachers and fellow students for his balanced personality and his truly great insight in mathematics and physics; he was equipped, moreover, with an exquisite sense of beauty. Both Hilbert and Klein were greatly impressed by the young man. One might say that he was, among the bright students drawn to Göttingen, Hilbert's favourite son. In 1913 Weyl was appointed in Zürich, where he was safe from the horrors and the inconveniences of the war. Already before the war he had won instant fame at the age of twenty eight with the above mentioned brilliant exposition of the theory of Riemann surfaces and its applications.

In 1918 he had further enhanced his status as a leading researcher and expositor, with two books: *The Continuum* and *Space, Time and Matter*.[44] The latter was an exposition of the theory of relativity, to which Weyl had significantly contributed, and the first one was the formulation and execution of a program for a version of constructive mathematics, to be precise, of predicative mathematics. Weyl had, like so many of his contemporaries, been worried by the paradoxes that threatened to paralyse mathematics. However, whereas most mathematicians considered the paradoxes, roughly, as a newspaper-reader considers an earthquake in a far away exotic

[40] Mulder (1917).

[41] *Die Idee der Riemannschen Flächen.*

[42] The two other events were Koebe's solution of the uniformisation problem and Hilbert's treatment of the Dirichlet Problem.

[43] Weyl to Klein, 6 May 1912.

[44] *Das Kontinuum; Raum, Zeit und Materie.*

country—an interesting phenomenon but too far away to worry about, Weyl felt that indeed everyday analysis (the theory of real functions) was jeopardised. We will return to Weyl's foundational work later in this chapter (cf. p. 308).

After the war Brouwer and Weyl met again in Switzerland; during the summer vacation Brouwer stayed in the Engadin, where Weyl visited him.[45]

With a little bit of luck the occasion could have been grander. Hilbert had visited Switzerland just before Brouwer, as it appears from a postcard of the latter to Hilbert, saying how sorry he was to have missed him. It would have been of decisive importance for the foundations of mathematics if the three of them could have sat down and discussed the new developments! Hilbert would have had the story of Brouwer's new intuitionism first hand, and it could have changed the course of the coming foundational conflict. These are history's little quirks.

For a cosmopolitan person like Brouwer it was a relief to be able to travel again, even though the circumstances were not wholly without risk. In Wiessing's autobiography there is a small passage referring to Brouwer's experience on the way back from Switzerland. Discussing the revolutionary events in Germany, Wiessing wrote 'Brouwer, the mathematician, returned yesterday from Switzerland. Everywhere along the Rhine he saw the revolutionaries with red bands around the arms and with guns in their hands, men who laid down the law.'[46]

The personal contact between Weyl and Brouwer must have opened Weyl's eyes to the deeper issues of constructive mathematics, for he immediately mastered the Brouwerian insights and started to present them in his own way. We will discuss Weyl's involvement in the intuitionistic reshaping of mathematics below, see Sect. 8.7. Brouwer was very pleased with Weyl's grasp of the issues and with his support.

At that time Weyl was a candidate for a chair in Berlin (cf. p. 291), and not much later Brouwer must have invited Weyl to Amsterdam. From the correspondence it becomes abundantly clear that Brouwer felt that he had found the right man to join him in Amsterdam. With a man like Weyl he was prepared to take on the mathematical establishment in an attempt to set mathematics straight! And thus he set himself to organise a chair for Weyl in Amsterdam.

The international contacts, including the crucially important visits to colleagues and universities, being resumed, Brouwer spent his next summer holiday after the war abroad once again; in August 1920 he breathed again the pure air of the Harz mountains. On this occasion he must have bought a house in Bad Harzburg to invest Cor Jongejan's money.[47] This may be inferred from the fact that in April he still stayed at the hotel Lindenhof in Bad Harzburg, whereas in August Bertus, Lize and Cor were living in their own villa, called *Friedwalt*, in Harzburg.[48]

---

[45] Cf. Brouwer to Fraenkel, 28 January 1927.

[46] Wiessing (1960), p. 33.

[47] Inherited from her father, see p. 249.

[48] The villa actually is still to be found in the Krodothal.

In a letter of 7 September 1920, written in Bad Harzburg, Brouwer discussed the position in Amsterdam with Weyl. The letter is significant, as it gives us a glimpse of Brouwer's views on the centre of mathematical activity that he had in mind.

> With respect to mathematics in Amsterdam, I am in no way pursuing the plan to organise there an intensive lecture and seminar activity, but only to bring together a circle of people whose mathematical papers are mainly stimulating and controlling additional phenomena of their general spiritual development, that is of people who feel themselves more or less the thinking organ of the community and who frankly put the immediately concrete academic teaching practice on the second place for this vocation. (I view, indeed, the mathematical drive for knowledge—which is basically distinct from the joy of the solving of mathematical problems—the hallmark of the state of mind which guarantees in the most diverse moral and practical domains a wide and free view, which considerably surpasses the common view.) I may add that we mathematicians in Amsterdam have won in the last years a very high measure of academic freedom, and make use of it in the above sense.
>
> Furthermore we are respected in our faculty (of the sciences), and our discipline is there regarded without scepticism. With the other faculties (maybe partially with the exception of the medical faculty) we have, to be sure, roughly the reputation of Bolsheviks.
>
> As to assistants, I have one,[49] who administrates the reading room and who has worked out some of my courses. My colleagues have none and wish none. You can certainly have one as soon as you wish. Apart from the salary (which, by the way, will probably be raised again in the near future) there are no lecture fees. You can of course, if you prefer so, live in a suburb, like me, if necessary in Zandvoort at the North Sea, 30 minutes by train from Amsterdam. Rent and taxes will together make an amount of over 1000 guilders and less than 2000 guilders; in the suburbs both are considerably less than those in the city. My colleagues De Vries (Vossiusstraat 39), or even better, Mannoury (Koninginneweg 192), who has four children at the school going age, will be able to inform you accurately about schools. The schools in town are excellent. I have heard less praise about the ones in the suburbs, but there they certainly are also tolerable. There is even a German school in Amsterdam; about its quality nothing is, however, known to me.
>
> As to your official language, you automatically have permission to teach for two years in a language other than Dutch; this permission is then, when necessary, extended year by year. Ehrenfest already lectured in Dutch in his second year; for Denjoy, however, it is already the fourth year that he lectures in French in Utrecht. At the end of this week I will again be in Holland and travel from there to Nauheim. So please write [...] to me in Laren. The courses start on the first of October.

[49] J.H. Schogt, later succeeded by Belinfante.

> Give my respect to your wife and accept with the greetings of my wife, a warm handshake from
>
> your
> Egbertus Brouwer

Weyl contemplated Brouwer's offer with more than routine interest. He had reached a point in his mathematical development which asked for radical decisions. He reported to the Board of the ETH in Zürich that

> At the meeting of the 'Naturforscher' at Nauheim, Professor Brouwer conveyed to me an offer from the University of Amsterdam. At the rejection of the offer from Berlin, Brouwer got permission for a sizeable extension of the Mathematics Department at the Amsterdam University; the position that was offered to me is a new chair in function theory, to be founded as a consequence of this plan. It is above all my close relation to Brouwer—we work jointly for a revolutionising and re-founding of analysis—which makes me consider the offer in earnest. I would in any case like to make myself familiar with the matters at the spot, before I make a decision, and thus I will travel this very vacation to Amsterdam.[50]

In spite of the attraction of a close co-operation with Brouwer, Weyl preferred to stay in Zürich, provided certain conditions were met. In negotiation with the authorities of the ETH Weyl indeed obtained all he wished.[51] In a letter of 20 January 1921 to the President he confirmed the rejection of the offer of Amsterdam. An important reason for declining the Amsterdam offer was Weyl's poor health. On account of his asthma he had to withdraw from time to time to the mountains with their clear air. As a matter of fact, the ETH granted him a regular vacation in the first months of the year.

Weyl announced his research plans, with a request for financial assistance. He hoped to be able to publish the first results soon: 'It concerns, on the one hand, impulses to a new foundation of the analysis of the infinite, the present foundation of which is in my opinion untenable; on the other hand, in close connection with that and in connection with the general theory of relativity, a clarification of the relation of the two basic notions with which modern physics operates, that of *field of action and of matter, the theories of which can at present by no means be joined together consistently.*'

To conclude our account of the postwar period of normalisation, let us return to the vacant chairs in Amsterdam, Berlin and Göttingen. Brouwer had turned down both offers, strengthening his position at home. Weyl, similarly had preferred to remain in Zürich. He rejected the offers from Berlin, Göttingen and Amsterdam. In December 1920 he informed Brouwer of his decision. Brouwer was sincerely disappointed; like Hilbert he was much taken with the virtually universal young

[50] Weyl to the President of the ETH, 27 September 1920.

[51] Cf. Frey and Stammbach (1992).

mathematician, who not only knew his way around in mathematics and physics, but also in philosophy, and who displayed an unbounded enthusiasm combined with a solid integrity.

> When your telegram arrived, my disappointment was very great, which surprised me and from which it really became clear to me how I would have loved to have you here.[52]

The letter makes it evident beyond doubt that the professional relation had been transformed into a mutual friendship, which may have gone through a long period of hibernation in the thirties and forties, but which was still strong enough to be rekindled at the end of Weyl's life.

Brouwer was for the moment at a loss. He wrote Weyl that he hoped to succeed in attracting 'another young personality with a really special reputation'. Bieberbach, he thought, would not qualify, 'as he is not sufficiently known outside the strict professional circle of mathematicians, and in addition he does not yet have a completely unchallenged name as an innovator'. For the chair offered to Weyl, Brouwer thought perhaps of Birkhoff, 'whom also the astronomers and the specialists in mechanics know, and who is especially considered as a star of the first magnitude, ...'

He ended his letter with a cordial

> Now my dear chap, here is to 1921 for you and your family. May the mountains bring you full health and vitality. I long now for a sojourn there: I do not feel well at all the last weeks, and every day, if I do not lay down for a few hours, I have a considerable temperature at night. In itself this does not worry me, because I have been through many such periods; should my situation remain the same much longer, then I will have to request a vacation, and I will have to go to Switzerland for some time. We would then see each other really soon.[53]

And thus ended Brouwer's first attempt to create a mathematics department of the highest magnitude.

## 8.6 Intuitionism and the *Begründungs*-Papers

It is about time to retrace our steps to the last year of the war; after a relatively quiet period of reflection on the foundations of mathematics, Brouwer returned to the revision of mathematics, a project that had withered somewhat after the surge of interest following the paradoxes and Zermelo's axiom of choice. The comparative rest of the period 1914–1918 had given him an opportunity to think over the problem of infinity and the nature of mathematical objects, and thus the foundations of mathematics once again became the prime target of Brouwer's efforts. He had discovered

[52]Brouwer to Weyl, 1 January 1921. ETH. HS.91:493.

[53]In March Brouwer asked for and obtained a three month sick leave.

that it made sense to consider infinite sequences of well-defined mathematical objects, such as natural numbers, which were *not* given by a law, and he had become convinced that these sequences, called *choice sequences*, were the answer to a host of questions in mathematics. The first observable result of these reflections was a substantial exposition of the new constructive mathematics: '*Founding mathematics independently of the logical theorem of the excluded middle. First part: general set theory*.'[54] This paper was published by the Academy in Amsterdam in 1918. It was followed by a second and third part, respectively in 1919 and 1923. This first paper (in German), in which the basics of Brouwer's intuitionistic mathematics was presented, was a purely *technical* paper; it contained no philosophical remarks. Neither, and this is surprising, did the paper contain the term 'intuitionism'! It is worth noting that in the 1920 paper *Intuitionistic Set Theory* Brouwer for the first time adopts the name 'intuitionism' for his new program. His earlier use of the term 'intuitionistic' covered, as we have seen, a more amorphous collection of principles and methods. It certainly includes the French (semi-) intuitionists. On some occasions, notably his inaugural address *Intuitionism and Formalism*, Brouwer called his mathematics 'neo-intuitionistic', to introduce his 'new, improved intuitionism'. 'Neo-intuitionism' was clearly no more than a convenient term for attracting the reader's attention; it never caught on and Brouwer obviously had no use for it. In his later publications he spoke of (the French) 'pre-intuitionism' and (his) 'intuitionism'. The term was survived, so to speak, in the influential book *Einleitung in die Mengenlehre* of A. Fraenkel.

The title of the *Begründungs*-papers may have misled readers who did not go on to read the body of the paper. In the first place, an uninformed reader might expect to find a Cantorian or perhaps axiomatic approach à la Zermelo to set theory. In the second place he might think that the author advocates some sort of mathematics based on a restricted logic. In both cases he would be wrong; the sets (*Mengen*) that Brouwer introduced were made up of choice sequences subjected to specific conditions, and logic does not figure at all in the papers (nor is the principle of the excluded third mentioned!). In Brouwer's terminology there lingered the old aim of presenting an alternative to Cantor's set theoretic universe. Already in the dissertation Brouwer had listed the possible kinds of sets, and now he was carrying out a similar project for his new intuitionism.

One gets the impression that Brouwer on the one hand wanted to lay down the principles for a conceptually sound mathematics, but on the other hand did not want to estrange the mathematical community. Perhaps for that reason he cautiously stuck to the positive aspects of intuitionism; in particular he did not repeat the dissertation's attacks on all and sundry. Nor did he present his counterexamples, based on unsolved problems. Compared to his missionary dissertation, the *Begründungs*-papers seemed a harmless intellectual exercise of another eccentric mathematician.

---

[54] *Begründung der Mengenlehre unabhängig vom logischen Satz vom ausgeschlossenen Dritten. Erster Teil, Allgemeine Mengenlehre*. Brouwer (1918a). We will refer to the papers of this series as the *Begründungs*-papers.

If the leading exponents of non-constructive abstract mathematics got to see the papers at all, it probably would not have cost them a minute of their sleep.

Yet, the very first paper contained a revolutionary principle that by itself refuted the practice of traditional mathematics (*classical mathematics*, as Brouwer started to call it). It seems highly unlikely, however, that anybody discovered the sting in Brouwer's new approach, since the papers were published as little separate monographs, and did not even appear in the Proceedings of the Academy, so potential readers were few. In short, the 'marketing' of Brouwer's new ideas was as low key and uncontroversial as only a learned treatise can be.

The first paper starts right away with the definition of the main concept, given in one sentence of 13 lines. In a later publication he repeated this definition and added dryly: *Dem Umstand dasz die Mengendefinition langatmig ist, ist leider nicht abzuhelfen.* (The circumstance that this definition of set is long-winded can unfortunately not be helped.)[55]

A *set* (*Menge*) is what nowadays is called a *spread*; the terminology is rather unfortunate, since Cantor had already fixed the meaning of the term set as 'collection to a whole $M$ of definite, well-differentiated objects $m$ of our intuition or our thought', which has become the standard usage in mathematics, and which was already firmly entrenched in Brouwer's days.

The confusion was further increased by introducing the term *Spezies* (species) for what we would call, following Cantor, set, in Brouwer's words, a property that a mathematical object may have.

In discussing Brouwer's intuitionistic mathematics we will, as a rule, stick to the modern terminology, that is we will use 'spread' and 'set' for *Menge* and *Spezies*. The definition of spread mystified almost all of Brouwer's contemporaries. Karl Menger in his 'My memories of L.E.J. Brouwer'[56] reported that set theoreticians like Hausdorff and Fraenkel failed to understand Brouwer's phraseology. Since the spread became the centre piece of Brouwer's mathematics, we will reproduce here the original definition:

> A *spread* is a *law*, according to which, if over and over an arbitrary number of the sequence $\mathbb{N}$ is chosen, every of these choices either generates a certain sign or nothing, or blocks the process and brings about the definitive destruction of the result, where for every $n$ after each unblocked sequence of $n-1$ choices, at least one number can be indicated, which, if it is chosen as the $n$th number, does not result in checking the process. Every sequence of signs that is generated in this way by the spread (which, thus can in general not be represented as finished) is called an element of the spread. We will denote the common mode of coming into existence of the elements of a spread $M$ also as *the spread* $M$.[57]

[55]Brouwer (1925a). In the first presentation the definition is given in *one* sentence running over *eleven* lines!

[56]Menger (1979).

[57]Eine *Menge* ist ein *Gesetz*, auf Grund dessen, wenn immer wieder ein willkürlicher Ziffernkomplex der Folge $\zeta$ gewählt wird, jede dieser Wahlen entweder ein bestimmtes Zeichen, oder nichts

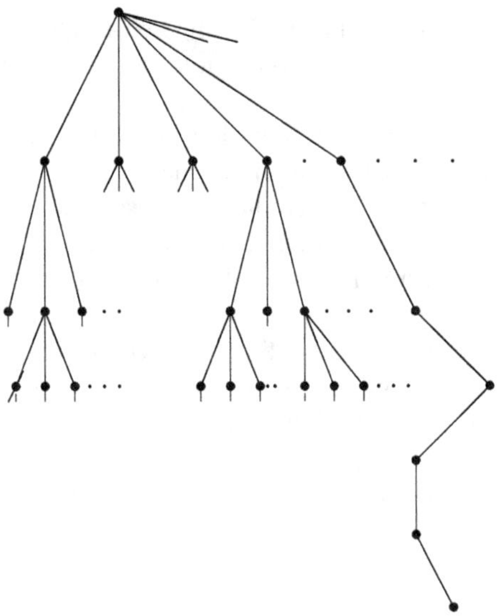

**Fig. 8.2** Underlying tree of a spread

The spreads and their elements make up the basic material of Brouwer's mathematics. It is convenient to visualize spreads by means of tree-pictures, Fig. 8.2.

The subsequent choices of natural numbers are indicated in the tree by the labels at the nodes. After each unchecked sequence of choices, at least one choice does not check the process, so below each node, at least one other node is eligible according to the spread (law). This means that, going down via the unchecked nodes, there are no finite (maximal) paths. In general, choice sequences are obtained as follows: unchecked choices generate signs; this means that after a finite number of choices $a_0, a_1, \ldots, a_n$, a particular object (sign) is generated; that is to say, we have an assignment of objects $s_{a_0,\ldots,a_n}$ to sequences $a_0, a_1, \ldots, a_n$. So the unchecked nodes have an extra label $s_{a_0\ldots,a_n}$. And the infinite sequences of these labels are the elements of the spread.

Imagine putting a cross at each checked node, then the *elements* of the spread can be visualised as the labelled infinite paths avoiding crossed nodes.

The simplest example of a spread that comes to mind is obtained by allowing at each step every natural number and by assigning the last number $a_n$ to the sequence $a_0, a_1, \ldots, a_n$. The result is the spread of all choice sequences of natural numbers. Of course, it is only a matter of convention to call the last mentioned sequences 'choice sequences of natural numbers'. One could also have considered the infinite

erzeugt, oder aber die Hemmung des Prozesses und die definitive Vernichtung seines Resultates herbeiführt, wobei für jedes $n$ nach jeder ungehemmten Folge von $n-1$ Wahlen wenigstens ein Ziffern-komplex angegeben werden kann, der wenn er als $n$-ter Ziffernkomplex gewählt wird, *nicht* die Hemmung des Prozesses herbeiführt. Jede in dieser Weise von der Menge erzeugte Zeichenfolge (welche also im allgemeinen nicht fertig darstellbar ist) heisst ein *Element der Menge.* Die gemeinsame Entstehungsart der Elemente einer Menge $M$ werden wir ebenfalls kurz als *die Menge* M bezeichnen.

sequences in the underlying tree as 'choice sequences of natural numbers'. The main idea is that one allows choice sequences to consist of elements of a previously given (or constructed) collection.

In the above definition the choice sequences are somewhat 'played down'; the spread appears as the primary notion, and the choice sequences behave as auxiliary objects. Since the choice sequences are the visible witnesses of a subjectivist approach to mathematics, Brouwer's particular presentation might well be the result of a deliberate policy to let sleeping dogs lie, that is not to invite the outcry and ridicule of a mathematical world that was undeniably cultivating a formalistic-positivistic (and definitely objectivistic) outlook.

The highlight of the first paper, although modestly hidden in the mass of new notions and techniques, is the proof of the non-denumerability of the universal spread, that is the spread in which no nodes are checked, and in which the assignment of signs to finite sequences of choices is the trivial one:

$$(a_0, a_1, \ldots, a_n) \longmapsto a_n.$$

This is the same spread which we called the spread of natural number choice sequences, and which may be denoted by $\mathbb{N}^{\mathbb{N}}$. In geometric terminology, the spread determined by this assignment is called the *Baire space*, a well-known topological structure.

Brouwer had used Cantor's diagonal argument in his 1915–1916 course on set theory, but he had abandoned it a year later in favour of his own new argument, which was unobtrusively slipped in amongst the other material.[58] In just ten lines a revolution was launched, which was to have marked consequences. Hardly anybody recognised the significance of the principle involved; it took a man like Hermann Weyl to recognise a revolution when he saw one.

Brouwer's argument runs as follows: suppose we have a function $F$ that assigns a natural number to each choice sequence of natural numbers. Then the number $F(\alpha)$ must be completely known after a suitable initial segment of $\alpha$ has been generated; and this initial segment must already determine $F(\alpha)$. But then all sequences $\beta$ with the same initial segment will be assigned the same value as $\alpha$. Hence $F$ cannot be one-one.

The initial segment-argument, used above, is Brouwer's continuity principle in a nutshell. Together with some other principles, it has enormous power.

Observe that the crucial condition is the totality of $F$, that is to say, $F$ must give an output for *all* inputs. A function like

$$F(\alpha) = \begin{cases} 0 & \text{if } \alpha \text{ is the constant 0 sequence} \\ 1 & \text{else} \end{cases}$$

does not qualify because no matter how many zeros we have chosen for $\alpha$ so far, future choices may yield a non-zero number. We just cannot tell for an arbitrary $\alpha$ whether it is eventually constant 0. Technically speaking, the identity relation between choice sequence is not decidable.

[58] Brouwer (1918a), p. 13, see also p. 238.

So we are not allowed to say that $F$ is defined for all $\alpha$. The intuitionist is not allowed to appeal to the higher authority of the principle of the excluded third, as the classical mathematician is.

The *Begründungs*-papers contained most of the tools required for a systematic practice of constructive mathematics, to mention just a few, the intuitionistic versions of:

- real numbers, the continuum;
- elementary topology (of the plane);
- located (*katalogisierte*) sets;
- measure.

Most, if not all, refinements are dictated by the modesty of logic under an intuitionist regime. In the *Unreliability* paper of 1908, Brouwer had drawn the consequences of his intuitionistic views; in particular he had indicated certain simple statements of the form 'there exists an object $x$ such that ...', one may think of 'there are infinitely many pairs of equal consecutive decimals of $\pi$', and 'there is a sequence $012\ldots9$ in the decimal expansion of $\pi$', which have neither been proved nor rejected.

For such statements $A$ there is no reason to claim the principle of the excluded third: $A$ or not $A$ (in symbols $A \vee \neg A$). Likewise there is no reason to accept that from the impossibility of the absence of a sequence $012\ldots9$ in $\pi$, the presence of such a sequence may be concluded, that is, the impossibility of the absence does not give slightest idea where this sequence might be among the decimals of $\pi$. Or, in general, $\neg\neg A \to A$, cannot be affirmed. The reader should realise that Brouwer did not say, in 1908, that $A \vee \neg A$ *cannot* be proved; he cautiously said that the principle is *as yet unreliable*.

In 1918 he had all the tools to go one step further and to show that $A \vee \neg A$ indeed cannot be proved. He drew this conclusion only much later; apparently even Brouwer could miss something that was staring him in the face.

The failing of the *double negation law* ($\neg\neg A \to A$) made it necessary to distinguish between a property and its double negation, for example, 'it is impossible that $a$ is not an element of $X$' is not the same as '$a$ is an element of $X$'. In symbols $\neg\neg a \in X$ is distinct from $a \in X$. So, for example, where we had the notion 'the sets $X$ and $Y$ are *identical*', that is '$\forall a(a \in X \leftrightarrow a \in Y)$' ($X$ and $Y$ have the same elements), Brouwer introduced a new notion '$X$ and $Y$ are *congruent*' if $\neg\exists a(a \in X \wedge a \notin Y) \wedge \neg\exists a(a \in Y \wedge a \notin X)$.[59]

The restrictions on the logic required a whole list of similar refinements, of which some never gained any practical significance, but others have become standard material in intuitionistic (but also in general constructive) mathematics.

Brouwer's practice of naming notions and objects is highly idiosyncratic. Many suggestive names have never been adopted. On the other hand, he doggedly used long descriptions of colourless names for key concepts. For example, he preferred

[59] A sniff of logic shows that this can be rewritten as $\forall a(a \notin X \leftrightarrow a \notin Y)$.

'absurdity of $A$' to 'not $A$' (let alone $\neg A$, $\sim A$, or $\overline{A}$). He never introduced names for the positive versions of 'non-empty' or 'distinct'. Notions like 'bar-induction' and 'continuity principle' fared no better.

## 8.7 And Brouwer—That Is the Revolution

At the turn of the century there was a flurry of foundational activity in mathematics. The paradoxes had just appeared, and a good part of the mathematical community was wondering if there was not something basically wrong with the subject. Zermelo's introduction of the axiom of choice added to the confusion: what were the legitimate objects of the mathematical universe? After the heyday of Frege's *Grundgesetze*, Russell's *Principia*, Hilbert's *Heidelberg Lecture*, Poincaré's predicativity and Brouwer's early intuitionism, the general attention turned towards other areas, and the 'foundations' were left to the individual scholar who was willing to spend his time on yesterday's fashion.

As we have seen, there were two well-respected mathematicians who were not satisfied with the general state of mathematics: Brouwer and Weyl. While Brouwer was rewriting mathematics in Blaricum, Weyl had embarked on a similar project, resulting in his monograph *Das Kontinuum*. Weyl was one of those exceptional characters who combined an impeccable technical know-how with deep philosophical insight. In Göttingen he had become Hilbert's star pupil; but he had also joined the circle around the philosopher Edmund Husserl, who had given a new impetus to the study of phenomenology. Like Brouwer, Weyl was in an excellent position to draw attention to the foundations of mathematics. His status as a mathematician was so unassailable that he could not be brushed aside as 'eccentric' or 'uninformed'.

The roads of Weyl and of Brouwer to an alternative mathematics were rather different. Whereas Brouwer was in a certain sense a radical, who wanted to guide mathematics back to its true sources (and thus to safeguard it from the excesses of Formalism and Cantorism), Weyl's first concern was to avoid the paradoxes that had appeared around the turn of the century. Comparing Weyl's and Brouwer's approaches to the foundations of mathematics, Weyl is the cautious revisionist, who wanted to find the safe kernel of traditional mathematics. Brouwer, like Samson, wanted to bring the whole building of mathematics down before erecting it again, whereas Weyl wanted a careful restoration.

Most mathematicians had their favourite villain, for some it was the comprehension principle, which caused the downfall of Frege by the hands of Bertrand Russell. For some it was the phenomenon of impredicative definitions that undermined mathematical practice. The early adherents of Cantor's set theory belonged to the first group; they mostly accepted Zermelo's solution, consisting of a well chosen axiomatisation. Poincaré and Russell were the exponents of the second group; they focused their attention on the phenomenon of 'impredicative definition'. Such a definition defines an object in terms of the totality to which it belongs. The familiar example of an impredicative concept is the supremum of a bounded set of real numbers. One has to know the set of upper bounds before one can determine the smallest among them; so the supremum, that is the least upper bound, presupposes the set of all upper bounds, of which it is itself an element. And how can one define something that has to be known before it can be defined? Impredicativity was at the bottom of the paradoxes of Richard and Berry, therefore both Poincaré and Russell each had proposed a ban on impredicative definitions (the *vicious circle principle*).

Weyl took the issue of predicativity seriously by constructing a continuum consisting of real numbers which were definable in a particular predicative way: 'arithmetically definable numbers'. In *The Continuum* he gave a systematic development of the theory of real numbers and their functions along the lines of arithmetic definability. He saved, so to speak, the continuum by restricting it. In this sense his method was a big step forward compared to the ideas of the French semi-intuitionists, who did not go into the details of what they meant by 'definable in finitely many words'. *The Continuum* was Weyl's coherent answer to the anarchy that was threatening the realm of mathematics. His conception was not a mere technicality for escaping unpleasant phenomena; it was based on the philosophical insights of phenomenology. Phenomenology was his explicit point of departure.[60]

Hermann Weyl's monograph was a gem, it was far beyond anything produced in the foundations of analysis, it thoroughly treated the mathematical-logical aspects, and actually dealt with issues of analysis. It could truly be called a mathematician's presentation, in contrast to, say, the works of Frege and Whitehead and Russell. *The Continuum* was, on his own view, a technical realisation of a phenomenological analysis of space (the real line). Weyl's monograph was considered to be rather forbidding, for example, Klein, thanking Weyl for a copy of *The Continuum*, wrote 'As you can imagine, it is difficult for me to make myself familiar with these abstract things, about which I would nonetheless like to have an opinion.'[61] The book was, as a matter of fact, much praised and cited, but it took the mathematical community some 60 years before it could and did read it properly. Only when Feferman, Kreisel and others started to investigate the foundational status of predicative mathematics was Weyl's creation recognised for its elegance, depth and daring. Weyl's monograph distinguished itself from Brouwer's contemporary papers by its philosophical excursions. Modern readers will view it as a pleasant and thorough exposition of the foundations of analysis, not in the last place because of Weyl's sincere effort to explain the philosophy behind his enterprise. In his own words: 'Although this publication aims above all at mathematical goals, I have not avoided *philosophical* questions, and I have not tried to dispose of them by that crude and superficial amalgamation of Sensualism and Formalism, which [combated with gratifying clarity by Frege in his *Grundgesetze der Arithmetik*] is still highly appreciated by mathematicians.'[62]

Weyl was not at all averse to radical positions and statements, even though he did not go all the way to a constructive mathematics. In particular he did not revise

[60]Cf. Weyl (1954).

[61]Klein to Weyl, 19 April 1918.

[62]Weyl (1918), p. IV.

the logic in *The Continuum*. He had independently reached a good many radical insights. Some of these have become famous in the wager between Weyl and his Zürich colleague Polya:

> within 20 years Polya and the majority of representative mathematicians will admit that the statements
>
> 1) *Every bounded set of reals has a precise supremum*
> 2) *Every infinite set of numbers contains a denumerable subset*
>
> contain totally vague concepts, such as 'number', 'set' and 'denumerable', and that therefore their truth or falsity has the same status as that of the main propositions of Hegel's natural philosophy.
>
> However, under a natural interpretation 1) and 2) will be seen to be false.[63]

Even Brouwer could not have outdone Weyl in provocation; his views were by no means identical with those of Weyl, but he heartily agreed with Weyl on the untenability of the classical principles.

When, after the long interruption of the war, Brouwer and Weyl finally met again in the Engadin in the summer vacation of 1919, they discussed the foundational problems of the day. Brouwer gave a private crash course on his new intuitionism, including the choice sequences. Weyl, as ever, was quick to grasp what was going on. He turned the new notions over in his mind, and soon he had mastered the topic. In the same year he gave a series of lectures with the title 'On the foundations of analysis' in Zürich.[64] The lectures contained, in fact, all the ideas of the subsequent paper; it is quite clear that Weyl had fully recognised the meaning of the new constructive views; from Gonseth's notes it appears that Polya heartily disagreed with Weyl's views, as illustrated by the following small discussion:

> *Polya*: You say that mathematical theorems should not only be true, but also be meaningful. What is meaningful?
>
> *Weyl*: That is a matter of honesty.
>
> *Polya*: It is error to mix philosophical statements in science. Weyl's continuum conception is emotion.
>
> *Weyl*: What Polya calls emotion and rhetoric, I call insight and truth; what he calls science, I call symbol pushing (*Buchstabenreiterei*). Polya's defence of set theory ([that] one may one day provide meaning to these formulations) is mysticism.
>
> To separate mathematics, as being formal, from spiritual life, kills it, turns it into a shell. To say that only the chess game is science, and that insight is not, *that* is a restriction.[65]

---

[63] 9 February 1918, cf. Polya (1972).

[64] 2, 9, 16 December in the seminar of Fueter.

[65] Polya had advanced the view that research in set theory should not be restricted.

Soon afterwards Weyl finished his beautiful provocative paper *Über die neue Grundlagenkrise der Mathematik*.[66] In May 1920 it was finished and a copy was sent to Brouwer:

> Zürich, 6 May 1920
> Dear Brouwer,
>
> Finally I have sent off the long promised [object] to you. It should not be viewed as a scientific publication, but as a propaganda pamphlet, thence the size. I hope that you will find it suitable for this purpose, and moreover suited to rouse the sleepers; that is why I want to publish it. I would be grateful for your criticism and comments. Did I enclose everything that you let me have only on loan? If not, please reclaim it; the lecture on *Formalism and Intuitionism*[67] was already in my possession in the old days; at that time I did not pay attention to it or understand it . . .

One can imagine how pleased Brouwer was with Weyl's conversion; here was one of the foremost members of the new generation of mathematicians, who was not only a master in almost all traditional topics in mathematics, but who had an exceptional insight in philosophical and foundational matters. The fact that Weyl was Hilbert's best student, and was viewed as the crown prince of mathematics in Göttingen, may have contributed to the importance of this first convert, but one should keep in mind that the *Grundlagenstreit* had not yet started and that the relations between Brouwer and Hilbert were still free from personal and scientific antagonism. As the enclosed manuscript showed, Weyl wholeheartedly gave up his own foundational program, trading his arithmetically definable sequences for Brouwer's choice sequences, and rejecting the principle of the excluded third (PEM). Only a rough draft, full of crossed-out parts, has survived of Brouwer's reaction. Nonetheless, one can fairly well guess Brouwer's views from this draft, and from a few remarks, inserted in pencil in the margin of Weyl's manuscript. The first lines of Brouwer's draft run

> Your wholehearted assistance has given me an infinite pleasure. The lecture of your manuscript was a continuous delight, and your explanation, it seems to me, will also be clear and convincing for the public . . . That we judge differently on some side issues will only stimulate the reader. However, you are completely right in the formulation of these differences of opinion; in the restriction of the objects of mathematics you are in fact more radical than I am; however, one cannot argue about this, these matters can only be decided by individual concentration.

Brouwer had carefully studied Weyl's manuscript and made some notes in the margin, but on the whole he was inclined to grant Weyl his own insights in the matter. He was critical here and there, for example, at the paragraph where Weyl

[66] Weyl (1921).

[67] Brouwer (1913b), note the 'Freudian' permutation.

seemed to pussyfoot in recognising the scope of the new notions. Weyl had, for example, stated that choice objects were alright as individuals, but that one could not (or should not) quantify over them, 'It should be stressed once more that certain individual functions of that kind occur from case to case in mathematics, that general theorems are, however, never asserted about them. The general formulation of this notion is therefore only required if one is giving a justification of the meaning and methods of mathematics; for mathematics itself, the subject matter of its theorems, it is never considered at all.' Brouwer not only disagreed with this view, but he also upbraided Weyl for being too cautious:

> It seems to me that the whole point of your paper is endangered by the end of the second paragraph of page 34.[68] After you have roused the sleeper, he will say to himself: 'So the author admits that the real mathematical theorems are not affected by his considerations? Then he should no longer disturb me!' and turns away and sleeps on. Thereby you do our cause injustice, for with the existence theorem of the accumulation point of an infinite point set, many classical existence theorems of a minimal function, and also the existential theorems of the geodetic line without the second differentiability condition, loses its justification!

Weyl distinguished in his paper two distinct views of the continuum, the *atomistic* one and the *continuous* one. In the first version the continuum is made up of individual real numbers which can be sharply distinguished. Weyl allowed in his program of *The Continuum* only specific (arithmetic) relations on natural numbers, and thus also restricting the class of Dedekind cuts. For this particular class of reals, Weyl accepted existence problems involving reals as meaningful:

> Only if we conceive the notion [of the continuum] in this way, which fixes and demarcates its extent, matters of existence concerning real numbers become meaningful. By this restriction of the notion, a bunch of individual points is picked from the flowing mash. The continuum is smashed to isolated elements, and the blending of all its parts is replaced by certain conceptual relations, based on the 'greater–smaller', between these elements. Therefore I speak of an atomistic conception of the continuum.

The arithmetically definable continuum of *The Continuum* has this atomistic character. Almost immediately Weyl added, however, that 'It has never been my opinion that the continuum given by intuition is a number system of Weyl (*Weylsches Zahlsystem*); rather that analysis merely needs such a system for its constructions and does not have to worry about the 'continuum' poured in between.' Indeed, already in *The Continuum*, Weyl pointed out that his atomistic version of analysis represented 'a *theory of the continuum* which has (over and above its logical correctness) to prove itself correct, much as a physical theory'.

After the discussion of the 'predicative', atomistic continuum, Weyl went on to give an exposition of Brouwer's continuum with choice sequences and choice real

[68] Weyl (1921), p. 66.

numbers. Weyl's style is beautiful and poetic, it is almost the opposite of Brouwer's style of presentation. The descriptions of the various new notions and the metaphorical presentation did more to popularise intuitionism than Brouwer's œuvre so far. Weyl introduced a number of slogans that caught the imagination of the otherwise not foundationally interested readers. Indeed, Weyl's 'New Crisis' reads as a manifesto to the mathematical community. It uses an evocative language with a good many explicit references to the political and economic turmoil of the post-war period. The opening sentences castigate the complacency of the mathematical community, which had not paid any serious attention to the potential dangers of the various paradoxes of Cantor, Russell, Richard, ...

> The antinomies of set theory are usually considered as border conflicts that concern only the remotest provinces of the mathematical empire and that can in no way imperil the inner solidity and security of the empire itself or of its genuine central areas. Almost all the explanations of these disturbances which were given by qualified sources (with the intention to deny them or to smooth them out), however, lack the character of a clear, self-evident conviction, born of a totally transparent evidence, but belong to that way of half to three-quarters attempts at self-deception that one so frequently comes across in political and philosophical thought. Indeed, every earnest and honest reflection must lead to the insight that the troubles in the borderland of mathematics must be judged as symptoms, in which what lies hidden at the centre of the superficially glittering and smooth activity comes to light—namely the inner instability of the foundations, upon which the structure of the empire rests.

The readers of 1920 were all too familiar with the phenomenon; the German State with its skirmishes in the Baltic and its political instability was a painful reminder of the self-deception of politics. Weyl used the political metaphor with great dexterity, he compared the classical use of existential statements with the use of paper money:

> The point of view sketched above[69] only expresses the meaning which the general and existential propositions in fact have for us. In its light mathematics appears as a tremendous 'paper economy'. Real value, comparable to food products in the national economy, has only the direct, downright singular; general and all the existential statements participate only indirectly. And yet we mathematicians seldom think of cashing in this 'paper money'! The existence theorem is not the valuable thing, but the construction carried out in the proof. Mathematics is, as Brouwer sometimes says, more activity than theory (*mehr ein Tun als eine Lehre*).[70]

And the final clarion blast of Weyl, one that fired the imagination and fed the wrath of many a practising mathematician, rang through the next decade: 'For this order cannot in itself be maintained, as I have now convinced myself, *and Brouwer—that is the revolution!*'

---

[69] That is, the construction of a specific object versus its non-effective existence.

[70] Cf. p. 101.

Weyl in his paper mainly addressed two points: the notion of the continuum, that is, the nature of the real numbers, and the meaning of logic, in particular the meaning of the quantifiers and the failure of the principle of the excluded third.[71] An analysis of Weyl's contributions is given elsewhere.[72] Here we will just look at the more superficial effects of the 'New Crisis' paper.

Real numbers are given by Weyl as infinite sequences of shrinking intervals, and thus the need arises to specify the notion of sequence further. Weyl allowed two sorts of sequences: those given by a law, standing for the individual points of the continuum, and the free choice sequences, which determine a continuum of 'becoming', or 'emerging', sequences (*werdende Folgen*),[73] that is, sequences of intervals that are freely chosen and hence cannot in a predetermined way point at an element of the continuum. In the most literal sense of the word, these sequences remain 'in statu nascendi' forever. In Weyl's words:

> It is a first basic insight of Brouwer that the sequence which is emerging (*werdend*) through free choice acts is a possible object of mathematical concept formation. Where the law $\varphi$, which determines a sequence up to infinity, represents the single real number, the 'choice sequence, restricted by no law in its freedom of development, represents the continuum.
>
> Brouwer's remark is simple but deep: here a 'continuum' arises in which, indeed, the single real numbers fit, but which itself does by no means dissolve into a set of completed existing real numbers, rather it is a medium of free becoming.

Weyl drew the conclusion that the true continuum consisted of the real numbers generated by choices and in addition those given by a law. Hence 'for all real numbers' meant for him 'for all choice real numbers', and existence could only mean 'there is a real number given by a law', for 'existence' must have the meaning 'I have a construction method here and now'. The consequence was that the negation of an existential statement over reals was not at all the same thing as the universal version of its negation. Hence there was not the faintest justification of the principle of the excluded third! But Weyl wanted more, he wanted to understand why the principle of the excluded middle failed even for such pedestrian objects as natural numbers. For statements like $A(n)$, $\exists n A(n)$ is established if one finds a number $n$, during the consecutive testing of $A(0), A(1), A(2), \ldots$. The negation would then come to 'after running through all natural numbers no instance $A(n)$ has been established', but Weyl correctly rejects this as nonsense, $\forall n \neg A(n)$ can only be true because it is part of the essence of the notion natural number. He concluded that these statements do not act as each other's negation. So this pleads against PEM. But, he continued,

---

[71] We will indiscriminately use the terms 'principle of the excluded third', 'principle of the excluded middle'. Brouwer preferred 'principium tertii exclusi'. The acronym used here is *PEM*.

[72] van Dalen (1995).

[73] The translation of the German *werdende Folge* is somewhat problematic. Various adjectives such as 'becoming', 'developing', 'emerging', and 'in statu nascendi' have been used in the literature.

> Although this speaks in favour of Brouwer, I am always thrown back to my old standpoint by the thought: if I run through the sequence of natural numbers and break off when I find a number with the property $A$, then this eventual breaking off does or does not occur; it is or is not the case, without wavering and without a third possibility [...] Finally I found for myself the magic word. An existential statement, for example, 'there is an even number', is not a judgement in the proper sense at all, which states a state of affairs; existential states of affairs are an empty invention of logicians.

Weyl termed such pseudo-statements 'judgement abstracts', to be compared to 'a piece of paper which announces the presence of a treasure, without giving away its place'. Similarly he provided a metaphor for the universal quantifier.

The 'New Crisis' paper is extraordinarily rich in ideas and claims; Weyl saw clearly what the Brouwerian revolution would bring. He drew some conclusions that were to be established a couple of years later by Brouwer. For example, he asserted the 'indecomposability' of the continuum (which includes the connectedness):

> 'A true continuum is simply something connected in itself and cannot be split in separated pieces; that contradicts its nature.'

This is the first enunciation of the unsplittability of the continuum. However, Weyl does not prove the principle, but rather asserts its plausibility on conceptual grounds. Similarly, the prize-winner—the continuity of real functions—is announced without a proof:

> 'Above all, however, there can be no other functions at all on the continuum than continuous functions.'

It is useful and necessary to correct the impression that Weyl thus proved the continuity of all real functions years before Brouwer. One should rather view Weyl's statement as an immediate corollary of his definition of real function. On his account an approximation of the value of a function was, by definition, determined by the approximation of the argument. And thus continuity was, so to speak, part of the definition of 'function'.

Another way to put it is: Weyl was convinced of the unsplittability of the continuum and the continuity of the real functions on the basis of a *phenomenological analysis*. Subsequently this was confirmed by Brouwer's *proof*.

We have devoted so much space to the 'New Crisis' paper of Weyl as it has become one of the landmarks of the foundational debate in the interbellum, known as the '*Grundlagenstreit*'. While Brouwer toiled on to provide an impeccable foundation to mathematics, producing scholarly but hardly exciting papers (it was as if Brouwer tried to avoid even the semblance of that circumlocution that is so characteristic of cranks and self-proclaimed missionaries), Weyl had sounded a trumpet blast that woke many to the serious problems posed by mathematics.

**Fig. 8.3** Group at the Nauheim Conference, 1920. *In front*: Hamburger; *first row* (l.t.r) Kerékjárto, Brouwer, Szász, Landau. *Second row*: Schur, Polya, Bessel-Hagen. [Brouwer archive]

## 8.8 Intuitionism, the Nauheim Conference

Brouwer, in the meantime, had put his foundational theories before the international mathematical forum at the *Naturforscherversammlung* at Nauheim, 1920. This meeting of the German scientists, which was more or less an act of defiance towards the Entente-dominated conference in Strasbourg, gathered the flower of German physics, mathematics and medicine. Brouwer gave his talk *Does every real number have a decimal expansion?*[74] in the mathematics section in the afternoon of 22 September 1920. Hermann Weyl was also present. He lectured in the joint mathematics–physics section on 'Electricity and Gravitation'[75] in the morning of the next day. The conference brought together a large number of, mostly German, mathematicians for the first time after the war, to mention some: Bernstein, Bessel-Hagen, Bieberbach, Brouwer, Engel, Fricke, Hausdorff, Hensel, Kerékjártó, Koebe, Landau, Mie, E. Noether, F. Noether, Polya, Radon, Schoenflies, Schur, Szasz, Weitzenböck, Weyl, and Wiener.

The list of speakers shows that the meeting was rather a 'family matter' for the Germans, no speakers were listed from the allied countries, and only three speakers from neutral countries: Brouwer (the Netherlands), Weyl and Polya (Switzerland). The title of Brouwer's talk, combined with Brouwer's reputation, was enough to

[74] *Besitzt jede reelle Zahl eine Dezimalbruchentwicklung?* (Brouwer 1921a).

[75] *Elektrizität und Gravitation*.

draw attention. How far the actual talk diverged from the written text of the subsequently published paper[76] is hard to say. On the one hand speakers usually try to enliven their exposition with didactic remarks, possibly elicited by questions from the audience. On the other hand, Brouwer was notorious for ignoring his audience.

So the talk could have contained elucidating comments, but it could equally well have been a scholarly recitation of the contents of the manuscript. Even now the paper does not strike the reader as 'accessible', but in 1920 it must have found few readers. In itself the paper is a small piece of art. It succinctly treats the fine structure of the continuum. Starting from a constructive version of Dedekind cuts, Brouwer considered classes of reals which were in particular ways determined by their position with respect to the rationals. For example, the reals with decimal expansions or continued fractions appear in a natural way.

The audience must have been puzzled. A contemporary reaction can be found in a letter from the mathematician Robert Fricke to his uncle Felix Klein, in which he reported on the highlights of the conference:[77]

> Of course, Brouwer has to be taken very seriously, but he essentially is incomprehensible. Landau proposed, in view of Brouwer's contribution, to introduce a section for pathological mathematics and to incorporate it in the medical section.

Whatever the audience might have thought of Brouwer's talk (which was clear, but not simple) one thing was clear: Brouwer had presented his program, and his criticism, in an international setting; he could no longer be ignored.

It is fairly evident that most observers considered what Brouwer had to offer a drastic austerity program. Indeed the Nauheim talk did not discuss the new notion of choice sequence, but remained rather neutral, and those who had not seen Brouwer's papers in the Amsterdam Proceedings, or the introductory paper *Intuitionistic Set Theory* (1919) in the home journal of the German mathematicians, would be rather confirmed in their opinion that constructivism only promised restrictions. The Set Theory paper, however, clearly spelt out the new notions of 'choice sequence' and 'spread'. But since Brouwer did not give applications, it is no wonder that the readers could not extract any benefits from the paper. It did, on the other hand, contain a criticism of Hilbert's dogma of the solvability of mathematical problems (that is, every mathematical problem can be solved or rejected). Brouwer had already in his 'unreliability' paper[78] identified the Principle of the Excluded Third and Hilbert's Dogma, adding that 'there was no hint of a proof' for either of them. Here he repeated the identification, remarking that PEM

> can be assigned none but a scholastic and heuristic value, such that theorems, in the proofs of which its applications cannot be avoided, lack every mathematical content.[79]

[76]Brouwer (1921a).

[77]Fricke to Klein, 28 September 1920.

[78]Brouwer (1908b).

[79]Brouwer (1919h).

Footnote 4 to this paper goes further into the general belief in the principle of the excluded third:

> In my opinion both the axiom of solvability and the principle of the excluded third are false, and the belief in these dogmas has been historically caused by the fact that people abstracted classical logic from the mathematics of subsets of a fixed finite set, and then ascribed to this logic an existence a priori, independent of mathematics, and that they finally wrongfully applied this logic to the mathematics of infinite sets, on the ground of this alleged *a priori*.

This rejection of the principle of the excluded third was the inevitable conclusion of the objections to 'logical arguments', put forward in Brouwer's letter to Korteweg of 23 January 1907, cf. p. 96. The above passage (which is repeated over and over again in later papers and talks) pin-points, so to speak, the disastrous consequences of the unjustified step from finite to infinite.

Although the paper showed a glimpse of the underlying philosophical motivations, Brouwer was reticent to provide full explanations. He admitted to have used 'old methods' in his 'philosophy-free' papers, but claimed at the same time to have endeavoured 'to deduce only such results, of which I could hope, that they could, after carrying out a systematic construction (*Aufbau*) of intuitionistic set theory, find possibly in a modified form a place in the new doctrine, and maintain a significance'.

It is, indeed, not hard to find places in the topological papers where proof by contradiction is used, or the second number class, but practice has born out Brouwer's claim to a surprising extent. Of course, one has to allow for suitable constructive modifications, see, for example, the intuitionistic version of the fixed point theorem, which Brouwer published more than forty years after the fixed point theorem itself.[80] The above claim was not an empty boast. Brouwer had enough visionary power to distinguish the constructive content of mathematical developments!

Before we leave the topic of intuitionism, let us see what the first assault at the fortress of mathematics had yielded. In the first two *Begründungs*-papers, dealing respectively with *General Set Theory* and *Point Sets*, a rich collection of tools was laid out. The first paper contained the basic concepts of *spread* (*Menge*) and *set* (*Spezies*), the various *inclusion* and *coincidence* relations for sets, the notion of *equivalence* (*gleichmächtigkeit*), and its weaker variants, notions of *finite* and *infinite*, various notions of *denumerability*, and the constructive theory of *order* and *well-ordering* (including *ordinals*).[81]

A number of these notions have reappeared in different guises at various places in later years, for example, the refinements of the notion of denumerable were to play an important role in recursion theory, where similar distinctions must be made. Some of the most common 'countability' notions are summed up (in modern language) as:

---

[80] Brouwer (1952a).

[81] N.B. Brouwer's ordinals were constructive, and hence countable.

**Fig. 8.4** Apartness for points

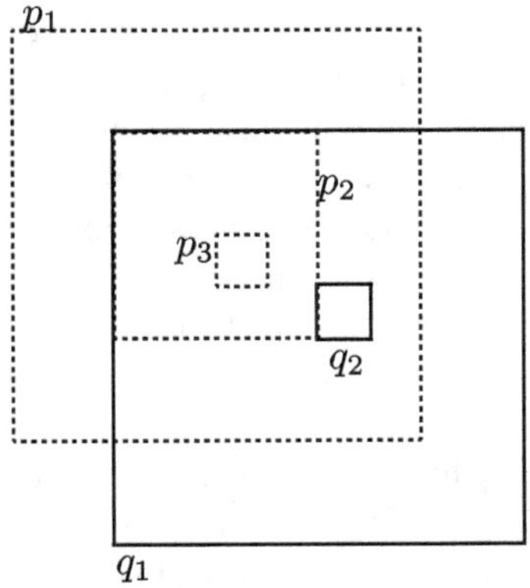


- S is *denumerable* if there is an injection $f : S \to \mathbb{N}$;
- S is *numerable* if it is denumerable, and if the range of $f$ is decidable (that is, '$\forall n(n \in ran f \vee n \notin ran f)$');
- S is *enumerable* if there is a surjection $f : \mathbb{N} \to S$.

For Brouwer's contemporaries, who were not in the least familiar with constructive mathematics, it must have been almost impossible to see that the distinctions made any sense. Even today it takes some time before students in an introductory course grasp the subtleties of constructive mathematics, so one can imagine that Brouwer's early papers, which were sparingly provided with didactic examples, found few readers!

The second paper introduced the basic concepts needed for elementary analysis and topology. It contained likewise a number of novelties:

- a constructive notion of 'point of the plane';
- a positive inequality relation: apartness (*örtlich verschieden*);
- limit point, boundary point, closure;
- located sets (*katalogisiert*), measurability, content.

Some of the above mentioned notions are fairly straightforward, that is, one has to make the proper constructive choices in the definitions. Some, on the other hand, are peculiar to constructive mathematics, and their significance disappears if one uses classical logic. One such notion is *apartness*; a number of definitions of this notion have been given in the past. We will sketch Brouwer's original definition here:

Consider the plane divided in squares $\kappa_1$ with integer corner points, and subdivide each square into four congruent squares $\kappa_2, \ldots$ A $\lambda_i$-square is a square consisting of four $\kappa_{i+1}$ squares. A point is defined as a choice sequence of $\lambda$-squares (that is $\lambda_i$-squares for certain $i$'s) such that each square lies properly inside its predecessor.

Now, let $P$ and $Q$ be defined by sequences of squares $(p_i)$, $(q_i)$, then $P$ is *apart* from $Q$ if some $p_i$ and $q_j$ do not meet, see Fig. 8.4.

We note in passing that Brouwer did not introduce a symbol for apartness, neither here, nor later.[82] Strictly speaking he could do without it, since in the presence of a metric, apartness is the same as being separated by a (strictly) positive distance.

The other notion that plays an essential role in analysis and topology is that of a '*located set*'. In general, subsets (say of the plane) can be rather elusive, for example, $\{x|A \vee \neg A\}$, where $A$ is some unsolved problem (e.g. the Riemann hypothesis),[83] has rather strange features; one cannot say that it contains a point, but it cannot be empty, for that would imply $\neg(A \vee \neg A)$, which is a contradiction.

The set $\{x|(x = 0 \wedge A) \vee (x = 1 \wedge \neg A)\}$ offers similar problems: one cannot say that either 0 or 1 belongs to it, but it cannot be empty. It is a subset of a finite set (namely of $\{0, 1\}$) but it cannot be said to be finite (for then it has exactly one element, and we should be able to tell which one).

Evidently, such sets make life miserable in mathematics. We want reasonably well-behaved sets, without requiring them to be decidable, because decidability is a scarce commodity in mathematics! Brouwer introduced *located sets* for the purpose.

Although there are now more convenient definitions, we will reproduce Brouwer's original definition: a point set $X$ is given by a set of sequences of $\lambda$-squares. It is no restriction to assume that all $n$-th elements of all sequences have the same length, say $2^{1-\mu_n}$. $X$ is called located (*katalogisiert*) if for any sequence $\sigma_1 < \sigma_2 < \sigma_3 < \cdots$ of natural numbers, and for each $n$, there is a finite set $S_n$ of squares $\lambda_{\mu_n}$ such that: (i) squares that do not belong to $S_n$ are not squares of points in $X$; and (ii) for each $m \geq 0$ and each square $\lambda_{\mu_n}$ of $S_n$, there is a square $\lambda_{\mu_n}$ of $X$, within a distance less than $2^{\mu_n}$ of $\lambda_{\mu_n}$.

This definition, popularly speaking, says that a located set can be arbitrarily well approximated by finite sets of squares. A somewhat more manageable definition is: $X$ is located if for any two squares $q_1$ and $q_2$, where $q_2$ is in the interior of $q_1$ and not touching its boundary, we have either $q_2 \cap X = \emptyset$ or $q_1 \cap X$ is inhabited.[84] Equivalently: $X$ is located if for each point $P$, the distance $d(P, X)$ exists.

## 8.9 The Failure of the Institute for Philosophy

Any reader who consults Brouwer's list of publications will be struck by the fact that this same man who stunned the mathematical world with his tremendous outpour of substantial papers before the war now took his time and carried out his revolution in instalments. One possible, and plausible, explanation is that the world had started taking its toll. Whereas during the years 1908–1913 Brouwer spent almost all his

---

[82] Heyting introduced in his dissertation (1925) $A\omega B$ (*A deviates from B*) for apartness. In Heyting (1936b) he replaced $\omega$ by $=||=$, and since Heyting (1941) the symbol # has generally been accepted. Warning: the Bishop school uses '$\neq$' instead.

[83] Some people worry about the moment this problem $A$ will be solved. There is no need to: there is an inexhaustible stock of unsolved problems; one can always take another one.

[84] A is *inhabited* if it contains an element, the positive analogue of non-empty.

time at research, now he not only had the duties that came with a professorship, but he had accumulated over the years a large number of extra activities. He was an editor of the *Mathematische Annalen*, a job he took extremely seriously. He studied the papers that were submitted to him thoroughly, corresponded with the authors, suggested and discussed improvements, etc. The time spent on editorial business was considerable, and Brouwer refused to cut a corner here and there; he had on principle all papers that were submitted to him copied (either by hand or typed), so that no author could pull the wool over his eyes. In the pre-Xerox days that was an awfully time-consuming procedure! Furthermore, he was seriously involved in the organisation and development of the philosophical activity known as the Signific Circle.

Generally speaking, Brouwer's lifestyle did not at all conform to the accepted idea of the single-minded researcher. He enjoyed his leisure, had an active social life, travelled a great deal in Europe, and corresponded with all and sundry. Cultural events also took up their share of the spare hours. He regularly frequented the concerts in the Concertgebouw, or even in The Hague, where his friend Peter van Anrooy was the conductor of the *Residentie Orchestra.* Indeed, the outburst of energy during his first topological period seems rather uncharacteristic!

All this time, while Brouwer carried out his professional duties, designed a new mathematics, refereed papers, and resumed his topology activities, he conscientiously played his role in the small group of (mostly amateur) philosophers, gathered around Van Eeden and Mannoury.

During the war, a good deal of activity had been going on at a national level. Now that the international contacts could be re-established, the thought about a Forte Kreis-like institution surfaced again.

In 1918 the future of Significs, as a scientific and an academic discipline, started to look somewhat rosier. In any case, the group started to attract some attention. Mannoury, who had been promoted to an ordinary chair (in analytic and descriptive geometry, mechanics and philosophy of mathematics) at the University of Amsterdam started a course in 'Mathematical Significs'. Frederik van Eeden, who had acquired a good deal of respect (and attracted a lot of abuse on the way!) was invited to give a talk on *Intuitive Significs* in the Auditorium of the University of Amsterdam. The small group consisting of Van Eeden, Brouwer, Mannoury, Henri Borel, and Jacob Israel de Haan got together regularly to discuss the business of the institute or to listen to an exposition of a fellow member. The company varied slightly, as new members were introduced from time to time. These new members usually left after some time, leaving behind a hard core comprising the above mentioned group. From 1919 onwards Father Jac. van Ginneken also attended the meetings.

Van Eeden's diary contains a number of short notes on the meetings of the group, accompanied by incidental comments on the individual members. He had slowly come to see himself as the centre of an exclusive intellectual universe and considered others as convenient and necessary supporters. There is no doubt that he was quite genuinely fond of his friends. Brouwer, especially, was close to his heart, although he was not blind to the shortcomings of the brilliant, impulsive mathematician. An exception was Henri Borel, who reciprocated the dark suspicions of Van Eeden.

The mutual dislike increased over the years, so that finally Borel refused to see Van Eeden any longer. In numerous letters to Brouwer, he complained about the perfidy of Van Eeden. The letters that have survived suggest that Borel and Brouwer formed a jolly couple; Borel addressed Brouwer as 'Joost' (for what reason is not clear) and signed as *Borreltje*, which in Dutch stands for a glass of *Jenever* (gin). Since only Borel's letters have been preserved, we do not know what Brouwer thought about all this jocularity. There is a strong suggestion, however, that the pair of them were not averse to an occasional binge and a bit of philandering.

On 3 January 1919, Joop de Haan left the group; he emigrated to Palestine. De Haan had lived a surprisingly active life so far. He was a talented poet and author, albeit of a somewhat shocking sort. Most of his acquaintances would agree that he was querulous and neurotic; Brouwer called him hysterical and unbalanced.[85] In spite of his pronounced sexual inclinations he married Johanna van Maarseveen, a medical doctor. After a period of rejection of the faith of his forefathers, he eventually became an orthodox Jew and a fervent Zionist. As a logical conclusion, he exchanged his country for Palestine, leaving his wife behind. After his emigration he remained faithful to the signific cause. As a correspondent for the newspaper *Het Handelsblad*, he wrote about a wide variety of topics, including, for example 'Hebraic Significs'. Once in Jerusalem, he discovered that the Zionist claim was not as unproblematic as he had thought. When he started to write critically about the Zionist program, he became the target of abuse and threats. His hardly veiled homosexual practice did not do much to endear him to his Jewish countrymen. Eventually he was executed on 30 June 1924 by a member of the Hagana.[86]

The International Institute for Philosophy never lived up to its name, as far as the predicate 'international' was concerned. A number of foreign scholars and artists were nominated for the membership of the International Academy for Practical Philosophy and Sociology,[87] but in the end only *one* foreign member joined the Academy (Eugen Ehrlich).

International traveling was in 1919 still a hazardous affair, trains were not for the faint-hearted, it took a compulsive traveller like Brouwer to visit the former Forte Kreis members in Berlin, and to re-establish the contacts that had laid dormant during the war. True, the Gutkinds had managed to come and visit Van Eeden in his Walden, but they had not met the whole significs group. In particular they had missed Brouwer. And now, prodded by Borel, who could not stop praising the charms of Lucia Gutkind, Brouwer travelled as a kind of emissary of the significists to Berlin. Borel would have come along, were it not for the miserable conditions of the German railways. He had heard about the horrors of wagons where one had to stand for 14 hours, where the toilets were blocked by masses of luggage, so that one could not pee for 12 to 14 hours.[88]

---

[85] Van Eeden diaries, 22 January 1919.

[86] The assassin was interviewed after 67 years in a Dutch television program on 29 October 1991.

[87] The Institute was the Foundation supporting the Academy. For the structure of the organisation, see p. 260 and Schmitz (1990a).

[88] Borel to Brouwer, 18 December 1919.

**Fig. 8.5** Erich and Lucy Gutkind. [Courtesy Library of the Amsterdam University]

The more adventurously inclined Brouwer survived the uncomfortable trip by train to Berlin; at his place of destination, still undecided about the Berlin chair (see p. 292), he asked Schoenflies to advise him on such mundane matters as housing and financial arrangements.[89] It appears that he was rather pessimistic about the general situation in Germany. In particular for a Dutchman, the economic perspective of life in Germany seemed rather bleak. Schoenflies' reply (if any) is unknown.

The planned visit to the Gutkinds proved more complicated than expected. The couple was away when Brouwer arrived. They wrote a worried letter to Borel, fearing that they had missed Brouwer altogether. In the end everything worked out, Brouwer was immediately taken in by Erich and Lucia. The Gutkinds wrote enthusiastically to Van Eeden, 'Brouwer is with us and he will bring you our greetings. We get along wonderfully and we are enormously pleased with him.' Mrs. Gutkind added 'It was a great pleasure to meet Brouwer, ...'[90] A comparison of the mystical writings of Brouwer and Gutkind easily convinces the reader that the two were of the same disposition. It is not surprising that they got along wonderfully; although the future drove them geographically apart (the Gutkinds emigrated, like so many, to America), the friendship closed in Berlin lasted for a lifetime.

Erich came from a Jewish family. He was deeply religious and had good contacts with the leading Jews in Berlin, among others Gustav Landauer, Martin Buber and Walter Benjamin. Both Gutkinds were deeply interested in spiritual and cultural matters. They offered the intellectual environment in which Brouwer felt comfortable, and which engendered a lasting spiritual relationship.

We have already seen that, philosophically speaking, Gutkind was not really close to Van Eeden. Brouwer clearly offered the reception and response that Gukind

[89]Brouwer to Schoenflies, 29 December 1919.

[90]Gutkind to Van Eeden, 15 January 1920.

had been waiting for. From Borel's letters, we can infer that Brouwer repeatedly visited the Gutkinds or met them somewhere in Germany, sometimes in the company of Borel. Reading Borel's letters one cannot suppress the thought that Brouwer must have made a strange impression on these travels. In one of Borel's letters we read, for example,[91] 'I just want to tell you that I have returned, and that I did indeed find a room in the hotel Windsor in Cologne. I got the impression that they held you for a burglar or a Bolshevik, for I came in with an enormous aplomb [...] and immediately got the 'only room still available', for only 22 German marks (*Reichsmarken*). [...] I had no more bad luck after you disappeared so mercilessly. You look too much like one of the murderers of Erzberger, or like a friend of Lenin or so.'[92]

The contacts between Brouwer and Borel were probably mostly based on the admiration of the latter for the former. Borel's letters are filled with plans for meetings and apologies for broken appointments. Among the assorted gossip, there are all kinds of complaints about people and things that did not conform to Borel's ideas. He was particularly critical of Van Eeden, whom he considered an enormous fake.

All the time, meetings of the signific group went on fairly regularly. The arrival of father Van Ginneken had introduced an element of Roman Catholic orthodoxy. This in itself did not cause problems, but it took all of Mannoury's intermediating capacities to keep the ensuing conceptual conflicts from flaring up.

In 1920 Barend Faddegon, a specialist in Sanscrit and Indian philosophy, joined the Institute, but by 1921 he had already left. A linguist, A. Verschuur, also had a short-lived interest in significs. He attended the meetings from 1921 to 1922.

The Institute for Philosophy did not flourish, as its founders had hoped; the international co-operation was not forthcoming, and in Holland the activities of the significists were watched with detached curiosity. The public at large mainly identified significs with Van Eeden, who was its main propagandist. But since Van Eeden's ideas were not representative, and were in due course coloured by his newly acquired sympathy for the Roman Catholic Church, his publications[93] did not do much for significs as a scientific enterprise.

In view of the failure to set significs on its way as a recognised discipline, the conspicuous lack of harmony, and, last but not least, the financial problems associated with the publication of the *Mededelingen* (Communications), the remaining members, Brouwer, Van Eeden, Van Ginneken[94] and Mannoury, decided to dissolve the Institute (23 March 1922) and the supporting association,[95] and to found a new institution, the *Signific Circle*. This circle was founded in an atmosphere of happy

[91] Borel to Brouwer, 20 September 1921.

[92] Brouwer was in a boyish way rather proud of the epithet 'Bolschevist', but it would go too far to attach any political significance to this fact. Let us add that there were some unconfirmed rumours that Brouwer met Lenin in Zürich.

[93] Cf. *Het roode lampje* (the red lamp), van Eeden (1921).

[94] Still a visitor and not a full member.

[95] Although legally it was only 'suspended'.

informality at Walden on 21 May 1922,—'At the lawn beneath the chestnut trees [...] we founded our Signific Circle.'[96]

And so the stage was prepared for the twenties. Brouwer had got himself involved in a large number of projects (or affairs, one could say): the intuitionistic revolution, significs, editorial business of the *Mathematische Annalen* and the struggle against the boycott of German scientists. One major event of the twenties was not foreseen by Brouwer: his renewed involvement in topology.

Altogether, the auspices for the decade were not favourable, although Brouwer might not have viewed it that way. But for a man with a full-time academic position, the punishment for spreading himself too thinly could and would prove disastrous.

---

[96]Diary Van Eeden, 22 May 1922.

# Chapter 9
# Politics and Mathematics

## 9.1 The *Conseil* and the Boycott of Germany

Resuming the pre-war routine might have been simple for scientists in the neutral countries; but for the citizens of the belligerent nations it turned out to be far from easy. The war had conjured up all those nationalistic feelings that before the outbreak of the conflict were quietly on their way out in the scientific community. In particular in mathematics, the otherwise abundant national differences, or animosities, had been virtually absent. Communication was free and open; international conferences, for example in Paris (1900), Heidelberg (1904), and Rome (1908), were the meeting places of a rich variety of mathematicians from all nations. The personal relations between the leading mathematicians had more or less guaranteed the absence of malicious or secretive manœuvres. The war had not, perhaps, changed the professional respect for colleagues across the dividing line of the Central Powers and the Entente, but it had fostered a collective hysteria that in some cases overruled the personal common sense of scientists. Although mathematics, in this respect, did not suffer as much as other disciplines, even in this privileged domain of quiet contemplation unrestrained emotions surfaced, as we will see in this chapter.

In spite of the relative sober-headedness of the scientific community, there had been nationalistic tendencies in certain scientific quarters. As a rule, the leading scientists who frequently travelled abroad and who cultivated an extensive circle of correspondents were fairly immune from bombastic nationalistic feelings, but the larger scientific community was not always invulnerable where the virus of nationalism was concerned. We have seen that high-minded thinkers like the members of the *Forte Kreis* reverted to a 'right or wrong, my country'-attitude as soon as the war broke out. How much less could one then expect from the rank-and-file intellectual? In Europe the *Wilhelminische* empire suffered in particular from the tragicomic illusion of grandeur; it cultivated a cultural missionary feeling to a dangerous degree. It is not hard to find traces of irritation with this German sense of superiority. For ex-

D. van Dalen, *L.E.J. Brouwer – Topologist, Intuitionist, Philosopher*,
DOI 10.1007/978-1-4471-4616-2_9, © Springer-Verlag London 2013

ample, one of France's top mathematicians, Emile Picard, published an essay with the title *The History of the Sciences and the Pretensions of German Science*.[1]

Where the exhibitions of nationalistic pride had so far been just an annoying or ridiculous phenomenon, which one could shrug off as immature behaviour, the outbreak of the war suddenly transformed the latent or not so latent national feelings into a dangerous mixture of aggression and self-pity. And so we see that, while the world at large showed every inclination to pin the blame for the war on Germany, the German scientists and intellectuals closed ranks and protested as one man against the real and imagined prejudices in the foreign press reports and in foreign political reactions. The violation of Belgium's neutrality, the sack of Leuven, and the barbaric treatment of Belgian and French civilians had indeed provided the international press with an abundance of material to write about.[2]

The German scientific community, following the general opinion in the country, reacted to the adverse publicity with a flurry of declarations. The German university teachers published an *Erklärung* on 7 September 1914. It was followed on 4 October 1914 by the fateful *Aufruf an die Kulturwelt* (Appeal to the Cultural World).[3] This declaration, known as the 'Declaration of the 93', was the source of a good deal of negative publicity. Brouwer had criticised it after its publication in the Academy meeting, cf. p. 245. The list of signatories of this document is a dazzling display of the flower of German science. Many of the leading scientists of the beginning of the century had lent their names for the defence of the national honour. Outside Germany, the declaration met with incredulity and indignation. The scientists who had signed were viewed with suspicion and contempt. We may note that among the large number of prominent physicists, chemists and representatives of the humanities who had signed the declaration, there was only one mathematician: Felix Klein. As so often happens, some were cajoled into signing and some had not even seen the statement they were signing.[4]

The scholars in the allied camp were not slow to publish *their* views; a group of English scholars replied to the 'Call' in an open letter to the Times of 21 October. Members of the Oxford Faculty of Modern History published a scholarly monograph 'Why we are at war'.

Whereas the war had a sobering effect on many of the directly concerned, in particular the young men in the trenches, many German scholars stuck to their nationalistic position. Throughout the war a strong conservative and annexionistic section of

[1] *L'Histoire des Sciences et les Prétentions de la Science Allemande, Revue des Deux-Mondes*, 1915. Cf. Picard (1922).

[2] See Tuchman (1962).

[3] For these and other details the reader is referred to Schroeder-Gudehus (1966) and Kellermann (1915).

[4] Klein, for example, had not seen the text and was under the impression that he supported a 'protest against foreign lies'. Cf. Schroeder-Gudehus (1966), p. 76.

the academic community squarely supported the war efforts and ideology, opposed only by a small band of scientists.[5]

It is a well-known fact that wars and conflicts tend to radicalise the various positions, and that had the unfortunate effect of hardening the determinedness of the victors to 'settle the bill'.

Indeed, at the end of the war there was a feverish activity at the side of the allied scientists. Not surprisingly, the most vehement denunciation of the Central Powers and their scientists came from the French and the Belgians, who had suffered horrific civilian and material losses, quite apart from their military losses. On various occasions the French and Belgian scientists vented their bitterness. Some of these statements have been preserved in the records of academic institutions and in newspapers. The leading and well-respected French mathematicians Emil Picard and Paul Painlevé[6] figured prominently on the post-war scientific-political scene. They spoke for a vast majority of Belgian and French scientists, who had experienced the war in person or in their families.

Picard, as the permanent Secretary of the Académie des Sciences, sternly rejected all contacts with German colleagues: 'The Academy judges that personal relations between scientists of the two groups of belligerents are impossible until the reparations and expiations, which have been made necessary by the crimes that have placed the Empires of the Central Powers under the ban of humanity, allow them to again take part in the concert of the civilised nations' (31 September 1918).

At the meeting of the Academy of 21 October 1918 he added: 'In order to restore the trust, without which fruitful collaboration is impossible, the central empires must renounce the political methods, the application of which has led to the atrocities that have filled the civilised nations with indignation.' Painlevé, doubtlessly, expressed the feelings of his colleagues and his countrymen, when on 2 December 1918, before that same audience, he expressed his views that the past war presented the picture of a desperate life-and-death duel between two concepts of civilisation, without even the possibility of dialogue. 'It was about learning if science will be a means for liberating and ennobling mankind, or an instrument for its enslaving.'

He denounced the German scientific community in a fierce indictment: '..., at the other side of the Rhine, science was a gigantic enterprise, where a whole nation

[5]There is a considerable literature on the role of German scholars during the war, see for example Schwabe (1969), Graf von Krockow (1990).

[6]Painlevé (1863–1933) had a double career as a mathematician and a politician. In 1910 he was elected as a deputy for the 5th 'Arondissement', and in 1916 he was re-elected. He was minister of education (*Ministre de l'Instruction et de Invention intéressant la Défense National*) from 1915–1916, minister of war (1917) and in 1917 he became prime minister. He was president of the Chamber of Deputies in 1924 and in 1925 he again was prime minister. In 1925 he was, for an interim period, minister of finance, followed by a longer period as minister of war (1925 to 1929) and finally he served as minister of aviation (1930–1933). He was, in the meantime, incredibly active in innumerable projects, for example, at the request of the Chinese Government, he reorganised the Chinese railway system (1920). Where Painlevé was, so to speak, politically predestined to fight German influence, Picard had suffered a personal loss during the war: his son fell at the front.

stubbornly continued to manufacture with a patient servility the most formidable killing machine that ever had existed.'[7]

One should of course not construe Painlevé's statements as a call for pacifism or non-violence. On the contrary, he was one of the advocates of the enlisting of science in the cause of warfare. At the session of the Academy of 7 January 1918, he elaborated on the contributions of his learned colleagues and their students toward the eventual defeat of the enemy. 'Your mission', he said, 'is the research of scientific truth, on which neither time, death, nor the human passions have any hold. Under the worst upheaval your reason will not deviate from its inflexible rules. But in the beleaguered city of Syracuse, Archimedes applied the rigorous correctness of geometry to the construction of gigantic catapults; who is therefore the scholar whose spirit will remain deaf to the call of his country (*patrie*) in danger.'

No doubt Painlevé would have viewed the military scientific projects on the other side of the Rhine in a totally different light. He would probably have argued that there is no symmetry in the situation of the aggressor and the victim.

For the French and Belgians the remedy was quite obvious: never again should German scientists be misused by national and military authorities. Since, in the instance of the Great War, if you did not live in Germany or Austria, there was a good reason for blaming the Central Powers, it escaped the politicians and the general public that the involvement of science in military projects was a general issue, not restricted to the Central Powers. The warning voices of those thinkers or politicians who managed to rise above a narrow nationalist viewpoint, Bertrand Russell and Romain Rolland, for example, were easily drowned.

The deep bitterness of the victorious but ravaged nations can be observed in many official documents and in the general press. The scholars of the city of Lille, for example, announced that they would not take part in any activity involving contact with Germans, stating that '... one is forced to recognise that in a very general way, with only very rare exceptions, the German heart is closed to noble, generous or human feelings'.

As soon as the arms had been laid down, the Entente scientists started to formulate rules for a new Science organisation, in which there was to be no place for the Germans. From 9 to 11 October 1918 representatives from Great Britain, France, the USA, Belgium, Japan, Serbia, Brazil, Italy and Portugal met at the *Conference of Inter-allied Scientific Academies* in London. The 'London resolutions' aimed at an organisation of international science, while at the same time isolating Germany and Austria. The next meetings of the Conference were in Paris in November 1918, and in Brussels, 18–28 July 1919.[8] The general atmosphere of these meetings was not encouraging for open international co-operation; the French and Belgians were firmly resolved to keep the Germans out, and they competently and effectively manipulated the proceedings to that effect. Even though the official wordings were 'matter of fact', a good deal of sabre-rattling went on in and out of the conference

[7] C.R. 167 (1918), p. 800.

[8] One can find some records of the post-war scientific organisation in *Ens. Math.* 20 (1918) and 21 (1920).

halls. Generally speaking, the leading personalities representing the scientific community of the Entente nations saw themselves as the counterparts of the political representatives at the Peace Conference at Paris. They did not hesitate to let political motives prevail over the scientific ones.

The minutes of some Entente academies contained language that would embarrass most of the present readers. The victors were quite frank about their goals, and they did not mince their words. Only academics from Entente countries participated in the discussions: 'Already now new societies—recognised as useful for the progress and applications of science—will be founded by the states that are at war with the Middle European Powers, possibly with the admission of the neutrals.'[9] Proposals for excluding the Germans were circulating, e.g. 'the Middle Europeans should be forced by the peace treaty to withdraw from all international scientific bodies'.[10]

A large-scale boycott of German scientists was envisaged. The Belgian astronomer Lecointe had already, in 1917, formulated a number of points in his memorandum for the Belgian government. In order to keep wavering colleagues in line, he proposed a black list of those who failed to observe the boycott rules, accompanied by punishments such as exclusion from public office and striking from honour lists.[11]

The influential Picard had also called for an exclusion of the 'nation that had placed itself beyond humanity', with the words 'There is too much blood, and there are too many crimes, separating us.'[12]

The meeting in Brussels saw the founding of the *Conseil International de Recherches*, with member countries Belgium, Brazil, USA, United Kingdom, Australia, Canada, New Zealand, South Africa, Italy, Japan, Poland, Portugal, Romania, and Serbia. Countries other than the Central Powers were allowed to join, provided they subscribed to the rules of the London Resolution, and in particular joined the boycott of Germany.

The practical consequences of the allied activities in the field of science were varied, and differed from discipline to discipline. Some Associations were rather strict. For instance at the International Tuberculosis Conference in Lausanne (Switzerland) German was not admitted as a conference language, so that the Basel newspaper drily observed that it was made impossible for the delegates from the German speaking part of Switzerland to speak their own mother tongue in their own country.[13]

The actual boycott of the Central Powers concentrated on a number of issues: exclusion of Germany from international conferences, a ban on the German language

---

[9] Ac. Royale de Belgique, Bulletin de la classe des sciences, 1919, p. 63.

[10] Picard, cf. C.R. 21 October 1918, vol. 107, p. 570.

[11] Schroeder-Gudehus (1966), p. 107.

[12] Ibid.

[13] Basler Nachrichten, 17 August 1924.

in scientific discourse, a reallocation of central bureaux, institutes and councils to countries of the allied part of the world and the termination of the German monopoly of bibliographical and review journals.

As was to be expected, the boycott was answered by German countermeasures; a number of organisations took care of the interests of German scientists and scientific institutions. Among those, the *Reichszentrale für naturwissenschaftliche Berichterstattung* was a very effective one. It was led by Karl Kerkhof, a civil servant from the Bureau of Measures and Weights. Kerkhof organised the flow of scientific information into Germany, and the converse flow of information from Germany to other countries. Thus the Germans actively organised their own counter-boycott in response to the Entente boycott.

This short excursion into post-war politics of science may help the reader to understand some of the events Brouwer got mixed up in. At the end of the war Brouwer was equally disposed towards the Allies and the Germans: his letters to Hilbert, Klein and Denjoy show that he wished both sides well. And the visit that he made with his friend Frederik van Eeden to the American Consul (cf. p. 280), on which occasion he advocated a council of scientists as the platform for peace talks, clearly demonstrated his trust in the rationality and goodwill of his fellow scientists. Although Brouwer's relations with his French colleagues had been somewhat less intense than those with his German colleagues, and in spite of his brush with Lebesgue, he fostered no animosity towards the French. It seems plausible that he kept himself informed of the developments at the Paris and Brussels meetings, which would not have been difficult, as the information would anyhow reach the Dutch Academy; and there is no doubt that he violently disagreed with the intended boycott of Germany. He soon found an occasion to demonstrate his feelings.

The *Conseil International des Recherches* had set itself the goal of creating new international organisations for the various branches of science, free from German influence and German members. The mathematicians had not been among the first to comply with the wishes of the leaders of the *Conseil International des Recherches*, who tended to reign over the scientific world in the style of the old absolute monarchs. In the absence of an international body, the French national committee of mathematics, presided over by the inflexible Picard, had taken the initiative by organising, in September 1920, an international conference of mathematics in Strasbourg. Brouwer was approached by the secretary of the committee, G. Koenigs, with the request to provide a list of Dutch mathematicians. The choice of Strasbourg could not very well have been a coincidence; the city had repeatedly changed hands between the French and the Germans. In 1871 the Germans had annexed the town, and in 1918 Strasbourg and the Alsace changed hands again. The organisation of an international conference in a contested (or newly acquired) town has always been an excellent means of demonstrating the new status quo, and of rubbing the adversaries' noses in the dirt.

## 9.2 The Nauheim Conference and Intuitionism

The Germans, barred as they were from this international meeting, organised their own conference at exactly the same time in Nauheim. The *Naturforscherversammlung* at Nauheim opened on 20 September 1920, two days before the conference in Strasbourg. The organisation of a congress of the size of the *Naturforscherversammlung* was a major operation so shortly after the war. Nauheim, a modest spa, north of Frankfurt a.M., had volunteered to house the participants (free of charge!)[14] and to provide meals at reasonable prices. The town and its inhabitants took great pains to make the 2500 participants welcome. Indeed, the visitors could easily have imagined themselves in a pre-war country, with pre-war conditions! The Nauheim Congress is memorable mainly for the confrontation between the proponents and opponents of the young theory of relativity, culminating in a debate at the end of the Relativity Theory session, between two Nobel Prize winners, Einstein and Lenard. The topic of the theory of relativity was a hot item in Germany; it had met with fierce opposition from conservative quarters, and the debate had attracted interest beyond the world of scholars. Einstein had already been the focal point of a tumultuous meeting in Berlin, and the Nauheim meeting became a battle ground for the various undercurrents in physics, with a considerable political-cultural bias.

Compared to this spectacular event in the centre of the congress, the mathematical section was solid but not surprising. As we have seen, Brouwer presented his talk with the provocative title '*Does every real number have a decimal expansion?*';[15] it treated the fine structure of the reals in an intuitionistic setting. Just like his earlier papers this was a scholarly exposition with hardly a controversial note, cf. p. 315.

At the time of the Nauheim Conference, Brouwer had already published several papers on his new intuitionism. Most of these were scholarly and dry: it seems almost as if Brouwer was avoiding controversy at all cost. There is one mild exception: the paper *Intuitionistic Set Theory* of 1920, published in the *Jahresbericht der Deutschen Mathematiker Vereinigung*. In this paper Brouwer gave an exposition of the principles and ideas of intuitionism and also discussed his objections to the current views on the foundations of mathematics. Hilbert is mentioned in connection with his 'Axiom of the solvability of each problem', known as Hilbert's dogma. Brouwer pointed out that Hilbert's dogma and the principle of the excluded third were equivalent—as he had claimed already since 1908, and that they were obviously false.

In view of later developments it is, of course, of some importance whether Brouwer did or did not provoke the formalists through personal attacks or biased statements. The modern reader will agree that even the 'Intuitionistic Set Theory'—the only pre-*Grundlagenstreit* paper that mentioned Hilbert—is a model of objectivity and courteous academic presentation.

---

[14] The reader is referred to Forman (1986) for more detailed information on the Nauheim Conference.

[15] The positive answer is usually taken for granted; even Brouwer saw fairly late that there was a problem. It was as late as 1919 that he explicitly asked himself if $\pi$ had a decimal expansion.

The reference to Hilbert was neither provocative, nor aggressive, just factual. His comments on the principle of the excluded third, on the other hand, might have offended its staunch believers.

One might well ask if the Nauheim Lecture was the opening of the foundational dispute between Hilbert and Brouwer, now known as the *Grundlagenstreit*.

In all fairness one cannot but conclude that the answer is 'no'. The Nauheim lecture (or at least its published version) is a scholarly exposition that would not excite the most sensitive natures. There is no mention of either Hilbert or the excluded third. Of course, one does not know what Brouwer actually said in his lecture, but even if he did elaborate the criticism of traditional mathematics, verbal comments are soon erased from the collective memory. It is not clear what the impact of Brouwer's lecture was. As we have seen in Fricke's letter (p. 317) the more jocular members of the audience did not see it as a serious challenge.

One would guess that Weyl, who at that time had adopted Brouwer's program, listened with approval, but apart from Fricke's reaction, little is known about the impact of Brouwer's message.

As we have seen in the preceding chapter, the situation radically changed when Hermann Weyl threw in his lot with Brouwer; the study of Brouwer's foundational papers, and doubtless the personal conversations, had convinced Weyl that his own program did not touch the heart of the problems in the foundations of mathematics. Weyl was not the person to practise his mathematical–philosophical convictions quietly in a corner. Often, when a new insight was mastered, he presented the outcome in grand and sweeping formulations to the scientific world.

In the provoking paper *The New Crisis in the Foundations of Mathematics*,[16] Weyl forcefully expounded the intuitionistic principles and criticism. The paper was widely read and some of its slogans became the watchwords of the twenties and thirties, cited by friend and foe. After a try-out at the mathematics colloquium at the ETH in Zürich on 2, 9 and 16 December 1919, he lectured on the topic in Hamburg on 18 July 1920.

Weyl published his 'New Crisis' paper in the *Mathematische Zeitschrift*, thus, no doubt with good reasons, bypassing the *Mathematische Annalen*, where Hilbert might possibly have raised objections, which would have, if not stopped, at least delayed the publication. One should note, however, that the paper, although written in a provocative style, did not mention Hilbert or his formalism. Weyl altogether avoided personal attacks. Nonetheless there could be no doubt as to his allegiance; he mentioned with approval, for example, Kronecker's views. The 'New Crisis' paper captured the imagination of the contemporary mathematician because of its imaginative use of language and its provocative claims. It was to dominate the discussion for some time to come.

Weyl's exclamation, '*Brouwer—that is the revolution!*', became a rallying cry for some, and a threat to others. Weyl's battle cries were to haunt the mathematical community in Germany for years to come, and doubtless they angered Hilbert, who

[16] Weyl (1921).

saw his favourite student desert him for a dangerous eccentric. Weyl did by no means intend to give up mathematics for the study of its foundations. His interests were too wide ranging to keep him fettered to one particular topic for a long period. Already at the Nauheim Conference, he lectured on mathematical physics, and his greatest work in mathematics was still to come, but not in the foundations of mathematics.

The intuitionistic ideas were certainly not ignored, but on the whole poorly understood. As Weyl had put it, it was considered more as 'interesting news from the border provinces', than as fascinating and worrying developments one had to learn about. In Göttingen things were taken seriously, in so far that Bernays and Courant gave talks at the Göttingen Mathematical Society on 1 and 8 February 1921, 'On the new arithmetical theories of Weyl and Brouwer'.[17] One may note here the adjective 'arithmetic'. From Brouwer's and Weyl's papers it should have been clear that the 'new' theories were not arithmetical; somehow it took a long time before even prominent mathematicians understood the true nature of Brouwer's program!

The general misconception, which survived into the late thirties, was that Brouwer (and hence Weyl) was a constructivist of the kind that only recognised the finite combinatorial part of mathematics. Even before Brouwer's new program, this would have been a doubtful proposition, but after 1918 nobody could have defended that view. Judging from the remarks in papers from the period, one can but guess how the general mathematical public got its information on Brouwer's ideas. The reception of Brouwer's ideas must have been slow; one explanation is that, before 1918, hardly any material was in print in the internationally accepted journals. Up to that time the dissertation and its 'appendix', the unreliability paper, gave the most extensive exposition of the foundational views, albeit in Dutch. Furthermore there were the Rome paper on the possible cardinalities and the translation of the inaugural address. In addition there was the review of Schoenflies' new version of the *Bericht*. One had to be a discerning reader to draw the right conclusion from the material! It was certainly tempting to view Brouwer simply as another Kronecker, that is, a constructivist with a bias towards the discrete. Although Brouwer was quite explicit in his dissertation about the unique irreducible nature of the continuum, it was not easy to explain the significance for ordinary mathematics. It is therefore not surprising that the more abstract, infinitistic features of his mathematics escaped a lot of readers. Even Bernstein, who was a personal friend of Brouwer, classified Brouwer, together with Poincaré, Richard, Borel and Lindelöf, as a finitist.[18]

One wonders, was Brouwer just poorly understood, or did he keep his foundational convictions to himself before 1918? The latter seems improbable in view of Brouwer's personality. We know at least that Hilbert was the one-man audience of an exposition on the foundations of mathematics (including the language levels) in 1909 at their joint walks in the dunes of Scheveningen, cf. p. 127.

---

[17]One would guess that Bernays' views on Brouwer and Weyl were along the lines of his paper *On Hilbert's thoughts on the founding of mathematics*, JDMV 1921.

[18]Bernstein (1919), *Cantor's set theory and finitism*, JDMV 1919.

## 9.3 The Denjoy Conflict

Soon after the Nauheim Conference, Brouwer was plunged into a political debate that a more cautious man would have sought to avoid. After the conference, Blumenthal visited Brouwer in Laren and used the occasion not only to discuss general matters, but also to ask the expert opinion of the Utrecht Professor Arnaud Denjoy on a paper submitted for the *Mathematische Annalen*, of which Blumenthal was the managing editor. The first symptoms of a new conflict, which was to drag on, and which was mostly fought by only two combatants, announced itself in a letter from Denjoy to Blumenthal—*chez M. Brouwer, à Laren*.[19]

In this letter Denjoy explained to Blumenthal that in spite of his fond recollection of their contacts during the Rome Conference in 1908, the time for a renewal of personal relations had not yet come. Instead of a brief refusal, the letter contained a list of arguments for avoiding contacts with the former enemy:

> The visible reasons for a renewal of the conflict between our two countries are from gone. One must have seen the devastation of certain regions in the North and North-East of France and measured the amount of work and expenses necessary for rebuilding to realise that the people of France, heavily reduced as they are in their means of production, will not consent in assuming that task alone, and in exonerating yesterday's enemies, less tested than they are.
>
> In case Germany would rise to escape her obligations and France would have to resort to force in order to submit her, the initiative, taken already by a French scholar, to ignore prematurely all reservations with regard to a German colleague, that position, taken in an offhand manner, would be regarded as thoughtless and irresponsible.

As soon as the French government thought that it had received from the German government guarantees of goodwill, and had judged them sufficient to elevate Germany again to the rank of normal nations, Denjoy would have no basic objections to renewing his relationships with those German scholars who had not displayed provoking behaviour by clamourous publicity. Until that moment, however, Denjoy chose to stick to the instructions that the policy of his country dictated to him.

The letter of Denjoy illustrates the bitterness that prevailed after the war. At the time of the writing of the letter, the peace treaty with Germany had been concluded in Versailles, but that had not automatically restored the pre-war situation. The size of the reparation payments had been set in June, but the actual implementation of the payments was to give rise to endless trouble, including a number of occupations of German territory by the French and Belgians, intended to guarantee the payments. It is an accepted historical fact that some of the allied nations displayed an understandable but short-sighted vindictiveness towards the former adversary. As a partial explanation, it should be pointed out that the Belgians and the French had borne the brunt of the almost apocalyptic destruction and killing of the Great War.

[19] Denjoy to Blumenthal, 4 October 1920.

**Fig. 9.1** Arnaud Denjoy. [Courtesy Bernard Denjoy]

Denjoy had adopted a formal position; he stated that his behaviour was dictated by the official French position. Today that seems rather harsh, but one cannot ignore the atmosphere of the Great War and its emotional impact. One wonders if Denjoy himself was a wholehearted supporter of the strict measures of the French government, or whether he was hiding behind a convenient excuse? Whatever his views were, he soon had no choice but to stick to his guns.

Brouwer was no sooner informed of this cold rejection of Blumenthal's request than he took the side of the insulted party. Characteristically, he did not write an avuncular letter which might have had a soothing influence, but instead dispatched a coolly argued indictment of the misguided behaviour of a professor in the service of the Kingdom of the Netherlands. In view of the friendly relationship between Brouwer and Denjoy, one would perhaps have expected an attempt to arrive at an understanding that would allow for some sort of compromise. But such a thing was not in line with Brouwer's strict code of behaviour. Misconduct of this sort could not be tolerated!

To avoid possible misunderstanding, it should be pointed out that Brouwer did not engage in this somewhat quixotic conflict because he had a bone to pick with Denjoy or because there was some hidden animosity. The correspondence prior to the Blumenthal incident had been cordial; nothing had foreshadowed this abrupt stern reaction. As a matter of fact Brouwer explicitly mentioned in a later letter to the Minister of Education that he had never detected any chauvinistic tendencies in Denjoy, and that, on the basis of personal experience, a civil reception of Blumenthal had been taken for granted.[20]

The relationship between Brouwer and Denjoy had been one of mutual respect and understanding. Denjoy appreciated Brouwer's topological fame, and Brouwer had enthusiastically welcomed the young analyst in Dutch mathematics. The correspondence preceding the conflict showed all the signs of a beginning friendship. The

[20] Cf. page 15 of an open letter of Brouwer to the Minister of Education, 27 September 1922.

nationality of Denjoy played no role at all. In general Brouwer treated his German colleagues and Denjoy on strictly the same footing.

On the occasion of the Versailles peace treaty Brouwer had not only sent letters to Klein and Hilbert, but also to Denjoy, who had acknowledged this the same day: 'Infinitely many thanks for your sympathetic thought to congratulate me with the great events which at the moment take place in the history of my country. [ . . . ]'

Denjoy's policy was not without its nuances. On one occasion he expressed his regret at not being able to meet Carathéodory, when Brouwer told him of Carathéodory's stay in Holland.[21] He excused himself on the grounds of an overloaded schedule. He remembered Carathéodory well from the Rome Conference, he said, which he had attended with Montel and other colleagues. It is, of course, possible that Denjoy did not wish to meet Carathéodory, but did not want to give offence either. Anyway, Brouwer could not have foreseen Denjoy's reaction to Blumenthal's request. In the above mentioned letter to the Minister of Education, Brouwer gave a complete exposition of the whole affair, explaining that Blumenthal had been staying with him in October 1920 for the purpose of discussing business matters of the *Mathematische Annalen*. Blumenthal had a manuscript of H. Hake, which touched on Denjoy's area of research; he felt that it would be good to consult Denjoy, but apparently he was hesitant, on account 'of the position adopted by several French mathematicians after the fall of Germany with respect to German colleagues'. Brouwer advised Blumenthal to contact Denjoy, as Denjoy in his role of Dutch civil servant and ordinary member of the Dutch Royal Academy, was to be considered as a Dutch rather than French mathematician, and hence would not harbour any grudge against Blumenthal. Furthermore, the position of Denjoy in the Kingdom of the Netherlands (as a professor, he had been appointed directly by Her Majesty the Queen; professors were so-called *Kroon docenten*, Crown-teachers) would in any case guarantee a courteous treatment of Blumenthal. In particular one might expect so in view of the rule of the Academy which said that 'the Academy was intended as a means of union between the scientists of the Netherlands and those of other countries'. Blumenthal, thus encouraged, had approached Denjoy, with the above result.

The possibility that Denjoy had a personal grudge against Blumenthal seems highly unlikely; Blumenthal was generally well liked, and no personal feud between the two men at that moment is known. Blumenthal had served in the army, but it is hard to say whether Denjoy was aware of that fact.

Denjoy's reaction to Blumenthal's request must probably be seen in the light of the enhanced national consciousness at a time of international conflict. This, in combination with the strict guidelines of the *Conseil* and the leading French mathematicians, may have forced him to do things that one would regret in a more sober mood.

When, so unfortunately, Blumenthal was rudely rebuffed, Brouwer's never failing sense of justice dictated the logical next move in this sad tragicomedy. He wrote a letter to Denjoy stating that:

[21] Denjoy to Brouwer, 14 October 1919.

> You will no doubt have realised the consequences of this incident for our mutual relationship, since the duties of hospitality oblige me to consider behaviour towards my guest as touching me personally.[22]

He added that his opinion of the political task of scholars ('in particular of ourselves, academicians of the neutral countries') was diametrically opposed to that of Denjoy.

This short note opened the sluices of Denjoy's eloquence, almost as if he had been waiting to lecture the Dutch on their neutrality. Three days later he spelled out his views on the rights and duties of hospitality, but above all on the behaviour of the Dutch scientists. After rejecting Brouwer's view on the rules of behaviour towards someone's guests, he went on to say that if Brouwer had written a letter to him from Nauheim (the place of 'the congress organised by the Germans and coinciding with the one in Strasbourg, which was of no less scientific interest than the first mentioned') he would have replied in his usual cordial way, but that the German hosts would have been mistaken to think that these feelings could be assumed to extend to them.

The letter goes on to state, in a fairly moderate tone, that once the suspicions against the Germans had laid themselves to rest, Brouwer's efforts for a reconciliation would be welcome, but that he would not find things all that easy. Moreover, he warned Brouwer not to try to impose his sympathies for persons of one of the sides on those of the other side. 'The French', he said, 'have no taste for orders, not to give and even less to take'. Denjoy repeated that once the suspicion against the Germans was officially lifted, Brouwer's attempts to re-unite the former enemies was perfectly in order. He was quite willing, for example, to put his reprints at the disposal of German scientists,[23] but he thought it was premature to try to bring authors belonging to the two nations together.

Denjoy had in particular been upset by Brouwer's phrase 'we, academicians of the neutral countries'; he discerned a tendency to subject him, as a member of the Dutch Academy, to obligations incompatible with his French citizenship, and this he resolutely declined to accept. Even stronger, if there had been obligations of the sort Brouwer suggested (or worse), then the choice between the membership of the Academy and his simple citizenship of France would not have been difficult.

'The general opinion', he wrote, 'lends to scientists, more perhaps than to the majority of individuals, businessmen, for example, a kind of national character which must make those scholars very circumspect, when it is a matter of lending their persons to conciliatory actions which could be criticised by sensible patriots'.

Denjoy's letter is typical for the curious situation in which many scholars found themselves after the war. They were obliged to follow the official directions in so far as international scientific contacts were concerned. What seems surprising, in the case of Denjoy, is that there is no trace of regret at the situation; he seems heartily to support the French position on contacts with colleagues from the Central Powers. One would expect at least some signs of reticence, where personal and scientific

[22] Brouwer to Denjoy, 17 October 1920.

[23] As he already had done at Brouwer's request.

matters are concerned. Let us, however, not judge too hastily here: Denjoy may have had private reasons for his attitude concerning the Germans, or he may have simply been swept along by the popular sentiment after the capitulation.

The above letter is in perfect accord with the official position of the French, as laid down and pronounced at the meetings of the *Conseil International de Recherches*: no contact with Germans and Austrians, until satisfactory promises have been given and reparations made. It lacks the harsh rhetoric of, for example, Picard, but it also lacks personal feelings towards those German colleagues that Denjoy must have known personally.

Brouwer's reply, a week later, was brief.[24] He did not see the purpose of a continuation of the discussion, for example, on hospitality. He could not resist pointing out that the hospitality of the conference in Strasbourg was offered him, because of the coincidence of his place of birth. As to the role of academics of neutral countries, he quoted the minutes of the particular meeting of the Academy (31 October 1914), at which Brouwer had commented on the role of scientists in the context of the war, in response to the 'Declaration of the 93'. He merely wished to draw Denjoy's attention to the rules for ordinary members of the Academy (in contrast to those for corresponding or foreign members) which stipulated that the Academy was to be a *means for uniting scholars of the Netherlands and those of other countries.* According to Brouwer, Blumenthal's scientific questions to Denjoy were justified on the grounds of this rule. He was surprised, he wrote, by Denjoy's interpretation of the rules regarding ordinary members of the Dutch Academy, but he wished to abstain from comments, since 'in the end it only concerned Denjoy and the Dutch government'. The letters so far showed some emotions, but no hostility. The flowery declarations of respect were certainly not insincere. Denjoy had ended his letter of 20 October 1920 with the words: 'There should not be any doubt that the sentiments expressed in this letter do not diminish in the least the great regard with which the mathematicians consider you and with which I associate myself without reserve.' Brouwer answered in kind: 'It is superfluous indeed, my dear colleague, to say that on the one hand I infinitely regret the circumstances that put a distance between me and a man of your value, and on the other hand that the circumstances diminish in no way the feelings of respect I have for you.'

Denjoy swiftly hit back, as if stung by a viper.[25] He was not, he said, going to consult the rules of the Academy; he would be surprised if the Academy forced its members to do the things Brouwer had mentioned. The Dutch government would not chase a Frenchman from its Academy just because he had declined to receive a German, if only because an incident of this sort would give too much pleasure to the enemies of the good relations between France and the Netherlands. The letter also contained a curious passage which hinted at a poorly veiled disapproval of the position of a small nation that did not take part in the war, that did not suffer, yet ventured to express its opinion in international affairs touching on the consequences of the war.

---

[24] Brouwer to Denjoy, 27 October 1920.

[25] Denjoy to Brouwer, 29 October 1920.

> French public opinion, which is all that matters to me, my letter to Mr Blumenthal has clarified that point, is already not too well disposed towards Holland. It is felt too clearly that certain people here would have seen the disappearance of France and her civilisation as a minor accident. They would not deplore it if the world had become German. The aggression of 1914, four years of German crimes, on land as well as on sea, all that would be no more than peccadilloes. It is in nobody's interest to confirm the belief of my compatriots, which for that matter is not exact, that all Dutch people think this way.

The last part of the letter refers to something that apparently was incomprehensible to Denjoy, that is, Brouwer's reference to the 'accident of his birth place'. Denjoy had hit on a problematic issue: almost all speakers mean by the phrase 'coincidence of my birthplace' something like, 'if I had been born in China, I would have been a loyal Chinese, but since I was born in Greece, I am a loyal Greek'. From Brouwer's lips it meant something else; as we have seen in the case of his review of Van Eeden's anti-nationalism brochure, Brouwer felt himself a true internationalist. Nationalism, for him, was one of the basic red herrings of social training. Possibly, Brouwer and Denjoy had discussed this topic before, or else Denjoy was an extremely keen reader and observer. Denjoy sketched in most eloquent words the ties that bind people to their nation, and that made him a Frenchman for better or worse. The eulogy has a sincerity and beauty that official addresses usually lack, and so I cannot resist the temptation to reproduce a translation here.

> Your letter shows me that you acknowledge only a vague attachment to Holland, created by the accident of place of birth. Are you not exaggerating your indifference? If the Belgians or the English had just invaded your country, and had pillaged and destroyed the wealthiest region, from Rotterdam to Amsterdam, and killed 300,000 young men, maybe you would have felt enough aversion towards the aggressor to make you feel Dutch.
>
> Your obligations towards what happens to be the place of your birth do not allow you to visit a conference at Strasbourg but they do allow you to visit one at Nauheim. I would have understood if you declined both the German and the French invitation. Your duties towards Holland entail rigours and accommodations that strike me as strange.
>
> Except for the kinds of countries you mentioned, one recognises also the country of affinity, a category the existence of which you will find difficult to challenge.
>
> But above all you forget the country of nationality. To belong to a nation implies charges but also advantages. Any man should see it as an honour, and for any man it is also wise, to be attached to a people under all circumstances.
>
> I dare to congratulate myself for being able to reunite in one country those of nationality, of affinity and of birth.
>
> It is no coincidence that my origins are in Gascogne. Given the fact that for many generations all my ancestors have been living in that corner of France, it would have been against all the odds had I been born elsewhere.

> I have no hesitations in feeling myself a member, and a very humble member, of one family, together with all those who have made my language incomparably superior to any other because of its rigour, its precision, its immaterial and energetic vigour. That language is perfectly apt to give expression to certain spiritual meanings I see in myself. And it does not easily lend itself to translate confused mental dispositions that my nature dislikes but in which many a foreign soul finds pleasure.
>
> I can recognise myself in the aversion of French intellect from vainglory, from charlatanry and from appeals to superficial curiosity.
>
> Among the dominating traits that are most characteristic for the French people is that I quite enjoy to rebel with all that is in me against characters opposed to mine.
>
> I know of no people with a greater inclination to criticise themselves and greater aversion from admiring themselves.
>
> There are no others on whom arguments of noblesse have more effect and contemptible reasons less impact. They are not like those to go to war hoping to come back rich.
>
> All these affinities determine my impression that I am not a Frenchman by accident.
>
> Your respect touches me, but I have never asked for it. Less respect for me, and less antipathy for my country would be more to my satisfaction

The above letter is an example of the patriotism that was fostered by the war, and that was—contrary to what Denjoy might have thought—not a French prerogative. Some of the claims made by Denjoy were traditional exaggerations that one should not take seriously; the superiority, for example, of the national language has been claimed by a rich variety of nations. There is, however, an undertone that seems to accuse Brouwer and the Dutch of an anti-French disposition. There seems to be little evidence for such a claim; until 1920 Brouwer published in the *Comptes Rendus*,[26] and his personal relationships with French mathematicians were affected by neither the war, nor the brief Lebesgue conflict. The circumstance that his contacts after 1920 were largely German can be explained by the fact that the developments in the foundations largely by-passed France. Privately, he was a frequent visitor to France. As for the Dutch, there was indeed a certain amount of pro-German feeling among the population, but this was certainly not a dominant feature.

Brouwer considered the above letter totally unsatisfactory, he could not share Denjoy's views, and he could not find a coherent answer to his remark concerning the rules of the Academy. He returned the letter[27] with a few lines:

> Dear Sir,
>
> The lines below, which I am truly unable to grasp, I do not wish to keep. I urgently beg you to direct no more letters to me.

[26]Brouwer only resumed his habit of submitting papers to the *Comptes Rendus* in 1950.

[27]But not before making some copies.

In the meantime be assured of my sincere esteem.

L.E.J. Brouwer

In the typed copy of the letter which is in the Brouwer archive, Brouwer had added in the margin a remark to the effect that Denjoy had misinterpreted his reference to the relationship between the Academy and its members. Indeed Denjoy's answer is hard to understand; Brouwer had intended to say that the relationship between a member and the Academy only concerned the member and the government of the Netherlands. He wrote in the margin

> This quotation is incorrect and must *perhaps* explain the—for me—incomprehensible sequel. If you really think that the government of the Netherlands has nothing to do with the way in which rules, made by her, are interpreted by the civil servant concerned, then I will not quarrel with you. What I wanted to say, was no more than that apart from the civil servant concerned and the government, *certainly no third party* should be involved, and you apparently agree to that.

So Brouwer wanted to say that the relationship between Denjoy and the Dutch Academy (viz. the government) did not concern the French government.

This was indeed the end of the exchange of letters between Brouwer and Denjoy; it was not, however, the end of the affair. Brouwer felt that a regular member of the Dutch Academy, irrespective of his nationality, had to observe the rules. He feverishly tried to get the Academy on his side. All through the year 1921 he compiled a Denjoy file, and hoped to activate the Academy or even better, the Minister. He did not stand quite alone in this respect. Apparently there was a growing irritation with the demeanour of Denjoy, and Brouwer, already displeased in 1920 with the fact that Denjoy could still not lecture in Dutch after four years, decided to force the language issue in the Academy.

He must have been rather pessimistic about the outcome of this particular action when he wrote in a letter of 23 September 1921 to various colleagues asking them to sign an enclosed letter protesting the behaviour of Denjoy. He told his fellow members that Denjoy considered himself rather an *agent de pénétration pacifique* in the service of the French government, than their colleague. If the addressees found the letter unfriendly with respect to Denjoy, he added, they should realise that Denjoy in his function in Holland refused to obey the rules of the Academy, but on the contrary accepted instructions issued by the government in Paris.

Brouwer had no illusions about the reaction of the Academy 'which is already halfway in the hands of French imperialism'. He just wanted to encourage 'those who found the responsibility for this new mocking of our native country too heavy, to remove it by means of this protest'.

Hk. de Vries, who was one of the recipients of the letter, was in fact upset by its sharp tone. Like Brouwer, he disliked the way Denjoy operated in Holland; he spoke of 'that Frenchman, who despises everything here, except our money (*onze rijksdaalders*)'. He pointed out to Brouwer that the legal basis for an action against

Denjoy's choice of language was extremely narrow. De Vries' advice had little effect, Brouwer just carried on, regardless of the consequences and the remote possibility of success.

In a letter to the chairman of the physics section of the Academy, Brouwer denied the personal nature of the exchange. In his opinion Denjoy's invectives were more directed at the institution behind Brouwer (that is, the Academy and the international community of scientists). Anyway, he said Denjoy blew his horn *pour la Galerie Parisienne*.[28]

The *Algemeen Handelsblad* of 29 October reported the meeting of the physics and mathematics section of the Academy and repeated verbatim a letter of Brouwer to the section's board. The letter explained the reason for Brouwer's absence from this particular meeting with the words:

> Since for part of the agenda of the ordinary meeting of the section, according to the convocation, a language other than Dutch has been chosen as a medium, and, in the opinion of the undersigned, in public meetings of mandatories of the Dutch Government no departure from the official language of the Kingdom of the Netherlands should be allowed without previously obtained permission, it is impossible for the undersigned to attend the present meeting.

Brouwer had already, before the meeting, informed F.A.F.C. Went, the chairman, of his intention. Should Denjoy again address the meeting in French, then he, Brouwer, would demand that the above message of his be read in the meeting. And to make sure that his gesture would not go unobserved, he informed Went that the press had received a copy in advance. In itself this was no breach of confidence, as the press had free access to the documents of the public meeting anyway. He wrote to Went that

> my respect for the Dutch Government and for the national character of the Academy forbids me to take part in sanctioning by my presence an improper act, such as the use of a foreign language by an official appointed four years ago is in my eyes. I will give a copy to the press.

A month later the *Algemeen Handelsblad* needed no prompting by Brouwer to report the meeting.

> *Royal Academy of Sciences*
>
> In the present meeting of the section of the Academy for mathematics and physics, professor Brouwer, after the reading of the minutes, asked for the reading of a note which he had sent to the Section.
>
> The chairman, professor Went, said that the particular note had been discussed in the extra-ordinary meeting.
>
> Professor Brouwer pointed out that the letter was directed to the ordinary meeting, and that he maintained his request to have it read. Only one mem-

[28] Brouwer to Went, 13 November 1921.

ber[29] seconded professor Brouwer, upon which he demanded permission to make a statement.

The secretary, professor Bolk, pointed out that professor Brouwer's desire for publicity had already been fulfilled, because he had published the note in the *Algemeen Handelsblad*.[30]

Thereupon professor Brouwer requested that it should be recorded in the minutes and in the report of the ordinary meeting, that he had been refused the right to read a message to the ordinary meeting.

Thus the matter was closed.

Faced with the flat refusal of the Academy to hear his case, Brouwer decided to go directly to the press; he sent a letter to the editor of the *Algemeen Handelsblad*, dated 27 November 1921, explaining to the public the issue of the 'official language'. He made it clear that he wished to be absolved from sharing the responsibility for allowing the unauthorised use of foreign languages in meetings of the Academy. The refusal of the meeting to take notice of this abuse, 'which constituted with respect to the undersigned a violation of the elementary rights of a minority', could only leave the readers of the reports of the Academy with the incorrect impression that the national character of the Academy could be disregarded without protest.

On 19 December Brouwer again sent a letter to the editor of the *Algemeen Handelsblad*; this time to correct an inaccuracy that had slipped in. He had erroneously claimed in his earlier letter to the editor that the readers would be informed of the lapse of the Academy in the minutes of the meeting of 29 October. According to recent new regulations the records of meetings, etc. would, however, be published in the yearbook. The above mentioned incorrect impression, caused by 'the strangulation of the protest of the undersigned', would thus be noted 'by the international readership' at a later moment.

The effect of Brouwer's actions must have been limited; even at the best of times the Dutch were not very sensitive to national honour; anything that smelt of pompousness was easily and happily ignored.

In the meantime the fears of the board of Utrecht University, that Denjoy would consider Utrecht an intermediate station in his career, turned out to be well-founded. In 1922 he left Holland to become *Maître de Conférences* in Paris, and so the person who had caused the conflict no longer figured in Dutch mathematics.

Denjoy's mathematical influence in Holland had been stimulating indeed. He gave courses and seminars which were attended by students from all over the country. He even went so far as to inquire with his colleagues whether they had any clever students who might be interested in working for a Ph.D. thesis with him! Brouwer sent his own student Belinfante to Utrecht to attend Denjoy's courses. Van der Corput and T.J. Boks were Denjoy's assistants; and Boks, H. Looman and J.

[29] Hk. de Vries.

[30] Strictly speaking this was not the case. Brouwer's note was printed as a letter to the editor, but reproduced within the report of the meeting of 29 October 1921.

**Fig. 9.2** Open letter to the minister of Education [Brouwer archive]

*Uebersetzung.*
*Als Manuskript gedruckt.*

**Unberücksichtigt gebliebenes Schreiben von Dr. L. E. J. Brouwer**

**an**

**Seine Exzellenz den Minister für Unterricht, Kunst und Wissenschaft**

**vom 27. September 1922.**

***Symptomatisches zu einer Gefährdung der niederländischen Staatshoheit.***

Ridder were his Ph.D. students in Utrecht. There is no evidence that Denjoy ever reconsidered the matter after the last letter was exchanged. Not so, however, Brouwer. When it became known that Denjoy was about to leave and thereby automatically be made a corresponding member of the Academy, Brouwer made a final attempt to bring him to justice. He wrote a long letter to the Minister of Education, Arts and Sciences, accompanied by a complete dossier.[31]

This letter summed up the events, starting from the request for advice by Blumenthal. Somewhat bitterly, Brouwer pointed out that Denjoy's role in the matter resembled more 'that of an agent of the French Government in a territory under French sovereignty, than that of a Dutch civil servant in an independent Kingdom of The Netherlands'. On the basis of Denjoy's interpretation of the rules of the Academy and their applicability, and the real or seeming disregard for the obligations of Academy members with respect to the Dutch Government, Brouwer concluded that the automatic transfer of the ordinary membership to a corresponding membership should not be granted, at least until Denjoy should have given the Dutch Government the necessary explanation and satisfaction. He expressed his hope that the Minister would act accordingly. The Minister did what a politician automatically does: he asked the Academy for advice; the Academy appointed a committee, which duly reported, through H.A. Lorentz, its chairman, the findings to the Academy on 27 January 1923. The report stated that Denjoy had, in his dealings with Blumenthal, been guided by his feelings as a Frenchman, rather than by the circumstance that he was a member of the Academy.

> It would certainly have pleased us if it had been otherwise and thus could have led to some reconciliation. All the same, we are of the opinion that the recovery of good relationships between scholars of mutually estranged countries, which we hope to be the case in the future, would not be furthered if we

[31] Brouwer to Minister of Education, 27 September 1922.

acted with regard to Mr. Denjoy, according to the ideas of Mr. Brouwer, and demanded an explanation.

The Academy advised the Minister accordingly. This left Brouwer far from satisfied. In another letter to the Minister he once more explained the issue, which he considered grossly misrepresented in the committee report.[32] With admirable perspicuity (which would have found a more rewarding goal in mathematical research) he pointed out that the treatment of Blumenthal was completely irrelevant; what mattered was Denjoy's attitude with respect to the Academy, a member of the Academy, the Dutch Government and the Dutch people. He, understandably, did not share the views of the committee, and suggested that the Minister ask for the desired explanation, and terminate Denjoy's membership in case of refusal. The dismissal of members of the Academy, by the way, was far from simple (it even required the Queen's signature). The Minister, in view of the modest importance of the problem, simply followed the advice of the Academy.

The Academy must have been rather displeased that Brouwer, in spite of the report of the committee, had thought it necessary to approach the Minister directly. At the meeting of 27 January, Brouwer defended his action in the (closed) meeting of the Academy.

This was the end of this rather quixotic episode in Brouwer's career; his emotional reaction to the insult of Blumenthal, coupled with his indignation at the treatment of the defeated Central Powers by, of all people, his fellow scientists, had carried him further than he might have foreseen. More than the earlier conflicts, this affair gave him the reputation of quarrelsomeness. He even went so far as to have his petition to the Minister printed at his own cost, including a German translation, with the ominous subtitle 'Symptomatic observations on the jeopardising of the Sovereignty of The Netherlands.'[33]

The whole Denjoy affair and the way it was handled by the Academy left Brouwer emotionally exhausted. It incapacitated him for months; in a letter to Mrs. Ehrenfest he confessed that '... since the Dutch Academy of Science has been pulling the loot wagon of the Parisian Shylock gang as the $n$-th yoke of oxen (and allows the members who do not swallow this humiliation to be abused by the Shylock lackeys) I have fallen into such a state of disillusion and apathy that most of the incoming letters are not answered by me'.[34]

Denjoy's career did not suffer through this intermezzo; perhaps, on the contrary, it was furthered. Among the many honours that befell him during his career, there was the *Legion d'honneur*, awarded in 1920.[35] The motivation contained the following passage: 'Scientist of very great value. Was sent as a professor to the University of Utrecht; has rendered through the success of his teaching the greatest services

[32]Brouwer to Minister, 12 February 1923.

[33]*Symptomatisches zur Gefährdung der Niederländischen Staatshoheit.*

[34]Brouwer to Mrs. Ehrenfest, 26 April 1922.

[35]*Chévalier*—1920, *Officier*—1935.

to the French propaganda. Was torpedoed twice while rejoining his post ...'[36] In a way the award confirmed the worst suspicions of Brouwer: the French government acted as if Denjoy was posted in the Netherlands, much as a soldier or diplomat is sent to a foreign outpost. The officials simply chose to ignore the fact that Denjoy had gone through an ordinary application procedure, and was appointed as an ordinary civil servant by the Dutch authorities. Perhaps the wording was chosen so as to teach the Dutch a lesson; it seems more likely that the French authorities could not see much difference between posting a professor in the Netherlands and a diplomat somewhere in Central Africa.

Brouwer was not intrinsically, or by disposition, a German supporter. There is little doubt that had the war taken another course and ended with victory of the Central powers, the peace conditions would have been equally harsh and short-sighted. In that case Brouwer would undoubtedly have campaigned for the other side. As it was, the present constellation of the international powers and executives tended to drive Brouwer more and more into the arms of the underdog—and of the German nationalists.

While Brouwer was conducting his campaign against Denjoy, his national mathematics association, the *Wiskundig Genootschap*, was wooed from its neutral position by the *Conseil International de Recherches.* The *Wiskundig Genootschap* was invited in April 1921 to join the *Conseil*; on the annual meeting on the thirtieth a letter to that effect from the *Conseil*, written by G. Koenigs, was read to the members. And when the meeting showed itself not averse to joining the *Conseil*, the board was authorised to settle the matter. And so that same afternoon, the board took steps to join the *Conseil*; from the minutes it appears that the motives for this hasty act were not wholly idealistic. There was a certain fear that the *Conseil*, that is to say, the French, would revise the international allocation of the review journals after the eclipse of the Germans, and the Dutch published a mathematical review journal, the *Revues Semestrielles*, which was eyed by some parties abroad with a keen interest. The society argued that nobody would rob a member of the *Conseil* of its review journal. And so the secretary duly wrote a letter of application on 11 May.

Brouwer's opposition was conspicuously absent. Why? The most likely explanation is that he was not aware of the goings-on. He had been on a holiday trip to Italy in April, where he visited Assisi and Florence with his wife. Right after the Italian tour he must have gone on to Germany, for it appears from a letter from Brouwer to Schoenflies that he was in the Harz in the beginning of May. Brouwer's presence in the Harz is not as surprising as it seems; already at an early age his health problems had been the cause of parental concern, and he had been an incidental visitor to health spas, among them the institution of Dr. Just in Jungborn, not far from Harzburg. Indeed some of the pre-war letters to Hilbert were dispatched from Just's Institution. After the war he resumed his old habit, and the letter to Schoenflies in effect was written in Just's institution, where Brouwer had sought refuge for

[36]Extract of the citation: *Savant de très grande valeur. Envoyé comme professeur à l'université d'Utrecht, a, par le succès de son enseignement, rendé les plus grands services à la propagande Francaise. A été torpillé deux fois en rejoignant son poste ...* (2 October 1920).

health reasons. One would be hard pressed to find a more idyllic and healthy area than the Harz, which had an additional advantage—it was close to Göttingen, and many prominent mathematicians spent their vacations or weekends in the Harz, for example Cantor, like Klein, was a habitué of Hahnenklee.

For the Mathematical Society, Brouwer's absence meant a quick decision and no painful debate; no one, and certainly not the board, would have looked forward to doing battle with Brouwer, who was a fearsome opponent and who could, like the legendary warriors of the past, cut down a numerous opposition single-handedly.

It is thus one of the minor ironies of history that the Dutch Mathematical Society, which counted among its members the most determined neutral opponent of the *Conseil International de Recherches*, inconspicuously joined it. One is reminded of decisions taken 'when father is away'!

The innocent cause of the Denjoy affair, Otto Blumenthal, had quietly disappeared from the scene. Brouwer, who was sincerely fond of Blumenthal, did some lobbying for him when Bieberbach left Frankfurt for Berlin. In glowing words he recommended Blumenthal to Schoenflies, as a man who in matters of 'universal mathematical knowledge, working capacity, helpfulness, and on top of that honesty and decency hardly finds his equal among our colleagues'.[37] He praised Blumenthal lavishly for carrying on the management of the *Mathematische Annalen*, which owed its prominence largely, according to Brouwer, to Blumenthal's unselfish efforts.

> That Klein and Hilbert have, nonetheless, never helped him to a university chair, I can only explain by Machiavelli's maxim: 'The first duty of kings is ingratitude', in addition to which the excessive modesty of Blumenthal (who never blew his own horn) has played a role.

Schoenflies' reply was sympathetic, but he regretted that he could not do much; Bieberbach had built up an excellent reputation as a Ph.D. adviser, with a wealth of scientific inspiration, and the faculty wished to find a comparable man. Max Dehn eventually became Bieberbach's successor in Frankfurt, and Blumenthal remained in Aachen. Brouwer was, of course, disappointed; he wrote that Blumenthal had exercised a considerable influence on himself: 'many of my papers I would never have written without him'.[38]

## 9.4 Weitzenböck's Appointment in Amsterdam

Gradually, the spirited and brilliant mathematician–philosopher Brouwer had become entangled in the web of academic traditions. Had he been primarily a researcher in the pre-war years, but by the time the war ended, Brouwer had, whether he liked it or not, become a part of the establishment, with all the obligations that

[37] Brouwer to Schoenflies, 17 January 1921. Schoenflies was Rector in Frankfurt at the time.

[38] Brouwer to Schoenflies, 14 May 1921.

eventually reduce a free man to a slave of the circumstances. During the years 1915 and 1916 he had served the Dutch Mathematical Society as a chairman, to be succeeded by his close friend and ally Mannoury. The Amsterdam Academy and the International Academy for Philosophy also laid a claim to his time and energy. And, of course, the faculty required his attention; one cannot escape the impression that Brouwer took these institutions somewhat more seriously than was good for him and for them.

Like so many theoretical disciplines, mathematics had no place of its own, just a faculty room in the central building of the Oudemanhuispoort. For Brouwer, who was familiar with the facilities in other places, this was far from satisfactory, but official signs of discontent have not been preserved in the minutes of the faculty. Brouwer was surprisingly faithful in attending the faculty meetings, but before the twenties he was not conspicuous. His activities were rather low-key: the minutes record his ideas and comments sparingly. For example, after Korteweg's retirement Brouwer and Hk. de Vries advocated to change the vacancy of the extra-ordinary chair into a lecturer's position (24 October 1918) and in February 1919 he is on record as requesting a typewriter for the use of the library assistant. The offers from Göttingen and Berlin were of course major events in the otherwise quiet faculty life. The minutes faithfully record the 'feelings of joy that Mr. Brouwer has resisted the lure of foreign parts' (17 March 1920).

In spite of his tendency to lose his audience in long monologues, Brouwer's teaching was appreciated by his students. The course material was always impeccably presented.

The records of Brouwer's courses show that his classes were well attended.[39] Some of his early followers already appear before 1920: the roll of the course 'Elementary Mechanics', 1915/16, contains the name of Maurits Belinfante, and in 1918–1919 the name of Arend Heyting appears on the list of the course 'non-Euclidean and axiomatic geometry'. Both were to become the first intuitionistic followers of Brouwer. They had been preceded by Brouwer's first Ph.D. student, B.P. Haalmeijer, who wrote a dissertation on a topological topic.[40]

Brouwer used the prestige of foreign offers of such considerable weight to bargain with the Curators of the University for substantial concessions. The permission to hire Hermann Weyl must have been one of them, but when Weyl declined the offer, Brouwer did not press his luck and settled for a lecturer. The minutes of the meeting of 23 February 1921 record that Brouwer had found a suitable candidate, Dr. Roland Weitzenböck, ordinary professor in number theory and the theory of invariants at Graz (Austria). The choice of Weitzenböck is difficult to explain. Brouwer knew many excellent mathematicians abroad, and there seemed no special reason to prefer Weitzenböck. At least it shows that Brouwer wished to widen the

[39] Only a few of the attendance lists were included in Brouwer's private files. There is no central registration of student attendance.

[40] *Bijdragen tot de theorie der elementairoppervlakken* (Contributions to the theory of elementary surfaces), 28 November 1917.

spectrum of expertise in Amsterdam: the new candidate brought new blood and new topics.

Weitzenböck was born 26 May 1885 in Kremsmünster (Austria). He studied at the Military Academy in Mödling and obtained his doctor's degree in Vienna. From 1911 to 1912, he studied in Göttingen and he wrote his *Habilitationsschrift* in Vienna. Subsequently, he taught mathematics in Graz, and got an offer from Prague just before the war. After the war he was appointed in Prague as an extra-ordinary professor. After being promoted to ordinary professor, he got an offer from Graz, where he was teaching at the Institute of Technology. He was still in Graz when the Amsterdam Faculty showed an interest in him. Although he was in active military service during the war, he had kept up his mathematical production; he was a regular speaker at mathematics meetings. He must have impressed Brouwer by his specific expertise. Brouwer and Weitzenböck already knew each other in 1912, as confirmed by a postcard from Brouwer in Göttingen to the Dutch mathematician P. Mulder.

Evil tongues had it that Brouwer had an ulterior motive for Weitzenböck's appointment: to annoy and keep out J.A. Schouten (who had a chair in Delft at the Institute of Technology). Weitzenböck's appointment followed on 11 October 1921 and since then he remained at Amsterdam.

The appointment of Weitzenböck was no routine matter; Brouwer and the faculty had many hurdles to take—couldn't the faculty find a Dutch candidate, at an earlier occasion two Dutchmen had been mentioned, would a lecturer's position do, etc. In the end the Curators gave in. De Vries was so bold as to suggest that hiring Weitzenböck would be a step in the direction of their common goal: 'to make Amsterdam the mathematical centre of the Netherlands'.

On 12 May they submitted Weitzenböck's name to the City Council for an appointment, and on 6 June the Rector was able to inform the Academic Senate that Weitzenböck would be appointed as a lecturer. The final appointment followed on 11 October of the same year. This appointment was, for the time being, the end of the expansion desired by Brouwer.

## 9.5 Kohnstamm and the Philosophy of Science Curriculum

The Amsterdam Faculty had boasted an excellent staff of professors; J.D. van der Waals, P. Zeeman, Hugo de Vries and D.J. Korteweg were the early stars, and Brouwer was a worthy second-generation member. Among the members of the faculty there was a former student and associate of J.D. van der Waals,[41] Ph. Kohnstamm, co-author with Van der Waals of a textbook on Thermodynamics. Kohnstamm was appointed in 1908 as the successor of Van der Waals, but gradually his interest in physics was overshadowed by other interests. We have met his name before, he was the editor of the *Tijdschrift voor Wijsbegeerte*,[42] who had handled Brouwer's 'Unreliability'-paper in 1908. Kohnstamm's comments at that time

[41] The Nobel prize winner, not to be confused with his son of the same name.

[42] Journal for Philosophy.

would have done little to increase Brouwer's respect for his colleague's philosophical insights (cf. p. 105).

Kohnstamm, a man of considerable charisma, eventually gave up his chair in physics, and accepted a special chair for pedagogy in Amsterdam.[43] In his function as a professor in the faculty of the exact sciences, he actively promoted the study of philosophy in the faculty of mathematics and physics. At that time philosophy, as a major subject, belonged to the literary faculty, and Kohnstamm tried to get philosophy, in particular the part that was oriented towards the sciences, accepted as one of the majors in the science faculty. He felt that the philosophy of the sciences had come of age, and that it was time to incorporate it into the academic curriculum; the proper place for the philosophies of the various disciplines seemed to him the corresponding faculty (called 'the faculty of mathematics and physics', but also containing chemistry, biology, astronomy, ...). Brouwer was fervently opposed to the whole enterprise, he was (correctly, one would guess) worried about the dilution of the content of the traditional curriculum.

The procedure for such a change in the academic curriculum was long and wearisome, but apparently Kohnstamm knew how to speed up the official machinery, and so we see that in a surprisingly short time the proposal reached the Academic Statutes.

The matter was brought up in the fall of 1919, when the official body for discussing matters of curriculum went over the major subjects of the faculties.

Kohnstamm handled the whole affair as a born politician; at the meeting of the faculties in Utrecht he was one of the spokesmen of the Amsterdam faculty. The faculty had bound him to a negative advice, to be conveyed to the other delegates. In the general confusion, when the meeting discussed this minor detail of 'philosophy in the science faculty', Kohnstamm managed to promote the introduction of the subject! The mix-up resulted in a general impression that philosophy had been accepted. When it finally dawned that there had not been a majority for philosophy, the positive recommendation had already been sent to the Minister.

Brouwer violently opposed Kohnstamm's move: he agitated where he could, in the faculty and in public, and indeed in January 1921 the united faculties had decided to advise against the proposal, but by a polished manoeuvre, stretching logic to its breaking point. Kohnstamm convinced the delegates that a vote against the proposal automatically implied the rejection of an older rule, which formulated the teaching qualification of students in philosophy. In this tragicomedy of error and calculation, the proposal eventually reached the Minister. In spite of Brouwer's attempts to change the mind of Minister De Visser—he enlisted the help of his fellow significist Van Ginneken—the study of philosophy was added to the science cur-

[43]There were three kinds of professorships: the ordinary one, which was a full time job, the extra ordinary one, which was usually a part-time appointment for some specialism, and finally the special one, which usually was a small part-time affair with scant remuneration, paid by some society for the furthering of the interest in ... These societies ranged from religious groups to para-psychologists.

riculum. The Minister was only vaguely aware of the purport of the proposal. He introduced the plans in parliament[44] with the words:

> For I have learned that lately a scientific (*natuurwetenschappelijk*) discipline has come up in the sciences, especially in physics, which, if I may say so, wants to see, more than before, that emphasis is laid on the psyche of plants and animals. And which, to stick to today's terminology, tries rather to do justice to a more idealistic direction than to the biased materialistic direction. As a couple of representatives of that movement here in our country, I would like to mention: Prof. Kohnstamm at Amsterdam and Prof. Buytendijk at the Free University.[45]

It is clear that the Minister was thinking of something rather different from what we nowadays call the philosophy of the sciences: he probably thought of something related to psychology (as the names of Kohnstamm and Buytendijk suggest). He probably thought that anything opposed to the evil of materialism was automatically a good thing.

Brouwer was not inclined to let the Minister get away with his curious motivation. As it happened, he had just joined the editorial board of a new literary magazine, *De Nieuwe Kroniek*.[46] This provided an excellent opportunity to write a biting comment on the proposal to adopt 'philosophy' as a major in the Science faculty. It is not known how Brouwer got into the literary–artistic circle that ran the magazine, but he clearly had no objections to try his hand at this unexpected opportunity.

In a few lines Brouwer took the plan and its motivation to pieces; he forcefully argued that there was no such thing as an independent philosophy curriculum (*leervak*) in mathematics and physics.

> As all measures that are impervious to reasonable realisation, the one of the minister will only have effect as a pretext for abuses; in this case the more so because of the subsidiary meaning of the word 'philosophy' in the mouth of eloquent superficiality, which is the expression of its reluctance to submit to any control of intelligence or factual knowledge.

Even less appreciation was shown for the minister's amateur opinion, or for his choice of representatives:

> And it is incomprehensible that the minister, who has no expertise in the field, and who should find it more important to inspire confidence than astonishment, felt that it sufficed to communicate his personal motives, and did not feel obliged to name the authorities, on the excellent authority of whom he has ignored the unanimous advice which he had received from the competent and involved faculties. The mentioning of the names of the gentlemen Kohnstamm and Buytendijk has no significance in this context, since Mr. Buytendijk is a

[44]Meeting of Parliament, 25 May 1921.

[45]The Calvinist university at Amsterdam.

[46]The New Chronicle.

> medical man, who never showed any mathematical or physical competence, whereas Mr. Kohnstamm, to be sure, has a place among the members of the faculties of mathematics and physics as an extraordinary professor, but who does not belong there, to those whose opinion one can make prevail over that of a large majority of colleagues, without affronting the latter and disgracing oneself.[47]

Objectively and factually, Brouwer was completely right. There was no urgent need to start a philosophical education in the science faculties, and there were (apart from Brouwer and Mannoury) no acceptable candidates for such a project. One may well surmise that Kohnstamm was promoting a private pet project, and that Brouwer saw through the manoeuvre. He was furious when Kohnstamm succeeded. At one instance he snapped at Kohnstamm, when the matter was discussed in the faculty: 'In this matter you are an ignoramus, and you have to abstain from an opinion!' Again, Brouwer's cool arguments were a front for highly emotional outbursts. Van Eeden recorded in his diary (15 June 1920) that 'Bertus was again fierce in sympathy and antipathy. Furious at Kohnstamm, ...'

Kohnstamm defended himself in an educational magazine with arguments that avoided the heart of the issue, and that breathed an atmosphere of reasonable reflection, suggesting that Brouwer was an isolated malcontent. This battle was lost by Brouwer, mostly because his opponent, Kohnstamm, had his campaign smoothly organised, and it seems likely that as a philosopher of the traditional kind, he was better able to mobilise the sympathy of the average philosophy amateur than the forbidding Brouwer, who did not spare his readers.

## 9.6 The New Chronicle

Brouwer's temporary platform, *The New Chronicle*, was a magazine that brought together artists and scientists. The editors were Brouwer, Frans Coenen (an author and journalist), F. Fisher, J.F. Staal (an architect) and Matthijs Vermeulen (a composer).

Brouwer's contributions to this journal were modest. He probably attended the meetings of the editorial board. He was, as a matter of fact, surprisingly conscientious in matters of that sort. He did not belong to the class of the perpetual absent. But in writing, there is little to call attention to; apart from two articles dealing with the philosophy issue and Kohnstamm, he contributed a short review of a Dutch translation of the Russian book of statutes of labour.

In this review Brouwer discussed the following principles of the Soviet law:

1. Only the state has the right of inheritance.
2. Man and woman have the same legal position.
3. All healthy individuals from 26 to 50 years of age are, by authority of the state, entitled to labour, in accordance to their ability, and to a recompense which guarantees their welfare.

[47] Brouwer (1922b).

4. All other individuals (including children) will be cared for by the state, even without their labour.

He accepted the first two without hesitation. The third and fourth principles he met, however, with considerably reservation. Brouwer's commentary is rather penetrating, especially if one takes into account the fact that in the first years after the Russian revolution intellectuals were inclined to give the new rulers the benefit of the doubt in almost every way. Keeping in mind the ultimate collapse of the Soviet experiment, which was viewed as the new paradise by the young idealists in the Western world, it is rewarding to reread Brouwer:

> The first two principles mean doing away with decayed institutions, obscuring selection and family life, and can hence be unconditionally welcomed with approval. The enforcement of the third and fourth principles, however, will then only meet their goals (putting an end to the needlessly cruel pauperisation of the socially weak) if with respect to the connected abolition of private property of real estate, means of production and transport, the utmost reservation is practised. For, one of the deepest rooted desires of civilised man aims at freedom, freedom of employment: unhindered unfolding of talents; private freedom: undisturbed possession of house and household goods, unhampered choice of company. And only the hope of freedom is capable of evoking in the individual that intense energy for work, through which alone a nation can gather riches, sufficient to bring affluence and civilisation to broad sections of the population. But hope for freedom, in a society without slavery, means hope for money, and hope for money is only possible in the case of a market for goods and labour which is at least partially free. By the total abolition of this market, i.e. the total militarising of society, the social efficiency of labour would be reduced to a minimum and universal poverty would be created. With respect to the equilibrium between state enterprise and private enterprise thus to be pursued, the following demands have to be made: In the first place, the higher revenues of private enterprise compared to that of state enterprise, should benefit the state not only indirectly, but also directly, and indeed by means of strongly progressive taxes. In the second place, luxury enterprises should be forbidden as long as a high general affluence has not been reached. In the third place, neither the wages in the state enterprises, nor the taxes on private enterprise, should be raised to such an extent that the possibility of existence of the latter is endangered. In particular, one should leave private trade the freedom of action, since only that can bring the goods in time and undamaged into the hands of the consumers. One should even keep up *dishonest* trade as an indispensable domain of escape for the activity of crooks and hustlers, whose perpetual birth in a normal percentage is after all guaranteed by the biological–statistical laws. Woe to the state that would block the possibility of trading to these gentlemen, and would force them to enter, all of them, politics!

Brouwer's review is interesting, as it shows that he was not blind to the idea of a reasonable state interference with society and trade, but he did not choose to forego his private philosophical views for a fashionable political idealism.

Extreme and unrealistic ideas were seriously considered by Brouwer. There is a passage in Van Eeden's diary in which Brouwer's views on social reform were recorded. According to Brouwer, two things were necessary for that purpose:

1. Abolishment of the anarchy in marriage and procreation, under the supervision of society.
2. Conferring a temporary value to money. Thus the value of a bank note is fixed for a limited period, after which it becomes worthless.

It certainly seems that ideas came first with Brouwer, the realizability was often left out of the considerations.

In view of the eventual decline and fall of the Soviet Empire, Brouwer's comments are surprisingly modern, but at the time of writing, his more progressive friends would not have found his views overly appealing. One might say that Brouwer's views were the paradigm of common sense, but one should also remember that in progressive circles common sense is often a scarce commodity.

*The New Chronicle* was a short lived magazine; it soon expired, thereby ending Brouwer's short fling with the literary world.

# Chapter 10
# The Breakthrough

## 10.1 The Signific Circle

Although the political affairs of the post-war period laid a disproportionate claim to his time, Brouwer still managed to think about more academic matters. There was his concern for the constructive content of mathematics and, on a smaller scale, his interest in the psycho-linguistic topics of significs.[1] Looking at the list of Brouwer's publications, one sees that an outburst of productivity in 1919 and 1920 was followed by a slack period, with no publications at all in 1922.

Brouwer blamed politics for the temporary lull in his scientific activity. In a postcard to Schoenflies of 1 September 1923, he wrote 'my own work has rested completely for three years, because my strength is almost fully taken up by the fight against our annexation by France, which is diligently promoted by the Lorentz clique'.

This claim may seem exaggerated, but we must realise that Brouwer was not a slick operator who could combine science and political manoeuvring without winking an eye. He was always emotionally involved in his fights against injustice. The conflicts sapped his energy to a dangerous degree; his work suffered perceptibly at times of crisis. So it is plausible that the Denjoy episode and the '*Conseil*' problems seriously affected his functioning as a mathematician. After Brouwer's letter to the minister (cf. p. 337) the Denjoy case was more or less closed, and there were no urgent extra-mathematical matters preventing Brouwer's return to his research.

This research consisted of his intuitionistic mathematics, but in addition there was his involvement in the signific enterprise. The latter was in the hands of a small group of Dutch scholars, the most prominent among them was the celebrated author–utopist–psychiatrist–linguist Frederik van Eeden, but the driving force behind the signific enterprise was Gerrit Mannoury, self-styled mathematician, philosopher and Marxist.

---

[1]The section on significs makes extensive use of Walter Schmitz' book, *De Hollandse Significa*, Schmitz (1990a). The reader will find a wealth of information in its pages.

D. van Dalen, *L.E.J. Brouwer – Topologist, Intuitionist, Philosopher*,
DOI 10.1007/978-1-4471-4616-2_10, © Springer-Verlag London 2013

Walden, the former colony (commune) of Van Eeden, was one of the favourite meeting places, the company also met at other places, such as the pharmacy of Brouwer's wife, Mannoury's place or Brouwer's house in Blaricum.

Schmitz gives in his book *De Hollandse Significa* a detailed report of the activities of the International Institute for Philosophy, its meetings and discussions, its principles and conflicts. For the present it may suffice to say that the discussion never produced the required consensus, at least not to a degree necessary for a program. Even Mannoury and Brouwer, the two members of the project who were close friends, differed on the topic of 'language'. Mannoury was keenly aware that Brouwer's concept of language reform was too specialised to attract supporters on a desired scale.

Things did not become simpler when a newcomer made his entrée. The old group, more or less the founding fathers of the International Institute for Philosophy, had been reinforced in November 1919 by the linguist Jacques van Ginneken, a powerful and charismatic man. Van Ginneken (1872–1945) was one of the shining examples of Roman Catholic emancipation, a man with an almost inexhaustible store of energy and a wealth of ideas and plans; he became a Jesuit in 1895 and, after writing a much admired dissertation on the foundations of psycho-linguistics, devoted himself to the conversion of the Netherlands to Catholicism. He gained prominence in the Catholic Netherlands where he was, among other things, the key figure in organising the mass movement *De Graal* (The Grail) for girls and women.

His authority in linguistics was recognised in 1923 when the newly founded Catholic University at Nijmegen made him a professor.

Right from his introduction into the company of significists, Father van Ginneken occupied an important, if somewhat puzzling, place in the group. He introduced an element of dogmatism into the discussion which had been absent so far. Whereas the older members, of course, had their convictions and did not lose an opportunity to defend them, their discussions were in general ruled to a high degree by the weight of arguments. Van Ginneken's Catholicism did not agree well with this tradition, but since his presence as an authority was appreciated, Mannoury had to exert himself in devising compromises. The very first act of the circle, the formulation of its basic principles, already called for such a compromise.

Back in 1919, Mannoury had put forward a system of language levels, which was adopted on 18 February of that year at the board meeting of the Institute for Philosophy.

The reader should not get any grand ideas at terms like 'Board of the Institute' and of the various committees mentioned in the prospectus of the Institute for Philosophy, since in practice the four friends, Borel, Brouwer, van Eeden and Mannoury, were the major actors who kept the enterprise going. The whole enterprise had something of an old-fashioned comedy, where the cast of a small travelling company had to play at least three parts each. So when we read that the board accepted Mannoury's language levels, we should not think of a proposal put on the agenda and subsequently put to a vote by the members. Everything was on a small, cosy scale. The four of them would spend months debating the issues involved, changing a little here, polishing a bit there, and finally reaching a satisfactory formulation.

The group had accepted the general premise that 'language is never able to represent or render any part of reality adequately, but that the so-called meaning of the words only depends on the effect which the speaker has in mind, or which the hearer undergoes'. The readers of Brouwer's *Life, Art and Mysticism* will not be surprised by this viewpoint. He must, however, allow for the fact that these ideas were, so to speak, in the air, in the words of the report: 'they were nowadays generally accepted'. The immediate task, formulated in the founding statement of 21 January 1919, was the compilation of one or more dictionaries for the 'ground words'. This should eventually lead to a clarification and purification of the exchange of thoughts.

The language levels were in this context an appropriate classification, intended to import some system into the amorphous domain of language. The opening statement was followed by a further specification, reproduced below.[2]

The content of the document is rather Brouwerian in tone and intention, but this is not particularly surprising in view of the relationship between Brouwer and Mannoury.

> *The distinction between the language levels with respect to social understanding.*
>
> The *ground language* need not presuppose any dynamic relationship of will to other individuals, and if it does, that relationship can be one of friendship as well as one of enmity.
>
> In the *mood language*, on the contrary, the mutual right to life of speaker and listener is recognised, though not further regulated, and hence loneliness is objected to, and enmity stylised.
>
> In *daily language*, a measure of unanimity of the speaker and listener is obtained, which excludes almost all militancy, and a strong limitation of the possibility of misunderstanding, by admitting to the understanding only those elements of life that are expressed in generally recognised human needs of a peripheral nature.
>
> In *scientific language*, this selection of elements of life has gone so far that only those are admitted which are inherent in the assumption of an objective outer world (common to all individuals).
>
> Finally, in the *symbol language*, only elements of life are considered which are covered by intellectual categories (common to all individuals) so that an almost complete exclusion of misunderstanding is reached.

The document was signed by the trustees of the International Institute for Philosophy: Mannoury, Brouwer, Borel and Van Eeden.

We have presented the formulation of the language levels in full, because they were adopted as a basis for further work. It is also a typical instance of the Mannoury–Brouwer approach to language. Further discussions of the language

[2] 18 February 1919. See Schmitz (1990a), p. 421.

levels is to be found in the document *Signific Language Research*, reproduced in Schmitz (1990a), p. 415.[3]

The discussions dragged on, and although some progress was made, the institute never became operational. In particular the lack of international response proved to be fatal. In passing it should be mentioned that the absence of financial support was another reason for putting the institute out of its misery. The brochures and mailings exhausted whatever slender means there had been. The institute in fact did not manage to rise above the level of a debating club of bright academics. The end of the project did not come as a surprise to its members.

When it was clear that the *International Institute for Philosophy* was not viable, it was dissolved by the remaining members on 23 February 1922. Not willing to give up their hopes for a signific revision of society and philosophy, the hard core, Brouwer, Van Eeden, Van Ginneken and Mannoury, almost immediately founded a new group: the *Signific Circle* (*Signifische Kring*). Mannoury was appointed chairman, and a little later, on 25 June Brouwer became its secretary.

The new group, called the *Signific Circle* at van Eeden's instigation, decided to present itself to the world in a common declaration of its founders.[4] The formulation of this declaration was no improvement over the founding declaration of the short lived Institute. As we all too often experience: too many cooks spoiled the meal. Comparing the new program to its predecessor, one notes that under the influence of Van Ginneken, the language levels lost their prominent role, and 'language act' had become the central term. Furthermore, Van Ginneken introduced statistical and experimental considerations into the program.

The new Circle was no longer after ultimate perfection and unanimity; therefore it was considered fitting to add the personal views of its members to the joint declaration. As to be expected, the formulation of the declaration involved a good deal of discussion, and gradually it became clear that Van Ginneken could not and would not endorse in all respects the common views that had been put forward in the past. Not surprisingly, his personal statement, submitted in the fall of 1922, contained a good measure of provocation. Considering himself the missionary soul of the emerging Catholic progressives, he did not try to hide his convictions and prejudices. Brouwer's expectation and hope that the significists would rise above their respective groups of reference as a 'neutral humanitarian community' did not meet with Van Ginneken's approval. On the contrary, he wrote to his fellow members that:

> The undersigned does not mind the inadequacy of language very much, as long as he directs himself to his equals, his close friends and relations, in short the members of his confessional group. And in this way the difference between me and my co-signatories has to be explained, I think, because I feel that affinity of soul in my private circle far stronger than they do.

[3] A detailed treatment of the history of significs would take us too far afield, so we will stick to the aspects that are somehow relevant for Brouwer's life and work. The book of Schmitz (on the basis of his Habilitationsschrift) is recommended for full information.

[4] Cf. Schmitz (1990a), p. 423.

This somewhat parochial and in a sense un-cooperative viewpoint irked Brouwer. On 24 November 1922 Brouwer confessed in a letter to Mannoury that he had given up all hope:

> In the contribution of Van Ginneken I read for the first time a formally pronounced, ruthless negation of the only thing that attracts me in significs: the hope on the creation of linguistic social reform tools, independent of all existing (and in my opinion mostly obsolete) groupings, and indeed by people who would rise in a neutral-humanitarian community above their respective groups. It is true that this view has been pushed more and more to the background, but I have patiently allowed this; in the first place counting on my learning capacity in the matter, and in the second place in the expectation that the community I hoped for would, in spite of all difficulties, eventually come about and function.
>
> This expectation I have to give up definitely, after the experience that one of my fellow members draws inspiration from the rejection of my (unchanged) principle. And the consequence of that can be no other than my resignation from our circle. I am even of the opinion that it would smell of dishonesty or lack of character if under these circumstances I took part in the publication of our joint manifesto, knowing that it will be followed by Van Ginneken's words.
>
> In spite of the above, I have the feeling that there is something that binds the four of us to each other, than to others, but *je ne sais quoi* seems not to be admitted to the domain of conscious reality.

Mannoury's superior statesmanship and his genuine personal warmth must have prevented the imminent collapse of the circle. Brouwer did not quit and he delivered his personal statement, be it with a considerable delay, on 25 March 1924. He stressed his fundamental views:

> For the undersigned, significs does not so much consist of a practice of linguistic criticism as:
>
> 1. The tracing of the affect-elements, into which the cause and action of words can be analysed. By this analysis, the affects that touch on the human understanding can be brought closer to control by conscience.
> 2. The creation of a new vocabulary, which also for the spiritual tendencies of life of man yields access to their reflective exchange of thoughts and, as a consequence, to their social organisation.
>
> For the realisation of the part of the program mentioned under 1, co-operation cannot be dispensed with: for, numerous affect complexes can only be analysed through the catalysing action of philosophical discussion between those not of the same mind.
>
> Also with respect to the creative labour mentioned under 2, I have for a long time believed in the great importance of co-operation. However, I have become more and more convinced that this higher task of significs can be accomplished only by the utmost concentration of mind of the individual.

Thus Brouwer remained true to his first conception. He was confident that via introspection and exchange of thought with, in particular, people of different socio-cultural convictions and backgrounds, the affects which underlie words could be traced and subsequently named, so that people could be more aware of them and that they could thus guide the conscience. His view, in fact, differed from those of the others (in particular Van Eeden and Mannoury) in that he assigned the social component of language a far more modest role.

The first practical enterprise which had to illustrate significs in action was the creation of an *Encyclopaedia of Significs*. The Circle composed a list of prospective authors: Faddegon, de Haan, Eugen Ehrlich, Révész, Bouman, Van der Plaats, Grünbaum, Bertrand Russell, Uhlenbeck, Husserl, Spengler, Peano, Couturat, Henri Borel, Brouwer, Van Eeden, Van Ginneken and Mannoury.

The Encyclopaedia project, formulated by Mannoury, meant a clean cut with the older ideas of social reform *à la* Van Eeden and Brouwer. As with so many plans of the significists, this one also came to nought. The envisaged authors, when approached, were not interested. After some more half-hearted initiatives, the Signific Circle, too, foundered. The last project, the publication of a survey of the various viewpoints, *The Signific Dialogues*, was undertaken in 1924. Its publication followed in 1937.[5]

Mannoury sadly accepted that the Signific Circle was doomed to sterility. In particular, Van Ginneken's objections to Mannoury and his communism played a role in the general ineffectiveness of the group. In 1924 Van Ginneken left the Signific Circle. He may have distrusted Mannoury's influence, as might be inferred from a letter of Van Ginneken to Van Eeden of 26 December 1924: 'Although the person of Mannoury remains very sympathetic to me, I feel that I have to break with him as a party member, since his principles force him to malign my principles.'[6] In view of the virulent anti-religious ideas of the more crude Marxists, van Ginneken may have had some grounds to keep his distance, but in the case of Mannoury he was sadly mistaken. Mannoury was, in spite of his occasional political rhetoric, an extremely tolerant person.

When the Signific Circle showed no prospect of healthy activity, Mannoury proposed in 1926 to dissolve it. This marked the end of an ambitious enterprise, one that failed on account of ordinary human weaknesses, and an unworldly approach to matters of organisation. The group had been too heterogeneous to succeed in the face of real differences of opinion. There have been attempts to compare the Signific Circle with the Vienna Circle, and indeed there are a number of similarities. But on the whole the Signific Circle lacked the coherence and academic homogeneity necessary for influencing the development of science and philosophy. In spite of their literary reputation, Van Eeden and Borel did not match the intellectual potential of Brouwer and Mannoury. Moreover, the goal that the significists had set themselves, a socio-linguistic reform, was so ambitious and vulnerable to squabble, that it was

[5] Brouwer et al. (1937a, 1937b, 1937c).

[6] Cf. Fontijn (1996), p. 550.

almost doomed to fail. Significs, as such, was promoted further by Mannoury, who gave numerous talks and published a number of books and papers, but even his efforts could not secure it a place among the popular philosophical topics, nor in the academic curriculum.[7]

Van Ginneken played more than one role in our history. He was also instrumental in the conversion of Van Eeden to Catholicism. His attempts to convert Borel and Brouwer did not meet the same success.[8] In his younger years Van Eeden had been rather critical of religion in general, but in his sixties, he changed his views and became, guided partly by Van Ginneken and his second wife, Truida, a faithful son of the Church.

Although there was no longer any social or political suppression of the Catholics, the church still had solid militant and expansionist ideas. It considered the conversion of Van Eeden as one of its spectacular 'catches' and decided to make the most of it. Van Eeden was baptised and received into the church on 18 February 1922, in the presence of Van Ginneken and Brouwer. At the occasion of the baptism, Van Eeden had to make a solemn retraction of most of his earlier publications, which were in part anti-religious. The books and plays that were not compatible with the views of the Catholic Church were displayed on a separate table. Van Eeden raised each book separately abjuring it. It made Brouwer feel sick to witness the procedure.[9]

Van Eeden's perception of Brouwer's reaction can be found in his diary (18 February 1922): 'Brouwer, too, was moved.' Reports of events of a personal nature have the inevitable tendency to contradict each other. Louise's recollection of the proceedings and Brouwer's reactions may very well be compatible with Van Eeden's observation.

Van Ginneken and Brouwer were present at the occasion, much to the indignation of Henri Borel, who wrote bitterly to Brouwer 'So you stood there as a friend and brother of that liar and faithless one, that fraud! [...]. Would Frederic Paulus now also condone the murder of Giordano Bruno and so many noble minds?'[10]

Van Eeden became 'more Catholic than the Pope', a phenomenon not uncommon with converts. He thoroughly reconsidered his views on religion and revoked part of his earlier views. A year after his baptism he was the star of a meeting, organised by the Catholic Church in the Concertgebouw of Amsterdam (22 March 1923) at which occasion he ardently testified to his new-found religion.

Van Eeden's conversion had not been the result of a sudden impulse; like a beleaguered fortress he had been made ready for the final attack. Truida, his second wife, had already embraced the Catholic religion two years earlier: she was received into the fold on 3 April 1920 and van Eeden's two children had been baptised on 23 August 1920. The influence of Truida on Van Eeden's decision had been considerable.

---

[7] As far as I know, only Mannoury, Van Dantzig, and Vuijsje taught courses in significs.

[8] Borel eventually became a Catholic at the end of his life.

[9] Interview, Louise Peijpers.

[10] Borel to Brouwer, 18 February 1922.

**Fig. 10.1** Louise as a novice. [Brouwer archive]

Brouwer's stepdaughter Louise Peijpers was baptised together with the children of Van Eeden. She had left home somewhere in the middle of the nineteen-tens[11] and entered a home in the Hague, run by Father van Ginneken. Under the guidance of Van Ginneken and his helpers, she became a devout Catholic, and expressed the wish to become a nun. Indeed, she entered a convent, but after a trial period her wish was denied. It appears that the strict rules and the obligatory menial duties did not agree with Louise's view of life. Eventually she became a Catholic, but the process took her more than eight years. According to Louise, Brouwer and Van Ginneken got to know each other through her.[12]

Later in life Louise became exceedingly religious, bordering on mania. She joined the church of the anti-Pope, and in her old age lived a secluded, pious life that was rather similar to that of mediaeval religious ladies, be it that she combined the love for the church with a deep and bitter hatred of her stepfather.

Before his conversion, Van Eeden had already occasionally gone into retreat in the St Paulus Abbey in Oisterwijk.[13] At his invitation, Brouwer accompanied him a couple of times. In fact, the Catholic practice of retreat may have appealed to Brouwer, because of the similarity with his own practice of mystical introspection. But there was certainly also a good deal of plain curiosity involved. Brouwer never missed an opportunity to attend mystical religious or pseudo-religious manifestations. For example, later in life he went with friends to hear Krishnamurti.[14]

The role of Van Eeden in Brouwer's life became smaller as Brouwer was more and more absorbed by his mathematical and academic duties. They continued to meet occasionally. The few cards and letters that were exchanged suggest a cordial relationship, but the close contact, the walks and visits, had gone.

---

[11] 'Ran away', as she put it.

[12] Interview, Louise Peijpers.

[13] A village in the province of North-Brabant.

[14] Interview, C. Emmer, one of Brouwer's family doctors.

In 1923 Van Eeden apparently entertained the idea that he might be considered a candidate for the Nobel Prize; he asked Brouwer to write a recommendation.[15] There is no trace of such a recommendation, so we cannot ascertain whether Brouwer complied with the request. In view of the standard procedures of the Nobel committee, Van Eeden's idea seems strange; it rather illustrates his self-esteem.

On Van Eeden's seventieth birthday, in 1930, a Liber Amicorum was presented to him, with contributions from friends. The list contains a fair number of prominent artists and scientists. There is no contribution from Brouwer, who is listed as one of the organisers.

Life at home was for Brouwer as hectic as ever during the early years of the twenties. He travelled, invited colleagues to stay, and carried out his extensive and complicated research and correspondence from the idyllic village Blaricum. Cor Jongejan had become a permanent member of the entourage, she acted as a general factotum, typing letters and manuscripts, copying large manuscripts which Brouwer handled for the *Mathematische Annalen*, went shopping, did jobs around the house, assisted Lize in the pharmacy, and at the same time was a companion to Brouwer. In these busy years Lize fell seriously ill. She turned out to have pernicious anaemia. In order to get the best possible prospect of recovery, she retired for some time to a health clinic in Antwerp.

## 10.2 Intuitionism—Principles for Choice Sequences

The scientific activity of Brouwer concentrated more and more on the further development of his new intuitionism. Of course, he still handled the topology papers for the *Mathematische Annalen* and therefore automatically kept himself informed of the progress in that area, but his personal research was completely devoted to foundational matters. He had set out no more, no less, to rebuild mathematics along constructive, that is, intuitionistic lines. In 1921 Hermann Weyl had taken his side and enthusiastically proclaimed the Brouwerian revolution, but, apart from a few isolated contributions, he hardly took part in the realisation of Brouwer's grand design. In practice, Brouwer had to reform mathematics single-handedly.

His basic papers of 1918 and 1919 had provided the basis for concrete mathematical theories, and in 1923 a substantial part of the intuitionistic theory of real valued functions was developed in the paper *The Founding of the theory of functions independent of the logical proposition of the excluded third. First part, continuity, measurability and derivability*.[16]

In the two earlier papers of the same series, the basic concepts, such as spread, real number, and function, had been introduced, and in addition Brouwer had added

[15] Van Eeden to Brouwer, 28 June 1923. Mannoury and Van Ginneken were also approached with the same request.

[16] *Begründung der Funktionenlehre unabhängig vom logischen Satz vom ausgeschlossenen Dritten. Erster Teil, Stetigkeit, Messbarkeit, Derivierbarkeit*, Brouwer (1923a).

to the arsenal of constructive mathematics the fundamental tool of *finitary spread* (*finite Menge*, later called *fan*) which played the role of the constructive substitute for compact sets. Basically, a fan is a set of choice sequences over a finitely branching tree. By reducing questions about sets to questions about finitely branching spreads (fans), Brouwer intended to exploit more fully the intuitionistic viewpoint. A major advance in the theory of real functions was presented right at the beginning of the paper on real functions. The first three theorems showed beyond a shadow of a doubt that intuitionistic analysis was not just the poor sister of classical analysis.

1. *A continuous function whose domain coincides with a finite point-spread is uniformly continuous.*
2. *Every function whose domain coincides with a closed located point set is uniformly continuous.*
3. *Every full function is uniformly continuous.*[17]

Although at first sight it is marvellous that all real functions are continuous, even an intuitionist would like to have things like step functions. Mathematical practice just begs for functions of that sort. Brouwer realised this, and he earnestly tried to find intuitionistic versions of discontinuous functions, which, of course, could not be everywhere defined. He did not find a wholly satisfactory solution, as one can read in his paper on the Domain of Functions.[18] Brouwer's letter of 1920 to Hermann Weyl confirms his pre-occupation with the problem of discontinuous functions.[19]

Brouwer soon realised that his continuity theorem required a more substantial proof than the one given in the 1923 paper. In March 1924 he submitted a new paper to the Academy with the title, *Proof that every full function is uniformly continuous*.[20] This paper, and its subsequent improvements, contained the major breakthrough of the new intuitionism. Until this time intuitionistic mathematics had been a high-spirited adventure, but with little to show—except new notions. An outsider could thus easily have convinced himself that Brouwer's program was purely negative: a banning of certain principles (notably the principle of the excluded third). It is true that Brouwer's constructivism of the first period, before the full-scale adoption of choice sequences, mainly held the ideas and tools for straightforward constructive mathematics. Brouwer's second Ph.D. student, Belinfante, wrote a dissertation in this vein on a constructive theory of series.[21] It was followed by a long series of papers on constructive function theory.

In 1924 Brouwer had reached the watershed; he could show the mathematical world results that proudly set intuitionism aside from classical mathematics. In other words, Brouwer started to produce theorems that classical mathematicians could not

---

[17]In this paper Brouwer restricts his attention to functions on [0, 1]; 'full' means 'everywhere defined on [0, 1]'. In general, 'full' stands for 'everywhere defined on the natural domain'.

[18]Brouwer (1927b).

[19]Cf. van Dalen (1995).

[20]*Beweis dass jede volle Funktion gleichmässig stetig ist*, Brouwer (1924a).

[21]Belinfante (1923).

**Fig. 10.2** Spread

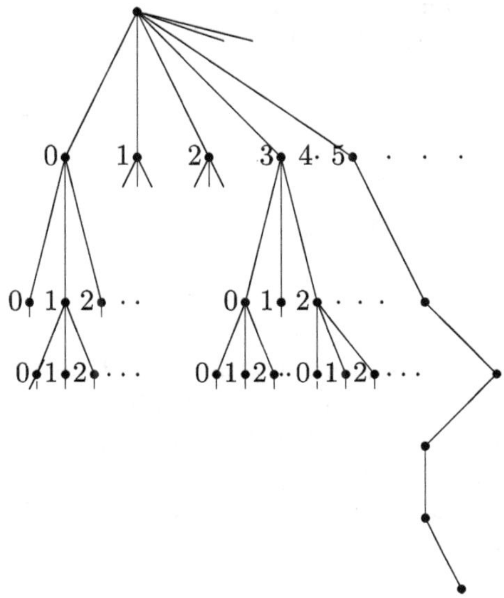

possibly prove. From now on classical and intuitionistic mathematics were recognised to be mutually incompatible. Brouwer used to express this as 'classical mathematics is contradictory'. The impact on classical mathematicians was negligible; even if they had been able to appreciate the finer details, they would have shrugged their shoulders. Why should Goliath pay heed to David?

In the following pages I will attempt to give a home-and-garden view of the ideas involved in the proof of the continuity theorems.

Since we are dealing with choice sequences of natural numbers, we have to introduce some notations for choice sequences and their (finite!) initial segments. The latter are nothing but finite sequences of natural numbers. We will use Kleene's notation, which is more flexible and convenient than Brouwer's.

We consider the universal tree of all finite sequences of natural numbers. Sequences are denoted by bold face numerals, $\mathbf{n} = \langle n_0, \dots, n_{k-1} \rangle$, and numerical functions by Greek symbols, $\alpha, \beta, \dots$ An initial segment $\langle \alpha(0), \dots, \alpha(k-1) \rangle$ of $\alpha$ is denoted by $\overline{\alpha}(k)$, in particular $\overline{\alpha}(0) = \langle \rangle$ is the empty sequence. The functions $\alpha$ can be viewed as infinite paths in the tree (that is, infinite sequences of natural numbers): in other words, Brouwer's choice sequences, see Fig. 10.2.

Now suppose that we are given a set $S$ of nodes of the tree (that is to say, a set of finite sequences) such that each infinite path contains at least one element of $S$. One can say, picturesquely, that $S$ cuts off all the infinite paths. The question is now: what do we know about such a set? The classically trained mathematician is thoroughly familiar with this question: in the case of 0–1-sequences, the infinite paths are the points of the *Cantor Space*. The points of the set $S$ can be viewed as an open covering of the Cantor space, and by compactness we know that a finite subset of $S$ already suffices to cut off all infinite paths. In analysis (infinitesimal calculus) the phenomenon goes by the name of the *Heine–Borel theorem*.

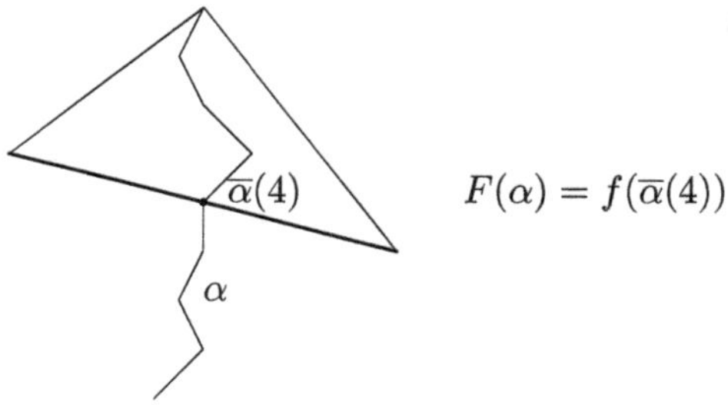


**Fig. 10.3** Value of $F(\alpha)$ determined at the bar by $f$

For Brouwer it was a matter of great urgency to find 'nice' subsets of $S$ that already suffice to cut off all infinite paths. The structure of the functions from the universal spread (that is the spread of all infinite paths through the universal tree) to natural numbers depended on the answer. The link between 'function from choice sequences to natural numbers' and 'path-cutting sets' is provided by Brouwer's continuity principle (cf. p. 240). This principle states that for a function $F : \mathbb{N}^{\mathbb{N}} \to \mathbb{N}$ (where $\mathbb{N}^{\mathbb{N}}$ stands for the set of all choice sequences of natural numbers) there is a set of nodes $B$ such that for each $\alpha$ there is an initial segment $\overline{\alpha}(k)$ in $B$; the values of $F$ are determined by these initial segments in $B$. A set like $B$ is called a *bar*, and the paths are *barred* by $B$. In logical notation:

$$\forall \alpha \exists k \big(\overline{\alpha}(k) \in B\big)$$

This bar is used to determine the outputs of $F$, that is, $F(\alpha)$ is determined by the node $\overline{\alpha}(k)$ of $B$, see Fig. 10.3. To be precise, there is a function $f$, acting on nodes, such that $F(\alpha) = f(\overline{\alpha}(k))$.

The behaviour of the function $F$ is thus determined by $B$ and $f$. It was Brouwer's *tour de force* to determine the structure of the bars $B$.[22]

There are no notes, no letters, no pieces of scratch paper that shed any light on the process of this brilliant achievement. In the case of some of his other discoveries there are traces of his preparatory research. (The Analysis Situs paper was the result of painstaking exploration of point set topology, made necessary by Schoenflies' *Bericht* with its mistakes.) In the case of the recognition of the legitimacy of choice sequences, we know of his wavering, his return to point set theory. But in this case there are no indications of any research, of any experiments. It is not unlikely that the insight came in a flash. Louise Peijpers recalled that at one particular occasion when Brouwer was working in the garden, he suddenly saw the solution to one of his mathematical problems at the moment that the sounds of a street organ that played on the road reached him. It is not known which problem it was, but it is said to be one of his major achievements. The story illustrates that Brouwer was no stranger to these sudden flashes of illumination.

[22]This is not an obvious generalisation of the compactness of Cantor space. For in that case we find that a finite subset of B will do. But the generalisation from finite to infinite is far from unique.

One thing is certain: Brouwer had the tools necessary for the job. In the first of the *Begründungs papers* he had devoted many pages to well-ordered sets. And well-ordering was the key to the problem of characterising bars.

In classical mathematics a set is well-ordered if it is ordered, and every non-empty subset has a first element. In intuitionistic mathematics this notion is no good, for even the set $\{0, 1, 2, 3\}$ is not well-ordered according to this definition.

One can see this by considering the set $A = \{n | (n = 0 \wedge R) \vee (n = 1 \wedge \neg R)\}$, where $R$ is an open problem, for example the Riemann Hypothesis. Clearly $A \subseteq \{0, 1, 2, 3\}$; and if $A = \emptyset$, then $0 \notin A$, so $\neg R$. Furthermore $1 \notin A$, so $\neg\neg R$. But this is impossible, hence $A \neq \emptyset$. Now if a number $n$ belongs to $A$, then it has to be 0 or 1; i.e. $R$ or $\neg R$ has to hold, which is unknown at the moment. Hence we cannot claim the existence of a least element in $A$, and this means that $\{0, 1, 2, 3\}$ is not well-ordered. Clearly this is not what we had in mind!

Brouwer recognised the problem and decided to use another, more constructive, definition. He went back to Cantor's first ideas about well-ordering and defined well-ordered sets as those sets that could be obtained by means of the following process: start with a one-element set and make new well-ordered sets by putting together finite or denumerable sequences of (smaller) well-ordered sets.

This process is perfectly constructive, and it yields many well-ordered sets. Here are some examples:

1. $\{a_0\}$
2. $\{a_0, a_1, a_2\}$ obtained from $\{a_0\}$, $\{a_1\}$, $\{a_2\}$ by putting them in a linear order
3. $\{a_0, a_1, a_2, a_3, \ldots, a_n, \ldots\}$ obtained from the denumerably many sets $\{a_i\}$
4. $\{a_0, a_1, a_2, \ldots, b_0, b_1, b_2, \ldots\}$ obtained from the two well-ordered sets $\{a_0, a_1, a_2, \ldots\}$ and $\{b_0, b_1, b_2, \ldots\}$
5. $\{a_0, a_1, a_2, \ldots, b_0\}$ obtained from $\{a_0, a_1, a_2, \ldots\}$ and $\{b_0\}$

As the reader will see, the actual nature of the elements $a_i, b_j$ is irrelevant. We could just as well have used dots:

| | | |
|---|---|---|
| (1) | $\bullet$ | $(= 1)$ |
| (2) | $\bullet\bullet\bullet$ | $(= 3)$ |
| (3) | $\bullet\bullet\bullet\cdots\cdots$ | $(= \omega)$ |
| (4) | $\bullet\bullet\bullet\cdots\cdots \quad \bullet\bullet\bullet\cdots\cdots$ | $(= \omega + \omega = \omega \cdot 2)$ |
| (5) | $\bullet\bullet\bullet\cdots\cdots \quad \bullet$ | $(= \omega + 1)$ |

Traditionally, these well-ordered sets are called *ordinals* or *ordinal numbers*, and the operation of putting together ordinals is thought of as a sum. So

$$3 = 1 + 1 + 1,$$

$$\omega = 1 + 1 + 1 + \cdots + 1 + \cdots \quad (\textstyle\sum a_i, \text{ with } a_i = 1).$$

Note that if we use the dot notation, both $\omega$ and $1 + \omega$ look alike:

$$\omega = \bullet\bullet\bullet\cdots, \qquad 1 + \omega = \bullet \quad \bullet\bullet\bullet\cdots$$

**Fig. 10.4** Ordinal trees

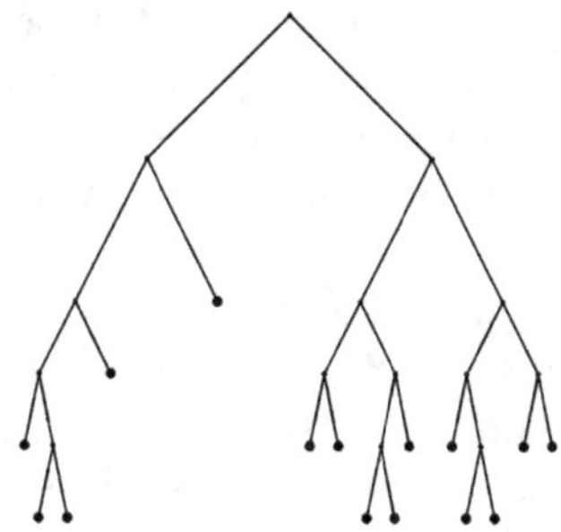

**Fig. 10.5** A bar in a fan

Hence we identify them.[23] There are specific rules in set theory about how to identify and order ordinals. Here it will suffice to consider them in our naive fashion.

Brouwer noted that the ordinals matched certain trees, in accordance with their construction process. The set 1 above corresponds to a tree with one node, and a composite ordinal corresponds to all the corresponding trees lined up under a single top node.

Let us call the corresponding trees, as shown in Fig. 10.4, *ordinal trees*.

Under the construction process different trees may belong to the same ordinal, for example $1+2$ and 3 both correspond to the same ordinal, but that is not a serious problem. The intuitive idea of the ordering of the ordinal is to run through the bottom nodes from left to right. From a mathematical point of view the ordinals and their ordinal trees are the same things in different disguises. Let us, for convenience, call such a bar, consisting of the bottom nodes of an ordinal tree, an *ordinal bar*.[24]

Now Brouwer's astonishing insight was that *a bar always contains a complete set of bottom nodes of an ordinal tree*.[25]

The above insight goes by the name *Bar Theorem*. It reflects rather bleakly on Brouwer's didactic gifts that this result remained nameless until after 1945. He gave it its name in his lecture at the Canadian conference talk in 1953.

He did, however, give a name to a special case of the theorem. Suppose we deal with a *fan*, that is, with a finitely branching tree, then the ordinal trees will all be finite, and hence, if in a fan all paths are cut off by a bar, this bar has to be *finite*! In particular one can determine a longest node in the bar, which means that there is an upper bound to the length of the nodes on top of the bar, see Fig. 10.5.

---

[23] Note that $\omega + 1 \neq 1 + \omega$; addition of ordinals is therefore not commutative.

[24] In his original proofs Brouwer considered thin bars, i.e. bars in which no two nodes are comparable (one above the other).

[25] There are actually some conditions to be put on the bar, such as monotonicity or decidability, cf. Troelstra and van Dalen (1988a).

**Fig. 10.6** Bar under **n**

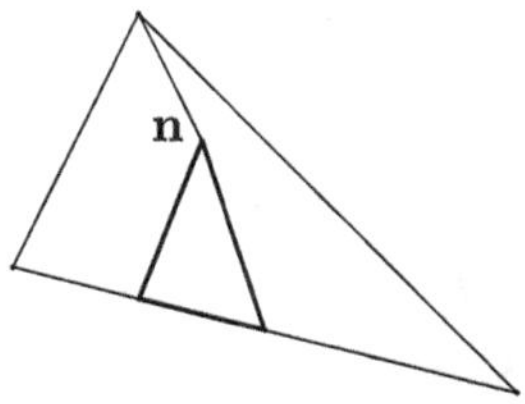


Let us give this theorem a more formal formulation. We use $B(n)$ for '$n$ is on the bar $B$'. Then 'every path $\alpha$ is cut off by $B$' is rendered as

$$\forall\alpha\exists x B\big(\overline{\alpha}(x)\big)$$

The conclusion above says that there is a maximal length $z$ such that each path is cut off at the latest at length $z$. This reads $\exists z\forall\alpha\exists x \leq z B(\overline{\alpha}(x))$.

Now the famous *fan theorem* is

$$\forall\alpha\exists x B\big(\overline{\alpha}(x)\big) \to \exists z\forall\alpha\exists x \leq z B\big(\overline{\alpha}(x)\big)$$

The fan theorem is an immediate corollary of the bar theorem. Brouwer was rightfully proud of the fan theorem, which was the main tool in his subsequent papers. In recognition of its importance, Brouwer gave it a name, albeit a rather uninspired one: *the fundamental property of fans* (*Haupteigenschaft der finiten Mengen*). The name *fan theorem* appears for the first time in his post-war publications.

It is an interesting historical curiosity that the fan theorem preceded its better known contrapositive: König's infinity lemma.[26] This lemma says that a fan with infinitely many nodes contains an infinite path. The infinity lemma is not constructively valid. An interesting observation is that this refutes the popular idea that constructive proofs and theorems are always posterior upgradings of classical theorems.

The proof of the bar theorem (as we will call it) was a complicated and mysterious affair. It is to be doubted that at the time anybody but Brouwer understood it, partly because few mathematicians were willing to devote much time to a project that met blank incomprehension in professional circles, but also because the only competent fellow revolutionary, Hermann Weyl, had left constructive mathematics.

Brouwer's proofs rest on an analysis of the data that go into the theorem, in this case a proof $\Pi$ that each choice sequence (path) hits the bar. Brouwer wanted to apply some form of induction to obtain his result, so he introduced auxiliary subproofs $\Pi_{\mathbf{n}}$ of $\Pi$ showing that 'all choice sequences through the node **n** hit the bar'.

The gist of the argument is that we can conclude from the proof $\Pi_{\mathbf{n}}$ that the little bar below **n** is an ordinal bar; having (by induction hypothesis) a suitable collection of **m**'s available for which there is a proof $\pi_{\mathbf{m}}$. We may 'pull up' this knowledge until finally $\Pi_{\langle\rangle}$ (that is $\Pi$ itself) is reached. Then the whole bar is an ordinal bar.

Now two things are clear:

[26]König (1926).

**Fig. 10.7** Joining bars

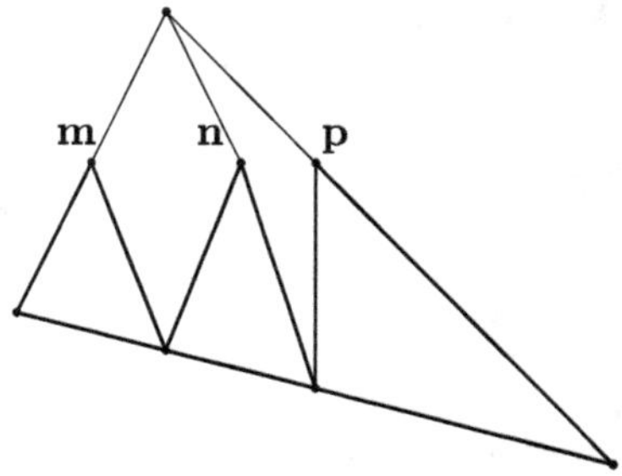


(1) If **n** is in the bar then the little bar below **n**, that is, a bar consisting of one node, is an ordinal bar (corresponding to the ordinal 1).
(2) If all nodes immediately below **n** have ordinal bars below them, then **n** itself has an ordinal bar below it.

So, starting at the bottom, one travels all the way up in the tree until the top node is reached. And then we have established that the bar is an ordinal bar.

The above sketch does not do justice to Brouwer's argument; rather, it gives the reader an impression of the underlying idea. The 'pulling up' trick in the proof is a special way of carrying out transfinite induction; it now goes by the name of *Bar Induction*.[27]

Since Brouwer was involved in actual constructive mathematics—not axiomatic reasoning on the basis of some principles (in this case the *Continuity Principle* and *Bar Induction*) he had to justify his argument conceptually. So he presented his argument as a constructive procedure based on the given data. Thus he had to analyse the given proof that each choice sequence hits the bar. Doing so he isolated the basic proof steps (1) and (2) above and argued that each such proof would be replaced by a canonical proof consisting of basic steps only. This canonical proof then enabled him to use bar induction.

In the early literature there are no comments on this proof. It is safe to assume that few, if any, of the experts in the foundations of mathematics had grasped Brouwer's argument. In the thirties some of the members of the Göttingen school, in particular Gentzen and Bernays, started to consider the fan theorem, but rather the result than the proof.[28]

Brouwer returned to the bar theorem in the same year, with a couple of further comments on the proof. In 1926 he submitted an extended version of his results to the *Mathematische Annalen* as a contribution to the Riemann volume (at the occasion of the hundredth anniversary of Riemann's birthday). This paper *On the domains of functions*[29] contained, in addition to the updated proof of the basic the-

[27]Cf. van Atten (2003), Kleene and Vesley (1965), Kreisel and Troelstra (1970), Sundholm and van Atten (2008), Troelstra and van Dalen (1988a).

[28]Gentzen, being a specialist in countable ordinals, must have recognised the significance of Brouwer's ordinals and ordinal trees. His first consistency proof of arithmetic was based on Brouwer's ideas, cf. Gentzen (1969).

[29]*Über Definitionsbereiche von Funktionen*, Brouwer (1927b).

orem, a number of foundational remarks. One of the more surprising statements concerns the nature of intuitionistic mathematical proof. In footnote 8, Brouwer casually remarked that 'intuitionistic proofs were constructed by means of the two generating operations from nil-elements [that is, immediately given insights] and elementary inferences, given immediately by intuition'. Thus, according to Brouwer, proofs are well-ordered sets of steps. In particular, he admitted infinite proofs. This looks stranger than it is, for 'infinite' means 'intuitionistically infinite'. And in that sense infinite remains potential.

In all of Brouwer's papers the bar theorem and the fan theorem were immediately followed by the *continuity theorem*, which states that 'all real functions are locally uniformly continuous'. Here, too, Brouwer could not fall back on older classical proofs, for the simple reason that this theorem was classically false. Brouwer proved the theorem by reducing functions on the reals to functions on choice sequences of nested intervals. The proof, once you have seen it, is not difficult, but it requires a certain degree of maturity to put it together for the first time.

It is fair to say that Brouwer's experience in topology and analysis gave him a head-start in constructive mathematics; especially in his particular infinitary constructive mathematics that dealt with infinite objects, such as choice sequences. Brouwer's career as an intuitionist was indeed intimately connected with his career as a topologist. A person without this penchant for the infinite and for topology would probably have embraced a more combinatorial or finitist constructivism, like Kronecker's.

Brouwer's own view of his breakthrough can be gathered from the titles of the first papers; he considered the local uniform continuity of the real functions the highlight of the new intuitionism. In this instance he showed a keen appreciation of the advertisement value of his work: the statement 'all real functions are continuous' was far more likely to shock the reader than some scholarly title, involving choice sequences or transfinite induction. The reader expecting provoking details must, however, have been seriously disappointed!

The 'domains of functions' paper also contained, as the title implicitly promised, a study of intuitionistic versions of discontinuous functions which are total in the classical sense. The results of this part are rather inconclusive. Brouwer notes that neither the measure theoretic nor the logical approaches are satisfactory.

The paper contains one real curiosity: in the introductory part the notion '*negative continuity*' is introduced, presumably as a didactic tool to introduce the reader to the new results. A function is called negatively continuous if it has no discontinuities.[30] A direct argument shows that full[31] functions are negatively continuous. The theorem can be seen as a best result in the absence of the continuity principle; it shows that even in a weaker intuitionism, functions have to behave well.

He must have discovered this lapse fairly soon, for in his Berlin lectures, which he gave in the first months of 1927, he had changed the presentation. There he

---

[30] In Brouwer's definition: if the values for a positively converging sequence converge negatively to the value of the limit of the arguments.

[31] i.e. everywhere defined.

showed that the existence of discontinuous functions implied a certain omniscience principle.[32]

Apart from the didactic reason for incorporating the 'negative continuity theorem', there may have been a more worldly one: to put on record that Brouwer had conjectured the continuity theorem before anybody else.[33] The 1927 paper contains a passage to that effect.

> On the basis of this theorem, mentioned often since 1918 in lectures and in conversations, which is an immediate consequence of the intuitionistic viewpoint, the conjecture of the validity of the following theorem, claiming much more, has become plausible. I succeeded with its proof, however, much later.[34]

It is no exaggeration to put the ideas that led to the continuity theorem in the same class as the basic ideas of Brouwer's new topology (for example the mapping degree). These two complexes of ideas are definitely the high water marks in Brouwer's career.[35]

One would expect that a similar period of feverish activity with its outpouring of papers would follow. Alas, the times had changed: Brouwer was no longer the unfettered researcher of the early years. Moreover, he was slowed down by developments in topology which called for his attention (cf. Chap. 11), and by the *Grundlagenstreit*, which started to assume unpleasant proportions (cf. Chap. 14).

Whatever the cause may have been, it is a fact that the breakthrough was not followed by a flurry of new results. The published fall-out in this period of the new insights is restricted to a few papers: on intuitionistic dimension theory and on an intuitionistic form of the Heine–Borel theorem.[36]

The first paper gave a metric characterisation of intuitionistic compact spaces and introduced intuitionistically the notion of dimension. It reproved his 1913 result, $\mathbb{R}^n$ is $n$-dimensional, in an intuitionistic setting. The second paper contained the correct intuitionistic formulation of the Heine–Borel theorem and its proof. The Heine–Borel paper is a clear instance of progress in the intuitionistic program; as late as 1923 Brouwer had branded the general Heine–Borel theorem (in its classical formulation) as false, but within three years he presented an intuitionistically correct version.[37]

---

[32]Omniscience principles are certain non-constructive statements; the prime example is the principle of the excluded third. Brouwer introduced these principles in the twenties, cf. Brouwer (1992). The term 'omniscience principle' was introduced by Bishop.

[33]Cf. p. 238.

[34]Brouwer (1927b), p. 62.

[35]Freudenthal referred to the pair of innovations, in an obituary of Brouwer, as 'the man who went through the wall twice with his head'.

[36]Brouwer (1926b, 1926c).

[37]The classical Heine–Borel theorem says that if a bounded closed subset of $\mathbb{R}$ has a covering by a collection of arbitrary open sets, then it is already covered by a finite sub-collection. Brouwer added an extra condition: the given set should also be located (cf. p. 320).

This seems a very meagre harvest for such a powerful new method. The situation is, however, not as bad as it looks. The Berlin Lectures presented a few more applications:

- the indecomposability (*Unzerlegbarkeit*) of the continuum;
- the natural ordering of the continuum is the only ordering of the continuum;
- the existence of the supremum of a continuous function on $[0, 1]$.

The indecomposability of $\mathbb{R}$ is a most surprising property of the reals; it says that one cannot split $\mathbb{R}$ into two disjoint sets.[38] It is the mathematical expression of the intuitive conviction that $\mathbb{R}$ is extremely closely knit. For example, the pair of sets $(-\infty, 0]$ and $(0, +\infty)$ is not a decomposition, since we are missing out on those points for which it is unknown whether they are on the left or on the right of 0; so $(-\infty, 0] \cup (0, +\infty)$ is a subset of $\mathbb{R}$ but not the whole of $\mathbb{R}$. Thus Brouwer's indecomposability theorem can be viewed as the mathematical expression of this surprising coherence of the continuum.

The continuum may be compared to those lines produced by syrup on a pancake; one cannot split a syrup line in two parts without losing some particles. Something will always stick to the knife.

In classical mathematics the splitting of the continuum is the result of a logical trick: the principle of the excluded third (PEM). This principle tells us, for example that each real number is either rational or irrational, although we have not the faintest idea how to test whether a real number is rational. On the strength of the PEM the classical mathematician claims that $\mathbb{R}$ is split into the rationals and irrationals. To pursue the metaphor: the classical continuum can be viewed as the frozen intuitionistic continuum; all points are nicely fixed in place, and there are no more uncertainties like 'where exactly is this point?'.

The indecomposability of the continuum was proved by Brouwer directly from the fan theorem in his Berlin Lectures; it was mentioned in his *Intuitionistic Reflections on Formalism* and again in *The Structure of the Continuum*.[39] In the first mentioned paper Brouwer used the indecomposability property to refute the generalised version of the principle of the excluded third: $\neg\forall x \in \mathbb{R}(x \in \mathbb{Q} \vee x \notin \mathbb{Q})$. The indecomposability is, of course, an immediate corollary of the continuity theorem.[40]

Let $\mathbb{R} = A \cup B$, with $A \cap B = \emptyset$. Define

$$f(x) = \begin{cases} 0 & \text{if } x \in A \\ 1 & \text{if } x \in B \end{cases}$$

Then $f$ is clearly a total function, and hence it must be continuous. But there are only two continuous functions with values 0 or 1, namely the constant 0 and the

---

[38] More precisely: if $\mathbb{R} = A \cup B$ and $A \cap B = \emptyset$, then $A = \mathbb{R}$ or $A = \emptyset$.

[39] *Intuitionistische Betrachtungen über den Formalismus*, Brouwer (1928a), *Die Struktur des Kontinuums*, Brouwer (1930).

[40] Cf. Heyting (1956).

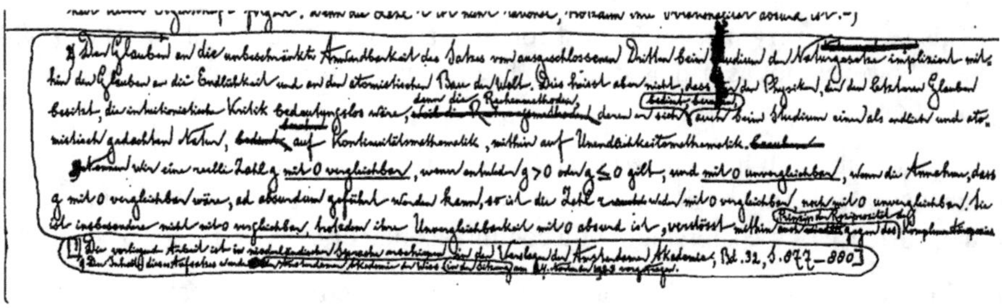

**Fig. 10.8** A draft of the indecomposability of the continuum. [Brouwer archive]

constant 1 function. If $f$ is the first one, then, by definition, $\mathbb{R} = A$; in the second case $\mathbb{R} = B$.

Brouwer never published a proof of the theorem or referred to it in print after the above mentioned papers. This may not have been intentional; he planned to publish the Berlin Lectures, but somehow the project aborted. The manuscript was found among Brouwer's papers, and it was finally published in 1992. His Cambridge Lectures also contain a proof of the indecomposability theorem.

The extra power and insight that the new methods had put at Brouwer's disposal should ideally have been the starting point for a real school of intuitionistic mathematics. However, as much as Brouwer would have liked to give intuitionism its rightful place in the world of mathematics, he lacked the leadership qualities and ambition of, say, Hilbert. There was enough of the mystic left in him to resist the temptation of a career as theorem prover or super manager. Instead of devoting all his efforts to the promotion of the true conceptual mathematics, he divided his time and energy between travelling, talking, quarrelling, lecturing, keeping an eye on the pharmacy, and a hundred more things.

## 10.3 Intuitionism in the *Mathematische Annalen*

Brouwer was not idle, nor did he neglect his intuitionistic research, but the driving impetus was missing. Nonetheless, he was deadly serious about promoting intuitionism. He carefully upgraded his *Begründungs* papers and submitted a series of three basic papers to the *Mathematische Annalen* under the title *On the foundations of intuitionistic mathematics*.[41] In the absence of any records it is hard to say who handled the papers, and who accepted them. It is not impossible that Brouwer himself had a hand in the procedure, but on the other hand he did not need tricks of that

[41] *Zur Begründung der intuitionistischen Mathematik*, Brouwer (1925a, 1926a, 1927a), submitted 20 June 1924, 14 March 1925, 28 November 1925. Note that there is a certain ambiguity in the title. It could be translated as 'On the foundations ...' or as 'On the founding of ...'. In view of the fact that Brouwer opted for 'foundations' (*grondslagen*) in the title of his dissertation, I have stuck to 'foundations'.

sort; his name was sufficient guarantee for quality. Even Hilbert could not have objected to these papers of his adversary; they were totally devoid of provocation. As usual, the papers were a paragon of respectable mathematics. The negative aspect was wholly absent. No names were mentioned; they refer to neither the principle of the excluded middle, nor classical mathematics. Brouwer presented his material with a superior disregard for the competing, and indeed prevailing, view. A prisoner, serving his years in isolation and being presented by a charitable institution with a copy of Brouwer's three expository papers, would not even have been aware that there was an alternative mathematics.

The new series lent a large measure of respectability to intuitionism; the general opinion of the *Mathematische Annalen* was so high that mediocre authors or impostors were not supposed to get access to its pages. The series also showed that, at a time when Hilbert was still issuing new programmatic statements, Brouwer had already worked out a substantial piece of mathematics.

The papers were an improvement upon the original papers published by the Amsterdam Academy.[42] In the intervening years, some loose ends had been discovered and corrected; in addition, Brouwer's views on the subject had ripened.

Brouwer did carry on his program, but in a rather eclectic way. The post-'continuity' papers deal with a number of topics that rather varied in importance. There was, for example, a paper that further refined the various equality and inequality relations between sets.[43] The results are instructive in so far as they illustrate the complications one may expect when dealing with sets; they also show that the intricacies of intuitionistic logic were no obstacle for Brouwer. To be fair, it must be said that the problems were somewhat esoteric, out of the mainstream. At the same time the contents must have provided a powerful deterrent for the curious among the mathematical community. Putting these concoctions of negations and double negations in the shopwindow was not the right policy to attract customers.

Together with his Ph.D. student B. de Loor, he published an intuitionistic proof of the fundamental theorem of algebra,[44] and a few months later he added the finishing touch to the fundamental theorem by proving the general case.

The fundamental theorem of algebra is an old family heirloom of mathematics. It says, roughly, that the complex numbers are rich enough to solve their own equations. It is well-known that the simpler number systems have their limitations when it comes to solving equations. Linear equations can be solved in the rational numbers, but not always in the integers; for instance $2x = 1$ has no integer solution. In the reals some equations, but not all, of higher degree have roots. Equations of odd degree always have a solution; but $x^2 + 1 = 0$ has no real solution. In order to solve this equation, the complex numbers, with the new number $i$ ($= \sqrt{-1}$) were introduced. Here the story stops: every algebraic equation $x^n + a_{n-1}x^{n-1} + \cdots + a_1 x + a_0 = 0$, with complex coefficients, has a complex

[42]Brouwer (1918a, 1919a, 1923a).

[43]Brouwer (1923d, 1924c, 1925d, 1927d).

[44]Brouwer and de Loor (1924).

solution. Gauss was the first person to prove the fundamental theorem of algebra, after a number of prominent predecessors had failed.

The fundamental theorem of algebra is one of the most famous *existence theorems*. It states that there is a number $\alpha$ such that $\alpha^n + a_{n-1}\alpha^{n-1} + \cdots + a_0 = 0$.

The usual proofs one finds in the textbooks proceed by contradiction; hence from an intuitionistic point of view they only show that it is impossible that there is no such $\alpha$. Therefore it was a matter of some urgency to give an effective procedure for constructing a root of the equation. This was accomplished in 1924 by Brouwer and De Loor and in the same year, independently, by Hermann Weyl.[45]

Brouwer saw, of course, that the theorem, as formulated above, was not fully general; it was based on the assumption that the leading coefficient was 1. Classically this is not a problem. If the leading coefficient $a_n$ is distinct from 0, one just divides all coefficients by $a_n$. Intuitionistically this does not work, since it may be unknown whether $a_n$ is *apart* from 0. The improved version of the fundamental theorem asserts that $a_n x^n + \cdots + a_1 x + a_0 = 0$ has a solution if at least one of the coefficients $a_n, \ldots, a_1$ is apart from 0. A matching characterisation in terms of a factorisation in linear factors was also provided.[46]

There are just two more papers of Brouwer to be mentioned. One of them is a refined analysis of the nature of spreads,[47] and the other contains a marvellous theorem on the intuitionistic theory of ordering.

Order is one of the basic tools of mathematics; without it, life would be very difficult, if not impossible. Now, the basic structures in mathematics—the integers, the reals—have a natural ordering, which in classical mathematics satisfies the *trichotomy law*: $x < y \vee x = y \vee y < x$.

In intuitionistic mathematics this law is valid for the natural numbers, the integers and the rationals, but, as Brouwer showed in 1923,[48] the law fails for the real numbers. He formulated this as 'the continuum is not ordered'. In his writings he insisted that orderings should be *total*, in the sense that for distinct elements, $a$ and $b$, either $a < b$ or $b < a$ holds.[49] His well-known counterexamples show that this is impossible for the ordering of the reals. Nowadays in constructive mathematics, one drops the requirement of totality, and adopts some weaker laws.[50] Somehow,

---

[45] Weyl (1924).

[46] *Intuitionistische Ergänzung des Fundamentalsatzes der Algebra*, Brouwer (1924b), see also van Dalen (1985), Troelstra and van Dalen (1988b), p. 434. The story does not end here. A more refined analysis in sheave models showed that without the hidden assumption of the countable axiom of choice the fundamental theorem fails, cf. Fourman and Hyland (1979), Troelstra and van Dalen (1988b), p. 794.

[47] Brouwer (1928d).

[48] Brouwer (1923c).

[49] In formal notation: $a \neq b \rightarrow a < b \vee b < a$ (note that $a \neq b$ stands for $\neg a = b$). This is, intuitionistically, not the same as the trichotomy property.

[50] In particular the trichotomy law is replaced by the following law $a < b \rightarrow c < b \vee a < c$.

Heyting deserves the credit for cleaning up the admittedly confused situation in the case of ordering, as left by Brouwer.

Brouwer had not yet made the final step to free himself from the time-honoured traditions of classical mathematics. Either unconsciously or expressly, he wished to stay as close as possible to the classical notions (at least in a number of instances). So, when he had observed that the relation $<$ fell far short of its classical counterpart, he decided to introduce a finer relation.

In order to give an idea of the intricacies, we will briefly look at the ordering of the reals.

Assuming that we have already ordered the rationals (which is unproblematic, as $a = b$ and $a < b$ are decidable for rational $a$ and $b$) and assuming that the real numbers $a$ and $b$ are given by Cauchy sequences,[51] we let $a < b$ stand for $b - a > 0$, that is

$$\exists k \exists n \forall p \big(b_{n+p} - a_{n+p} > 2^{-k}\big)$$

In other words, if 'eventually' the $b_i$'s are larger than the $a_i$'s plus a fixed positive rational number.

For this order one easily shows the following:[52]

$$a = b \rightarrow \neg a < b$$

$$a < b \rightarrow a \neq b$$

$$a < b \wedge b < c \rightarrow a < c$$

$$a < b \rightarrow c < b \vee a < c$$

Writing $a \leq b$ for $\neg b < a$ (not to be confused with the classical $a < b \vee a = b$) we see that the following holds:

$$a = b \rightarrow a \leq b$$

$$a \leq b \wedge b \leq a \rightarrow a = b$$

But we still do not get $a \leq b \vee b \leq a$.

Let us call $<$ the *natural ordering* on $\mathbb{R}$. Brouwer's examples showed that $<$ is far from trichotomous. He reacted to this phenomenon by introducing a finer ordering:

$$a \prec b := \neg a > b \wedge a \neq b$$

or in the above notation: $a \leq b \wedge a \neq b$.[53]

He called the new relation $\prec$ a *virtual ordering* and his treatment of ordering in the *Mathematische Annalen*[54] is based on this notion.

---

[51] Sequences of rationals which satisfy the Cauchy criterion: $\forall k \exists n \forall p (|a_n - a_{n+p}| < 2^{-k})$. Observe that in constructive mathematics one works with effectively given $\epsilon$ (and $\delta$). This explains why we have $< 2^{-k}$ instead of $< \epsilon$.

[52] Cf. Troelstra and van Dalen (1988a), p. 256.

[53] In Brouwer's notation, the natural ordering was denoted by $\circ\!\!<$, and the double negation ordering by $<$. It is not difficult to see that $a \prec b \leftrightarrow \neg\neg a < b$.

[54] Brouwer (1926a), p. 453.

Since the trichotomy law did not hold for virtual orderings,[55] the question arose if there is an even finer ordering (that is, an ordering defined for more pairs) which is trichotomous.

Brouwer settled the problem in a beautiful note *Virtual and inextensible order*,[56] which was published not in the *Mathematische Annalen*, but in the *Journal für Reine und Angewandte Mathematik*, also known as *Crelle's Journal*, or even as '*Crelle*'.[57] In this paper he introduced the notion of *inextensible order*, which is an order relation on a set with the property that; if for $a, b$ the relation $a < b$ is consistent with the given ordering and equality on the set, then $a < b$ is already the case in the given ordering ('what is not forbidden must happen'); similarly for $a = b$. Nowadays one would rather call this a *maximal ordering*.

The main result of the paper is a gem: an ordering is virtual if and only if it is maximal. This fact, which is rather isolated in the body of Brouwer's œuvre, is one of the encouraging events in the development of intuitionistic mathematics. Newcomers to the field usually start to feel that the stern conditions of intuitionism prevent almost any regularity or structure but for the most trivial situation. Indeed, this was the general feeling in the mathematical world: no doubt, intuitionism is a commendable, and perhaps the right, approach to mathematics, but its rules are so strict that one can only expect scattered *ad hoc* results. Finsler, in his inaugural address in Cologne in 1922, spoke for the majority of his colleagues:[58]

> Such [undecidability] assumptions may by themselves lead to very interesting investigations; an *exact* science cannot very well be based on them; apart from the great complications this would bring, many of the best secured results would have to be abandoned.

Brouwer's former ally, Hermann Weyl, also had 'second thoughts'. His initial enthusiasm of the 'New Crisis' paper:

> The new conception, as one sees, brings very far-reaching restrictions in comparison with the generality that virtually disappears into vagueness, to which an analysis, developed so far, has made us accustomed the last few decades. We must again learn modesty. We wanted to storm the heavens, and we have only piled cloud on cloud that cannot support anybody who tries in earnest to stand on them. That what is saved could at first sight appear so insignificant, that the possibility of analysis is questionable—this pessimism is, however, unfounded, [...]

---

[55] Cf. Brouwer (1926a), p. 455.

[56] *Virtuelle Ordnung und unerweiterbare Ordnung.*

[57] Brouwer (1927c). In view of Brouwer's long association with the *Mathematische Annalen*, the choice of journal is not quite obvious. It is possible that the relationship with Hilbert caused Brouwer to look for another place to publish, but that seems an unlikely motive. After all, the paper is of the inoffensive, scholarly sort. It is more plausible that Brouwer was invited to submit a paper to *Crelle* on the occasion of its centenary jubilee.

[58] Finsler (1925).

gradually gave way to more pessimistic views:

> Mathematics attains with Brouwer the highest intuitive clarity; his doctrine is idealism in mathematics analysed to the extreme. But with pain the mathematician sees the larger part of his towering theories fall apart. [...]
>
> Brouwer succeeds in developing the beginnings of analysis in a natural way, retaining much closer contact with intuition than was achieved before. But, progressing to higher and more general theories, one cannot deny that finally a hardly bearable awkwardness results from the fact that the simple principles of classical logic are not applicable. And with pain the mathematician sees that the larger part of his tower, which he thought was joined with strong blocks, dissolves into smoke.[59]

One can imagine that where Weyl's frustration was of the charitable variety, the majority of practising mathematicians would have little patience with a doctrine that promised hard labour and fewer results. The theorem about inextensible orders demonstrates, however, convincingly that there is more structure and regularity in intuitionistic mathematics than meets the eye. Clearly, intuitionistic mathematics is *not*, as many thought it to be, the loosely connected remnant of results of the powerful traditional mathematics. There is a good deal of structure, albeit more sophisticated than what people were used to.

The feature that makes this paper on virtual order stand out in Brouwer's work is that it is a general model-theoretic result. Most of the time, Brouwer's papers dealt with concrete, real life situations; in this case Brouwer considered arbitrary ordered sets. The 'virtual = inextensible' result is a striking piece of model theory *avant la lettre*.[60]

Heyting reported in the Collected Works (p. 596) that Brouwer was not quite certain of the correctness of the proof of the theorem. There is a note of 1933 in which he reconsiders the proof. One of the confusing features of Brouwer's paper was his convoluted use of language; one has to find a suitable way of reading the paper. Martino has, however, found a clean interpretation that completely vindicates the theorem.[61]

Brouwer's mathematical output in the twenties contains a few more interesting items, such as an intuitionistic proof of the Jordan curve theorem[62] and a number of results in the Berlin Lectures. The proof of the Jordan curve theorem was in itself not such a miracle, but it was important to Brouwer for a couple of reasons. It confirmed Brouwer's deep-rooted feelings that a considerable part of topology was constructive after all, once one found the proper formulations and proofs. Moreover, it showed that intuitionism did not reduce mathematics to abject poverty, as so many thought.

---

[59] Weyl (1928).

[60] Cf. Martino (1988).

[61] Ibid., cf. Brouwer (1992), pp. 10, 11.

[62] Brouwer (1925b).

The Berlin Lectures, in addition to the positive achievements, contained a list of counterexamples to well-known classical theorems, such as the existence of a maximum of a (uniformly continuous) function on [0, 1], the Bolzano–Weierstrass theorem, the intermediate value theorem, the incidence axioms of projective geometry and the fixed point theorem. In each case Brouwer indicated, however, an intuitionistic version, mostly in an $\epsilon$-formulation. For example, for a topological mapping $f$ of the unit disk into itself there is for each $\epsilon > 0$ a point $x$ at a distance less than $\epsilon$ from $f(x)$.

## 10.4 Beyond Brouwerian Counterexamples

There is one particular innovation in Brouwer's writings of the twenties that deserves special attention: *the fleeing properties* and *the pendulum number* (*Pendelzahl*).

Recall that in Brouwerian (counter-) examples certain sequences are constructed on the basis of some algorithm and a criterion. Here is such an example:

Let $R(n)$ stand for 'the decimal digits of $\pi$ at places $n-9$ through $n$ are all 9'. For each $n$ we can test $R(n)$ by simply calculating the digits of $\pi$ up to the $n$-th place;[63] we can thus effectively check if the last 10 decimals are nines.

Now we use $R(n)$ to define a real number as follows:

$$a_n = \begin{cases} (-2)^{-n} & \text{if } \forall k \leq n \neg R(k) \\ (-2)^{-k} & \text{if } k \leq u \text{ and } R(k) \text{ and } \forall p < k \neg R(p) \end{cases}$$

So the defining sequence oscillates until a string of 10 nines is met, then it becomes constant.

In other words, we write down simultaneously the decimal expansion of $\pi$ and powers of $(-2)^{-1}$.

$$\begin{array}{llllll} \pi = 3,1 & 4 & 1 & 5 & 9 & \cdots\cdots \\ \quad\ \ | & | & | & | & | & \\ a_n:\ -2^{-1}, & 2^{-2}, & -2^{-3}, & 2^{-4}, & -2^{-5}, & \cdots\cdots \end{array}$$

$$\begin{array}{lllllll} \cdots & \overbrace{9 \quad\quad 9 \quad\quad 9 \quad \cdots\cdots \quad 9} & & 9 & & 2 & \cdots \\ & | \qquad | \qquad\ | \qquad\qquad\quad | & & | & & | & \\ \cdots & (-2)^{-k}, (-2)^{-k-1}, (-2)^{-k-2} \cdots\cdots (-2)^{-k-9}, & & (-2)^{-k-9}, & & (-2)^{-k-9} & \cdots \end{array}$$

[63] That this can actually be done, was shown in Brouwer (1921a). Most mathematicians at the time would have considered this completely superfluous on abstract grounds. The intuitionist, however, needs a proof. Brouwer had jotted in the margin of his private copy of 'Mathematics, Truth and Reality' the question 'does that exist?', at the place where the decimal expansion of $\pi$ was exploited. Note, by the way, that the precise number of digits in the sequence figuring in the Brouwerian counterexamples is immaterial. One can always take a very large string.

It is not hard to see that $a_n$ converges to a real number, say $a$. This $a$ can be used to refute some classical laws; for example, $a = 0 \vee a \neq 0$ cannot be asserted, for that would mean that *at present* we have a proof that either there will never be a sequence of 10 nines, or that it is impossible that there is no such sequence. But we have no evidence supporting this, so we have no right to assert that $a = 0 \vee a \neq 0$.

Neither can we assert that $a < 0 \vee a = 0 \vee a > 0$. Indeed, for $A$ to be negative or positive, we have to indicate a place where there are 10 consecutive nines in the expansion of $\pi$, which we so far have not done; for $a$ to be zero we have to show that there is no such sequence, which we have not done either.

The special feature of this example is that it is completely algorithmic, so the number $a$ is constructed by a law.

After Brouwer had come to the conclusion that choice objects are not only legitimate but even indispensable, he had reached the point of view that the distinction between classical and intuitionistic real numbers was that classical real numbers were completely fixed in advance, that is, given by a law. This is a difficult point, for it raises questions about lawlike sequences. What does one mean by a law, and would the classical mathematician be prepared to give such a law? The term 'predeterminate' seems more apt in this case. The classical reals are fixed by the magic of the principle of the excluded third, not by honest calculation.

Intuitionistic reals could, on the other hand, also be given by a choice process. In the Berlin Lectures and the Vienna Lectures, he called the continuum of all choice reals the *full continuum*, and the continuum of all lawlike reals the *reduced continuum*. Both forms of the continuum were treated side by side.

The introduction of the reduced continuum was not a passing fad. Brouwer kept the distinction up to the end of his career.[64] In Brouwer (1954b) Brouwer called the lawlike part of the continuum the *classical continuum*, thus affirming the view that classical reals could be considered from an intuitionistic viewpoint. He did not add any comments, but we may be fairly certain that he merely wanted to express his conviction that choice phenomena did not belong to classical mathematics, not that the lawlike reals have classical properties. Roughly speaking, one might say that Brouwer wanted to indicate certain reals that rather resembled the classical reals.

The early Brouwerian counterexamples thus referred to the reduced continuum. In the late twenties Brouwer cast his counterexamples in a somewhat more abstract form. Instead of fixing the property $R(n)$ he only required that

(i) $\forall n(R(n) \vee \neg R(n))$ (that is $R(n)$ is decidable); and
(ii) neither $\exists n R(n)$ nor $\forall n \neg R(n)$ has been shown.

For such an $R$ he carried out similar constructions as above; Brouwer considered the construction sufficiently important to give names to the key notions: the least number $n$ (if any) for which $R(n)$ holds is the *critical number* of $R$. $R$ itself was called a *fleeing property*, and the real number constructed by means of $R$ carried the evocative name of *pendulum number*.[65] Wittgenstein, who attended Brouwer's

[64] Cf. Brouwer (1952b), Brouwer (1981), p. 41.

[65] *Pendelzahl* of $R$, cf. Brouwer (1929a).

Vienna Lectures, refused to admit the pendulum number as a proper mathematical object.[66]

Brouwer wanted, however, to move on to non-lawlike Brouwerian counterexamples. He did so in his Berlin Lectures.[67] The method was later called the method of the *creating subject*,[68] but until 1948 it remained nameless. We will illustrate the new version of Brouwerian counterexamples:

Let $A$ be a mathematical statement for which neither $A$ nor $\neg A$ has been proved. Define

$$a_n = \begin{cases} (-2)^{-n} & \text{as long as neither} A \text{ nor } \neg A \text{ has been established} \\ (-2)^{-k} & \text{if } k \leq n \text{ and between the choice of } a_{k-1} \text{ and } a_k \\ & \text{either } A \text{ or } \neg A \text{ has been established.} \end{cases}$$

Clearly, the sequence $(a_n)$ converges to a real number, say $a$. Again we cannot assert that $a < 0 \vee a = 0 \vee a > 0$.

The above technique is taken from the Berlin Lectures (p. 31). It produces, like the example based on $\pi$, a weak counterexample. Brouwer returned to the counterexamples based on the creating subject only after the Second World War; by then he had found a way to strengthen the method so that actual contradictions could be produced, instead of mere 'it cannot be shown that'. The remarkable fact remains that Brouwer waited for 20 years before he published these 'extended counterexamples'.[69]

In his postwar publications he clarified the ideas behind the procedure. The example above, reformulated in the terminology of 1948, runs as follows:

$$a_n = \begin{cases} 2^{-n} & \text{as long as the creating subject has experienced neither} \\ & \text{the truth nor the falsity of } A \\ 2^{-k} & \text{if } k \leq n \text{ and the creating subject has experienced the} \\ & \text{truth of } A \text{ or of } \neg A \text{ between the choice of } a_{k-1} \text{ and } a_k. \end{cases}$$

What had to be read between the lines in 1927 is spelled out in 1948 in provoking detail. The full extent of Brouwer's subjective approach to mathematics is hidden in this procedure. The creating subject produces choices depending on its mental experiences with respect to a particular statement. It was as far as Brouwer could go in his intuitionism.

There are few reports on Brouwer's private reactions to all these new developments of the twenties. It is beyond doubt that he was proud of his fan theorem and

[66]Cf. Wittgenstein (1984).

[67]Brouwer (1992), p. 31.

[68]In the literature, it is known as the *creating subject*. However, Brouwer's naming seems much more apt, keeping in mind the intended meaning.

[69]We will return to this method at the proper place to show how Brouwer cleverly exploited the logical possibilities of the creating subject.

the continuity theorem, but perhaps the discovery of the creating subject and its possibilities did not leave him untouched either. There is a letter from the significist Henri Borel to Brouwer, right after the Berlin Lectures, in which he mentioned that "Gutkind wrote me that 'you had made an enormous discovery', which 'had to do with the foundations of logic' ".[70] Taking into account the timing of the statement, there are two candidates for this discovery: the 'virtual–inextensible' theorem, or the creating subject. (To be fair, it could have been the continuity theorem or the indecomposability of the continuum just as well, but they belong rather to mathematics than to logic.) Under the circumstances, the creating subject would be a good candidate for this discovery, assuming that Brouwer had referred to a recent event. Whichever may have been the alleged discovery, we may rest assured that Brouwer considered the creating subject as an important step in the further development of intuitionism. In general, Brouwer was not given to bouts of self-congratulation, so we may safely conclude in this case that Brouwer had expressed his own true view on the recent developments in intuitionism.[71]

The investigations into more general choice sequences, depending on the subjective behaviour of the individual mathematician, probably did not yield satisfactory results, for Brouwer did not publish any of the material that he lectured on in Berlin until after World War II.[72] It is, of course, possible that Brouwer anticipated an outcry over this openly subjective mathematics and that, fearing a reversal of the generally tolerant opinion of the mathematical world, he suppressed the material for strategic reasons. This does not seem plausible; like the late Samson, he was willing to face the Philistines at any time and any place. It seems more likely that he was not satisfied with the results obtained so far. Indeed, the strong results of the post-war years were well within his reach, and we cannot exclude that he was in the possession of the decisive results, and that he lost interest for reasons that had little to do with mathematics but, as we will see, everything with the power structure in the mathematical community.

## 10.5 Fraenkel's Role in Intuitionism

Brouwer's intuitionistic mathematics found by no means the response that was necessary for a wholesale adoption. As he directed his missionary activities mainly towards the German mathematicians, it is in Germany that one has to look for the first signs of appreciation. The first convert was Hermann Weyl, but a second one was not easily forthcoming. No matter how one thought about the foundations of

[70] Borel to Brouwer, 26 March 1927.

[71] One cannot exclude the possibility that Brouwer referred to the bar theorem and the fan theorem. This certainly changed the face of logic, be it that Kleene and Kreisel had to act as midwives some forty years later.

[72] With the exception of his Vienna lectures, Brouwer 1929, but there the uninitiated reader could easily miss its significance.

mathematics, giving up the existing practice for an arduous life of hard labour in the house of intuitionism was going too far for most mathematicians.[73]

Intuitionism found a sympathetic reception with the German mathematician Adolph Fraenkel,[74] a specialist in the young discipline of set theory. Fraenkel was a man with vivid foundational interests. He had written a famous monograph *Introduction to Set Theory*[75] (1919) while serving in the German army during World War I. His name lives on in Zermelo–Fraenkel set theory, an extension of Zermelo's set theory. Fraenkel (and independently Skolem) had added a powerful existence axiom that allowed the introduction of larger sets and infinities. Furthermore, he was the first person to attack successfully the independence problem for the axiom of choice; the Fraenkel–Mostowski models are the lasting witnesses of his pioneering work.

Fraenkel's wife came from Holland, and the couple cultivated strong ties with that country. While staying with his in-laws in 1921, Fraenkel visited Brouwer in Laren, and a cordial relationship was established. He used Brouwer's library and attended his lectures in Amsterdam during the break between the (German) winter and summer term of 1923. The discussions with Brouwer stimulated his interest in intuitionism:

> It was, among other things, very interesting for me to observe the fresh life of intuitionism, already pronounced dead from many quarters; inside me it still ferments with these questions.[76]

At that time Fraenkel was preparing a new and extended edition of his *Introduction to Set Theory*. He had discussed the book with Brouwer, who had read the proofs and commented on the material (probably the chapter on intuitionism, but most likely the entire text, for Brouwer's interest in set theory had not waned). Brouwer and his wife visited the Fraenkels in Marburg during the Annual Meeting of the German Mathematical Society in 1923. When the second edition of Fraenkel's book appeared, Mrs. Fraenkel personally presented a copy to Brouwer, and Brouwer gratefully acknowledged its receipt, expressing his expectation that 'The book will exert [...] in very wide circles, both of mathematically and epistemological interested readers, an intensive and beneficial effect.'[77]

Fraenkel was a brilliant expositor, and so, when he gave a series of lectures on set theory in Kiel in 1925, the publisher Teubner was only too happy to publish the lectures as a book with the title *Ten Lectures on the Foundations of Set Theory*.[78] In 1926 the proof sheets had already been sent to the author, who once more asked

---

[73]The reception of Brouwer's intuitionism in the crucial years before 1930 is the topic of the dissertation of Dennis Hesseling (Hesseling 1999).

[74]Who later in life, for obvious reasons, changed his name into Abraham.

[75]*Einleitung in die Mengenlehre*.

[76]Fraenkel to Brouwer, 18 April 1923.

[77]Brouwer to Fraenkel, 15 December 1923.

[78]*Zehn Vorlesungen über die Grundlegung der Mengenlehre*.

**Fig. 10.9** A.A. Fraenkel. [Courtesy Mrs. M. Fraenkel]

Brouwer to go over the proofs and to suggest changes where necessary. And here, suddenly, a conflict arose. When Fraenkel, after three weeks, told Brouwer that there was no more time to make changes, the latter reacted as if stung by a bee:

> What kind of a wizard do you think I am, that I could in the middle of term, with my time almost completely occupied otherwise, study a book of more than 100 pages so thoroughly that I could bear the responsibility for suggestions of changes!

He went on to say that on account of this disastrous haste, the only thing left for him to do was to write a critical review '... where I would indeed (in particular with respect to intuitionism) have to correct a great deal, but maybe it is really good that I have an occasion to dispose of the false information on intuitionism that is offered to the public'.

Brouwer's wrath was mainly caused by the sections about intuitionism. Regardless of a possible revision, he demanded some immediate changes, among others the insertion of a list of intuitionistic publications, '... among which there are, after all, so far, the only publications existing in the field of intuitionism, apart from Heyting's dissertation, that "do not just talk, but also create" '.[79] In the absence of Fraenkel's original text, it is hard to say what exactly incited Brouwer's anger, but given the correspondence and the final text, one can make an educated guess. Fraenkel, in line with the prevailing tradition (encouraged by Hilbert) had probably described intuitionism as a more or less continuous foundational program, initiated by Kronecker, carried on by Poincaré, and taken over by Brouwer. Commentators found it often difficult to distinguish Kronecker's program, which was in the most literal sense an arithmetisation of mathematics (in particular of analysis and algebra) from Brouwer's intuitionism.

It was even harder for the general mathematician to keep Brouwer's intuitionism and the French semi-intuitionists apart. The latter group was far from homogeneous;

[79] Brouwer to Fraenkel, 21 December 1926.

it held strongly diverging views on foundational matters. It was not uncommon to present Brouwer's intuitionism as a hybrid offshoot of the programs of Kronecker and the French semi-intuitionists. The popular view of Brouwer's program was however mistaken: the influence of Kronecker on Brouwer is negligible. Neither in Brouwer's dissertation nor in his inaugural address is Kronecker mentioned. In the Berlin Lectures there is a reference to Kronecker in the historic introduction, but in the Vienna Lectures his name does not appear. Taking into account Brouwer's basic ideas, one has to admit that the programs of Kronecker and Brouwer had preciously little in common. The resemblance of the two programs was largely accepted on Hilbert's authority.

As for the French intuitionists, one would be hard pressed to extract a coherent program from their publications. Their unreserved acceptance of classical logic made them and Brouwer strange bedfellows. It is quite possible that Fraenkel had, under the influence of the prevailing opinion, fallen a victim to similar crude simplifications.

It is not easy to interpret Brouwer's comments and complaints; some of the objectionable expressions had probably been deleted by Fraenkel in reaction to the arguments of Brouwer. There is, for example, a passage in Brouwer's letter which suggests that Fraenkel had lamented the lack of consistency in Brouwer's nomenclature in the treatment of the Cantor–Bendixson theorem (called 'the fundamental theorem of Cantor'): 'every closed set is the disjoint sum of a countable set and a perfect set'. In the second *Begründungs*-paper[80] Brouwer had introduced a number of constructive notions in the topology of the plane, partly for the purpose of proving an intuitionistic version of the Cantor–Bendixson theorem. One of those was an analogue of the derived set of a set, yielding the notion of *Abbrechbare Punktmenge* (deconstructible point spread). The resulting version of the Cantor–Bendixson theorem was accordingly formulated in terms of deconstructible spreads. In *Intuitionistic Set Theory*, however, Brouwer declared that in the *Begründungs*–paper 'the consequences of intuition were not completely clear in my mind'; for that reason he felt obliged to revise his earlier statements.

One can safely say that it was not a trivial matter to keep track of some of Brouwer's notions and theorems. For example, in the review of Schoenflies' *Bericht*, Brouwer had pronounced the Cantor–Bendixson theorem trivially true.[81] But in his *Intuitionistic Set Theory*, he retracted his words and declared that it was a task for intuitionistic mathematics to classify the sets for which the theorem holds.

Fraenkel, not surprisingly, was struck, it seems, by the evolution of the Cantor–Bendixson theorem in intuitionism. His manuscript probably contained some statement to the effect that this was the consequence of a 'gradual refining' of the definition. In Brouwer's eyes this was such a flagrant misconception of the development of intuitionism that he immediately set out to lecture Fraenkel:[82]

---

[80]Brouwer (1919a).

[81]Brouwer (1914).

[82]Brouwer to Fraenkel, 12 January 1927.

Dear Fraenkel,

That Cantor's fundamental theorem is obvious for totally *deconstructible* sets and false for general point sets has nothing to do with a 'gradual refinement' of the fundamental notions, but only with the fact that the intuitive initial construction of mathematics (which, where it occurs with my predecessors, nowhere exceeds the denumerable) was conceived by me at first (1907) as totally *Abbrechbare*, finitely spread; next as a totally *Abbrechbare* (not necessarily finitely) spread, and finally as a spread *tout court*; but in the introductory phase it was always denoted as just 'spread'. One cannot keep introducing new notions; therefore I have always called my intuitionistic basic construction of mathematics, when it required an extension, 'spread',[83] just as before.

Indeed, a couple of months ago, another such extension became necessary, as you can read in my note *Intuitionistic introduction of the notion of dimension*. Also, after this extension, some so far 'obvious' theorems will turn out to be 'false', without lending warnings of yours, as in the footnote under consideration, the least justification. Should you stick to this insulting and hollow insinuation, even after my urgent request and urgent advice to strike them out, then the competent reader (as I claim to be myself) can only read this as a declaration of war on me. I ask myself in vain what grounds I may have given you. Forgive me that I write sharply and clearly, but I will have to do this also in public; and then it should not be said that I have not called your attention to the consequences of the above-mentioned statement of yours, and warned you.

With friendly greetings, your Brouwer.

It is difficult to explain this sudden outburst of Brouwer. Of course, he was already irritated by the undue haste of the publisher, but his extremely defensive reaction must have had deeper grounds.

Before we go further into the remaining part of the correspondence let us point out that Brouwer did not harbour any grudge against Fraenkel. On the contrary, he had been impressed, and somewhat flattered, by the young man's interest in his foundational views. Moreover he could not have failed to notice that Fraenkel possessed the valuable gift that he himself lacked, or maybe did not have the patience to cultivate: the gift of exposition. A man like Fraenkel could bring intuitionism to the attention of the mathematical community in a way which was much more suited to rouse interest and sympathy than Brouwer's uncompromising, stern papers. At the time of the correspondence, the development of intuitionism was in a critical phase; solid and competent mathematicians were more and more inclined 'to give intuitionism a chance'. But this required from Brouwer more than just being right, it required a measure of convincing that was alien to Brouwer's way of thinking. The process of patient indoctrination did not appeal at all to him. He was quite willing to lecture on intuitionistic mathematics and to show other mathematicians

[83] i.e. *Menge*.

where they went wrong, but he had no taste for the more didactic–propagandistic side. Much later Heyting attempted to regain the lost ground, but even he could not elevate intuitionistic mathematics above the level of an interesting specialism in the corner of dungeons of the mathematics castle. In the twenties there was, however, still hope that intuitionistic mathematics might become an alternative to the existing practice, and Fraenkel was more than anybody else at that time equipped, and mentally suited, to influence the mathematical opinion. Hence Brouwer's high hopes of the influence of Fraenkel's monographs.

As we have seen, the relationship between Brouwer and Fraenkel had been warm and unproblematic. Fraenkel admired Brouwer, and Brouwer valued the opinion of the bright younger man. Under the circumstances Fraenkel did not know what to make of this stern letter. He must have ascribed it to some accidental irritation, and written to Brouwer in that spirit. In his letter he probably commented, again, on Brouwer's 'dynamical' presentation of the Cantor–Bendixson theorem, and told Brouwer that such change of notions 'easily led to mistakes'. This was more than Brouwer, who was in the middle of his course on intuitionism in Berlin, could bear; he answered with a long letter full of bitter complaints.[84] In spite of his utter disappointment at Fraenkel's reaction, he had managed to convince the publisher Teubner that more time should be allowed for the proof reading and the corrections. As Brouwer's list of complaints is rather significant for his state of mind at the time of writing, let us have a look at some of them.

He started by urgently requesting Fraenkel not to assist in the 'expropriation, which the German-speaking review journals practised on him, by making me share with Poincaré, Kronecker and Weyl what is my exclusive personal and spiritual property'. It is quite true that his commentators always referred to his work in connection with the aforementioned three persons; for instance, Hilbert with his immense authority suggested that Brouwer was something like a mathematical reincarnation of Kronecker. In order to straighten out these misunderstandings, Brouwer sent Fraenkel a German copy of a paper which he had written for the *Revue de Métaphysique et de Morale* (but which never appeared) in which he presented his view on the status of intuitionism.[85]

He went on to dictate to Fraenkel a few corrections, which were partly adopted. The reader who knows Brouwer's style will be amused by these sudden interruptions of Fraenkel's narrative by solemn Brouwerian formulations.[86] In addition he asked Fraenkel to insert a number of bibliographic references (all to himself).

Finally he proposed to replace 'opinion of the radical intuitionist' by 'opinion of Brouwer' (p. 156 top); for, he explained, 'this opinion is, even when it has been repeated since then by others, nonetheless my spiritual property'.

The letter ended on a sad, almost pathetic note:

---

[84]Brouwer to Fraenkel, 28 January 1927.

[85]No manuscript of the paper has been found.

[86]See Fraenkel (1927), p. 35, starting from 'Die zweite'; p. 48, line −8 to −3. p. 50, line −9 to −3 (changing 'Urintuition' to 'Intuition'). Cf. van Dalen (2000).

> According to a statement of Schopenhauer, there will be practised against each innovator, by the automatically appearing opposition, at first the strategy of (factual) ignoring (*totschweigen*), and after the failure of this strategy, that of priority theft. Should this also apply in my case, I am convinced that you do not belong to my enemies, that on the contrary you harbour the wish—and after learning the above—will help to make the above-mentioned strategy against me as little successful as possible. Finally I beg you to believe that the purely factual content of this letter is accompanied only by benevolent and friendly feelings towards you.[87]

It is most likely indeed that Brouwer sincerely wished to avoid a fight. His relationship with the Fraenkels was friendly enough, and although Fraenkel may have had difficulties grasping Brouwer's unusual theories, he definitely was a scholar of recognised integrity.

There is one more letter from Brouwer to Fraenkel, which shows that their relationship had survived this emotional storm. Fraenkel produced many more books after the 'Ten Lectures on Set Theory' and each of them contained an exposition of intuitionism. It is a fact that Fraenkel's books introduced many readers all over the world to the idea that there was more in this world than set theory. At the end of his life, he was even preparing, together with Bar-Hillel, a new edition of his last version of *Foundations of Set Theory*. Above all, Fraenkel should be praised because he conscientiously strove to incorporate the basic ideas of mathematics in his books, including Brouwer's ideas on intuitionism.

## 10.6 Heyting's First Contributions

The reconstruction of 'everyday' mathematics, that is geometry, algebra, probability theory, topology and so on, was a project that figured on Brouwer's agenda, but he could not really bring himself to the toil of routine labour. What he urgently needed was a group of young disciples who could and would revise mathematics in a systematic way. In this respect he was only marginally successful. Although his classes on foundations of mathematics were well attended, few students were willing to devote themselves to the task that Brouwer envisaged. There were two notable exceptions: M.J. Belinfante and Arend Heyting. The first one completed his doctoral examination (M.Sc.) on 16 February 1921, and the second on 22 May 1922. Belinfante took up intuitionistic mathematics and published a series of papers on the theory of infinite series and on the theory of complex functions. He was Brouwer's assistant from 1921 to 1925. In 1925 he became a *privaat docent* at the University of Amsterdam, where he lectured on intuitionistic topics. Those who attended his lectures described them as exceedingly dull and clerically precise. During the Second World War, Belinfante, a Jew, was deported. Euwe, in his Obituary, lamented his

[87] Brouwer to Fraenkel, 18 January 1927.

**Fig. 10.10** Arend Heyting. [Courtesy Lien Heyting]

tragic fate: 'Belinfante had more opportunities to bring himself to safety than many others, for as a Portuguese Jew he was in a preferential position for some time, so that he had ample opportunity to prepare thoroughly his disappearance.[88] But Belinfante was not the kind of person to do so. Being a hundred percent correct and reliable person, he had difficulties—even in this case—to distrust given promises. Advice from friends was of no influence: he stuck to his decision and only remarked that others had so much less than he had. He even wanted to help! A big child, but at the same time a great man!' He ended his life in Theresienstadt. The physicist A. Pais was in the same category. He was fortunate enough to escape the fate of some of his fellow Portuguese Jews, who had been given a so-called reprieve.[89]

Heyting, the second student who embraced constructive mathematics, became the best-known intuitionist after Brouwer. When Heyting expressed his wish to engage in research in intuitionism, Brouwer suggested the investigation of the axiomatisation of intuitionistic projective geometry as a suitable topic for a dissertation. Although this may seem a strange topic for an intuitionist (for it deals more with the linguistic description of a part of mathematics than with the subject matter proper) it was a plausible one. Hilbert's monumental *Foundations of Geometry* had set the tone in geometry, and Brouwer had been lecturing on that topic right from the beginning of his career; a tradition he pursued until the forties. It does not seem to have worried Brouwer that this Ph.D. student made axiomatic investigations. He was sufficiently open-minded to recognise the practical and theoretical value of the axiomatic method as such, as long as one refrained from using unwarranted logical principles or tools and from drawing unwarranted conclusions. After all, he offi-

[88]The Dutch term for 'go into hiding' was 'onderduiken'.

[89]The story of the Portuguese Jews is complicated. Arguments were offered to show that they were not Jewish in the normal sense. Part of the group was given permission to stay in Holland for the time being. Eventually they were transported to Theresienstadt, where some managed to survive the war, but a large number was sent to Auschwitz, with the expected fatal consequences, cf. Presser (1965).

cially held the chair in (among other things) axiomatics! Of course, he denied the axiomatic method any 'creative' role.

In line with the tradition of the day, Heyting left the university after passing his final examinations. He became a high school teacher in Enschede, a town close to the German border. Notwithstanding his duties as a full-time teacher (luxuries such as research grants or part-time jobs were unheard of) he managed to complete his dissertation in three years. Heyting conducted his research in relative isolation; for some time he had no contact with his Ph.D. adviser. According to Heyting, Brouwer usually left his few Ph.D. students largely to their own resources. When he did not hear from his student Heyting, he simply assumed that he had given up. And so one day Heyting found out to his surprise that Brouwer had assigned the same research topic to someone else! When he wrote to Brouwer about this, Brouwer replied 'I have not heard from you, so I thought that nothing had come of it.'[90] Heyting, however, persevered, and finished his dissertation three years after his final examination; on 27 May 1925 he defended his dissertation in the central auditorium of the University at the traditional *Oudemanhuispoort* before the Faculty of Mathematics and Physics.

Heyting's dissertation *Intuitionistic Axiomatics of Projective Geometry* was a beautiful piece of work indeed. He was the first to explore the inaccessible intricacies of intuitionistic axiom systems. In particular, he had to solve the far from trivial problems of formalising the specific positive notions that made intuitionism stand out from classical practice. For geometry one needs a relation like Brouwer's 'apartness' (*örtlich verschieden*) in order to give a reasonable version of the familiar axioms. For example, 'through any two distinct points there passes exactly one line'; this axiom cannot be copied literally, because one cannot determine the requested line if the points are not clearly distinguishable. To be more precise, if we cannot put a positive distance between two points, then we have no means to find the direction of the line passing through the two points.[91] Similarly, there has to be a positive equivalent of the notion 'a point is not on a line'. In analytic geometry one can reduce apartness for points to apartness for real numbers, but in axiomatic geometry one is not allowed to use number systems; one has to find intrinsic properties of 'apart'!

Heyting, with more sense for notational convenience, introduced a symbol for apartness: #. He also formulated three simple axioms for the notion of apartness that have ever since been an indispensable part of intuitionistic practice:

(i) $a \# b \rightarrow b \# a$
(ii) $a = b \leftrightarrow \neg a \# b$
(iii) $a \# b \rightarrow c \# a \vee c \# b$

The first axiom is rather obvious, but the second and the third captured the essence of 'being apart from'. The third axiom says that 'apart' is a global property: if $a$ and $b$

[90] Oral communication, A. Heyting.

[91] Think of the co-ordinate version of projective geometry: it is not sufficient for the solving of two linear equations that the determinant is distinct from 0. It has to be apart from 0!

are far apart, then any $c$ is far apart from at least one of them. Phrasing it negatively: two things cannot be far apart if there is something close to both of them.[92] The second axiom relates '=' and '#'. It gives a very specific meaning to 'apart', not just as the negation of 'close to', but (in terms of 'distance') as 'having at least a distance $2^{-n}$ for a suitable $n$'. Brouwer had indeed shown earlier that in the case of the points in the plane, 'not apart' coincides with 'is equal to'. Once he had found the correct formulations, Heyting developed a true mastery in the manipulation of the axiomatic machinery. His dissertation is full of clever axiomatic constructive arguments.

In the same period a Ph.D. student from abroad, B. de Loor, had been attracted by Brouwer's fame. He had come to Amsterdam from South Africa to study with Brouwer, and in 1925 he defended his dissertation, *The fundamental theorem of algebra*. We have already seen that Brouwer and De Loor published an intuitionistically correct version of the fundamental theorem of algebra in 1924. On the basis of this experience De Loor was expected to compose a dissertation. In 1925 the relationship between Brouwer and De Loor suddenly turned sour. The latter was offered the position of lecturer in Pretoria (South Africa). There was, however, a condition attached to the offer: a doctorate was required. So, to oblige him, a few corners were rounded: in exchange for the acknowledgement of the fact that most of the work for the thesis was done by Brouwer and Belinfante, he would be awarded the doctorate 'on grounds of sufficient assimilation by him of a dissertation appearing under his name, but completely composed by others'. The faculty would then graciously grant the doctor's title 'in the interest of cultural relations between the Netherlands and South Africa'. When the dissertation was presented, to the annoyance of Brouwer, no such acknowledgement was to be found. When at the end of the Ph.D. ceremony it was Brouwer's duty to deliver a short laudation on the work of the young doctor, he publicly castigated De Loor for his breach of academic mores, an understandable but slightly curious action since he had entered into the unusual deal in the first place.[93]

From a mathematical point of view Brouwer's program was showing some real progress: hard theorems were proved. The general conviction that intuitionism was a synonym for poverty was firmly refuted (although the mathematical community at large never bothered to base its opinion on these facts). Objectively speaking, the prospects for intuitionism were fairly rosy.

[92]The formulation is un-intuitionistic, but it may suggest the background of the property: the contrapositive of (iii) is the ordinary transitivity of equality.

[93]Minutes of the faculty meeting of 11 March 1925.

# Chapter 11
# The Fathers of Dimension

## 11.1 The Two Russians

For the next, and final, episode in Brouwer's topological career we have to go back to 1923. Now forty-two years old, Brouwer had acquired a reputation in mathematics that few could match. With his glorious past he had become in a way the wizard of topology, a man with an apparent direct access to the mysteries in which topology had been, and still was, veiled. That this man had given up his first place in topology, to save mathematics from the sterility that was threatening it, only gave greater intensity to his aureole of unselfish redeemer of mathematics. Like Moses he gave up his prominent role at the court of mathematics, in order to lead his people out of the affliction of Egypt into the land of intuitionism flowing with milk and honey.

That few were prepared to give up the 'flesh pots of Egypt', for a prolonged journey through the desert, did not matter. The reputation of radical reformers often ignores this realistic point.

Brouwer's critique of existing mathematics, his acknowledgement of fundamental uncertainty, fitted well in the mood of the times. The world—and in particular Germany—had lost its point of reference; the war had left behind a new awareness of social imperfections and of the inadequacies of the institutions of the older generation. In the wake of the war an exodus of royalty had taken place. The central powers were reduced in size, stability, and international status. Germany, the primary home of mathematical foundational activity, and the country where Brouwer's ideas could count on vivid interest, was in an exceptional state of confusion. The republic was struggling for recognition, for authority, and for survival. And the academic world was not being very sympathetic towards the new state. The consequences of a lost war, multiplied by the treaty of Versailles, were considerable; there was the trauma of the loss of national identity, and the all too apparent infringement of the national integrity. The French occupied the Ruhr in January. There was the galloping inflation—from day to day the mark sunk further; on July 25 the exchange rate had dropped to a terrifying $1 to 600,000 Mark. And the political instability was disturbing—a certain Hitler was putting his party organisation in order, and in November his Coupe in Munich just barely miscarried.

D. van Dalen, *L.E.J. Brouwer – Topologist, Intuitionist, Philosopher*,
DOI 10.1007/978-1-4471-4616-2_11, © Springer-Verlag London 2013

In this uncertain world thousands of scientists worked, and tried to live under the notice 'business as usual'. This was not Brouwer's style, he was heavily involved in the struggle for the rehabilitation of German scientists and scientific organisation, and he quietly carried on his research for a rebuilding of mathematics under the laws of his intuitionism. The year 1923 saw the last of his efforts to finish off the Denjoy affair. He had started a new life after the war, and the rebuilding of mathematics along constructive lines occupied him almost exclusively. But in spite of Brouwer's farewell to topology, other than as a subject of editorial and general interest, the topic was not to leave him alone. The twenties brought him a certain degree of pleasure as it furnished the recognition that was due to him for his penetrating insights, but it also plunged him into one of the most bitter fights of his career. It all had to do with a minor slip of the pen (*Schreibfehler*) in 1913.

In 1923 the German Mathematics Society, *DMV*, held its annual meeting from 20 to 25 September 1923 in Marburg. The meetings of the *DMV* mirrored in a way the state of mathematics; new trends became visible in the list of topics. The 1923 meeting showed a pronounced presence of topology, represented by the speakers Alexandrov, Urysohn, Schoenflies, and Furch. There were furthermore many mathematicians among the participants with a first-hand expertise in topology, such as Bieberbach, Brouwer, Carathéodory, Dehn, Reidemeister, Schreier, Vietoris, and Wilson. The post-war development of topology had persisted, and finally the pioneering works of Brouwer found their worthy successors. Two young Russians, Alexandrov and Urysohn, had taken up the thread of Brouwerian style topology, and they shared a strong and natural talent for the subject. Both owed their topological education to Egorov and Lusin. In 1913 Alexandrov, the scion of an upper middle class family, enrolled at Moscow University at the age of seventeen. Having been taught by an outstanding mathematics teacher, he already knew most of the first year mathematics; he started to read the mathematics texts for himself, and good fortune brought him in contact with the writings of Cantor. From that moment, he was under the spell of set theory and topology. One of his fellow students was M. Suslin, who left his stamp on significant parts of set theory although he died of typhus, only 25 years old, in 1919. Alexandrov took Egorov as his guide in mathematics, who introduced him to such topics as measure theory, Hilbert spaces, and Baire's discontinuous functions. By 1915 Alexandrov had, in answer to a problem set by Lusin, determined the cardinality of Borel sets, and introduced the operation $A$ (called thus after its inventor by Suslin). Lusin later called sets obtained by the $A$-operation analytic sets, claiming that '$A$' stood for 'analytic'.[1]

At the time of Alexandrov's first results in descriptive set theory (comprising the theory of Borel sets, analytic sets, the Baire hierarchy, Borel measure, etc.) the young student Urysohn entered the stage. Urysohn was two years younger than Alexandrov, but he soon proved a match for him in topology.

The mathematical career of the two was interrupted for a few years by various kinds of activities (for example, Alexandrov spent a great deal of time in artistic

[1]Cf. Alexandrov (1979).

circles), and of course by the revolution. Fortunately both survived the period of civil war, and in 1920 they returned to the study of mathematics in Moscow. From then on things went fast.

In 1921 Urysohn finished his dissertation and was subsequently appointed as a lecturer. At the instigation of Egorov he turned towards topology. The problems that Egorov pointed out to him were right in the heartland of contemporary topology: what is, topologically speaking, a curve or a surface? The question may seem strange, in particular since 'everybody knew what a curve (surface) was'. But the usual definitions usually described curves by equations, and thus made use of notions which were not prima facie topological. What one needed was a definition that used only topological notions, something like 'a curve is a set such that ... ', where in the definition only notions from (basic) topology, such as point, neighbourhood, continuous, ..., are used.

In the course of his investigations Urysohn came upon the notion of dimension. In view of the topological nature of curves and surfaces, this was not so unexpected, already Euclid had stated that 'the extremities of surfaces are lines'.

In Alexandrov's memoirs there is a description of the birth of Urysohn's dimension theory:

> One morning in August [1921], Urysohn and I were both at the Burkovo dacha and went to swim in the Klyaz'ma. During our bath Urysohn told me about his definition of dimension at which he had just arrived and then began to expound at great length the basic propositions of dimension theory. I was thus present at the conception of one of the finest chapters in topology: Urysohn's dimension theory.[2]

Soon Alexandrov and Urysohn started to put their ideas on paper, the first result being their famous '*mémoire sur les espaces topologiques compacts*', published many years later by the Dutch Academy of Sciences.[3]

The two young mathematicians were sufficiently enterprising to see that beyond Moscow there was a whole world waiting to be discovered. And where better could gifted beginners go than to Göttingen, the capital of the mathematical world? One might object that beginning topologists would do well to go to Amsterdam and sit at the feet of the old wizard of the subject. But Brouwer had more or less said farewell to the subject; although he had published some 18 papers on topology after the war, this activity was more an aftershock of a big earthquake, than the beginning of a new life. Moreover, Brouwer was a loner, and in Göttingen Alexandrov and Urysohn would find a whole group of eager researchers. There might have been other options—Paris, Berlin, but in the mathematical culture of the time Göttingen was the perfect choice. It was a mathematical wonderland for them. During the summer semester of 1923 they met all the great mathematicians; Klein, Hilbert, Courant, Emmy Noether, .... Indeed, Göttingen went out of its way to welcome these first two Soviet mathematicians that had crossed the border.

---

[2] Alexandrov (1979), p. 296.

[3] Alexandroff and Urysohn (1929).

Alexandrov, in his memoirs, describes his experiences; he went to the lectures of Hilbert, Courant, Landau and—best of all—Emmy Noether. Hilbert was lecturing on intuitive geometry. His lectures were presented in 'an inspiring way, with a large number of individual remarks, always interesting, often witty, and sometimes profound'.[4] It is interesting to read that Alexandrov was not impressed by the presentation itself—'Hilbert spoke badly and could not even draw the simplest figure. Once he wanted to draw an ordinary rectangular parallelepiped. He tried to do so without success, and finally he turned angrily on his assistant (that was Bernays that summer). Bernays got up and (also without much sparkle) drew the ill-starred parallelepiped.' When Hilbert, with Cohn-Vossen, prepared the publication of these lectures, he asked Alexandrov to provide an appendix on topology. But Alexandrov did not oblige. Instead, he wrote a short monograph '*Einfachste Grundbegriffe der Topologie*' (The simplest basic notions of topology), which became as popular as the Hilbert–Cohn–Vossen book.

In Göttingen was Emmy Noether, the greatest influence on Alexandrov and Urysohn. She was already acknowledged as the mother of the new algebra, and she was lecturing on the theory of ideals. This term was the beginning of a lasting friendship.

Urysohn and Alexandrov gave a number of talks on topology in Göttingen, and Hilbert was sufficiently impressed to invite them to submit some papers to the *Mathematische Annalen*. They did so, but Ostrowski, who was asked to handle the papers, failed to take the appropriate steps to get them published. Apparently he considered them unsuitable for the *Annalen*, and so they just gathered dust for a year. When Emmy Noether, in 1924, discovered that the papers had been disregarded, she told Hilbert and insisted that the manuscripts should be sent right away to Brouwer for referee reports. Ostrowski was castigated by Hilbert and the papers were published without delay. It is not clear if Brouwer's help was called in after all.[5]

In August 1923 Urysohn and Alexandrov made a trip to Norway; it was the end of term, they went on a long walking tour along the coast and fiords. In order to save their shoes, they walked barefoot when possible.

## 11.2 The Definition of Dimension

They returned to Göttingen and ended their visit to German at the meeting of the *DMV* in Marburg. There Alexandrov and Urysohn lectured respectively on 'Investigations from the theory of point sets' and 'Theory of general Cantorian curves'.[6] Urysohn had in Moscow continued his research in dimension theory, and the results were presented by Maurice Fréchet to the French Academy for the publication in the

[4] Alexandrov (1979), p. 298.

[5] Alexandrov (1979), p. 300. Blumenthal apparently proposed that Brouwer should check the proof sheets after the papers were accepted, cf. p. 413.

[6] *Jahresber. d. Deutschen Math. Vereinigung* 1924, pp. 68, 69 (italics).

*Comptes Rendus*, where they appeared in 1922 and 1923.[7] During his stay in Göttingen someone must have told Urysohn about an earlier paper of Brouwer on the same subject. He duly looked up Brouwer's 1913 paper on natural dimension, read it, and discovered a mistake. He mentioned this in his lecture in Marburg, which was not attended by Brouwer; of course the news reached Brouwer, be it that Urysohn approached him privately, or that a participant at the meeting told him. Brouwer was shocked—a ten-year-old basic paper, and nobody so far had found anything wrong with it! Here Brouwer was confronted with the consequences of his decision to publish his dimension paper in '*Crelle*' and not in the *Mathematische Annalen*: the paper had been systematically overlooked. In fact Brouwer's results were not used for further research, or only discussed in passing in seminars.

Brouwer, who had not thought about the dimension notion since the publication in 1913, asked Urysohn to send him an exposition of the alleged mistake.

At the Marburg meeting Brouwer lectured in the 'foundations' section. He gave a talk on the negative consequences of his intuitionistic program, 'The role of the principle of the excluded middle in mathematics, in particular in the theory of functions'.[8] This talk had already been presented at the Flemish Congress for Science and Medicine in Antwerp, one month before, where for the first time Brouwer demonstrated his Brouwerian counterexamples in public.[9]

Brouwer's talk was wedged in between those of Behmann (Algebra of logic and decision problems) and Fraenkel (New ideas on the founding of analysis and set theory). The *Jahresbericht* of 1923 contained brief reports of the lectures at the Marburg meeting. At the end of the report on Behmann's lecture, an objection of Brouwer to the use of 'etc.' and 'finite number' is mentioned. The speaker and the chairman (Schoenflies), the report goes on, refuted these objections, as the notions played no role in the lecture. Brouwer was not altogether pleased with this detail. In the next volume he protested in a letter to the editor that his remarks had been misinterpreted.[10] One may guess that the remarks were provoked by Behmann's somewhat irrelevant statement that 'Thanks to the efforts of the symbolic logicians (Frege, Peano, Russell) we now know that, in the first place, the whole of mathematics is represented as a collection of purely logical facts, ...'. In fact, Brouwer only wanted to point out that the contents of Behmann's lecture had no foundational consequences, precisely because of the use of the above notions. And, so Brouwer concluded, his remarks were refuted neither in the lecture, nor by the speaker or the chairman. Was the matter important enough for a reaction, one wonders. Brouwer, in any case, thought so; he could not appreciate the implicit suggestion that his contribution to the discussion was silly and irrelevant.

---

[7]Urysohn (1922, 1923).

[8]*Die Rolle des Satzes vom ausgeschlossenen Dritten in der Mathematik, insbesondere in der Funktionentheorie.* Brouwer (1923a).

[9]See p. 442. In a sense the Unreliability paper of 1908 contained Brouwerian counterexamples. The application of Brouwer's technique to realistic mathematical problems followed in 1923.

[10]Brouwer (1925c).

In order to put the topic of this section in the proper perspective we now have to retrace our steps to the beginning of the twentieth century.[11]

The grand master of mathematics, Poincaré, had in the nineteenth century dominated the development of mathematics, his innovations in many areas have largely determined the shape of mathematics as we know it. But not only was he a great and deep scholar, he also had an admirable gift for popularisation, a very rare talent indeed. He published a large number of essays, mostly in the *Revue de Métaphysique et de Morale*, that were subsequently collected into small paperbacks. These books were an immense success. What made these books so extra-ordinary was that Poincaré explained things that could be understood by high school students, but that were equally well of use to professional mathematicians and physicists, who would draw inspiration from Poincaré's ideas and views.

Brouwer, for one, had been a devoted reader of Poincaré's philosophical essays, as is testified by his dissertation, where Poincaré is quoted ten times. This influence of Poincaré is best seen in the section on mathematics and the world (i.e. physics) and that on mathematics and logic.[12]

The topic of the present chapter, dimension, was also presented by Poincaré in one of his semi-popular expositions. He had already considered the question of the dimension of our (physical) space in 1895; that particular approach was based on the group theoretical approach of Sophus Lie, in the tradition of the *Riemann–Helmholtz Raumproblem*. For our account of the dimension theory this approach is not really relevant; much more so is a renewed attack of Poincaré on the question 'why has space three dimensions?' From 1903 onwards the idea of 'cut' appeared in his considerations. Some similar ideas had already been expressed by Euclid, who stated that a point is the end of a line, a line the boundary of a plane, and a plane the boundary of a solid body. Roughly speaking, we assign the dimension 0 to points and the dimension 1 to a line because if we cut it in a point we get two pieces. Similarly, if you cut a plane along a line, you get two pieces, etc. In 1903 Poincaré basically restricts these arguments to physical space, but in 1912 in the paper 'Why has space 3 dimensions?'[13] he considers mathematical space and its objects, and states his famous definition:

> A continuum has $n$ dimensions when it is possible to divide it into several parts by means of one or more cuts which are themselves continua of $n - 1$ dimensions. The continuum of $n$ dimensions is thus defined by the continuum of $n - 1$ dimensions; this is a definition by recursion.

The 'one or more cuts' are necessary, a circle for example, which one would definitely wish to be 1-dimensional, needs two cuts before two pieces are obtained.

---

[11]My account of the dimension episode makes use of original documents, of Freudenthal's comments in the Collected Works of Brouwer, and of the outstanding papers of Dale Johnson, Johnson (1979, 1981). The reader who wishes to learn more about the topic is urged to consult these publications.

[12]Cf. p. 77.

[13]*Dernière Pensées*, Poincaré (1912), p. 488.

Poincaré's paper is a beauty, full of sound insight into geometry and physics. As he died that same year, it is idle speculation to guess what Poincaré would have done with this new purely mathematical definition. It is worth noting that at the time that Poincaré submitted his paper to the *Revue de Métaphysique et de Morale*, Brouwer's proof of the invariance of dimension had already been published, and that Lebesgue's second paper on that topic had already been presented to the *Académie des Sciences*.[14]

In Poincaré's paper there is no mention of this fundamental fact, as Dale Johnson puts it: 'The attitudes of Poincaré and Brouwer, the two greatest topologists at the beginning of our century, towards the problem of proving dimensional invariance were very different. While the former apparently thought that this problem was not very important, the latter thought that it was highly important, urgently requiring solution.'[15] The importance of the invariance problem is indeed a consequence of one's faith in the correct choice of notions. If one believes that dimension is essentially a topological notion, i.e. independent of our knowledge of the real line, plane, space, etc., and that topology is the science that studies properties invariant under topological transformations (i.e. continuous transformation without cutting or pasting), then Brouwer's invariance theorem is the supreme test for the notion of dimension. On the other hand, if one thinks that it is too early to say whether we have the right notions, and a certain amount of adjustment is still to be expected, then the invariance of dimension is nice, but not the last word. Perhaps this was roughly what distinguished Poincaré and Brouwer.

Before Poincaré's last paper, Frederich Riesz and René Baire had already investigated the topic of dimension, however, without success.[16] As their work is not relevant for our story, we will move on to Brouwer.[17]

The birth of modern dimension theory took place in 1913 in one of Brouwer's last papers before the First World War. This paper, with the cryptic title 'On the natural notion of dimension', was published in the *Journal für die reine und angewandte Mathematik*, also known as *Crelle's journal*, or simply '*Crelle*' (after its founder). The choice of this journal remains a matter of conjecture; although it was one of the better journals, it certainly did not cater to the specialists in topology. And, as a matter of fact, Brouwer's paper was totally overlooked by all and sundry. Why, is hard to say. If only Brouwer would have sent his paper to the *Mathematische Annalen* the history of dimension theory might have taken a totally different course. It is difficult to ascertain why Brouwer sent it to *Crelle* instead of the *Mathematische Annalen*; one reason might be that Brouwer wanted to avoid a leak that would give away the result, and might cause more priority problems. The reader may recall that

---

[14] 27.III.1911.

[15] Johnson (1981), p. 105.

[16] Cf. Johnson (1981).

[17] There are two more interesting definitions, Fréchet's *dimension types* and Hausdorff's dimension. They are interesting, and up to a point, fruitful notions, but they fall short of being 'natural'. Hence we will not consider them (cf. Johnson 1981). Note, however, that Hausdorff's dimension has suddenly become important in the theory of fractals.

**Fig. 11.1** Separation

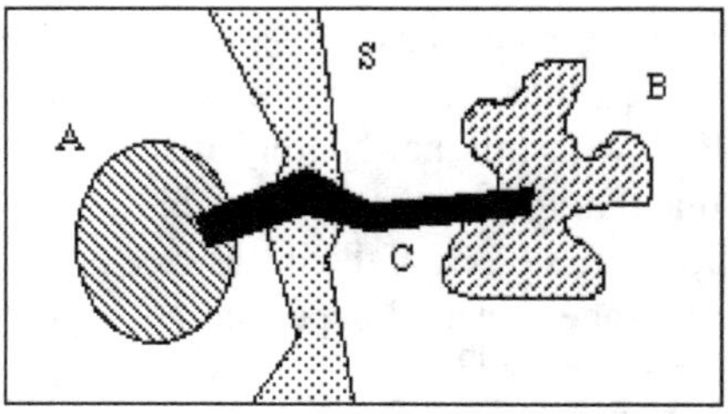


in 1913 Lebesgue had not yet produced a proof of his paving principle, although he had informed Brouwer by mail of a new proof, to be published in the *Bulletin de la Société Mathématique de France*.[18] So if Brouwer's proof of the principle appeared in 1913, one could say in all fairness that Lebesgue had been given ample time to make good his claims.

A letter from Hellinger, editor of Crelle, has been preserved, in which Brouwer is promised immediate publication.[19] So, maybe Brouwer was in a hurry to get the paper published before others would catch on to Poincaré's ideas.

Brouwer's paper started with a discussion of Poincaré's proposal. The most obvious shortcoming was that some objects were assigned a dimension that one clearly would not want them to have, e.g. consider a double cone (like a diabolo), one cut of just one point separates it into two parts (to be precise, the removal of one point—a 0-dimensional set—produces two parts), so it should have a dimension 1. *Quod non.*

Furthermore, Poincaré had used the word 'continuum' in his definition. In order to develop the notion of dimension 'out of nothing' Brouwer therefore replaced the word 'continuum' by 'normal set in the sense of Fréchet'; which is a set with the property that any two of its points are contained in a closed connected subset.

Brouwer also objected to the use of 'one or more' in Poincaré's definition, pointing out that a finite number of $m$-dimensional manifolds can be joined to form an $(m + p)$-dimensional manifold.[20]

He then proceeded by giving a definition of separation: a set $S$ separates two sets $A$ and $B$ if every closed connected set[21] $C$ meeting $A$ and $B$ intersects $S$. In the definition $S$, $A$ and $B$ are closed subsets of a set $P$.

Next he went on to define '$n$-dimensional':

> The expression: '$P$ has the general dimension degree $n$', in which $n$ denotes an arbitrary natural number, now will express that for every choice of

[18]Footnote in Brouwer (1913a), p. 151. Lebesgue's proof appeared in 1921.

[19]Hellinger to Brouwer, 21.XI.1912.

[20]The simplest example is: the sets of the rationals and of the irrationals are 0-dimensional, but their union—the real line—is 1-dimensional.

[21]Brouwer was using the modern notion of 'connected'. This was not an obvious thing in 1913, and questions about the use of the notion have been raised by Urysohn and Menger. Whenever in the sequel 'connected' is used in the text without additional specifications, the modern notion is meant.

> [closed sets] $A$ and $B$ there exists a separating set $S$, which has the general dimension degree $n-1$, but not for every choice of $A$ and $B$ there exists a separating set which has a general dimension degree less than $n-1$. Furthermore the expression: '$P$ has general dimension degree zero, resp. infinite' means that $P$ has no part that is a continuum, resp. that for $P$ neither zero, nor any natural number can be found as its general dimension degree.[22]

Brouwer's definition is, like Poincaré's one, inductive ('recurrent' as he called it), it avoids the 'double cone'-difficulty because all points (subsets $A$ and $B$) are treated on an equal footing.

It is not too far fetched to guess that Poincaré, could he have returned to the topic, would have hit on the same or a similar improvement. Speculations of this sort are always tricky; it is nonetheless pertinent to point out that Brouwer's geometrical intuition was better adapted to the infinite variety of possible pathologies than Poincaré's intuition, which felt more comfortable with the natural aspects of geometry. So Brouwer may have been in a better position to handle the intricacies of dimension.

Brouwer, of course, realised that it is one thing to introduce a new general notion, but that the proof of the pudding is in the fact that the notion extended the old one, so he went on to put this definition to the test. That is to say, he set out to prove that the traditional dimension of our space is indeed 3 according to the new definition (or rather that $\mathbb{R}^n$ has general dimension degree $n$).

An interesting detail, more of a methodological than of a topological nature, is that Brouwer rephrased the definition in the form of a two-person game. Let the players be called $\forall$ and $\exists$.[23] The players move alternately, and we will denote the moves by $i_\forall$, $i_\exists$.

Here are the moves of both players:

$1_\forall$—$\forall$ chooses two disjoint closed subsets $A_1$ and $B_1$.
$1_\exists$—$\exists$ chooses a closed separating set $S_1$.
$2_\forall$—$\forall$ chooses two disjoint closed subsets $A_2$ and $B_2$ of $S_1$.
$2_\exists$—$\exists$ chooses a closed separating set $S_2$ in $S_1$. Etc.

$\exists$ wins if a separating set $S_h$ is reached which does not contain a continuum (we would say, is totally disconnected); otherwise $\forall$ wins. We say that $\exists$ has a winning strategy if, no matter what the moves of $\forall$ are, $\exists$ wins. If $\exists$ has a winning strategy such that each game (sequence of moves) ends after at most $n$ moves (of each) then the original set has dimension at most $n$. If there is no winning strategy for $\exists$ to win in at most $n-1$ moves, then the dimension is exactly $n$ (i.e. $\forall$ can choose his moves in such a way that $n$ moves are required for $\exists$ to win). Brouwer's reformulation of

---

[22] We have replaced Brouwer's $\pi$, $\rho$, $\rho'$, $\pi_1$ by $P$, $A$, $B$, $S$. Note that 0 is not considered a natural number. The problematic and controversial aspects of this definition will be considered in detail in the sequel (see p. 408 ff.).

[23] *Abélard* and *Éloise* in Wilfrid Hodges' terminology. The presentation of the game is somewhat updated.

the definition of dimension in terms of a game is remarkable because it is (to my knowledge) the first such reformulation of an inductive definition.[24]

Here is an example of how the game works for the dumb-bell:

Two solid balls are connected by a line. If ∀ chooses closed sets in the left and right hand balls, then ∃ can separate them by a point, however, ∀ can play a better game—if he chooses two points in the right hand ball, then ∃ has to separate them by, for example, a plane section, then ∀ can choose two points in this disc, which ∃ will separate by a line, and finally ∀ may choose two points in the line, which ∃ separates by a point. This game takes 3 steps, and indeed, ∃ has a winning strategy for a 3-step game, but not for a 2-step game, so the general dimension degree is 3.

Far from considering the game definition an exotic toy, Brouwer used it in the following soundness proof.

Brouwer had already observed that the new notion was no good unless one could prove the *dimension theorem*,[25] which states that the $n$-dimensional Euclidean space has general dimension degree $n$. He had immediately realised that here Lebesgue's tiling principle was the thing to use, and so he wrote down a proof, which was—by the way—short and elegant. He used the occasion to indicate in a footnote the gap in Lebesgue's 1911 proof, and to give a slick proof of the tiling principle by means of his mapping degree. In a footnote he pointed out that, since the dimension (dimension degree, as he called it) was a topological invariant, the invariance of dimension now became an immediate corollary.

At this point the matter rested for some 10 years, until, in 1923, Brouwer to his surprise met a young man who had independently discovered the definition of dimension, and Urysohn was equally surprised to find out that some one (and not just any old topologist) had long ago given a definition of dimension. There must have been some measure of consolation in his quick discovery of an 'irreparable mistake', as he called it.

Brouwer might have switched from topology to intuitionism, but that did not mean that he was willing to acquiesce to the criticism of Urysohn. Being very meticulous in his work, Brouwer was not prone to serious errors, so he felt that he was basically right. It could not be denied that Urysohn's remarks seemed to imply that Brouwer's claim to the right notion of dimension, including the 'soundness proof', was forfeited. So a reaction was in order. As soon as he returned from the Marburg meeting, he consulted his notes, and almost immediately sent a card to Urysohn from Zandvoort, where he spent a short holiday at the seaside. Only Brouwer's draft remains,—Urysohn never received the card.[26] It ran as follows:

> Saturday 29.IX.23 from Zandvoort, pension John Bückmann, written to Dr. P. Urysohn, Mathematisches Seminar der Universität, Moskau, concerning the

---

[24]Cf. Aczel (1977).

[25]Brouwer (1913a), p. 148.

[26]He replied (27.XII.1923) that he blamed the loss of this card on the university rather than on the Russian mail. Letters addressed to his house had always reached him, but recently three letters, addressed to the university, had gone astray.

note in pencil (at the separation definition) in the margin of my personal copy (*hand exemplaar*) of *Über den natürlichen Dimensionsbegriff*. This pencilled note, that clarifies everything, must date back many years; it is very well possible, that it has been made as a result of a remark by a colleague (in that case probably Weyl, Gross or Rosenthal). I shall try to ascertain this and also check if the note has not already been added as an Erratum to a later publication.

(signed) L.E.J. Brouwer.

Notwithstanding, his conviction that—in view of the correction—his dimension notion was impeccable, he was somewhat worried about Urysohn's handling of the matter. After all, a beginning mathematician does not every day catch a famous mathematician nodding. Brouwer was himself no stranger to this vice. A letter to Bieberbach, one of the editors of the *Jahresberichte*,[27] which routinely published reports of the annual meeting of the German mathematicians, sheds light on Brouwer's worries:

......Furthermore I would like to request the following (which I beg you to pass on to Gutzmer, in case it concerns not you but Gutzmer): In Marburg a Russian, Mr. Urysohn from Moscow, gave a talk (which I did not attend), in which he presented my general dimension theory as being untenable on principle. This certainly was unjustified, for my paper in question has, in the course of the years, not only been studied and checked in numerous seminars, but also by very critical and penetrating mathematicians like Weyl, Rosenthal, Birkhoff, Veblen and Alexander,[28] without yielding anything but a few omissions, such as occur in any paper, that don't damage the body in the least and the correction of which occurs automatically to the reader: and so the paper has so far always been cited without objections, and now Urysohn doubtlessly has misunderstood it.—Now, if the mentioned Marburg talk or a report of it should be submitted to the *Jahresberichte*, I would like to ask and to advise you to submit the manuscript in question to my inspection (taking into account the above mentioned, and to avoid unpleasant polemics) before sending it to the printer.

It is likely that the letter to Bieberbach predates the card to Urysohn, otherwise Brouwer would probably have mentioned the 'minor correction in the margin' referred to in the card to Urysohn. Brouwer always maintained that the mistake discovered by Urysohn was a minor slip of the pen, one of those little details that any serious reader would automatically correct. In view of the fact that—in spite of all foundational preoccupations—topology had become second nature to him, it is quite plausible that Brouwer was somewhat amazed at the fuss made about such a little gap in the definition. The above letter also shows that in Brouwer's view

[27] Only Brouwer's handwritten and undated copy survives, but it must have been written between the meeting at Marburg and Urysohn's letter with details.

[28] Even though the paper was apparently discussed in certain circles, it found no follow up in the mathematical literature.

the natural dimension paper was not an overlooked item. He had of course sent out reprints to his brethren in topology, and the above cited mathematicians may very well have read the paper, but grosso modo it remained relatively unnoticed until Urysohn brought it to the attention of qualified mathematicians. From a letter to Schoenflies it appears that Brouwer had inserted the corrections in handwriting in the reprints that he had sent out, but which he seemed unable to recover when they would have been helpful.

Anyway, it is evident that right after the Marburg conference Brouwer could not quite recollect the precise details as to what happened with the 1913 paper and its corrections. As we will see, he soon found enough evidence to convince himself that he had not overlooked the point raised by Urysohn. Bieberbach apparently complied with Brouwer's request and sent him Urysohn's manuscript and Brouwer did not find any objectionable remarks in it (the published account of the talk, anyway, contained no reference to Brouwer's oversight)[29] for he wrote to Bieberbach:

> Enclosed, Urysohn's report is returned, with warm thanks. My name is not mentioned in the citations. I hope to have a discussion with Mr. Urysohn (who was in Marburg, probably under the fresh influence of the smear campaign (*Hetze*) led by Hilbert against me), for the time being I would have only one request to you to inform me, should Mr. Urysohn in a possible correction of his report insert my name afterwards, ......

Clearly Brouwer did not wish to take risks; as we have already seen, he was somewhat over-sensitive to real or imagined slurs on his professional status, and he did not want to be ridiculed by a newcomer on account of an ordinary lapse in exactness. He wanted to be prepared if it should come to a public debate, and so without qualms he used his status to get first hand information on a possibly looming battle.

It must have been a relief for him to find out that Urysohn was not after a cheap victory; to some it might have been tempting to correct the high and mighty Brouwer, but Urysohn had no inclinations of the sort, nor did he need the doubtful publicity involved in a bit of one-upmanship.

Brouwer may have been right about Urysohn being somewhat prejudiced against him; the remark on Hilbert's campaign against himself is evidence that is not easy to overlook. Brouwer was all too familiar with the goings on at the major universities in Germany (in particular at Göttingen), to remain unaware of an unkind reception of his foundational views. It is unlikely that it was 'all in the mind'.

Urysohn, who had not received Brouwer's card of 29 September, reacted to Brouwer's request for details, which Brouwer had made in Marburg. In a letter of October 24, in which he clearly and precisely pointed out where Brouwer's proof of the 'dimension theorem' failed: 'In Marburg you have called upon me to impart to you the objections which I have made in my lecture to your proof in Crelle's journal.' Urysohn's letter gives a perspicuous and to the point exposition of his objections. It contained a clever counter example to

---

[29]JDMV 1923, p. 69 italics.

some claims in Brouwer's proof, Brouwer quoted it in his note in the Proceedings of the Amsterdam Academy.[30] Urysohn's letter would have alarmed anybody less well-versed in topology than Brouwer. Criticising Brouwer's definition of 'connectedness', Urysohn warned, 'I will show below that your proof remains defective under *any* definition of these notions.' The conclusion being that (i) the notion of dimension was not properly defined, (ii) whatever the corrections were, the proof of 'dimension degree of $\mathbb{R}^n$ is $n$' could not be saved.

Brouwer must have been in some confusion how to react. He finally replied in December, apologising for the delay and explaining that he had hoped to get a reaction to his card of September, so that an exchange of views might be more fruitful. He inquired if Urysohn had received the card at all, and he closed the letter on an unusually complimentary note:

> I have the impression that the method sketched there[31] leads to an extremely valuable insight into the structure of continua, after trivialities have been published by the dozen in this field during the last years.

For those readers who think that Brouwer is laying it on a bit thick, it must be pointed out that Urysohn was indeed an exceptionally talented mathematician; during the short period of his mathematical activity (only three years) he produced a wealth of results and insights. To this day his name is connected with important notions and results in topology. Brouwer simply recognised a young genius when he met one (or read his work).

In the ensuing correspondence Brouwer and Urysohn promised to exchange reprints; it is characteristic of the hardships of the time that Urysohn could not afford to pay the price for the usual number of reprints of the *Comptes Rendus* notes, so he promised Brouwer to copy them for him by hand.[32]

In a letter of 22 January 1924,[33] Brouwer set himself to discuss the dimension paper in some detail. He began by repeating the content of the lost card:

> After returning home from Marburg, your objection raised there became immediately clear to me by consulting my personal copy of the paper *Ueber den natürlichen Dimensionsbegriff*, where I have an old note in the margin of p. 147, lines 17–20, according to which this passage is 'to be brought into conformity with page 150 at *)'.[34] It was this note in the margin that my card, mailed from Zandvoort, referred to.

He then entered into Urysohn's objections; in the first place he remarked that the notion of domain (Gebiet)[35] in his older papers had been used in a way that fixed

---

[30] Brouwer (1924e).

[31] i.e. in the *Comptes Rendus* notes of 1922.

[32] The copies are still in the Brouwer Archive.

[33] Brouwer's copy (handwritten).

[34] '*mit S. 150 bei *) in Uebereinstimmung zu bringen*'.

[35] A domain is an open, connected set, see e.g. Brouwer (1910a), p. 169.

the meaning in the natural dimension paper of 1913, in the second place, he agreed that the key to the gap in the proof was the separation definition. Here Brouwer recounted the history of the paper and its sequel:

> As to the origin of the oversight (*Versehen*) of p. 147,[36] my records of that time make it probable that in the manuscript of the paper there had originally not been an explicit definition of separation, as for example in my paper 'Proof of the invariance of dimension', published in Math. Ann. 71, and that such a definition was inserted rather thoughtlessly, after a co-reader of the proofs had pointed out the absence to me. When, not long after the publication of the paper, the oversight became clear, a quick correction must have remained forthcoming, because I expected soon the publication of the paper, on the same topic, promised by Lebesgue and mentioned on p. 151, and I was convinced that this paper would require a rejoinder, which would accommodate in a natural way the required correction. When subsequently the paper promised by Lebesgue kept us waiting year after year, the matter disappeared in the course of the years gradually from the realm of my thoughts, and without your interpellation I would perhaps never have thought of it again. I have now also, as a result of your remarks, studied the statement of Lebesgue, published with a delay of 10 years (and not, as agreed, in the Bull. de la Soc. Math., but in the second volume of Fundamenta Mathematicae, and seen that it, exactly as I already expected 10 years ago, calls for a rejoinder on my part, because the proof of Lebesgue of the auxiliary theorem of p. 150 of 'On the natural notion of dimension' only offers a botched up form of my proof of the same theorem. I hope that this rejoinder will appear soon. It will contain (mentioning your priority) at the same time the correction of my old slip. [...]
>
> To be sure, my own investigations are oriented since some years in another direction, but my interest in topology has remained and I consider you as one of the few who could here open up new perspectives.

Within a month Brouwer mailed a manuscript containing the correction to Urysohn.[37] What, in fact, was this mysterious 'oversight'? This is spelled out in Brouwer's revised paper,[38] which provided an argument based on a seemingly trivial correction. The cause of all problems and confusion was the erroneous insertion of the adjective 'closed' in the separation definition.

Consider three disjoint closed sets $A$, $B$ and $S$. $S$ is said to separate $A$ and $B$ if every closed connected set meeting $A$ and $B$ also meets $S$. By deleting the last occurrence of 'closed' the standard separation notion is obtained. What Brouwer claimed in the above letter was that the separation definition was added at the proof reading stage, and that the insertion of 'closed' was a slip made in haste. We will

[36]Brouwer (1923e).

[37]Brouwer to Urysohn, 19.II.1924.

[38]Brouwer (1923e).

return to this slip later. In the published corrections Brouwer used a somewhat different version of separation: $S$ separates $A$ and $B$ if $S$ determines a domain containing $A$, but not $B$.

Both corrections were noted by Urysohn in his letter of 24 October. In fact, the modern reader will find Urysohn's formulation of the separation definition more congenial.

The old definition stated that any closed connected set 'running from $A$ to $B$' must intersect $S$ ($A$ and $B$ disjoint). By dropping the 'closed' from the definition $S$ has to block more 'connecting sets', so the notion of separation becomes stronger. In other words, it becomes easier for $\exists$ to win the dimension game. We therefore speak of a weaker dimension.

Indeed, Urysohn's example (reproduced in Brouwer 1924e) exhibits a separating set that meets all closed connected 'intersecting sets', but not all 'connecting sets' which are merely connected.[39]

The alternative separation definition does not need this barrier between $A$ and $B$: $A$ and $B$ are separated if there is an open connected set $S$ such that $A \subset S$ and $B \cap S = \emptyset$.[40]

This letter is followed by a long silence on the part of Urysohn. So long indeed that Brouwer started to worry, did Urysohn get the letter, was Urysohn not convinced? One should keep in mind that Brouwer's reputation was at stake. He had become the paragon of exactness, as one may infer from a letter from Carathéodory to Hilbert in 1912: 'You know how often, if you make an exception for Brouwer, in this part of mathematics people are sinning.'[41] With his high standards of exactness, he did not hesitate to publish his own corrections. But it was more than he could bear to seeing the crowning paper of his topological days reduced to a 'nice try'. The more so, as he was certain that the criticised point was no more than a slip of the pen, a detail that every well-informed reader would correct in passing. When Urysohn's reaction remained forthcoming, he took a minor preventive action. At the Marburg meeting Urysohn had made it known that a comprehensive treatment of the notion of dimension (and much more) was to appear in Fundamenta, so any criticism that had been communicated privately to Brouwer could be expected to be made public in this paper. And so Brouwer wrote to the editor of *Fundamenta Mathematicae*, Sierpinski, that the substantial paper, submitted by Urysohn might contain a criticism of Brouwer's dimension paper that was based on a 'small isolated error' in the definition of a technical notion, and that a correction of this definition answered all objections of Urysohn.[42] He conjectured that the unreliable

[39]Cf. Johnson (1981), p. 173.

[40]In Brouwer's formulation '$S$ separates $A$ and $B$ if $S$ determines an open connected set $S'$ such that $A \subseteq S'$ and $B \cap S' = \emptyset$'. The modern formulation runs as follows: '$S$ separates $A$ and $B$ if the complement of $S$ is the disjoint union of two open sets $A'$ and $B'$, such that $A \subseteq A'$ and $B \subseteq B'$'. Cf. Hurewicz and Wallman (1948).

[41]Carathéodory was discussing topological papers of his for the *Annalen*. 5.V.1912.

[42]In the letter to Sierpinski (25.III.1924), Brouwer cited a specific place in his paper on the Jordan curve theorem (Brouwer 1910a, p. 170, line 9), where one of the basic terms is introduced. Its role becomes clear in his correction, e.g. Brouwer (1924h).

Russian postal system might be the reason for the breakdown of the communication between Urysohn and him,

> Being convinced that the editorial board of the Fundamenta Mathematicae, just like me, wants to prevent avoidable polemics, I am informing you for the present that, except for the above mentioned correction, my memoir 'Ueber den natürlichen Dimensionsbegriff' is perfectly in order. I hope therefore that you will be so kind to see that the work of Urysohn, which you must publish, contains no unfounded criticism.

It turned out, however, that Brouwer's intervention with Sierpinski was not necessary. Urysohn's reply showed that he took Brouwer's point:[43]

> I admit that your new definition[44] is the only one that is connected with the remaining contents.[45]

He added that he failed to see why Brouwer called this notion of separation 'the usual one',[46] for in publications before the natural dimension paper separation was only used for manifolds—'where the two existing notions are identical'.

The matter may appear somewhat confusing, but one can visualise a separation of (say) two points $x$ and $y$ by a circle, and more generally by a set that is connected. Now 'the more connected' this set is, the harder it is to get from $x$ to $y$. So if you cannot get from $x$ to $y$ past a weakly connected set (a thin wall) you can certainly not get from $x$ to $y$ past a strongly connected set (a thick wall) and so if two sets (or points) are already separated by a weakly connected set, they are strongly separated.

Without following Urysohn's criticism in detail let us remark that the crucial point was the notion of *separation* and hence also that of *connected set*. As the episode of Urysohn's critique (and the later controversy with Menger) rests on the question 'was Brouwer aware of the correct notion of connectedness?', we will take a somewhat closer look at the matter.[47]

The first idea that comes to mind when considering connected sets in the plane is that of 'arcwise' connectedness, i.e. a set is (arcwise) connected if any two of its points can be connected by an arc or a path (i.e. continuous image of the closed interval [0, 1]). This idea goes back to Weierstrass, who used connection by a polygonal line. The next step towards a more abstract notion was taken by Jordan, who defined a connected closed set as one that could not be split into two non-empty disjoint closed sets.[48]

Schoenflies adopted the same definition in his *Bericht* of 1908 (without the 'non-empty' clause), and he gave separate definitions for open and closed connected sets.

---

[43] Brouwer to Sierpinski, 28.III.1924. 'After waiting for more than three months I finally received a sign of life from Urysohn, which makes my letter of 25 March superfluous.'

[44] Of separation.

[45] Urysohn to Brouwer, 20.III.1924.

[46] Brouwer (1923e), footnote 11.

[47] We follow Freudenthal's exposition in Brouwer (1976).

[48] Jordan (1893), p. 25.

As late as in 1918 Carathéodory followed this example in his lectures on real functions.

The present notion, which says that a set $A$ in a topological space is connected if it cannot be written as the disjoint union of two open (or closed) sets in the relative topology of $A$, was formulated by Lennes,[49] and (almost certainly) independently by Brouwer in the same year.[50] There is also an elegant completely general definition of connectedness in Brouwer's letter to Engel.[51]

Hausdorff's book,[52] the bible for whole generations of topologists, spelled out the modern definition, and so the notion is generally ascribed to him, but Brouwer would consistently refer to Lennes.[53] In a private note (in the handwriting of Cor Jongejan) he remarked:

> I only started to quote Lennes after I observed that the definition of connectedness which was in the old days considered obvious, was blown up into a discovery by a later generation.

Anyway, there is no doubt that Brouwer knew the general notion of 'connectedness'. When Karl Menger suggested in the late twenties that Brouwer in 1913 did not know Lennes' paper (and hence the above notion), Brouwer even produced letters showing that he had refereed the paper of Lennes that was a sequel to the 1911 paper.[54]

Actually, virtually unknown to the topologists of the day, Frigyes (Friederich) Riesz had already given the correct definition in a long paper, which was unfortunately published in an inaccessible place (1906).[55] So, as far as one can judge, he should also be credited with the modern notion.

The notion of connectedness was really important for the definition of dimension, since both the notions of 'separation' and 'domain' were based on it, as Urysohn had seen at once.

In his manuscript of the revised dimension paper Brouwer had explained to Urysohn that he had indeed meant by 'connected' and 'domain' the correct notions all the time, and that the fateful word *abgeschlossen* (closed) was a slip of the pen, and that he always had the intended concept of separation in mind, cf. p. 409.

For a punctilious man like Brouwer the matter of the oversight presented an awkward situation, to say the least. He must have felt that outsiders would and could

---

[49] Lennes (1911). We know that Brouwer had seen Lennes' definition in the subsequent publication of Lennes in 1912.

[50] Brouwer (1911d), p. 308. 'Eine innerhalb $\kappa$ abgeschlossene Punktmenge [$\pi$] soll *zusammenhängend* heißen, wenn sie nicht in zwei innerhalb $\kappa$ abgeschlossene [(oder relative abgeschlossene)] Teilmengen zerlegen lässt.' The insertions are Brouwer's, made in the margin of his private copy.

[51] A set $X$ is connected if for any assignment of neighbourhoods $U_p$ to points $p$ in $X$ and for any two points $p$ and $q$ there is a finite sequence $p = p_0 = \cdots = p_n = q$, with $p_i \in U_{p_{i+1}}$. Brouwer to Engel, 9.III.1912, CW II, p. 149.

[52] *Grundzüge der Mengenlehre*, 1914.

[53] Cf. Freudenthal in Brouwer (1976), p. 487.

[54] Blumenthal to Brouwer 3.II.1912, 12.II.1912, cf. Brouwer (1976), p. 487.

[55] Riesz (1960).

consider his claims as 'fiddling the books of history'. Freudenthal, in his edition of Brouwer's topological work, has carefully sorted out the evidence, and although Brouwer's case would probably have a hard time in a court of law (for after all, a devious person could fake the historic evidence), it looks strong enough to be adopted as the correct one.

In the first place, Brouwer had corrected his reprints in handwriting by inserting 'connected in the sense of Lennes' in the margin of the definition in question. In his private copy the word *abgeschlossen* is struck out with pencil, and commented 'to be deleted conform footnote *) p. 150' and finally, Brouwer had added a note in the proofs of Schoenflies' new edition of the 'Bericht' (1913) on p. 382: 'similarly the investigations of Brouwer in Math. Ann. 70, p. 161–165 and Journ. f. Math. 142, 146–152 (in the latter one the word 'closed' on p. 147, line 18 has to be deleted, according to a communication of Brouwer)'. That particular part of the proof sheets of Schoenflies' Bericht is still in the Brouwer Archive. It shows a handwriting that fits his style of writing in the early teens of the century. The new edition of Schoenflies' *Bericht* did not contain this footnote. The reason for this is unclear, perhaps Schoenflies thought the note irrelevant (and he might have a point), perhaps the printer missed it. It had escaped Brouwer's eye, for only after Urysohn's intervention did he find out that it had not been adopted.

Urysohn, evidently, was sufficiently satisfied to drop the matter and to accept Brouwer's views and hence his priority for the definition of dimension. It should be pointed out that Brouwer's and Urysohn's definitions are in general not equivalent. There is also a technical distinction: Brouwer gave a *global* definition of dimension, i.e. for a space in its totality, whereas Urysohn gave a global and a *local* one (where dimension is defined in a point $x$, i.e. by applying the separation procedure in arbitrarily small neighbourhoods of $x$).[56]

The, for Brouwer so desirable, peace was suddenly disrupted when on June 21 a registered letter from Göttingen was forwarded to Brouwer, who was at that moment conducting gymnasium examinations in the country (the traditional, time consuming but useful, voluntary task of Dutch professors). It contained an urgent message from Urysohn, and a proof sheet of a paper that Urysohn had submitted to Hilbert for the *Annalen*. It was entitled '*Über den natürlichen Dimensionsbegriff*', and contained an exposition of the mistake in Brouwer's eleven year old paper with the same title. The note leaves the impression that Brouwer had utterly failed the goal he had set himself. Hilbert must certainly have felt a certain quiet amusement, seeing his (by now) arch-enemy stumble, especially where Brouwer had always shown little patience with the weaknesses of others, such as Lebesgue, Engel and Schoenflies.

But it is always easier to attack the weaknesses of a person you don't know personally, than those of a man you have come to appreciate and like, so Urysohn, somewhat embarrassed, wrote Brouwer that he had already submitted this small note to the *Annalen* in July (thus before the Marburg meeting), and since then had completely forgotten about it. So when he suddenly received the proofs he was at a loss,

[56]Cf. Johnson (1981), p. 178.

> It is completely unclear to me what I should do with it, perhaps the 'supplement added in the proof' will satisfy you.

The supplement ran as follows:

> In the preceding claims I have obviously based myself on the assumption that one sticks to the definition of dimension of Vol.142 of Crelle's Journal. Now, Mr. Brouwer has since then published a correction,[57] where he indeed changes the notion of separation on which the notion of dimension is based. *In that way* the proof has been corrected completely, and I would like to stress particularly that, as I have been informed, the necessity of such a modification of the definition of separation had been known already for a long time to Mr. Brouwer, and that its publication has not been carried out due to a lapse. Nonetheless, I believe that the above lines may be useful, since Mr. Brouwer, in his Correction, has not indicated *why* the old definition has to be rejected.
>
> Göttingen, 21 June 1924.

Brouwer acted immediately, he elaborately explained to Urysohn that publication would be unwise,[58]

> Many thanks for sending me the proofs of your forgotten small note for the Annalen and for seeking my advice on it. It is my opinion, that in both our interests, the publication of this note should absolutely be refrained from. For the publication of an oversight, which had escaped the notice of scholar B by scholar A, is only then compatible with the dignity of scholars, if either the oversight can only be grasped by means of an extensive exposition of new discoveries of A, or if all consultation between both parties involved has become impossible (e.g. on political grounds or because of the death of B). In each other case such a publication creates a suspicion that either A allows himself to be carried away by imprudent ambition, resp. deliberately wants to offend B, or that B did not want to acknowledge his oversight to A, resp. has refused public acknowledgement, at least to fullest extent. Fortunately, none of the above circumstances exists in the present case, rather the contrary, in every respect.

'It would be useful', he continued, 'to publish your counter example, but that would find a natural place in a note in which I am going to exhibit the evidence of my early correction'. He added that he hoped that Urysohn would agree with the retraction of the note, saying that the matter with the Editorial board and the publisher of the *Mathematische Annalen* could easily be arranged by himself, being after all an editor.

Urysohn's letter and the proof sheet may possibly not have caught Brouwer unawares. It is not unlikely that Alexandrov and Urysohn's manuscripts were, after all, passed on to Brouwer for advice.

---

[57] Brouwer (1923e, 1924h).

[58] Letter of 24.VI.1924.

Blumenthal had written on June 14 that he was sending Brouwer proof sheets of a paper of Urysohn and Alexander plus three more papers that were already in print. Which papers were involved is not known. The small note mentioned above should also have passed through Brouwer's hands, since Brouwer contacted Blumenthal, before he had received Urysohn's letter, asking permission to deal with Alexandrov and Urysohn directly in the name of the editorial board. Blumenthal replied on 14 June that the decision was not his; Hilbert had to be consulted about the matter. On the same day, in a letter, he had suggested that Hilbert authorise Brouwer to discuss the matter of the dimension note directly with Urysohn, and that on account of the alleged error the paper might after all have to be rejected. He also suggested that it would be wise to send Brouwer the already accepted manuscripts of Alexandrov and Urysohn so that he could if necessary make corrections—'for this single example makes me apprehensive'. In his letter to Brouwer he carefully added that Brouwer could only correspond with Urysohn in his own name; since he was personally involved in the matter of the definition of dimension, it would not be fitting to act in an editorial capacity. Hilbert's reaction has not been preserved, but it is likely that he gave Blumenthal his fiat. So when Brouwer wrote to Urysohn, he was not violating any editorial guidelines. He must have felt confident—on account of Urysohn's letter—that an agreement would be forthcoming; indeed he informed Blumenthal that Urysohn was about to agree that his paper was not suitable for publication. Should some more persuasion be necessary, he wrote, then Blumenthal could tell Urysohn that Brouwer would pay the bill for the now useless proofs, and that the editorial board 'would obviously follow in this matter the judgement of the only involved and only qualified member'. We see that Brouwer was not totally adverse to some muscle-flexing, the cause justified this, as Brouwer had already refuted Urysohn's claim that Brouwer's mistake was beyond repair; moreover Urysohn had accepted Brouwer's explanations. In Urysohn's words, 'I have received your card of the twenty second and your letter of the twenty fourth of this month; I am very grateful to you that you have given me so kindly and extensively your opinion. I completely agree with everything that you have written.' It would certainly have been a serious blot on Brouwer's topological record if Urysohn had been right. So this episode in the history of dimension ended well. It remains a bit of a mystery why Hilbert did not assign the papers of Alexandrov and Urysohn to Brouwer right away. One would expect that in a well-run editorial board manuscripts would be handled by the recognised experts, and certainly in a tricky subject like topology.

In the Brouwer–Urysohn correspondence there is a particular interesting letter, which plays no role in the discussion, but which sheds more light on the confusion of notions. On 14 June Brouwer wrote:

> Perhaps the enclosed variant to the passage in Crelle's journal 42, between p. 149, l.2 bottom and p. 150, line 10 bottom, by means of which the proof is adapted to the separation definition found on p. 147, (thus without the deletion of the word 'closed') will interest you. [...]
>
> This variant, which I recently found among my papers from the years 1912 and 1914, is most probably communicated in the correspondence that I pur-

> sued at the time with Schoenflies, Gross and others about, among other things, dimension. I will see if perhaps the other parties have filed this correspondence better than I have. My own interest has been diverted for a full nine years from these topics, and as an archivist I have always been a failure. I consider, by the way, as before, the separation definition without the word 'closed' as the proper and more fruitful one for dimension theory.

The enclosed part of a proof was submitted two days later to the *Mathematische Zeitschrift*, and published that same year.

The upshot of this message was that Brouwer had at an earlier period investigated another notion of separation (weak separation), which led to another (strong) concept of dimension. Freudenthal, who had no access to the Brouwer–Urysohn correspondence, noted in his comments on the dimension episode, 'The fine distinction between weak and strong connexion may appear as an a posteriori implantation—such subtleties in the stone age of topology! Yet, Brouwer was subtle.'[59] The above letter proved Freudenthal right. From the argument of the paper in Crelle and Brouwer's use of the correct notion of connectedness, one may well conclude that he indeed had the correct (weak) dimension notion in mind, and the 'slip' was really a slip. The fact that Brouwer discussed the above strong dimension is probably a partial explanation of the confusion, and also a demonstration of his undiminished topological powers. Both dimension notions are mentioned again in Brouwer (1924e) and Brouwer (1928f).

Urysohn did not react to Brouwer's alternative notion (at least no written evidence is extant).[60]

When Urysohn received Brouwer's letter concerning the note for the *Mathematische Annalen*, he suddenly recalled that in the manuscript of his paper '*Mémoire sur les multiplicités Cantoriennes*', submitted to the Fundamenta Mathematicae, a similar reference to Brouwer occurred, and so he hurried to ask Sierpinski to replace the reference by a revised one.[61] The original reference to Brouwer was stern in tone:

> I lately learned about a paper of M. Brouwer *(Über den natürlichen Dimensionsbegriff)*, ... where he proposes to solve this question by a method which seems (at first sight, at least) very close to mine. Now, the proof of M. Brouwer contains an error which, it seems to me, cannot be corrected, and which undermines all his results. I refer for the details to a supplement to be found at the end of the first part of the present mémoire.[62]

[59]Brouwer (1976), p. 551.

[60]For a thorough analysis of the various notions the reader is referred to Freudenthal's commentary in the Collected Works II, p. 548 ff. and Johnson (1981), p. 171 ff.

[61]This request was made in a letter mailed from Göttingen, so it was not written under the supervision of Brouwer. Urysohn to Sierpinski, 27.VI.1924.

[62]Copy in the Menger Archive.

Urysohn pointed out to Sierpinski[63] that although the remark was justified with respect to the original paper, it would be out of place after Brouwer's correction.[64]

> Analyzing my 'Brouillon', I have seen that in the introduction to my 'Mémoire sur les multiplicités Cantoriennes' I have written about Brouwer's 'Natural notion of dimension' an observation, in which I wrote roughly the following: 'Now the proof of this theorem contains an error which, it seems to me, cannot be corrected.' Although this observation is justified with respect to the old formulation of Brouwer's paper, in his correction[65] he changes the *definition of the notion of dimension*, it seems to me that the publication of my observation would not be appropriate. Therefore I permit myself to beg you to modify this observation: if possible to replace it by the observation below, if it is already impossible on technical grounds, to delete it at least altogether. Here is the text of the desired observation:
>
> "the memoir was already finished when I got to know the paper '*Über den natürlichen Dimensionsbegriff*', published by Brouwer in 1913 in the Journal für [die reine und angewandte] Mathematik (v.142, p.146). I hope to return to the definition of Brouwer and mine."

The letter must have reached Sierpinski, since Menger at some later date acquired a handwritten copy. The first proofs of Urysohn's paper do not contain the desired correction, so either Sierpinski thought the matter of marginal importance, or the proofs were ready before he could inform the printer.

The correspondence between Brouwer and Urysohn (often in combination with Alexandrov) continued right up to the death of Urysohn. The two Russians left Göttingen first for Bonn to visit Hausdorff. After repeated fruitless visits to the Dutch consulate in Cologne, they finally got a visa for an extended trip to Holland and continued their journey. The two stayed in Blaricum with Brouwer. From this visit there is a pictorial witness: a photograph of Brouwer sitting between his two young friends in the garden.[66]

Not surprisingly, a great deal of time was spent on discussing the work of Brouwer and of Urysohn on dimension theory. One of the topics, of course, was the above mentioned mémoire.

Brouwer was so impressed by the two young visitors that he insisted that they should come back in October and spend the year in Holland.

Unfortunately Brouwer had a previous engagement in Göttingen,[67] so that he had to leave the two Russians behind. Urysohn and Alexandrov tried to make the best of it, and they went as proper tourists with Cor Jongejan and a lady friend of

---

[63]Urysohn to Sierpinski, 27.VI.1924 (Menger Archive).

[64]Brouwer (1923e).

[65]Brouwer (1924e).

[66]See Collected Works II, p. 453.

[67]He was invited for talk on the foundations of mathematics; it was one of the exchanges in the Brouwer–Hilbert debate, commonly known as the *Grundlagenstreit*.

**Fig. 11.2** Alexandrov, Brouwer, and Urysohn in the garden at Brouwer's hut (1924) [Brouwer archive]

hers to Amsterdam. From Amsterdam they sent a postcard to Brouwer, lamenting his absence. The Russian must have made a somewhat curious impression on the Dutch, as appears from the fact that they were kicked out of the Rijksmuseum on the grounds that they did not wear a jacket.

From Amsterdam they moved on to Paris where they met Fréchet. Their next stop was in Brittany, in Batz, a small fishing village at the end of the world. Urysohn reported their experiences in a letter:[68]

> *Hochgeehrter und lieber Herr Professor,*
>
> Only now we finally got down to writing a letter. In Paris we walked around every day from 9 in the morning until 10 at night[69]—for apart from the city and the museums there was still the police headquarters, which gave us troubles, and the German consulate, where we asked for a transit visa for the return journey etc. After four days we got so tired[70] that we made the decision to postpone the continuation of Paris until the return journey (Urysohn), resp. until eternity (Alexandrov). We have come here the day before yester-

[68]Urysohn and Alexandrov to Brouwer, 29.VII.1924.

[69]With the greatest pains: Paris is even more horrible than I could ever have thought [Urysohn's note].

[70]And Alexandrov had cursed so much and has become so unbearable [Urysohn's note].

day, and it took us a whole day before we could find a quiet place at the coast. [.........] With our best wishes for you and the two ladies

Paul Urysohn Paul Alexandrov
Le Batz (Loire Inférieure), Pension de famille 'Le Val Renaud'.

Alexandrov and Urysohn had in the meantime contemplated Brouwer's invitation to spend a year in Amsterdam. Being exceptionally bright, they could have spent a year almost anywhere, in particular Göttingen must have been on their mind. For any young mathematician that place offered glimpses of mathematical heaven. It says something for Brouwer that they decided to come to Amsterdam. Their curricula vitae and a formal letter of acceptance were enclosed with their letter.

Brouwer answered by return post, reporting about the reception of his talk,[71] and asking Urysohn to add his signature to the preceding letter.

The letter was immediately answered by Urysohn and Alexandrov, and the missing signature was provided.[72] The letter is written in a jocular tone, it closes with the promise to tell Brouwer all about France at the meeting of the German Mathematics Society in Innsbruck.

A week later they sent a picture postcard from the Pointe du Raz,—'Many warm greetings from the place here at the picture, which we climbed and swam around in all directions.'

Returning from Cap Finisterre to Batz they moved into a small cottage at the sea shore, where they alternatingly pursued their research and went swimming. On August 15 and 16 the sea started to get rougher and 'our swims became more and more interesting', as Alexandrov put it. On the 17th Urysohn had thought out the proof of his famous metrisation theorem and he managed to finish to write down the first page of the paper '*Zum Metrizationsproblem*'. At 5 o'clock the two went down for their customary swim.

The dramatic consequences were recorded in Alexandrov's 'Pages from an autobiography':[73]

> When we got into the water, a kind of uneasiness rose up within us; I not only felt it myself, but I also saw it clearly in Pavel. If only I had said, 'maybe we shouldn't swim today?' But I said nothing ... After a moment's hesitation, we plunged into a not very large shore wave and swam some distance into the open sea. However, the very next sensation that reached my consciousness was one of something indescribably huge, which suddenly grabbed me, and this sensation was accompanied by the rather absurd but quite precisely formulated thought: had this wave come to me all the way from Venezuela to no useful purpose here? A moment later I came to myself on the shore, which was covered with small stones—it was the shore of a bay, separated from the open sea by two rocks between which we had had to swim as we made our

[71] See p. 446.

[72] Alexandrov and Urysohn to Brouwer, 4.VIII.1924.

[73] Alexandrov (1980).

way to the open sea. I had been thrown over by a wave, right across these rocks and the bay. When I was on my feet, I looked out to the sea and saw Pavel at those same rocks already in the bay, in a half-sitting position. I immediately swam up to him. At that time I saw a large group of people on the shore. (It was a Sunday, and many people from various places had come to Batz to admire the sea.) After swimming to Pavel, I put my right arm around him above his waist, and with my left arm and my legs I began to paddle to shore with all my might. This was difficult, but no one came to my assistance. Finally, when I was already quite near the shore, someone threw me a rope, but within a few moments I reached land. Then eye-witnesses told me that the same great wave that had thrown me across the bay had struck Uryson's head against one of the two rocks and after that he had begun to roll helplessly on the waves in the bay.

When I pulled Pavel to the shore and felt the warmth of his body in my hand, I was in no doubt that he was alive. Some people then ran up to him, and began to do something to him, obviously artificial respiration. Among these people, there happened to be, as I was later told, a doctor, who apparently directed the attempts at life-saving. I do not know and did not know then how long they continued, it seemed like quite a long time. In any case, after some time I asked the doctor what the condition of the victim was and what further measures he proposed undertaking. To this the doctor replied 'What do you want that I should do with a corpse?'.[74]

As I now remember, the only thought that entered my mind when I heard these words was that the word 'fasse' is the 'présent de subjonctif' form of the verb 'faire' and that our French teacher at my school had often asked us for this form and for the subjunctive in general.

Some more time passed, and I went into my room and finally dressed. (Until then I had remained in my swimming clothes.) Pavel Urysohn lay on his bed covered by a sheet. There were flowers at the head of the bed. It was here that I thought for the first time about what had happened. All my experiences, all my consciousness, with such distinctness and clarity. All this merged into a single awareness of how good, how exceptionally good, things had been for each of us, only about an hour ago.

And the sea raged. Its roaring, its crashing, its bubbling, seemed to fill everything. The next day, I sent telegrams to Brouwer, and to my brother Mikhail Sergeevich, in Moscow, whom I asked to tell the Urysohn family about what had happened. That same evening I received in reply a telegram from Brouwer with the words 'Appelez-moi où vous voulez'. I asked Brouwer to come to Göttingen, where I planned to stop for a few days on my way to Moscow.

The funeral was on 19 August. In the belief that it would accord with the wishes of Urysohn's father, I asked a rabbi to perform the funeral rites. As far as the funeral itself is concerned, I remember the huge number of people who

[74] 'Que voulez vous que je fasse avec un cadavre?'

> came to it, the pile of living flowers on the new grave, and the noise of the sea, which could be heard even in the cemetery. On 20 August I left Batz and after stopping in Paris for a day I arrived in Göttingen on the 22nd, where Brouwer, Courant and Emmy Noether awaited me. Hilbert and Klein asked me to come and see them. This was my last meeting with Klein. He died in the summer of the following year.

The death of Urysohn was a shock for everybody who had known him. Many mathematicians who had met him during his short presence on the European scene of mathematics sent their condolences to Alexandrov.

Brouwer, in particular, suffered from the terrible blow. He had in their all too brief acquaintance come to see Urysohn as a precious gift from heaven, as the predestined person to carry on topology from where he had left it.

The little time Alexandrov, Urysohn and Brouwer were together was spent with intense discussions on Urysohn's recent work (and undoubtedly Alexandrov's work as well). They discussed in particular Urysohn's dimension theory and his big memoir for the Fundamenta. Knowing Brouwer's principles where historical credit and detail were concerned, it was no surprise that considerable attention was paid to these aspects of the paper. The conversations confirmed Brouwer's impression, created by the correspondence; here was the spiritual son he had not even dared to hope for. Indeed, all reports of Urysohn's personality bear witness to a genuine geniality and to an exceptional, sympathetic character.

Long after Urysohn's death, Brouwer expressed his feeling in a moving letter to Paul's father.[75]

> Dear Mr. Urysohn,
>
> I thank you for your kind and trusting letter. I believe I can understand your feelings, precisely because of the deep and almost mystical impression Paul left in my mind. He must have united in a rare way the best from you and from his mother's nature, and indeed in such a mature way, that already during his lifetime his soul almost seemed to dream and to float over the earth. This particular impression put itself, while we were together, alongside my admiration for his considerable mathematical achievements, which I felt for a long time. In particular for the powerful and surprising new life that he infused into the scientific subject topology, cultivated in the past by me.
>
> It almost has the appearance as if there were a transcendental causal connection between his superterrestial state of mind and his short meteor-like corporal existence, and as if death had been for him more a waking up than a falling asleep.
>
> May you, heavily tried father, in awareness and in certainty of the epic beauty of his short earthly course of life, find the solace that I wish you so much with all my heart.

[75]Brouwer to Urysohn Sr., 14.II.1925.

In case you should want to go this summer to the resting-place of your dear deceased, and I can be of any assistance in making this journey possible, please dispose yourself of me without any reservation.

I remain in warmest sympathy
Yours sincerely
L.E.J. Brouwer

Although the paths of Urysohn and Brouwer met only briefly, Urysohn made a lasting impression and Brouwer with his characteristic impetuosity set himself the task to protect the memory and reputation of Urysohn.

## 11.3 The Viennese Connection

At the time that Urysohn was working out his ideas on dimension, another young man was turning his mind to the same problems. Karl Menger, an Austrian student and son of the famed economist Carl Menger, had the good fortune to be in his first year at the University of Vienna when Hans Hahn joined the faculty.[76] Hahn, born in 1879 in Vienna, studied mathematics in Strasbourg, München and Vienna, where he got his Ph.D. in 1902. After a sequence of short appointments at and visits to Göttingen, Vienna, Innsbruck and Czernowitz he was drafted into the army, where he sustained serious injuries and was subsequently decorated and discharged in 1916.

After a five year term as a professor in Bonn, he returned to Vienna where he stayed until his death in 1934.

Hahn was a versatile mathematician with wide ranging interests. He published in a great number of areas, among others variational calculus, function theory, set theory and topology, real functions, Fourier series and foundations of mathematics. His book '*Reelle Funktionen*' became the standard work for a generation of mathematicians. Hahn's philosophical interests led him to take part in some of the sessions of the Vienna Circle. One of his first actions in Vienna was the introduction of a mathematics seminar for students. The first topic was the theory of curves. Menger fell immediately for this fascinating subject. He started to work out his ideas and within a week presented Hahn with a proposed solution. Encouraged by Hahn, he pursued his ideas further and in June 1921 he handed in a short note with the basic notions of curves, including the notion of one-dimensionality.

Overtaxing himself in a situation where the consequences of the war were still noticeable in the form of lack of heating and proper food, he fell ill and had to spend more than three semesters in a sanatorium. Nonetheless he finished his first paper at the end of 1921. In the preceding fall of 1921 Menger felt so worried about the state of his health and the prospects for future research and publications that he decided to deposit his main results with the Austrian Academy, remembering that his father

[76] Mayrhofer (1934).

once told him that one of the roles of academies was to act as trustees of people's ideas by accepting sealed envelopes and testifying later to their contents and the date of their deposition.[77] Menger's document contained the discussion of a number of basic concepts of topology, their definitions and properties.[78]

A copy, authenticated by a notary, is in the Brouwer Archive. The document contained under the heading I. *On the dimensionality of Continua (Draft)*[79] the following items:

§1. The notion of curve
§2. Theorems about curves
§3. The notion of surface
§4. A continuum is called $n$-dimensional if $D$ is $(n-1)$-dimensional[80]

Menger's definition of dimension, although equivalent to the one of Urysohn had a very appealing form. It did not make use of any form of separation, but directly made the induction step by reducing a neighbourhood to its boundary. Here is Menger's definition: A set is at most $n$-dimensional if any of its points has arbitrary small neighbourhoods with at most $n-1$-dimensional boundary. It is a routine matter to define 'has dimension $n$'.

The note ends with the remark that the definition of curve, surface and dimension were already given by him in April 1921, as could be confirmed by Hans Hahn and Menger's fellow student, Otto Schreier.

The paper 'On the dimensionality of point sets', that was to become the first published account of Menger's theory of curves and dimension, was submitted to the *Monatshefte für Mathematik und Physik*, the Austrian journal for mathematics and physics, and it was read by Hans Hahn, one of the editors and definitely the most competent referee one could wish for. Hahn discovered a mistake which Menger could not repair (for good reasons: Sierpinski had already provided a counter-example), and although Menger in a letter to Hahn[81] suggested a solution by changing the definition of 0-dimensional, the paper was not published in its present form.

At the time of the writing of his paper, Menger was still unaware of Brouwer's dimension paper of 1913. This explains why Menger erroneously ascribed to Brouwer the viewpoint that 'dimensionality is a property belonging to certain point sets derived from the possibility to map them one-one and continuous onto certain other ones'. In the days before the general notion of dimension was made explicit, geometric dimension was considered only in the traditional setting of Euclidean spaces,

---

[77] Menger (1979), p. 251.

[78] Sealed document No. 778 (1921). Subsequently published in Menger 1929, part 1.

[79] *Über die Dimensionalität von Kontinuen* ('*Zur Theorie der Punktmengen*' in Menger 1929, the change of title is not explained).

[80] $D$ is assumed to be the boundary of the intersection of a sufficiently small neighbourhood with the continuum.

[81] Menger to Hahn, 15.II.1922.

and Brouwer's invariance of dimension must be viewed in this light. Menger's above characterisation would rather apply to Baire's work.

Menger's poor health kept him in the sanatorium until April 1923; roughly at that time he became aware of Brouwer's 1913 paper.[82] Later in the year he found out that Urysohn (who was virtually unknown at the time) had presented a talk at the Marburg conference in September. He hastened to get hold of the Comptes Rendus notes and was just in time to insert a reference to Urysohn in the proofs of his first published paper.

Upon publication of this paper 'On the dimensionality of point sets, part I', he mailed reprints to various mathematicians, and got in exchange a number of reprints from Fréchet, Brouwer, and the Warsaw school. Brouwer obviously was interested because of the close similarity with his own and Urysohn's work.

The first contact between Brouwer and Menger must have been in late February early March 1924, for there is a letter of March 12 in which Menger thanks Brouwer for sending him a reprint of the 1913 paper, adding that:

> In 1921, when I tried to define the notion of curve and of dimension, I was in the first year of my study at the university and I was not aware at all of your paper, dear Professor, in the *Journal f. d. reine und angew. Math.* 142, in which the definition is essentially anticipated.

In June 1924 Brouwer had communicated his own paper 'Remarks on the notion of natural dimension' to the Amsterdam Academy, in which he took the Menger–Urysohn definition into account and showed that it was equivalent to the natural dimension for locally compact metric spaces, a result that was already mentioned in Urysohn's letter of 24.X.1923, as Brouwer pointed out in a footnote. Brouwer sent a copy to Menger, who replied:[83]

> I cannot, dear Professor, thank you enough for your attention that you paid to my little note, and also for your kind letter and the mailing of your paper which I had already read with great interest in the Proceedings of 28.6.1924.

He went on to point out that for wider classes of spaces the $MU$- and $N$-dimension were not equivalent.[84]

Menger's interests were by no means restricted to topology, and he shared an interest with his teacher in foundational matters. He told Brouwer that:

> Recently I had to present a talk on research on the foundation of mathematics in a privatissimum of the epistomologist Prof. Schlick. It may well have been the first time here in Vienna that an extensive presentation of intuitionism was offered. The report was followed by a long discussion. It would be very

[82] In my description of the role of Menger in the development of dimension theory, I am relying on Menger's reminiscences, Menger (1979), the surviving correspondence and printed material.

[83] Menger to Brouwer, 13.XI.1924.

[84] $MU$-dimension = Menger–Urysohn dimension, $N$-dimension = natural (i.e. Brouwer's) dimension.

> fortunate for me to obtain in a few months time instruction from your lectures on these fundamental questions which touch me in my innermost being.

Fate had plunged Brouwer once more into topology; it definitely was not by his wish that he had to divide his attention between his mission in life—intuitionism—and his one-time love—topology. But the sudden appearance, first of Urysohn and Alexandrov and then of Menger left him no choice but to return to the fields of his first glory and to attend to some unfinished business.

## 11.4 The Scientific Legacy of Urysohn

During his life, cut short by the tragic accident, Urysohn had been incredibly creative and productive. The papers and manuscripts of the topological genius allowed Alexandrov to publish two volumes of collected works of almost one thousand pages.[85] At the time of his death only a small part had appeared in print. We have already come across Urysohn's *Comptes Rendus* notes, and the manuscripts that he had submitted to the *Mathematische Annalen* (p. 399). In addition, at the time of Urysohn's death there was a substantial manuscript in the hands of Sierpinski, the editor in chief of *Fundamenta Mathematicae*.

The personal and scientific impact of Urysohn on Brouwer had been tremendous—a brilliant scientist and a soul mate who had miraculously appeared. Terrible as the shock of Urysohn's death must have been for Brouwer, it was nothing compared with its impact on Alexandrov. He had lost his closest friend, a fellow mathematician who shared his soul with him. Reading the surviving correspondence one gets the impression that Alexandrov had lost the will to live. In one of Brouwer's characteristic, deeply empathetic letters he encouraged and comforted the despairing young man,[86]

> I have received both letters, and I am in my thoughts continually with you. Yet I would not pray, in accordance with your statement, that you will not have a long life. In the first place, as it is not because of objective events, but only for the illumination of our sense of duty and for the sake of strength to bear the trials that are imposed on us, that we may pray. In the second place, because our earthly existence was given us only for the purification of our soul from the original sin of fear and desire, and that it is only in accordance with the time required for the satisfaction of *this* purpose, that the span of life of the righteous is measured.
>
> Just because of that the death of the righteous person has for himself always the characteristic of fulfilment, a release and a salvation, and we should after his death offer him further our love, but not our compassion, in particular not when his transition to death was a light one.

---

[85] Urysohn (1951).

[86] Brouwer to Alexandrov, 31.VIII.1924.

> And for the mourning surviving friends and relatives the following holds: each sorrow has for the heart that feels it, its ennobling meaning and in the days of sorrow it is often easier than in the days of joy, to become aware of God's presence, because sorrow, to be born in peace, forces dematerialisation. May this also be so for you!

Almost immediately after the heartbreaking message from Alexandrov, Brouwer had decided to see that justice was done to Urysohn's unpublished work. For a man with his meticulous editing habits, this was not just a matter of sending manuscripts to the printer after a cursory inspection of the text. He was known for his punctilious treatment of papers, his own as well as those of others. The page proofs of his own papers that have been preserved show a rich variety of corrections, concerning both the formulation and the mathematical content. Mostly these corrections were of a cosmetic nature, improving the text. It was only natural that he and Alexandrov should act together in this enterprise. Both were more than competent where the content of the papers of Urysohn was concerned; Alexandrov wholeheartedly shared Brouwer's sentiment, for him this work of love for his deceased friend was a matter of course. As guardians of Urysohn's scientific estate there was much to do, sorting out the manuscripts and representing the scientific interests of the deceased. Their first challenge was the monograph-sized paper for the *Fundamenta Mathematica*, the so-called '*memoir*'. In a way this was a touchy project, mainly because the two first-rate topologists who had taken it upon themselves to supervise the editing of this fascinating manuscript were in their own research so close to the subject matter that extra care had to be exercised to keep the necessary distance. For Brouwer the matter was particularly sensitive, as Urysohn's original text contained a rather unflattering characterisation of Brouwer's dimension paper. Brouwer was, however, not the man to adopt the safe procedure, and to back out. He had no doubt at all that he was the right man for the task.

Urysohn's criticism of Brouwer's dimension paper was concentrated in a particular footnote which explicitly claimed that Brouwer in 1913 had failed to establish the right notion of dimension (see p. 415). Urysohn had written the memoir unaware of Brouwer's earlier definition of dimension. He discovered Brouwer's paper just in time to insert a last minute footnote, as appears from a letter to Sierpinski.[87]

The first page proofs of the memoir still contained the old footnote, see p. 415, but the next proof took Urysohn's wishes into account. The formulation, however, is different from the one in Urysohn's letter; a reference to Menger is appended,

> I mention moreover the work of Menger (Monatshefte für Math. u. Phys. 23 (1923)), where the point of departure is more or less the same as the one of my notes cited in the present memoir, and which has only come to my notice in the spring of 1924.

Indeed Urysohn learned about Menger's work, as appears from a letter written in March 1924,[88]

[87] Urysohn to Sierpinski, 27.VI.1924.

[88] Urysohn to Menger, 22.III.1924.

> As it appears, we have almost simultaneously found the definition of dimension: mine I have found in July 1921, presented to the Moscow Mathematical Society in October 1921. Two announcements have appeared in the Paris Comptes Rendus (1922, Vol. 175, pp. 440, 481); the first part of my major publication is at present in print at the Fundamenta Mathematicae.

This, apparently, was the only contact between the two new masters of dimension theory.

As we have seen, p. 409, Brouwer had already contacted Sierpinski about the memoir. In March, uncertain of Urysohn's views, Brouwer had approached the editor in chief of the *Fundamenta Mathematicae*, asking him to keep an eye on possibly unjustified comments on Brouwer's definition of dimension—an unmistakable reference to the footnote. A possible argument between Brouwer and Urysohn had been smoothly averted, when Urysohn had accepted Brouwer's explanation. Unfortunately Urysohn had failed to correct his comments in the memoir, and he had informed Sierpinski in haste, scarcely a month before his visit to Brouwer, that a correction of the footnote was in order.

Once Alexandrov and Brouwer had agreed to edit Urysohn's scientific estate, interested parties had to be informed. In the case of the *Mathematische Annalen* there was no real problem. Brouwer was the editor who handled the papers of Alexandrov and Urysohn, and it was completely natural that the matter could be settled by Brouwer and Alexandrov. As for the *Fundamenta Mathematicae*, the editor had to be informed of the new arrangement concerning the corrections of the proofs. On 11 September Alexandrov wrote to Sierpinski that Brouwer was so favourably impressed with Urysohn that all correspondence on the subject of Urysohn's papers would be handled jointly by the two of them.[89]

It seems that Sierpinski was pleased to accept the arrangement, for Brouwer referred in a letter to information from Sierpinski:[90] the memoir was scheduled to appear before the autumn of 1925 in two volumes of *Fundamenta Mathematicae*, the introduction and the first two chapters in Volume 7, and the rest in Volume 8. He advised Alexandrov to leave the matter alone for the moment. But he also foresaw the possibility of difficulties,

> In the meantime it seems that we should, alas, take into account the possibility that Kuratowski will have on his own authority declared the introduction of Paul's memoir ready for printing (although such an act without the authority of Paul's heirs should appear incomprehensible to me), that thus about note 3, which criticises me, nothing could be done. Finally, the proofs that were sent to me do not look as if they are ready for printing at all; they still contain many annoying printing errors.

Almost all of the following correspondence is concerned with the proofs and corrections. There seemed to have been one long struggle to get the proof sheets,

[89] Alexandrov to Sierpinski, 11.IX.1924.

[90] Brouwer to Alexandrov, 24.X.1924.

to convince Sierpinski to grant more time to insert corrections, both of a material character and of a purely typesetting nature. Sierpinski was probably not happy at all to have two excellent topologists go over the proofs with a fine-tooth comb, and so we see a running battle over access to proofs, the right to insert corrections and the like.

A week after Brouwer's letter Alexandrov tried to get Sierpinski's co-operation,[91]

> Shortly before his death Paul Urysohn has talked at length with Brouwer about his own papers and in particular his *memoir*. In the course of the exchange various questions concerning the final formulation were discussed and joint decisions were taken accordingly. Therefore it is very important that the proof sheets of the memoir, until the moment that they are marked with the decision 'imprimatur!', are read by Professor Brouwer, so that, if it were only in appendices, we can enter such changes as Paul Urysohn would have entered himself, had he been alive.

Without speculating on the possible state of mind of a deceased person, one can say that, if anybody, Alexandrov knew Urysohn's mind; he had worked with him over the past years, and he was in an excellent position to understand Urysohn's work, not just the polished results but also the thoughts and deliberations that had gone before. So it is not unreasonable to trust him in this matter. As for Brouwer, he had a singular gift to fathom a kindred spirit. So there was no pair better equipped for this task than Alexandrov and Brouwer.

The task of supervising the editing of the *memoir* was not an easy one. Printing was in an advanced stage, and apparently the proof reading was done by local staff in Warsaw, so it seemed almost hopeless to expect that the publisher would welcome more proofreading, and hence more corrections.

A week later Alexandrov, who had returned to Moscow, where he was living with Urysohn's family (and actually was thrown out of the apartment together with them), dropped a harmless looking bombshell. He wrote Brouwer that he was about to approach Sierpinski, and he included the text of the letter.[92] In this letter he confessed a painful oversight: although he and Urysohn had together gone over the final version of the text, they had forgotten to change the expression 'se propose a résoudre' to 'résout', see p. 415. As the old formulation contained 'a certain offensive meaning' for Brouwer, a meaning which had escaped them at the time, he had begged Sierpinski to change the formulation 'as required by scientific truth and by the wishes of the author as expressed in his last days'.

In his letter to Brouwer, Alexandrov cautiously expressed his hope that things would work out, 'I believe that the revision of the flawed footnote 7) in the introduction is not totally hopeless.' Brouwer could not draw much comfort from this kind but rather lame reassurance. Some two weeks later he himself wrote to Sierpinski; thanking him for his co-operation, he went on to say that then were a num-

[91] Alexandrov to Sierpinski, 20.X.1924.

[92] Alexandrov to Brouwer, 20.X.1924.

ber of corrections and emendations, made necessary by the exchange of ideas between Urysohn, Alexandrov and Brouwer, shortly before the death of Urysohn—'in a form that only the author could have recognised [. . .], a form which Alexandrov and I have determined in accordance with the last notes and the last verbal statements of the deceased, for which we accept full responsibility'. He added that one of the corrections concerned himself, and that Alexandrov had written about it.

A complicating factor on Brouwer's side was that he was still recovering from a serious illness. In October he mentioned his health problems, and Alexandrov was worried accordingly.

Brouwer's health was again in a poor state, in letters to Alexandrov cod liver oil is mentioned, and a recommendation of the family doctor to spend the Christmas vacation in the Engadin (Switzerland). Needless to say that Brouwer refused to go, there were too many things he had to do. His ill health no doubt was real, but keeping Brouwer's medical history in mind, one can hardly doubt that Urysohn's death and the subsequent pressure of Urysohn's posthumous publications made things worse. It looked as if his not uncommon nervous breakdowns had surfaced again.

On the thirteenth of October Brouwer replied to an earlier letter of Alexandrov saying that he was still in bed. 'My recuperation progresses smoothly, but not very fast, and my doctor declares moreover that I have to be careful for a long time and that I will have to spare myself. I have not been able to extract a diagnosis with a scientific medical name: he spoke of "flu with complications".'

In a touching letter Alexandrov replied, expressing his worries that the mysterious illness might have been a flu combined with pneumonia, which in his opinion could easily develop into tuberculosis. 'I am very much worried about your weak lungs .... You should therefore not only take care, but also make your food as intensive as possible, to put the whole organism in order, otherwise it will not work. You should take great quantities of butter, eggs, milk, cream etc., and *cocoa*, I think that now meat is indispensable to you. .... Please keep me continuously informed of the news of your health.'[93] The last part of the advice was not self-evident for a vegetarian like Brouwer! Alexandrov's advice to take cod-liver oil ('oleum jecoris Aselli') was, however, dead on target, Brouwer adored any medication that came straight from Mother Nature's drugstore.

Alexandrov's letter also contained a German translation of the poem 'Worüber singt der Wind?' of Alexander Block, a poem that summed up the sad emotions of Alexandrov.

> Warte, mein alter Freund, und dulde, dulde,
> Das Dulden wird nicht lange dauern, nur schlafe fester;
> Es wird ja alles doch vergehn,
> Es wird ja niemand was verstehn,
> Weder dich verstehn, noch mich,
> Noch das, was von dem Winde
> Uns vorgesungen ist...

[93] Alexandrov to Brouwer, 20.X.1924.

> So much was sung us by the wind. It was really as if between the lines of life a divine song was sung, intertwined. I recall how we landed at 4 o'clock in the morning for the first time at the Norwegian coast at Christiaansand. The sun had not yet risen, we stood in this cool morning air and we gazed at the rocks of the shore, at the motionless, totally quiet sea, at the blazing red sky. So we stood for a long time, silent, until the sun had come up, and still longer. We did not speak a word, but we both felt in the same way the eternal infinity of nature. But even in daily life, in the most ordinary aspects, this other thing was present, that what we felt and understood, and what separated us. This haze of eternity ... it surrounded our joy, thus perhaps it lasted so briefly ...

It was actually neither the first nor the last time that 'tuberculosis' was mentioned, nor the first time that Brouwer's vegetarian diet was deplored. On 16 February 1921 Van Eeden wrote in his diary: '—I went home with Brouwer. He said that he had been ill, and I understood that it was the beginning of tuberculosis. I loved him so much, he was friendly and so warm-hearted. If only he would eat meat.' Van Eeden's fears may have been somewhat exaggerated, but he was a medical man and he definitely knew how a tuberculosis patient looked! The conclusion one may safely draw is that Brouwer's ascetic way of life and the nervous tensions were taking a heavy toll.

On 20 October Brouwer wrote Alexandrov that his health was improving and that he hoped to resume his lecturing the next Friday. 'I feel as if I have come slowly to the light from a dark abyss', and on 3 December he sent a card to Alexandrov with a reassuring message, 'My health is now really satisfactory; I live indeed most carefully (the cod liver oil is faithfully taken).' That Christmas Brouwer stayed home, his family doctor had wanted to send him to Engadin, but Brouwer argued successfully that the backlog of tasks he would find on returning from Switzerland was a greater danger for his health than a Christmas in Laren. Brouwer was grateful for the attention of his young friend. There was also good news concerning the grant. The Rockefeller Foundation had approved the arrangements for Alexandrov's stay in Amsterdam, where he would receive $180 per month from the grant.

The exchange of letters must have been quite frequent in the fall, although only a few have survived. There is a note of Brouwer's in which a few points of a letter of 7 November to Alexandrov are written down; one of these says 'vision in my dream of Urysohn: "Yes, yes, yes, yes,—yes, of course" about my plan to have a new typesetting of the imperfect text'. Far from being a cheap trick to impress Alexandrov, it shows that the memoir occupied a prominent place in his thoughts.

A few day earlier he had written to Sierpinski, stressing the importance of allowing more time for a proper proof reading.[94] After thanking him for the proofs of the introduction, he expressed his hope that

> there would possibly be time to make changes and corrections in the introduction, that had become necessary by the exchange of ideas and that had

[94]Brouwer to Sierpinski, 4.XI.1924.

> taken place in the last weeks of the life of the author, between him, his friend Alexandrov and me. In this hope I will send you shortly the corrected proofs of the introduction in the form that only the author could recognise at the day of his death, a form which Alexandrov and I have established from the last notes and the last verbal statements of the deceased, and for which we accept the full responsibility.
>
> Among the changes there is one which concerns me personally, and about which Alexandrov has written you a letter, which he will send through me as an intermediary and which I will get registered myself to-day.
>
> In the proofs which I will return to you I have done my best to correct typographic errors which were left in fairly large numbers and some of which are not all that innocent.

On 17 November Brouwer again turned to Sierpinski, explaining that certain corrections could only be made in the final proof.[95] There were for example, he wrote, 'passages [...] thoughtlessly written at the time, and recognised as absurd by the author shortly before his death—which, if they were printed in their actual form, would gravely damage his scientific memory ...'. For this reason he begged Sierpinski to suspend all printing, to send all proof sheets to him, and to have new proof sheets made after his corrections, including those pages which had gone through their final printing. Any necessary costs would be gladly paid by him. One would not have to be extremely sensitive to resent a letter of this sort. No editor likes to be told in a patronising way what to do. The next day Brouwer reported to Alexandrov that he had written in unmistakable terms to Sierpinski about the poor editing and correcting. He urged Alexandrov to support him. Thus Alexandrov duly informed Sierpinski of his surprise and displeasure, to find out that the introduction and the first chapter had been printed *without enabling him to read* the text. The fact that Kuratowski had read the text did not mean that there was therefore no reason for Brouwer and himself to read the proofs. He forcefully appealed to Sierpinski to redress the situation.[96]

Sierpinski reacted swiftly and firmly.[97] Urysohn, he wrote, had asked for one correction (see p. 416) *if possible*, and he clearly did not expect to see any proofs. 'Therefore the editor of the Fundamenta Mathematicae considered himself entitled to start the printing of the memoir of Paul Urysohn, without sending any proofs, except for that single correction. Paul Urysohn never proposed any more modifications of his memoir.'

What had started as a simple duty to honour Urysohn's memory, and a service to a well-known mathematical journal, thus started to take the form of another conflict. That was exactly how Brouwer saw it.[98]

---

[95] Brouwer to Sierpinski, 17.XI.1924.

[96] Alexandrov to Sierpinski, 24.XI.1924.

[97] Sierpinski to Brouwer, 25.XI.1924.

[98] Brouwer to Alexandrov, 21.XI.1924.

> Now it also occurs to me, that for the fortifying of our position (it indeed is like a real war; our position assumes the role of a fortified camp, in the wall of which, as a consequence of your fateful letter to Sierpinski, a breach was shot, which we have to close with all our might) it were very important to add the following passage …

In view of the position taken by Sierpinski, Brouwer thought it best that Alexandrov would make it clear to Sierpinski that Brouwer and he would take no important decision without mutual consultation, and that hence it was obvious that Alexandrov had to see the proofs. Moreover, Sierpinski's claim that Urysohn had not asked for proofs seemed in conflict with the fact that the first galley proof of the memoir had reached Urysohn in Le Batz. Anyway, would any author of a memoir of more than three hundred pages voluntarily forego the proof reading?

Should it come to the worst, then Brouwer could, as an 'ultimate medicamentum heroicum, withdraw in name of the family, the paper from the Fundamenta'. That would probably teach the editors a lesson.

In December another complicating factor arose—Brouwer, with his refined instinct for possibly hidden insults, had discovered that the uncorrected version of the manuscript implied a reading with drastic personal consequences. Namely, Urysohn had characterised Brouwer's notion of separation as 'the old notion of cut' (*coupure*). This, he was certain, would immediately tell every reader that Brouwer (i) missed the essential point in his definition of dimension, and (ii) tried to cover up his lapse. To what extent Brouwer's fears were justified is hard to say. Urysohn's formulation was unfortunate, and went back to the period before his exchange with Brouwer, and indeed an acute reader might doubt Brouwer's explanation. So, Brouwer foresaw a disaster, unless he could repair the formulation. In view of the fact that Urysohn had fully accepted Brouwer's view, and in view of the discussions in Blaricum (where, one may be certain, Brouwer once more pointed out the true history of his oversight), there is little doubt that the unfortunate passage was overlooked in July 1924. But for Brouwer the dimension matter began to border on an obsession. In his letter to Alexandrov he forcefully argued his case.[99] He spoke of 'Sierpinski and Kuratowski tainting Paul and the two of us'. Somehow the matter was settled, for on 15 January 1925 he wrote 'The peace with the Warsaw people, also towards me, has been completely re-established.'

Peace or no peace, the end of the problems had not been reached. On February 17 Brouwer informed Alexandrov that a new disaster had taken place. Brouwer had returned the galleys 8 and 9 to Fundamenta Mathematicae, but the printing shop had in the meantime, 'on account of the small stock of type (just enough for a few galley sheets)' destroyed the type for galley 8, 'as Sierpinski informed me in a further most polite and apologetic letter'. So galley 8 had to be set again, and—whereas galley 9 was impeccably set—'contained, of course, again the normal quantity of Kuratowski errors'.

[99] Brouwer to Alexandrov, 24.XII.1924.

The last surviving letter to Sierpinski on the matter of editing Urysohn's memoir is the one of May 24, 1925. In it Brouwer conveyed the wish of the family of Urysohn to preserve the original manuscript of chapters 4 through 6, which they considered 'the culminating part of the scientific activity of the deceased', as 'a relic'. So he begged Sierpinski, in the name of the Urysohn family, to cede that part of the manuscript. A copy 'carefully made under my direction (in which moreover some posthumous wishes of the deceased were taken into account)' was offered in exchange. The correspondence does not say if Sierpinski honoured the request of the family.[100]

Urysohn's memoir appeared without further problems in the Fundamenta Mathematicae. It was a substantial piece of topology, all together 236 pages; the paper dealt with a rich variety of topics, including dimension theory and indecomposable continua. The modern reader, accustomed to Bourbaki-style presentation would find the paper a bit long winded here and there, but this opus of the twenty-five-year-old topologist contains the fundamental theorems of dimension theory.

The editing of Urysohn's memoir took a heavy toll on Brouwer, both emotionally and clerically. In addition to this, Brouwer was also handling the newer papers of Alexandrov and Urysohn for the *Mathematische Annalen*, complete with correcting formulations, streamlining proofs, etc. All of this took place at the time that he was occupied with the preparation and publication of his finest intuitionistic work: the bar theorem, fan theorem and the continuity theorem.

Did Brouwer actually 'falsify the text', as Menger later suggested? It is highly unlikely; after all he did not act without Alexandrov's approval, and Alexandrov had been present and taken part in the discussion between Brouwer and Urysohn in July 1924. Those who have known Alexandrov know that he would not be participant in shady practices of this sort, and he was certainly not one to be pressured into such an act.

It is an altogether different question whether it was wise to take part in the editing. Although his motives were totally unselfish, and based on his admiration and love for this sudden topological prodigy, less kind tongues could easily spin a tale of text manipulation. It is a pity that Brouwer and Alexandrov did not add an editorial comment to the Memoir.

Brouwer's factual influence on the text is hard to assess. With his editorial routine, he probably did some polishing of the formulations, and added an occasional useful reference. The most damaging accusation would of course be that of systematic self-promotion. Without access to the original manuscript there is little one can say. Brouwer is mentioned a number of times in footnotes (re references), but these are all harmless 'useful information for the reader'. There is no reference beyond generally recognised facts. Yet, the editing of the memoir would become the subject of some most unpleasant altercations.

The correspondence with Alexandrov contains regular references to a planned stay in Amsterdam. In the letter of February 17 arrangements for Alexandrov's

[100]So far no copies of the manuscript in whatever form have been found.

stay in Amsterdam are mentioned. The Rockefeller Foundation had put a grant at Alexandrov's disposal and it had taken some ingenuity to get permission for this stay from the Soviet authorities. Brouwer left no stone unturned, wrote letters and asked his friend and colleague Mannoury to write a letter to the Soviet authorities. Mannoury duly did so, he asked the Volkscommissariat for education to grant Alexandrov permission to come to Holland to collaborate with Brouwer. He signed his letter as 'Professor of Mathematics at the University of Amsterdam' and 'Member of the K.P.H.' (Communist Party of Holland).[101]

Brouwer had also enlisted the help of his colleague Pannekoek, the astronomer, who had an international reputation as a Marxist theoretician. Pannekoek, in turn had asked a Dutch engineer, Rutgers, to put pressure on the authorities.[102] The organisation of Alexandrov's stay in Amsterdam began to take shape. For Brouwer it was important to arrange the details, as he wanted to make the most of not only Alexandrov's presence, but also that of a number of other visitors, to wit Menger, Vietoris, Kerékjártó and Wilson.[103] A special set theoretical lecture series would only be worthwhile if all the visitors could attend. Brouwer strongly advised his visitors to come and live in Laren or Blaricum, he preferred to have them around, so that he could organise meetings and seminars at home, without travelling to Amsterdam. In his letter of March 15, 1925 he discussed the housing arrangements with Alexandrov; it had apparently been agreed that Alexandrov was going to live in Laren. Was it not a good idea, he asked, if Menger also found himself rooms in Laren, or the neighbouring towns of Hilversum or Bussum. 'I hesitate to make this suggestion myself to him, because I would accept thus certain imponderable obligations, from which I want to safeguard myself as long as I don't know Menger personally.' Alexandrov would not be bound by any obligations if he arranged temporary housing for Menger. Brouwer would, of course, if necessary look for suitable rooms for Menger. One should not read too much into these lines, Brouwer had always been extremely sensitive in his contacts, being completely in the dark with respect to Menger, he felt he could not commit himself as he did towards Alexandrov.

Menger, in fact, had already contacted Alexandrov; in an undated letter from Alexandrov to Menger an earlier letter of Menger is mentioned. Alexandrov's letter must have been written in the first months of 1925 (perhaps even earlier). It is an enthusiastic welcoming of a fellow topologist, expressing admiration for the parallel efforts of Menger and his friend Urysohn in the area of dimension theory—'of course you will be cited in the 'Mémoire sur les multiplicités Cantoriennes', and the relevant footnote has already been sent to the editors; Brouwer and I are completely of the same opinion on this matter'. He looked forward to work with Menger in Brouwer's seminar, believing that this would be useful to both of them. In fact, he bade Menger to come to Holland a bit later in order to make their stays overlap for a

[101] Mannoury to Lunatscharsky, 4.XI.1924.

[102] Rutgers had offered his expert services to the Soviet government, and in time he became an appreciated courier of Lenin, with quite significant assignments for the promotion of the world revolution.

[103] Wilfrid Wilson was a British topologist who took his doctorate with Brouwer in 1928.

longer period. He looked forward, he wrote, to make the acquaintance with Menger from whom he expected many interesting things.

Menger had entered Brouwer's life through correspondence, and Brouwer had immediately recognised Menger's contribution to dimension theory. Indeed, in his sequel to the dimension paper of 1913[104] he referred to Menger (1923) and spoke of the *Menger–Urysohn dimension* ($MU$-dimension).

In 1925 things were brightening up for Brouwer and his Urysohn project. Alexandrov had already informed him in January that the peace with the Warsaw mathematicians had been completely restored. Of course, Brouwer had enough to grumble about, but he was no longer hampered by editorial interference. In the spring his topological admirers started to arrive, and the stage was set for a period of intensive study and research in modern topology.

[104] Brouwer (1924e). Brouwer (1924g) is an abridged version in which Menger is not mentioned.

# Chapter 12
# Progress, Recognition, and Frictions

> If I yet have the time and energy, I will myself show the mathematical world that not only geometry but also arithmetic can point the way to analysis, and certainly a more rigorous one. If I cannot get this done, then they who come after me, will do it, and they will also recognise the incorrectness of all those inferences with which the at present so-called analysis operates.
>
> L. Kronecker

## 12.1 The First Skirmishes in the Foundational Conflict

To an outsider the foundational situation in mathematics after the crisis of the set theoretic *paradoxes* may have seemed quiet and unproblematic. It was almost as if a moratorium had been called on foundational research. In the early years of the century great men had proposed various remedies for the sudden outbreak of paradoxitis in the wake of Cantor's monumental innovations. Poincaré, Russell, Hilbert, Brouwer each had their ideas for saving mathematics. In this line Brouwer came last, and after his contribution to the foundational debate no new enterprises were undertaken until after the First World War. It would be a mistake, however, to assume that everyone had returned to 'business as usual'. Hermann Weyl's concern about the foundations, for example, had been (re)awakened in 1910, as appears from a letter to his Dutch friend Mulder, 'lately I have been thinking a lot about the foundations of set theory, and there I arrive at views that deviate rather from those of Zermelo, and which are in a sense close to the, here much ridiculed, position of Borel–Poincaré'.[1]

Hilbert had not given up his interest in the foundations, and he lectured regularly on the foundations of mathematics.[2] He also had a number of researchers around him, who worked on various issues. The best known collaborator of the early years was Ernst Zermelo, a singularly gifted mathematician, who is now mostly known

[1]Weyl to Mulder, 24.VII.1910.

[2]See Sieg (1999).

D. van Dalen, *L.E.J. Brouwer – Topologist, Intuitionist, Philosopher*,
DOI 10.1007/978-1-4471-4616-2_12, © Springer-Verlag London 2013

for his proof of the well-ordering theorem and the axiomatisation of set theory. was in Göttingen until he left in 1910 for Zürich. Under Hilbert's influence the philosophers Edmund Husserl and Leonard Nelson were also appointed for some time in Göttingen. Hilbert managed to get a steady stream of young researchers to work on his ideas. Among these Paul Bernays became his principal assistant for his own research. Bernays was born in London in 1888; he possessed many talents, among which those for music (he was an accomplished pianist), mathematics and philosophy stand out. After studying in Berlin and Göttingen, he obtained his doctorate in 1912 from Edmund Landau. In the same year he got his 'habilitation' under Zermelo. He followed Zermelo to Zürich, where he stayed until Hilbert called him in 1917 to Göttingen. As a matter of fact, it was Polya who, on a walk in the mountains, had called Hilbert's attention to the promising young man. A most fruitful collaboration followed, without Bernays' competent and patient toil, Hilbert's program might not have materialised.

Suddenly, after the First World War, the foundations of mathematics came into focus again. In 1918 Hermann Weyl published his monograph *Das Kontinuum*, Hilbert's Axiomatic thinking (*Axiomatisches Denken*) appeared in the *Mathematische Annalen*, and Brouwer had launched his new constructive program in the first *Begründungs*-paper. Hilbert's paper contained the text of a lecture in Zürich the year before; it was one of those expository masterpieces for which Hilbert was famous. It consisted of a brief tour of the foundations, with views of mathematical problem areas, including physics. The significance and scope of the axiomatic method was discussed in broad terms (still a long way from the later proof theory). The paper was free of personal attacks, it contained just a slightly depreciative observation on Kronecker and Poincaré.[3] It contains an interesting change of view on 'Hilbert's dogma'. He mentioned 'the problem of solvability in principle of each mathematical problem', whereas in his famous Paris Lecture of 1900 spoke of 'this axiom of the solvability of every problem'.[4]

Brouwer continued his *Begründungs*-series with the second paper, which dealt with the topology of the plane and the theory of measure.[5] Just like its predecessor it was dry as dust, scholarly to the point of dull. In retrospect, one might say that Brouwer carefully steered away from any possible controversy. How far he followed a deliberate policy will remain a matter of conjecture; did he fear to upset Hilbert, whom he greatly admired? Generally speaking Brouwer was well-informed about the mood in Göttingen, and it is not wholly impossible that he foresaw Hilbert's disapproval. He had already disappointed Hilbert by turning down the offer of a chair in Göttingen (cf. Sect. 8.4), and it is not certain how this influenced Hilbert's view of his Dutch admirer. One should bear in mind, however, that it was not a grave sin to turn down a flattering offer in order to cash in at one's own university. Hilbert was

[3]Hilbert claimed that Kronecker rejected set theory on the grounds of the paradox of the set of all sets. This is rather a misjudgement of Kronecker's motives, moreover Kronecker died before the paradoxes surfaced.

[4]Cf. Gray (2000), p. 248. Brouwer was indeed quick to point out this shift, Brouwer (1919h).

[5]Brouwer (1919a).

himself no stranger to the game, in 1919 he turned down an offer from the university at Bern 'after the ministry has granted all my wishes'.[6] Nonetheless, it is one thing to improve one's own position through 'position bargaining', and another thing to be used in the game. Whatever Brouwer had in mind, the result was a series of non-provocative, scholarly papers, hardly read by the mathematical community. The first paper that addressed a larger audience was 'Intuitionistic set theory', published in the *Jahresbericht* and in the proceedings of the Dutch academy.[7] The paper contained a number of non-technical, conceptual remarks that put some meat onto the skeleton of his intuitionistic mathematics. In particular he equated Hilbert's dogma with the principle of the excluded third, and gave an argument for their rejection. Hilbert is explicitly mentioned in the paper. None of this seemed to provoke Hilbert, but when in 1921 (cf. Sect. 8.8) Hermann Weyl sided in his spectacular paper 'On the new foundational crisis of mathematics' with Brouwer, it was more than Hilbert could take. Weyl's paper is one of the all-time beauties; formulated in a sparkling, literary style, it gave convincing reasons for joining Brouwer's intuitionism. None of this could be called derogative or insulting. Perhaps this was one of the things that irked Hilbert: not only had his prize student left him for another, but he had done so with great charm and wit![8]

Yet, the developments could not have taken Hilbert by complete surprise. Weyl had addressed the Göttingen Mathematical Society on May 11 of 1920. It is true that Hilbert had missed the lecture because he was not informed in time. But he must have heard what Weyl had to say. Indeed he had ordered Bernays and Kneser to draft an extensive report. As he wrote to Weyl, 'there are all kinds of similar lines of thought, as I also have pursued in my lecture courses of the last semesters, even though our fundamental tendencies seem to be very different. I am very curious about the development of your ideas'. There is no trace here of exasperation, on the contrary, Hilbert seemed somewhat pleased with some similarities that he noticed.[9] Admittedly, there is a slight possibility that Weyl's talk dealt with his older ideas. The title of the talk was 'The Continuum', so he might have kept his powder dry, and stuck to the less controversial material. This, however, seems highly unlikely. Weyl was at that time fully committed to Brouwer's program and he had already lectured on the topic in Zürich, and prepared the manuscript of the New Crisis-paper. Perhaps

[6]There are certain indications that Hilbert's independent position during the war had made him no friends among his colleagues. Born wrote in a letter to Hilbert (27.VIII.1919) that he guessed that Hilbert's action was largely determined by his relationship with the 'Göttingen' colleagues, who mostly subscribed to the *Alldeutschtum*. In view of the treatment by this group during the war, Hilbert's threat to leave Göttingen might not have been empty talk.

[7]Brouwer (1919h, 1922a).

[8]See Sect. 8.8.

[9]Since there is no elaboration of this remark, one can only make an educated guess as to what these similarities were. A possible candidate could be Weyl's treatment of quantified statements as 'judgement abstracts'. In Hilbert's writings a similar phenomenon surfaces in the distinction between finitary, meaningful statements and quantified, ideal statements, cf. Hilbert (1926), p. 174. In Hilbert's papers no reference to Weyl's ideas is to be found. There is, however, a reference in Hilbert and Bernays (1934, 1939).

**Fig. 12.1** Hibert and Weyl [Courtesy Niedersächsische Staats- und Universitätsbibliothek Göttingen]

his talk in Göttingen did not contain the provocative statements that are so characteristic for the idealistic revolutionary in the prime of life. Some time after this letter, the daring paper of Weyl appeared, and Hilbert apparently realised that there was a challenger at the gate. He quickly took the offensive and used an invitation of the University of Copenhagen to lecture, among other subjects, on the foundations of mathematics. On 18 February Niels Bohr, Hjelmslev, and four other members of the faculty had submitted a proposal to award David Hilbert an honorary doctorate. The academic machinery handled the matter quickly and efficiently, and on March 16 the King authorised the faculty to award the doctorate. At a meeting of the faculty on that same day the faculty carried through Hilbert's honorary doctorate ceremony. With characteristic modesty the great man described his work with the words, 'My scientific work is in the field of algebraic invariants, higher arithmetic and analysis; it occupies itself moreover with the foundations of geometry and the axiomatisation of the physical and mathematical sciences.'[10] The doctorate was not the sole purpose of Hilbert's visit, he also gave three lectures on the following topics: 'Nature and mathematical cognition', 'Axiomatic theory and consistency', and a discussion session.[11] Later that year Hilbert repeated the lectures on the foundations in Ham-

[10]Festschrift udgivet af Københavns Universitet. November 1921, p. 114.

[11]*Natur und mathematisches Erkennen* (March 14), *Axiomenlehre und Widerspruchsfreiheit* (15 and 17 March). I am indebted to Sigurd Elkjaer for the information and documentation on Hilbert's Copenhagen lectures and doctorate.

burg on July 25–27.[12] The Copenhagen and Hamburg lectures were enthusiastically received by eager audiences. It must indeed have been fascinating to hear a living legend speak about matters of life and death for mathematics.

It is, however, a dangerous matter to publish a lecture for a mixed audience as a specialist paper. One tends to slip in details and side remarks that did well before an sympathetic audience, but that do not lend themselves for a scholarly readership. Maybe Hilbert did not realise this, maybe he hoped that readers in their study would react as the audience in the lecture hall. Whatever he may have had in mind, the modern reader is struck by the querulous pedantic style. It is worth noting that all Hilbert's papers, but one, between 1918 and 1931 were reports of lectures. And indeed all of these published lectures have this same emotional, partly indignant, partly sulking, but always aggressive tone.

His colleagues and students must have been surprised to see the brilliant expositor, who had enthralled the mathematical world with his epochal contributions, turn into an aggressive, argumentative preacher. Hilbert always had a reputation for a certain brusque wit; visitors to the Göttingen mathematics seminar were often confronted with Hilbert's caustic Baltic irony, in which tactfulness took second place to truth. But in the foundational lectures the wit was bordering on rudeness. Hilbert's counter-revolution opened in 1921, with his Copenhagen-Hamburg lectures and the subsequent report 'New founding of mathematics'.[13] This opening volley starts with five pages of criticism of the revolutionaries Weyl and Brouwer, with additional shots at Poincaré and Kronecker. As to be expected in a talk for a non-specialised audience, none of the points of critique are elaborated or convincing. Weyl's predicative approach, including his objections to the vicious circle, is answered perfunctorily: Weyl's introduction of the vicious circle in mathematics is waved aside as artificial: there is no evidence that the class of sets of natural numbers is susceptible to the same paradoxical phenomena as the universe of sets. And, understandably, Hilbert appeals to the axiomatic method: the continuum can be introduced axiomatically. Instead of offering the required consistency proof (which, by the way, would have to reckon with Weyl's objection), Hilbert notes that this axiomatic approach corresponds roughly to our intuition. As he was to find out in his subsequent talks and papers, the consistency of analysis was not the answer to Weyl's observations, but the problem itself. Weyl, he said, when speaking of 'the inner instability of the foundations, upon which the structure of the empire rests'[14] and the 'impending dissolution of the State of analysis', is seeing ghosts.

Hilbert's view of the work of Weyl and Brouwer is simple and drastic,

> What Weyl and Brouwer do, amounts in principle to striding along the former path of Kronecker: they try to provide a foundation of mathematics by jettisoning everything that seems inconvenient to them and by setting up a dictatorship of prohibitions à la Kronecker. This means, however, to chop up

[12] See Hesseling (2002) for details.

[13] Hilbert (1922).

[14] Cf. p. 313.

> and mutilate our science, and we risk to lose a large part of our most valuable treasures if we follow such reformers. Weyl and Brouwer outlaw the general notion of irrational number, of function, even of the number theoretic functions, the Cantorian numbers of higher number classes, etc.; ... and even the logical Tertium non datur.[15]

Serious allegations indeed. But to what extent did Hilbert's accusations do justice to the revolutionary duo? He could point to Weyl's paper and its restrictions on quantification,[16] but certainly not to Brouwer's foundational views. Indeed a large part of Brouwer's first *Begründungs*-paper dealt with Cantor's second number class, that is to say with Brouwer's constructive version of it. Nor did Brouwer's or Weyl's approach rule out the totality of irrational numbers (in Brouwer's terminology 'a species'). Reading the philosophical part of the Hamburg lecture, one is tempted to think that Hilbert rather felt a duty to answer the challenges of Hermann Weyl, than that he deemed a mathematical, foundational analysis of the papers of Brouwer and Weyl worthwhile. His formulations betray a degree of vexation that is unusual in mathematical discourse. The fact that two leading mathematicians, one of them his favourite student, dared challenge the mathematical tradition, including his own views, must have shocked him. The tone of this paper, and most of the later ones, can hardly be explained by the issue itself. In mathematics it is an old custom to refute mathematical claims by mathematical arguments. Blasting opponents is and was certainly not common practice. There must have been a powerful emotional urge behind Hilbert's reaction, partly the result of his impression that the barbarians were at the gates, partly the resentment of the older, established professor who is confronted with insubordination and desertion. In his words,

> Brouwer is not, as Weyl thinks, the revolution, but only the repetition of an attempted coup (*Putsch*) by old means, which in its day had been attempted with more vigour, but which failed utterly, and which now, where the power of the state is so well armed and strengthened by Frege, Dedekind and Cantor, is all the more doomed to failure from the beginning.

The modern reader may think the talk about revolutions and coups (*Putsch*) an amusing exercise in metaphorical language, but for the German readers it was not so in 1922. The German nation was thoroughly unstable and coups were part of the harsh reality. In 1920 Berlin had its '*Kapp Putsch*' and Munich had been the scene of a left-radical revolution in 1919, the same year that the Spartacist coup took place in Berlin. So Hilbert's rhetoric was not the abstract reference of an intellectual to some misfortune of the old Romans, but the voice of the statesman-soldier who, in desperate times, follows in Cincinnati's footsteps. The reader should, by the way, keep in mind that Hilbert's political metaphor was a reaction to Weyl's 'New Crisis'-paper, in which the revolution was proclaimed.[17] Not even Kronecker and Poincaré

---

[15] Hilbert (1922), p. 159.

[16] Cf. p. 314 ff.

[17] Cf. pp. 313, 313.

were spared Hilbert's displeasure, 'Kronecker coined the slogan: the Lord created the integers, everything else is the work of man. Accordingly, he outlawed—the classical prohibiting dictator—what was no integer for him.' Poincaré gets off with a lighter reprimand, but Hilbert remained adamant: 'His objection that this principle [i.e. complete induction] could not be proved, then by complete induction, is not justified and will be refuted by my theory.'[18]

It tells something about Hilbert's faith in his new program that he was willing to make these strong claims. On the one hand, it shows that he trusted the tools that he was designing, on the other hand a certain naiveté cannot be denied. To be fair, there was little or no experience in the more sophisticated parts of logic; before Gödel hardly anybody had an idea about the limits of what was provable. Of course, Poincaré and Brouwer had opposed Hilbert's claims, but rather on philosophical than on technical grounds.

The remainder of Hilbert's paper consisted of a general exposition of his program. He sketched how a formalisation of number theory would be obtained. There is an explicit formulation of the separation of the material into (i) mathematics proper (*die eigentliche Mathematik*), which is a collection of provable formulas (in the formal system), and (ii) *metamathematics*, which consists of contentual inferences about mathematics proper, and which served to show the consistency of it. The term metamathematics appears here for the first time in print. And a few pages earlier the term 'proof theory' (*Beweistheorie*) was coined for the study of formal proofs in a formal system, an important step, as it explicitly mentioned formal proofs as objects of study.

Before moving on, let us note that Hilbert mentioned in his attacks Weyl and Brouwer in that order. In view of Weyl's New Crisis-paper he may have viewed Weyl as the ringleader even though Weyl gave the credit for the revolution to Brouwer.

With the lectures and papers of Weyl and Hilbert the *Grundlagenstreit* (foundational conflict) had started in earnest.

Weyl had certainly provoked his old teacher, and the mathematical world saw Hilbert's counter offensive as the proper reprimand for a person who dared to challenge the supreme commander of mathematics.

Neither Brouwer nor Weyl answered Hilbert's attack in kind, both men revered the grand old man (who, by the way, was only sixty) too deeply to upbraid him for a spell of impetuosity.

The discussion immediately drew the attention of the mathematical community, and the words 'revolution' and 'Putsch' became bywords. Adolph Kneser, for example, quoted Hilbert with apparent approval in his paper on Kronecker:[19] 'As a matter of fact, as Hilbert has said very aptly referring to a similar new movement, it was not a revolution, but a coup (*Putsch*) which had to collapse under the general strike of the analytical workers (*analytische Arbeiter*).' Kneser apparently expected a similar breakdown for the intuitionistic revolution.

[18]Cf. Poincaré (1905), Les derniers efforts des Logisticiens, Poincaré (1908) (livre II, Chaps. IV, V), Brouwer (1907), p. 172.

[19]Kneser (1925).

In 1923 Brouwer broke with his habit of producing scholarly, technical expositions. He published a paper in the *Jahresbericht*, which was the result of lectures at conferences. He lectured, first in Antwerp at the Flemish conference of science in August, and a month later at the annual meeting of the DMV in Marburg, on September 21, 'On the significance of the theorem of the excluded third in mathematics, in particular in function theory.'[20]

A paper with the same title appeared in the *Journal für die reine und angewandte Mathematik*, the journal that also had published Brouwer's dimension paper, and also in Dutch in the *Wis- en Natuurkundig Tijdschrift*; Brouwer must have had his reasons for not submitting it to his favourite journal, the *Annalen*. In the first section the principle of the excluded third[21] is discussed. For finite systems, Brouwer acknowledged the applicability of PEM: for properties deduced with the help of PEM 'always the certainty exists that, given a sufficient amount of time, one can arrive at the empirical confirmation'.[22]

Now, from the fact that many objects and mechanisms of the outer world can be viewed as 'finite discrete-system for certain known parts ruled by certain laws of temporal concatenation', one may conclude that the laws of logic, including PEM, are applicable, notwithstanding the fact that 'a complete empirical confirmation is mostly a priori materially impossible'. In view of this experience, logic and PEM were granted an a priori character. One particular historical consequence of this phenomenon was that PEM was applied in all of mathematics, without taking into account that the results did in general defy theoretical or empirical confirmation. Already in his 'Intuitionistic set theory' of 1919,[23] Brouwer had observed that PEM, as well as Hilbert's dogma, was an unjustified generalisation of the mathematics of finite systems. The present paper contained a more elaborate foundational analysis of this phenomenon.

At the end of the first section Brouwer very briefly mentioned the formalist consistency program. Considering Brouwer's earlier criticism, along the lines of Poincaré, the comments are surprisingly mild: one need not despair of reaching the ultimate goal of a consistency proof. He added, however, that such a proof would not amount to much:

> nothing of mathematical value will thus be gained: an incorrect theory, even if it cannot be inhibited by any contradiction that would refute it, is none the less incorrect, just as a criminal policy is none the less criminal even if it cannot be inhibited by a court that would curb it.[24]

[20] Brouwer (1923b, 1923f), translated in van Heijenoort (1967).

[21] Abbreviated as PEM, *principle of the excluded middle*.

[22] Brouwer (1923f), p. 2.

[23] Brouwer (1919h).

[24] In Fraenkel (1923), p. 239, Brouwer is quoted slightly differently: 'For intuitionists, however, 'consistency' has by no means the same meaning as 'existence', no more than 'a crime which' cannot be detected by prescribed means of investigation stops being a crime.'

A rather mild *badinage* that would not upset any modern reader. But did it upset Hilbert (if he saw it at all)? It would not have done much to endear Brouwer to him. In fact, Brouwer had formulated his objections to the formalist program more pointedly in his dissertation:[25] 'from the consistency of the axioms the existence of the accompanying mathematical system does not automatically follow', and 'it has not been proved anywhere that if a finite number has to satisfy a system of conditions, for which the consistency can be proved, that this number indeed exists'.

The paper does not, however, contain any personal attacks, in particular, Hilbert's name is not mentioned. The significance of PEM for mathematical practice is shown by means of 'Brouwerian counterexamples', that is, examples showing that the validity of some mathematical property would imply the solvability of a particular open problem for which no evidence exists. The standard open problem is the presence of a specific finite sequence of decimals in the expansion of $\pi$: 'if in the decimal expansion a sequence 0123456789 occurs, then...'[26]

Brouwer listed a number of classical theorems which could not be accepted in the intuitionist framework:

- The ordering of the continuum has the trichotomy property.[27]
- Every set is either finite or infinite.
- A continuous function on a closed interval has a maximum.
- The theorem of Heine–Borel.
- The theorem of Bolzano–Weierstrass.

Brouwer's lecture in Marburg was attended by a good many mathematicians, and the audience must have been surprised, to say the least, by the consequences of Brouwer's criticism.

Hilbert's Hamburg paper will have reassured most mathematicians that nothing was wrong with classical mathematics that could not easily be set right. The general impression was that Brouwer and Weyl would not recover from this blow, as is implicit in Fraenkel's letter to Brouwer of April 18, 1923: 'It was, among other things, most interesting for me to observe the active life of intuitionism which had already been pronounced dead in many quarters.'[28] It is no exaggeration to say that Fraenkel's books and papers were most influential in disseminating the facts about progress in the foundations of mathematics, in particular he reported in a neutral way on what was going on in intuitionism. The expanded version of his famous 'Introduction to set theory'[29] of 1919 was under preparation when Fraenkel met Brouwer,

---

[25]Brouwer (1907), p. 138 (footnote), pp. 141, 142.

[26]Cf. I.10.4, p. 392. Brouwer had picked a hard, although elementary, problem. Only in 1998 was the sequence 0123456789 found in the decimal expansion of $\pi$, Borwein (1998). Needless to say that the demise of a single counterexample does not weaken the intuitionist cause—there are enough new counterexamples.

[27]The trichotomy property asserts the comparability of any two elements: $x \leq y \vee y \leq x$. Brouwer's favourite formulation was 'the continuum cannot be ordered', Brouwer (1950, 1992).

[28]Cf. p. 386; see van Dalen (2000) for more on Fraenkel and intuitionism.

[29]*Einleitung in die Mengenlehre*, Fraenkel (1919).

and Brouwer was sufficiently impressed to assist Fraenkel with the proof reading of the second edition (1923). Fraenkel's exposition helped to put oil on the troubled waters. On the one hand it took intuitionism seriously and lent it the respectability that Hilbert's talk had tried to deny it, on the other hand it put the reader at ease—this revolution was not the only way to save mathematics. The mathematician Konrad Knopp voiced a widespread feeling in a letter to Fraenkel, 'I am glad to see in it a confirmation of my feeling that the 'destruction of the foundations' by Brouwer is by no means so disastrous as it seemed to be.'[30] Fraenkel himself had not given up his neutrality, he was well aware of the foundational virtues of Brouwer's intuitionism, but, like his fellow mathematicians, he was loath to exchange the fleshpots of Egypt for the desert of intuitionism:[31]

> After this excursion to philosophical grounds, we must go on to another group of dangerous revolutionaries, which is exclusively made up of mathematicians.
>
> This group is more dangerous, because the attack is carried out with much sharper, partly mathematically finely whetted arms, but also insofar as here it is not just a small border correction, aimed at the exposed province of set theory, at the cost of the mathematical empire, but the attack is carried into domains of this empire that are most flourishing and imagined as most safe.
>
> If it ultimately succeeds, then there will remain of present-day mathematics, apart from tightly bordered impregnable areas (in particular arithmetic in the narrow sense), only ruins, from which indeed through the labour of generations new, somewhat inhabitable (and not comparable in comfort to the old one) housing can be constructed.

Fraenkel's appreciation for Brouwer's ideas competed with an understandable attachment to his first love—Cantorian set theory. The author is, he wrote, 'emotionally and, in view of the factual successes of science, convinced that the far reaching intuitionistic amputations are not necessary', at the same time he recognised 'the present justification of a part of the intuitionistic objections'.[32]

After Hermann Weyl, Fraenkel was probably the best informed man where Brouwer's intuitionism was concerned. He had in fact attended Brouwer's lectures in Amsterdam, and discussed the foundational problems with Brouwer in private.[33] He was certainly not afraid to speak his mind, as one can see in his discussion of Hilbert's second paper,[34] where he points out that a particular argument of Hilbert, pertaining to PEM, 'touches on the final part of an older essay of Brouwer'.[35]

---

[30] Knopp to Fraenkel, 2.I.1924.

[31] Fraenkel (1923), p. 164.

[32] Fraenkel (1923), p. 173.

[33] See p. 386.

[34] Hilbert (1923).

[35] Fraenkel (1923), p. 239.

If there were mathematicians who thought that Hilbert had silenced the opposition, Brouwer's Marburg Lecture shattered that hope. Indeed, Brouwer pursued in relative quiet the founding of his intuitionistic mathematics. In Amsterdam he was virtually on his own. His first intuitionistic Ph.D. student, Maurits Belinfante, had defended his dissertation 'On infinite series'[36] on December 12, 1923. This dissertation treated the theory of infinite series from the intuitionistic point of view.[37] Brouwer developed in 1924 his new insights further in a number of papers.[38] In 1924 the relationship of Hilbert and Brouwer had not reached the point where contact became impossible. No direct comments of Brouwer on the situation have been preserved; he had no difficulty acknowledging the greatness of the mathematician who had stimulated his research and recognised his qualities; he bore his dissatisfaction in silence. As Weitzenböck wrote to Hermann Weyl, 'As far as Hilbert's matters are concerned, Brouwer for the time being shrugs his shoulders. I believe that he is somewhat annoyed that Hilbert ignores his things.'[39] Slowly and imperceptible the old relationship, which was characterised by Brouwer's outspoken admiration for the great mathematician, and the appreciation of an established authority for a gifted younger colleague, began to assume the Saul-David pattern. Brouwer, who was generally well-informed about the atmosphere in Göttingen, could not dismiss the rumours that his intuitionistic program estranged him from Hilbert. In a letter to Bieberbach Brouwer spoke of the 'smear campaign conducted by Hilbert' against him.[40]

Brouwer had occasion to feel personally the winter of Hilbert's discontent when he was invited to give a lecture in Göttingen before the Mathematical Society on July 22, 1924. The lecture was probably a sequel to his Marburg lecture, it bore the title 'Consequences of the intuitionistic position in mathematics'.[41] Brouwer addressed a large audience; Hilbert, against his custom, preserved an icy silence during the lecture, but when Brouwer had finished, he rose saying 'With your methods most of the results of modern mathematics would have to be abandoned, and to me the important thing is not to get fewer results but to get more results.'[42] He then sat down to an enthusiastic applause. And that was all Hilbert contributed to the discussion. Another member of the audience attacked Brouwer's constructive position on existence with the words 'You say that we can't *know* whether in the decimal representation of $\pi$ ten nine's occur in succession—maybe we can't know, but God knows.' Objections of this sort, which illustrate the fact that education and

[36]Belinfante (1923).

[37]Belinfante had not made use of Brouwer's new results, such as the fan theorem and the continuity theorem. From a modern point of view, one could say that he belonged to the Bishop school *avant la lettre*.

[38]Brouwer and de Loor (1924), Brouwer (1924a, 1924b, 1924c, 1924d).

[39]Weitzenböck to Weyl, 16.IV.1923.

[40]Brouwer to Bieberbach, 1924, undated. See p. 406.

[41]*Konsequenzen des intuitionistischen Standpunktes in der Mathematik.*

[42]Reid (1970), p. 184.

old habits make it easy to miss the point, were quite commonplace in the early days of intuitionism; obviously, Brouwer had a ready reply, 'I do not have a pipeline to God'. The occasion left Brouwer with a poor impression; after the talk he visited the seventy-five-year old Felix Klein and had a long talk with him.[43] In a letter of August 1 he wrote to Alexandrov 'In Göttingen there was a lot of interest for my subjects—I had an audience of roughly 150 persons, but the discussion remained rather superficial. [...] With Klein I had a discussion of several hours. He is completely lucid and (much more than the rest of the Göttingers) mathematically and personally broadly oriented and interested.'[44] One may read a mild disappointment in this letter. Already people were taking sides in the formalist-intuitionist discussion. Constance Reid reports that the majority of the Göttingen mathematicians did not see the point of Brouwer's criticism. In the words of Hans Lewy,

> It seems that there are some mathematicians who lack a sense of humour or have an over-swollen conscience. What Hilbert expressed there seems exactly right to me. If we have to go through so much trouble as Brouwer says, then nobody will want to be a mathematician. After all, it is a human activity. Until Brouwer can produce a contradiction in classical mathematics, nobody is going to listen to him.
>
> That is the way, in my opinion, that logic has developed. One has accepted principles until such time as one notices that they may lead to contradiction and then he has modified them. I think this is the way it will always be. There may be lots of contradictions hidden somewhere; and as soon as they appear, all mathematicians will wish to have them eliminated. But until then we will continue to accept those principles that advance us most speedily.[45]

Lewy's view was doubtlessly the prevailing one among working mathematicians, for whom foundational considerations were a luxury option. Only the most crude criterion, viz. consistency, appealed to them.

The finer arguments of Brouwer and Weyl, and even of their icon, Hilbert, must have been wasted on the mathematicians of the day. The modern distinctions based on effectiveness, complexity, degree of abstraction and the like, of which we know so much more than our forebears, have equipped us with a sensitivity for the finer traits of mathematics that one could not expect in the twenties. It is all the more surprising that mathematicians with an extraordinary vision, such as Weyl and Brouwer, could see beyond the first necessities. Nonetheless, Hilbert's 'we want more theorems' has a hollow ring that even at that time must have struck the more perceptive minds as dubious. After all, one does not want just *more* theorems, but rather good theorems, beautiful theorems, useful theorems. ... The history of mathematics abounds with disputes over the right theories, the right theorems, ... To be

---

[43]Since Blumenthal, Brouwer and Klein were at the time entangled in the unpleasant Mohrmann affair, there was no shortage of topics, see also p. 583.

[44]In the same letter he told of his strongly favourable impression of Kneser and Neugebauer.

[45]Reid (1970), p. 184.

fair, Hilbert would probably agree, but his reaction shows that nice catch phrases often get the better at public occasions.

The foundational discussion, soon to become a conflict, certainly had its serious observers, such as Bernays and Fraenkel, but there was also a measure of loyalty involved of the sort 'right or wrong, my country'. Hilbert, in particular, could inspire a large measure of loyalty; Brouwer was by nature a loner; his following remained modest to the point of marginal.

In the middle of Brouwer's hectic life, being occupied with his new intuitionism, the *Grundlagenstreit*, and dimension theory, a tragic event took place in his personal life. His close, and perhaps only, intimate friend, Carel Adama van Scheltema died. In March Carel had a fall that did not seem serious, but in April he suddenly suffered from dizzy spells, and on 6 May 1924 he closed his eyes after days of excruciating pain. A neglected concussion was conjectured to be the cause. On 9 May the socialist poet prince was carried to his grave in an impressive ceremony.[46]

Brouwer had lost not just a friend, but his spiritual brother, the man who had watched over Brouwer's mind, who had been his mentor and who, in spite of a difference in political outlook, had a magical and mystical rapport with him. In a terse homage he remembered his friend:

> It was the reflection in your eyes, the inflowing grip of your hand, the warm engulfing of your voice, the peaty smell of your overcoat. It was the wild wealth of your dream life, the confusing abundance of your imagination.
>
> But around you roams the power of determinacy that cannot be dismissed, which you sensed and had to acknowledge; and you were determined, and you wanted to understand, and to be a figure. You have understood much and you have become a figure. And a part of your flourishing rhythm became generally accepted.[47]

## 12.2 Consolidation and Entrenchment

The second half of the twenties was probably the most dramatic period of Brouwer's life, it contained episodes and moments of great exhilaration, but it ended in gloom and disaster. The threads of his various activities all came together in this short period, and the sheer complexity of it would have been enough to undo a lesser spirit. As it was, Brouwer had to pay the price at the end—the juggler had too many glittering balls in the air. We will deal with Brouwer's activities one-by-one, but the reader has to keep in mind that Brouwer was all the time playing simultaneously at a large number of boards.

There was his intuitionism that had to be fostered, topology that required his attention, the editorial business of the *Mathematische Annalen*, the fight against the

[46] Drost (1952), p. 26.

[47] Brouwer (1929b).

boycott of Germany, faculty politics, teaching, the *Grundlagenstreit*, the management and administration of his wife's pharmacy, significs, private life, etc.

During the twenties a steady stream of papers of Brouwer,[48] and later also of Heyting, started to build up a body of intuitionistic mathematics. Of course, this body could in no way match the existing body of traditional mathematics, but that was a point that did not worry Brouwer. In the first place, this was just 'work to be done', and in the second place large parts of traditional mathematics could, with proper care, be 'rescued'. Heyting's thesis and the function theoretic work of Belinfante were unmistakably the first steps towards a rebuilding of mathematics along intuitionistic lines, and one had to hope and trust that more would follow. When Brouwer delivered the laudation at the occasion of Heyting's Ph.D. defence in 1925, he asserted that the day had come that intuitionistic mathematics had proved to be viable, and to encompass the central parts of mathematics. He was convinced that his mathematics would eventually be largely accepted.[49]

Brouwer's personal research was more along fundamental lines. In particular he sought to fathom the nature and the consequences of the new notions that he had introduced. He had come to see the incredible complexity of the notion of choice sequence, and tried to analyse the notion further.

In his first *Begründungs*-paper the notion of choice sequence was left almost completely open. He spoke of 'choices of arbitrary digit-complexes'.[50] It may seem surprising that Brouwer used linguistic entities, such as 'sign', 'digit complexes', 'numeral' in his definition, as mathematics was for him independent of language. It was only in his 1947 paper 'Guidelines of intuitionistic mathematics' that he spelled out his intention: 'the word *symbol* (*Zeichen*)... must be understood in this definition in the sense of *mental symbols*, consisting in previously obtained mathematical objects of our thinking (*denkbaarheden*).' In his post-war publications[51] Brouwer spoke of 'sequence of natural numbers'.

Little to nothing, however, was said in 1918 about the choice process and its limitations. Only in 1925 did Brouwer add some comments, albeit only in a footnote:[52] 'Including the character of its freedom of continuation, which can narrow down after each choice (possibly until its complete determination, each time, however, according to a spread law).' By itself this footnote is not all that illuminating, after all a spread law, say the underlying tree of the spread, can at a certain point exhibit a path that has no branchings. This was not, however, what Brouwer had in mind. He wanted to allow the possibility that, although many numerical choices could be made, the subject might just stick to a law.

---

[48] 1921—1; 1923—4; 1924—7, 1925—2; 1926—2; 1927—3; 1928—2; 1929—1 (not counting the multiple versions).

[49] Interview, A. Heyting.

[50] *Ziffernkomplex*. In Brouwer (1925a) Brouwer used '*Nummer*'.

[51] e.g. Brouwer (1954a).

[52] Brouwer (1925a).

In one of his reprints a large number of corrections and additions has been inserted. One of these reads 'The *arbitrariness* of these "restricting-clauses" associated to a finite choice sequence, under preservation of the possibility of continuation, assigns to this choice sequence, and hence also to all its continuations, a new arbitrariness.'

One might well wonder what good higher order restrictions may do, and what kind of restrictions Brouwer had in mind. So far, we know only one example, it occurs in a letter of June 26, 1924 to Heyting: the completely free choice sequences.[53] In that case there is a second order restriction: there will be no future restrictions on the numerical choices.[54] The matter of the correct version of the notion of choice sequence kept Brouwer occupied; in 1942 the higher-order restrictions were explicitly mentioned in print,[55] but subsequently doubts set in; in the Cambridge Lectures[56] the higher order restrictions are still mentioned, but with a caveat: 'But at present the author is inclined to think this admission[57] superfluous and perhaps leading to needless complications.'[58] And a few years later he made a complete U-turn:

> However, this admission is not justified by close introspection, and moreover would endanger the simplicity and vigour of further developments.[59]

So as far as higher-order restrictions are concerned, Brouwer had in 1952 made the full circle.

Intuitionism was viewed by some as a threat to mathematics, an infectious disease that had to be stopped. Now, almost a century later, one has difficulty understanding all the excitement. In order to get a realistic view of the 'strategic' situation in the twenties, we have to review the development of foundational ideas and methods very briefly. As pointed out before, hardly anybody was aware of an intuitionistic program before 1920.[60] Only after Brouwer's talks at meetings (in particular of the German Mathematics Society) and Hermann Weyl's 'New Crisis' paper in 1921 did people start to take intuitionism seriously. Some considered it as a phenomenon that merited attention, simply because honest foundational ideas, such as those of Kronecker, Cantor, Poincaré, Russell, Brouwer and Weyl held promises or warnings for mathematics. Some considered it as an idiosyncrasy of a few mathematicians with a hang-up about paradoxes and the like, and, finally, some considered

---

[53] Strictly speaking one particular second-order restriction had been implicitly recognised all along, namely the restriction of all future choices to a law.

[54] These sequences were later independently introduced by Kreisel as 'lawless sequences', cf. Kreisel (1968). See also Troelstra (1982).

[55] Brouwer (1942a).

[56] Cambridge Lectures 1946–1950, Brouwer (1981).

[57] Of higher-order restriction.

[58] Brouwer (1981), p. 13.

[59] Brouwer (1952b), p. 142. Indeed, higher-order restrictions in general destroy not only the simplicity, but put the continuity principle in danger. See van Atten and van Dalen (2002).

[60] See Hesseling (1999).

Brouwer and Weyl a danger for mathematics and mathematicians. Among the latter Hilbert figured prominently. Already in the late nineteenth century Hilbert had run into opposition to his spectacular innovations. He had shown by abstract means that a system of invariants had a finite basis (nowadays known in algebra as *Hilbert's basis theorem*), in such a way that the number of elements could not actually be determined. The grand old man of constructivism, Kronecker, had been sternly critical of abstract mathematics, including Hilbert's contributions. Kronecker actually formulated a program for a reduction of mathematics (that is to say, algebra, number theory and analysis) to natural numbers. Kronecker's undisputed authority as a mathematician lent extra force to his attacks at 'modern' mathematics, and his shadow fell over mathematics long after his death. In a personal interview with Hilbert, the older man had told him outright that everything in algebra could be replaced by 'the discrete and singular', a remark that Hilbert never forgot.

In continuation of his beautiful work in geometry, Hilbert had started to study the axiomatisation of arithmetic (and set theory) in the hope and expectation to obtain consistency results that would remove the last doubts from mathematics. His first basic paper in this area was presented at the International Conference of Mathematicians in Heidelberg, 1904. It contained a, not wholly flawless, axiomatic study of the natural numbers and a consistency proof for a small fragment of arithmetic. The actual details of the system and the proof are not that important, what really counts is that Hilbert gave a combinatorial, metamathematical, consistency proof, that is, he applied mathematical techniques to a formal system. This paper[61] remained an isolated effort to carry the consistency tradition over from geometry to, say, arithmetic and analysis. As we have seen it came under a heavy attack in Brouwer's dissertation, but it remains to be seen to what extent Hilbert was aware of the fact. It is a matter of speculation whether he had a copy of Brouwer's dissertation (written in Dutch). Perhaps Brouwer sent him a copy at the time that he was writing his papers on Lie groups, or maybe after his walks with Hilbert in the dunes at Scheveningen,[62] where, as he wrote in his paper 'Intuitionistic reflections on formalism',[63] 'An oral communication of the first insight [the necessity to handle the contentual study of formal systems by means of 'the intuitionistic mathematics of the set of natural numbers'] to Hilbert had taken place in the fall of 1909 in several discussions'. The researches of Sieg[64] have shown that Hilbert did not abandon the foundations of mathematics after 1904. He lectured repeatedly on foundational topics in the years before he started to present his new program in public. An account of the development of his ideas over the years would provide a welcome historical clarification.

Interest in foundational matters had markedly increased in the post-war years. In addition to the fundamental papers of Brouwer, Hilbert and Weyl there had already been a number of more or less expository papers on foundational topics, but

[61] Hilbert 1904, reprinted in the later versions of *Grundlagen der Geometrie.*

[62] See p. 125

[63] Brouwer (1928b).

[64] Cf. Sieg (1999, 2000).

generally speaking these were not terribly exciting, or even well-informed. For example, Felix Bernstein extensively discussed 'Finitism in Mathematics', and lumped Poincaré, Richard, Borel, Lindelöf and Brouwer together as finitists.[65]

After the first manifestations of Brouwer and Weyl, Hilbert had returned to his formalist founding of mathematics. A perspicuous formulation of Hilbert's new ideas was given by Paul Bernays at the annual meeting of the German Mathematical Society in September 1921.[66] He reported on the conceptual progress that Hilbert had made since 1904. In the 1904 talk no serious attention was paid to the more refined metamathematical problems, e.g. a syntactical consistency proof was carried out by (a tacit) induction, without any discussion whether this constituted a reduction to a more basic theory. In the meantime, possibly recalling the discussions with Brouwer, Hilbert had grasped the significance of the meta-mathematics, i.e. the mathematics employed in the consistency proofs. The problem was to justify (or to show the consistency) of a mathematics dealing with infinite objects, such as sets of natural numbers, real numbers, and the like. These objects and their theories were called, for obvious reasons, transcendent. Now Hilbert proposed to handle the mathematics of these transcendent theories by simple means, 'to justify those transcendent assumptions in such a way that only *primitive intuitive insights* (*Erkenntnisse*) *will be employed*'.[67] In this way 'A sharply outlined, comprehensible program has taken the place of the earlier, indeed obscure, indications.'[68]

Bernays quite correctly observed that if one wants to base metamathematics on primitive intuitive notions and objects, one has to take into account Brouwer's considerations, and to question the applicability of PEM. Hilbert, said Bernays, wanted to exploit the constructive features of mathematics rather than give them up,—'More fundamental than the contact with symbolic logic, is the fact of Hilbert's enterprise touching the constructive theories of Weyl and Brouwer.'[69]

The paper of Bernays is a cogent and objective exposition of what was to become 'Hilbert's program'. The problems of the program are duly mentioned, and clearly explained. For example the problem that a consistency proof will doubtlessly require mathematical induction. But in contrast to Poincaré and Brouwer he does not view that as an insurmountable obstacle. With Hilbert, Bernays is convinced that one only needs a weak form of induction. That is to say, one where only simple combinatorial properties come into play, such as 'if the sign + occurs in a concrete proof, then one finds, reading through the proof, a place where it occurs for the first time'. There seemed to be a feeling that if the subject matter is simple (here sequences of symbols), then one only needs simple applications of induction, generally speaking, a simple question will find a simple solution. Indeed, the whole 'pre-Gödel' era of

[65] Bernstein (1919), pp. 63–78.

[66] Bernays (1922), submitted 13.X.1921. Translations of many of the key publications of this period can be found in Mancosu (1998). See also van Heijenoort (1967).

[67] Bernays (1922), p. 11.

[68] Ibid. p. 10.

[69] Ibid. p. 15.

Hilbert's program is characterised by a contagious optimism: just a few technical problems to be solved and we are done. In retrospect one can only conclude that Hilbert and his collaborators had an overly rosy view of the project.

It is curious that Hilbert could not see, even for the sake of the argument,[70] the point of Brouwer's program; after all, one would expect that a mathematician should have no difficulty appreciating that questionable means of proof could not yield reliable results. Admittedly, it was not easy to unravel the foundational ideas behind Brouwer's early papers, but Weyl's presentation should have done much to compensate for that shortcoming. Moreover, one would expect some discussions on the topic in Hilbert's Göttingen. But with Weyl in Zürich and Brouwer in Amsterdam, who would have the courage and insight to play the devil's advocate? Bernays could have done it, but he was not the person to stand up to Hilbert, even as an exercise in purely academic argument.

The whole conflict that started with Weyl's manifesto and Hilbert's reply is marked by an unwillingness of both parties, Brouwer and Hilbert, to contemplate the virtues of the other side; from the people involved only Hermann Weyl managed to take a moderate position between the parties. While Brouwer considered Hilbert's formalism an enterprise doomed to failure, even if it succeeded (see p. 442), Hilbert saw Weyl and Brouwer as iconoclasts of the worst kind.

Somehow Hilbert managed to miss the point of the new intuitionistic mathematics. To mention two of Hilbert's examples—the status of the irrationals and of the number theoretic functions—it is certainly true that in the first intuitionistic period, beginning with the dissertation, the irrational numbers were not always recognised as a set, but from 1919 onwards the irrationals had found a place in constructive mathematics.[71] The rejection of the general notion of number theoretic function too was a thing of the past. Brouwer's choice sequences were precisely designed to allow the set (spread) of *all* numerical functions (infinite number sequences). Of course, many a traditional theorem failed to make it to the intuitionistic practice, and this was what annoyed Hilbert. Brouwer, for his part, had once and for all abjured the axiomatic method; that is to say, in its extended sense, not as a convenient tool for daily practice, but as a basic tool for capturing and creating the true mathematics. He had summed up his objections in his dissertation and they had not been met by the proponents of axiomatics, not even by the best principled ones, including Hilbert.[72] And so he saw no reason to change his views. From the available material one may infer that Brouwer was reasonably well informed about the views and prospects of the formalists, in the dissertation he showed a profound knowledge of the contemporary foundational situation (with the exception of Frege's work) and in

[70]That is to say, Bernays pointed out in his writings, cf. Bernays (1922), that Hilbert did not lack in appreciation for constructive metamathematics. For some reason Hilbert just did not see, or did not wish to see, his intellectual debt to Brouwer and Weyl.

[71]The terminology is anachronistic; Brouwer started to use the name 'intuitionism' for his program after World War I.

[72]Brouwer (1907), Chap. 3.

the twenties the real formalist innovations were rather of a technical than philosophical nature—no reason for Brouwer to contemplate drastic changes in his position.

The *Grundlagenstreit* is a typical example of one of those scientific tragicomedies, where both parties run around the stage, shouting but not listening. In particular Hilbert acted as a headmaster dealing with two protesters who threaten to set the building on fire. Indeed, Hilbert viewed Weyl and Brouwer as real threats to the survival of mathematics; being the leader of contemporary mathematics, he felt it his personal duty to save mathematics and to thwart the efforts of the evil pair. And he made this perfectly clear in his lectures and publications (cf. p. 439).

From 1922 onwards Hilbert and Brouwer exchanged regular blows like legendary knights of the past. In 1922 Hilbert addressed the *Deutsche Naturforscher Gesellschaft* in Leipzig with a paper titled 'The logical foundations of mathematics'.[73] This paper constituted real progress with respect to the preceding one; it contains the idea to reduce mathematics (arithmetic) to a small fragment, i.e. finitistic mathematics. Furthermore it introduced the $\tau$-operator, a forerunner of the $\varepsilon$-operator. This paper was totally uncontroversial, it abstained from diatribes against third parties.

While Hilbert was defending mathematics in his lectures and papers, a good deal of the painstaking research in proof theory was done in Göttingen by his modest, even shy, assistant, Paul Bernays. From 1917 onwards Bernays had worked in Göttingen on various topics in mathematics. But gradually his sole task became the study of Hilbert's metamathematics. It is no exaggeration to state that without him (or a suitable replacement) the project would have come to nought, the intricacies of logic and proof theory, which nowadays are more or less routine, had to be explored, and all consequences had to be painstakingly checked. Eventually Bernays moved to Zürich and became his own master, somewhat to his surprise, but in the twenties he was the quiet man behind Hilbert's grand schemes. The relation between the two was unmistakably that of master and servant. The historian of mathematics, Dirk Jan Struik, told an anecdote illustrating the relation between Hilbert and Bernays; Struik was once spending some time in Göttingen with a Rockefeller stipend, and one day he asked Bernays to accompany him on a Sunday morning walk, 'All right', said Bernays, 'but I first have to get permission from *Herr Geheimrat*'.[74] The collaboration between Hilbert and Bernays culminated in a two volume treatise, the famous *Foundations of Mathematics*,[75] finished after Bernays had to leave the country in 1933.

Paul Bernays was indeed one of the most saintly persons one can imagine, he was wholly incapable of harming anybody—a man of an absolute and unassailable integrity. During the Brouwer–Hilbert conflict he managed to preserve good relations with both adversaries, and also in later years Brouwer and he were on quite cordial terms. There is no doubt that Bernays appreciated the force of the arguments from

---

[73] Hilbert (1922).

[74] Cf. Rowe (1989).

[75] *Grundlagen der Mathematik I* (1934), *II* (1939).

both sides, unhindered as he was by personal antagonism, or a 'save mathematics'-complex.

Both parties in the conflict were in 1922 busily investigating their respective projects. Brouwer had not yet achieved his great breakthrough, but he was making an inventory of the damage sustained by classical mathematics. He knew that a good part of traditional mathematics had to be revised and that another part had to be thrown overboard. From 1921 onwards he had lectured on the theory of real functions (a practice he kept up until 1929), and this provided ample opportunity to test out the destructive consequences of his new intuitionism.

In spite of Hilbert's attacks there was not yet an open conflict. Actually the two adversaries, Brouwer and Hilbert, hardly ever met during the foundational debate. Whereas in certain disciplines and certain debates the opponents got together to try to decide the matter in a personal exchange of thought (think of the Bohr-Einstein debate), Brouwer and Hilbert probably did not see what could be gained from a personal discussion. The *Grundlagenstreit* was conducted like some sort of a match, the players would in turn address some conference or seminar, present the progress in their own project and return to their respective desks.

We have already seen that Brouwer unfurled his banner on German soil in 1923 at the Marburg meeting of the DMV.[76] And a year later he addressed the Göttingen mathematicians.[77] Hilbert had lectured on his new program in Copenhagen (1921), Hamburg (1921) and Leipzig (1922).

The next episode in the *Grundlagenstreit* took place in Münster; the *Westfälische Mathematische Gesellschaft an der Wilhelm-Universität* organised a Weierstrass-week from June 2–6, 1925. The meeting offered a select group of speakers: Mittag-Leffler, Bieberbach, Hilbert, Perron, Koebe, Knopp and Weyl.[78]

Hilbert presented at the Weierstrass meeting his famous talk 'On the infinite and the founding of mathematics',[79] a lecture that presented his progress in proof theory and which also contained an attempt to solve the continuum problem. Again he used the occasion to point out the futility of the objections from various quarters. He pointed out that he was going to provide the ultimate security of the mathematical method and he demonstrated (probably genuine) astonishment at 'the old objections, that were believed to have been renounced long ago, which reappeared in a new cloak. So at present something like this is put forward: even if the introduction of a notion were possible without danger, i.e. without yielding contradictions, and this could be shown, then this would not yet establish its justification.' One hears a faint echo of Brouwer's words. But where Brouwer saw the justification of theories

[76]See p. 442. Brouwer's Nauheim lecture predates the *Grundlagenstreit*; it is by all standards a totally innocuous academic treatise on effective procedures, without programmatic claims. Nobody could possibly take exception to it. Therefore I will not view it as part of the foundational conflict.

[77]See p. 445.

[78]Curiously enough Brouwer was in Münster on June 1 and 2. At least that was what Brouwer announced in a letter to Gerda Holdert, 14.V.1925: 'I have to be in Münster on June 1, 2.' He may or may not have gone to Münster after all, but it remains a curious coincidence.

[79]*Über das Unendliche und die Begründung der Mathematik.*

in mathematical constructions, Hilbert saw it in the ultimate success of the practice of these notions. 'Indeed, success is necessary, it is here the highest authority (*Instanz*), to which everybody bows.'[80] The clarification of the notion of infinite was for Hilbert something that transcended the sciences by far: it was necessary for 'the honour of human understanding itself'. Indeed, Hilbert shows himself here fully modern, possibly more progressive than his younger audience; he was prepared to uphold set theory to the ultimate consequences, 'From this paradise that Cantor gave us, nobody will be able to expel us.'[81]

The comments on the notion of infinite are quite balanced, and give credit to the two basic notions: potential and actual infinite. One can even hear an echo of Brouwer: 'in analysis we have only to deal with the infinitely small and the infinitely large as limit notions, as something in growth, arising, generated, (*etwas Werdendem, Entstehendem, Erzeugtem*)...'[82]

There is also a rather remarkable reprimand for the finite-number-of-steps people: 'The mathematical literature finds itself, if one takes notice of it, inundated by absurdities and lack of thought, mostly due to the infinite. Thus, for example, if in the sense of a restrictive condition it is required that in the exact mathematics only a finite number of steps are allowed in a proof—as if somebody ever could succeed in carrying out infinitely many steps.' No doubt Hilbert meant well, but this example looks somewhat like the teacher who scolds a pupil who has just given the right answer, because he might have thought of the wrong answer.

The paper was, on the whole, rather more expository than polemic, it did not call the adversaries by name (Weyl, anyway, no longer figured in the debate) and it radiated a spirit of confidence and enthusiasm, as for example expressed by the words 'And what we experienced twice, once when it concerned the paradoxes of the infinitesimal calculus, and then when it concerned the paradoxes of set theory, that cannot happen to us a third time and will never happen again.' It is a spirit of progress and of the ultimate security of mathematics that is within reach: 'Our proof theory which is sketched here, can not only secure the foundations of the mathematical sciences, but I believe that it opens up a route by which general problems of a fundamental nature that fall under the domain of mathematical thought can be treated, that could not be got at in the old days.' Hilbert saw mathematics already as the supreme judge of fundamental issues, and, he added, 'Even the claims of the recent so-called "intuitionism"—modest as they may be—will in my opinion get their authorisation from this tribunal.' The mathematical-logical content was also more substantial than that of earlier papers.

The paper, again the text of a lecture, whatever its shortcomings, still leaves the reader with a strong impression even today. It is the inspiring talk of a general to his troops in the face of a dangerous enemy. It radiates strong confidence and it contains

---

[80] Hilbert (1926), p. 163.

[81] Ibid. p. 170.

[82] The fact that Hilbert accepted the modest constructive approach: potential infinite for *cardinalities* in the style of Brouwer 1907, of course does not imply that he would accept, or even contemplate, choice sequences.

not just the pep talk, but an outline for concrete foundational work and progress. If one would like to make a critical remark at all, it would be that as before the content consists of sketches, expectations and promises. After this paper the exchanges were suspended for a while.

Hilbert's paper is indeed rich in ideas, it contains for example the notion of 'finite attitude of thought' (*finite Einstellung des Denkens*). The idea had crossed Hilbert's mind already in 'The logical foundations of mathematics' (1922); it is explicitly mentioned in his next paper:[83]

> Our thinking is finitary;[84] while we are thinking, a finitary process takes place. This truth, acting by itself, is more or less used in my proof theory in such a way that the experience of this contradiction should automatically bring with it the choice from among the infinitely many objects. In my proof theory it is therefore not claimed that the selection of an object among infinitely many can always be carried out, but rather that one can act, without the risk of an error, as if the choice had been made.

It is worthwhile noting that Hilbert did not make excessive claims about the effectiveness of this finitary attitude.

In 'On the infinite' the fundamental ideas of the notion of 'finitary' is further worked out. Finitary arithmetical statements are for example $2 + 5 = 7$ or $12 > 3$, and even $n + m = m + n$.

The latter statement roughly means, if we take for $m$ and $n$ concrete numbers, then a finitary check will establish the truth. However, $\exists x A(x)$ for some finitary statement $A(x)$, for example, is no longer finitary. It involves an infinitary operation, namely unbounded search. A finite set can be checked element by element, albeit highly idealised. But for an infinite set no such good fortune can be expected. One consequence of this insight is that finitary statements cannot be negated, or more precisely: if $A$ is finitary, then $\neg A$ need not be so. Hilbert formulated the case of the existential statement as 'In general, an existential statement of the form: there is a number with such and such a property, has from the finitary point of view only a meaning as a *partial statement*'. There are striking parallels between Hilbert's ideas about finitary statements and Hermann Weyl's earlier 'judgement abstracts'. More historical research will be required to clear up the relation. In the sense of 'who published first', Weyl's priority is clear, but it is very well possible that Hilbert had already been discussing the issue in his lectures. Although Weyl's New Crisis-paper was known to Hilbert, there is no reference to Weyl in Hilbert's infinity-paper.

Hilbert concluded quite correctly that if we restrict ourselves to the realm of finitary statements—'as we should'—then 'those laws of logic, which mankind has been using all the time, since it started to think, and which indeed Aristotle has taught, are not valid'. But, he continued, 'one could set oneself to formulate valid logical laws for the domain of finitary statements; but that would not serve us, as

---

[83] Hilbert (1923), p. 160.

[84] Hilbert used the German '*finit*' instead of '*endlich*'. Following Kleene, the term finitary has been adopted in the foundations of mathematics.

indeed we do not want to forego the use of the simple laws of Aristotelian logic, and nobody, though he might speak with the tongues of angels, will keep man from negating arbitrary statements, to formulate partial judgements and to apply the Tertium non datur'.

It is apparent that Hilbert had moved quite a bit in the direction of the intuitionists; given the basic finitary nature of the operation of our thinking, he agreed that the logic of finitary statements is really the appropriate one. In this sense he actually was a sub-intuitionist. However, he took the pragmatic view of mathematics, and sought to salvage the rest by logic. The solution he had in mind is imaginative, he compared the situation to the old geometrical practice of interpreting the projective plane (or space) in terms of the affine one. Given the affine plane, one defines points at infinity in terms of pencils of parallel lines. The old points of the affine plane are called *real points* and the new points *ideal points*.

In this sense one obtains ideal statements from the finitary ones by unrestricted quantification. Hilbert then went on to formulate his new version of the consistency proof:

> For there is one condition, just one, but indeed an absolutely necessary one, with which the method of the ideal elements is tied up, and this is the consistency proof: the extension by means of the adjunction of ideals is in fact only permissible if it causes no contradiction in the old domain, that is, if the relations for the old objects that result at the elimination of ideal objects are always valid in the old domain.[85]

The great innovation of the paper was the introduction of the choice operator, $\varepsilon$, the embodiment of the 'thus far in the mathematical literature most challenged so-called axiom of choice'. The $\varepsilon$-axiom is in Hilbert's formulation $A(a) \rightarrow A(\varepsilon A)$. It in fact eliminates the $\exists$-quantifier: $\exists x A(x) \rightarrow A(\varepsilon A)$. This indeed was a brilliant stroke, by getting rid of the quantifiers one could concentrate on much simpler formulas. However, the complications now shifted to the $\varepsilon$-operator, and these difficulties were just as complicated as the old ones. This was something that further research would indeed show. In sophisticated areas, such as proof theory, free lunches are rare.

In September Hilbert repeated the same lecture in Copenhagen, where it found universal acclaim. Courant reported that 'your Münster lecture in Copenhagen came as a real bombshell and was received from all quarters with the greatest enthusiasm'. He added, 'It did me also an exceptionally great pleasure to see that finally the outsiders begin to understand that your theory saves the situation (*dass mit Ihrer Theorie das erlösende Wort gesprochen ist*). I believe that your lecture will really have an enormous influence in the most diverse directions.'[86]

---

[85] Hilbert (1926), p. 179. A more modern version would run roughly as follows: if a finitary statement is provable in the extended system, then it is already provable in the finitary system (the extension is *conservative*).

[86] Courant to Hilbert, 10.IX.1925.

## 12.3 The Riemann Volume

In the middle of the *Grundlagenstreit* there is an intermezzo that has nothing to do with foundational issues, but which widened the gulf separating Brouwer and Hilbert. The issue involved the old bogeyman, the *Conseil*, and the boycott of German scholars.

Mathematics knows a number of men of unusual genius, whose life and career was tragically cut short, at an early age. Among these was the nineteenth century mathematical genius, Bernhard Riemann. He died of consumption at the age of forty, at the height of his mathematical career. It is no exaggeration to say that his ideas transformed function theory, differential geometry and topology. He also had a tremendous and inspiring influence in the foundations of mathematics through his monograph *Ueber die Hypothesen welche der Geometrie zu Grunde liegen* (1867), which dominated the discussions in geometry far into the twentieth century under the name of the *Riemann–Helmholtz Raumproblem*.

As a tribute to Riemann's genius, the *Mathematische Annalen* had decided to dedicate a volume to Riemann's memory at the occasion of the commemoration of his hundredth birthday in 1926. The usual plans were made by the editorial board, the main issue being whom to invite for a contribution. This seemingly innocuous question raised problems that from our vantage point seem irrational and childish. In 1924, when the matter was raised, German scientists were still the outcasts of the international scientific community and in reaction to the official boycott proclaimed and enforced by the *Conseil international des Recherches*, (cf. Chap. 9) the German scientists had answered with a counter boycott. Professional organisations were wary, to say the least, of contacts with parties from the Allied countries. Efficient as ever, a network of organisations and persons had been put together for the purpose of protecting and furthering the interests of German science.

There were a number of organisations that occupied themselves with the promotion and protection of German science and scientists; the two that concern us here were the *Reichszentrale für naturwissenschaftliche Berichterstattung* under Karl Kerkhof and the *Notgemeinschaft der Deutschen Wissenschaft*, directed by F. Schmidt-Ott. The *Notgemeinschaft*, founded on 30 October 1920, was a corporation of the five academies of the *Kaiser-Wilhelm Gesellschaft* and the universities, institutes for technology, etc.[87] It was essentially a scientific and not a political enterprise. All the same it was a reaction to the boycott, and its members may have fostered anti-*Conseil* sentiments, but its main business was the support of actual scientific activities. It is interesting to note that the two deputy presidents were chosen from the exact sciences: W. von Dyck, the mathematician from Munich, and Fritz Haber, the famous chemist from Berlin. The *Reichszentrale für naturwissenschaftliche Berichterstattung*, an official organisation for the promotion and support of the exact sciences had a strong political role, that is to say, it occupied itself intensely with the foreign relations of German science.

---

[87]Cf. Schroeder-Gudehus (1966), p. 199, Schmidt-Ott (1952).

The head of this bureau was Dr. Kerkhof, a man with an impressive international network.[88] In the years of the Weimar Republic, resentment ran high in certain academic circles; almost every occasion was used to indulge in self-pity and in anti-*Conseil* sentiment. We have seen that the *Conseil* and its various branches did not excel in the virtue of brotherly love, and they abhorred the sinner just as much as his sin. Even worse, the allied wrath was equally aimed at sinners and innocents. In 1924 the recollections of the war and the subsequent peace treaty of Versailles was fresh enough to turn the average German into a nationalist. Under these circumstances the matter of inviting foreign scholars to contribute to a Riemann volume became an apple of discord.

Brouwer, when invited to submit a paper for the Riemann volume, replied that he would be happy to do so, he added that he agreed with the proposed list of contributors, with one exception: Painlevé.[89] This was not a wholesale refusal of French contributors, indeed the second Frenchman on the list, Hadamard, was perfectly acceptable to Brouwer. To underscore his objections he went on to list a selection of Painlevé's offensive characterisations of Germans and more in particular German scientists.[90] In his eyes Germany would make itself the laughing stock of the world if the man who, in his functions of president of the *Académie des Sciences* had heaped such insults on German science, were invited to contribute to the Riemann volume. He invited Blumenthal to send copies of his Painlevé quotes to the editorial board, so that none of the members would even contemplate the possibility of this invitation.

Blumenthal, a man of peace, did not know how to handle this letter, so he asked another managing editor, Carathéodory, to reply to Brouwer.[91] Carathéodory admitted in his letter to Brouwer[92] that he shared Brouwer's opinion, although he wondered 'if the nonsense, that was put together in all countries, should be recalled all the time, then it would not have been necessary to stop the shooting'. He did not see why the Riemann volume would need any French contributors at all, but if one wanted a French author, then Painlevé was precisely the man to invite. Only a man with his mathematical, but even more, political reputation could take part in the Riemann volume without getting the yapping pack of barbarians after him.[93] Painlevé seemed to have mellowed considerably towards the old enemy, as was reported by the physicist Nernst, who had told that Painlevé had not so long ago offered to lecture in Berlin. Even the minister of foreign affairs had welcomed Painlevé, but some

[88] See p. 332.

[89] Brouwer to Blumenthal, 1.XI.1924.

[90] Cf. p. 329.

[91] The editorial board contained two kinds of editors: the proper editors, listed on the cover as '*Herausgeber*' (publishers), and associated editors, listed under the heading '*unter Mitwirkung von*' (with the co-operation of). When relevant, we will call the first kind 'managing editors' and the second kind 'associate editors'.

[92] Carathéodory to Brouwer, 6.XI.1924.

[93] '*ohne Gefahr zu laufen, von der ganzen Meute der Banausen angebellt zu werden.*'

Berlin colleagues had successfully resisted this overture. 'You see that, at least according to this report, Painlevé seems to have forgotten the words that you charge him with.'

Blumenthal was in an awkward predicament, it is more than likely that he was not a free agent in this matter, and that he was acting under instructions of Hilbert. So he could not just drop the matter. Given the sensitivity of the matter he turned to Einstein, who was also a managing editor.[94] Explaining the situation he suggested a solution along the lines of Carathéodory. Painlevé, he thought, was no longer an active mathematician and the invitation would be just pro forma. But with Painlevé's moral blessing one could safely invite Hadamard. In view of Brouwer's hard-line views, he added, the managing editors, Hilbert, Carathéodory, Einstein and Blumenthal, would have to take a decision in the matter. Not wishing to upset the sensitive Brouwer, he wondered if Einstein could give his opinion on the proposed procedure. Einstein was perfectly willing to act in this matter on behalf of the *Mathematische Annalen*. 'Painlevé', he wrote, 'is a man who has for years been active in cleaning up the spiritual atmosphere. He may have erred; yet he is of a sort of which we unfortunately have none, or not enough.'[95] In his opinion Carathéodory was quite right, one could only approach other French mathematicians through Painlevé. In the case of Hadamard such a detour was certainly advisable, 'for Hadamard is timid and something of a Germanophobe, not at all with a free view in matters that are far from his science'. He promised to approach Painlevé via Langevin.

Blumenthal had, however, been too optimistic, thinking that only Brouwer needed pacifying. In January Bieberbach (one of the associate editors) sent him a letter with a formal motion to renounce a decision taken by the editorial board, when meeting at the occasion of the annual meeting of the German mathematicians in Innsbruck in 1924.[96] This so-called Innsbruck resolution (*Innsbrucker Beschluss*) is referred to at a few places, it probably dealt with the organisation of the Riemann volume.

After reading Brouwer's note, Bieberbach had come to the conclusion that an invitation of Painlevé was out of the question. A day later Otto Hölder joined the dissidents; the matter had more or less escaped his attention, he wrote, but even without reading the statements of Painlevé, he had strong objections to the participation of French or Belgian mathematicians.[97] Somewhat later Walter von Dyck from München sent a violent protest to his fellow editors. Einstein spoke of 'a rabid letter from München, containing a protest against the participation of Frenchmen in the *Festschrift* [i.e. the Riemann volume].... Who is the author of this letter anyway?'[98] After quoting some objectionable statements of Painlevé, von Dyck had concluded that 'I cannot imagine, although Brouwer reports to the contrary, that

---

[94]Blumenthal to Einstein, 15.XII.1924.

[95]Einstein to Blumenthal, 16.XII.1924.

[96]Bieberbach to Blumenthal, 12.I.1925.

[97]Hölder to Blumenthal, 13.I.1925.

[98]Einstein to Blumenthal, 20.I.1925.

after learning this monstrous insult, anyone would contemplate the contribution of Painlevé even for a moment!' Not content with barring Painlevé from the Riemann volume, he went even further: foreigners could only be admitted if there 'was absolute proof of their pro-German attitude and activity'.[99] The letter was dated '*am Tage der Gründung des Deutschen Reichs, 18 Januar 1925*' (on the day of the founding of the German empire). Von Dyck was not alone in his outrage over the anti-German feelings and actions following the war, sentiments of this sort were not all that exceptional. In a covering letter to Bieberbach, von Dyck expressed his vexation over the procedure for the Riemann volume, and over the *Annalen* in general. In particular he wondered about the role of Einstein, who 'at that memorable stay in Paris had declared that he was not a German!' He added that anyway 'one believes rather to be in Hungary, Poland, Russia and Bulgaria, reading the names of those who publish nowadays in the Annalen'. This side remark is a small illustration of the frustration of an older scholar who, quite apart from his nationalistic feelings, finds it difficult to accept the natural influx of talent from rising nations. On this point Brouwer was without prejudice, as we have seen he warmly encouraged the young topologists from Russia and Poland to publish in the *Annalen*.

Von Dyck's letter was not the only worry for Blumenthal, Carathéodory also had unpleasant news for Blumenthal: his letter had never reached Brouwer, as he found out from Brouwer a few days ago.[100] So now the whole matter had gone wrong. He added that he found 'the solution that Frenchmen will not be invited at all, quite good. I only objected to trying to invite the co-operation of the French before contacting Painlevé. It seems to me, however, that the resumption of normal relations should be left to a later generation.'

It was Bieberbach who suggested a compromise. Although he has gone down in history as an evil man,[101] it should not be forgotten that he was an active member of the mathematical community, with a talent for organising and with excellent social graces. He certainly was not after a big conflict in the *Mathematische Annalen*. So his efforts in the present matter should be viewed as sincerely directed at the interest of the *Mathematische Annalen*. Since Einstein had learned through Langevin that Painlevé had reacted favourably towards a co-operation in the matter of the Riemann volume, the following approach seemed promising: Einstein could gratefully acknowledge Langevin's assistance, and ask him to see to it that suitable contributions would reach him in time. No special invitations would be sent out, and thus a special invitation to Painlevé could be avoided. Bieberbach guessed that the editors might agree to this.[102]

Blumenthal apparently was happy to adopt Bieberbach's proposal, for he wrote on January 31 to Einstein that the Bieberbach proposal could avoid a conflict in the editorial board. If Einstein agreed, he could submit the plan to Hilbert. On the 19th

---

[99] '*wenn absolute Beweise deutschfreundlicher Gesinnung und Betätigung vorliegen*'.

[100] Carathéodory to Blumenthal, 20.I.1925.

[101] Cf. p. 627.

[102] Bieberbach to Blumenthal, 23.I.1925.

of February Hilbert informed the editors of the *Mathematische Annalen* of the unanimous decision of the managing editors. Basically the Bieberbach plan was adopted; to avoid misunderstanding he added that foreign mathematicians who submitted contributions, thus indicating their interest and position, should not be excluded on grounds of their nationality.

This almost settled the matter, had not Bieberbach called Hilbert's attention to a few awkward points.[103] In view of the fact that a number of editors were opposed to French participation, he asked Hilbert to have the decision approved by not just the managing editors, but by all editors—'I must fear that otherwise this matter will seriously interfere with the otherwise good co-operation'. The implicit purpose of his proposal had been to discourage French submissions, and possible French authors, as far as possible. Furthermore the blanket invitation of French mathematicians via Langevin would result in an unequal treatment of German and French mathematicians, for in Germany only a few mathematicians were selected. He had discussed the point with Einstein, and the latter was prepared to explain the predicament to Langevin, and to inform him that French contributions were not welcome. So it would probably be best to accept Einstein's offer.

It seems rather uncharacteristic for Einstein to suggest such an inelegant rebuff of his French colleagues. Whatever Einstein may have had in mind did not matter, for exclusion of the French was completely unacceptable for Hilbert. He wrote in no uncertain terms to Einstein and Bieberbach, declaring that the managing editors were after all responsible for the policy of the *Mathematische Annalen*, so it was only proper that they should take the decision. A unanimous decision of all editors would have been even better, but was not to be expected. He did see the point of the invitation to the French mathematicians. There would probably not be that many papers submitted by Frenchmen. Moreover, in the end it was the board of managing editors that decided which papers to accept. For Painlevé and Hadamard, of course, an exception should be allowed. Both had 'given spontaneous and clear evidence that they respect us and want to co-operate with us'. They should not be turned down on account of their nationality. '*This decision is for me irrevocable, and I would like to ask you to act accordingly*' (underlined by Hilbert in the letter to Einstein), and to Bieberbach he added that the decision was in the interest of Germany at large and of the German mathematicians.

The decision of the managing editors had effectively ended the discussion. When the volume appeared there were no French contributors; was there still an aversion to German scientists and their institutions? The volume contained only 14 papers and there were 9 nationalities involved. Four editors published in the volume: Brouwer, Carathéodory, Einstein and Hilbert. The affair did not attract much attention, it remained in the seclusion of the editorial board, where it had come up.

Brouwer had more or less triggered the crisis, and probably made certain that the problem was taken seriously. Hilbert gave in when confronted with a compromise organised by Bieberbach and Blumenthal. The whole affair could hardly have endeared Brouwer to Hilbert, indeed later developments confirmed that Hilbert could

---

[103] Bieberbach to Hilbert, 25.II.1925.

neither forget, nor forgive Brouwer's interference.[104] The Riemann volume could have served to bring a rapprochement between the French and the Germans, or at least it could have broken the unity of the anti-German front in France. It was an opportunity missed, but it probably was too early. More time was required.

## 12.4 International Relations

Brouwer's involvement in the discussion on the Riemann volume was in the end a corollary of his struggle for the re-integration of German scientists into the international community. In chapter 9 we have seen that Brouwer went to great lengths to set a real or imagined injustice right. His crusade against Denjoy was carried out with his usual tenacity, but without tangible result. There are no documents to trace Brouwer's anti-*Conseil* activities, but we may be convinced that like a slumbering volcano, Brouwer's resistance went on under the surface. After the policy for the Riemann volume was determined, the *Conseil* again asked Brouwer's attention. He had already been in contact with Karl Kerkhof and his *Reichszentrale für naturwissenschaftliche Berichterstattung*. The year 1925 was going to be a crucial year for German science; there were signs that the wind was changing, some members of the *Conseil* had enough of the French-Belgian supremacy and the rumours of an action to invite Germany to join gathered momentum. Kerkhof was apparently mustering his forces.

Brouwer's role in the affair of the Riemann volume is just one illustration of his implacable opposition to the *Conseil* and its boycott of German scientists. He had fought Denjoy, and when his own academy and his mathematical society had compromised themselves by joining the *Conseil* and the *Union mathématique*, he had conducted a campaign to get the *Wiskundig Genootschap* to sever its ties with the *Union*. In the meeting of February 24, 1923 Brouwer, Mannoury and Hk. de Vries tabled a motion to the effect that the *Wiskundig Genootschap* withdraw from the International Mathematical Union. Brouwer argued that by remaining in the *Union*, the *Wiskundig Genootschap* had taken sides in a political issue, namely the exclusion of Germany. Even England, he added, had remained aloof from the *Union*, so not even all allied nations had joined. To illustrate the lack of principles and coherent policy of the French, Brouwer pointed out that in 1908, at the Rome Conference, the French had made fun of international co-operation in mathematics, claiming that mathematics was a matter of a purely private nature.

His former teacher, Korteweg, took issue with Brouwer's claims. In his opinion the *Wiskundig Genootschap* would make a political statement exactly by terminating its membership of the *Union* on the grounds of the boycott of Germany. There was no evidence, he said, that the *Union* had been founded to keep the Germans out; 'Being prepared to co-operate to the same degree with all nations, ignoring

---

[104]Cf. p. 551.

the discord that may exist between nations, he deemed the right neutrality.'[105] The chairman, Wolff, recalled the arguments of the *Wiskundig Genootschap* for joining the *Union*. These were not political, but rather financial. It was hoped at the time that the *Union* would take over the *Revues Semestrielles*, the Dutch review journal, or at least share the financial burden. Although he could see the point of Brouwer's proposal, he thought it wiser not to act hastily. And so the matter was postponed to the next meeting.

In the November meeting the matter was again postponed. Brouwer, fearing that the issue would not get proper attention, then called a special meeting of the *Wiskundig Genootschap* on January 26, 1924. At the meeting Mannoury acted as the spokesman of the *Union* critics. Since the situation had not changed, he said, the original proposal could be maintained. He advocated a 'gentle solution'; it was best to tiptoe quietly out of the bad company in which the *Wiskundig Genootschap* had fallen. Brouwer added that the policy of the *Union* with respect to the exclusion of the Germans had not changed. 'The leaders of the *Conseil*', he said, 'were fierce haters of Germans'. The proposal of Brouwer, Mannoury and De Vries was strongly advised against by Korteweg. He pointed out that so far no country had left the *Union*. That the English mathematicians had not joined the *Union* was correct, he said, but it was deplored by many individual mathematicians. In fact the London Mathematical Society had turned down the invitation to join. Thus the United Kingdom was represented in the *Union* by a committee of the Royal Society. It was obvious that so soon after the war direct contact between the French and the Germans was not possible, but he saw signs of change. He indeed expected that in the meeting of the *Conseil* in 1925 the rules would be changed in favour of the admission of the Germans. And if this were not the case, he added, it should seriously be considered to get out of the *Conseil*. Brouwer immediately countered that the admission of the Germans in 1925 was irrelevant, 'On the German side the unanimous opinion is that no German will wish to join this *Union*.' Mannoury also took part in the discussion, noting that the Dutch mathematicians should not have assisted in the immoral act of excluding the Germans; they were in principle opposed to the exclusion, but had hoped for a change. That was wrong, and invited revenge. The error, he urged, should be corrected as soon as possible. After some more discussion a vote was taken on Korteweg's motion not to discuss the possible resignation of the *Wiskundig Genootschap* before the next meeting of the *Conseil*. The motion was carried with seven votes to three.

This small episode illustrates the irresoluteness of official organisations. Although there was a general feeling that the international situation in science was unsatisfactory, the tendency was not to rock the boat. It showed, by the way, that Brouwer's influence had its limits. Even in his own professional organisation he could not have his way. This may have induced him to look elsewhere, mainly in Germany, for a resistance to the *Conseil* and its subsidiaries.

History does not simply side with either Brouwer or Korteweg. The latter knew what he was talking about; being close to Lorentz, he was probably reasonably well-

[105] Minutes of the meeting of the *Wiskundig Genootschap* of 24.II.1923.

informed about the *Conseil* and the undercurrent of discontent about the inflexible position of France. Lorentz was a true internationalist, and his many personal relations with scholars all over the world guaranteed a perfect understanding of the need for international co-operation. He was recognised by scientists and politicians alike as a man of high moral values without prejudices.

When Einstein in 1923, two months after the occupation of the Ruhr area by the French, gave up his membership of the *Commission internationale de Coopération intellectuelle*, Lorentz was appointed in his place.[106]

In 1925 the scientific world was already attempting a cautious return to a more normal state of relations in certain quarters. In particular Painlevé showed himself inclined to bridge the existing gap between the French and the Germans. On the German side Fritz Haber was paving the way for a reconciliation. These developments raised hopes that the *Conseil* might revoke the clause excluding the Germans. And so Lorentz proposed on July 8, 1925 at the meeting of the *Conseil*, on behalf of the Dutch, Danish and Norwegian delegates, to drop the exclusion rule. Six years after the war, he said, it is time to make scientific research really international. He urged the meeting not to forget that even during the war there had been German scientists who showed goodwill towards the enemy. The nominally international organisation should now become a real international one. If the meeting could not adopt a truly international attitude, he feared for the future of the organisation. Lorentz' call and the implicit warning were countered by the delegate of Belgium. He rejected the proposal as premature and dangerous for the internal peace and the indispensable tranquillity of the atmosphere. Lorentz' proposal was defeated, as it did not get the obligatory two thirds of the votes. It was a fateful result; in a sense it had been the last chance to turn the *Conseil* into a beneficial organisation. We now know that not many years were left before a new and far more deadly danger announced itself.

It is one of the quirks of history that the French and the Belgians tended to consider Lorentz too soft on the Germans, whereas the German nationalists viewed him as a friend of the French. The German historian Karo did not hesitate to suggest that Lorentz was rewarded for his pro-French sympathies with a *légion d'honeur*.[107] Even Brouwer, who admired Lorentz, wondered in a letter to the editor in a newspaper how Lorentz could reconcile his internationalist principles with this decoration in view of the French inspired anti-international policy. Commenting on Lorentz' reply to the French ambassador, 'that science will after all put an end to the dissension that nowadays is manifest in mankind', Brouwer sourly remarked that 'Mr. Lorentz is waiting the further auspicious task, to point with the full authority, which he enjoys among Dutch scientists, the KNAW in the first place at the proper limits

[106]In 1924 Einstein joined the C.C.I., an organisation of the League of Nations, after all. The larger part of my information on the scientific organisations after the war is drawn from Schröder-Gudehus' books, Schroeder-Gudehus (1966, 1978), which are highly recommended for their treatment and the use of sources.

[107]Karo (1926). It is worth noting that Karo was a Jew, and that he emigrated to the USA during the Nazi period. The protest against the boycott of German science and scientists was supported by a large section of the German academic community.

of Christian charity, which encompasses the person who has gone astray, but not his aberration, in the second place at the motto of our country'.[108]

Emil Picard, the imperturbable president of the *Conseil* failed to see the omens. 'Six years is a long time for some nations', he said in his closing speech, 'who think that the forgetting can now commence. To others this time span seems too brief, to cover so much abomination and crime with the veil of oblivion,—all the more if not the least repentance is shown over it.' It is worth noting that the two leading mathematicians, Picard and Painlevé, who had started as implacable enemies of the Germans, had gone their separate ways. Picard had remained unrelenting, but Painlevé had realised that hatred and punishment were not a suitable basis for a flourishing community of scientists. Painlevé had even openly remarked in a telegram that there had been attempts to influence 'intransigent scholarly specimen, in particular Picard'.[109]

The outcome of the meeting of the *Conseil* had proved Korteweg wrong, and vindicated Brouwer. This had no practical consequences, the *Wiskundig Genootschap* did not break with the *Union*. Brouwer, however, must have felt that the regular means for a rehabilitation of German science had been exhausted. He turned towards the Germans and German organisations that fought the boycott with varying success. He was in fairly close contact with Karl Kerkhof and provided him with material and advice. At one occasion Kerkhof asked for references about Blumenthal, who had been recommended to him as a particularly well-informed man where international relations were concerned.[110] However, he had his doubts whether Blumenthal had the necessary understanding for the dignity of the German scientists. For he noted that Blumenthal had, in spite of a severe rebuke of Denjoy, see p. 336, tried to establish again and again friendly relations with French scholars who had harshly attacked German science. Brouwer replied that he felt that as an editor of the *Mathematische Annalen* he was not in a position to provide information on (or rather against) a fellow editor. Kerkhof would do better to ask Bieberbach, he added. In a separate confidential note, however, he agreed with Kerkhof that Blumenthal would be the worst possible authority in this matter. He added cynically that Blumenthal's behaviour seemed to justify the French when they accused the German scientists of servility.[111]

Before long, at the German side an intolerance had emerged that matched that of the allied side at the time of the founding of the *Conseil* (cf. p. 331). Brouwer, who somehow had become more radically anti-*Conseil* than most of his German colleagues,[112] supported Kerkhof's schemes with information and documents. He did not lapse into the mistake of holding all Frenchmen responsible for the sins of the *Conseil*, but he fought the latter with determination.

---

[108]The motto is: *Je maintiendrai.*

[109]Schroeder-Gudehus (1966), p. 248.

[110]Kerkhof to Brouwer, 4.V.1925.

[111]Brouwer to Kerkhof, 11.V.1925.

[112]Witness the remark of Sommerfeld in a letter to Blumenthal (26.I.1925): 'I can but experience it as humiliating that Brouwer feels more German than we do.'

In the KNAW Brouwer had not played a role in the pro- and contra-*Conseil* movements. When the Academy joined the *Conseil*, there had been a fierce opposition of various factions and individual members. In the meeting of January 25, 1919 Brouwer had suggested a formal way out of the membership invitation of the *Conseil*, by pointing out that the old associations still existed, and that co-operation with those should be maintained. The members of the Academy from Groningen were likewise opposed to the *Conseil*. They submitted a proposal to found an independent '*Conseil*' of neutral nations, which could act as an intermediary between scholars of all nations. And somewhat later the group sent a note to the Academy, pointing out that the proposal of the board to join the *Conseil* was in fact 'a polite and humble request' to the allied academies to be allowed to co-operate in an organisation that 'on the ground of the exclusion of Germany and Austria, is and will be of a pronounced hostile nature to the Central academies'. The Academy nonetheless did join the *Conseil* and this was, for the astronomer Kapteyn, a reason never again to attend Academy meetings. The philosopher Heymans went so far as to resign from the Academy. In due time peace was restored, but there were some attempts to dislodge the KNAW from the *Conseil*, or at least to admit Germany and Austria. In 1922 Cornelis van Vollenhove, the law professor, proposed in the meeting of the *Conseil* to drop the exclusion clause. This motion was easily defeated. After the rejection of Lorentz' proposal in 1925, there was another anti-*Conseil* action in the meeting of the KNAW of September 25; Brouwer, J.F. van Bemmelen and J. Versluys condemned the *Conseil* and advised the KNAW to withdraw from it. The attempt failed.

Brouwer's observation that no German scholar would wish to join the *Conseil* had been somewhat sweeping, but he did have a point. The more nationalistic scholars had, no doubt influenced by Kerkhof's organisation, decided that they would rather not join a company which characterised them as brutal, slavish killers. The insults addressed to the collective German scholars were still ringing in the ears of the sulking scientists, who were no more familiar with forgive-and-forget than their French colleagues. And so gradually out of the boycott a counter-boycott was born.

An influential group was actively campaigning against membership of the *Conseil*. Many pamphlets and brochures appeared that sketched in vivid notes the offensive treatment of German scientists. Already in 1923 the *Hochschulverband* had issued a list of five instructions on how to cope with international contacts.[113] The instructions were a perfect counterpart of the rules and instructions of the *Conseil*. Here are some passages:

> —The German Universities (*Hochschulen*) and individual German scholars should on principle and carefully avoid everything that could in some way be interpreted as an attempted approach to the Entente.
>
> —[with respect to private contacts] Contacts via the backstairs should be turned down as unworthy with the request that first the slanderous state-

[113] Cf. Schroeder-Gudehus (1966), p. 207.

ments[114] should be publicly withdrawn in the same solemn formulation in which they were enounced.

—Any co-operation with France and Belgium should under all circumstances be out of the question.

The radically anti-French section of German science had started to organise itself, and their statements did not give much hope for a compromise. Professor Georg Karo compiled a long list of anti-German measures and statements in his 'The spiritual war against Germany'.[115] There is no denying that the Germans were sinned against, but they certainly gave as much as they received. Karo spoke for a sizeable number of his colleagues when he concluded 'We are not served by a solemn retraction of the boycott, as long as the system of regulation of the scientific world that was founded together with it, remains. Basic dignity forbids us any rapprochement with the *Conseil Internationale de recherche*, which was born out of hate and contempt for us, even when it would be reorganised.'

Brouwer did play a modest role in the anti-boycott, e.g. he once ordered (and presumably, distributed) 100 copies of Karo's monograph. He would have subscribed to some of the texts of the anti-*Conseil* movement, but certainly not all of them. He cultivated, for example, contacts with French and Belgian colleagues. But he shared the radical sentiment that demanded apologies for the post-war insults. His role was probably no secret in Germany. Kerkhof made use of Brouwer's disposition and Planck, for example, asked Brouwer to find out certain disputed facts about the *Conseil*. Apart from his role in the Riemann volume, Brouwer's influence was marginal. His big moment came at the time of the Bologna conference.

## 12.5 The Dutch Topological School

In 1925 Brouwer was back in topology, not that he prepared new publications or continued his research, but the encounter with Urysohn and Alexandrov had reawakened his interest on a more contemplative level; he realised that a new generation of topologists had come forward, and he liked what he saw. In particular Urysohn's work was after his own heart, large parts could be seen as a continuation of his innovating papers. As always, there were the dangers of priority matters, but as we have seen, Urysohn and Brouwer came to a perfect agreement, and Urysohn must have been an unusually engaging and mature person. During his short sojourn in the mathematical community in Germany and France, he had succeeded in making friends everywhere. In particular, he and Alexandrov were highly regarded in Göttingen, where the tragic death of Urysohn was felt more deeply than one would generally have expected from such a meteor-like appearance in the mathematical firmament. Emmy Noether, Bernays, Felix Bernstein, Hausdorff, Fréchet, each and

[114] Cf. p. 329.

[115] *Der geistige Krieg gegen Deutschland*, Karo (1926).

**Fig. 12.2** Cor Jongejan with cat in Blaricum [Brouwer archive]

every one of them felt the loss of this brilliant but modest young man.[116] Brouwer, who had a singular gift to reach the soul of another person—if he cared to do so—had held long and penetrating conversations with Urysohn during the period of their brief encounter in Blaricum. He had almost instantaneously developed an immense liking and appreciation for the young man. The sudden loss of the young genius who had so briefly entered his life brought him immeasurable desolation. Perhaps he cared all the more for the remaining young Russian, whom he subsequently invited to Amsterdam.

In 1925 Amsterdam suddenly became the centre of topology in Europe, and Brouwer was tacitly recognised as the founding father of the 'new topology'. Menger had arrived in Holland at the end of March and Alexandrov in May, in the autumn Vietoris also joined the group and Newman came over for a short visit. Furthermore Wilson, a student of Brouwer, regularly took part in the activities.

An important source of information on the period described here are the published memories of Alexandrov and Menger.[117]

As customary in those days, professors worked largely at home, unless they had their laboratories at the university, and they only turned up for lectures and faculty meetings.

It was Brouwer's custom to work at home in Blaricum, and he more or less insisted that his visitors and collaborators should be within easy reach. This implied that they had to find lodgings in 't Gooi. For Alexandrov this was no problem, on the contrary, there was a remarkable affinity between his and Brouwer's lifestyle. Both liked the outdoors, swimming, hiking, etc., although Brouwer no longer went in for

[116] As testified by letters in the Alexandrov archive.

[117] Alexandrov (1980), p. 322 ff., Menger (1979), p. 241, ff.

excessive physical exercises. It was no problem to find lodgings for Alexandrov in Laren in the house of a friend of Cor Jongejan, and something similar was envisaged for Menger, cf. p. 433.

Brouwer's way of life had not altered since he came to live in Blaricum, his wife still attended the pharmacy and during the week lived mostly in Amsterdam. Brouwer's home was run by Cor Jongejan, who not only took care of the house, but also acted as his secretary and friend. She assisted Brouwer in keeping order in his files, typing letters—sometimes writing them by hand, copying manuscripts that were submitted to the *Proceedings* of the Academy or to the *Mathematische Annalen*, etc. Eventually Cor got the proper recognition for her secretarial duties: she was appointed as an assistant in 1925. So far Belinfante had filled the position of assistant, and in the academic year 1925/26 Brouwer managed to get a record number of assistants: Belinfante, Menger (first assistant), Alexandrov (first assistant), Vietoris (first assistant), and C. Jongejan (adjoint assistant). Belinfante had been admitted to the university as a 'privaat docent', and in this capacity he taught his specialties: intuitionistic analysis, the theory of infinite series, and complex function theory. In order to make a living he also taught in high schools.

One would expect that by now Brouwer's fame would attract the brighter students in mathematics. This was only partly the case. Brouwer did not care for public relations, he taught and operated in a rather low key. Most students did not even know that their local mathematics professor was a topological genius, let alone a man suspected of overthrowing the mathematical world. It was more or less a fortunate coincidence that in the early twenties one of the brightest young men of the day came to study mathematics in Amsterdam: Bartel Leendert van der Waerden.

Van der Waerden enrolled in Amsterdam in 1919. The study of mathematics was for him the proverbial 'piece of cake'. Reminiscensing about his studies, he said: 'I heard Brouwer's lectures, together with Max Euwe and Lucas Smid.[118] The three of us listened to the lectures, which were very difficult, he treated the integration theory of Lebesgue along intuitionistic lines, and that works. It was very curious, Brouwer never paid any attention to the audience. All the time he gazed at a point on the opposite wall. He lived in Laren, rather isolated; he came roughly twice a week to lecture in Amsterdam.'[119]

Brouwer's commuting was rather time consuming, but it had an attraction for him: he could meet acquaintances, students, or colleagues on the tram (or bus, or train, as it happened to be). In winter, when darkness fell early, Brouwer as a rule carried a storm lantern, as the streets in Laren-Blaricum were not all that well-lit. His neighbour, Mr. Crèvecœur met him once on the bus. Brouwer tapped him on the shoulder and said, 'Mr. Crèvecœur I see you smoke, and I do not. You must have matches, could you light my lantern?' Arriving in Blaricum, Brouwer accompanied

[118] Smid became a specialist in insurance mathematics. Max Euwe won in 1935 the world title in chess. Later he became a professor in computer science.

[119] Brouwer actually lived most of the time in Blaricum, but right at the border with Laren. This border has moved back and forth in the course of time. Sometimes he rented a house in Laren or in Blaricum. A large part of his letters bears the name of Laren, later letters show Blaricum.

**Fig. 12.3** Van der Waerden and Emmy Noether in Göttingen. [Courtesy Alexandrov archive]

Crèvecœur home, but being in a good mood, he talked and talked, explaining how one could determine the surface of a ball by emptying it. Crèvecœur had heard quite enough, so that when they reached the house of Crèvecœur, he said at his door, 'I am tired, now I want to go to bed.' Brouwer would not hear of it, 'Come along to my house, and we can talk a bit more.' So they went from house to house a couple of times, until Brouwer suddenly said, 'I go home, good night', leaving Crèvecœur in the dark.

Van de Waerden did not find it easy to talk to the professor, 'He immediately departed after the lecture, so that it was very difficult to make contact with Brouwer.' Van der Waerden meticulously took notes in class, and usually that was enough to master all of the material; Brouwer's class was an exception. Van der Waerden recalled that 'at night he actually had to think over the material for half an hour and then he had in the end understood it'.[120] It was not until he visited Göttingen that he heard about Brouwer's topological work. Students in Amsterdam were on the whole unaware that their department housed one of the great topologists of the century. Van der Waerden was an extremely bright student, and he was well aware of this fact.

He made his presence in class known through bright and sometimes irreverent remarks. Being quick and sharp (much more so than most of his professors) he could make life miserable for the poor teachers in front of the blackboard. During the, rather mediocre, lectures of Van der Waals Jr. he could suddenly, with his characteristic stutter, call out: 'Professor, what kind of nonsense are you writing down now?' He did not pull such tricks during Brouwer's lectures, but he was one of the few who dared to ask questions. This was no easy matter, Brouwer lectured in a very precise, concentrated manner. He was completely wrapped up in himself and reconstructed the material as he went, he presented his mathematics in a long monologue, any interruption was apt to break the line of thought. Normally the students accepted

[120]Interview Van der Waerden.

this phenomenon as part of the unavoidable hardships of university life, but at one occasion they conquered their awe and sent a delegation to Brouwer. Admitted into Brouwer's office, the spokesman solemnly brought his message, 'Professor, we have difficulty following your lecture, you go so fast.' Brouwer thought for a moment and replied 'I see, well, I will try to lecture as slowly as the gentlemen think.'[121]

When Van der Waerden gathered all his courage and asked Brouwer a question in class, Brouwer replied politely, but later he sent his assistant to Van der Waerden with the message that further interruptions would not be appreciated.[122]

One would think that such a bright student was a man after Brouwer's heart. The truth is that Brouwer had no affinity with Van der Waerden's mathematics, furthermore, Brouwer wanted to be left alone to do his own mathematics. A clever young man who would interrupt his own contemplations with bright remarks and questions was the last thing in the world he wished for. He certainly appreciated Van der Waerden's mathematical gifts, and he went out of his way to get him a Rockefeller stipend for Göttingen. Given Van der Waerden's algebraic interests, the person to take care of him was Emmy Noether. Once in Göttingen, under Emmy's wings, Van der Waerden became a leading algebraist. Emmy was very pleased with the young Dutchman, 'That Van der Waerden would give us much pleasure was correctly foreseen by you. The paper he submitted in August to the *Annalen* is most excellent (Zeros of polynomial ideals)...', she wrote to Brouwer.[123] Notwithstanding his popularity in Göttingen, Van der Waerden came back to Amsterdam for his doctor's degree. Perhaps he would have liked Brouwer as a Ph.D. adviser, but Brouwer systematically discouraged students from writing a dissertation under his supervision. Brouwer was not interested in the honour, pleasure and toil of the Ph.D. advisor role. Whenever a student turned to him for a Ph.D. project, he referred him to Weitzenböck or Hk. de Vries. Weitzenböck was regarded as a leading specialist in the theory of invariants, and quite a number of Ph.D. theses were written under his supervision. Max Euwe was one of his Ph.D. students, and so was Griss, the founder of negationless mathematics. Hk. de Vries had a totally different personality, he was a cultured man, deeply in love with mathematics, but his admiration for the subject of his love did not express itself in deep or original contributions. He was well aware of the fact that depth and originality were not his strong points; he could jokingly refer to his colleague Brouwer, whom he admired, as a 'point collector' (*puntverzamelaar*). Even his lectures, although enthusiastically presented, were somewhat superficial. The students loved him and his courses; the love for his subject is illustrated by the following story. At the end of one of his lectures, he was so excited, that he left the room with a resounding 'long live mathematics'. He wrote

[121]Oral communication F. Kuiper.

[122]Max Euwe told a different story. When Van der Waerden asked his question, Brouwer replied sternly and then concluded: 'Mr. Van der Waerden, I advise you to study the material completely before you start a discussion.' The story went from student generation to student generation. Freudenthal's version was that 'Van der Waerden thought to have found a mistake. But Brouwer was completely right.'

[123]Emmy Noether to Brouwer, 14.XI.1925.

quite a number of attractive textbooks, more praised for the style and presentation, than for depth or precision, including books on the history of mathematics. In spite of his modest reputation as a creative mathematician, he was the man who inspired generations of students. 'If you had heard De Vries' lectures, you would forever be under the spell of mathematics', said Euwe. It was De Vries who took the role of Ph.D. adviser of the young Bartel upon himself. The topic of Van der Waerden's dissertation was enumerative geometry,[124] a subject that had once been treated in a monograph by De Vries himself.[125] Van der Waerden's dissertation earned him instant fame in the world of algebraic geometers for its importance as a solid basis of the subject. Although later in life Brouwer and Van der Waerden drifted apart, their relationship in the early days was unproblematic. Van der Waerden expressed in his dissertation his intellectual debt with the words, 'You, learned Brouwer, I thank you for your lectures, which went up to the ultimate limit of exactness, and for your vigorous help, amply shown to me on various occasions.'

The obligatory list of theses, which were part and parcel of dissertations in Holland, dealt with a rich variety of topics; one of these treated a traditional source of confusion in intuitionistic mathematics (and logic): negation. It read, 'It is recommendable to use the word *not* in intuitionistic mathematics exclusively with the meaning *impossible*, and no longer with the meaning *there is no reason to assume that.*' This thesis demonstrated Van der Waerden's familiarity with the strong intuitionistic negation, and it embodied an attempt to do away with the confusion between, e.g. 'the Goldbach conjecture is false' and 'we have no proof of the Goldbach conjecture'. However, the weak negation is firmly entrenched in daily language, and its abolishing would make conversation most awkward. Brouwer, when questioning the candidate at the public defence of his dissertation, said as much by pointing out that 'you are not a doctor, but that does not mean that you will never be one'.

In mathematics Van der Waerden was easily recognised as an outstanding scholar, but in the 'real world' he apparently did not make such a strong impression. When Van der Waerden spent his period of military service at the naval base in Den Helder, a town at the northern tip of North-Holland, his Ph.D. adviser visited him one day. He said later that the commander was not impressed by the young man, 'he is a nice guy, but not very bright'.

After getting his doctorate in 1926, Van der Waerden became an assistant in Hamburg, and in 1927 he became a *Privatdozent* in Göttingen. A year later he accepted a chair in Groningen, where he wrote his famous *Moderne Algebra.* A few years later he moved on to Leipzig, where he held a chair from 1931 till 1944.[126]

Alexandrov saw a great deal of Brouwer, together they edited Urysohn's papers, which often came to reconstructing the text on the basis of Urysohn's notes.

---

[124] 'The algebraic foundations of the geometry of number', [i.e. enumerative geometry], van der Waerden (1926).

[125] de Vries (1936).

[126] For biographical information on Van der Waerden, see Springer (1997) and the papers in *Nieuw Archief voor Wiskunde*, 1994(12) no. 3.

Brouwer conscientiously read Alexandrov's drafts and made small editorial corrections, when necessary. They often got together in Brouwer's cottage to work and talk. Alexandrov was a man with an engaging personality, a ready wit and of a cheerful disposition. Of course, the loss of his close friend and co-topologist, Urysohn, had thrown him off-balance, but his subsequent stays in Blaricum and in Göttingen did much to revive his spirits. Alexandrov, on his wit and graciousness made quite an impression with the female inhabitants of 't Gooi, including Cor Jongejan. Brouwer even imagined a real threat to his relationship with Cor. Gerda Holdert, the wife of Van Eeden's former secretary and member of Brouwer's inner circle, reported that he once asked her to keep an eye on the two.[127] Alexandrov used the first half of July to be with his old love: the sea. He made a walking tour along the seacoast and settled for a fortnight in Katwijk, at that time a small fishing town, which in summer hosted a good number of holidaymakers. In his autobiographical notes he mentioned that Princess Juliana, the future queen, stayed in Katwijk during that period.

Alexandrov spent the rest of July 1925 in Göttingen, where he presented, at Hilbert's request, a survey of topology. In August he returned to the fateful Batz where he was joined for a month by Urysohn's father. In the month of September he set out on a walking tour in the Pyrenees. At the end of October he returned again to Batz, where Brouwer joined him to spend the first two weeks of November together. They returned together to Blaricum and almost immediately the topological activities started.

Menger had also left Holland for the summer. Staying in Heidelberg, a telegram that was forwarded to him by Brouwer brought him the sad news of his mother's death. Menger was heart-broken, in the depth of his suffering he lamented,[128]

> I thank you most warmly for your kind letters of condolence with the most terrible blow of fate which has struck me. What I lost in my dear mother, I cannot sketch in words; her goodness of heart was unbounded. And to the pain about what has been taken from me, the unspeakable thought is added, that she, who as long as I live has done and sacrificed so infinitely much for me, died precisely now, when a more quiet evening of her life, which she looked forward to with pleasure and was still capable enjoying, had begun. She followed from afar everything concerning me, grateful in particular to you, for all the support you gave me. That news was her last enjoyment.

After hurrying back to Vienna, Menger suffered a severe nervous breakdown. As a result he had to take a cure of several weeks. His thoughts wandered to the tragic fate of Urysohn, 'Again and again I thought in this period of the poor Urysohn, and wished that I would have perished instead of him. Only the thought, that I may not destroy that what my beloved mother had built with her life's effort, gives me now the will, to regain, if possible, my health and then to achieve something.'

[127] Oral communication Gerda Holdert.

[128] Menger to Brouwer, 3.VII.1925.

Brouwer, who considered his topological visitors more his foster children than just a group of scientists, 'postdocs' we would say nowadays, on their way to outstanding careers, sympathised with Menger's loss in its full tragic degree. He replied on July 8 with a deeply felt, caring letter; he was indeed relieved to see that Menger had survived the shock, because during Menger's stay in Holland Brouwer had surmised

> to what extent the sphere of your mother has radiated through your life. I also sensed the extent of your loss, and expected the crisis that the sudden emptiness, and the sudden necessity, to adopt another spiritual mode of breathing, would bring forth in you. But after this first crisis has been borne, I am certain that you will find the required concentration and religious devotion to find yourself a way through, and that the certainty of the dear deceased's wish, aimed at these, as well as the thoughtful memory of her that will be with you, will help you in this.

Being uncertain how well-off Menger was financially, he offered him an assistantship in Amsterdam, as he was not certain how long it would take to wind up the details of the Rockefeller stipend.

Menger managed, nonetheless, to carry on his mathematical activities and on August 16 he could send Brouwer a copy of the manuscript of a large expository paper on dimension theory, to be submitted to the *Jahresbericht*; in the accompanying letter he asked Brouwer's permission to dedicate a separate publication as a small monograph (in fact off-prints with a special cover) to him, 'Dedicated to L.E.J. Brouwer in deep reverence.'[129] In the same letter another key figure from dimension theory is introduced: Witold Hurewicz. Hurewicz was born in 1904 in Lodz as the son of a rich industrialist; he studied in Vienna where he became informally Menger's student. Menger promised to send Brouwer a copy of Hurewicz' recent manuscript on dimension theory, which was subsequently published in the *Mathematische Annalen*.

The letter also shows that Menger was by no means a narrow specialist. There is a passage in the letter that shows Menger's occupation with graph theory; he in fact believed that he had found the key to a proof of the four-colour problem.

In Menger's autobiographical sketch 'My memories of L.E.J. Brouwer',[130] a penetrating description of Brouwer and his *entourage* is given; Menger showed himself a keen observer—one cannot but deplore that he did not go on and write down the total story of his life. In a way his fate is typical of the middle European scientist of the pre-war era, highly successful, versatile, cultured, with many contacts, who through a forced emigration suddenly became separated from his scientific-cultural background and lived on in relative obscurity (he would never reach the heights of his European period again).

Coming to Blaricum, he met Brouwer for the first time in person. As he remembered the first impressions,

---

[129] *L.E.J. Brouwer in tiefer Verehrung gewidmet.*

[130] Menger (1979).

> he looked older than his 44 years. His figure was lank and youthfully agile; but his hollow-cheeked face, faintly resembling Julius Caesar's, was extremely nervous with many lines that perpetually moved changing his expression from one moment to the next.
>
> His features seemed incessantly to reflect intense inner reactions to what he was hearing and seeing, just as outward intensity in speech and movement in action was the hallmark of his personality.[131]

The picture is wholly accurate, Brouwer had an unusually expressive face and his movements have been described as agile and almost fluttering. He could come down the aisle of the great auditorium like a gigantic bird, his large delicate hands flapping like wings. Menger was struck by Brouwer's fluency in various languages:

> Whichever the language, Brouwer talked very fast and with great precision, though often in long and involved sentences. This he did also in lectures, where he went into minute details. But he delivered them with an impressive emphasis.

If Menger had expected to find a quiet scholar, devoting his time to the creation and dissemination of knowledge, he was in for a surprise. Brouwer was largely occupied with organisational matters, non-mathematical activities and innumerable legal cases. For example, at that time Brouwer still spent a great deal of time fighting the *Conseil International de Recherche* (as we will see later).

Soon after his arrival Menger accepted Brouwer's offer of an assistant position; the duties were fairly light, varying from reading manuscripts that Brouwer had to referee for the Proceedings of the Academy or the *Mathematische Annalen*, lecturing, etc. Both Alexandrov and Menger shared the normal teaching duties, be it that they taught their own speciality. Alexandrov gave a course on general topology, and Menger a course on dimension theory. Brouwer conscientiously attended these courses and took part in the lively discussions that followed. Brouwer himself lectured on intuitionistic mathematics and on continuum mechanics. In the first one he covered a lot of ground, in particular he treated topology from an intuitionistic point of view, for example the theorem of Jordan. The second one also contained a fair amount of topology, as Alexandrov formulated it 'much homology theory (in the specifically Brouwerian manner, with pseudo-manifolds and so-called 'fragments' instead of cycles and chains.)'[132]

Menger was in close contact with Brouwer:

> I frequently saw Brouwer alone and came to know more about his habits and tastes. I never found him depressed, but I rarely saw him laugh or display signs of a sense of humour. He usually manifested a high-strung personality—least tense when listening to or talking about music. He loved classical music and frequently attended concerts, especially of Beethoven's works, about

[131] Here and in the following pages I quote from Menger (1979).

[132] Alexandrov (1969), p. 117.

> which he spoke beautifully. In painting he seemed to favour the style of the Renaissance and was fond of portraits of that period. He looked with obvious distaste at the expressionistic graphic with social themes by the Dutch and Flemish artists of the 1920s (which I greatly admired)—even at Masereel's marvellous novels in woodcuts. They lacked the kind of beauty he seemed to expect from art. He also looked for comeliness in handwriting and developed in his middle years an aesthetic style of writing. But even more interesting were his ways of crossing out passages in manuscripts. In order to delete an entire paragraph he would draw many diagonal lines in both directions forming networks of perfect regularity; some had wider meshes, others were quite narrow grills. Single words or sentences he blocked out with solid black lines or rectangles. Some corrected manuscripts of his were quite picturesque (in a remotely Mondrian-like way). The handwriting was also of interest to him in letters that he received; but there, his chief concern was the consistency of style of writing in a person's consecutive letters.

Alexandrov, too, mentioned Brouwer's love for music; he regularly accompanied Brouwer and Cor Jongejan to the Sunday afternoon concerts at the Concertgebouw where the great Mengelberg conducted. They also saw Stravinsky conduct his *Sacre du printemps*. Quite often important visitors of Brouwer joined them for the concerts. Brouwer was for the greater part of his life a familiar figure at the Sunday concerts, the traditional meeting place of the Dutch music lovers. Brouwer, although not consciously drawing attention to himself, was a conspicuous figure in the Concertgebouw, one could not help noticing this extra-ordinary person.

Even though Brouwer was no longer actively involved in topological research, Menger warmly appreciated Brouwer's interest in his work:

> In the late spring and the fall of 1925, I reported to him my latest results almost every week and immensely enjoyed these occasions. He never formulated problems to me; but he listened with an eagerness that I found very stimulating. Through the perpetually changing expression of his face he had the rare ability to be, as it were, even silent with intensity. He also showed concern for my then still unstable health and expressed warm sympathy when I suffered a bereavement. He warned me against overworking. 'A mathematician's mind is like a fruit tree', he once said to me. 'Not all blossoms turn into fruit; and if one year they all do, then the tree bears no fruit the following year.' I revered him with a deep affection during those months.
>
> Brouwer regarded mathematical creation as a decidedly youthful activity. 'Even Gauss', he once remarked, 'spent his later years on boring numerical computations'. I observed that according to what I had heard, Gauss enjoyed numerical computations from childhood on, all his life. 'Mathematicians' old age is sad', Brouwer answered sombrely, 'Gauss' old age was sad, too.' It was the only time that I ever discovered in him something like gloom. Only much later did I realise that these remarks were the key to understanding Brouwer's personality, especially several traits of his that I was soon to discover and which (together with some of his political views) I would find incompatible with the idealised picture that in youthful enthusiasm I had formed of him.

Emmy Noether also joined the group for a brief period, she spent a month in Blaricum staying over for Christmas and New Year. In his paper 'Topology in and around Holland in the years 1920–1930'[133] Alexandrov reports about this visit, 'Right then Emmy Noether formed the opinion that group theory is the proper foundation for combinatorial topology, that in particular the numerical invariants—the Betti numbers and the torsion numbers—must be replaced by homology *groups*.' Alexandrov recalled 'a dinner at Brouwer's in her honour during which she explained the definition of the Betti groups of complexes, which spread around quickly and completely transformed the whole of topology'. Back in Göttingen she gave a lecture on 'the algebraisation' of topology at a meeting of the Göttingen mathematics society, of which the dinner crash-course in Blaricum had been a preview.[134]

This may be a good point to dispel a widespread misconception about Brouwer's views on women and society. In his student days he had aired some radical opinions on the matter, but his actual policy in his private and professional life was in such a striking contrast, that one is inclined to attribute his early statements to a provocative student mentality and to a genuine mystical introversion. Brouwer did, however, enjoy and appreciate female company until the end of his life, and not only as the charming creatures that spread happiness and light, but equally in professional circles. In science it did not matter a bit to him what the gender of a scholar was, an interesting and active mind counted more than conventional distinctions. Emmy Noether, in particular, was highly regarded by him, and her arrival in Blaricum in a period during which the *Grundlagenstreit* had already dangerously soured the relations between Hilbert and Brouwer is a telling proof that she reciprocated Brouwer's regards. As a matter of fact, their acquaintance went back to before 1914; in answer to a postcard from Carathéodory and Brouwer[135] she wrote 'I also recall with pleasure the days in Karlsruhe, and I hope that there will soon be another mathematics conference where we can meet. Won't you come soon to Göttingen and give a lecture?'

In November 1925 she had approached Brouwer about the extension of Alexandrov's Rockefeller grant for a stay in Göttingen. She had come to appreciate Alexandrov and Urysohn so much during their stay in 1923, that she had invited Alexandrov to Göttingen. Alexandrov was, understandably, gratified by the invitation, but under no circumstance did he want to antagonise Brouwer, so he discussed the matter with Brouwer, who did not object, although he remarked to Emmy Noether[136] that 'In my opinion it would be an excellent choice of the Georgia Augusta[137] if it could get Alexandrov a temporary teaching position (*Lehrauftrag*) for Göttingen. [...] I would even prefer to secure Alexandrov for a longer period in Holland, or even keep him here permanently, ...'

---

[133] *Die Topologie in und um Holland in den Jahren 1920–1930*, Alexandrov (1969).

[134] Cf. MacLane (1981).

[135] E. Noether to Brouwer, 7.IX.1919. For the Karlsruhe meeting, see p. 175 ff.

[136] Brouwer to E. Noether, 21.XI.1925.

[137] The University of Göttingen.

Alexandrov, who like so many young mathematicians had fallen for the Göttingen Goddess of modern algebra, apparently knew the world well enough to realise that his attachment to Brouwer would not be considered a recommendation in certain places. So when corresponding with Emmy Noether, he begged her to consider his letter as strictly personal,[138]

> I would not like that my purely moral dependence on Brouwer in these and many other matters, which I take freely upon me, and which therefore does not clash with my personal and scientific freedom, would lead to any misunderstandings at all.

For some time the paradise-like situation lasted in the topological family but soon tensions built up. It is hard to put a finger on the exact cause, probably a number of factors played a role. In a sense Alexandrov and Menger were, temporarily at least, Brouwer's topological sons—Vietoris was also a member of the topological family but he seemed to have had a knack for keeping out of trouble, indeed for remaining unaware of frictions around him. To extend the metaphor a bit farther, Alexandrov strongly resembled the beloved brother of Urysohn, whose premature death cast such a deep shadow over his close associates. This introduced an undesirable asymmetry in the family, and one must fear that Brouwer did not obey the golden rule of all parents: not only love your children equally well, but make certain that they feel equally well-treated. The relation between Alexandrov and Menger, with Brouwer in the background, is the classical story of two brothers, the charming, intelligent, sport-loving boy adored by all friends of the family, and the awkward boy who cannot tell his jokes properly and who is not applauded for the same performance as his brother.

Indeed, where Alexandrov was the accepted darling of any mathematical community, Menger confused his fellow men with his statements that may have been intended as jokes, but certainly did not enhance his popularity. The more active and outspoken scientists usually generate anecdotes that in one way or another illustrate their personality; Menger was no exception. A single one may serve as an example; on one occasion, he refused to take the thermometer[139] from a nurse with the words: 'Do you know who I am, I am the Napoleon of mathematics.'[140] Also his worries about priority (something he shared with Brouwer!)—in particular the story about the deposited manuscript—were the subject of comments in the circle of his colleagues. Alexandrov, when discussing a particular meeting of the Berlin Seminar on topology in Göttingen, jokingly remarked that Menger would certainly go with three notaries to the patent office, and get himself an 'original cover protected by law for his course notebook, where there would be two forms of this cover: one exclusively for national, and the other one for international use, as is customary with famous laxatives and the like'.[141] The difficulty with anecdotes is that they are not fair, they

[138] Alexandrov to E. Noether, 11.XI.1925.

[139] Rectal, as the hospital tradition was.

[140] Oral communication, Mrs. J.F. Heyting-van Anrooy.

[141] Alexandrov to Hopf, 10.IV.1927.

blow up a particular feature, not unlike a caricature. The anecdotes about Menger tend to highlight his tendency for an inflated feeling of importance. Those who have known Menger agree that this was only one side of his personality, nonetheless it did not make contacts with him any easier. As to the matter of the deposited manuscript, one should bear in mind that this was the story of a 20 year old boy, seriously ill and in confusion, acting on the advice of his 80 year old father.

Altogether, the combination of characters in Blaricum was one that spelled trouble,—and that would have thrilled an Agatha Christie.

The atmosphere was gradually changing in the Blaricum centre. The exact cause is a matter of some uncertainty. There are two main views: those of Menger and of Brouwer. The older of the two was at the zenith of his career, respected, and occasionally feared in the scientific world. The younger one was making his way in the mathematical world; he was already basking in the gratifying admiration of the topological world. Brouwer, from his place at the mathematical Olympus, had no intention to suffer any belittling attacks or comments, and Menger, with his not inconsiderable self-esteem, was inclined to see conspiracies to do him out of his rightful place at the top. From these two perspectives conflict material was abundantly available. Brouwer had seen enough of the mathematical world to know the value of correct citations and references; he therefore insisted on appropriate references in matters of dimension theory. Basically this came down to the references to the three founding fathers: Brouwer himself, Urysohn and Menger. He quite correctly took the position that as far as the new dimension theory was concerned, the priority lay with Urysohn, albeit that Menger's ideas were conceived at roughly the same time.[142] The independence of the work of Urysohn and Menger was never disputed.

During Menger's first year in Amsterdam the 'historical' aspects of dimension theory must have been the subject of the conversation in the group of topologists. After all, Menger was teaching a course on dimension theory, and the other members of the group would attend whenever possible. We may assume therefore that questions were asked and replies given, perhaps even debated with some fervour. Menger was not the man to stand aside, and let Alexandrov and Brouwer dominate the discussion. Having a strong interest in the proper distribution of credits, Menger in the end set out to present the historical development of his dimension theoretic ideas in their proper order. This required a complete set of manuscripts and notes that would count as objective evidence. This evidence had to be collected in Vienna. On April 10 Menger could inform Brouwer that 'the priority matter is now completely resolved'.[143] The material had been kept in a safe, and he promised to hand

[142]We recall that Urysohn discussed his dimension theory for the first time (with Alexandrov) in August 1921 (cf. p. 397 and Johnson 1981, p. 228). He lectured on the topic in Moscow in the academic year 1921/22, and submitted three printed notes on the subject to the Moscow Mathematical Society. The first internationally accessible version appeared in the *Comptes Rendus* in 1922. Menger developed his ideas on dimension theory between April 1921 ands February 1922. An account was submitted to Hahn in November 1922, and was withdrawn after Hahn discovered a mistake in it. Menger's first publication followed in December 1923, Menger (1923).

[143]Menger to Brouwer, 10.IV.1926.

over the whole collection (four items) to Brouwer. The matter had put tremendous pressure on Menger's mental state, 'Had I not intended to put the documents, dear Professor, into your hands, for what you have written about the theory, and what you have done for me,—I could not have suffered what I had to live through.' A nervous collapse indeed followed. There was also corroboration of Menger's assertions by third parties; Otto Schreier, a close friend of Menger, went through his notes to ascertain the dates and events that were relevant,[144] and Menger's teacher, Hahn, gave an account of Menger's work on dimension theory.[145]

It did not take Menger long to complete the survey of his dimension theoretic work; on 29 May Brouwer submitted a paper entitled 'On the genesis of my papers on dimension and curve theory'.[146] Three years later, when the dimension conflict was at its height, Brouwer recalled that he had to use all his persuasive powers 'to get him [Menger] to write down only verifiable [*beweisbar*] matters with an indication of the proof in his historiography; and to accept personally and openly the responsibility for historical views, for which only he possessed the documents, therefore neither to demand publication of these views from less expert friends, nor to convey these, instead of explicitly, only implicitly as a consequence of irrelevant marginal facts'.[147]

Menger's genesis-paper later became a subject of controversy; Brouwer had insisted on a precise record of the mentioned manuscripts and papers. He therefore demanded that those documents that could be exhibited be referred to as 'still available' (*noch vorhanden*).

> The refutation of my assertion that the MS of February 1922 could not "be found" remains a second difficult point (also in the new version of Menger). For when Menger during his stay in Amsterdam handed over his old documents for publication in the Amsterdam Proceedings, I initially had objections to submitting the text in which he refers to the MS of February 1922, without prior inspection. Only after my repeated requests to Menger (also in the presence of other interested parties) had failed, I have agreed to the publication of the text in such a form that a distinction was made between 'still available' and 'not available' documents, and that the MS of February 1922 would be classified in the latter category. Even when I subsequently had accepted a very moderate and almost veiled form of this distinction, it was nonetheless maintained with respect to content, and expressed unmistakably in the particular publication.

The manuscript of Menger's February 1922 note was indeed not available at the time of the genesis-paper, but a correction of the note was produced. Freudenthal,

---

[144] Schreier to Menger, 7.IV.1926.

[145] Hahn to Brouwer, 10.IV.1926. It is not clear whether Menger or Brouwer asked for this information.

[146] Menger (1926).

[147] Brouwer to Hahn, 22.X.1929.

in his edition of Brouwer's topological work, commented, 'It is hard to understand why the *correction of a paper of 1921 is published in 1926 without the paper itself*, and only the careful reader of Menger (1926) will discover that the mark 'noch vorhanden' is to be associated with the correction, and not with the paper itself, which, indeed, was not available when Brouwer asked for it in 1926'.[148] Brouwer's device was so subtle indeed, that hardly anybody could have observed the above effect. In particular Menger suffered no loss of face. But one may well conjecture that the discussions that preceded the publication did little to endear Menger to Brouwer.

The composition of the group of topologists changed towards the end of the academic year 1925/26. Alexandrov left in the spring of 1926 for Göttingen, and Hurewicz took his place; Brouwer procured a Rockefeller stipend for Hurewicz, who stayed in Amsterdam until he left in 1936 for the Institute of Advanced Study, and subsequently, Brown University and MIT.

In August 1926 there was a sudden outbreak of unpleasantness; the event that triggered Menger's displeasure was objectively speaking wholly insignificant, but it is characteristic for the changed mood. One day, Menger had as usual visited Brouwer in his home, and when parting Brouwer asked him to come round in a few days to discuss the dedication of Menger's dimension *Bericht*. And so Menger duly went to Brouwer's house twice during the next two days, but he found Brouwer out. The next four days he was confined to his bed with a flu. As soon as he recovered, he went to Brouwer's house, where he found Cor Jongejan, who told him that Brouwer had left the country for a longer period.[149] She added that no message was sent to him, as Brouwer assumed that Menger had left. Menger flew into a temper. He returned home and immediately composed an angry letter. 'I must tell you, professor, that I have for the first time in my life heard such a comment which contains such an imputation of lack of character, education and manners.'[150] It is a fact that a well bred young scholar from Vienna would consider it unthinkable to fail to keep a social obligation towards his host, and thus Menger's dismay can be imagined. It would probably go too far to assume that Brouwer thought Menger guilty of a breach of good manners; Menger probably read too much in the event.

Having vented his feelings, Menger turned to a matter that urgently needed attention: the dedication of the Dimension Report. He suggested the formulation

| | |
|---|---|
| | Herrn L.E.J. Brouwer |
| entweder: | dem grossen Förderer der Topologie |
| oder: | dem bahnbrechenden Bearbeiter der Topologie |
| | zugeeignet. |

| | |
|---|---|
| | Dedicated to L.E.J. Brouwer, |
| either: | the great promoter of topology, |
| or: | the pioneering researcher of topology. |

[148]Brouwer (1976), p. 567.

[149]Brouwer was off to Batz, to meet Alexandrov.

[150]Menger to Brouwer, 19.VIII.1926.

It is curious that a simple matter like the formulation of a dedication had become the subject of a serious exchange of thought. Brouwer, with his penchant for refined and precise formulations, may have stressed the importance of the right words, or Menger may have wished to make it clear that he was praising a giant of the *past*. Whatever considerations may have played a role, one does not need a great deal of imagination to have second thoughts about the dedication; true—Brouwer was promoting topology even now, and he had done pioneering work, but all the same—the formulation would equally well suit a civil servant who had promoted, say, the export of tulips, and '*Bearbeiter*' sounds like a person who is sorting out an existing field. The proposed formulations definitely had a lame ring. In view of the later developments and of Menger's report of his Amsterdam experiences, it does not seem far fetched that the dedication was a well meant, but calculated, present. Menger could not simply have dropped the dedication after first asking permission to dedicate the *Bericht* to Brouwer. When the dimension report appeared the special offprints carried a simple dedication: '*Herrn* L.E.J. BROUWER *zugeeignet*', and Brouwer's copy carried the handwritten dedication 'Dedicated to you by the author in sincere appreciation of your work and in gratitude.'[151]

Much later Brouwer mentioned this dedication to Hahn;[152] the topic was Menger's alleged disparaging view of Brouwer's activity in dimension theory. On Hahn's assurance that Menger did not in the least question Brouwer's quality in that respect, Brouwer dryly pointed out that Menger 'would thus only confirm a conviction that he continually stated in the past, which was expressed most clearly when he wanted to dedicate his dimension-Bericht to me. I had at first to decline such a dedication—which Menger had originally had in mind as "in admiration and gratitude"—in this form, but which I was prepared to accept if it would bear on my quality as creator of the notion of dimension. Only after Menger had declared his agreement with this view (which was in accordance with the formulation of his Bericht) in the presence of others, did I accept this dedication.'

In the same letter Menger thanked Brouwer for the honour of being his assistant for a year, but he bade Brouwer at the same time not to extend the assistantship for another year. This, he added, he was certain would also 'satisfy your own wishes'. One does not have to be an accomplished psychologist to read between the lines, that the two men had become estranged. Even the last line, in which Menger bade Brouwer to visit him when in Vienna, could not take away that impression.

In contrast to the Brouwer–Hilbert conflict, where no relevant correspondence is available at all (and it is doubtful if there was any), there is quite a bit of written evidence in the Brouwer–Menger relationship. In reconstructing the development, one has to take the correspondence seriously, later reflections could easily have been coloured by rationalisation after the facts, and by subsequent experiences. There is a curious discrepancy between the accounts of Menger and Brouwer; we will have to look carefully into the actual exchange of letters in the twenties.

---

[151] *In oprechte waardering van Uw werk en in dankbaarheid U opgedragen van den schrijver.* Menger had for all practical purposes mastered the Dutch language.

[152] Brouwer to Hahn, 27.VIII.1929.

Menger, the other party, was not blessed with social graces. The few surviving reports on him show him an introverted person, with a vastly exaggerated feeling of his own importance. Alexandrov, in a letter to Hopf of 23.XII.1926, reported that 'Menger seems, as before, to remain Brouwer's assistant. He will not live in Laren or Blaricum, but in Amsterdam. The wagging tongues of the Laren-Blaricum ladies say that Menger had not found in any house in Laren-Blaricum (according to his pretensions) a fitting admiration, and since he had fallen out with all the housewives, none of them seems to be willing to forward his mail.' The specialist who at the time was called in to treat Menger for a disorder of the lungs had without much ado diagnosed Menger's affliction as the result of a pathological inflation of his ego, and prescribed a treatment accordingly.[153]

It would go too far to say that bringing two characters like Brouwer and Menger together was asking for problems, but those who knew both persons were aware of the volatile situation. A careless remark or a conflicting view could easily blow the delicate balance to pieces. Let there be no misunderstanding, there was no lack of appreciation between the two men. Menger revered Brouwer as a profound scholar and a keeper of the Holy Grail of mathematical wisdom, and Brouwer recognised in Menger a gifted thinker and a man with a precious geometrical spirit.

Reading Menger's memoir of his Dutch intermezzo, one wonders why, after all, he stayed on for another year. Perhaps Brouwer managed to convince him that he was an appreciated member of the mathematical group in Amsterdam, perhaps Menger took another look at the situation, and decided that one more year in Amsterdam would be alright.

During this summer of 1926 Menger set himself, among other things, the task of digesting Brouwer's courses on intuitionism which he had attended in the spring term. It suddenly struck him that Brouwer's spreads (and fans) were nothing but the analytic sets from descriptive set theory.[154] The reason that nobody had seen this before was, he guessed, that the set theoreticians found Brouwer's intuitionistic papers unreadable, whereas those who could understand Brouwer's intuitionism had no taste for descriptive set theory.[155]

Menger's discussion of the analogy between spreads and analytic sets in 1928 did not draw the attention it deserved. His paper has been overlooked by set theorists and intuitionists alike; Menger justifiably complained about this.[156] At the publication of the paper Heyting told Menger that he hoped that it would help people to understand the Brouwerian terminology, and indeed, Hausdorff confessed that now he understood finally what Brouwer's '*Menge*' was about.

---

[153]Brouwer to Hahn, 22.X.1929.

[154]Menger (1979), p. 246.

[155]One can sympathise with Menger, although my experiences were the other way round. When I first learned about analytic sets, I thought, 'Why, these are spreads!'

[156]Menger (1979), pp. 86, 246.

The problem with Menger's paper[157] was that it pointed out superficial similarities, obscuring the finer points which were at the bottom of the differences. In particular, the peculiarities of Brouwer's spreads derived from the choice sequences that they were made of, and their properties were (at least in Brouwer's approach) best seen through functions defined on spreads. It would be unfair to blame Menger for this, because it took the foundational community as a whole a long time to see Brouwer's point. Brouwer had seen Menger's manuscript while still in preparation, this is confirmed by the note in Brouwer's handwriting, which served as a draft for a letter to Menger. In the note he comments on the text, one particular remark is 'Delete remark about Hurewicz, because this observation has been made by many, namely so far by every one who got to know my 'Punktmengen' [spreads] and also the Alexandrov, or Souslin [Brouwer's spelling] sets.'[158] From this draft we learn that Hurewicz had also been shown (or perhaps, had noted) the correspondence, and that Brouwer considered the observed similarity a matter of common knowledge, in other words, that it was not all that unnoticed. Since the name of Hurewicz does not occur in the paper, one may conclude that Menger saw and adopted Brouwer's advice. The note further implies that Menger's observation was not the novelty Menger considered it to be. Heyting, in his review,[159] noted that Menger presented a historical discussion of the external features of intuitionism, without however doing justice to the conceptual content, in particular Brouwer's contributions. Menger's analogy is mentioned with the warning that the analogy is essentially a classical phenomenon. In his monograph *Mathematische Grundlagenforschung. Intuitionismus. Beweistheorie*, Heyting observed somewhat uncharitably that the interest for intuitionists was negligible, as Menger's paper made free use of the principle of the excluded third. After that, the paper seems to have dropped out of sight. Menger probably viewed the neglect of his contribution as part of an intuitionistic conspiracy, but in all fairness, it is hard to see what the analogy could contribute more than a didactical device. Thus it would be fair to say that Menger's paper was likely to promote as much misunderstanding as enlightenment. Menger was certainly interested in the aspects of constructive mathematics, but somehow he missed the point of Brouwer's intuitionism, and by the time the theory of recursive functions entered the stage, he had lost interest in the subject. It is therefore not surprising that Menger's papers on constructive mathematics play only a modest role in the literature.

The academic year 1926/27 showed, at least on the surface, no signs of conflict. Menger had decided to stay on in Amsterdam, but he no longer felt comfortable. In his 'My memories of L.E.J. Brouwer' he reported an increasing awareness of Brouwer's pre-occupation with matters of reference and credit. Attempts to discuss points brought up by Brouwer only led to sharp altercations. So he was relieved

---

[157] Menger (1928a), '*Über Verzweigungsmengen*', translated in Menger (1979) as 'An intuitionistic-formalistic dictionary of set theory'. See also Menger (2003), pp. 3–22.

[158] See Menger (2003), p. 4.

[159] Heyting (1931b).

when Hahn offered him a position in Vienna. The actual situation in Amsterdam is hard to assess. Menger's feelings towards Brouwer had changed from admiration to aversion and distrust. Menger was not willing to take second place to anyone, and the intermezzo of the genesis-paper (see p. 481) must have been an embarrassing occasion. Even though all face saving steps were taken, the fact that one had to render an account of one's past research to the world at large was not flattering. Being inclined to find his surroundings at fault rather than contemplate the possibility of a weakness on his side, he did not take it well to be on the defensive. Brouwer no doubt saw the strong as well as the weak points of Menger, and he felt it his responsibility to steer him in the right direction, in particular to instruct him in the subtle art of crediting. The surprising fact is that Brouwer's letters remained friendly in tone; there are no hidden reproaches 'between the lines'. Menger's ambitions had by no means escaped him, as a letter to Hahn shows.[160] Looking back at Menger's stay in Holland he remarked that 'Menger has in the past in Laren expressed himself in the same insulting way with the same injustice as now about me, about anyone whom he felt to be in his way to a lightning world-fame'. In short, Brouwer saw a gifted young man with megalomaniac ambition, and Menger perceived Brouwer as a scheming member of the old generation who wanted credit for some old, imperfect ideas.

Nonetheless, he wanted to conduct topological research in Amsterdam on a 'business as usual' basis.

The relation between Menger and Brouwer gradually worsened, and Menger reported in his 'My memories of L.E.J. Brouwer' that Brouwer insisted on certain changes in Menger's manuscripts and became incensed when Menger tried to argue with him. By themselves these interventions were not strange, Brouwer always inserted comments and corrections in papers that he was handling for the *Annalen* or the Proceedings of the KNAW. It is in fact a normal editorial procedure. Although Menger does not say so, his account seems to suggest that Brouwer's interference was not quite on the level. The available information does not seem to bear this out, it is more likely that Menger disliked being bossed by anybody.

If there were any concrete examples of Brouwer's intervention beyond matters of formulation or mathematical detail, Brouwer's own dimension papers or Urysohn's would most likely be involved. In fact, there is such an example in Brouwer's letter of 2 November 1924, before the two had met. Menger had referred to Brouwer's definition of dimension in a formulation that suggested that only after a subsequent correction could Brouwer's dimension be considered as such, and only then could be compared to the definition of Menger. In other words, before the correction there was no definition at all. Brouwer's viewpoint, as we have seen, was that the definition was correct up to a slip of the pen. He had convinced Urysohn of his view, and he encouraged Menger to take the same view. Menger did indeed adopt Brouwer's formulation, but that did not prevent him from sticking to his convic-

[160] Brouwer to Hahn, 22.X.1929.

tion that Brouwer had missed the right definition, as appears from his later publications.[161]

Menger conjectured[162] that Brouwer was no longer satisfied with being associated with the dimension *concept*, but wanted to be considered the founder of dimension *theory*.

If Menger was right—and the evidence is not quite conclusive—then this was something that emerged as a reaction to certain developments. In Brouwer's papers up to 1924, the phrase 'dimension theory' does not occur. The term appears only in his historical comments on dimension theory (Brouwer 1928f, see below). This paper, which basically is a critique of Menger's book *Dimensionstheorie*, contains a number of explicit statements on this issue.

The paper opens with 'I have founded dimension theory...', and on the same page Brouwer explains his position with respect to the further exploitation of the notion:

> In my cited paper of 1913 I have restricted myself to the founding of dimension theory and abstained from the publication of further dimension theoretic developments, on the one hand because with the justification theorem[163] the posed epistemological goal had been attained, on the other hand because for the subsequent considerations (in the first instance those which are grouped around the sum theorem and the decomposition theorem) an intuitionistic realisation, unlike that for the justification theorem, was not plausible.

This argument may have carried a limited weight in 1928, but the modern mathematician will agree that once a proper definition of dimension is available, the immediate consequences are not all that surprising, and a superior topologist like Brouwer would certainly have seen the first steps. However, his dimension paper appeared at a time when his interest in topology was already waning, while at the same time his foundational interests had become stronger. He may very well have sensed the difficulties offered by, for example, the continuum in an intuitionistic setting. Modern developments in this field have fully vindicated Brouwer's views.[164]

It should be noted that at no time during his life did Brouwer feel inclined to exhaust the consequences of his ideas. In almost all cases he proved the fundamental results and left the exploitation of his ideas to others. The first topological period

---

[161] For comparison, here are the two formulations: Menger—'*in einem allerdings weniger bekannten kurzen Aufsatz (Crelle Journ. 142, S.146–152) eine Definition n-dimensionaler Kontinua gegeben, die nach Korrektur (Amsterdamer Akademieber. XXVI, 1923) mit unserer Definition des n-dimensionalen Kontinuums äquivalent ist*'; Brouwer—'*in einem allerdings weniger bekannten kurzen Aufsatz (Journ.f.Math. 142, S.146–152; vgl. auch die Korrektur eines daselbst befindlichen Schreibfehlers in den Amsterdamer Proceedings 26, S.796) eine Definition n-dimensionaler Kontinua gegeben, die mit unserem Definition des n-dimensionalen Kontinuums äquivalent ist.*' The reader will appreciate the difference in suggestive force.

[162] Menger (1979), p. 247.

[163] i.e. the theorem that $\mathbb{R}^n$ has natural dimension $n$.

[164] Cf. van Dalen (1999a), in which it is shown that, e.g., the irrationals (the complement of the rationals) are indecomposable, and hence 1-dimensional.

was in a way an exception. There it was the challenge of Lebesgue and Koebe that produced such a wealth of results.

So why did Brouwer, apparently, object to a modest role of the inventor of the notion of dimension? The most plausible answer seems to be that somehow Menger claimed the ownership of dimension theory for himself, and that he exuded, or even explicitly stated this belief. Tact was not Menger's strongest point, and a discerning man like Brouwer could easily have felt slighted, not to mention the belittling of Urysohn's role. In cases like this both sides tend to overreact and escalation sets in.

Menger was overly sensitive where his rights were concerned, and even before the Brouwer–Menger conflict flared up, he was inclined to see conspiracies to rob him of his well-deserved credit. Alexandrov was at one point his *bête noire*. In a letter of February 21, 1926 Menger wrote to Brouwer, begging for support,

> ... I would have liked to say a few words about my own business. You had, dear professor, dropped the last few days in conversation comments about my dimension theory papers that have deeply saddened me, because I believe not to deserve them. I can imagine how Alexandrov, during the last weeks of his stay here, has worked under full pressure to put the finishing touch to what he has been trying systematically and with all means since a year: to discredit me and my papers in your eyes. Please, do not lend him your ear. If I should feel that he succeeds in this plan, I would consider myself robbed from my total present existence.

In the absence of corroborating material there is little one can say about the matter. It is true that Alexandrov took a jocular view of Menger's preoccupation with priority rights, but he was not likely to go beyond a few quips. Another Russian author, Tumarkin, had also entered the field of dimension theory, and in his papers he had overlooked the work of Menger. Since Tumarkin was a protégé of Alexandrov, Menger blamed Alexandrov for the lapse. Alexandrov duly apologised, accepted full responsibility for the omission[165] and promised a printed recognition of Menger's priority. In a letter to Brouwer, Menger drew his attention to Tumarkin's paper, but insisted that he should not get annoyed, 'the matter is by no means so important that it is worth half an hour of your time'.

Roughly at that time Brouwer must have decided that the priority claims of Menger should be settled once and for all. The result was Menger's 'genesis'-paper, discussed above.

With Menger safely in Vienna, the friction had disappeared, and both parties could heave a sigh of relief. Most of the information on the increasing friction between Menger and Brouwer comes from Menger's reminiscences. The curious fact is that, if one reads the letters from Brouwer to Menger, the situation does not look as bleak as Menger makes it appear. Up to 1928 the letters from Brouwer are cordial and contain no accusations or attacks. There are no signs that Brouwer considered Menger as an (even potential) adversary. Perhaps Brouwer considered the dimension

[165] Alexandrov to Menger, 1925 (no precise date known).

matter settled after Menger's account of the genesis of his ideas on the dimension notion. As we will see, he was well aware of Menger's personal streaks, but he probably accepted those, as one accepts medical complications in a person—things that could be straightened out in the long run. In all, he considered Menger as a bright and promising postdoc, and treated him as a scientific collaborator. Once having accepted Menger as a student and fellow researcher, he felt it his responsibility to guide Menger both in purely scientific matters, and in the *mores* of the world of scholars. In these respects he fully recognised the strong and the weak points of his assistant. It would be absolutely incompatible with Brouwer's personality to detest Menger, and write kind letters at the same time.

In Vienna Menger became the key figure in the mathematical community. He continued his own research in topology, initiated the new field of metrical (distance) geometry, published on a wide spectrum of topics and organised the influential Mathematical Colloquium in Vienna. For the foundations of mathematics his role as scientific promotor and protector of Gödel was particularly significant.[166]

The peace between Brouwer and Menger was now and then interrupted by a small salvo, for example when Brouwer sent the proofs of Tumarkin's paper 'On the dimension of non-closed sets'[167] for comment, Menger sharply protested against the violation of his priority; he (too) politely informed the editors (*die verehrliche Redaktion*) that publication in the present form might make a rejoinder necessary. One wonders how Menger judged the situation; Freudenthal stated that 'Correspondence up to 1928 shows Menger attached to Brouwer',[168] but letters are not so easy to judge—what is old fashioned civil politeness, and what is sincerity? Menger could with some justification think that Brouwer was after Menger's priority rights. After all, if a manuscript of Tumarkin could reach the stage of proof sheets, then Brouwer had failed to protect his (Menger's) rights.

Another source of discontent was the choice of authors (editors) of the prestigious Encyclopaedia of Mathematical Sciences; Tietze and Vietoris had been asked to prepare a chapter on 'Relations between the various branches of topology'.[169] Brouwer informed Menger about the contribution and promised to remind the authors to send the proofs, but Menger was in no mood to lend a hand to a project that had been arranged behind his back. He was already aware of the Tietze–Vietoris project, as he had met Vietoris in the Vienna seminar. The authors had added insult to injury by categorically refusing to show the proofs to Menger, only Rosenthal and Kneser were allowed to see (and read) the proofs! Menger, who had definite views on the pecking order in topology, commented angrily:[170]

---

[166]Cf. Menger (1994), p. 200.

[167]Tumarkin (1926), Menger to Brouwer, 8.IV.1927.

[168]CW II, p. 564.

[169]*Enzyklopädie der mathematischen Wissenschaften, Dritter Band, Geometrie, Ch. 13.* Enzyklopädie der mathematischen Wissenschaften (submitted 15.X.1929). In fact Klein had asked Brouwer to engage Vietoris to write the chapter. Alexandrov and Menger had opposed the idea (interview Vietoris).

[170]Menger to Brouwer, 17.I.1928.

> If the authors expect more help from those two gentlemen than from me, let them believe it. [...] That I must fear under the present circumstances, also in view of Vietoris' total ignorance of the fundamental theorems of dimension theory (published in 1926), to have serious objections against this chapter, is clear indeed. I suppose, of course, that you will have arranged the formulation of the word of thanks to you in such a way that it will still be possible, in view of the highest regard for you, to formulate possible objections against the trash of Tietze–Vietoris.

The Encyclopaedia chapter was, when it appeared, not the shocking piece of injustice or ignorance that Menger had feared, it was loaded (as usual) with references, and Menger got his fair share.

The above letter also contained a cryptic message: you will soon get a letter from Hahn and Ehrenhaft. The announced letter was part of an attempt to invite Brouwer to Vienna for a talk on the foundations of mathematics. Before we can discuss the 'Vienna connection' we shall have to go back and retrace our steps to the *Grundlagenstreit* and the reception of intuitionism.

The developments of topology until the middle twenties were adequately summed up by Georg Feigl in the *Jahresbericht*, a perspicuous survey of topology,[171] in which Brouwer got full recognition for his contributions to the field. Together with Poincaré and Fréchet he was listed as one of the founders of modern topology. Feigl, by the way, had no problems in giving Brouwer credit for making 'the first and most important step towards dimension theory', by providing an internal topological definition of dimension.

[171] *Geschichtliche Entwicklung der Topologie*, Feigl (1928).

# Chapter 13
# From Berlin to Vienna

## 13.1 More Intuitionism

More or less against his will, Brouwer had to prolong his stay in the fair country of topology. The Russians, and subsequently the Austrians Menger and Vietoris, had re-awakened his interest in the area. The sudden influx of gifted young mathematicians had turned Amsterdam into a centre for topology, and it remained so until the second world war. Being the leading authority on topology on the editorial board of the *Mathematische Annalen*, Brouwer was the obvious person to handle the steady stream of pioneering papers. Moreover, he submitted a large number of topological papers to the KNAW. But although he invested a great deal of time in his activity as editor and supervisor, his main research efforts were undeniably directed at the development of his intuitionistic program. He was sincerely convinced that this was the new mathematics, that, if it would not drive out the traditional mathematics, it would at least be recognised as a viable alternative.

Brouwer had organised his campaign along two lines; there was his massive exposition of the basic parts of intuitionistic mathematics in the *Mathematische Annalen*, '*Zur Begründung der intuitionistischen Mathematik I, II, III*',[1] and there were quick, effective expeditions into the territories of the mathematical empire.

His great expositions were basically an expansion and revision of the old *Begründungs*-papers of 1918 and 1919. New insights were incorporated, old notions were refined, but the content was not unfamiliar to those who had followed Brouwer from the beginning of his second intuitionistic program. It is no secret that few had read Brouwer's earlier expositions, and we can say that this new series offered in fact a first large scale international dissemination of the intuitionistic ideas.[2]

[1] Brouwer (1925a, 1926a, 1927a).

[2] The reception of the papers was, judging from the review in the *Fortschritte*, rather nondescript. The reviewer, Arthur Rosenthal, one of Brouwer's topology friends, restricted himself to a few lines, referring to his earlier review of the old *Begründungs*-papers. But even that review of more than two pages did little to lift the veil of mystery that covered Brouwer's writings (*Jahrbuch der Fortschritte der Mathematik* 47, p. 171, 51, p. 164).

D. van Dalen, *L.E.J. Brouwer – Topologist, Intuitionist, Philosopher*,
DOI 10.1007/978-1-4471-4616-2_13, © Springer-Verlag London 2013

The material of the three parts of the 'Founding of intuitionistic mathematics' was not a mere update and rearrangement of the first *Begründungs*-series, e.g. the notions of spread (*Menge*) and choice sequence were extended to allow higher-order restrictions (cf. p. 449). In the second paper, which treated the notion of order, Brouwer shifted his attention to the so-called *virtual ordering* (cf. p. 379), which, roughly speaking, extended the natural ordering by a double negation. Brouwer chose to make his ordering relation as strong as possible. Indeed, his paper 'Virtual ordering and inextensible ordering'[3] established that the virtual ordering is the best one can get. Virtual order on the continuum just orders 'better' than the natural order. History has shown Brouwer wrong in his preference for more order, in fact the order which is the traditional companion of apartness (i.e. satisfying $a \# b \Leftrightarrow a < b \vee b < a$) turns out to be the most useful notion for practical constructive mathematics. From 1925 onwards, Brouwer gave virtual order a prominent place in his expositions. In Brouwer (1949a) he used the term 'negative order' for 'virtual order', perhaps a better, albeit less colourful, name. He also gave an exposition of his analysis of the fine structure of ordering, as introduced in his paper 'Does every real number have a decimal expansion?'[4] The third paper in the series treated the notion of well-ordering, much along the same lines as Brouwer (1919a).

The shorter papers convincingly demonstrated that intuitionistic mathematics could indeed handle certain topics that were generally seen as test cases for a mature mathematics. Brouwer dealt successfully with the fundamental theorem of algebra,[5] Heine–Borel,[6] the Jordan theorem for the plane,[7] intuitionistic metric spaces and the definition and justification of the notion of dimension.[8]

The last mentioned paper makes good Brouwer's claim that his pioneering work in dimension theory was basically intuitionistically correct. Of course, a number of intuitionistic refinements had to be introduced, but the results of his 1913 paper could clearly be upheld.

The paper that is most often quoted for Brouwer's proof of the bar theorem, the fan theorem and the continuity theorem, is his contribution to the Riemann volume of the *Mathematische Annalen*, 'On the domains of functions'.[9] The title of the paper referred to a serious problem that is characteristic for intuitionistic mathematics:[10] if all total functions on the continuum are continuous, how does one handle

---

[3] *Virtuelle und unerweiterbare Ordnung*, Brouwer (1927c). A proof of the main theorem of this paper already occurs in Brouwer's classroom notes of 1925.

[4] Cf. p. 316 and Brouwer (1921a).

[5] Cf. p. 377 and Brouwer and de Loor (1924), Brouwer (1924b).

[6] Brouwer (1925b).

[7] Ibid.

[8] Brouwer (1926b).

[9] Brouwer (1927b), also cf. pp. 372, 373.

[10] Of course, also for later branches of constructive mathematics.

discontinuous functions? Clearly discontinuous functions can only exist on domains smaller than the continuum. For example, the classically total function

$$f(x) = \begin{cases} 0 & \text{if } x = 0 \\ \frac{1}{x} & \text{else} \end{cases}$$

can be mimicked intuitionistically by

$$f(x) = \begin{cases} 0 & \text{if } x = 0 \\ \frac{1}{x} & \text{if } x \# 0 \end{cases}$$

Its domain is, however, a proper subset of $\mathbb{R}$; namely $\{x \in \mathbb{R} | x = 0 \vee x \# 0\}$. The last section of the paper dealt with the following important problem: what are the proper intuitionistic analogues of the classical total discontinuous functions. After a number of case studies, Brouwer offered the suggestion that 'Only those subsets of the unit continuum should be admitted as pseudo-full domains, which are in the first place congruent to the unit continuum, and which are measurable for all measures of the unit continuum and have measure 1.'[11]

The paper is as scholarly as one could wish; there is just one footnote that betrays the tensions of the times. In footnote 8 (p. 64) Brouwer pointed out that 'canonical' proofs are well-ordered objects (in the intuitionistic sense), to be distinguished from their 'finite necessarily inadequate' linguistic representation. Summing up his objections to the formalist program, Brouwer continued:

> The preceding remark contains my main argument against the claims of Hilbert's metamathematics; a second argument is this, that the settling of the problem of reliability of the principle of the excluded middle (taken, by the way, from intuitionism) is sought by Hilbert in a vicious circle; for if one wishes to found the correctness of this principle by means of a proof of its consistency, the principle of the reciprocity of the complementary set, thus of the principle of the excluded middle itself, is implicitly presupposed.

It is worthwhile to point out that Brouwer's position on infinite (well-ordered) proofs predates that of Zermelo and Hilbert.[12]

## 13.2 Feelings of Crisis and German Science

The First World War had left Germany in a crisis, not only socially and economically, but also spiritually. The newspapers and the army bulletins had given the citizen the general impression that Germany was fighting a just war, and that the military situation was not unfavourable. The sudden armistice, and the eventual peace of Versailles, left the Germans with a serious trauma. In a surprisingly short time the empire was transformed into a republic, and suddenly Germany was confronted

[11] 'Congruent' means roughly 'identical up to a double negation'.

[12] Zermelo (1929), Hilbert (1930).

with uprisings and revolutionary movements. Although the new republic managed to restore a measure of order, the average German citizen felt betrayed and insulted. The scientists were not excepted from the general feeling of malaise. They were, in a sense, worse off than the general citizen; they suffered not only from the disastrous economic situation, but they were also systematically excluded from international contacts; officially (although not always in practice) they were made the pariahs of the western scientific world. The German scientific community saw itself treated in a harsh and, for them, incomprehensible way. For example

- the Germans were excluded from 129 of the 195 international conferences between 1920 and 1924.
- the role of the German language in international publications was severely reduced, from 40% in 1909 to 27% in 1929.
- the German monopoly in review journals was broken up, some new review journals went so far as to refuse the review of papers in the German language.
- personal contacts with scholars from the *Conseil*-associated countries were forbidden.[13]

In this climate the German intellectual community was thrown back onto itself. It saw itself excluded from the international exchange of ideas and scholars, for which it was famous. The intolerance of the scientific boycott and the general humiliation provided a fertile soil for a rich variety of nationalistic movements under the collective name of *Deutschnationalen*.

A considerable part of the German scientific establishment was far from happy with the new political institutions that replaced the old Wilhelminian order, for example the physician Müller, at the Nauheim conference,[14] sketched a black picture of the status of science and scientists:

> ... until now it was the custom at our congresses that at the inaugural session we greeted with veneration the Kaiser and the local prince, in whom we saw the embodiment of our country.—That we can now no longer do!—[mournful pause]—But is it not our duty, gratefully to recall the support which Germany's princes have accorded science and especially the natural sciences?... Monarchies care for the sciences and honour their significant scholars.—Republics support instruction and leave... concern for the sciences largely to the initiative of private individuals... Revolutions, however, destroy, letting a Pavlov starve and beheading Lavoisier.[15]

And although mathematics has a reputation for unworldliness, it was equally well influenced by the post-war malaise. Weyl's proclamation of the revolution in mathematics was far from an abstract pun of an occupant of the ivory tower, but a symptom

[13]Cf. the Denjoy affair, p. 336 ff.

[14]Cf. p. 316 ff.

[15]Forman (1986), see also p. 316 ff.

in the wider context of the German state of society. It was indeed a rather geographically restricted crisis that concerned the German (and German speaking) mathematicians. The threats to mathematics were not purely abstract; the post-war authorities contemplated for example a reduction in the secondary and higher education mathematics and science curriculum. The newly founded National union of German mathematical associations and organisation[16] warned 'At the current revision of the school systems there are forces active which aim at a reduction or elimination of the mathematical curriculum. These attempts find a well-prepared ground in the hostile mood, that certain world-wide distributed works of the *belles lettres*, in spite of the weakness of their arguments, have been able to generate against mathematics.'[17] It is not difficult to guess the 'world-wide distributed'-source of irritation: Spengler's book, *Der Untergang des Abendlandes*, was a best-seller; how serious it was taken may be illustrated by the fact that the mathematician Hessenberg explicitly warned in his inaugural address of 1921 for this 'danger to the youth'.

It is hard for the present reader to imagine how influential Spengler's *Decline* was; the intellectual community in the western world indulged in apocalyptic visions of a fall of western civilisation. The book was found in almost any bookcase, it was widely discussed and quoted, until it quietly became invisible during the Nazi period. Although most mathematicians saw through the flimsiness of the arguments of the culture-pessimists, some were left with an uneasy feeling—was something wrong after all? For example, Richard von Mises, the eminent applied mathematician, thought to discern a 'crisis in mechanics'. In general there was a certain tendency to counter the cultural pessimism by a measure of accommodation, e.g. by stressing or acknowledging the cultural roots of mathematics. The optimism of the natural sciences, the belief in progress, based on the spectacular and solid advances of the past, had become suspect.

The mathematician Timerding, also at the Nauheim conference, remarked that 'We feel in Nature a spiritual essence... in its core incomprehensible.... The transcendent... has thus finally appeared even in scientific research, and therewith optimism is destroyed forever.' The new passwords for science in the Weimar Republic had become 'irrationality', and 'individualism', 'causality' and 'determinism' were out.

Under these circumstances it is hardly surprising that Brouwer's foundational program caught the attention, it offered a scientific basis for an anti-deterministic, man-based mathematics. It remains, of course, questionable in how far the participants in the broader foundational discussions were informed about Brouwer's underlying philosophy. Before 1928 no complete exposition of Brouwer's philosophical views was available; the Dutch publications, *Life, Art and Mysticism*, the dissertation, and 'The Unreliability of Logic' could hardly be considered to be accessible to the German public.[18] The first of these publications was not even known in Holland,

[16]*Reichsverband deutscher mathematischer Gesellschaften und Vereine.*

[17]For further information and literature, cf. Mehrtens (1984).

[18]Brouwer (1905, 1907, 1908b).

and the latter two would have found few readers outside Holland. In all likelihood, those who appealed to intuitionism as a basic philosophy for a 'return to intuition' (*zurück zur Anschauung*), based their beliefs, if not on Weyl's 'New Crisis' paper, or maybe a lecture of Brouwer himself, on the seductive call of 'intuition' (*Anschauung*).

The friction (to put it mildly) between 'modern abstract' mathematics, including formalism, and conceptual, intuitive mathematics worried a great many mathematicians and in the discussion of the twenties many viewpoints will draw the attention of the patient spectator. The well-known mathematician Ludwig Bieberbach, who took in his inaugural address of 1913 a strikingly formalistic position, defended in the twenties his view that the true spirit of mathematics was that of the *Anschauung* (intuition), and he praised Felix Klein for promoting the practice of mathematics along the lines of geometric tradition. He strongly condemned Hilbert's formalism for severing the ties between intuition and application on the one side and mathematics on the other side.

In a talk in 1926 (unpublished), he said:

> ... the time will not be far off... where its overrating [i.e. of formalism] will fade away, and the catastrophic consequences that follow from it, namely the ignoring of the problems of the concrete reality, will belong to the past.

Intuitionism appeared to him 'a breath of spring air' and 'the escape from a nightmare'. To avoid any misconceptions, it should be pointed out that at that time there were no political overtones. Like so many, Bieberbach was seriously worried about the health of mathematics.

There is no reason to associate the views of Bieberbach with his later political activities. Even a man as mild as Blumenthal spoke out at the DMV meeting at Bad Kissingen (1927) against the new fashion, 'The misfortune is our cowardice. It is far easier to occupy oneself with ideals, groups and groupoids, and to cocoon in a system of abstract axioms, than to tackle the really important, concrete problems of mathematics and its applications.'[19]

Traditionally there had been two main centres for mathematics in Germany: Berlin and Göttingen. In the nineteenth century Berlin was the more prestigious of the two; although Göttingen certainly had its star scientists, the Berlin establishment had a larger mass and more famous names. The balance started to shift towards Göttingen at the end of the nineteenth century. Felix Klein moved to Göttingen and made it a base for both pure and applied mathematics. After his exploits in group theory, geometry and automorphic functions, he dedicated himself to the organisation of a department for the study and advancement of high school mathematics. Being virtually in charge of the mathematics department, his view in matters of appointments carried a considerable weight. His choice of colleagues could hardly have been better, among his first appointees were Minkowski and Hilbert. Both of them were top mathematicians, and in a short time they made Göttingen the

[19] van der Waerden (1928).

Mecca of European mathematics. Hilbert's extraordinary qualities were publicly recognised when he was invited to give a large survey talk at the 1900 mathematical congress in Paris,—resulting in the famous list of Hilbert's problems. Since then a long series of present and future leading mathematicians made their name in Göttingen, some stayed on, but most accepted positions at other universities, where they set an example to students and staff.

Berlin, of course, played its part in the mathematical culture of the early twentieth century, but Göttingen was undeniably the place that set the tone for the new mathematics. Among the mathematicians of that period a certain division of loyalties could be discerned, not in any serious way, but enough to be quietly pleased if either Göttingen or Berlin scored a point. In this atmosphere of friendly (sometimes not so friendly) competition the Berlin mathematicians more or less sided with Brouwer against Hilbert. It was therefore not surprising that efforts were made to get Brouwer to Berlin, if not as a full professor, at least then as a visitor. Erhardt Schmidt had campaigned for a chair for Brouwer and when nothing came of it, he may have opted for a visiting appointment. Already in June 1921 the faculty in Berlin had proposed Brouwer as a visiting professor for the winter term, but that time things did not work out. It is likely that Bieberbach renewed the efforts to get Brouwer to Berlin as a visitor, and indeed, in January 1927 Brouwer, after obtaining a leave from the University of Amsterdam, made his triumphant entrance into the Berlin mathematical community. He had asked the curators in Amsterdam permission to teach a course on intuitionistic mathematics. In his opinion this was 'in the interest of science and the University of Amsterdam'.[20] He started his course on January 15.

## 13.3 The Berlin Lectures

Contrary to what most people thought in the later years of the century, intuitionism had a definite appeal; serious mathematicians were inclined to agree that something was wrong in mathematics and that Brouwer and Weyl had a point. Some even went so far as to use intuitionistic methods in their elementary teaching. Loewner and Study, for example, taught at one time their first course in analysis (or calculus) according to the intuitionistic principles.[21] Loewner's constructive inclinations are also mentioned in Lipman Bers' introduction to Loewner (1988). In his Berlin period Loewner became very much impressed by the intuitionist critique of classical mathematics. He once taught a calculus course using only constructive proofs and, as he recalled many years later, concluded the first term by proving that a uniformly continuous function, defined on a bounded closed interval, is bounded. Loewner did not remain an intuitionist, but did retain a strong preference for constructive proof.

[20] Brouwer to Curators of the UVA, 10 January 1927.

[21] According to Freudenthal, Loewner's approach was far too complicated—'In short, a catastrophe', see Freudenthal (1987a).

It was not only among the established mathematicians that intuitionism found a certain acceptance, the more adventurous among the students, taking their cue from Hermann Weyl, fervently supported the revolution in mathematics. The actual appearance of the great revolutionary in the lecture halls in Berlin caused a furore. Somewhat to the surprise of the audience, the Dutch revolutionary was a soft spoken man with a gift for long, intricate sentences, and flowery language; a man with a quick wit and steel blue twinkling eyes. A Merlin rather than a Robespierre. The lecture hall was filled till the last seat, and intuitionism and foundations became the talk of the town. Even the newspapers followed the events with interest. The followers of Brouwer, proudly adopting Hilbert's taunt from the 1922 talk in Hamburg, called themselves *Putschists*, and the mathematician-poet Hubert Cremer rhymed in his Carmina Mathematica

Und wird mir das ganze
Getu hier zu trist,
Dann kauf ich mir'ne Kanone
Und werde Putschist.

Ja klassisch da schließen s'
Mit falschem Genie;
Sie mag net die andern,
Drum mag's also mi.

Wir Putschisten aber sagen,
Des stimmt net deswegen,
Denn sie braucht nämlich leider überhaupt
kein net z' mógen!

Ach, zwischen "Sie liebt mich"
Und "Sie liebt mich nicht"
Da gibt's noch ein Drittes,
Die alte Geschicht.

Und grade das Dritte,
Des gibt mir an Riß,
Man kann's net entscheiden,
Man weiß's halt net g'wiß!

The lectures were attended by a mixed audience, consisting of students, professional mathematicians and interested laymen.

Among the listeners there was a young student, Hans Freudenthal, a man who was going to play an important role in Brouwer's life and in Dutch mathematics. Freudenthal had privately studied Brouwer's work; in the first place, of course, his topology, but also his foundational work. Reichenbach, the young mathematician-philosopher, was also drawn to Brouwer's lectures. Another noteworthy member of the audience was the young André Weil, who later in life became a leading number theorist, and one of the star mathematicians of the Princeton Institute for Advanced Study. Weil had received his mathematical training at the *École Normale Supérieure* in Paris. Having a strong mind of his own, he had decided to visit a number of prominent mathematics departments in Europe, including some of the German ones. That he was breaking the boycott decreed after the First World War did not worry him overmuch.

Weil spent some time in Göttingen, and subsequently visited Frankfurt and Berlin. In his memoirs he related his encounter with Brouwer.[22] No comments on the subject matter of Brouwer's course are to be found, but Weil recalled how he found himself after one of the lectures next to Brouwer in the café, where the *Nachsitzung*[23] was held. He told Brouwer that he had agreed to visit Mittag-Leffler in Stockholm, and had promised to assist him in a certain project. He confessed that he was not looking forward to discussions with the older man who had little to offer in the way of actual mathematics. He would indeed have been happy to get out of his obligation. 'Nothing easier than that', Brouwer immediately replied, 'just pick a fight with him.'[24] It must be admitted that Brouwer sometimes made use of this strategy—it saved a lot of time.

Freudenthal followed Brouwer's lectures with a keen interest. He had already become acquainted with the ideas and practice of intuitionistic mathematics through Loewner's course on differential and integral calculus. Furthermore Freudenthal had read and studied a great deal of philosophy and history. So after Brouwer's lectures he could pose sensible questions, which betrayed a more than average intelligence and intellectual curiosity. He corresponded with Brouwer, and asked for reprints of his intuitionistic papers. Brouwer was indeed so impressed by the young man, that when Freudenthal had obtained his doctor's degree in Berlin, Brouwer offered him a position as an assistant in Amsterdam.

One should not get, by the way, the impression that Brouwer was quietly sitting in Berlin, giving his courses and discussing mathematics. In February we find him in Vienna where he had business to conduct. There is a card to Alexandrov,[25]— 'Tomorrow I dine with Wirtinger, Ehrenhaft, Hahn, Vietoris and Loewy. In Berlin the colleagues are very good to me and my lectures are well attended.' And Cor Jongejan, who accompanied Brouwer to Vienna, had added in a corner 'Brouwer is very much lionised. He drags me along everywhere. Now a diner jacket will be bought.'

At home the old fears revived, would Holland lose Brouwer to Berlin? Henri Borel wrote: 'Have you said farewell to the Amsterdam University, and accepted a chair in Berlin? Is your beautiful house now for always deserted?'[26] And Frederik van Eeden, his ally from the days of the signific circle, complained that he could not miss Brouwer for three months.[27]

---

[22] Weil (1991).

[23] Informal meeting after a lecture.

[24] *Verkrachen Sie sich mit ihm.*

[25] Brouwer to Alexandrov, 3.II.1927.

[26] Henri Borel to Brouwer, 19.I.1927.

[27] Frederik Van Eeden to Brouwer, 16.I.1927.

Among Brouwer's correspondence there was a letter of historical interest. His old teacher and friend, Gerrit Mannoury, had composed a question for the annual Prize Contests of the Dutch Mathematical Society, which had Brouwer's intuitionism for a subject,[28]

> *Draft Prize Question*
> Although Brouwerian set theory is essentially not to be identified with the conclusions which can formally be drawn from some pasigraphy,[29] certain regularities can be observed in the language with which Brouwer accompanies his mathematical intuition, which can be collected into a formalist-mathematical system. It is asked, 1. to formulate such a system and to indicate the deviations of the formalism following from that system and Brouwer's theories, 2. to investigate if by means of a (formal) exchange of the principium tertii exclusi and the principium contradictionis an associated dual system can be derived. [...].

This is the influential prize question that was to be answered by Arend Heyting. According to the rules contestants had to submit their essays anonymously, with only a motto for identification. After the evaluation of the essays the committee would solemnly open the envelope with the winners motto, and destroy the other envelopes. Heyting had chosen the telling motto 'Stones for bread'.

The interest of Brouwer in real estate, which we have noted before, must have extended far over the borders of his native country. We have seen that Cor Jongejan bought a house in Bad Harzburg, no doubt on Brouwer's advice.

For some reason he bought at some time in the twenties a house in Berlin-Zehlendorf (perhaps he considered settling in Berlin, but it could also have been an investment), from 1921 to 1928 there is a considerable amount of correspondence with real estate agencies and the city housing office. At the time of the economic slump, the house proved an awkward liability. Brouwer spent a great deal of time on his Berlin house. The house had to be let, there were repairs, the rent had to be collected, and in the end he had to find a buyer and arrange a sale. It is a saddening thought that so much time, money and ingenuity was wasted on—money!

At all times Brouwer remained a faithful health adept; for a part his interest went back to his youth, when he was seriously bothered by health problems, for another part it suited his way of life. It was this preoccupation with health and healthy living that brought him to the Harz. He had found an establishment with a big reputation in a small place, not far from the town of Harzburg, *Rudolph Justs Kuranstalt Jungborn*.[30] It was a Spa in a small place, Stapelburg, with a sanatorium 'with natural cures; associated school of vision'. Dr. Just's institution had all the features that appealed to Brouwer, and that may have set the pattern for Brouwer's private health

---

[28] Mannoury to Brouwer, 26.I.1927.

[29] A somewhat outdated term for 'symbolic (universal) language', often used by Mannoury.

[30] Just had opened the Spa Jungborn in 1896. The German vicar Felke followed Just's example and opened in 1915 his own Spa, organised in accordance with Just's ideas. Brouwer referred to Felke in his 'Life, Art, and Mysticism'.

**Fig. 13.1** Cor Jongejan, Lize and Bertus in the Harz at the Brocken mountain [Brouwer archive]

practice at home. There was a large central building surrounded by a large number of open air cabins and huts (*Lufthäuschen*). One got up at sunrise, exercised in the natural state, played ball games, had mud baths and open air cures—separated for both sexes. There were diet and fasting cures of a wide variety. There was in addition a special School of Vision, following the ideas of the American doctor W.H. Bates, which practised a better vision method under the slogan 'Learn to see again. Away with glasses.' We may safely assume that the Spa influenced Brouwer's way of life to a large extent, the general principles were largely in harmony with Brouwer's ideas. He often visited the Spa, sometimes accompanied by Lize, who shared in the health practices of Bertus. The buildings survived the war, but as the property was almost exactly on the border of the two Germany's, most of it was torn down, and now nothing remains of it.

**Fig. 13.2** The house at Bad Harzburg [Photo Dokie van Dalen]

During his stay in Berlin, Brouwer renewed old friendships and made some new ones. One name has already come up in connection with Alexandrov, that of Heinz Hopf, a man thirteen years younger than Brouwer. He was born and raised in Breslau (Wroclaw) where he had also started his university studies. Called into military service, he happened, during a short leave, to hear the set theory course of Erhardt Schmidt (the man who, according to Weil, was as quick and sharp as Hadamard) just when he was treating Brouwer's proof of the invariance of dimension by means of the mapping degree. This determined the course of his mathematical career. In 1920 when Schmidt moved to Berlin, Hopf went with him. Brouwer came in contact with Hopf when he handled the latter's paper 'Vektorfelder in $n$-dimensionalen Mannigfaltigkeiten' for the *Mathematische Annalen*. Hopf's paper dealt with matters that were close to Brouwer's own work; as a matter of fact Brouwer informed Hopf about the connection between the work of Hadamard and Brouwer with respect to developments in 1909/10.[31]

Brouwer was so pleased with the work of the young author that he expressed in this letter his hope to meet the author in person. This was the beginning of a lifelong friendship that survived all storms. Before Brouwer came to Berlin, Hopf had already visited Brouwer in Amsterdam. The second topologist who was influenced by Brouwer was Georg Feigl. Both Hopf and Feigl were teachers of Freudenthal. Indeed all the Berlin topologists or analysts more or less adhered to Brouwer-style topology.

[31] Brouwer to Hopf, 21.XII.1925. Cf. p. 154 ff.

There is a perspicuous survey of the topology of the twenties,[32] from Feigl's hand, published in the Jahresbericht in 1927, in which Brouwer got full recognition for his contributions to the field. Together with Poincaré and Fréchet he was listed as one of the founders of modern topology. Feigl, by the way, had no problems in giving Brouwer credit for making 'the first and most important step towards dimension theory', by providing an internal topological definition of dimension. At the same time he acknowledged the dimension theory of Urysohn and Menger. Feigl's survey represented the general view of the mathematical community. For a short while it looked as if the dimension discussion had reached a happy ending.

Hopf had of course attended Brouwer's lectures, and could therefore be expected to appreciate some of the finer points. After he had finished his lectures, Brouwer wrote Hopf a card commenting on the topic of his final lecture; Brouwer had presented as the grand apotheosis of his lecture series a counterexample to his fixed point theorem, an act that never ceased to baffle people. How could one prove such a marvellous theorem and subsequently renounce it?[33] Of course, there is a simple explanation: the theorem is false from a constructive point of view, and it happens that there is a perfectly good substitute for the fixed point theorem. The above mentioned postcard elaborates the intuitionistic viewpoint:[34]

> Dear Hopf,
>
> It occurred to me that I owe you and Feigl an additional statement as to what I said about the fixed point theorem, and I would like to pay this debt before my departure. When I said that the classical fixed point theorems could not be saved intuitionistically as fixed point theorems, I did not, by any means, want to say that these theorems would not admit an interpretation from an intuitionistic point of view, which remains valid there too. On the contrary: the classical theorem that the transformation $\tau$ of the compact space $R$ (which we assume to be metric) exhibits a fixed point, has the intuitionistically correct meaning that for every $\varepsilon > 0$ a point $P$ of $R$ can be determined that is less than $\varepsilon$ apart from its image. And the classical theorem that the transformation $\tau$ of $R$ yields $n$ distinct fixed points has the meaning, which remains intuitionistically correct, that there is an $a > 0$ with the property that for each $\varepsilon > 0$, $n$ points $P_1, \ldots, P_n$ can be determined which all are less than $\varepsilon$ apart from their images, and each two of which have a distance $> a$. These theorems are, however, not fixed point theorems, for one has no means to indicate, i.e. to approximate, a fixed point.
>
> Please also show this card to Feigl. It is meant for both of you.

---

[32] *Geschichtliche Entwicklung der Topologie*, Feigl (1928).

[33] There were more rumours about Brouwer's behaviour and personality than about most mathematicians, some of them absolutely unfounded. I recall that in the sixties a leading mathematician gave a talk at a monthly meeting of the Dutch Math. Soc., who at one point remarked 'Brouwer no longer believed in the real numbers, did he?'

[34] Brouwer to Hopf, 8.III.1927.

The content of this card did not find its way into the mathematical literature. It is another example of the Brouwer's reticence to publish his results. His courses often contained new material; Brouwer was well aware of the novelty value of the material, and in some instances he required the students to sign a statement to the effect that they would not publish (parts of) the material of the course. It was common knowledge that a book published by Haalmeijer and Schogt (the former Brouwer's first Ph.D. student and the latter his first assistant) used material from Brouwer's early courses on point set theory in their book on set theory,[35] a fact that rather displeased Brouwer. In a letter to his publisher Noordhoff of 10 October 1929 he mentioned the matter, 'I don't consider this matter as of great importance, but nonetheless I think that you should be informed about it, if only to prevent as much as possible repetitions of such matters.' Quite often Brouwer assigned some student the task to prepare a set of notes of his course. In the case of the Berlin lectures he could not leave the composition of a transcript of the lectures to students, but he managed to get assistance in transferring his handwritten notes into typescript. The task was carried out by Eva Wernicke, a most competent secretary. Eva went about her job with great enthusiasm, and in the process the author fell completely under her spell, and she under his. One can, looking at the remaining correspondence, easily imagine what the mutual appeal was. For her, Brouwer was the great man with startling unorthodox ideas, who had completely preserved his own boyish charm and sparkling originality, whereas she (to judge from the letters) was a lively person with a fair knowledge of philosophy, literature and with a lively sense of humour. She not only prepared his manuscript, but also—after his departure—looked after his business interests. She talked to the representative of the publisher De Gruyter and tried to convince Brouwer of the necessity to postpone no longer a book on intuitionism (to be published in the series *Göschen Lehrbücher*, which also included Hausdorff's *Set theory*). If she had been on the pay-roll of De Gruyter, she could not have been more persuasive:

> *Please don't get a fright!* This idea, that finally a *book* of yours would come into being (which would doubtlessly be snatched from your hands), is worth not only to be 'discussed', but to be *realised*.

She even offered to come to Holland to take notes,

> Dear Brouwer, don't cry the moment you are requested to write a book. This is not a matter for you to produce new scientific material, but to collect together and present what is already there. So it is basically about you, scraping your publications up-till now from their dungeon in the wooden cottage, spreading them on a big table and then generating the linking text. I am at your disposition for this job.[36]

Not only did she decipher Brouwer's manuscripts and type a first version of the envisaged book, but she also negotiated the royalties with the publishing house.

---

[35] Haalmeijer and Schogt (1927).

[36] Eva Wernicke to Brouwer, 30.III.1927.

Eva took Brouwer's cause firmly in hand and did her utmost to coax Brouwer into producing a text for a book. She must have been an exceptionally humorous and charming lady, who indeed could twist Brouwer around her little finger. In a mixture of admiration and mockery, she could quote tongue in cheek the newspapers and the general opinion to him 'I have read in the newspaper about your 'crown of fame'. Great man I congratulate you and all yours. I enjoy reading something like that, in particular in the present which otherwise manifests itself so plebeian. I hope that the beautiful sentence of the '*selten geworden Menschen*' (Man who has become rare) applied to you...'. Eva had completely mastered the refined art of appeasing Brouwer; with a mixture of flattery, nonsense and appreciation she arrived at results that would have baffled his friends. Her letters would open with 'Verehrter und liebster Professor, *Freund und Übermensch*', or '*Teuerste Lucian*' ('Lucian' for 'Luitzen'), or just '*Lieber Professor*'. When Brouwer sent a worried telegram about the typescripts that had not arrived in time, she filled half a page with references to German craftsmen, Strindberg, class hate, and the underworld, before she came to the point. One could almost say that she had taken a leaf out of Tucholsky's book.

The reader should not get the impression of a secretive affair; Eva had met Corrie and Lize in Blaricum, and she was adopted as a friend of the family. Greetings were exchanged and plans for visits to Blaricum were discussed. Eva not only assisted Brouwer with his manuscript, she also represented his interests as a Berlin landlord. As to be expected, his house in Berlin proved to be something of a disaster,—taxes, rent protection, realtors, .... A steady stream of letters on the matter of the house in Berlin passed from Berlin to Blaricum and vice versa. Real estate and land property played an important role in Brouwer's whole life; old contracts, insurance policies, contractor's bills of all sorts are to be found in Brouwer's archive. There is, for example, a reference in one of the letters to Eva to a piece of meadow in Harzburg, left over from the earlier property—as so often the taxes required attention.

Although in July 1928 Brouwer announced that 'the Berlin lectures will appear in print soon',[37] the book did not materialise after all; Brouwer kept revising the typescript until the middle thirties. One guesses that his perfectionism played a major role in the failure to deliver a book on the subject of his heart. Moreover, after 1928 Brouwer was so disillusioned that he lost interest in the project. The manuscript in the end was published posthumously together with parts of a book on real functions (style Baire-Borel) that Brouwer was working on in the twenties.[38]

The material of the Berlin Lectures has a certain overlap with the series of papers in the *Mathematische Annalen*, but it touched more on everyday mathematics. It contained an extensive analysis of the continuum, not quite identical with that of the *Mathematische Annalen* presentation. A number of the results is based on the idea of the *creating subject*, which appeared in print only after the Second World War. He showed furthermore that if the continuum can be ordered at all, the or-

[37]Brouwer to Heyting, 17.VII.1928.

[38]Brouwer (1992).

**Fig. 13.3** Brouwer in front of his Berlin house (Zehlenberg). [ETH-Bibliothek Zürich, Bildarchiv]

dering has to coincide with the natural ordering.[39] Since the latter is not an order in Brouwer's sense, there is no ordering whatsoever. This material was not published until 1950.[40] The final chapter of the lectures consisted of applications and counterexamples; culminating in a counterexample to the fixed point theorem, plus a formulation of the intuitionistic version of a special case of the theorem: For a topological transformation of a circular disk $\kappa$, for each $\varepsilon > 0$ a point $P$ of $\kappa$ can be found whose distance to its image is less than $\varepsilon$. A similar theorem for the sphere appeared in print in 1952.[41]

The notes of the Berlin lectures did not contain a proof of the theorem 'virtual ordering = inextensible ordering', neither did the *Mathematische Annalen* presentation. One wonders if Brouwer thought the material too sophisticated, or did he have

[39] A partially ordered set is *ordered* under Brouwer's definition if $a \neq b \rightarrow a < b \vee b < a$, this is weaker than the trichotomy property. Brouwer (1992), p. 50.

[40] Brouwer (1950).

[41] Brouwer (1952a).

second thoughts? The theorem is mentioned, so he had no doubts about the fact itself. In Amsterdam he had indeed presented a proof in his course on 'Intuitionistic Order' (1925). A polished proof was published in the year of the Berlin lectures in *Crelle*, the same journal that had published his dimension paper. One cannot think of a convincing reason to prefer Crelle over the *Mathematische Annalen*. Brouwer was probably invited to contribute to the jubilee volume of Crelle's centenary.[42] Whatever the reason was, the paper suffered from a definite lack of visibility.

There is a puzzling letter from Henri Borel, which quotes Gutkind saying that: 'you have made an enormous discovery which influences the foundations of mathematics'.[43] Neither in Brouwer's papers, nor his private notes, is such a discovery recorded, assuming that Gutkind referred to a recent event. The basic facts of the 'Domains of functions'-paper were already established earlier, and the 'virtual = inextensible'-theorem was new and surprising, but it was not going to rock the foundational world. It is just possible that Brouwer had discussed the idea and the use of the creating subject as a tool to study the full continuum.

The outstanding event of 1927 remained the Berlin course on intuitionism. Brouwer had every reason to view the future with confidence, Berlin had received him in grand style and intuitionism was on its way to recognition. After a period of friction between formalism and intuitionism, a modest place under the sun for intuitionism seemed a realistic option. Brouwer's lecture series was not just an internal mathematical matter, it was received in Berlin as a cultural phenomenon. So much so that the newspapers thought it worthwhile to solicit contributions, including from Brouwer, to illustrate the foundational debate in mathematics. Brouwer made it clear to Eva Wernicke that he would do no such thing as long as the newspaper offered him a fee that—according to the prevailing standards—was rather that of 'the better science journalist' than of a 'foreign expert'.[44] Neither did he feel like publishing a collection of his old papers, the time did not seem ripe, and he would certainly not do so in Berlin as long as:

> they offered second class fees, and the academy in this city did not consider him worthy of its membership (but they did so for a mediocre loudmouth, and also a counterfeiter and plagiariser, like Koebe, who is younger than I am, and moreover thanks part of his fame to a theft practised on me, a circumstance which was known to several members of the Berlin Academy before his appointment was announced).[45]

The last part of the quotation shows that even after all those years he had not succeeded in putting the old insults of the Koebe-conflict behind him, and that, although he could be magnanimous and warm hearted, there were moments he could not govern his sharp tongue.

---

[42]*Jubiläumsband I. Aus Anlass des 100 jährigen Bestehens.*

[43]Henri Borel to Brouwer, 26.III.1927.

[44]Brouwer to Eva Wernicke, 22.XII.1927.

[45]For the Koebe-affair, see p. 180 ff.

Alexandrov, who had created his own topological kingdom in Moscow, kept up his contact with Brouwer, and in recognition of Brouwer's role in topology he saw to it that Brouwer was made an honorary member of the Moscow topological society. In his letter of February 15, 1927, he informed Brouwer that both Hopf and he were appointed honorary members. Hopf was the leading Berlin topologist at the time, and Alexandrov had met him at one of his many stays in Germany. Between the two a close friendship developed. They were eventually to write a textbook of topology that dominated topology for years on end. The letters of Alexandrov to Hopf are little gems of impish humour, Alexandrov being an expert in spotting and exploiting the little absurd aspects of the world. Here, as a modest example, is a comment on fixed points: 'Thus I can with the same conviction with which Brouwer speaks of the failing of all marriages, state my point of view, which is beyond any doubt, that the method for proving all kinds of fixed point theorems is that of yours and Brouwer (and not of Lefschetz).'[46]

Brouwer's intuitionism was attracting attention not only in Germany, the country with the oldest foundational tradition, but also in the Soviet Union. One of the brightest Russian mathematicians of that period, the young Kolmogorov, had in 1925 published a paper on intuitionistic logic that showed a sharp insight into the impact of intuitionism on logic. His brilliant analysis predates Heyting's formalisation, but since the paper was in Russian, it unfortunately went unnoticed in the West. On wonders if he sent a reprint to Brouwer? There was another Russian mathematician, one with a philosophical turn of mind, who was attracted by the constructive aspects of Brouwer's intuitionistic mathematics, Samuel Osipovich Shatunovsky. Shatunovsky was already in his late sixties when he decided to inform Brouwer about his own ideas on constructive mathematics. In the Brouwer archive there is a translation of a text of Shatunovsky on constructive mathematics, in fact the introduction to his dissertation. Shatunovsky was a mathematician with original ideas on various subjects. He suffered from the fact that he did not have a school certificate, so that he could not enter upon a formal university career. He attended courses at a number of institutes in St. Petersburg, he heard for example Chebyshev's lectures. After a stay in Switzerland, he found recognition in Odessa, where he spent the rest of life. There is a scathing reference to Shatunovsky in a letter from Alexandrov, 'You give me your opinion on the nonsensical paper of Shatunovsky. As this opinion was always privately mine too, I saw no reason to make somehow a statement about it, and I have restricted myself to pass this on to Mrs. Ehrenfest in order to inform her.'[47] Cor Jongejan sent Mrs. Ehrenfest a German translation with Brouwer's compliments, and that was the end.[48]

The name of Shatunovsky later crops up in a letter to Veblen of 2 April 1932, Alexandrov—probably in answer to a question of the former—described him as

---

[46]Alexandrov to Hopf, 15.II.1927.

[47]Alexandrov to Brouwer, 15.III.1927. Shatunovsky may have approached Mrs. Ehrenfest for an introduction to Brouwer.

[48]Brouwer to Mrs. Ehrenfest, 3.VII.1927.

a very gifted mathematician, a precursor of Brouwer and Moore. As a constructivist Shatunovsky was closer to Kronecker than to Brouwer, who rejected him. Shatunovsky, Alexandrov wrote, was lacking in mathematical culture—a gifted solitary eccentric![49] According to Alexandrov, Shatunovsky paid Brouwer a visit, which turned out a disappointment for both.

Apart from the Berlin Lectures, there is little to report about Brouwer's scientific activities in 1927. The paper that contained a substantial contribution to the intuitionistic program, 'On the domains of functions', had already been submitted in 1926, and the 'virtual order'-paper in *Crelle* was an imaginative piece of work, but not really in the centre of the intuitionistic program.[50]

Although the *Grundlagenstreit* was in full swing, Brouwer chose not to take part. But at the end of the year he for once spelt out his views on Hilbert's program. Perhaps he could not resist the challenge laid down in Hilbert's 'On the infinite'. Hilbert's conviction that classical mathematicians were not going to be cast out of Cantor's paradise was one thing, but that the leading formalist judged the mathematical progress of the 'recent so-called "Intuitionism"' to be modest, went too far. To straighten out the issue, he presented in December 1927 his paper 'Intuitionistic reflections on formalism' to the Dutch Academy.[51] We will return to this and other papers in a later section.

Brouwer was now forty six years old, he had seen the light and the shadows of life, but so far he had observed the human tragedies of others. This year fate struck closer to him, he lost his mother. On May 3 she died of a neglected stomach ulcer.[52] Brouwer was a curious mixture of family man and a hermit guarding his privacy. Whereas we know that the relationship with his father was tense, to say the least, little is known of his relationship with his mother. She was a strict but fair woman, educating her sons came easier to her than cuddling. When she died she had saved out of the meagre salary of a schoolmaster the tidy sum of 23,000 guilders.[53] There are no documents or oral communications shedding light on the impact on Brouwer of the death of his mother. Mourning was a very private matter to him.

Busy as he was, carrying the banner of the intuitionistic revolution into the capital of Germany, Brouwer found time to indulge in his dabbling in real estate in Blaricum. His neighbour, Mr. Crone, contemplated the sale of a few lots and the villa 'De Pimpernel'. Months were spent in negotiations and in the end Brouwer became the owner of a pleasant villa, bordering on his hut and grounds.

As a result he owned a considerable piece of land in Laren-Blaricum, covered with the typical trees and shrubs of the sandy soil. Brouwer occupied himself quite intensely with financial affairs. He negotiated loans, mortgages, bought land, and

[49] A gifted '*Eigenbrötler*'.

[50] Cf. p. 379.

[51] The paper was also published in the reports of the Berlin Academy, Brouwer (1928c).

[52] Lize to Louise, 27.XI.1949. Lize spoke of 'a nitwit of a doctor'.

[53] This was indeed an incredible feat, the savings amounted to more than twice the annual income of a full professor.

**Fig. 13.4** The villa 'De Pimpernel' [Photo Dokie van Dalen]

also tried his hand in investments. In 1927 he made a minor, and relatively insignificant investment—he bought obligations emitted by the *Sodalitas Oblatorum Ordinis Sancti Benedicti*. Innocuous as this may seem, the 'Sodalitas affair' was to haunt Brouwer for the better part of the thirties, as we will see.

On Brouwer's grounds there was a number of small cabins, constructions of diverse kinds, some of them reminders of professor van Rees' colony of Christian Anarchists.[54] In 1925 a modest prefab cottage, called the *Padox* after its factory, was erected on the grounds; it was usually occupied by visitors, and occasionally it was let. Although he did not have a car (and Cor Jongejan got hers only after the war), a garage was added to the grounds. It was a small building with a high roof; according to the neighbours, it was mainly built to block their view of Brouwer's property. In practice the garage was used as an extra cabin for housing visitors (girl friends as the neighbour said).[55]

By now Brouwer was one of the icons of the Amsterdam University. His fame was well established, the offers from Berlin and Göttingen had not gone unnoticed. His authority in the senate of the university had probably something to do with his status in science, but it was also to no small degree based on his verbal virtuosity. He could convince the Senate of almost anything by his convoluted but fascinating arguments. Heyting, who had occasion to observe Brouwer's behaviour in the faculty

[54]Cf. p. 59.

[55]Interview, H.J. Looman.

**Fig. 13.5** The Padox, a prefabricated cottage. [Photo Dokie van Dalen]

and senate later in life, recalled that 'even in meetings, Brouwer spoke in complete sentences; also at meetings he often uttered nonsense, but impeccably formulated and motivated, so that one had to think hard to find the error'. In the minutes of the faculty or the Senate Brouwer does not figure prominently, but one should not conclude that therefore his role was negligible.

In 1927 family and faculty matters got mixed up; a vacancy had occurred when the geologist-palaeontologist Dubois retired, and the faculty was looking for a geologist. In May the Curators recorded in the minutes that 'H.A. Brouwer had not yet put his desiderata on paper', so negotiations with Brouwer's brother had been going on. The matter dragged on and in the minutes of September 10, we read that L.E.J. Brouwer had pointed out that all the geologists in The Dutch Indies were foreigners 'because there is no training here'. He urged the University to take the matter in hand. The faculty and the board of the University must have been satisfied with the qualities of Aldert Brouwer, because his appointment followed in the beginning of 1928.[56]

Intuitionism had gradually become an accepted, be it controversial, part of mathematical life. In the twenties an interest arose in its logical aspects. As we have

[56] Aldert had been very successful in his profession. After his Ph.D. in 1910 (Delft) he worked as a geologist in the Dutch Indies (now Indonesia); in 1918 he got a chair in Delft, and in 1925 he was an exchange professor in Michigan. Utrecht made him an extraordinary professor in 1925. The Royal Academy in Amsterdam elected him a member in 1922. His chair in Amsterdam was in 'general and practical petrology'.

seen, Kolmogorov analysed intuitionistic logic in a masterly paper.[57] By means of a syntactical translation he also established the embedding of classical logic in intuitionistic logic. Here Kolmogorov not only anticipated Heyting's full system of intuitionistic logic, but also Gödel and Gentzen by providing the first example of a *translation*. This so-called Kolmogorov translation basically adds double negations to every step of the construction of a proposition.[58]

Already in 1926 Heinrich Scholz, the German theologian who had turned logician and philosopher, followed Fraenkel's example and visited Brouwer to discuss foundational matters; although this did not result in any significant progress in intuitionistic logic, it showed that logicians of an independent mind started to visit the founder of the subject. Intuitionistic logic gave rise to an impressive body of technical research, but it also tended to confuse the lesser minds. An example is the long co-operation of two Belgian mathematicians Barzin and Errera. Already in 1925 they published a paper on Brouwerian logic which contained a not uncommon misconception. In the early thirties Heyting engaged in a discussion with the two; we will return to the matter, see p. 608. In 1927 Alonzo Church conducted a correspondence with Barzin and Errera, without convincing them.[59]

The greatest blow for Brouwer in 1927 was, in a sense, the plan of the City of Amsterdam to expropriate his wife's pharmacy at the Overtoom.[60] The pharmacy was left over from the medical practice of Lize's father. After his death in 1880, Lize's mother carried on the pharmacy with the help of a so-called *provisor*, that is, a certified manager. Lize later studied pharmacy and took charge of the pharmacy in 1908 after she got her degree. The building and the practice had been bought by Brouwer from his mother-in-law in 1905 for an annuity of 1100 guilders. Brouwer had developed an almost emotional relation with the pharmacy, it had become an important part of his life. He often handled the administration, he personally dealt with the tax office, helped to select staff, etc.

And now the City demanded the building and the grounds for development. In Brouwer's eyes this was not only sacrilege, but also a base attack at his wife's livelihood. He, of course, rejected the City's offer of 40,432 Dutch guilders for the property.[61] He contested the expropriation procedure, and in due course found himself in the district court at the Prinsengracht where he pleaded his own case on November 22. When his objections were turned down, he tried to get a better deal. He argued that the pharmacy had at that time 2142 clients, and resettling the pharmacy meant

[57] He actually formalised a fragment of full intuitionistic predicate logic, cf. van Heijenoort (1967), p. 414, van Dalen (2005).

[58] Cf. Plisko (1988b, 1988a), Troelstra and van Dalen (1988a), p. 59; Troelstra (1978); van Heijenoort (1967). Another famous Russian, Molotov, also proposed a translation, albeit in a completely different setting. According to Leonard Mosley in *Dulles*, 1978, Molotov once during a negotiation said 'I propose that we insert a "not" before every verb in the text'.

[59] Cf. Church (1928).

[60] Lize ran the pharmacy, but Brouwer was formally the owner. See pp. 53 ff., 195 ff.

[61] There were extensive grounds behind the building, stretching all the way into what is now the *Vondelpark*.

a serious loss of clients plus a stiff competition at another location. So he requested a higher compensation, or a location at the Overtoom not too far from the existing shop. The negotiations lasted until the end of 1928, but Brouwer lost on both counts, and so the laborious process of relocating the pharmacy had to be started. The pharmacy at the Overtoom was demolished in 1929.

Politically speaking, Holland was a quiet island in the turmoil of the twenties. Brouwer's active role in those years was mostly in the world of scholars, where the fight between the old adversaries was fought around the *Conseil International de Recherche*. By 1927 the die-hards on the allied side were losing ground. One decisive battle was still looming; the abolition or the reshaping of the Mathematical Union, a subsidiary of the Conseil. That battle was going to be fought in all fierceness in 1928.

Brouwer's close friend, Gerrit Mannoury, was active in national and international politics. He was a staunch Marxist and member of the Communist Party of Holland, K.P.H. He had joined the SDAP, Social Democratic Labour Party, in 1909, and followed it when it changed its name to *Kommunistische Partij Holland*. In the party organs and pamphlets he published innumerable expositions of the Marxist doctrine for the workers. Mannoury was a rather exceptional person, he maintained the highest possible standard of moral and intellectual integrity in the middle of the infighting of his political friends. Far from sticking devotedly to the party line, he had already in 1925 spoken out against the persecution of Trotski. Mannoury had the curious mixture of brotherly love for mankind and Marxist class war that one could find in socialists and communists in the early part of the twentieth century. In his many publications in *De Tribune*, the communist newspaper, one recognises both the spirit of the pietist preacher and the Marxist reformer. In fact, Mannoury was one of the kindest men on earth and his rhetoric must be viewed in the light of the vocabulary of the time. Nonetheless it is surprising to find him defending the old 'you cannot make an omelette without breaking eggs' doctrine, as he did when one of Brouwer's old fellow students Bonger attacked the communists for using of the Sacco-Vanzetti case for political advantage.[62] He had written in the *Handelsblad* 'Even worse than the anarchists are the communists in this case. If anybody in this matter should keep silent with a blush on his cheek, they should, in view of all the injustice that the Russian henchmen bring over their political opponents—when does a worldwide protest begin against that?—the significance of the American affair becomes negligible. He who commits himself injustice or explains it away, and complains about what somebody else may have done, only makes himself ridiculous and despicable.' Mannoury felt that he had to defend the Soviet Union, he replied in *De Tribune*: 'Soviet-Russia is a bastion that communism has raised against the violence of its opponents, and that therefore needs itself violence for its defence. In the first period of the revolution this struggle of violence took the form of an overt war, of late the enemy has taken recourse to hidden violence of murder attempts and

[62]Bonger was a through and through socialist. Later he became a violent opponent of national socialism, and at the German invasion of Holland he committed suicide rather than fall in the hands of his arch enemies.

conspiracies; and in the face of that violence, in my firm conviction, terror, that is summary justice in all its horrors, is necessary and just. But I am also convinced that it is possible, in the middle of those horrors and violence, never to lose sight of the lofty standard of humaneness and justice...' And yet, Mannoury was a man who would and could not 'harm a fly', as the saying goes. It is greatly to Mannoury's credit that he did not hesitate to expose crash cases of party injustice.[63]

In 1928 Mannoury found himself in the centre of a party conflict. Stalin had banned Trotski to Alma Mata; this was in Mannoury's eyes an act that no true communist should condone. He thus sent on January 17 a protest telegram to Stalin (with a copy to Trotski). The Dutch comrades could, as to be expected, not appreciate such a gesture, as Mannoury was to find out in due time. He was indeed expelled from the party on May 20, 1929. Not wishing to accept this injustice, he carried a determined defence to all corners of the party, including the Moscow organs—to no avail.

In December 1927 Brouwer fell once more victim of the flu; in his letter to Eva Wernicke of January 24, 1928 he spoke of 'a bad case of flu, at which roughly 12 days ago my right eye has gone blind'. The ophthalmologist[64] mentioned a haemorrhage in the optic nerve. So far, he wrote, there had been no recovery, but the specialist was optimistic. In the meantime Brouwer trained himself to see with one eye only, and he hoped to be able to travel to Vienna in a few weeks time, albeit with a dark patch over one eye. The correspondence with Eva, of which no letters later than April 30, 1928 have been preserved, contain little glimpses of Brouwer's affection—'Again I am not writing anything personal. You know however, also without words, how my relationship with you is', (22.XII.1927), or 'I admire you, I thank you and I embrace you' (24.I.1928). The main part of the letters, however, deals with Brouwer's house in Berlin. There are all kinds of instructions for Eva concerning the real estate agents, legal advisers, the tax collector, tenants, prospective buyers, etc.

In February the topology school bore another fruit, the young Englishman, Wilfrid Wilson, who had come to study with Brouwer, defended his dissertation on the tenth in the Dutch tradition. His thesis *Mappings of spaces* was a generalisation and extension of Brouwer's work.[65]

## 13.4 The Vienna Lectures

The big event of the early months of 1928 was undoubtedly Brouwer's visit to Vienna. He had been invited by the 'Committee for the organisation of guest lectures of foreign scholars in the exact sciences'[66] to give an exposition of his intuitionistic mathematics and philosophy. The members, Ernst Egger, Felix Ehrenhaft, Hans

---

[63]Cf. Mannoury (1930), e.g. the executions following the assassination of Wojkof.

[64]Prof. Van der Hoeve, the same person who met Urysohn and Alexandrov in Norway.

[65]*Afbeelding van Ruimten*, Wilson (1928), Brouwer (1924f).

[66]*Komitee zur Veranstaltung von Gastvorträge ausländischer Gelehrter der exacten Wissenschaften*.

Hahn, Anton Lampa and Rudolph Wegscheider, envisaged a promotion of the exact sciences through the invitation of prominent scientists. The choice of Brouwer was not all that surprising, in Vienna there was considerable interest in the sciences in general, but in particular in the philosophy of the sciences. There was the Schlick circle, and later the Wiener Kreis. Furthermore, at least two of the members, Hahn and Ehrenhaft, were close friends of Brouwer. Brouwer, by the way, had the privilege of opening the series of lectures. He gave two lectures, 'Mathematics, Science and Language' and 'The Structure of the Continuum'.[67] The first one on March 10, and the second one on March 14. Both lectures contained, among other things, material that roughly speaking was conceived at the time of Brouwer's dissertation (1907), which had never been presented to the public at large. A first version appeared in Brouwer's dissertation, where, however, a substantial part was suppressed by his Ph.D. adviser, Korteweg.[68] This time Brouwer presented a concise survey of his views on the role of the individual. The first lecture presented Brouwer's views on the genesis of the basic entities of the subject's inner and outer world. This exposition was most helpful indeed, as in the dissertation the central intuition more or less seemed to come out of the blue.

I will quote a few parts from the lectures.

> Mathematics, science and language constitute the main functions of the activity of mankind, by means of which it dominates nature and maintains order in its midst. These functions find their origin in three forms of action of the will to live of the individual.[69] 1. mathematical attention, 2. mathematical abstraction, and 3. the imposing of will by means of sounds.
>
> The *mathematical attention* comes about as an act of will serving the instinct of self-preservation of the individual in two phases: the *temporal disposition* and the *causal disposition*.[70] The first one is nothing but the intellectual principal phenomenon of the falling apart of a moment of life into two qualitatively distinct things, of which the one is experienced as giving way to the other, and nonetheless is experienced as preserved by the act of memory.
>
> At the same time the split moment of life is separated from the Ego and shifted to a world of its own, called the world of perception [*Anschauungswelt*].

Once this sequence of the manifestation of two phenomena is recognised, one can easily create longer finite sequences. These sequences are then subjected to the *causal attention*:

---

[67] *Mathematik, Wissenschaft und Sprache*, Brouwer (1929a) and *Die Struktur des Continuums*, Brouwer (1930), p. 83.

[68] Cf. p. 92 ff.

[69] In German Brouwer uses 'einzelner Mensch'. We will use 'individual'. It is important to note that in this paper Brouwer does not take a solipsist position.

[70] In Brouwer (1933a) Brouwer uses '*tijdsgewaarwording*' and '*causale aandacht*', translated as 'perception of time' and 'causal attention'.

> Now the causal disposition consists of the act of the will to 'identify' certain of these temporal sequences of phenomena which extend over the past as well as the future. Thereby comes into being a common substratum of these identified sequences, which is denoted by a 'causal sequence'.

These causal sequences form the building material of the universe. In particular the stable, permanent causal sequences are called 'objects'.

The mathematical disposition finds its sole justification, according to Brouwer, in what in *Life, Art and Mysticism* was called 'the jump from end to means'.[71] In the Vienna lectures Brouwer spoke of the *mathematical act*. This act consists of the following: if we know that a particular desirable phenomenon occurs in a causal sequence, and the phenomenon cannot directly be attained, then one may try to realise some intermediate event in the sequence, in the expectation that the causal sequence will (more or less) automatically take us to the desired phenomenon. Here the means is substituted for the end.

As Brouwer points out, the efficacy of the mathematical act[72] is greatly enhanced if 'external' interference can be blocked. We may add that if all interference is eliminated, so that the end follows with absolute certainty, algorithmically so to speak, from the means, a prototype of the constructive interpretation implication is obtained.

The paper furthermore contains a fairly detailed treatment of the role of language and communication, including the moral techniques and principles necessary for an organisation of human society.

Brouwer's views on language and communication had not changed much since the early days:

> Now there is, however, for the transmission of will, in particular for the transmission of will accomplished by language, neither exactness nor certainty. And this state of affairs unrestrictedly remains the case, if the transmission of will concerns the construction of purely mathematical systems [...]. This situation cannot be fixed by the procedure, that one, as the formalist school does, objects the mathematical language itself (i.e. the symbol system that serves to evoke purely mathematical constructions in other people) to mathematical contemplation, and lends it by means of an overhaul the precision and stability of a material instrument or a phenomenon of the exact science, and communicates about it in a language of the second order or a metalanguage.

After pointing out that none of the above can exclude the possibility of misunderstanding, Brouwer passed on to a discussion of the misplaced trust in logical principles.

The final section dealt with intuitionistic mathematics and its consequences. The novelty here, which probably escaped the audience, was a presentation of his method

[71] See pp. 66, 88 ff. and Brouwer (1905), p. 19.

[72] Which much later is called the 'cunning act', Brouwer (1948b).

of the creating subject *avant la lettre*. That is to say, Brouwer introduced a method for counterexamples, which went beyond the Brouwerian counterexamples. One could, however, easily miss this point, as Brouwer did not stress the new methodology. The Berlin lectures were clearer in this respect. The two terms that are remembered from this exposition are 'fleeing property' [*fliehender Eigenschaft*] and 'pendulum number' [*Pendelzahl*]. The section closes with the remark that the extended principle of the excluded third has contradictory instances, the example being 'all real numbers are either rational or irrational'.[73]

There is a report of Brouwer's lecture by Karl Menger in his memoirs,[74]

> While none of the other members of the Circle knew Brouwer as a lecturer, I had often heard him speak in Amsterdam. So, thinking of the conversation with Schlick a few weeks before, I said to Waismann "Why don't you invite Wittgenstein to these lectures? Brouwer is a stimulating speaker and his topics may arouse Wittgenstein's interest." Waismann reflected for a split second and said "That is a very good idea. I will speak to Feigl. Perhaps we can induce Wittgenstein to attend." Two days before Brouwer's lecture, Waismann told me that Wittgenstein would be present.
>
> Hahn, who was to introduce Brouwer, was notified by Waismann or Feigl when Wittgenstein entered the auditorium. (Like myself, he had until then seen Wittgenstein only in a photograph, which incidentally proves that the latter had never attended a meeting of the Circle in its first years.) From a distance I watched Hahn walking down the aisle to introduce himself and to welcome the guest. Wittgenstein thanked him with an abstract smile and eyes focused at infinity, and took a seat in the fifth row or so.
>
> I always tried to avoid making the acquaintance of someone who appeared to be not interested in making mine and thus stayed far away. But I was curious to see how the guest would behave during the lecture which he attended, probably unbeknownst to him, at my suggestion. So I took a seat two rows behind him and well to his left. Motionless from beginning to end, Wittgenstein looked at the speaker with a slightly startled expression, at first, which later gave way to a faint smile of enjoyment.

Feigl described Wittgenstein's reaction after Brouwer's lecture as follows: 'He became extremely voluble and began sketching ideas that were the beginnings of his later writings. That evening marked the return of Wittgenstein to strong philosophical interests and activities.' The opinions on Brouwer's influence on Wittgenstein

---

[73]The terminology may seem confusing. Brouwer had observed that PEM itself is not contradictory. For consider some open problem, e.g. the Riemann hypothesis $R$. Then if '$R$ or not-$R$' were contradictory, both $R$ and not-$R$ would yield contradictions, So both not-$R$ and not-not-$R$ would be true, which is impossible. Hence the negation of PEM for a single statement is consistent. The same argument works for finitely many statements, $R_1 \wedge R_2 \wedge \cdots \wedge R_n$. However for an infinite set of statements (the extended case of PEM) this is no longer the case. Consider 'for all real numbers $x$, $x$ is rational or $x$ is irrational'. This cannot be true, because it would yield a decomposition of the continuum, which is impossible, cf. p. 375.

[74]Menger (1994), Chap. X.

diverge. Brouwer's views on communication must have appealed to Wittgenstein, and his later use of language games seems an adequate consequence of these views. In mathematics there are certain traces of Brouwer's influence, e.g. Wittgenstein's use of the decimal expansion of $\pi$ to illustrate some points concerning real numbers. It should be noted, however, that in the printed version of the Vienna talks the decimal expansion of $\pi$ does not occur. Brouwer may of course have used the example in the talk, or Wittgenstein could have read Brouwer's earlier papers, or—for that matter—even Borel.[75]

Four days later Brouwer gave his talk 'The structure of the continuum', which was intended for a mathematical audience. The paper has a historical introduction which is not quite identical with the introduction to the Berlin Lectures, nor with the historical sketch Brouwer wrote in connection with Fraenkel's '*Ten Lectures on Set Theory*'.[76] The old-intuitionistic school of, in particular, Poincaré and Borel, is criticised quite properly for freely using classical logic; it only practices a certain caution in the presence of impredicative definitions.

The paper concentrates on a refined analysis of the continuum with respect to the standard notions, such as order, separability, compactness and the like. A special feature which the paper shares with the Berlin lectures is the separate treatment of the full continuum (i.e. as based on choice sequences) and the reduced continuum (as based on lawlike sequences). This corresponds perfectly to the techniques of the creating subject and that of the Brouwerian counterexamples.[77] Whereas the Berlin lectures anticipate the presentation of the basis of intuitionistic mathematics in the form of the first and second act of intuitionism (creation of natural numbers and creation of choice sequences), the choice sequences as such are played down in 'The structure of the continuum'. The term '*Wahlfolge*' (choice sequence) occurs only once, and in one of Brouwer's private copies it is replaced by '*Folge von solchen Wahlen*' (sequence of such choices), but though the word hardly occurs, the notion plays its role all right. The representation of the (unit) continuum by a fan demonstrates the basic role of choice sequences. One might wonder what happened to the 'ur-intuition' of the dissertation of 1907. Under influence of Brouwer's later expositions the impression became prevalent that the ur-intuition was just a beginners device, a thing of the past. In the paper 'The structure of the continuum' the intuitive continuum, including the ur-intuition, is however once more given its prominent place. A private note, acting as a reminder, reads, 'In the continuum lecture, add at the end of section I, that the continuum is thus given by means of the *ur-intuition* just as with Kant and Schopenhauer.' Section I indeed pays at the end generously attention to this topic.

Wittgenstein, in his *Philosophical Remarks* (1930), discusses a large number of issues pertinent to, or derived from Brouwer's lecture. He energetically rejects the legitimacy of choice sequences as objects of mathematics. E.g. when discussing

[75]Brouwer (1908b, 1923f), Borel (1908a).

[76]Brouwer (1923f, 1992).

[77]Cf. van Dalen (1999a).

the unending choice process of 0 and 1 in the generation of a decimal expansion, he concludes 'That, what in the process of throwing a die is arithmetic, is not the factual result, but the infinite non-decidedness. But that indeed does not *determine* a number.'[78]

In the same volume, no. 173, Wittgenstein comes very close to Hilbert's 'there is no ignorabimus'.[79] After conceding that Brouwer was right in stating that the properties of the pendulum number are not compatible with the principle of the excluded third, he continued 'For if the question whether a statement is true or false is a priori undecidable, then the statement loses its meaning and exactly for that reason the laws of logic lose their validity.' So the pendulum number and choice sequences in general are for Wittgenstein, on the basis of meaning considerations, not a part of mathematics.

On the other hand there are points with a certain Brouwerian ring, e.g. no. 159 says: 'One cannot describe mathematics, but only practice it.' Compare this to Weyl's rendering of Brouwer's viewpoint as 'Mathematics is an activity, not a theory.'[80] The use Wittgenstein makes of the decimal expansion of $\pi$ is closer to Borel's examples. In 'Les paradoxes de la théorie des ensembles', Borel draws attention to phenomena, by mathematicians usually viewed as 'ill-considered'; here is an example: consider the decimal expansion of $\pi$ and replace all digits 7 by 8, and vice versa. Question: is the resulting real number algebraic? In general there is little one can say about these problems, they excellently illustrate the fact that at a very elementary level hard questions can be posed. Brouwer uses the decimals of $\pi$ also for the illustration of unsolved problems, but his trade mark is the occurrence of certain sequences. Wittgenstein, in his '*Philosophical Remarks*' adopts the 'switch of digits' trick (no. 182 ff.). Since Wittgenstein was parsimonious with references, it is not obvious what his source was. Given the date and place, it is however likely that Brouwer was his inspiration (his mentioning of the 'pendulum number' also points in the direction of Brouwer).

It would be interesting to know with whom Brouwer met in Vienna. He certainly met Hahn and Ehrenhaft, Wirtinger and Menger. But did he meet the Viennese sphinx Ludwig Wittgenstein or the young genius Kurt Gödel, or the other members of the Vienna Circle?

As to Wittgenstein, we have some definite, albeit scarce, information from Roy Finch. Finch met Brouwer in 1953 when Brouwer was doing the grand tour of the United States.[81] In a conversation Brouwer mentioned 'that he had a private all-day meeting with Wittgenstein on an island, during which they discussed Brouwer's lecture which had made such a strong impression on Wittgenstein'. Finch assumed that this island was off the coast of Holland, but apparently no records of the meeting have survived. A small Danube island would fit the bill as well.

---

[78]Wittgenstein (1984), no. 179.

[79]Wittgenstein's arguments, however, should be distinguished from those of Hilbert, Brouwer, Heyting, and Martin-Löf.

[80]pp. 101, 313.

[81]Cf. p. 795. Finch to Van Dalen, 10.X.1990.

Gödel and Brouwer did probably not meet; at that time Gödel would not so easily have approached a famous visitor, moreover his name had still to be made. There is, however, a letter from Gödel to Menger of April 21, 1972, in which he wrote 'I was never introduced to Wittgenstein and have never spoken a word with him. I only saw him once in my life when he attended a lecture, I think it was Brouwer's.' With a little goodwill one might infer that Gödel attended Brouwer's lecture.[82]

Gödel was quick to understand Brouwer. There are some notes in Carnap's diary which shed some light: 'Gödel says: when I want to follow strictly the constructivist point of view, then I have either to reject the principle of the excluded third (because it is not the case that either $p$ or $\neg p$ is provable), or on the other hand to presuppose a complete (decidable) logic. That seems correct.' 'Gödel on the inexhaustibility of mathematics. It was Brouwer's Vienna Lecture that gave the first impulse to these thoughts.'[83] Gödel's subsequent research beautifully illustrates how to use the new tools of logic to obtain conceptually significant insights.

In the early part of 1928 Holland was honoured by a visit of one of the most prominent European philosophers, Edmund Husserl. Husserl visited Holland in April at the invitation of the Dutch philosopher Pos.[84] Husserl was on the whole rather pleased with the occasion, in a letter to Heidegger he wrote 'The most interesting events were the long conversations in Amsterdam with Brouwer, who made a most significant impression on me, that of a completely original, radically honest, real, completely modern man.'[85] There is also a reference to this visit in the letter of Brouwer to Eva Wernicke of April 30:

'Here, at the moment, Husserl is going around, I am also strongly attracted. He thinks Miss Gawehn the most intelligent person he has come across in Holland.'[86] The brief passage makes clear that Brouwer was as strongly impressed by Husserl as Husserl by him.

Apparently Husserl had met in Amsterdam a gifted young lady, who impressed him by her intelligence. Since she did not make a reputation in philosophy one may well wonder who she was and what became of her. Irmgard Gawehn was born in 1900 in Memel, where she went to school. She studied in Heidelberg between 1920 and 1924, interrupted by a summer semester in Göttingen. She specialised in topology under Rosenthal, who became her Ph.D. adviser. She obtained her doctorate in 1931.[87] The dissertation *Über unberandete 2-dimensionale Mannigfaltigkeiten* was

---

[82]See Köhler (1991). There is however a letter from Gödel to the American Philosophical Society of 19.I.1967 in which he says 'I have seen Brouwer only on one occasion, in 1953, when he came to Princeton for a brief visit'. If 'seen' means 'met' there is no contradiction.

[83]Gödel über Unerschöpflichkeit der Mathematik. Erst danach Brouwer Wiener Vortrag zu diesen Gedanken angeregt worden'. Carnap 14.II.1928 and 23.XII.1929. See also Wang (1987), p. 50.

[84]The *Amsterdamer Vorträge* were published in Husserl (1997).

[85]Husserl to Heidegger, 5.V.1928, Husserl (1994).

[86]Hier tummelt sich jetzt Husserl herum, wobei ich stark mit herangezogen werde. Er findet Frä ulein Gawehn den intelligensten Kopf, der er in Holland angetroffen hat.

[87]The oral exam was in 1925. Rosenthal and Liebmann examined her in mathematics, Lenard in physics and Jaspers in philosophy.

accepted by the *Mathematische Annalen*, where it was published in 1927. Brouwer handled the paper; the correspondence concerning her paper was put in Menger's hands.

Irmgard was already a legend in Heidelberg when she finished her studies. She had the reputation of a first-class topologist. With this reputation she went to Berlin to pursue her career further. Freudenthal, who studied in Berlin at that time recalled that everybody was excited to get such a wonderful topologist in Berlin. However, when she gave on June 25, 1926 her first colloquium talk, *Set theoretic definition of 2-dimensional manifolds without bounds*, it turned out to be bordering on the incomprehensible. Whatever knowledge she might have had she could not get across.

Soon she discovered, however, that her real interest was philosophy, and she built herself a reputation as a philosopher. But again, she could not really make a career in philosophy. In a letter to Hopf of April 9, 1927 Brouwer asked Hopf to keep an eye on Irmgard, and to see that her philosophical paper got published quickly; so there was, or there seemed to be, some publishable philosophical activity.

In order to provide Irmgard (who was always referred to in a formal manner as 'Fräulein Gawehn') with a basis for a secure income, Brouwer had already suggested that she should take a '*Staatsexamen*', that would give her a licence to teach at gymnasia or high schools. The exam out of the way, Brouwer would apply for a Rockefeller stipend for her to study with him in Amsterdam. When she took the advice, it turned out to be a bitter disappointment. The stumbling block was the oral examination, the committee would ask questions on all kinds of topics, and she simply could not convince the examiners of her mathematical knowledge.

So she came to Amsterdam without her *Staatsdiploma*. There she became Brouwer's assistant from 1927 through 1930. And once in Amsterdam she surprised everybody by showing a striking lack of ability both as a topologist and a philosopher. Brouwer apparently kept her on for light work. The veil over the enigma Gawehn was lifted by Rosenthal, who later told Freudenthal that Irmgard was for whatever reason the object of admiration of the male student population. In Heidelberg she found a topologist as a lover, and the man wrote her dissertation. In Berlin a philosopher demonstrated his love for her in a similar way. When left to her own resources, she could not cope. Hence the disaster of the state examination. Brouwer, who only knew her dissertation, was not aware of her history. In Amsterdam she had also a number of lovers, whom she invariably dumped after some time. But finally one of her suitors dumped her, and according to the local gossip, that was more than she could bear. She mentally caved in and spent the rest of her life in a mental institution in Holland. Cor Jongejan and Lize visited her regularly until she died, after the Second World War.

The story, tragic as it is, illustrates that even in mathematics, the inner sanctum of reason, not all is what it appears to be, and that famous philosophers may be fooled.

It never became clear who wrote the thesis in topology.

The University of Oslo commemorated in 1929 the centennial of the death of Niels Henrik Abel and at the occasion the mathematics faculty awarded a large number of honorary doctorates to the most prominent mathematicians of the day. Brouwer received an invitation for the festivities with the announcement that the university proposed to appoint him as an honorary doctor.[88] Brouwer gladly accepted the honour, but fate decided against him: he wrote on April 4 that he could not attend the ceremony because of an illness. The list of honorary doctors was impressive, it contained the names of Engel, Fueter, Hadamard, Hardy, Hensel, Juel, Landau, Lindelöf, Painlevé, Phragmèn, Pincherle, Takagi, de la Vallée-Poussin, Veblen and Weyl. On the day itself, however, only four of them showed up.

## 13.5 Other Activities

With the founding of his new intuitionism, the topological school, the *Grundlagenstreit*, the fight against the *Conseil*, and his normal teaching, Brouwer had enough on his hands to be fully occupied. Nonetheless he found time for activities outside the academic sphere. The signific enterprise, in a sense, was one of those side activities. The signific circle has sometimes been compared to the Vienna Circle, and certainly it had for some time the potential of growing into an influential centre. There was, however, no driving power behind significs. The leaders, Brouwer and Mannoury, were not ambitious enough to get significs on the map. That is to say, the members of the circle indulged in grand schemes, such as the encyclopaedia project and the attracting of influential foreign experts, but somehow it never came to anything. Moreover, the only one who was willing to devote his scientific career to significs was Mannoury. Neither Brouwer, nor Borel, Van Ginneken or Van Eeden, were willing to give up their own practice, be it scientific, social or artistic. But Mannoury, successful as he might have been as a teacher, never managed to get influential philosophers to join his enterprise.

When in 1924 the circle was running into problems (cf. p. 362), Mannoury seriously considered its dissolution. After some discussion, it was decided to keep the circle going as long as a glimmer of hope was left. The members of the circle set themselves the goal of publishing the *Signific Dialogues*, a report of some of the discussions of the circle. Even this proved to be problematic. The circle met two more times, on 12 November 1925 and 2 December 1926.[89] After that the circle just faded away. The *Signific Dialogues* eventually appeared in 1937.[90] In the late thirties a new organisation for significs was founded; Brouwer only took part in this as an interested party.

By the time Brouwer got involved in his other schemes, significs was a thing of the past. What was left, was his friendship with Mannoury and Borel. The friendship with Mannoury lasted; it was not at the same level as the friendship with Scheltema, but there was an immense appreciation and trust between the two. Brouwer was well aware of the destructive element in his relations with fellow human beings, he once remarked to Mannoury, 'Gerrit, you bring out the good from people. I bring out the

[88] Rector Oslo University to Brouwer, 19.II.1929.

[89] For details, see Schmitz (1990a), 6.3.

[90] Brouwer et al. (1937a, 1937b, 1937c, 1939).

bad.'[91] Indeed, Mannoury had the makings of a saint, he was highly valued by his colleagues, and revered by his students. And his role in politics had shown him to be a man of principle.

The other friend that the signific episode had left him was Henri Borel. After the foundering of the circle the two remained in contact; there are a number of letters from Borel to Brouwer preserved. They show Borel as a tragic figure, unable to find a proper place in life. When he died he left a large number of books and brochures behind, and curiously enough, in spite of his vehement criticism of Van Eeden for turning to Catholicism, he embraced the faith of the Mother Church in the end.

A considerable part of Brouwer's time was taken up by matters concerning his wife's pharmacy. The original building and grounds at the Overtoom had been expropriated by the City of Amsterdam. In spite of his protests, Brouwer had to relocate the pharmacy. He found a suitable location at the Amstelveenseweg 21-23 in a street which is a continuation of the Overtoom. After some effective bargaining he bought the house, at number 21-23, for 44,500 guilders, which was a substantial sum at the time.[92] Even if the city compensated Brouwer for a comparable sum (the original offer was 40,342 guilders), Brouwer had to furnish quite a bit of capital. In fact the cost of converting the building came to an extra 1500 guilders. It may be said that Brouwer remodelled the building in style. The shop was panelled in 'Slavonic oak', the central part was designed in the style of an old library, complete with a gallery and fitting staircases. In matters of furnishing the pharmacies and providing the necessary equipment Brouwer kept rather high standards; in the case of the Amstelveenseweg pharmacy, for example, he engaged one of the better architects.

The second and third floor of the building were occupied by tenants. On the first floor lived Louise, the daughter of Lize, and Brouwer had his study on the same floor. His study contained a grand piano, and Mr. Guasco, the son of a tenant, remembered that usually when Brouwer returned from his lectures, 'the heavy tones of the overture Egmont resounded through the house'. The Guasco's lived on the second floor, and the third floor was let to a music teacher. Louise, in spite of her preoccupation with religion and the eternal, had conceived a personal interest in the man. As she could hear all the traffic on the staircase, she might present herself as if by chance in the open doorway when the teacher happened to pass and invite him in for a neighbourly chat. Eventually the poor man tried to sneak out on stockinged feet, but even that could not always mislead Louise's sharp ears.[93] At roughly the same time Brouwer acquired two more premises, not far from the shop at the Amstelveenseweg. He bought a shop at the Surinameplein[94] and rented a wooden temporary building at Surinamestraat number 12.[95] The premises at the Amstelveenseweg and those at the Surinamestraat and Surinameplein were separated by a canal. It seems that Brouwer had counted on a future bridge in front of

[91] Oral communication, C. Vuysje-Mannoury.

[92] He got a mortgage of 29,000 guilders.

[93] Interview R.A.F. Guasco.

[94] Suriname place or square.

[95] The transaction was concluded on 2 April 1930.

**Fig. 13.6** Silver wedding of the Brouwer couple [Brouwer archive]

his Amstelveenseweg pharmacy, which would make it more attractive for clients. When the city decided to build the bridge some distance to the North he was rather angry, but though he contemplated legal action, there was little he could do.

There is a tendency to equate 'owning a pharmacy' and 'being rich', but that did not apply to Brouwer's pharmacy; it was on the whole rather poorly managed. Somehow Brouwer could not operate in a businesslike manner where the pharmacy was concerned. It was overstaffed, and, since Brouwer and his wife took a somewhat paternalistic view of the assistants, they were almost appointed for life. In the end the shop started to look like a home for the elderly pharmacy assistants. Brouwer took an active part in the running of the pharmacy, he carried out administrative duties, prepared the tax forms, dealt with the municipal bureaus, and represented the pharmacy at the national organisation of pharmacists. At the meetings of the latter he could act in a rather overbearing manner, until finally he was barred from the meetings, on the grounds that he was not a pharmacist.

Family life remained as hectic as before. Lize usually spent the weekdays in Amsterdam, where she slept over the pharmacy. This had the advantage that she could keep an eye on Louise. In spite of the fact that Louise was not accepted in the convent as a novice, she remained a devoutly Catholic, and—unknown to Brouwer—Lize was baptised as a Catholic at roughly the same time of Van Eeden's conversion (cf. p. 363). The ceremony took place on 24 December 1924 in the St. Dominicus Church in Amsterdam. Neither her husband nor her friends were aware of her conversion, it was a secret between her and Louise.

In 1927 Lize spent some time in Antwerp in a paramedical institute of some sort, probably halfway between a boarding house and a health centre. The scarce information seems to indicate that she suffered from pernicious anaemia.[96]

[96]Lize to Louise, 17.V.1935.

# Chapter 14
# The Three Battles

> If mathematics would for the sake of its safety seriously retire to this status of pure game, it would no longer be a determining factor in the history of mind.
>
> H. Weyl

## 14.1 The *Grundlagenstreit*

We left the foundational debate in a fairly peaceful state. Hilbert's talk at the Weierstrass conference in 1925 was rather moderate in tone; there were a few references to dissident views, but more admonishing than belligerent in tone. Apart from a repeat of the Münster lecture in Copenhagen no further major activities took place that year. In November Hilbert was diagnosed with pernicious anaemia; he had not been well for some time, but the actual cause was not discovered earlier. At that time the disease was more often than not fatal. The immediate consequence was a serious decline in working power, and the disease took its toll on Hilbert's intellectual activities. Hilbert, nonetheless, tried to carry on his research as usual.

The years 1925–1926 were for Brouwer mainly filled with duties that had little to do with research. He spent time editing the Urysohn memoir for Fundamenta, and the necessary correspondence with Sierpinski and Alexandrov.

As an editor of the *Annalen* he was also very much occupied with the new influx of topological papers of, among others, Alexandrov, Urysohn, Menger, and Tumarkin. In between he found time to work on his contribution for the Riemann volume.

The foundational debate was watched with interest in mathematical circles, but only a handful of mathematicians actually took a position. If they did, it was mostly on the formalist side. The choice was, more often than not, determined by feelings of loyalty—any student or alumnus of the Göttingen university fostered 'right-or-wrong my teacher' feelings.[1] Another argument for siding with the formalists was

[1]Cf. Lietzmann (1942).

D. van Dalen, *L.E.J. Brouwer – Topologist, Intuitionist, Philosopher*,
DOI 10.1007/978-1-4471-4616-2_14, © Springer-Verlag London 2013

of a pragmatic nature—life with PEM is easier than without it. It is, however, fair to say that the majority of the mathematical community deplored the hostilities of the last few years. A perfect occasion for a peaceful exchange of ideas arose when in the summer of 1926 Brouwer once more visited Göttingen.

Alexandrov, who was already in Göttingen, later told the charming story of how the two opponents came to lay down their arms.[2] Brouwer, he wrote, was immediately at home, as in the old days, in the circle of mathematicians. He was warmly welcomed in the intimate circle around Courant and Emmy Noether. The ill-feelings and suspicions having melted like snow for the sun, plans were made to get Brouwer and Hilbert together again. One evening a group of mathematicians, including Hilbert and Brouwer, was invited to Emmy Noether's apartment. And so Brouwer, Hilbert, Courant, Landau, Alexandrov, Hopf and a few other young mathematicians found themselves at the table under the roof of Emmy's friendly quarters (Landau used to question the validity of Euler's polyhedron theorem for this room). It fell to Alexandrov to start a conversation that would break the ice between Hilbert and Brouwer. Alexandrov's solution was as ingenious as simple: what better expedient was there to bring two persons together, than the mentioning of a third person both enjoyed criticising. Alexandrov cleverly introduced the acknowledged paragon of vanity Paul Koebe (in his paper he discretely referred to the well-known function theorist of Luckenwalde). The trick worked better than could be expected, it did not take long before Brouwer and Hilbert were outbidding each other in criticism of poor Koebe, 'at the same time they were nodding more and more friendly at each other, until finally they completely agreed in a mutual toasting'. The thus established peace lasted for the duration of Brouwer's visit. Even his lecture in the Göttingen mathematical society on 22 July 1926, 'On everywhere- and seemingly everywhere defined functions',[3] could not disrupt the peace. There are no notes of the lecture, but one may safely assume that it contained material from Brouwer's paper for the Riemann volume.

After his stay in Göttingen, Brouwer pilgrimaged in the company of Alexandrov to Batz to commemorate the lamented Urysohn. The topological fire in Brouwer had apparently not been extinguished completely, for Alexandrov wrote excitedly to Hopf that 'Brouwer has proved a topological theorem'.[4]

Less than a week later Brouwer was to be found at one of those meetings where he could indulge in the spiritual side of life. He attended a Sufi-meeting in France with a lady named 'Mies'. 'I am undergoing in humility the first impacts of the Sufi-order' he wrote to Bertha Adama van Scheltema, the sister of his deceased friend.[5] Brouwer was no stranger to spiritual intermezzo's of this kind. It is not easy to guess what interested him more, the spiritual experience and reflection offered, or the attraction of the colourful participants.

---

[2] Alexandrov (1969).

[3] *Überall und scheinbar überall definierte Funktionen.*

[4] Alexandrov to Hopf, 12.VIII.1926.

[5] Brouwer to B. Adama van Scheltema, 22.VIII.1926.

Not much later Brouwer turned up at the Düsseldorf meeting of the DMV, where he delivered on 23 September his talk 'On domains of functions'.[6]

Hilbert, in the meantime, was very poorly. He had his ups and downs, but generally speaking there was not much improvement.

In Amsterdam the faculty did something that certainly must have pleased Brouwer, and one is tempted to see his hand in it. In the meeting of October 24 the faculty decided no longer to admit philosophy as a major in the curriculum. Four years ago his colleague Kohnstamm had managed by clever manoeuvring to get philosophy accepted as a major, and Brouwer, in spite of his sharp intelligence, had lost that battle.[7]

The next significant event in the *Grundlagenstreit* was Hilbert's second talk in Hamburg in July 1927. His health had somewhat improved, but it was a tired old man who appeared at the rostrum. Nonetheless, the event was a tremendous success. Courant, who accompanied Hilbert to Hamburg, reported to Springer, 'The days in Hamburg with Hilbert were in every respect most satisfying. Hilbert's condition was so good, that he could speak for an audience of over 100 with a soft and tired voice, it is true, but with intense temperamental outbursts and very impressive, and that he even took part in a big festive dinner in the evening of the second day. The whole occasion was a great triumph for him and did him psychologically exceptionally good, without harm to the body. Objectively, his blood picture has strongly improved and now our greatest hope is a liver cure. It will be, however, hardly possible to get Hilbert to eat liver in the required large quantities, but we have cabled to Harvard medical school, and we will probably soon get the liver preparation produced there, from which we expect much.'[8] The liver cure started not long after the lecture, followed indeed by the American preparation. The influence of American colleagues and students had certainly been helpful to introduce Hilbert to this newly discovered treatment.[9] This was by no means the end of the story; the right dose had to be found, and annoying and painful symptoms reared their heads. It was a surprising proof of the success of the treatment—and the determination of Hilbert—that he was again one of the main speakers at the annual meeting of the DMV in Leipzig. His topic was 'The axiom of choice in mathematics'. The title is somewhat misleading, as 'choice' referred here to the $\varepsilon$-operator.

The Hamburg lecture of 1927 was, like its predecessor of 1921, published in the *Abhandlungen des Hamburgischen Seminar*, this time with the simple title *The foundations of mathematics.*[10]

The paper was in content a leisurely exposition of the Infinity-paper, a didactical introduction to the proof theory of arithmetic and analysis, laced with some popular foundational motivation, and the usual pot-shots at dissenters. Again, we must bear

---

[6] *Über Definitionsbereiche von Funktionen.*

[7] Cf. p. 352 ff.

[8] Courant to Springer, 2.VIII.1927.

[9] See Reid (1970), Chap. 21.

[10] *Die Grundlagen der Mathematik*, Hilbert (1928).

in mind that the paper should be excused for its heavy handed propaganda, since it was doubtlessly the sort of language that could and would rouse an audience. One cannot read the paper without being touched by the belief of the author in his philosophical credo, and by his fighting spirit. Here was a man who had discovered the final answer to the foundational problems of mathematics. The wisdom of publishing all these belligerent remarks may be doubted. As a rule they reflect on the author, and moreover, they stand in the way of a proper scientific discussion. In particular, the formulations used by Hilbert tended to demonise the opponent. One might almost conclude that Hilbert did not want a *discussion*, he wanted to *win*.

As in earlier publications, Hilbert was determined not to give up the *Tertium non datur*. After pointing out that from a finitist point of view the negation of statements like $\forall x A(x)$, where each instance $A(n)$ can be finitistically checked, is problematic, he continues:

> But we cannot give up the use of the applications of the Tertium non datur or any of the other laws of Aristotelian logic which are expressed in our axioms, as the construction of analysis is impossible without them.

It may be remarked here that Hilbert was too pessimistic about a Tertium non datur-free mathematics. In due time work in the intuitionistic school and above all the results of the school of Erret Bishop were to give a powerful impetus to constructive mathematics by actually rebuilding large parts of analysis in a constructive manner. He showed that 'constructive' does not stand for 'ugly and cumbersome'. He showed convincingly that constructive mathematics possesses the same elegance as classical mathematics.[11] We know now that (i) a substantial part of analysis can be developed in the intuitionistic frame work, (ii) the proofs are not as cumbersome as people in Hilbert's day expected, and (iii) that the mathematical problems arising in the Brouwer–Bishop tradition are interesting and rewarding. Of course, it would be anachronistic to expect Hilbert to see this point. Hilbert's solution was exemplary for the methods of the working mathematician. If you cannot solve the equation $x^2 + 1 = 0$, extend your system with $i$ $(= \sqrt{-1})$. So if you cannot negate and quantify at the same time, add the ideal statements (which allow negation and quantification), and add Tertium non datur. It is like the extension of an algebraic structure. And then, of course, show the consistency.

After a quick survey of some of the tools and methods of proof theory, Hilbert returned to answer the various objections that had been raised, and which he found 'one and all as unjustified as possible'. He started by reconsidering Poincaré's objections against his consistency proofs that intended to safeguard induction by applying induction. This time Hilbert elaborated his defence, as promised in 1922, cf. p. 440; Poincaré, he said, had not understood the distinction between conceptual induction and formalised induction. On this point Hilbert's position was weak, to say the least. Unless he managed to pull off the finitist consistency proof that he had announced, he could not uphold his own view. Brouwer and Poincaré had a better intuition on

[11] See Bishop (1967) and Bishop and Bridges (1985).

this point,[12] Hilbert saw Poincaré's negative comments as a consequence of his pronounced prejudice against Cantor's theory. The argument does not convince, why should a mistrust in set theory lead to a rejection of a combinatorial practice such as proof theory?

On the whole Hilbert was not impressed by what he read in the foundational literature. Most of it he considered as backward, as if belonging to the pre-Cantor period. Since Hilbert's yardstick was calibrated by the continuum hypothesis, Hilbert's dogma, 'consistency $\Leftrightarrow$ existence', and the like, he was by definition right. But if one is willing to allow other yardsticks, no less significant, but based on alternative principles, then Brouwer's work could not be written off as obsolete nineteenth century stuff.

However, Hilbert clearly saw Brouwer as his prominent opponent, and thus he spent the rest of the paper giving him a thorough roasting. As he put it: 'By far the largest amount of space in the present literature on the foundations of mathematics is taken up by the doctrine that Brouwer has formulated and called intuitionism. Not out of an inclination for polemics, but to express my views clearly, and to prevent misconceptions about my own theory, I have to go further into certain statements of Brouwer.'[13]

There are actually two main points that Hilbert singled out for comment: the meaning of existence and the Tertium non datur. 'Brouwer declares existential statements by themselves to be one and all meaningless, insofar they do not at the same time contain the construction of object that is asserted to exist, for worthless scrip: it is through them that mathematics degenerates into a game.'[14] Hilbert's main defence against this was based on the $\varepsilon$-operator. What Hilbert exactly wished to achieve is not clear, in the following argument the $\varepsilon$-operator plays no role. A possible option might be the use of the $\varepsilon$-operator to obtain a term that can be seen as the solution promised by $\exists x A(x)$, namely $\varepsilon_{x+2=7}$ is the solution of the equation $x + 2 = 7$.[15] What is left is a passionate plea for pure existence proofs on general grounds, they do away with lengthy computations, they provide real insight, .... No mathematician will disagree, but the fact remains that Hilbert evades the issue: can the promise $\exists x A(x)$ be fulfilled? Instead Hilbert stuck to his lofty views: 'Pure existence theorems have thus in fact been the most important landmarks in the historical development of our science'. 'But', he added, 'such considerations do not trouble the devout intuitionist'. As to the game, so deprecated by Brouwer, it is Hilbert's opinion that the formula game reflects exactly the *technique of our thinking*. And indeed, 'the basic idea of my proof theory is nothing else but to describe the activity of our in-

[12]Cf. Brouwer (1913b), p. 88.

[13]Ibid, p. 77.

[14]The reference to 'worthless scrip' suggests that Hilbert has Weyl's New Crisis-paper in mind. But it covers Brouwer's views equally well.

[15]It is simple to prove $\exists x(x + 2 = 7)$.

tellect, to make a protocol about the rules according to which our thinking actually proceeds'.[16]

The defence of the Tertium non datur is, so to speak, left to the $\varepsilon$-axiom, from which it follows; the informal argument for the Tertium non datur comes down to the bottom-line: 'it has never yet caused the slightest error'. In the end Hilbert's arguments are pragmatic, and emotional, as the following quote shows:

> Taking this Tertium non datur from the mathematician would be the same as, say, denying the astronomer his telescope, and the boxer the use of his fists. The ban of existence theorems and the Tertium non datur roughly boils down to renouncing the science of mathematics as a whole.

The last part reflects the then prevailing pessimism with respect to intuitionistic mathematics. One simply could not see how the essential parts of mathematics could be salvaged. Even those who were sympathetic to Brouwer's cause felt discouraged by the complications that had to be faced. Abraham Fraenkel saw the intuitionistic landscape as a bleak place; he feared that after the intuitionistic revolution, only ruins would remind the passing traveller of the former splendour of the architecture of mathematics. Perhaps some corners of the old buildings, e.g. arithmetic, might be saved, cf. p. 444.

Hermann Weyl was no less pessimistic:[17]

> Mathematics attains with Brouwer the highest intuitive clarity. He succeeds in developing the beginnings of analysis in a natural way, retaining contact with intuition much closer than before. But one can not deny that, progressing to higher and more general theories, the fact that the simple principles of classical logic are not applicable finally results in a hardly bearable awkwardness. And with pain the mathematician sees that the larger part of his tower, which he thought to be joined from strong blocks, dissolves in smoke.

Understandably, Hilbert, who did not particularly love the competition, shared these bleak views:

> For what can the wretched remains, the few incomplete and unrelated isolated results, that the intuitionists have obtained without the use of logical $\varepsilon$-axiom mean, compared to the immense expanse of modern mathematics!

There is little doubt that Hilbert's assessment was right, but what was his point? After all, any new discipline, say group theory, had to start from scratch. Around the turn of the last century abstract group theory was not everywhere received with open arms. How could this new subject of group theory ever hope to catch up with the enormous mass of techniques and results of, say, projective geometry? If one takes into account that intuitionism aimed at rebuilding mathematics, then it is evident that

[16]A most interesting claim. Given a number of extra specifications, there are good arguments for Hilbert's claim when made for the system of Gentzen.

[17]Weyl (1928).

Hilbert was premature. Moreover, his own program would, under the above criteria, be disqualified as well.

Hilbert's ultimate disproval of Brouwer's foundational tenets is expressed in an observation following his own credo, 'The fact is that thinking proceeds parallel to speaking and writing, by forming and concatenating sentences.' From Brouwer's early writings, and from the Vienna lectures, we know that this belief is diametrically opposed to Brouwer's view on language and communication. Quite consistently, Hilbert continues, 'after all, it is the task of science to liberate us from arbitrariness, sentiment and habit, and to protect us from the subjectivism, which has made itself noticeable in Kronecker's views, and which, as it seems to me, finds it apex in intuitionism'. This is again one of those populist phrases that go well with almost any audience. One may be almost certain that Hilbert's statement was interpreted as 'Kronecker and Brouwer act without objective justification, they ban what they dislike'. It is more than likely that even Hilbert intended this interpretation, because otherwise he would not lump Kronecker and Brouwer together. Kronecker, so to speak, was the super objectivist, his rejection of such things as arbitrary sets and functions, which are ultimately a matter of belief, makes him a stricter rationalist-objectivist than even Hilbert. Brouwer, on the other hand, accepted such subjective objects as choice sequences, that by their nature escaped description and communication.[18] This could rightly be seen as subjectivism in mathematics, and Brouwer would insist on doing so. But here Hilbert's attitude seems somewhat unreasonable, for would it not be a matter of progress if such notions could be handled mathematically?

The critique of Brouwer and his intuitionism found an apotheosis in the following, undeniably subjective, expression of exasperation:

> Under the circumstances, I am baffled that a mathematician doubts the strict validity of the Tertium non datur. I am stunned even more about the fact that, as it seems a whole community of mathematicians has nowadays constituted itself, that is doing the same. I am most astonished about the fact that also even in the circle of mathematicians, the power of suggestion of a single, high spirited and imaginative man, can exert the most improbable and eccentric influences.

Strong language, understandable in the euphoria of an exciting lecture, but out of place in a paper in a respectable scientific journal.

After this emotional intermezzo Hilbert returned to business, he sketched some progress made by Ackermann, and closed with the optimistic statement that the success of his program was just round the corner, 'the remaining problem is just to carry out a purely mathematical finiteness proof'. His conclusion was therefore, 'mathematics is a science without presuppositions. For its foundation I need neither God, as Kronecker does, nor a special faculty of our intellect attuned to the principle of

[18] Cf. van Dalen (1999c).

complete induction, nor Brouwer's ur-intuition, nor, finally like Russell and Whitehead, axioms of infinity, reducibility, and completeness, which are in fact actual, contentual assumptions that cannot be compensated by consistency proofs.'

It is interesting to read these proclamations, in view of the lessons Gödel taught us about Hilbert's program. Nonetheless, the certainty and enthusiasm of Hilbert must have been infectious, as Courant's report indicates.

One can only speculate what made Hilbert so angry, after all in August 1926 he and Brouwer had made their peace, and since then not that much had happened. Or had Brouwer's spectacular success in Berlin upset Hilbert? Had he become convinced that Berlin was a hotbed of intuitionists—it is possible, where else would he have conjured up 'a whole community of mathematicians' abjuring the Tertium non datur? In fact, one may doubt if the number of practising intuitionists would surpass five. If anything, the cause of Hilbert's anger and anxiety must have been psychological. Here was a younger man, with an impeccable scientific past, who with great tenacity kept pointing out the Achilles heel of the formalist program, and who was not daunted by the displeasure of the reigning sovereign of mathematics. And worse, a man who had even inspired the defection of Hilbert's star student, Hermann Weyl. Hilbert was reputed to take a common sense position in political matters, to steer away from his conservative colleagues, and to accept the republic for what it was. He was, however, not able to practice the same common sense with respect to himself. He could not forget that he was the famous professor who was always right; indeed his students usually went out of their way to spare the great man embarrassments or inconvenience. So when this Dutchman offered him admiration for his mathematical achievements and for his role as a promoter of mathematics and mathematicians, but declined to drop his foundational program, he committed something comparable to high treason. Finally, Hilbert's mental-physical situation was put under pressure by the persistence of his pernicious anaemia. In all, there were enough factors present to worry even an experienced diplomat. A diplomat can, however, consult advisers, but Hilbert had isolated himself from potential advisors on foundational matters (and not only those). Advice was to be given, not to be taken. Under the circumstances, there was probably no one to question the wisdom of such a crude course. One might almost think that Hilbert was provoking an open conflict, which would allow him to sever all ties with Brouwer. Whether he was enough of a Bismarck to hatch such a plot is of course questionable. From Courant's report to Springer, we may believe that the Hamburg audience applauded the address of the Grand Master; the provocative and aggressive character could, however, not have escaped anyone in the audience. Whereas some saw Hilbert's talk as a triumph for mathematics, Hermann Weyl did not share that view. Following Hilbert's lecture he made a number of remarks, published in the same volume of the *Abhandlungen aus dem mathematischen Seminar der Hamburgischen Universität*.[19]

Weyl took his place at the lectern as if he were the counsel for the defence, who saw that his client had just been sentenced in absentia. 'Allow me to say a few words

[19]Weyl (1928), van Heijenoort (1967).

in defence of intuitionism', he opened. Mathematics, he said, was considered a system of contentual, meaningful and evident truths. This was the generally accepted platform before the advent of Hilbert's proof theory. Poincaré took that position, and Brouwer seconded him. 'But he was the first to see exactly and in its full extent how it had in fact transgressed the limits of contentual thought. I think that we are all indebted to him for this recognition of the limits of contentual thought.' Hilbert too recognised these limits in his metamathematical considerations, and he accepted that there was nothing artificial about these limits. 'Thus', he went on, 'it does not seem strange to me that Brouwer's ideas found a resonance (*Gefolgschaft*); his point of view followed necessarily from a thesis that was shared by all mathematicians before the formulation of Hilbert's formal approach, and from a fundamental new, indubitable logical insight, that was recognised by Hilbert too'. The consequences of these insights led, however, to widely differing reactions. Brouwer rebuilt mathematics from the inside, putting up with the limitations; Hilbert opted, however, in Weyl's words, for a wholesale reinterpretation of the meaning of classical mathematics, that is the formalisation. At this point Weyl made a gesture that caused some sensation, 'Also in the epistemological appraisal of the thus created new situation, nothing separates me from Hilbert, as I am happy to acknowledge.' Did this mean the definite farewell to intuitionism? That seems too hasty a conclusion, for we have to keep in mind that it was the same Weyl who wrote in that year,

> The ice field was broken into floes, and now the element of the flowing will soon completely be master over the rigid. L.E.J. Brouwer designs—and this is an accomplishment of the greatest epistemological importance—an exact mathematical theory of the continuum, which conceives it not as a rigid being, but as a medium of free becoming.[20]

The recognition of the epistemological basis of a system does not automatically lead to the rejection of another system which is compatible with it. In fact, one can very well be an intuitionist and practise proof theory. One could even, as Weyl seems to suggest, share the epistemology of the proof theorist. Let us consider one of the major points of difference, the interpretation of existential statements. Even the success of Hilbert's program did not provide an existential truth, like $\exists x A(x)$, with a construction of an $a$ such that $A(a)$. The commitment of proof theory *à la* Hilbert is to finitary statements, not to all real statements. The latter would only be consistent.

It looks from our point of view more as if Weyl was granting Hilbert his project. For a discerning man like Weyl must had seen that the goal of the program was still not reached, and a few private calculations would have convinced him that this was a treacherous swamp to enter. So he might have, quite reasonably, accepted the philosophical part of Hilbert's program, without committing himself to either proof

---

[20]'*Die Eisdecke war in Schollen zerborsten, und jetzt ward das Element des Flieszenden bald vollendes Herr über das Feste. L.E.J. Brouwer entwirft—und das ist eine Leistung von der gröszten Erkenntnistheoretischen Tragweite—eine strenge mathematische Theorie des Kontinuums, die es nicht als starres Sein, sondern als Medium freien Werdens faszt.*' Weyl (1928).

theory or intuitionism. Weyl's career, his work and personality show, I think, that he in the first place was a 'working mathematician' with an interest that could not be pinned down to one project or doctrine. There are enough statements to be found in his later writings that make it clear beyond doubt that basically Brouwer was right. For example, in his review of *The philosophy of Bertrand Russell*:[21]

> Brouwer made it clear beyond a doubt that there is no evidence supporting the belief in the existential character of the totality of all natural numbers, and hence the principle of the excluded middle...

And in *The Open World*:[22]

> If mathematics would seriously retire to this status of pure game for the sake of its safety, it would no longer be a determining factor in the history of mind. (p. 77)
>
> If mathematics is taken by itself, one should restrict oneself with Brouwer to the intuitively recognisable truths and consider the infinite only as en open field of possibilities; nothing compels us to go farther. But in the natural sciences we are in contact with a sphere which is impervious to intuitive evidence; here cognition necessarily becomes symbolic construction. (p. 82)

In 'David Hilbert and his mathematical work' he returned to the issue:[23]

> L.E.J. Brouwer by his intuitionism had opened our eyes and made us see how far generally accepted mathematics goes beyond such statements as can claim real meaning and truth founded on evidence. I regret that in his opposition to Brouwer, Hilbert never openly acknowledged the profound debts which he, as well as all other mathematicians, owe Brouwer for this revelation.

Although Weyl was reluctantly prepared to live in a world where Hilbert's formalism prevailed, one in which a doctrinaire meaninglessness should be the ultimate scientific moral, he notes with sincere regret that 'If Hilbert's view prevails over intuitionism, as appears to be the case, *then I see in this a decisive defeat of the philosophical attitude of pure phenomenology*, which thus proves to be insufficient for the understanding of creative science even in the area of cognition that is most primal and most readily open to evidence-mathematics'.

Why was Weyl so defeatist? To be honest, there was no reason to give up hope so soon. After all, Hilbert's program had no palpable results to show. Most of the steps on the road to success had yet to come. There is a well-known quote of Weyl that may provide a hint: 'My work has always tried to unite the true with the beautiful and when I had to choose one or the other, I usually chose the beautiful.'

Hilbert was a master of sweeping presentations, but his technical evidence was not always solid. As Von Neumann remarked in a letter to Carnap, 'There are many

[21] Weyl (1946), p. 275.

[22] Weyl (1932).

[23] Weyl (1944), p. 157.

programmatic publications of Hilbert, in which Hilbert states that something is being proven or almost proven, for which this is not even approximately the case.'[24] It could not have escaped Weyl that the grand old man of Göttingen had enough faith in himself to confuse wish and fact. And even if Hilbert would succeed, why give up 'meaning' for the evidence of the table of multiplication (Kreisel)? It is more plausible that Weyl's motivations were of an altogether different sort. Let us not forget that here was his revered teacher, who was in a state which could very well be terminal. And circumstances like that tend to inspire a certain compliance. Moreover, if one shares the intuitionistic conviction that there is a real, sound mathematics, created by ourselves, then it is a matter of magnanimity to grant the pleasure of 'playing the game', which formalism professes. In short, Hilbert could argue, if Brouwer is right, he can prove the consistency I am looking for, so why worry? For Hermann Weyl this would have been too cheap a solution of a deep problem; but who knows, he may have had his reasons to prefer a compromising policy. We will probably never know. Whatever the meaning of Weyl's epistemological move was, it was a serious blow for intuitionism. There are no comments known of Brouwer on the above. He never lost his esteem for Weyl, and kept his New Crisis paper on the reading lists until the fifties.

Brouwer's reaction to the Hamburg lecture is not known, although it can easily be guessed. Apparently, Hilbert's infinity paper was already as much as he could take, for he decided to write down his version of the foundational conflict. On December 17, 1927 he submitted his paper of 'Intuitionistic reflections on formalism', together with a paper on an embedding of spreads[25] to the KNAW. Two months later the same paper was submitted at the Prussian Academy in Berlin.

The paper lists all the relevant articles that had appeared so far. The last paper of Hilbert on this list is the infinity paper, so one may reasonably conclude that he had not yet seen the Hamburg paper when he submitted his manuscript. Compared to the rather excited paper of his adversary, Brouwer's Reflections-paper is moderate in tone. It opens with what could be consider a peace proposal:

> The disagreement of which is correct, the formalistic way of founding mathematics anew or the intuitionistic way of reconstructing it, will vanish, and the choice between the two activities be reduced to a matter of taste, as soon as the following insights, which pertain primarily to formalism but were first formulated in the intuitionistic literature, are generally accepted. The acceptance of these insights is only a question of time, since they are the results of pure reflection and hence contain no disputable element, so that anyone who has once understood them must accept them. Two of the four insights have so far been understood and accepted in the formalistic literature. When the same state of affairs has been reached with respect to the other two, it will mean the end of the controversy concerning the foundations of mathematics.

[24]Von Neumann to Carnap, 7.VI.1931.

[25]Brouwer (1928b, 1928d), the translations below are Stefan Mengelberg's, van Heijenoort (1967).

Brouwer shows himself optimistic, not to say certain. The acceptance of his points was just a matter of time, he said, as it is only a matter of reflection; the points contain no disputable elements, and everybody who has understood them must accept them. It is striking indeed that Brouwer had given up the radical condemnation of formalism of his early years. It is not unlikely that in the changing world of mathematics, he started to see that logic and formalisation at best offered certain benefits after all, and at worst, were harmless. After all, his own student had designed a logical codification for intuitionistic mathematics, and he, Brouwer, had approved the project.

After the brief introduction, which ended with a prediction of the end of the *Grundlagenstreit*, Brouwer formulated his four insights.

> First insight. *The differentiation, among the formalistic endeavours, between a construction of the 'inventory of mathematical formulas' (formalistic view of mathematics) and an intuitive (contentual) theory of the laws of this construction, as well as the recognition of the fact that for the latter theory the intuitionistic mathematics of the set of natural numbers is indispensable.*
>
> Second insight. *The rejection of the thoughtless use of the logical principle of excluded middle, as well as the recognition, first, of the fact that the investigation of the question why the principle mentioned is justified and to what extent is valid constitutes an essential object of research in the foundations of mathematics, and, second, of the fact that in intuitive (contentual) mathematics this principle is valid only for finite systems.*
>
> Third insight. *The identification of the principle of excluded middle with the principle of the solvability of every mathematical problem.*
>
> Fourth insight. *The recognition of the fact that the (contentual) justification of formalistic mathematics by means of the proof of its consistency contains a vicious circle, since this justification rests upon the (contentual) correctness of the proposition that from the consistency of a proposition the correctness of the proposition follows, that is, upon the (contentual) correctness of the principle of excluded middle.*

The list is followed by a critical discussion, in which Brouwer precisely checks the provenance of these insights.

> The first insight is still lacking in Hilbert (1905) (see in particular Sect. V, pp. 184–185, which is in contradiction with it). After it had been strongly prepared by Poincaré, it first appears in the literature in Brouwer (1907), where on pp. 173–174 the parts of formalist mathematics mentioned above are distinguished by the terms *mathematical language* and *mathematics of the second order*, and where the intuitive character of the latter part is emphasised.[2)] In Hilbert (1922) (...), under the name *metamathematics*, mathematics of the second order broke through in the formalist literature. The claim of the formalist school to have reduced intuitionism to absurdity by means of this insight, borrowed from intuitionism, cannot very well be taken seriously.
>
> [2)] [Brouwer's note] An oral discussion of the first insight took place in several conversations I had with Hilbert in the autumn of 1909.

We have seen that in the fall of 1909 Brouwer and Hilbert met in Scheveningen (cf. p. 125). On that occasion Brouwer discussed the various levels of language, logic and mathematics from his dissertation with Hilbert.

The precise distinction *formal, meaningless mathematics—contentual mathematics (metamathematics)* is only much later introduced in Hilbert's first Hamburg paper (1922). There is no reference in any of Hilbert's papers to Brouwer's role in clearing up the confusion about the two levels. Perhaps Hilbert had completely forgotten about the exchange in 1909. Courant confirmed what was common knowledge at the time, that Hilbert had great difficulty in keeping track of the information that he picked up. Courant mentions a few instances, which concern substantial research, they are by no means small, neglected corners of mathematics—'so indeed, Hilbert's theory of integral equations, one of his greatest achievements, was triggered by a bad memory'.[26] Whatever was behind Hilbert's unreliable memory, one should not forget that at the time of the *Grundlagenstreit* Hilbert had passed his sixtieth year, and he was at the height of his career. The success and admiration had left its trace in the form of a '*roi soleil*' mentality, it would not do to remind the king of such matters of credit. As usual, there is hardly any reason for a scientists to suppress reference to his sources, and certainly not for a successful mathematician like Hilbert. It is well known that some scholars only need the slightest hint, and they will build a beautiful theory. Others excel in originality, but have no urge to carry out the follow up. It is the co-operation of the two kinds that often produces striking results. We have seen that in the case of dimension theory, Brouwer was satisfied to work out the underlying idea so far that its significance was beyond doubt. It took a man like Menger (and had he lived, Urysohn) to work out all the interesting consequences. In the present case of the mathematics and language levels, one is hard pressed to find a reason for Hilbert to deny Brouwer his credit. Hilbert's grand design for proof theory was there for all the world to see; the most plausible explanation is that Hilbert simply forgot about Brouwer and his levels. That still leaves the question of why Hilbert did not acknowledge Brouwer's contribution after the subtle hint in the Reflections-paper. At worst, he could have said, 'Brouwer independently arrived at the idea of the mathematics levels'.

Brouwer definitely had not forgotten, and this time he decided to mention the fact; the above mentioned footnote serves the purpose. In the main text there is a hidden allusion to this matter, the two level distinction occurs in the dissertation where 'the mentioned parts of formalistic mathematics are distinguished as *mathematical language* and second-order mathematics, and where the intuitive character of the latter part is emphasised. It broke through in the formalist literature with the name *metamathematics* for second-order mathematics in the first Hamburg lecture.' In Brouwer's archive there is a small note to the effect that the 'Hilbert of 1900 did not feel the contradictory consequence of the lack of the intuiting of metamathematics, of which I freed him only in 1909 on a walk in the dunes, whereupon he changed my description "mathematical language of the second order" to "meta-

[26]Courant (1981), p. 161.

mathematics" '.[27] Clearly, in the absence of written evidence or witnesses, it is hard to prove that, on this walk in the dunes, the topic of formalisation and levels was discussed. However, Brouwer was a scrupulous man when it came to historical facts; there is no reason to doubt the 'discussion in the dunes', and it is most plausible that, apart from Lie groups and topology, foundational matters also came up. And what is more plausible than that Brouwer turned to the criticism formulated in his dissertation. Hilbert never reacted to Brouwer's statement. This may or may not be significant. One would guess, however, that a public reference to a poor credit policy would not have amused him. Hilbert had a reputation for a certain cavalier attitude with respect to credits. Courant, who belonged to Hilbert's inner circle, lists in his *Reminiscences* a number of instances where Hilbert completely forgot where he picked up what.[28] Being a man of undisputed technical capacities, he could easily work out a topic for himself on information that reached him in conversation. Courant cites F. Schur and Fredholm as two fellow mathematicians who fell victim to Hilbert's absent-mindedness; there were undoubtedly many more, but apparently nobody went as far as Brouwer to mention the matter of intellectual debt in public. Courant was acutely aware of Hilbert's practice, and he observed the influence on his circle of students, 'it did create in Hilbert's students and assistants a feeling of neglect. A certain duty exists, after all, for a scientist to pay attention to others and to give them credit. The Göttingen group was famous for a lack of feeling of responsibility in this respect. We used to call this process—learning something, forgetting where you learned it, then perhaps doing it better yourself, and publishing without quoting correctly—the process of "nostrification". This was a very important concept in Göttingen.'[29]

For the remaining three insights Brouwer also analysed Hilbert's papers, and showed that in each case an acceptance of the insight by the formalists came *after* the intuitionist allegation. It is interesting to see that usually Hilbert, after some rearguard skirmishes, adopted Brouwer's views. It is certainly strange that Hilbert never bothered to take Brouwer's arguments seriously in public.

Number three on the list is a bit surprising, as Hilbert's dogma is a rather marginal matter. One of those slogans that belong rather to popular lectures than to science proper. That Hilbert consistently missed the point is not all that surprising, as one must view the dogma, just as 'PEM', with an intuitionistic eye. Something that Hilbert could not, or would not, do, not even for the sake of argument.

Number four was, of course, the heart of the matter. Accepting it would mean for Hilbert to declare his program bankrupt. It took the soft spoken Gödel, and an equally withdrawn Gentzen, to present Hilbert with a fait accompli.

After analysing his four insights, Brouwer devoted the second and last section to the consistency of various forms of PEM, but not before answering politely, but firmly, Hilbert's snubs:

---

[27] The note is not dated, but it is almost certainly written in his later years. It also mentions Russell, Kant and Mannoury.

[28] Courant (1981), p. 160.

[29] Courant (1981).

> According to what precedes, formalism has received nothing but benefactions from intuitionism and may expect further benefactions. The formalistic school should therefore accord some recognition to intuitionism, instead of polemicising against it in sneering tones while not even observing proper mention of authorship. Moreover, the formalistic school should ponder the fact that in the framework of formalism *nothing* of mathematics proper has been secured up to now (since, after all, the meta-mathematical proof of the consistency of the axiom system is lacking, now as before), where intuitionism, on the basis of its constructive definition of spread[30] and the fundamental property it has exhibited for finitary sets[31] has already erected anew several of the theories of mathematics proper in unshakable certainty. If, therefore, the formalistic school, according to its utterance in Hilbert (1926), p. 180, has detected modesty on the part of intuitionism, it should seize the occasion not to lag behind intuitionism with respect to this virtue.

One can see that Hilbert's qualification of the intuitionistic efforts as 'modest', or worse, 'wretched remnants', rankled.

We know that Brouwer was no stranger to emotional outbursts, but he always kept his public reactions under control. The above quotation is an illustration, there is no name-calling, just a summing up of some facts, and a mild admonition.

The paper contained a novelty, namely the strong refutation of the principle of the excluded middle. In fact this was an immediate corollary of his theorem that the continuum is indecomposable. Hence the statement 'very real is either rational or irrational' is contradictory. The indecomposability was proved directly in the Berlin lectures, but it is obviously an immediate corollary of the continuity theorem.

The *Grundlagenstreit* was warming up, yet nobody could foresee that it would soon be over. Brouwer used his Vienna lectures to lay a coherent philosophical picture before the world; two talks without antagonistic elements. Hilbert was preparing his grand lecture for the Bologna conference. A storm was brewing, and when it broke, its rage threatened to sweep away substantial parts of the castle of the German mathematical community.

## 14.2 The Bologna Conference

After the war the mathematicians had twice been called to an international meeting. One in 1920 in Strasbourg, and the next in 1924 in Toronto. The atmosphere in Strasbourg had been one of exhilaration, of victory. The vanquished, not allowed to attend, had met in Bad Nauheim licking their wounds. The Toronto meeting was still out of bounds for the Germans; certainly there had been an attempt to put an end to the boycott, but the motion for readmitting the Germans was not even tabled by the *Union internationale de Mathématique*, it was simply passed on to the *Conseil*.

---

[30]Cf. p. 304 ff.

[31]Cf. p. 371 ff.

After the defeat of the Dutch-Danish-Norwegian proposal to drop the exclusion clause from the statutes of the *Conseil*,[32] a substantial part of the mathematical community wished to see that the next international mathematics conference truly deserved that name. The Union had already decided on the place: the international mathematics conference was to be held in 1928 in the venerable university of Bologna. And so the various mathematics societies duly advertised the conference. The German mathematics society, DMV, had inserted the flyer announcing the meeting in its journal, the *Jahresbericht der Deutschen Mathematischen Vereinigung*. The DMV traditionally informed their members of relevant events, and it clearly saw no objection to honour the old policy of free information. The more radical adversaries in Germany, however, viewed an announcement of the organisation that did not recognise them, in their own periodical, as a case of adding insult to injury. Brouwer, the implacable foe of the *Conseil* and its satellite, the Union, made himself the spokesman of the *Deutschnationale* opposition. In January, Brouwer wrote to Bieberbach about possible actions against the Union. Apparently, he had approached Bieberbach, as the editor in chief of the *Jahresbericht*. Bieberbach, in his reply of January 20, informed him that as the managing editor of the *Jahresbericht*, he could not possibly publish a comment of Brouwer on the invitation of the Union. 'It would create a novum if we wished to accept political arguments in the *Jahresbericht*, thus we have avoided, because of the political side of the matter, to speak of the planned conference in Bologna.' Brouwer, never to be caught out, agreed that his comments on the *Conseil* were only in form, but not in content, more political than the invitation for the Bologna conference.[33] His account, he said, would expose the hidden content of this invitation. Anyway, if the Jahresbericht would not publish his note, an enclosed leaflet would do. And indeed, a leaflet was produced, in which Brouwer reminded the readers of the motivation and formulations of the advocates of the boycott of German scientists.[34] He quoted the statements made by Painlevé in the *Académie des Sciences*.[35] These statements belonged to the emotional atmosphere of the war and its aftermath, and as such they were understandable, but they were not well-suited to promote peace and cooperation in the scientific world. Indeed, they were nothing less than a wholesale insult and condemnation of all German scholars. Painlevé's conciliatory moves in the affair of the Riemann volume of the *Annalen*, and his general role in the scientific community, make it clear that he was no longer the rabid anti-German of 1918.[36] But a public retraction of everything said and done in the early years after the war was probably asking too much. Nonetheless this was what the German nationalists among the mathematicians insisted upon.

Brouwer, in his leaflet, left it as a—none too subtle—suggestion: '... the readers of the *Jahresbericht* may contemplate in how participation in the planned congress

[32]Cf. Sect. 9.1.

[33]Brouwer to Bieberbach, 23.I.1928.

[34]Dated March 1923 and addressed to the members of the DMV.

[35]Cf. p. 329.

[36]Cf. Schroeder-Gudehus (1966), pp. 248, 250.

is possible, without mocking the memory of Gausz and Riemann, the humanitarian character of the science of mathematics, and the independence of the human spirit.'

Not content with a passive resignation, Brouwer, possibly after consultation with like-minded Germans, decided to visit the Italian organisers in person, in the hope that the role of the Union could be reduced to naught. The physicist Sommerfeld wrote Brouwer in March that the new invitation for the Bologna congress did not mention the name of the *Conseil* or similar organisations. He expected that Brouwer would find an open ear in Italy.[37] Some time in late March or in April Brouwer visited Pincherle, the president of the organising committee, and Levi-Civita. The discussions were friendly, and for both sides most satisfactory. From the side of the organisers, it was agreed that the conference would not be a Union-conference, but an international congress under the aegis of the University of Bologna. Brouwer could thus return home with good news: the Union was sidetracked.

In the German mathematical community the neutral mathematicians and the anti-*Conseil* group had strong feelings about the conference. The neutrals advocated participation in Bologna, and the anti's were, in spite of the Italian initiative, still wary. At the annual meeting in Bad Kissingen the DMV had decided not to send representatives to the Bologna conference, but it had at the same time sent out the parole to its members to attend in as large numbers as possible.[38] The Göttingers largely supported the Bologna conference, and the Berlin group contained some outspoken anti's. It was no secret that a certain amount of friction existed between the Berlin group and the Göttingen group. Although there was no lack of co-operation between the mathematicians of both groups (and it should be kept in mind that Berlin and Göttingen were icons for many mathematicians all over Germany), there was a certain tendency to accuse the Göttingen group of smugness. Vice versa there was little feeling of antagonism, a Göttingen mathematician was usually well aware of the true or imagined superiority over sister universities.

Pincherle, in his role of president of the congress, had to face one unpleasant duty: to inform Picard, the president of the *Conseil* and the Union of the decisions of the Bologna committee. And he proved himself up to the occasion. On June 8 he sent a long letter to Picard—'Monsieur et illustre Maître!' He respectfully, but firmly, explained that one could no longer stick to the format of the Strasbourg and Toronto Conferences. If one wishes, he wrote, to re-establish between the scholars who practice the most pure of all sciences, the good relations which are so necessary for their progress, one should leave formal considerations behind. The organisers felt that they had to shape the congress in such a way that this meeting of all mathematicians would be possible. And so he informed Picard of the decision to make the congress an Italian matter, independent of the Union. Of course, he added, the Union will still have its business meeting during the conference, where they could fix their policy for the future. He hoped, nonetheless, that the great Emil Picard would be there to open the lecture series of the conference.

---

[37] Sommerfeld to Brouwer, 24.III.1928.

[38] Perron to Bieberbach, 6.V.1928. Perron inquired if Bieberbach would go to Bologna and give a talk.

Picard, confronted with a fait accompli, replied that his quality of president of the *Conseil* did not permit him to attend the congress.

In view of the pleasing initiative of Pincherle, one would guess that the problem, at least in Germany, had been solved to everyone's satisfaction. Perhaps a number of French mathematicians would feel hurt, but the international character of the congress was at least saved. Yet, in Germany new problems and discussions arose. Bieberbach, in reply to a circular letter of Geheimrat Ziehen, the rector of the university at Halle, summed up the touchy issues that still rankled.[39] He pointed out that the information on the Bologna congress was far from consistent. Indeed, there must have been some mix-up affecting the invitations. Some mentioned the involvement of the Union, others did not; by way of illustration—Brouwer obtained a personal, albeit printed, invitation from Peano on July 8, in the usual flattering wording, full of misprints, and mentioning explicitly on the third line '*Sous les auspices de l'Union Mathématique Internationale*'.

Bieberbach's list of grievances contained more: the French invitation listed the Bologna congress as the eighth international mathematics congress, thus elevating the Strasbourg and Toronto congresses, from which the Germans were excluded, to a status of internationality they did, in his eyes, not deserve. As far as he knew the effects from a certain neutral side had resulted in the fobbing off in the form that in the latest circular the Union was no longer explicitly mentioned. 'A diplomatic feat, by which the real state of things was even more underlined.'

Furthermore, he took exception to the tactless plans for an excursion to the power station at the Ledrosse in South Tirol, which had changed hands after Versailles.

The call of the rector of the University at Halle to the universities and academies, to send representatives to Bologna, amazed Bieberbach. He could not see a justification in customs of the past. The Prussian Academy had left the decision to attend congresses of the *Conseil* to the individual scientists, but advised a large measure of restraint. 'Only if this congress, compromised beyond rescue, essentially suffers damage through the lack of German participation, can we expect with certainty that the next international mathematics congress will be really international, without involvement of the Union, either that this totally superfluous and ineffective organisation cleanses itself by extricating itself from the *Conseil*, or by a complete reorganisation satisfying the German wishes.' Here was a concrete goal; whereas Brouwer vented his moral objections against the *Conseil* and the Union, Bieberbach saw a solution to the present discontent in mathematical circles, and not only in Germany. The breaking up of the Union and the *Conseil* could start in Bologna, he thought. Ziehen must have informed the leading mathematicians of Bieberbach's letter, an unfortunate and rash action. For the letter was a personal reaction, and Bieberbach had not foreseen that Ziehen would circulate it.[40] Had he known the use Ziehen was going to make of his letter, he would probably have been more guarded in formulation and content. Bieberbach had not expected any such action from Ziehen, as

[39] Bieberbach to Ziehen, 18.VI.1928.

[40] Cf. Bieberbach to Courant, 15.VII.1928.

the *Rektorenkonferenz* and the *Hochschulverband* had agreed that all foreign affairs would be handled by the latter.

The Bieberbach–Ziehen exchange was not entirely ignored: there was at least one sharp and aggressive reaction, it came from the grand-master of German mathematics. Hilbert wrote an angry letter to the German rectors and directors of mathematical seminars, resolutely repudiating Bieberbach's views: 'An die Herren Rektoren der deutschen Hochschulen und die Leiter der mathematischen Seminare. *Betrifft den Internationalen Mathematiker-Kongress in Bologna*'.[41]

Bieberbach, he said, was ill informed, and citing rumours. The congress had officially severed its ties with the Union; agitating against the Italian organisation and boycotting the congress would alienate friendly colleagues and organisations, and could only be seen as a slap in the face of the Italians.

The letter ended with an explicit call to the German colleagues, 'It would be in the interest of German science and German prestige to wish most urgently, that no university and no academy will desist from accepting the official invitation to participate in a friendly manner.'

The letter clearly outlined Hilbert's political position; he did not speak about the past insults, but stuck to the possibility offered by the Italian organisation. At the same time he put Bieberbach in his place, and this caused—as to be expected—some hard feelings.

For many German mathematicians Hilbert's action was the ardently awaited signal. Thus Perron wrote to Landau 'Finally the circular of Hilbert, for which I had been waiting passionately, has arrived.'[42]

Hilbert, in a manner of speaking, had viewed the discussion and the arguments concerning the Bologna congress from the hight of his Olympus, blurring the finer details. He was probably quite right to do so, because an overly discerning gaze is no asset if one wants to reach a compromise. But it should be noted that in this way he did not do justice to Bieberbach's observations. Small wonder that Bieberbach replied publicly to Hilbert's attack. Since his '*hochverehrter Lehrer, Herr Geheimrat Hilbert*' had severely reprimanded him, he felt obliged to defend himself. The letter is a clever balancing act, between not insulting Hilbert and sticking to his guns. On the whole, one must say that, viewed as an academic exercise Bieberbach performed most credibly. E.g. in countering Hilbert's argument, that Picard had felt offended after Pincherle's letter, he remarked that this was beside the point, 'for our aim is only the satisfaction of our own value, not the degree of irritation of Mr. Picard'. He deplored that Hilbert, by using the word 'denunciated', had introduced a false note into the discussion.[43] His veneration for his teacher and the level of courtesy in a discussion would forbid him to follow suit.

---

[41]Hilbert to rectors of German universities, 29.VI.1928.

[42]Perron to Landau, 2.VII.1928.

[43]*...Dieses Schreiben nimmt in schroffster Weise Stellung gegen den diesen Herbst in Bologna stattfindenden Internationalen Mathematiker-Kongress, denunziert den Kongress als eine Veranstaltung des conseil internationale de recherches und rät daher dringend an zu einer Ablehnung der an die Universitäten usw. ergangener Einladung.*

There must have been an enormous amount of correspondence concerning the Bologna congress, most of it will have found its way to the wastepaper basket, but some of the letters of the influential mathematicians have been preserved. There are basically two kinds of views, one, like Hilbert's, shows itself satisfied with the Italian results, another one, like that of Bieberbach and Brouwer, considers the concessions promising but not conclusive. Von Mises, for example, in his letter to Courant,[44] agrees with the latter that 'we should have nothing to do with the Union'. But he disagreed with Courant (and the Göttingen mathematicians)[45] about the real content of the Italian communications. The role of the Union is left completely vague, there is no evidence that the Union is put out of action. For one thing, Pincherle, who became president of the organising committee in his function of chairman of the Union, should have abdicated in the latter function.

From von Mises' letter we learn that Brouwer had not aimed, in his discussion with Pincherle, at a complete separation between congress and Union, but rather at a more modest goal: the appointment of an independent international committee that would after the conference decide on the future conferences, so that there would be a guarantee that these would no longer be congresses of the Union. That much was promised to him. The main thing was, he argued, to prevent that the Union, 'which at this moment is fighting its final battle', could present the Bologna conference as 'sufficiently international', for then 'the Union, this pronounced *political organisation*, would still exist in four years time, and all the present conflicts would have to be repeated'. He concluded that the only definite solution could be reached through a wholesale ignoring of the conference by the German mathematicians.

By now, in the early days of July, all positions had been taken up. It was unlikely that any German mathematician would change his views. The remaining correspondence can be viewed as damage control. Courant, who had become Hilbert's main lieutenant, tried to pacify Bieberbach in a letter of July 10. Hilbert had acted in such a radical manner, as he had promised to give a talk in Bologna, and believed that the political issues had been straightened out. To his surprise, the letter from Halle, followed by Bieberbach's letter, had given substance to the idea that going to Bologna was un-German. Hence his sharp reaction, which was in no way intended personally. As Courant put it, 'Hilbert did not want to react once more to your reply, he had, however, expressly assured me how little from his side a personal insult was intended.' One might wonder why Hilbert did not send a short note to that effect to Bieberbach, but one has to keep in mind that he still suffered from his pernicious anaemia. He probably tried to avoid all action that was not strictly necessary. Of course, there is also the possibility that the king of mathematics does not apologise to a former student, who happened to be a Berlin professor.

Courant went on to explain that Hilbert had been seriously annoyed by the Riemann affair: 'Hilbert had felt deeply insulted when, as a consequence of Brouwer's

[44] Von Mises to Courant, 7.VII.1928.

[45] He referred tongue in cheek to the Hilbert–Bieberbach discussion, noting that 'it is a good illustration of your remark that inside the mathematics society no controversies exist'.

exhortation and, as Hilbert thinks, with your co-operation the collaboration, desired by Hilbert, of French mathematicians to the Riemann volume, was blocked.' Courant's, no doubt kindly intended, attempt at reconciliation met with doubt and scorn. Bieberbach replied five days later that he assumed that Courant did not agree with the use of the term 'denunciate' to describe his action. 'I presume that this is solely on Hilbert's conto.'[46] Courant was completely wrong, Bieberbach added, in his view of Brouwer's role in the Riemann affair. Bieberbach's letter shows some of the discontent one would find in Germany over the high-handed actions of Hilbert.

> As far as this matter of the Riemann-volume is concerned, there was, as you know, a discussion about it in Innsbruck.[47] And I have made use of my right to ventilate my opinion. I have never committed myself to voice only opinions that pleased Hilbert; and I would find it foolish if such a demand would seriously be made to me. Also in this matter I have not unfolded any agitation, as I have better things to do. I have, only when my opinion was asked, stuck to my point of view.

Bieberbach also felt that Hilbert had not done his homework, but nonetheless accused Bieberbach of 'using secondary information'. Here Bieberbach had a point, right from the beginning he had been involved in the discussions concerning the Bologna congress. He was well-informed about all the negotiations. Thus it is not surprising that he was upset, if not offended, by the amateurish diplomacy of the Göttingers, 'Rather one could reproach the Hilbert clique, that they attach decisive importance (*Bedeutung*) to ad hoc formulations, e.g. like the letter of Bohr, without taking into account the previous history and the laborious correspondence over many years.'

Bieberbach's letter also provides some useful information on Brouwer's role. From the point of view of Göttingen it seemed as if there was a huge conspiracy with Bieberbach, Brouwer and Von Mises as the main characters. But actually, Brouwer's actions in Italy were his own, Bieberbach only learned about them from Von Mises. The only significant contact about the Bologna matter concerned Brouwer's letter to the editor of the Jahresbericht (see p. 542). As to Brouwer's involvement in the Bologna affair, Courant had confirmed that Hilbert's thoughts went beyond just suspicions: 'The idea that Brouwer's disposition was actually normative for the colleagues in Berlin was moreover confirmed by reports of Von Mises; and I can only say that Hilbert rejected most passionately this interference and this playing the judge.'

In his reply Bieberbach soberly asked 'Why, by the way, is Brouwer's interest in the matter an 'outside interference', but in contrast, that of Bohr and Hardy not?' As Bieberbach wrote, Hilbert had already at the time of the Bad Kissingen meeting, thus before the conflict arose, put himself forward as a lecturer at the conference. This certainly was an extremely visible sign from the German side of the wish for

[46] Bieberbach to Courant, 15.VII.1928.

[47] Apparently a meeting of the editors of the M.A. editors at the time of the annual meeting of the DMV.

**Fig. 14.1** Ludwig Bieberbach. [Courtesy U. Bieberbach]

appeasement, in particular since it came from the greatest mathematician, at a time when the congress was still a full-blooded Union congress. He, Bieberbach, had acquiesced, as one might assume that few Germans would attend. Hilbert's role as a symbol of German good will was part of Bieberbach's master plan. He had not dreamed that 'Hilbert was essentially so little independent, that he demanded a general participation, and considered the alternative as an insult for his person.' In short, Bieberbach was not subdued by the carrot-and-stick letter of Courant.

In another letter, to the applied mathematician Walther, Bieberbach again analysed the key statements concerning the separation of Union and congress, pointing out the various readings and glaring inconsistencies (e.g. in June Bortolotti had told Bieberbach that Picard's objection to the procedure was that the invitations for the conference were not offered for inspection to the *Conseil*, and that the rector of the University of Bologna had sent them out). He had seen too many diplomatic formulations not to be sceptical. 'But', he wrote, 'mundus vult decipi and who wants to go to Bologna, and thinks that he has to let his glory shine in the beams of the international sun, to him such formulations are *Butter aufs Brot*'. The correspondence between Courant and Bieberbach went on for another round. Since only Bieberbach's letter has been preserved, one has to guess the content of Courant's letter. Bieberbach, in his vehement protest against Courant's claims, maintained that Brouwer's disposition had not been of decisive influence on him. It had not exercised any influence at all, he wrote. Hilbert's conciliatory move towards Bieberbach must again have been brought up by Courant. But Bieberbach remained adamant. Hilbert had publicly insulted Bieberbach—even after Bieberbach's reaction of July 3, Hilbert's circular had been distributed further; for example to the Prussian Academy. Instead of authorising Courant to make friendly noises, he should have stated his apology before the same Forum that had witnessed the insult. But, he went on, 'in view of the circumstance, that Hilbert is substantially older than I am, and that he is my teacher, I will be satisfied if you will get yourself an authorisation from Hilbert, to state to me that he regrets and retracts the expression 'denunciate''.

Since the negotiations concerning the Bologna congress were now all water under the bridge, he advocated a composed conduct of the German delegation in Bologna: 'It is more important to prevent that in Bologna the Germans present there sneer altogether too much at those who stayed away to paint a picture of their own excellence at the expense of German prestige and unity. Perhaps you should just think about it if you could not contemplate this idea. So far, everything that comes from Göttingen has a ring of such disdain for dissenting Germans, that I am really worried how it will be in Bologna, if this mood comes to an eruption, just as it is said to have burst out in Göttingen before foreigners.'

Brouwer's role had become marginal, after his negotiations in Rome and Bologna, he carried on his correspondence, but no longer played a role. He once more sent out his leaflet with the unfortunate statements of Painlevé (August).

The unity of the foreign advocates of ending the boycott of Germans had begun to crumble. Harald Bohr had, after an exchange of letters with Pincherle, come to the conclusion that the Italians were offering a really international meeting where all nationalities participated with equal rights. He saw the Union as definitely defeated—'we, the internationally minded, (i.e. people like you, Hardy etc.) have in fact won so completely, that from my point of view it would be neither natural, nor advantageous, if now the congress would be blown up by the opponents of the Union. The relatively minor questions and formalities should not create a rift between us.'[48]

Hardy also wrote to Brouwer about the conference boycott; he had, like Brouwer, been a staunch enemy of the *Conseil*—'I detest it, and have never had anything to do with it.' But—contrary to Brouwer—he hoped 'that the Germans would, on the whole, adopt the policy of being more magnanimous than they could reasonably be expected to be'. In short, he shared Bohr's view. Having fought the *Conseil* tooth and nail, with Brouwer, Bohr and others, he could not share Brouwer's adamant views, as appears from the closing hints of his letter: 'Finally, if you are to demand that everybody should formally retract all the imbecilities which have been uttered during the war, then assuredly there will never be any Congress of any kind until everybody born before 1914 is dead. And I should hope that, whatever happens at Bologna, it will at any rate be enough to make everybody realise that the era of imbecility is passed.'

So this was the end of the road, a small group of die-hards refused to go to Bologna, so close to Canossa, as Bieberbach had remarked. A sizeable contingent of the German mathematical community did accept the assurances of the organisers, and participated. There was also a number of mathematicians that did not go, but for other than political reasons (e.g. Hardy and Menger).

The conference became a huge success. The mathematician Härlen reported to Brouwer that Pincherle had opened the congress by declaring that the Italians were

[48]Bohr to Brouwer, 3.VII.1928.

the only organisers of the congress, and on many occasions it was proclaimed 'that this was the first truly international congress after the war'.[49]

Three weeks later Härlen sent a more complete report to Brouwer. The German aspects were quite satisfactorily reckoned with, he wrote, there were for instance a large number of German posters instructing the participants, more than English or French ones. The congress organisation clearly had done its best not to offend the Germans. In particular there were quite a number of apologetic remarks about the recent past. Pincherle was elected with general approval as president of the meeting. The choice of the vice presidents was judicious: Hadamard, Hilbert ('exceptionally strongly applauded, very striking'), Fehr (Switzerland), Young (UK), Birkhoff (USA), Bohr (Scandinavia), Rey Pastor (Spain & South America), Sierpinski (Poland), and Lusin (USSR).

The series of lectures was opened by Hilbert. Constance Reid described the occasion in her biography of Hilbert: 'At the opening session, as the Germans came into an international meeting, for the first time since the war, the delegates saw a familiar figure, more frail than they remembered, marching at their head. For a few minutes there was not a sound in the hall. Then spontaneously, every person present rose and applauded.'[50] Härlen also described this event:

> ... the first lecture of Hilbert, who was greeted with a storm of applause. Numerous repetitions; power of concentration clearly much hampered by physical suffering. Content familiar by latest publications. Strong applause—Also a strong applause greeted Hadamard, whose lecture was also very good in presentation much more effective than Hilbert's. At Hadamard's lecture the applause much stronger than before. In Hilbert's case the applause was meant for the person, in Hadamard's case also for the lecture.

It should probably not surprise us that the Italian government made the most of this meeting. At the opening the Podesta greeted the participants in the name of 'the fascist city, that was happy to show the foreign visitors the accomplishment of the fascists'. The participants were given tags with bands in the Italian colours and at breakfast there were little Italian flags for the participants to wear. This unnecessary and somewhat childish nationalism did not exactly please all participants.

The Union was active after all, it distributed flyers with invitations to join its business meeting. At this meeting Pincherle announced that Switzerland had offered to host the next congress. In his closing address the Union did not figure at all. In preceding private sessions invitations from Prague and the Netherlands had reached the meeting. Switzerland had declined, in spite of pressure from all sides. The Dutch professor Schouten had telegraphed after the meeting to the Dutch government, and he had obtained a positive reaction. The next day in informal discussions objections against the Netherlands were raised on the grounds of Brouwer's actions. After some more pressure Switzerland came back on its refusal, and the Netherlands and Prague withdrew their offers.

[49] Härlen to Brouwer, 6.IX.1928.

[50] Reid (1970), p. 188.

After the congress there were some retrospective discussions, but on the whole the hopes were high that from now on the boycott would be a thing of the past.

Brouwer may not have been successful in discouraging the Germans to participate, but he could well be satisfied with the final outcome. If one tries to give a reason for the uncompromising attitude of the anti-Unionists, one should look at the past history of the *Conseil* and the Union. All attempts to drop the exclusion clauses had been rudely blocked by Picard and his followers. In certain quarters there was absolutely no faith in the will or capability to accept the enemy of the past as the colleague of today. Hence the uncompromising stance of people like Bieberbach, Brouwer and Von Mises. On the other side there was a more optimistically minded group, led by the Göttingers, who believed that a compromise at the Bologna meeting would pave the way for a restoration of the old free co-operation. History has proved the Göttingers right, but in 1928 the future still offered two options.

Looking back at Brouwer's role in the Bologna affair, one is almost surprised by his innocence. For him it was a fight between good and evil, between a closed shop of *Conseil*-connected scholars and institutions, and the free world of science. His attacks were conspicuous for the quotations of his adversaries, not for his own strong language. The most effective action he undertook was his personal visit to Rome and Bologna, and the negotiations with his Italian colleagues. It probably was the turning point in the action for detaching the congress from the Union. His pamphlet might be considered a clever attempt to keep the German mathematicians away from Bologna, but it is doubtful if it changed the view of those who had not made up their minds to stay home. For a successful action of that sort, one needs an organisation and strongly motivated activist-supporters. None of this was within Brouwer's province. He had always been a loner, and even the sympathy of a number of German colleagues was not more than the support of kind well-wishers. Against a network like that of the Göttingen mathematicians, an unworldly Amsterdam professor stood no chance. Brouwer, like the legendary Roland, was famous for his lost battles.

Brouwer's actions and pamphlets were personal in the sense that he referred to certain mathematicians in the higher ranks of the *Conseil*, but no attacks at German colleagues were launched. Hilbert on the other hand personally attacked Bieberbach, and there is a private note of his, probably of this period, in which he poured his wrath over Brouwer:

> *Erpressertum.*
> In Germany a political blackmail of the worst kind has come up. You are not a German, not worthy a German birth, if you don't speak and act as I tell you now. It is very easy to get rid of these blackmailers. One has but to ask, how long they have been in the German trenches. Unfortunately German mathematicians have fallen victim of this blackmail, for example Bieberbach. Brouwer has known to make use of this state of the Germans, and without having been active himself in the German trenches, all the more to work towards inciting and to cause discord among the Germans, in order to pose as the master of German mathematics. With complete success. He will not succeed in this for the second time.

One should of course keep in mind that Hilbert's health was very poor at the time, and this may well have influenced his judgement and emotional balance. A far more fateful manifestation of Hilbert's illness would in the aftermath of the Bologna congress bring German mathematics to the brink of 'civil war'.

## 14.3 The War of the Frogs and the Mice

Who will rid me of this meddlesome priest

Henry II

The dreaded and perilous pernicious anaemia had not left Hilbert alone. After a modest improvement in January 1928, Hilbert was plagued again by its various symptoms. In this condition he had a heart attack in the weeks before he set off for Bologna.[51] So when he appeared in Bologna, he was physically in a poor state. His triumphant reception and the success of his lecture must have given him the strength to make it through the congress, but after the congress he completely broke down. He was taken to a sanatorium in Luzern where he spent a full five weeks. His situation was for a longer period so desperate that he saw himself at death's doorstep. It was this unfortunate and sad combination of events that lay at the root of the following tragic history of a conflict in the German mathematical world.

On 27 October 1928, a curious telegram was delivered to Brouwer, a telegram that was to plunge him into a conflict that for some months threatened to split the German mathematical community. This telegram set into motion a train of events that was to lead to the end of Brouwer's involvement in the affairs of German mathematicians and indirectly to the conclusion of the *Grundlagenstreit*. The telegram was dispatched in Berlin, and it read:

> Professor Brouwer, Laren N.H. Please do not undertake anything before you have talked to Carathéodory who must inform you of an unknown fact of the greatest consequence. The matter is totally different from what you might believe on the grounds of the letters received. Carathéodory is coming to Amsterdam on Monday.
>
> Erhard Schmidt

Following Schmidt's instructions, Brouwer, puzzled as he was, went about his daily routines. Two registered letters that had arrived at the post office added to the mystery. Brouwer collected these letters from Göttingen and waited for the arrival of Constantin Carathéodory. The letters were still unopened when Carathéodory arrived in Laren on the thirtieth of October. Carathéodory's visit figures prominently in the history that is to follow.

In order to appreciate the full dramatic magnitude of the following history, one must keep in mind that Brouwer was on friendly terms with all the actors in this

[51] Courant to Springer, 3.X.1928.

**Fig. 14.2** Constantin Carathéodory. [Courtesy M. Georgiadou]

small drama, with the possible exception of David Hilbert. Some of them were even intimate friends of his, for example Carathéodory and Otto Blumenthal.[52]

Carathéodory found himself in the embarrassing position of being the messenger of offensive news, in which he was involved against his will. The first letter, he explained to Brouwer, should have carried more signatures, or at least Blumenthal's signature. Carathéodory's name was used in a manner not in accordance with the facts, although he would not disown the letter should Brouwer open it. Finally, the sender of the letter would probably seriously regret his action within a couple of weeks. The second letter was written by Carathéodory himself, although Blumenthal's name was on the envelope. He, Carathéodory, regretted the contents of the letter.

Thereupon Brouwer handed the second letter over to Carathéodory, who proceeded to relate the theme of the letters. The contents of the second can only be guessed, but the first letter can be quoted verbatim. It was written by Hilbert, and copies were sent to the other dramatis personae in the tragedy that was to follow.

Hilbert's letter was brief:

> Dear Colleague,
>
> Because it is not possible for me to co-operate with you, given the incompatibility of our views on fundamental matters, I have asked the members of the board of managing editors of the Mathematische Annalen for the authorisation, which was given to me by Blumenthal and Carathéodory, to inform you that henceforth we will forgo your co-operation in the editing of the Annalen and thus delete your name from the title page. And at the same time I thank you in the name of the editors of the Annalen for your past activities in the interest of our journal.
>
> Respectfully yours,
> D. Hilbert.

[52] A wealth of information on Carathéodory can found in the recent biography (Georgiadou 2004).

The meeting of the two old friends was painful and stormy; it broke up in confusion. Carathéodory left in despondency and Brouwer was dealt one of the roughest blows of his career.

Although Brouwer had kept himself completely under control during the visit, he suffered under a strong reaction. After the visit he was ill for a few days and ran a temperature.[53]

The *Mathematische Annalen* was the most prestigious mathematics journal at that time. It was founded in 1868 by A. Clebsch and C. Neumann. In 1920 it was taken over from the first publisher, Teubner, by Springer.

For a long period the names of Felix Klein and the *Mathematische Annalen* were inseparable. The authority of the journal was largely, if not exclusively, based on the mathematical fame and the management abilities of Klein. The success of Klein in building up the reputation of the *Annalen* was to no small degree the result of his choice of editors. The journal was run, on Klein's instigation, on a rather unusual basis; the editors formed a small exclusive society with a remarkably democratic practice. The board of editors met regularly to discuss the affairs of the journal and to talk mathematics. Klein did not use his immense status to give orders, but the editors implicitly recognised his authority.

Being an editor of the *Mathematische Annalen* was considered a token of recognition and an honour. Through the close connection of Klein—and after his resignation, of Hilbert—with the *Annalen*, the journal was considered, sometimes fondly, sometimes less than fondly, to be 'owned' by the Göttingen mathematicians.

Brouwer's association with the *Annalen* went back to 1915 and before, and was based on his expertise in geometry and topology. In 1915 his name appeared under the heading 'With the co-operation of'. Brouwer was an active editor indeed; he spent a great deal of time refereeing papers in a most meticulous way.

The status of the editorial board, in the sense of by-laws, was vague. The front page of the *Annalen* listed two groups of editors, one under the head *Unter Mitwirkung von* (with the co-operation of) and one under the head *Gegenwärtig herausgegeben von* (at present published by). I will refer to the members of those groups as associate editors and chief editors. The contract (25 February 1920) that was concluded between the publisher, Springer, and the editors (*Herausgeber*) Felix Klein, David Hilbert, Albert Einstein, and Otto Blumenthal speaks of *Redakteure*, but does not specify any details except that Blumenthal is designated as managing editor.

The loose formulation of the contract would prove to be a stumbling block in settling the conflict that was triggered by Hilbert's letter. At the time of Hilbert's letter the journal was published by David Hilbert, Albert Einstein, Otto Blumenthal and Constantin Carathéodory, with the co-operation of L. Bieberbach, H. Bohr, L.E.J. Brouwer, R. Courant, W. von Dyck, O. Hölder, T. von Kármán, and A. Sommerfeld. The daily affairs of the *Annalen* were managed by Blumenthal, but the chief authority, undeniably, was Hilbert.

[53] Gawehn to Von Mises, 9.XI.1928.

Given the status of the *Annalen*, any mathematician would be more than happy to join the editorial board, and once on the board, expected to remain there until a ripe old age. Two of the most prominent editors, Klein and Hilbert, served resp. from 1876 till 1924, and from 1901 till 1939. So Klein retired at 75, one year before his death, and Hilbert at 77, four years before he died. Under the circumstances, Brouwer, who was an active associate editor, could expect eventually to become an editor, and to remain so for years to come. Hilbert dashed these expectations with one stroke of the pen. Even if the procedure remained a secret, which was to be doubted, Brouwer would be the laughing stock of the mathematical world. The readers of the *Mathematische Annalen* would note that Brouwer's name had disappeared from the list, and they would draw their own conclusion. Had Hilbert contemplated these consequences when he signed the dismissal in an emotional impulse? And did he realise the enormity of the insult he was committing?

It is hard to imagine what Hilbert had expected; he could not have counted on a calm resigned acquiescence from the high-strung, emotional Brouwer. In Brouwer's eyes (and quite a few colleagues would have taken the same view) a dismissal from the *Annalen* board was a gross insult.

Carathéodory, in trying to win Brouwer's acquiescence, had apparently revealed only part of the underlying motive, as appears from Brouwer's letter of November 2 to Blumenthal,

> Furthermore Carathéodory informed me that the *Hauptredaktion* of the *Mathematische Annalen* intended (and felt legally in the position) to remove me from the editorial board of the Annalen. And only for the reason that Hilbert wished to remove me, and that the state of his health required to give in to him. Carathéodory begged me, out of compassion with Hilbert, who was in such a state that one could not hold him responsible for his behaviour, to accept this shocking injury in resignation and without resistance.

Hilbert himself made no secret of his motivation; in a letter of 15 October he asked Einstein for his permission (as a *Mitherausgeber*) to send a letter of dismissal (the draft to the chief editors did not contain any explanation) and added

> Just to forestall misunderstandings and further ado, which are totally superfluous under the present circumstances, I would like to point out that my decision—to belong under no circumstances to the same board of editors as Brouwer—is firm and unalterable. To explain my request I would like to put forward, briefly, the following:
>
> 1. Brouwer has, in particular by means of his final circular letter to German mathematicians before Bologna, insulted me and, as I believe, the majority of German mathematicians.
>
> 2. In particular because of his strikingly hostile position vis-a-vis sympathetic foreign mathematicians, he is, in particular in the present time, unsuitable to participate in the editing of the Mathematische Annalen.
>
> 3. I would like to keep, in the spirit of the founders of the Mathematische Annalen, Göttingen as the chief base of the Mathematische Annalen—Klein,

> who earlier than any of us realised the overall detrimental activity of Brouwer, would also agree with me.[54]

In a postscript he added: 'I myself have for three years been afflicted by a grave illness (pernicious anaemia); even though this disease has been taken its deadly sting by an American invention, I have been suffering badly from its symptoms.'

Clearly, Hilbert's position was that the chief editors could appoint or dismiss the associate editors. As such he needed the approval of Blumenthal, Carathéodory and Einstein. Blumenthal had complied with Hilbert's wishes, but for Carathéodory, the consent was problematic; apparently he did not wish to upset Hilbert by contradicting him, but neither did he want to authorise him to dismiss Brouwer. Hilbert may easily have mistaken Carathéodory's evasive attitude for an implicit approval. Carathéodory had landed in an awkward conflict between loyalty and fairness. He obviously tried hard to reach a compromise. In view of Hilbert's firmly fixed conviction, he accepted the unavoidable conclusion that Brouwer had to go; but at least Brouwer should go with honour.

Being caught in the middle, Carathéodory sought Einstein's advice. In a letter of 16 October he wrote 'It is my opinion that a letter, as conceived by Hilbert, cannot possibly be sent off.' He proposed, instead, to send a letter to Brouwer, explaining the situation and suggesting that Brouwer should voluntarily hand in his resignation. Thus a conflict would be avoided and one could do Brouwer's work justice: 'Brouwer is one of the foremost mathematicians of our time and of all the editors he has done most for the *Mathematische Annalen*.' The second letter we mentioned above must have been the realisation of Carathéodory's plan. Einstein answered, 'It would be best to ignore this Brouwer affair. I would not have thought that Hilbert was capable of such emotional outbursts.'[55] The managing editor, Blumenthal, must have experienced an even greater conflict of loyalties, being a close personal friend of Brouwer and the first Ph.D. student (1898) of Hilbert, whom he revered. Einstein did not give in to Hilbert's request. In his answer to Hilbert (19 October 1928) he wrote:[56]

> I consider him [Brouwer], with all due respect for his mind, a psychopath and it is my opinion that it is neither objectively justified nor appropriate to undertake anything against him. I would say: 'Sire, give him the liberty of a jester (*Narrenfreiheit*)!' If you cannot bring yourself to this, because his behaviour gets too much on your nerves, for God's sake do what you have to do. I, myself, cannot sign, for the above reasons such a letter.

Carathéodory, however, did not possess Einstein's strength to cut the knot, once the moral issue was decided. He was seriously troubled and could not let the matter rest. He again turned to Einstein:[57]

---

[54]Carathéodory had also received a copy, and probably Blumenthal as well.

[55]Einstein to Carathéodory, 19.X.1928.

[56]Einstein to Hilbert, 19.X.1928.

[57]Carathéodory to Einstein, 20.X.1928.

> ... your opinion would be the most sensible, if the situation would not be so hopelessly muddled. The fight over Bologna... seems to me a pretext for Hilbert's action. The true grounds are deeper—in part they go back for almost ten years.[58] Hilbert is of the opinion that after his death Brouwer will constitute a danger for the continued existence of the *Mathematische Annalen*. The worst thing is that while Hilbert imagines that he does not have much longer to live (...) he concentrates all his energy on this one matter, (...). This stubbornness, which is connected with his illness, is confronted by Brouwer's unpredictability... If Hilbert were in good health, one could find ways and means, but what should one do if one knows that every excitement is harmful and dangerous? Until now I got along very well with Brouwer; the picture you sketch of him seems me a bit distorted, but it would lead too far to discuss this here.

This letter made Einstein, who in all public matters practised a high standard of moral behaviour, realise that these were deep waters indeed:[59]

> I thought it was a matter of mutual quirk, not a planned action. Now I fear to become an accomplice to a proceeding that I cannot approve of, nor justify, because my name—by the way, totally unjustifiably—has found its way to the title page of the *Annalen*... My opinion, that Brouwer has a weakness, which is wholly reminiscent of the *Prozessbauern*,[60] is based on many isolated incidents. For the rest I not only respect him as a man with an extra-ordinarily sharp eye, but also as an honest man, and a man of character.
>
> While I beg you not to blame me for my stubbornness and while I assure you that I will never make use of the fact that it was *you* who informed me about this, I remain with warmest greetings.

From these letters, even before the real fight had started, it clearly appears that Einstein was firmly resolved to reserve his neutrality. Einstein had called Brouwer 'an involuntary champion of Lombroso's theory of the close relation between genius and insanity', but Einstein was well aware of Brouwer's greatness, and did not wish him to be victimised. It is not clear whether Einstein's opinion was based on personal observation or on hearsay; the two knew each other well enough, they had met in 1920 when Einstein visited the Academy in Amsterdam, and Einstein had stayed with Brouwer in Blaricum.[61] Furthermore they had probably met during some of the visits of Einstein to Leiden, and at meetings of the editorial board of the *Annalen*.

---

[58] An enigmatic remark. It is far from clear what Cara (as Carathéodory was called among friends) had in mind. The only possible point of friction around 1918 could have been the offer of a chair in Göttingen to Brouwer, who subsequently turned it down. But is hard to see how that could have bothered Hilbert.

[59] Einstein to Carathéodory, 23.X.1928.

[60] This probably refers to the troubles in Schleswig-Holstein during roughly the same period when farmers resisted the tax policies of the government. Hans Fallada has sketched the episode in his *Bauern, Bonzen und Bomben*.

[61] Oral communication, Mrs. F. Heyting-van Anrooy.

Einstein's somewhat crude characterisation of Brouwer in the letter of 19 October may also have been prompted by a wish to pacify the unstable Hilbert. There is no better remedy to calm a person down, than by outdoing him.

It did not take Brouwer long to react. Brouwer was a man of great sensitivity, and when emotionally excited he was frequently subject to nervous fits. In the days following Carathéodory's visit, Brouwer was actually physically incapacitated (see p. 554).

On 2 November Brouwer sent letters to Blumenthal and Carathéodory, from which only the copy of the first one is in the Brouwer archive—it contained a report of Carathéodory's visit. The letter stated that 'in calm deliberation a decision on Carathéodory's request was reached'.

The answer to Carathéodory, as reproduced in the letter to Blumenthal, was brief:

> Dear Colleague,
>
> After close consideration and extensive consultation I have to take the position that the request from you to me, to behave with respect to Hilbert as to one of unsound mind, qualifies for compliance only if it should reach me in writing from Mrs. Hilbert and Hilbert's physician.
>
> Yours
> L.E.J. Brouwer

This solution, although perhaps a clever move in a political game of chess or in a court of law, was of course totally unacceptable—even worse, it was a misjudgement of the situation. The prevailing view is that a gentleman rather suffers the accusations of an unaccountable person, than to mention this unaccountability. Cara was thoroughly upset; he reported the outcome of his trip to Blaricum in a letter to Courant;[62] the idea had been that by explaining that Hilbert had been acting under the pressure of his illness, the pill could be sugared, thus making it possible for Brouwer to withdraw voluntarily. Erhard Schmidt had thought that this would satisfy Brouwer. At first Brouwer was quite sensible, Cara wrote, and he promised to do nothing before he had talked to Schmidt. 'Unfortunately I have today received a totally absurd letter from Brouwer, so that my whole action seems to have fallen through.' If there was any bright spot at all, he said, it was that now the Berlin mathematicians would no longer unconditionally support Brouwer, and that would reduce the risk that the whole matter got into the open. Being a man of honour, Carathéodory added, 'However, after my mediation has so sadly failed, I must resign as soon as possible from the board of the *Annalen*.'

Blumenthal's first reaction was guarded, 'With Brouwer it's complete chaos, you will hear soon enough. Hilbert must not hear about Cara's trip.'[63]

Hilbert's brief note had triggered a development that would have greatly surprised him. He had dismissed an associate editor, and that was it. The idea that he

[62] Carathéodory to Courant, 3.XI.1928.

[63] Blumenthal to Courant, 4.XI.1928.

had to justify his decision would not have crossed his mind. For Brouwer, on the other hand, it was unthinkable that one could fire an editor just because of an 'incompatibility of fundamental views'. In particular, since the formulation allowed only one interpretation: 'no intuitionist on my board', Brouwer had every right to be upset. And so would every well meaning editor. For Brouwer, giving in meant swallowing a grave insult; for the Hilbert side, Brouwer's demand as formulated above, was equally unacceptable.

Hilbert, like one of his ideal statements, had in the mean time been eliminated from the discussion. For his protection, his friends and students had decided to avoid any excitement that could harm his precarious health. And so the defence of the old master was taken up by the younger Hilbertians. There must have been quite a bit of hurried consultation, which could not have been all that easy, as Hilbert's extremely negative view of Brouwer was certainly not universally shared. Most editors were of course aware of the skirmishes in the *Grundlagenstreit*, but they would not dream of considering Brouwer a poor editor or a political risk for the *Annalen*. In fact, Brouwer was on friendly terms with most German mathematicians, be it from Göttingen, or elsewhere. There were of course obvious exceptions, such as Koebe, but on the whole he was a welcome guest at any university.

The conflict had presented itself so suddenly and so totally unexpectedly to Brouwer that he failed to realise to what extent Hilbert saw him as a deadly danger for mathematics, and as the bane of the *Mathematische Annalen*. His belief that the announced dismissal was the whim of a sick and temporarily deranged man emerges from a letter he dispatched to Mrs. Hilbert three days later:

> I beg you, use your influence on your husband, so that he does not pursue what he has undertaken against me. Not because it is going to hurt him and me, but in the first place because it is wrong, and because in his heart he is too good for this. For the time being I have, of course, to defend myself, but I hope that it will be restricted to an incident within the board of editors of the *Annalen*, and that the outer world will not notice anything.

A copy of this letter went to Courant with a friendly note, asking him (among other things) to keep an eye on the matter: 'As a matter of course, I count especially on you to bring Hilbert to reason, and to make sure that a scandal will be avoided.'[64] Courant, after visiting Mrs. Hilbert, replied to Brouwer that Hilbert was in this matter under nobody's influence, and that it was impossible to exert any influence on him.[65] The reader should realise that there was no animosity between Courant and Brouwer. As the *Annalen* affair left in the end deep scars and lasting aversions, it is well to keep in mind that there were no hidden or open personal conflicts between Brouwer and the other members of the editorial board. It was not as if a bone had to be picked. In particular there was no bad blood between Brouwer and Courant,

[64] Brouwer to Courant, 6.XI.1928.

[65] Courant to Brouwer, 10.XI.1928.

the favoured assistant of Hilbert. Indeed sometime in the past Brouwer had warmly recommended Courant for a mathematics chair in Münster.[66]

Apart from Einstein, who kept a strict neutrality, all the editors (mostly reluctantly) did take sides—the majority with Hilbert. Hilbert himself, however, no longer took part in the conflict. His position was fixed once and for all, and in view of his illness the developments were as far as possible kept from him (e.g. Blumenthal stressed in a letter to Courant that Cara's attempted intervention with Brouwer should be kept secret from Hilbert).[67] One might wonder whether Brouwer, as a relative outsider (one of the three non-Germans among the editors), stood a chance from the beginning; his letter of November 2 to Carathéodory doubtlessly lost him a good deal of sympathy and proved a weapon to his opponents.

In a circular letter of 5 November 1928 Brouwer appealed directly to the publisher and the editors, thus widening the circle of persons involved. The letter was clearly addressed to all editors, both chief and associate. This, of course, widened the circle of the informed, and it would make it more difficult for the Hilbert side to sweep the matter under the carpet.

*To the publisher and the editors of the Mathematische Annalen.*

> From information communicated to me by one of the chief editors of the Mathematische Annalen at the occasion of a visit on 30-10-1928 I gather the following:
>
> 1. That during the last years, as a consequence of differences of opinion between me and Hilbert, which had nothing to do with the editing of the Mathematische Annalen (my turning down of the offer of a chair in Göttingen, conflict between formalism and intuitionism, difference in opinion concerning the moral position of the Bologna congress), Hilbert had developed a continuously increasing anger against me.
>
> 2. That lately Hilbert had repeatedly announced his intention to remove me from the board of editors of the Mathematische Annalen, and this with the argument that he could no longer 'co-operate' (*zusammenarbeiten*) with me.
>
> 3. That this argument was only a pretext, because in the editorial board of the Mathematische Annalen there has never been a co-operation between Hilbert and me (just as there has been no co-operation between me and various other editors). I have not even exchanged any letters with Hilbert since many years and that I have only superficially talked to him (the last time in July 1926).[68]
>
> 4. That the real grounds lie in the wish, dictated by Hilbert's anger, to harm and damage me in some way.
>
> 5. That the equal rights among the editors (repeatedly stressed by the editorial board within and outside the board)*) allow a fulfilment of Hilbert's will

[66]Cf. Freudenthal to Hopf, 22.I.1932. Freudenthal wrote that in Brouwer's opinion Hilbert was not pleased at all, feeling that Brouwer was trying to do him out of a valuable collaborator.

[67]Cf. p. 558.

[68]Cf. p. 528. The last preserved written communication was dated 20.VIII.1919.

only in so far that from the total board a majority should vote for my expulsion. That such a majority is scarcely to be thought of, since I belong to the most active members of the editorial board of the Mathematische Annalen, since no editor ever had the slightest objection against the manner in which I fulfil my editorial activities, and since my departure from the board, both for the future contents and for the future status of the Annalen, would mean a definite loss.

6. That, however, the often proclaimed equal rights, from the point of view of the chief editors, was only a mask, now to be thrown off. That as a matter of fact the chief editors wanted (and considered themselves legally competent) to take it upon themselves to remove me from the editorial board.

7. That Carathéodory and Blumenthal explain their co-operation in this undertaking by the fact that they estimate the advantages of it for Hilbert's state of health higher than my rights and honour and freedom of action (*Wirkungsmöglichkeiten*), and than the moral prestige and scientific contents of the Mathematische Annalen that are to be sacrificed.

I now appeal to your sense of chivalry and most of all to your respect for Felix Klein's memory, and I beg you to act in such a way that either the chief editors abandon this undertaking, or that the remaining editors split off and carry on the tradition of Klein in the management of the journal by themselves.

---

*) [Brouwer's note] (From the editorial obituary of Felix Klein, written by Carathéodory) 'He (Klein) has taken care that the various schools of mathematics were represented in the editorial board and that the editors operated with equal rights alongside of himself—He has (...) never heeded his own person, always had kept in view the goal to be achieved.' (From a letter from Blumenthal to me, 13-9-1927). 'I believe that you overestimate the meaning of the distinction between editors in large and small print. It seems to me that we all have equal rights. In particular we can speak for the *Annalenredaktion* if and only if we have made sure of the approval of the editors interested in the matter under consideration.—Although I too take the distinction between the two kinds of editors more to be typographical than factual (I make an exception for myself as managing editor), I understand your wish for a better typographical make up very well. You know that I personally warmly support it. However, we can for the time being, as long as Hilbert's health is in such a shaky state as it is now, change nothing in the editorial board. I thus cordially beg you to leave your wish for later. At the right moment I will certainly and gladly bring it out.

---

Laren, 5 November 1928
L.E.J. Brouwer

The above circular letter was dispatched at the same day as Brouwer's plea to Mrs. Hilbert; the two letters are in striking contrast. One letter is written on a conciliatory note, the other is a determined defence and closes with an unmistakable incitement to mutiny.

Blumenthal immediately took the matter in hand. He wrote to the publisher and the editors[69] to ignore the letter until he had prepared a rejoinder. The draft of the rejoinder was sent off to Courant on November 12, with instructions to wait for Carathéodory's approval and to send subsequently copies to Bieberbach, Hölder, von Dyck, Einstein and Springer. It appears from the accompanying letter that Carathéodory had already handed in his resignation, although he had given Blumenthal permission to postpone its announcement, so that it would not give food to the rumour that Carathéodory had turned against Hilbert.

Blumenthal, being the acting managing editor, had more or less assumed responsibility for the defence of Hilbert. It is hard to understand that he, who had been a close friend of Brouwer's, could from one day to the next turn against his former friend, and organise a campaign against him. It is not unlikely, however, that the Riemann affair, and the subsequent Bologna affair, had already introduced a measure of estrangement. It was of course well-known that Hilbert ran a tight ship, but it still comes as a bit of a shock to see that Blumenthal actually feared that Carathéodory would be banned from Hilbert's circle: 'Poor Cara has, in spite of his best intentions, got himself into a tight corner, and I don't know yet if Hilbert will break off relations with him.'[70]

In the meantime Brouwer had travelled to Berlin to talk the matter over with Erhard Schmidt, and to explain his position to the publisher Ferdinand Springer. Brouwer, accompanied by Bieberbach, called at the Berlin office of Springer. The visit is described in a memorandum *Aktennotiz* 'Unannounced and unexpected visit of Professor Bieberbach and Professor Brouwer' (13 November 1928). As Springer wrote, his first idea was to refuse to receive the gentlemen, but he then realised that a refusal would provide propaganda material for the opposition. Springer opened the discussion with the remark that he was firmly resolved not to get involved in the skirmishes and that he did not consider the *Annalen* the sole property of the Company (like other journals), but that the proper *Herausgeber*, Klein and Hilbert, had in a sense entrusted it to the publisher. Moreover he would choose Hilbert's side, out of friendship and admiration, if he would be forced to choose sides.

The unwelcome visitors then proceeded to inquire into the legal position of Hilbert, a topic that Springer was not prepared to discuss without the advice of his friends and which he could not enter into without consulting the contract.[71] Thence the two gentlemen proceeded to 'threaten to damage the *Annalen* and my business interests. Attacks on the publishing house, which could get the reputation of lack of national feelings among German mathematicians, could be expected.'

Springer took this stoically, and assured that he would know how to react to such statements, but that he would accept any negative effects without complaint.

---

[69]Blumenthal to the editors, 16.XI.1928.

[70]Blumenthal to Courant, 28.XI.1928.

[71]The contract had indeed a clause to the effect that changes in the editorial board required the assent of the publisher. The problem was of course how to interpret 'editor'.

The implicit threat was definitely in bad taste, not in the last place because the Springer family had Jewish ancestry. Bieberbach's later political views have gained a good measure of notoriety;[72] it certainly is true that already before the take-over of the Third Reich his views had grown more and more nationalistic. Brouwer's position on these matters was neither political, nor nationalistic, it was dictated by his extreme aversion of the scientific boycott of Germany. Nonetheless, the above lines show that, wittingly or not, he ran the risk of being associated with right wing Germans.

Thus rejected, Bieberbach and Brouwer asked if Springer could suggest a mediator, to which Springer answered that he was not sufficiently familiar with the personal information involved, but that two *deutschfreundliche* foreigners like Harald Bohr and G.H. Hardy might do.[73] Before leaving, Brouwer threatened to found a new journal with De Gruyter, and Bieberbach declared that he would resign from the board of editors if it definitively came to the exclusion of Brouwer. In a letter to Courant (13 November 1928) Springer dryly commented 'On the whole the founding of a new journal, wholly under Brouwer's supervision, would be the best solution to all difficulties.'[74] He also conveyed his impression of the visit: 'I would like to add that Brouwer, as a matter of fact, does make a scarcely pleasant (*unerfreulich*) impression. It seems, moreover, that he will carry the fight to the bitter end (*der Kampf bis aufs Messer führen wird*).'

In Aachen, Blumenthal was preparing his defence of the announced dismissal of Brouwer and, following an old strategic tradition, he took to the attack. After consulting Courant, Carathéodory and Bohr he drew up a kind of indictment. From a letter from Bohr and Courant to Blumenthal,[75] one may infer that the draft of 12 November was harder in tone and more comprehensive than the final version. There is mention of a detailed criticism of Brouwer's editorial activities and of matters of formulation ('... leave out capriciousness (*Schrullenhaftigkeit*)...').

Carathéodory remained an uncertain factor in the coming power play; Bohr and Courant realised this, and they therefore would prefer him to do his bit: 'We feel that Cara should make a stand himself because of the misuse Brouwer made of his kindness, respectively that he should explicitly authorise you to use defensive words.' A clever suggestion, but not exactly considerate. Bohr and Courant explicitly warned Blumenthal:

> To what extent Brouwer exploits without consideration every tactical advantage that is offered to him, and how dangerous his personal influence is

[72]Cf. Mehrtens (1987).

[73]This suggestion of the publisher encouraged the impression that the conflict had a political origin. Blumenthal complained to Courant (letter of 18 November 1928) '... the bad thing is, that Brouwer managed to move everything on to the political plane, just what Carathéodory thought he had prevented'. The idea of mediation was not pursued.

[74]Brouwer indeed founded a new journal, the *Compositio Mathematica*, with the Dutch publisher Noordhoff.

[75]Bohr and Courant to Blumenthal, 14.XI.1928.

(Bieberbach), can be seen from the enclosed notice which Springer has just sent us.[76]

The correspondence of Blumenthal, Bohr, and Courant shows an unlimited loyalty to Hilbert, which it would be unjust to ascribe to Hilbert's state of health alone. There is no doubt that Hilbert as a man and a scientist inspired a great deal of loyalty in others, let alone in his students. Sentences like 'We don't particularly have to stress that we are, like you, wholly on Hilbert's side, and also, when necessary, prepared for action',[77] illustrate the feeling among Hilbert's students.

A revised version of Blumenthal's letter is dated November 16, and it is this version that was in Brouwer's possession. It incorporated remarks of Bohr and Courant, but not yet those (at least not all of them) of Carathéodory. It contained a concise *resumé* of the affair so far, and proceeds to answer Brouwer's points (from the letter of 5 November 1928).

Blumenthal partly based his handling of the matter on correspondence, partly on conversations with Hilbert in Bologna. The contents of the latter conversation remain a matter of conjecture, but it may be guessed that in August at the conference Hilbert had made clear his objections to Brouwer—in particular after Brouwer's opposition to the German participation in the conference.

From Blumenthal's circular letter, the editors—and also Brouwer—learned the contents of Hilbert's letter of October 25.

In view of the importance of the letter, it is worthwhile to reproduce Blumenthal's letter here.

> To the publisher and the editors of the *Mathematische Annalen*.
>
> As manager of the board of the *Annalen*, I feel obliged to reply to Brouwer's circular letter to the publisher and the editors of the *Mathematische Annalen*. My exposition relies in part on letters of Hilbert, Carathéodory, and Brouwer, in part on an extensive discussion I had with Hilbert in Bologna.
>
> I would like to point out in advance that the formulation of Brouwer's letter is misleading: one can get from it the impression that the editor who visited Brouwer on October 30 (Carathéodory) formulated the statements 1–7. This is of course not the case for any of them, these are rather opinions that Brouwer formed himself.
>
> In the following I give a brief report of the developments, and react to the relevant points of Brouwer's letter.

Blumenthal proceeded to quote Hilbert's letter of 25 October to Brouwer in full (cf. p. 555) and continued

> Brouwer has not opened this letter, as I should note already here, and as I explain later. He is, however, informed by Carathéodory about its content, in

[76] The above mentioned *Aktennotiz*.

[77] Ibid.

particular also of the motivation for Hilbert's action given in the first sentence. Brouwer's points 2 and 3 refer to this. About this I have to say the following:

*On points 2 and 3.* Brouwer interprets the notion of co-operation in a literal sense (point 3). This is complete misapprehension of the true meaning. It is rather the case that Hilbert had become convinced that Brouwer's activity was detrimental for the *Annalen*, and that he could therefore no longer take the responsibility to act as a chief editor in an editorial board, to which Brouwer belonged. By no means does this concern a pretext.

*On points 1 and 4.* The motivations for Hilbert's way of operating, indicated by Brouwer, in these points is not correct. The motivation in point 4 is spiteful and therefore requires no refutation. Also the scientific differences with respect to the foundations, of which one could think, play no role. In particular it is not true what Brouwer seems to suggest in point 5, that the mathematical direction, represented by him, will in future get less opportunity to speak. Also Brouwer's circular letter before the Bologna congress, the expression of which Hilbert found insulting, has only in conjunction with other, perhaps more important, factors acted as a catalyst on his decision. The causes lie much deeper. I will give them in my formulation, but I am certain to get Hilbert's meaning precisely.

Felix Klein had, until his resignation from the editorial board, formed among us a kind of supervising body, that in difficult cases could be called in, or that acted on its own initiative, to support important decisions (for example the transfer of the *Annalen* to Springer Verlag), or to smooth disagreements inside the editorial board. It is good and necessary that in a numerous board like ours, there is such a supreme body available, that, relieved from the details of the management, keeps an eye on the general relations and feels responsible for these. After Klein's death, Hilbert had felt obliged to fulfil this position, and has already acted in this sense, and I for one have also personally always instinctively recognised him as such.

Hilbert saw in Brouwer a headstrong, unpredictable and domineering character. He had feared that, when he at some time should have left the board, Brouwer would bend it to his will, and he has judged this such a great danger for the *Annalen*, that he wanted to stand in his way as long as he still could do so. Probably under the influence of his renewed illness, he felt obliged in the interest of the *Annalen* to order Brouwer's exit from the board, and to tackle this measure immediately and with all energy.

Cara and I, who were associated with Brouwer in a long-standing friendship, had objectively to recognise Hilbert's objections to Brouwer's editorial activity.

True, Brouwer was a very conscientious and active editor, but he was quite difficult in his dealings with the managing editor and he subjected the authors to hardships that were hard to bear.

For example, manuscripts that were submitted for refereeing to him lay around for months, while in principle he had prepared a copy of each submitted paper (I recently had an example of this practice). Above all there is no doubt that Klein's premature resignation from the editorial board is to be traced back to Brouwer's rude behaviour (in a matter in which Brouwer was formally right). The further course of events has shown that Hilbert was even far more right than we thought at the time.

Since we could not reject the objective justification of Hilbert's point of view, and were confronted by his immutable will, we have given our permission for the removal of Brouwer from the editorial board.[78] We only wished—unjustifiedly, as I now realise—a milder form, in the sense that Brouwer should be prevailed upon to resign. Hilbert could not be induced to this procedure, so we finally, though reluctantly, have decided to give in to him (*den Weg freigeben*). Einstein did not comply with the argument that one should not take Brouwer's peculiarities seriously.

*Point 5 and 6.* In how far it was justified that the other editors were not first informed of Hilbert's plan, I don't want to go into here. Formally speaking the justification seems to be given by the distinction between '*Mitwirkenden*' and '*Herausgeber*' on the cover.

*The events after the dispatch of Hilbert's letter.*

On October 26 and 27 Cara and I were in Göttingen to discuss the situation. Subsequently Cara travelled in the interest of the matter to Berlin. Although he saw objectively that Brouwer's eviction was unavoidable, he decided in Berlin to make a last attempt, to settle the matter in an amicable sense, by weakening the categorical form of the expulsion. For that reason he came on the thirtieth of October to Laren, after Brouwer had been asked in advance, by telegram, not to take any steps until Cara's arrival. Since Brouwer had not opened Hilbert's letter, Cara informed him of the content (but not the formulation), and proposed him to resign of his own free will from the editorial board and to leave the letter unopened. He wanted to prevent that Brouwer would feel insulted by the form, and he felt justified to do so, as the rudeness seemed partially determined by Hilbert's ailing health.

He did not make it clear to Brouwer that in our opinion he had to leave the board, and bade him, out of consideration with Hilbert and his illness of that moment, to withdraw by himself. Brouwer reserved a decision until further calm deliberation. He had left Hilbert's letter unopened, and written to Cara on November 2 the following letter:

At this place Blumenthal inserted the text of letter of Brouwer of 2 November, see p. 558.

For this frightful and repulsive letter, which Brouwer also sent to me in copy, I can only offer the explanation that Brouwer (on purpose or inadver-

[78]Blumenthal is here less than truthful. Cara had not authorised Hilbert. He had probably tried to pacify Hilbert, who took this for the desired approval. Cara wrote, referring to Blumenthal's draft, 'In fact I have not given an authorisation in the legal sense in my letter to Hilbert.' (Carathéodory to Courant, 14.XI.1928.)

> tently) had put together from Cara's statements and entreats precisely the ugliest part. I must confess, and Cara has written me likewise, that I have been thoroughly deceived in Brouwer's character, and that Hilbert has known and judged him better than we did. I too am no longer in a position to co-operate further with the writer of this letter in the board, and I now side actively with Hilbert. I cannot understand that Brouwer after this letter can appeal to the chivalry of the editors and to the memory of Felix Klein.
>
> I beg the gentlemen either for a speedy reaction, or for their tacit consent, that from the next issue Brouwer's name is no longer on the cover, and that he no longer gets my Annalen information.

So far the 'case for the prosecution'. As an indictment the above letter did not make a convincing impression. It was written to refute Brouwer's points, and to justify Hilbert's decision. In neither was Blumenthal very successful. Admittedly he was in a difficult position, he had read Hilbert's letter to his fellow '*Herausgeber*', and that letter listed some concrete complaints. But he could not very well use these, as the letter was for the chief editors only, he could not ask Hilbert's permission to use the letter, because Hilbert was not supposed to know what was going on, and finally, Hilbert's complaints were very subjective, another editor would probably have seen the mentioned facts in a completely different light. Worse, these facts would have given ammunition to Brouwer.

Blumenthal indeed acted as if he had direct access to Hilbert's thoughts. For example, in the case of the refutation of Brouwer's points 1 and 4, he gave no evidence, he just denied Brouwer's statements. Looking at the evidence, cf. p. 551, one cannot but conclude that the *Grundlagenstreit*, the Bologna affair, and the Riemann affair brought out some uncontrolled emotions on Hilbert's side. Hilbert's references to intuitionism and Brouwer went beyond scholarly comment. The matter of the Göttingen chair is not as clear. Position bargaining was a normal thing in Germany, and that Hilbert's offer was used by Brouwer to improve his situation in Amsterdam would not have surprised anybody. And yet—might not Hilbert have taken Brouwer's decision ill? After all, he was planning to get the leading topologist for Göttingen, and perhaps, in a corner of his heart, he saw Brouwer as a useful addition to his foundational team—the discussion in 1909 might have left some memories. And there is Carathéodory's statement that Hilbert's objections go back ten years.

Considering all evidence, one would be inclined to side with Brouwer rather than Blumenthal. Point 4 is a debatable one. When not under the influence of a fatal disease, would Hilbert have had no wish to hurt Brouwer? From the height of his Olympus he might have acted objectively and impersonally: 'I cannot get along with this man, so he'd better go.' But perhaps Hilbert, when not on his Olympus, was enough of a man of flesh and blood to be susceptible to the 'I'll get you' mood. Whatever was the real state of affairs, the point is subjective, but precisely for that reason Blumenthal should not have dismissed the point so perfunctorily.

Hilbert, in the letter to his fellow chief editors, mentioned three points: (1) the insult implicit in Brouwer's Bologna circular letter; (2) Brouwer's anti-*Conseil* and anti-Union feelings as standing in the way of his editorial work; and (3) the *Annalen* should remain in Göttingen. The heart of Hilbert's argument lies in (2). For, one could hardly accept (1) as a ground for dismissal.[79] If Hilbert was so sensitive about insults, had he forgotten his words in the second Hamburg lecture? And (3) was rather irrelevant. Journals have no fixed abodes (not counting the proceedings of academies, and the like). Of course, it is a pleasure and an honour for a department to be almost a synonym for a prestigious journal, but all these things would pass one day. Whatever Hilbert's assessment may have been, there is not the slightest indication that Brouwer would have tried to move the *Annalen* elsewhere. Knowing Brouwer's attachment to Göttingen, he would have insisted on keeping the journal where it was.

Yet, even a man like Carathéodory, to whom the term 'the milk of human kindness' could justifiably be applied, had his reservations on this point. It tells something about the internal frictions in German mathematics, that this man, who obtained a doctorate and habilitation in Göttingen, and who was for some time Felix Klein's successor, was critical of the Göttingen imperialism. It should be added that he knew the mathematical world better than most, having studied in Berlin, where he briefly held a chair, and at the time of the *Annalen* conflict he was professor in Munich. Nonetheless, it might have come as a surprise to Courant when Carathéodory told him that Hilbert's claim was for him one of the grounds to resign from the *Mathematische Annalen*.

No, Hilbert saw problems with, in particular, French authors. Brouwer, if he were a chief editor, might, in his opinion, block a paper of Painlevé, or of Picard. Here he was mistaken; Brouwer could very well distinguish between an individual and an organisation. We may recall that Brouwer had lectured Denjoy at length for not observing this distinction.[80] His objections against an invitation of Painlevé to participate in the memorial volume of Riemann was based on the invitation and on the special occasion. As it was, Brouwer was only an associate editor with no influence on papers that were not refereed by him. No matter how one looks at this point, one cannot but conclude that Hilbert's fears were mostly the product of his unfortunate health situation.

Anybody with a cool mind could see that the Bologna conference had restored the international character of mathematics, and that the nationalistic differences would no longer play a role. This clearly had escaped Hilbert when he was in Bologna. Either he was fixed on the past, or he had an axe to grind.

A few editors responded to Blumenthal's letter in writing, but the majority remained silent. Only von Dyck, Hölder and Bieberbach[81] sent their comments. Von Dyck could 'neither justify Brouwer's views nor Hilbert's action' and he hoped that a peaceful solution could be found. Hölder was of the opinion that he could not approve of a removal of Brouwer by force (November 27).

Bieberbach's letter showed a thorough appreciation of the situation. And he at least, was willing to take up the case of the underdog. In view of his later political

[79]If anything, Brouwer's circular letter could been condemned rather as divisive, than as insulting.

[80]Cf. Sect. 9.3.

[81]Bieberbach to Blumenthal, 24.XI.1928.

extremism, one might be inclined to question the purity of his motives; however in the present letter there is no reason not to take his arguments at their face value. Like Brouwer, and probably the majority if not the whole of the editorial board, he contested the right of the *Herausgeber* to decide matters without the support of the majority of all the editors, let alone without consultation. Indeed this seems to be a shaky point in the whole procedure. Bieberbach referred to the 'Innsbruck resolution', which laid down that the chief editor had no claim to the dismissal of editors. 'And now', he said, 'the right to dismiss editors should follow from typographical characteristics. That would reduce the members of the editorial board to subalterns, who can be dismissed any day by the chief editors.'

As a matter of fact, the contract between Springer and the *Herausgeber*[82] is not very concrete in this particular point. It states: 'Changes in the membership of the editorial board require the approval of the publisher.' The correspondence makes it clear that Hilbert did not observe this rule. Bieberbach observed that a delay in handling papers cannot be taken seriously as grounds for dismissal: such things ought to be discussed in the annual meeting of the board.

He devoted a few lines to the procedure proposed by Blumenthal. Blumenthal had not asked for a vote on the expulsion of Brouwer, his plan was to decide the matter on the basis of individual reactions plus the 'silence lends consent'-principle, without even mentioning a time limit. Such a procedure contradicted the most elementary principles of justice, Bieberbach observed. The exclusion of an editor should be handled with extreme care. 'I would consider it correct', he continued, 'that in our circle the question of expulsion should only be made a point of discussion if such a sensational failure of one editor is the case, that one count on unanimous agreement, not, however, if beforehand a prominent member of the board of chief editors, such as Einstein, opposes the exclusion, ...'

As one of the main dissidents in the matter of the Bologna affair, Bieberbach added some 'scholastic' comments to Blumenthal's report of Hilbert's feelings.

Brouwer's pamphlet was directed at those visiting the Union-congress in Bologna. As Hilbert himself claimed, the Bologna conference was *not* a Union-congress, how could he be insulted?

Bieberbach found no difficulty in dissecting Blumenthal's case against Brouwer. He concluded that 'A dismissal without any notice of an editor, who is moreover a scientist of world fame, after thirteen years of diligent activity could only be justified by defamatory actions or so, not by incidents that only hold inconveniences for the editor in chief (*geschäftsführender Redakteur*)'.

Blumenthal had been making the most of Brouwer's, admittedly tactless, letter to Mrs. Hilbert. Bieberbach correctly spotted a serious flaw in Blumenthal's charge involving Brouwer's 'terrifying and repulsive' letter.

> Finally I hold it totally unjustified to concoct material against Brouwer from letters that he wrote after learning about the action that was mounted against himself. For it is morally impossible to use actions, to which a person

[82] 25.II.1920.

is driven in a fully understandable emotion over an injustice that is inflicted on him, afterwards as a justification of this injustice itself.

The point is well taken. It does not exonerate Brouwer, but it at least makes clear that to use it against Brouwer is in poor taste. Bieberbach explicitly stated that he would not support Brouwer's dismissal; on the contrary, he strongly sided with Brouwer, without, however, attacking Hilbert.

There is a certain tendency to dismiss Bieberbach's statements and opinions on the basis of his later political views and actions. Needless to say that this does not go well with rational reflection. Be that as it may, Bieberbach's letter painfully exposed the weaknesses in the case against Brouwer.

The publisher reacted in a cautious way. Springer thought that Brouwer was 'an embittered and malicious adversary', and that he should not receive a copy of the circular letter without the permission of the lawyer of the firm. Springer also concluded that the publisher should not state in writing that he officially agreed to Brouwer's dismissal, because it would imply a recognition of Brouwer's membership of the board of editors in the sense of the contract. In short, Springer abstained from voting on Blumenthal's proposal.

At this point the whole action against Brouwer seemed to have reached an anticlimax. One may surmise a good deal of activity in the camp of the Göttingers. The matter now began to take on national proportions. In view of the barely veiled animosity between Berlin and Göttingen, as the ultimate bastions of mathematics, and Brouwer's close connections with the Berlin mathematicians, the temporary leaders of the Göttingen group started to worry that the Brouwer dismissal might lead to an open rift between the two groups. This might even lead to further unpleasantness, as the main loyalties among German mathematicians were with Berlin or Göttingen; once a Göttinger, always a Göttinger, and the same for Berlin. It thus became a matter of some urgency to settle the Brouwer matter. Had Hilbert, much like the ghost of Hamlet's father, not continually hovered in the background, the editor's would have been able to find a compromise, so that Brouwer could stay on. Unfortunately the outcome was fixed in advance. Nonetheless the Hilbert party had its worries.

If only Einstein could be persuaded to give his assent...! Although it seemed doubtful that anyone could succeed where even Hilbert had failed, Max Born tried to convince Einstein to remain at least neutral. In a lengthy letter he summed up the arguments, sketched the mood, outlined the consequences. Hilbert, he said, was so ill that he wouldn't live much longer. Every emotion could do serious harm, and would shorten the time left to finish his work. 'What is more, he is full of a strong will to live, and he sees it as his task to carry out his new founding of mathematics, to which he has to devote himself with his last strength. His mind is clearer than ever, and the rumour, spread by Brouwer, that Hilbert were not completely *compos mentis* is an extraordinary heartlessness.' Born had spoken with Hilbert on the topic 'Brouwer', and Hilbert had declared that he saw Brouwer as an eccentric and unbalanced man. According to Born, Hilbert considered Brouwer's ultra-German behaviour in the Bologna matter a folly, 'but the dreadful thing was that the Berlin mathematicians fell for Brouwer's nonsense'. Born mentioned Bieberbach,

Von Mises and Schmidt as dupes of Brouwer, a claim that was not borne out by evidence, at least Bieberbach and Von Mises were anti-Bologna-Union on their own accord. Born had travelled with Von Mises in the USSR, and at one occasion Von Mises opened the conversation with, 'the Göttinger simply ran after Hilbert, who was no longer quite responsible for his actions' (*unzurechnungsfähig*). Born immediately broke off the conversation, since in his opinion Von Mises was too insignificant to have an opinion on Hilbert. As he pointed out, this was before the Bologna conference, thus before Hilbert's collapse. In every community there is invariably an amount of gossip floating around, in mathematics no less than in other subjects. It is more than likely that the rumours of Hilbert's disease, combined with his strange emotional behaviour at the Hamburg and Münster lectures, had encouraged speculation about his mental health. Thus von Mises' view may have displeased, but not surprised, Born.

In view of a meeting to be held at Springer's in Berlin, Born begged Einstein not to do anything that might harm Hilbert's interest. It was important that the chief editors should speak with one voice.

The pressure on Einstein was, from a strategic point of view, understandable. His immense scientific and moral prestige made him a key figure in any debate. If he could be persuaded to side with Hilbert the battle would be half won. In spite of personal pressure from Born (20 November 1928) on behalf of Hilbert, Einstein remained stubbornly neutral. In his letters to Born and to Brouwer and Blumenthal one may sense a measure of disgust behind a facade of raillery. In the letter to Born (November 27) the apt characterisation of '*Frosch-Mäusekrieg*' (war of the frogs and the mice) was introduced.[83] After declaring his strict neutrality he went on:

> If Hilbert's illness did not lend a tragic aspect, this ink war would for me be one of the most funny and successful farces performed by that sort of people who take themselves deadly seriously.
>
> Objectively, I might briefly point out that in my opinion there would have been more painless remedies against an overly large influence on the managing of the *Annalen* by the somewhat mad (*verrückt*) Brouwer, than eviction from the editorial board.
>
> This, however, I only say to you in private, and I do not intend to plunge as a champion into this frog-mice battle with another paper lance.

Einstein's letter to Brouwer and Blumenthal of November 25 is even more cutting and reproving.

> I am sorry that I got into this mathematical wolf-pack (*Wolfsherde*) like an innocent lamb. The sight of the scientific deeds of the men under consideration here impresses me with such cunning of the mind, that I cannot hope also in this extra-scientific matter to reach a somewhat correct judgement of them. Please allow me therefore to persist in my 'booh-nor-bah' (*Muh-noch-Mäh*)

[83] War of the frogs and the mice—a Greek play of unknown author; a late medieval German version, *Froschmeuseler*, is from the hand of Rollenhagen.

position and allow me to stick to my role of astounded contemporary. With best wishes for an ample continuation of this equally noble and important battle, I remain

Yours truly,
A. Einstein

The whole affair now rapidly reached a deadlock. A week before, Springer had, after seeking legal advice at Blumenthal's urging, written optimistically to Courant[84] that the legal adviser of the firm, E. Kalisher, was of the opinion that it would suffice that those of the four chief editors who did not want to advocate Brouwer's dismissal actively would abstain from voting, thus giving the remaining chief editors a free hand. Apparently Springer did not realise that since two editors with a high reputation had already decided not to support Hilbert, the solution, even if it was legally valid, would lack moral support. If this solution should turn out to raise difficulties within the editorial board, the publisher could still fire the whole editorial board and reappoint Hilbert and his supporters, so the advice ran. In the opinion of the legal adviser the publishing house was contractually bound to the chief editors (*Herausgeber*) only; there was no contract with the remaining editors.

Bieberbach's letter, mentioned above, apparently worried Carathéodory to the extent that he decided to ask a Munich colleague from the law faculty for advice. This advice from Müller-Erzbach plainly contradicted the advice from the Springer lawyer. It made clear that

(1) Brouwer and Springer-Verlagwere contractually bound since Brouwer had obtained a fee.
(2) Hilbert's letter was not legally binding.

Müller-Erzbach sketched three solutions to the problem:

(a) Springer dismisses Brouwer. A letter of dismissal should, however, contain appropriate grounds.
(b) The four chief editors and the publisher form a company (*Gesellschaft*) and dismiss Brouwer.
(c) A court of law could count the *Mitarbeiter* as editors. In that case the only way out would be to dissolve the total editorial board and to form a new one.

Carathéodory considered the first two suggestions inappropriate because it would not be fair to saddle Springer with the internal problems of the editors. Hence he recommended the third solution.[85] Here, for the first time, appeared the suggestion that was to be the basis of the eventual outcome of the dispute.

Hilbert, the main contestant in the *Annalen* affair, had quite sensibly withdrawn from the stage. The developments, had he known them, would certainly have done

[84]Springer to Courant, 17.I.1928.

[85]Cara to Blumenthal, 27.XI.1928.

his health no good. He had authorised Harald Bohr and Richard Courant to represent him legally in matters concerning the *Mathematische Annalen*. Thus the whole matter became more and more a shadow fight between Brouwer and an absentee.

At this point the dispute had reached an impasse. Although Springer upheld in a letter to Bieberbach the principle that the chief editors could dismiss any of the other editors, the impetus of the attack on Brouwer seemed to ebb away. A meeting between Carathéodory, Courant, Blumenthal and Springer had repeatedly to be postponed and finally had been cancelled.

Courant agreed with Carathéodory that the dissolution of the complete board would be a convenient way out;[86] however, it would require a voluntary action from the editors and the ultimate organisation of the editorial board should not have the character of a legal trick with the sole purpose of rendering Brouwer's opposition illusory.

Carathéodory, who, on the basis of Müller-Erzbach's information, had come to the conclusion that the original plan of Hilbert, even in a modified form, would not stand up in a court of law, expressed his willingness to assist 'out of devotion to Hilbert' in the liquidation of the affair, but quite firmly refused to be involved in the future organisation of the *Annalen*.

The reluctance of Carathéodory to be involved in the matter beyond the bare minimal efforts to satisfy Hilbert and spare Brouwer (his friend) is throughout understandable. As far as we can judge from the correspondence, only Blumenthal exhibited an unbroken fighting spirit. He realised, however, that his circular had not furthered an acceptable solution,[87] and he leaned towards alternative solutions. In particular, Blumenthal wrote, the time was favourable to Carathéodory's plan. The *Annalen* were completing their hundredth volume, and it would present a perfect occasion to open with volume 101 a 'new series' or 'second series' with a different organisation of the editorial board. But at the present time he was facing a dilemma. Because Hilbert's letter clearly had no legal status, Brouwer was still a *Mitarbeiter* and his name should appear on the cover of the issue that was to appear—this, however, conflicted with Hilbert's wishes. Could Bohr and Courant, as proxies of Hilbert, authorise him to print Brouwer's name on the cover? Otherwise the publication would have to be postponed. The authorisation probably was given.

It seems that Bohr had also put forward a solution to the affair. From the correspondence of Carathéodory and Bohr with Blumenthal, one gets the impression that Bohr's proposal was a slight variant of Carathéodory's suggestion. The main difference was that Bohr advocated a total reorganisation of the editorial board. In his proposal there would only remain *Herausgeber*, and no *Mitarbeiter*. So the solution would look like a fundamental change of policy, and hence it would no longer be recognisable as an act levelled against Brouwer.

Apparently Bohr envisaged Hilbert, Blumenthal, Hecke and Weyl as the members of a new board. And should Weyl decline, one might invite Toeplitz. Blumenthal questioned the wisdom of reinstating himself as an editor; it could easily be

[86] Courant to Carathéodory, 30.XI.1928.

[87] Blumenthal to Courant and Bohr, 4.XII.1928.

viewed as the old board of chief editors in disguise.[88] In his letter to Courant, the next day, he considered the dissolution of the editorial board at large as necessary, and he fully agreed that Hilbert should choose the new editors.

From then on things moved smoothly; Springer accepted the dissolution of the editorial board and agreed to enter into a contract with Hilbert on the subject of the reorganised *Annalen*. By and large only matters of formulation and legal points remained to be solved.

One might wonder where Brouwer was in all this—he was completely ignored. In a letter of November 30 to the editors and the publisher he confirmed the receipt of Blumenthal's indictment which had only just reached him. In a surprisingly mild reaction he merely asked the editors to reserve their judgement—blissfully unaware that nobody was going to cast a vote—for the composition of a defence would take some days.

Because the dissolution of the editorial board had to be a voluntary act, it was a matter of importance to get Einstein's concurrence. The contract of 1920 presented an elegant loophole that would allow both parties to settle the matter without breaking the rules. In Sect. 5 the clauses for termination of the contract were listed, and one of them stipulated that if the editors (*Redaktion*) renounced the contract, without a violation from the side of the publisher, the latter could continue the *Mathematische Annalen* at will.

Possibly Einstein's agreement could be dispensed with, but it is likely that a decision to ignore Einstein's vote would influence general opinion adversely; moreover, it would be wise to opt for a watertight procedure, as Brouwer would not hesitate to test the outcome in court.

So pressure was brought to bear on Einstein. James Franck, a physicist and a friend of Born, begged him to listen to the new plan. He stressed the political side of the issue, 'At this time, ..., whether the mathematicians split into factions or whether the affair is arranged smoothly, depends on your decision. It would almost be an inappropriate joke (*ein nicht all zu guter Witz*) if in this case you would be claimed for the nationalistic side' (undated). Franck was not the only person to discover a (real or imaginary) political aspect in the controversy at hand. Blumenthal had already complained to Courant (November 18) that Brouwer had managed to introduce the political element into the matter. Born also, in his letter to Einstein of November 11, tied the conflict to the political issue of the German nationalists and the animosity of Berlin vs. Göttingen.

The successful conclusion of the undertaking was conveyed to Springer by Courant. In his letter of December 15 he announced the co-operation of Einstein, Carathéodory, Blumenthal and Hilbert in the transition. At the same time he proposed that a new contract be made between Hilbert and the publisher, and that Hilbert get *carte blanche* for organising the editorial board. Blumenthal should be invited to continue his activity as managing editor and, according to Courant, he would probably accept. Also—and this is a surprising misjudgement of Einstein's

[88] Blumenthal to Bohr, 5.XII.1928.

mood—Courant thought that there was a 50% chance that Einstein would join the new board. As far as he himself was concerned, Courant thought it wiser to postpone his own introduction as an editor until the dust had settled (the matter apparently had been discussed earlier).

Courant had to work hard to prepare the various documents, to solicit comments, make changes, etc. The *Annalen*-new style would have one *Herausgeber*, Hilbert, and a variable (but small) number of *Mitarbeiter*. Hecke and Blumenthal were eventually chosen for the latter function.

The new arrangement promised a satisfactory end to the *Annalen* affair, but not everyone was happy. Blumenthal, for example, cautiously pointed out that Hilbert would become the only chief editor. If he intended any criticism, he was careful to leave it to Courant to read it between the lines.[89] Carathéodory on the other hand openly expressed his disappointment. He, too, deplored the end of the old regime. When confronted with Hecke's comments on the practice of the past (letter from Courant to Carathéodory, December 17): '... that Hecke, when he learned about the organisation of the editorial board and the competence of the *Beirat* [the advisory editors] grasped his head and judged a revision and a more strict organisation absolutely necessary.' Carathéodory heartily disagreed: 'For, Klein had organised the board of editors of the *Mathematische Annalen* in such a way that it formed really a kind of Academy, in which each member had the same rights as the others. That was in my opinion the main reason why the *Annalen* could claim to be the first mathematics journal in the world. Now it will become a journal like all other ones.'[90] It did not take Blumenthal long to recognise the negative sides of the new set up. On 2 February 1929 he sent out a note, 'On the future organisation of the Annalen', in which he drew the attention to the decline of the journal compared to other journals. Since the associated editors (*Nebenredaktion*) had been eliminated, one simply needed a larger staff: 'the increasing necessity of scientific advisers follows inevitably from the increasing specialisation'. In short, Blumenthal proposed to reinstate something like the old associate editors under a different name. In the same letter he broached the question of the successor of Hilbert, should he step down. One finds it difficult to reconcile this letter with the arguments that were put forward in favour of the solution to the conflict.

Finally Courant suggested that the publisher alone should inform all present editors of the collective resignation. With respect to Brouwer, he advised Springer to write a personal letter explaining the solution to the conflict, and to stress that he [Springer] would regret it if Brouwer were left with the impression that the whole affair would restrict his freedom of action, and that the publishing house would be at his disposal should he wish to report on his foundational views. It is not known whether this letter was ever written, but Courant's attitude certainly was statesmanlike and conciliatory.

The fact that the whole board was going to be dismissed, and that only Hilbert and Blumenthal were going to be reappointed, was an unpleasant message for most,

---

[89]Blumenthal to Courant, 16.XII.1928.

[90]Carathéodory to Courant, 19.XII.1928.

if not all, of the sitting editors, but in particular for the senior members Von Dyck and Hölder. Van Dyck had even at one time been a chief editor. It is greatly to Blumenthal's credit that he asked Courant to intervene with Hilbert, so that the latter would write a few nice words to the two; the pill needed a strong dose of sugaring, he said.

Like a good statesman, Courant realised that the past events carried the potential for a long period of friction. Of course he was aware of the reputation of arrogance of the Göttingen group, but he was sensible enough to see that there was life outside Göttingen. As he wrote to Carathéodory,[91] 'We should also think of the future relations between the German mathematicians. If a part of the colleagues does not learn to understand what really motivated Hilbert, then the vexation will not yield and can burst out here and there. If such a latent tension—that will not come from Hilbert's circle—is to be avoided for the future, then one must make use of the present moment to rob the matter of any unjustified ugly appearance and enter into a basis of mutual understanding and trust. It would be gratifying and comforting if you would help us to make all persons involved, in particular our Berlin colleagues, to adopt this position.'

Once the decision was taken, no time was wasted; after the routine legal consultations the publisher carried out the reorganisation and the editors were informed of the outcome (December 27). In spite of Courant's considerations mentioned above, the letter was signed by Hilbert and Springer. Brouwer, like everybody else, was thanked for his work and was given the right to a free copy of the future *Annalen* issues. The matter would have been over, were it not for some rumblings among the former editors and for a desperate but hopeless rearguard action of Brouwer.

Carathéodory had been considerably distressed during the whole affair; from the beginning he had been torn between his loyalty to Hilbert and his abhorrence of the injustice of Brouwer's dismissal. His efforts to mediate had only worsened the matter and the final solution was an immense relief to him. In a fit of despondency he wrote to Courant:[92] 'You cannot imagine how deeply worried I was during the last weeks. I envisioned the possibility that, after I had parted with Brouwer, the same thing would happen with all my other friends.' He had even considered accepting a chair at Stanford that was offered to him. In his answer Courant tried to set Carathéodory's mind at ease:[93] he believed that he had succeeded in convincing Hilbert that Carathéodory, in his position, could not have acted differently; the matter was settled now 'without fears of a residue of resentment on Hilbert's part'. Two days later he wrote that the night before he had discussed the whole matter with Hilbert, who had asked Courant to tell Carathéodory that 'he thinks that you would have done everything for him, as far as possible'. Hilbert was completely satisfied with the result of the undertaking, and in his opinion the *Annalen* were even better protected now than through his original dismissal of Brouwer '... and by and by it

[91] Courant to Carathéodory, 23.XII.1928.

[92] Carathéodory to Courant, 12.XII.1928.

[93] Courant to Carathéodory, 15.XII.1928.

has become completely clear to me that in fact no personal motives have inspired Hilbert's first step, ...' Carathéodory expressed his pleasure with Hilbert's views but he was not satisfied with Courant's evaluation of the motives behind Hilbert's move.[94] 'Now, he himself has given as the exclusive motive for his decision that he felt insulted by Brouwer; I would find it unworthy of him, to construe after the fact, that only impersonal motives had guided him.' This last remark could hardly be left unanswered by Courant. He had worked hard to pacify the participants in the affair, and here one of the former chief editors was lending support to the rumour that Hilbert was not completely devoid of some personal feelings of revenge. In an attempt to quench this source of dissent he and Bohr admonished Carathéodory. Courant calmly repeated his view and referred to Hilbert's personal statements that he 'fostered no personal feelings of hate, anger or insult against Brouwer'.[95] Even a bit of subtle pressure was brought to bear on Carathéodory: 'Our responsibility to Hilbert at this point is even greater, as he is not yet filled in on the development of the conflict; in particular he does not surmise your visit to Laren and the disconcerting report of it by Brouwer.'

Bohr was less subtle in his approach (same letter); if Carathéodory were not convinced of Hilbert's impersonal motives, he should ask Hilbert himself. 'For, that Hilbert—without being aware of it and without being able to defend himself—should first be considered 'of unsound mind' and then 'not to the point' (*unzurechnungsfähig... unsachlich*), that is a situation, that I, as a representative of Hilbert, cannot in the long run witness without action.' In spite of Bohr's sabre rattling Carathéodory stuck to his guns: 'To judge Hilbert's motives is a very complicated matter; I believe that I see through his motives because I have known his way of thinking for more than 25 years. It is true that the motivations that you indicate, and which H. also expounded in Bologna in discussion with Blumenthal, were there. The total complex of thoughts, that caused the explosion of feeling of October 15,[96] was much more complicated.'

Who was right, Courant and Bohr, or Carathéodory? The matter will probably never be completely settled. There is no doubt that the question of 'how to safeguard the Annalen from Brouwer's negative influence (real or imagined)' was uppermost in Hilbert's mind. But who is to say that no personal motives were involved? There are Hilbert's own statements to the effect that no personal grudge led to his action, e.g. to Blumenthal and Courant, but how much weight can be attached to them? In any case they contradict the letter of October 15.

Finally, there was the 'blackmail' note (see p. 551), which indicates a strong emotion and vexation, if not more. And if Hilbert had personal motives, so what? Would we think less of a person if he were not the cardboard saint that some would prefer him to be?

---

[94]Carathéodory to Courant, 19.XII.1928.

[95]Courant to Carathéodory, 23.XII.1928.

[96]Cf. letter to Einstein, October 15.

Were Courant and Bohr themselves all that certain about Hilbert's motives? This question will probably never be answered. The available correspondence is not really informative, the lack of personal motives is systematically given credence by quoting Hilbert. They may have realised this, when they wrote 'Thus it is nothing less than a reconstruction after the fact, if one stresses now at the liquidation these objective motives, although the first step of Hilbert, made under such singular circumstances, could perhaps create another impression.'

The whole problem seemed to have been settled satisfactorily. Hilbert, who was only partially informed of the goings on, wrote to Blumenthal 'a triumphant letter, that everything was glorious'.[97] Courant had written a conciliatory letter to Brouwer in which he expressed the hope that the solution to the matter satisfied Brouwer. He also wished to convince Brouwer that no personal motives had played a role in Hilbert's action, and definitely no motives 'whose existence were in conflict with the respect for your scientific or moral personality'.[98] Little did he know Brouwer!

As a matter of fact Brouwer launched another appeal to the publisher and the editors the same day Courant was offering Brouwer the 'forgive—and—forget' advice. Brouwer insisted that in the interest of mathematics the total editorial board of the *Mathematische Annalen* should remain in function. As he realised that a written defence from his hand would inevitably wreck the unity of the editors, he was willing to postpone such a letter; moreover, Carathéodory, in a letter of December 3, had promised him to do his utmost to find an acceptable solution, and had begged him to be patient for a couple of more weeks. Sommerfeld had also pressed Brouwer to wait for Carathéodory's intervention. The final solution, as formulated in the Hilbert–Springer letter, did *not* satisfy Brouwer. He recognised that the reorganisation of the *Annalen* was mostly, if not wholly, designed to get rid of him. Also, Brouwer had explicit views on the ideal organisation of the *Annalen*. In a circular letter (23 January 1929) to the editors, Blumenthal and Hilbert excluded, Brouwer rejected the final solution. According to him, the *Mathematische Annalen* were a spiritual heritage, a collective property of the total editorial board. The chief editors were, so to speak, appointed by free election and they were merely representatives *vis a vis* the mathematical world. Thus, Brouwer argued, the contractual rights of the chief editors were not a personal but an endowed good. Hilbert and Blumenthal, in his view, had abstracted this good from their principals, and hence were guilty of embezzlement, even if this could by sheer accident not be dealt with by law (the reader may hear a faint echo of Brouwer's objections to the consistency program, see p. 442 of Brouwer 1923c). Brouwer then proceeded to attack Blumenthal's role in the *Annalen*. He repeated Blumenthal's earlier views on the equal rights of all editors and referred to certain irregularities in the management of the *Annalen* in 1925, when Blumenthal had committed an exceptionally strong infringement.[99] Brouwer had only given up his plan to call a meeting of the collective

[97]Blumenthal to Courant, 31.XII.1928.

[98]Courant to Brouwer, 23.XII.1928.

[99]The German term is *'Übergriff'*, which could also be translated as 'impertinence'. In view of Blumenthal's reputation, that version seems implausible.

board to discuss Blumenthal's lapse, when Blumenthal made it clear that he planned to give up the position of managing editor no later than at the publication of volume 100. There is no information on this alleged intervention. Since 1925 was the year of the Riemann affair, it might have played a role.

There is concrete evidence that Blumenthal wanted to resign from the *Annalen* board. A stern letter from Hilbert to Blumenthal of November 18, 1925, opens with words that quelled all opposition: '*NEIN*, I do not agree at all with your plan.' Since the letter provides useful background information for the *Annalen* conflict, it is worthwhile to reproduce it here in part.

> Already by itself, I would at the moment indeed follow the tendency to reduce the number of members of the editorial board, instead of increase it—including the chief editors, where, by the way, Springer should also have his say. When Cara once wanted to resign, I have—but only for the case that you would insist on an immediate replacement—mentioned who is by far the most important and most generous mathematician of his generation, Hecke as his successor; you turned him down with the argument that you, as a substitute for me, needed a real worker and manager for yourself, and at that point you were right. And now Weyl should be such a person? Weyl, who has a purely academic and luxury professorship, and for the rest lives for his scientific and literary activity, and for his health! And such a man should from Zürich conduct the business of the Annalen! His name on the title page were nothing but an honour from us for his person. [...] Thus I come to my main motive for rejecting your plan: if you lay down the managing of the Annalen, I would like to get it to Göttingen.

Hilbert wanted in fact to strengthen in Göttingen the connections with physics. But, he continued:

> Fortunately you are for the time being prepared to conduct the management, and it is my sincerest and warmest wish that you will do this for a long time to come. But I recommend you—in particular in your own interest not to let yourself be distracted from the trusted practice of your management by external influences, and I recommend a smooth development *without* editorial meetings, *without* changing editors, and *without* formal innovations.

And that was the end, Blumenthal had to stay on, whether he liked it or not. So, maybe Brouwer should not blame him for hanging on. The letter sheds light on a few details that bear on the previous pages. Much of what went on in Göttingen and Berlin was of course circulating in mathematicians' gossip. The problem is that this is hard to trace, and that it is usually coloured one way or another.

The *Annalen* were settling down under the new regime and, due to a tactful handling of all publicity, the excitement in Germany was dying out, even as Courant wrote to Hecke, among the colleagues in Berlin—and Brouwer was completely ignored. After waiting for months—and probably realising that the battle was over and that everybody had gone home—he fired his parting shot, the letter of defence against Blumenthal's indictment of 16 November 1928. The letter is three and a half folio sheets long and contains a report of the events mentioned above, as experienced by Brouwer. The tone of the letter is bitter, Brouwer felt that he was let down by his supporters, 'To my astonishment and disappointment up to now no correction has followed from the other side, in spite of my challenge, of the false representations in Blumenthal's circular letter of 16.11.1928.' This applied most of all to Carathéodory, who had failed to straighten out Blumenthal's distorted interpretation of the facts about Carathéodory's visit. So there was no choice but to speak himself.

In the first place he denied Blumenthal's claim that Brouwer had substituted his own interpretation for Carathéodory's version of the developments leading to, and including, Hilbert's action. The views, he wrote, were not mine, but 'views that, during the aforementioned visit, came up between Carathéodory and me in mutual agreement, i.e. that were successively uttered by one of us and accepted by the other'. He also elaborated the grounds for not acquiescing in the dismissal. He had told Carathéodory

> I would consider a possible dismissal from the editorial board not only a revolting injustice, but also a serious damage to my possibility to function and, in the face of public opinion as an offending insult; that, if it really came to this unbelievable event, my honour and freedom of action could only be restored by the most extensive flight into public opinion.

At the end of the otherwise friendly visit the shadow returned, and the two parted shattered and in grief.

Reading Brouwer's report, one gets a gloomy impression of a meeting of two friends, confronted with a human tragedy they know they cannot prevent. It was an almost paradoxical situation, both knew that the request was an injustice, yet both knew that a decision had to be taken. In view of the role of Carathéodory's visit in the history of the conflict, Brouwer's report follows here:

> At his visit on 30.10.1928 Carathéodory informed me first of all, while the two letters lay unopened, that the 'fact of the greatest consequence, which was unknown to me' consisted of the following: recently the taking of a wrong medicine had brought out in Hilbert a situation of such a serious nature, that on the one hand 'he could no longer be taken seriously at all' (Carathéodory's words)[100] and that on the other hand the slightest opposition to his will could be fatal for him. In this situation Hilbert got the idea to remove me from the board of the Annalen, and wished to realise this idea by all means. It should be evident that the realisation of Hilbert's plans would mean a scandalous

[100][Brouwer's footnote] One might think for a moment that the communication of such utterances carried something incorrect, because one assumes with respect to these naturally an atmosphere of confidentiality. This assumption of confidence, however, and the solidarity which it presumes, in as far as it is not invalidated by the end of the conversation, is certainly not compatible with Carathéodory's later silence in the face of Blumenthal's later false representations. Moreover, even justifiable scruples must take second place in the present case, where a discussion of wicked defamation and isolation of status is concerned—just as in the case of the question of a witness in a court of law.

injustice. In order not to jeopardise Hilbert's life, he (Carathéodory) begged me to take no action against it, for the time being. Hopefully Hilbert would soon return to the right medicine, and as a consequence of the improvement of his condition, regain better views, before anything definite could happen.

One of the two unopened letters was Hilbert's. The message it contained, that Hilbert dismissed me as an editor, 'authorised by Blumenthal and Carathéodory', were not justified; for when he (Carathéodory), after his return from America, received a letter from Hilbert, asking for his authorisation, he had answered that he would on principle not oppose Hilbert, but that he would come to Göttingen to discuss the matter. Arriving in Göttingen he had learned from Blumenthal that Hilbert had dispatched the letter of dismissal, mentioning the above authorisation.[101] In the following conversation of half an hour with Hilbert, the matter was mentioned between them, as little as it was until today. As far as the second letter was concerned (with Blumenthal's name on the cover as sender), this was written by him (Carathéodory), in which he begged me to withdraw of my free will from the editorial board, in consideration of the situation of Hilbert's health. He now regretted, however, to have written this letter.

Subsequently I have returned the second letter unopened to Carathéodory, and declared that I would consider my possible removal from the board not only a scandalous injustice, but also a serious harming of my professional prospects, and in its public aspect, a despicable insult; that should it really come to this unheard of event, my honour and professional prospects could only be regained by means of the most far-reaching flight into publicity; as a consequence the crime practised against me would cause a public scandal—Carathéodory replied that he had been prepared for such a position of mine, that in his expectation the Annalen would meet its doom over the realisation of the plan hatched against me. And that he had already made the decision to resign from the editorial board, in which decision could for the time being—again out of consideration for Hilbert's health—not be carried out.

The further course of our conversation then came to the seven points in my circular letter of 5.11.1928.

Concerning the consideration with Hilbert's state of health, demanded by Carathéodory, I gave as my opinion that if there were an immediate mortal danger for Hilbert, it would be a crime to assist him in ending his life with a crime; on the other hand that unreasonable tolerance would possibly only increase his sensibility and lust for power in a manner that would endanger his happiness in life. I promised, however, to consult on these latter psychological questions a suitable acquaintance. In case after further reflection my position would not change, the acceptability of Carathéodory's request, to undertake for the time being nothing against Hilbert's plans, would for me be equivalent to the probability of a cancellation of these plans even without my active

[101]Only a thoroughly confused person would thus run the risk of a court case for falsification. If true, this supports Carathéodory's and Brouwer's view.

interference. The conversation closed with Carathéodory's repeated reference to Hilbert's terrible state, and the words that he (Carathéodory), under these circumstances, 'appealed to my mercy'.

During this conversation of two hours in the morning of October 30 Carathéodory's position was indeed that of a mediator, friend and ally, who counselled me on the possibilities and means to prevent a calamity. The discussion seemed concluded in full agreement, in spite of the temporary differences in our evaluation of details of the situation. Carathéodory stayed accordingly some more hours at my house with some guests that I had invited at the occasion of his visit; all guests had the impression of a perfect mood. Only at the parting, when I was again alone with Carathéodory, I mentioned a thought that came up at that moment, that, as Hilbert had survived Einstein's objections against his plans, he could also suffer without danger a repudiation of the authorisation, mentioned without justification in his letter to me. Only when I did not get from Carathéodory a reply to the question, but only exclamations such as 'What can one do' and 'I don't want to kill a person' (perhaps to be ascribed to the uneasiness of parting), amazement, uncertainty and irritation came up in me, which expressed themselves, under a complete change of mood at my side, in phrases like 'I cannot follow you any more', 'I consider this visit a farewell', and 'I am sorry for you'.

A fortnight later Brouwer visited Schmidt in Berlin. Schmidt's account roughly agreed with Carathéodory's, but he added one piece of information: Hilbert's anger was to a great extent caused by Brouwer's actions in the Riemann affair.

The second part of Brouwer's circular letter concerns Blumenthal and his indictment. He saw Blumenthal's hand in the action against him, for only Blumenthal had insight in the record of the individual editors. The objections listed by Blumenthal 'could only degenerate into anecdotes, if one would ascribe them to Hilbert. He counts already for years so little as editor, that for a regular handling of the business, it had even been proved to be dangerous to submit manuscripts to him. Accordingly Hilbert does not try to mention these grounds in his well-known letter of dismissal.' As a motive for Blumenthal's alleged action, Brouwer mentioned the promise to resign at the completion of Volume 100, and the admonishments he repeatedly directed to Blumenthal in relation with arbitrariness and damaging actions in general.

In defence of Blumenthal, it must be acknowledged that in his unshakable loyalty to Hilbert he would probably go very far, but he would never seek to protect himself. Indeed, there is convincing evidence that Blumenthal almost immediately regretted his actions against Brouwer. Blumenthal was a man of high integrity, but in this case his *Doctorvater* overruled his conscience. That Blumenthal mismanaged the editorial procedure from time to time was a recognised fact; the Lebesgue note of 1911 is an example. But that would not have been a reason for ousting Brouwer.

Brouwer's refutation of Blumenthal's four points is more interesting, as it provides relevant information:

Ad 1. There could very well be a reality corresponding to the word 'rude' (*schroff*), if the meaning of the word is fixed as follows: will for integrity

> (the duty of every human being), with in addition the will for clarity (destiny of the mathematicians).—These wills have manifested themselves with me if the honour and prestige of the Annalen was at stake. (There were, by the way, cases, where Blumenthal himself had called upon me.) Then neither the vanity of the authors, nor Blumenthal's wish to please, could be spared—When I have occasionally carried my will through against that of the managing editor [Blumenthal], the latter has indeed found no support with his colleagues in the board, or has had reasons not to look for it.

There is not much evidence in this case. There is hardly any correspondence between Brouwer and authors left, but there are notes that show that Blumenthal was quite happy to use Brouwer as a trouble-shooter. Furthermore, a certain natural wish to ingratiate in Blumenthal cannot be denied. This may not be a recommendation for a managing editor.

> Ad 2. The occasion Blumenthal hints at in his report on Klein's resignation can hardly be anything but the following: I had a discussion with Klein about a paper that I had already handled, the author of which had appealed to Klein, as chief editor, concerning the changes I had demanded, and he had made him his views plausible in a personal discussion. In the conversation with me, Klein then saw that the author was wrong (not formally, as Blumenthal suggests, but contentual), and that he could, as a consequence, not stick to the promise given to the author.[102] In the further course of his discussion Klein offered his view that the manner in which the chief editors were mentioned at the cover apparently gave the public a misleading impression, and he for himself, as far as it concerned his person, could not very well bear the responsibility for this impression.—Some time later he resigned as a chief editor.—Such a conduct does as much credit to Klein as little as it does for Hilbert that, with on his part, a much smaller contribution to the editorial activity than Klein made at the time of his resignation, the opportunity existed to exploit the internal weakness of his position for the external strengthening of it.
>
> Ad 3. As I devoted yearly some thousand hours, it is almost obvious that manuscripts that had come in were usually for months in my possession. Only the word 'lay around' (*Lagern*) is misleading, for never papers were temporarily forgotten, or even missing (as happened with Hilbert), but they were always the subject of the most intensive editorial activity, by which their content was as a rule considerably influenced. As I moreover kept only in extremely exceptional cases, in which great deficiencies were found, manuscripts beyond the normal printing time limit, the papers were much better stored with me, than that they would have 'laid around' at Blumenthal's.—Blumenthal was, by the way, until shortly of the opinion that my procedure was normal and scrupulous, otherwise he would not so often have appealed to me for refereeing, even of papers, for which the content of I could not in the least be considered an expert.

[102]The available evidence points at Mohrmann.

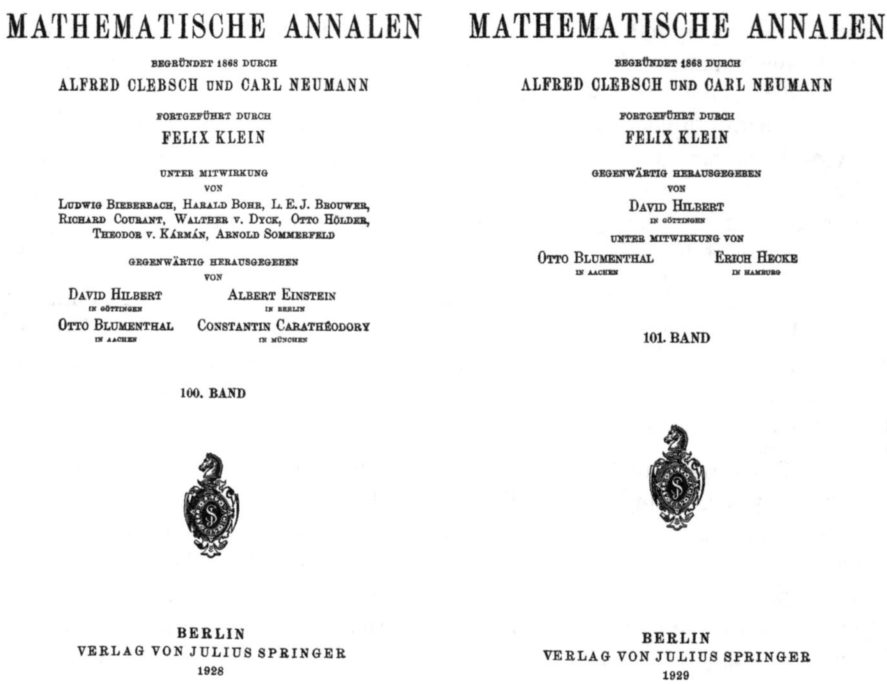

MATHEMATISCHE ANNALEN
BEGRÜNDET 1868 DURCH
ALFRED CLEBSCH UND CARL NEUMANN
FORTGEFÜHRT DURCH
FELIX KLEIN
UNTER MITWIRKUNG
VON
LUDWIG BIEBERBACH, HARALD BOHR, L. E. J. BROUWER,
RICHARD COURANT, WALTHER V. DYCK, OTTO HÖLDER,
THEODOR V. KÁRMÁN, ARNOLD SOMMERFELD
GEGENWÄRTIG HERAUSGEGEBEN
VON
DAVID HILBERT IN GÖTTINGEN
ALBERT EINSTEIN IN BERLIN
OTTO BLUMENTHAL IN AACHEN
CONSTANTIN CARATHÉODORY IN MÜNCHEN
100. BAND
BERLIN
VERLAG VON JULIUS SPRINGER
1928

MATHEMATISCHE ANNALEN
BEGRÜNDET 1868 DURCH
ALFRED CLEBSCH UND CARL NEUMANN
FORTGEFÜHRT DURCH
FELIX KLEIN
GEGENWÄRTIG HERAUSGEGEBEN
VON
DAVID HILBERT IN GÖTTINGEN
UNTER MITWIRKUNG VON
OTTO BLUMENTHAL IN AACHEN
ERICH HECKE IN HAMBURG
101. BAND
BERLIN
VERLAG VON JULIUS SPRINGER
1929

**Fig. 14.3** Mathematische Annalen, Volume 100, Issue 1, 1928 © and Mathematische Annalen, Volume 101, Issue 1, 1929 ©

Brouwer's reply is, at least to us, completely convincing. A modern editor would be happy indeed if his referees were as punctual as Brouwer. Brouwer in fact had a reputation of being a most scrupulous editor, he would—in cases where it made sense—rethink the content of the paper, which resulted in considerable, and sometimes essential improvements. He mentioned the copying procedure below, this may have been the expression of a certain caution, adopted over the years. This meant a considerable extra work load, borne by Cor Jongejan, who was paid an assistant's salary for these and similar jobs.

> Ad 4. Although Blumenthal knows to give an example of my 'principle' to obtain a copy of every manuscript coming in, and although I think that this act is by itself the elementary right of the refereeing editor, I have since many years done so only then, when a paper seemed indeed acceptable, but fit for publication only after rewriting or considerable emendations. Then I considered it a duty in the face mathematical history, and indeed because one must take into account the possibility of an unjustified reference to the date of submission.

The circular ended with:

> I challenge Blumenthal to make public the Annalen correspondence, in particular with the complete correspondence between him and me. I claim that these documents will really refute those accusations of him against me.

In an open discussion in a free meeting of editors, Brouwer would undoubtedly have scored, but the days of Klein's 'academy-concept' were over. The *Annalen* had become a one-man enterprise; this one man happened to be a man of great prestige, and an outstanding mathematician (although already then some of his colleagues would prefer the past tense), but a man with absolute power is not so easily corrected, as history has taught.

Whatever one may think of the crisis-management of Hilbert's lieutenants, Bohr and Courant, they had managed to outwit Brouwer. The whole matter was decided behind closed doors and the books were closed on Brouwer.

And so the curtain fell over one of the most tragic and completely unnecessary conflicts of twentieth century mathematics. It seems desirable to sum up the affair, but the motives of the invisible main actor were so confused that it is hard to come to a satisfactory conclusion. The available evidence tends to show that Hilbert's primary motivation was Brouwer's political activity and the Riemann affair. Although he nowhere explicitly says so, he felt it as a loss for the *Annalen* that French authors would keep their distance. The fear that was formulated in correspondence, that Brouwer would take over the *Annalen* after Hilbert's death, was rather a figment of Hilbert's imagination. It is not unlikely that Brouwer would have been pleased to move up to the chief editors, but that his influence would go any further than what a chief editor normally does, i.e. accept or reject papers, is not plausible. Even in his own faculty, his own mathematical association, his own academy, he had no excessive influence. So why would the *Annalen* be different? Some people have mentioned the danger that Brouwer would turn the *Annalen* into an intuitionistic bastion. The record shows that even his own journal, *Compositio Mathematica*, did not function as such.

There remains the matter of the *Grundlagenstreit*. Here we have to admit that Hilbert's emotions ran high. His aversion to this odious doctrine, with its by far too clever proponent—as witnessed, for example, by his refusal to quote Brouwer for any foundational contribution—certainly went beyond rational evaluation.

There is one more point that might have added to Hilbert's emotional outburst. David Rowe has pointed out that Brouwer's reminder of a forgotten credit, see p. 538, was the first public manifestation of this sort. There had been grumbling before, but no one had the courage to put it in print. Here we may very well have an instance of the old-time German professor, who could not forget such a slur.

Reactions on the *Annalen* affair are scarce. Of course the German mathematicians must have known that something was going on. A healthy dose of gossip is part of academic life. But it may be doubted that the facts were also known beyond the borders of Germany. There is for example a letter from Ehrenfest to Van der Waerden from that period:[103]

---

[103] Ehrenfest to Van der Waerden, 8.X.1928.

> It was by itself a great pleasure to be again in Göttingen, together with the wonderful Harald Bohr! In spite of all the sadness I actually had to laugh about this scandal bomb that exploded around the editors of the Mathematische Annalen. It looks almost as if the physicists are a more tolerant and humorous people than the mathematics-only people. They have such a terribly refined brain that it can tip over at any moment.

Alexandrov, when informed about the *Annalen* affair by Cor Jongejan, thought that 'such a dismissal is a terrible insult, and also an act that is hard to justify to the international mathematics'. But knowing both Brouwer and the Göttingers, he hesitated to take sides. 'Be that as it may, the whole matter is and remains most unpleasant, and it will hardly contribute to the moral prestige of the mathematicians in central Europe. But what I find most distressing, is that the painful, and for the collective mathematics in Germany damaging, antagonism Göttingen–Berlin will be further intensified.'[104]

A later letter of Alexandrov to Hopf suggests that there were rumours about a boycott of the *Annalen* in protest against Brouwer's dismissal. That seems, however, never to have been a serious plan.

Looking back at the *Mathematische Annalen* affair and the surviving correspondence, and keeping in mind that before the final blow-up all the persons concerned, with the possible exception of Hilbert, had been living in the general atmosphere of friendship that was so characteristic for the international brotherhood of mathematicians, one gets a strong impression of regret and reluctance at the side of the managing and associated editors. Leaving aside Hilbert and Blumenthal, nobody was out to hurt Brouwer. The majority of the editors had at one time or another more or less close connections with Brouwer, and all of them were aware that he was a strict, perhaps difficult, but not unfair, colleague. Bohr and Courant, Hilbert's two lieutenants, for example, were doing their utmost to reach a clean, objective solution; they did not muster the zeal of Blumenthal, but looked for a minimal damage compromise, in the hope that Brouwer would play along and be so magnanimous as to swallow the insult that no cosmetic procedure could obscure for him.

Blumenthal was an exception; there had apparently been frictions between him and Brouwer before 1928, but the main source of his vehement denouncement of Brouwer—so uncharacteristic for the man of peace he was—was his unerring loyalty to Hilbert. Later in life, according to his daughter, he sincerely regretted the consequences of his actions.

Hilbert, the leading actor in this almost Shakespearean drama of the 'Frogs and the Mice', only played a role at the beginning and the end of the conflict. After his initial appearance in the role of defender of the realm, he was whisked back stage, only to reappear in the finale for the ceremony of signing the letters that sealed the conclusion of the conflict. There are no known reports of comments or reminiscences of Hilbert that shed more light on the episode or on his motivations. It was as if for him the book on Brouwer was closed once and for all. As we have seen

[104] Alexandrov to Hopf, 6.XII.1928.

above, Hilbert told his closer associates at the time of the conflict that there was no personal element in his actions and statements, and most of his followers were happy to take his word for it. Nonetheless, one would probably do well to entertain a certain measure of doubt where motives and self-knowledge are concerned. A person who is the subject of unlimited admiration and trust is mostly not in a good position to do serious soul searching.

Brouwer himself refused to publish further in the *Mathematische Annalen* and he convinced Heyting to follow his example. Heyting's paper was the outcome of a prize problem set by the *Wiskundig Genootschap* in 1927; it was formulated by Mannoury, cf. p. 500, and obviously endorsed by Brouwer. It may come as a surprise, but Brouwer was highly pleased with Heyting's formalisation of intuitionistic logic, so much indeed that when Heyting had rewritten the material as a full blown paper, Brouwer expressed his appreciation in no uncertain terms, 'Your manuscript has interested me extraordinarily, and I am sorry that I have now to hasten to return it. In future I would appreciate it if you made a copy of your manuscripts before sending them to me, if at least you value a more than superficial reading. In the meantime I have already got such a high appreciation for your work, that I ask you to edit it in German for the Mathematische Annalen.'[105] At the time of writing Brouwer could not have foreseen what Hilbert had in store for him, but after the battle of the frogs and mice had run its course, he asked Heyting to withdraw his manuscript from the *Mathematische Annalen*. Two months later he wrote to Heyting that had hoped that Blumenthal and Hilbert would show repentance and mend their ways before the summer was out, but that he had now given up hope.[106] Anticipating Heyting's understanding, he had decided 'not to leave the manuscript in the hands of those who call themselves falsely editors of the Mathematische Annalen'. He proposed to submit the paper to the Prussian Academy, where Bieberbach could handle it. As a consequence Heyting's historic contribution to logic appeared in the reports of the academy, which did not have the distribution and visibility of the *Annalen* by a long chalk.

That Brouwer was seriously impressed by the advances made by his student may further be illustrated by his letter to Weyl. Brouwer was proposing Heyting for a chair in Utrecht, and he asked Weyl to support his proposal.

> In Utrecht there is an important vacancy list of the mathematics faculty: Barrau, H.J.E. Beth, Schaake (all three insignificant). My (alphabetical) list: Heyting, Hurewicz, Van der Waerden (respectively intuitionist, topologist, algebraist). Heyting and Van der Waerden are Dutch, Hurewicz (my assistant) though a Polish citizen, and educated in Moscow and Vienna, is already for a long time in Holland. In order to document my list with the minister, I need foreign testimonials. For Heyting (so far my only really gifted intuitionistic student) you are the only eligible author of such a testimonial. Such a testimonial should on the one hand stress the general importance of intuitionistic

[105]Brouwer to Heyting, 17.VII.1928.

[106]Brouwer to Heyting, 28.IX.1929.

> research in the present stage of development of mathematics (in Holland nobody outside Amsterdam believes that), on the other hand it should qualify Heyting's papers (which are enclosed) as epoch-making.[107]

The letter also shows that Brouwer had few illusions about the acceptance of his program in his own country; it must be said that he did little to convert his colleagues to his views. Brouwer was quite prepared to explain his views in lectures and papers, but he would not go round to practice empire building.

In 1929 there was another publication in the intuitionistic tradition: an intuitionistic analysis of the game of chess by Max Euwe.[108] It was a paper in which the game was viewed as a spread (i.e. a tree with the various positions as nodes). Euwe carried out precise constructive estimates of various classes of games, and considered the influence of the rules for draws. When he wrote his paper he was not aware of the earlier literature of Zermelo and Dénès König. Von Neumann called his attention to these papers, and in a letter to Brouwer Von Neumann sketched a classical approach to the mathematics of chess, pointing out that it could easily be constructivised.

## 14.4 The Endings of the *Grundlagenstreit*

The issues of the *Grundlagenstreit* were far from settled. In military terms, Brouwer had built an impregnable citadel, and Hilbert produced blue prints for an immense fortress.

Brouwer's intuitionism could only be defeated on philosophical grounds, his mathematical intuitionism consisted of honest mathematics, and there was no grand claim for which an account could be asked. Strictly speaking there was a master plan: the rebuilding of mathematics along intuitionistic lines. But that was an open-ended claim. One could not reject intuitionism, because it had, say, not yet proved the fixed point theorem for the unit square. In the first place, because the list of such challenges is endless, and none of them would be crucial. In the second place, the reply is flexible. In the case of the fixed point theorem, the intuitionist would say, I cannot give you the fixed-point theorem, because it is false in my mathematics, but I can give you our version, which is an $\varepsilon$-theorem, cf. p. 503.

In Hilbert's approach there was a crucial test: the consistency proof for, say, arithmetic by finitary means. In 1928 at the Bologna congress Hilbert 'saw the light at the end of the tunnel', that is to say, he had set up his machinery, and considered it just a matter of time before the remaining technicalities were carried out.

The *Annalen* affair rudely interrupted the conflict; Brouwer felt deeply insulted, and retired from the field. He did not give up his mathematics, but he simply became invisible. He no longer appeared at meetings of the DMV to report on intuitionistic mathematics. Even worse, he gave up publishing for a decade, there is evidence

---

107 Brouwer to Weyl, 16.II.1928.

108 Euwe (1929).

that he carried on his more reflective research, but the results remained restricted to such things as classroom notes, communications to Heyting and Freudenthal, etc. His withdrawal from the debate did not mean a capitulation, on the contrary, he was firmly convinced of the soundness and correctness of his approach.

Looking back, one has to come to the conclusion that there never had been a discussion between Brouwer and Hilbert on the essential points. One doubts if the two ever had a private conversation on the foundations, with the exception of the Scheveningen walks. It is very likely that Hilbert never read Brouwer's basic papers, such as the 'domains-of-functions'-paper (although he probably attended his lecture in Göttingen in 1926, and the paper was a contribution to the Riemann volume). Indeed, referring to Brouwer's publications, Bernays asserted in an interview, 'he has not read these things at all'.[109] All of Hilbert's attacks consisted of rather superficial comments on hearsay bits of Brouwer's repertoire. Brouwer, on the other hand, repeatedly put his finger on the crucial spots of Hilbert's program: (1) consistency of induction requires induction (siding with Poincaré); and (2) consistency does not prove existence. Hilbert was in fact encouraging work on (1), and he did not see the point of (2).

The historic facts were in fact all speaking for Hilbert; he had every right to be optimistic about the success of his program. There was considerable activity in his proof theory, there was significant work of Ackermann (1925, 1928), Herbrand (1928–1931), and von Neumann (1927).

Yet Hilbert's program foundered when in 1930 a young man with his soft voice announced that formalised classical mathematics was incomplete. This stunning event took place at the *Tagung für Erkenntnislehre der exakten Wissenschaften* in Königsberg. At this meeting (which was embedded in the Meeting of the Natural Sciences and Medicine)[110] an exchange of ideas between the 'big three' of the foundations of mathematics was arranged. Rudolph Carnap spoke for the logicists, Arend Heyting for the intuitionists, and Johann von Neumann for the formalists.

Reidemeister, who was in charge of this meeting, wrote to Heyting that neither Brouwer, Hilbert, nor Russell[111] were invited as speakers; this was a symposium where the younger researchers presented their work, so it seemed more appropriate not to invite the big names. Apparently Heyting was worried that Brouwer might feel passed over. Reidemeister also begged Heyting not to be upset that Hilbert was going to be present, he was invited by the *Naturforscher*. In view of Hilbert's recent lecture in Bologna, von Neumann was definitely a better choice. It was a meeting of the second generation, and as a consequence the tone was friendly and objective.

The surprise of the meeting was Gödel's contribution to the discussion at the end. He had already presented a talk which dealt with his completeness theorem. This solved one of Hilbert's 1928 problems. His contribution to the discussion at

[109] *Er hat die Sachen überhaupt nicht gelesen.* Interview, Bernays 18.VII.1977.

[110] To be precise, the 6th *Deutsche Physiker- und Mathematikertagung*, and the 91st *Versammlung der Gesellschaft Deutscher Naturforscher und Ärzte*. September 5–7, 1930 Königsberg. Proceedings published in *Erkenntnis* (23), 1931.

[111] Weyl (1946), p. 275.

the closing session, however, spelled disaster for Hilbert's program. It said, roughly, that there were simple statements, comparable to Goldbach's conjecture, that were neither provable nor refutable in classical mathematics. In symbols, statements $A$ such that $T \nvdash A$ and $T \nvdash \neg A$, where $T$ is a suitable mathematical theory, for example arithmetic.[112]

This was a dramatic moment—the end of a program, but what a magnificent end! Hilbert, however, missed it, as he was whisked off in a taxi to the radio studio, where he was expected for a broadcast of his '*Logik und Naturerkenntnis*' (Logic and the understanding of nature). According to contemporary sources, Hilbert learned about this stroke of fate only months later; no one had the courage to disappoint the old master. When he learnt about Gödel's work, he was angry.

Bernays had heard about Gödel's result through the grapevine. He wrote to Gödel, asking for advance information in the form of proof sheets.[113] One may thus take it that Hilbert was not aware of the results either. This would be consistent with the paper that Hilbert published after the Königsberg meeting, namely 'The Founding of elementary number theory' (submitted 12.12.1930).[114] In this paper Hilbert upheld his program as if nothing had happened. He did, however, introduce in his paper a novelty, the $\omega$-rule:

$$\frac{A(0), A(1), A(2), \ldots, A(n)\ldots}{\forall x A(x)}$$

In words: if you have derived $A(0), A(1), A(2), \ldots, A(n)\ldots$, then you have also derived $\forall x A(x)$. This could have been an answer to Gödel's theorem, but as he considered the $\omega$-rule to be a finitary rule (p. 456), the system would be finitary and fall under Gödel's theorem. There is no reference to Gödel in this paper, but that does not say much in view of Hilbert's reference practice. The paper was the report of a talk before the philosophical society in Hamburg, in December 1930.

Hilbert's next, and last, paper, 'Proof of the Tertium non datur' (submitted 17.7.1934), does not mention Gödel either, but as it is more modest in scope, one may guess that Hilbert revised his perspective.

Gödel published his epochal paper in the Austrian mathematics journal, the *Monatshefte für Mathematik und Physik*, in 1931. In the same year the *Erkenntnis* issue with the lectures of the Königsberg meeting appeared with a brief synopsis of his incompleteness results. This time it included the second incompleteness theorem, 'For a system in which all finitary forms of proof are formalised, a finitary consistency proof is not possible.'

Immediately after Gödel's talk John von Neumann had collared Gödel. Von Neumann, who was reputedly the fastest thinker of his generation, had right away seen what was going on, and he had realised that Gödel's argument could be 'internal-

[112] *Erkenntniss*, 1931, pp. 147, 148.

[113] Bernays to Gödel, 10.XII.1930, cf. Dawson (1997), p. 282.

[114] Hilbert (1930).

ized'. This then would yield the unprovability of the consistency of (say) arithmetic in arithmetic.[115]

Gödel's incompleteness theorems brought the second ending of the *Grundlagenstreit*. Where Hilbert had won the conflict in the social sense, he had lost it in the scientific sense.

What could be learned from the history of the *Grundlagenstreit*? In fact little, the whole matter was in the real sense a struggle of Titans. Both Brouwer and Hilbert were exceptionally gifted, be it that their characters were almost completely opposite. The older man was totally engrossed in mathematics—the paradigm of the German professor as the icon of traditional *Bürgertum*, the younger man equally infatuated with mathematics, but without the compulsion, who always was glad to change the three piece suit of the professor for the bohemian outfit that belonged to his free and artistic way of life, complete with numerous love affairs. The fact that their philosophies of mathematics were so different undoubtedly played a role, but should not be overestimated. Those who had followed Hilbert's evolution from the first steps in 1904 to the middle twenties must have seen that he sought safety for mathematics in the realm of the finitary, and that whatever intuitionism might be, it certainly encompassed Hilbert's finitary mathematics. Abraham Fraenkel, who had observed the developments closely, was quoted in the *Vossische Zeitung*, 'With charming wit Fraenkel called Hilbert the second intuitionist',[116] and in his 'Ten Lectures on Set Theory' Fraenkel noted that 'one could even call him an intuitionist'.[117]

Brouwer had come to the same conclusion in his 'Intuitionistic Reflections on Formalism'. If anything, Hilbert could rather accuse Brouwer of surpassing the limits of finitary mathematics. It is doubtful whether he was aware of this, although Bernays wrote him in 1925 that 'he had discovered a certain difference between the finitary position and that of Brouwer'.[118] In 1977 Bernays said that he had grasped the difference between 'finitistic' and 'intuitionistic' through the Gödel translation. Taking into account the cited letter, the plausible reading seems to be 'the Gödel translation confirmed my earlier view'.

As Brouwer had put it, after acknowledging a few foundational insights, the whole matter 'intuitionism—formalism' became a matter of taste. So why the incessant harping on Brouwer? Even in his last paper but one, Hilbert repeated his complaints: 'Nonetheless there are even today followers of Kronecker, who do not believe in the Tertium non datur: it is well by far one of the crassest disbeliefs that we find in the history of mankind.'[119] Since the paper has the character (and hence

---

[115]Gödel called this the Von Neumann conjecture: 'If there is a finitary consistency proof at all, then it can be formalised. Then Gödel's theorem yields the impossibility of a consistency proof at all' (15.1.1931). See also Dawson (1997), p. 68 ff.

[116]8.X.1929.

[117]Fraenkel (1927), p. 154.

[118]Bernays to Hilbert, 25.X.1925.

[119]Hilbert (1930).

the precision) of a popular lecture, one should not worry too much about the details, but here are a few noteworthy statements:

- The a priori is nothing more nor less than a basic disposition, which I would like to call the finitary disposition.
- Kronecker has clearly enunciated the view, and illustrated it by means of numerous examples, which nowadays coincides essentially with our finitary disposition.
- The problem of the foundations of mathematics is, as I believe, definitely dispelled by proof theory.

So we note that Hilbert's a priori is even more restricted than Brouwer's of 1907. Furthermore, Hilbert made after all these years his peace with Kronecker. In Hilbert's hindsight Kronecker was, foundationally speaking, a forerunner of himself, who did not go all the way. One guesses that for a similar rehabilitation of Brouwer it was too early. Moreover, Brouwer was still alive, so, in contrast to Kronecker, he could possibly protest.

All great men, with great ideas, attract criticism, one may think of Cantor, Einstein, Gödel. Sometimes (and perhaps often) the criticism is offered by persons halfway between crank and amateur, sometimes by colleagues, who lack the imagination or intuition to appreciate novel circumstances or principles.[120] Reading Hilbert's foundational papers, one gets a strong impression that he did not suffer the criticism (of fools?) gladly. In the majority of these papers he complained about all those people who held obscure views and could not see the point of his program. His final reaction was, 'The critic of my theory should point to me exactly the place where my alleged error is to be found. Otherwise I decline to check his line of thought.'[121] A definite exasperation speaks from these lines, but was it reasonable? Assume someone would claim to have proved that after $10^{10^{10}}$ there are no more primes. Would he not rather say, go and study your Euclid, than check the long and intricate proof. The ultimate consequence of the above viewpoint would be that Hilbert had to reject Gödel's criticism of his program, for Gödel did not carry out a proof check of Hilbert's (non-existent!) proof. Apparently, in his emotions, he allowed himself to say things that he could not possibly mean.

The reader may have got a somewhat negative impression of Hilbert's proof theory. This should be corrected at once, it is a fact that Hilbert gave the world a new discipline of great finesse; after Gentzen added beauty and structure to it, which it was so sadly lacking, it became an elegant and powerful mathematical tool. In modern logic it occupies an important place, being the tool *par excellence* to bring out the finer details of complexity and structure. It also happens to be the part of logic that is of considerable interest to computer science. That Hilbert saw it primarily as a prop for settling all fundamental questions about the foundations of mathematics was simply a matter of a lack of experience in topics of this complexity and intricacy. It was a gift from Hilbert to mathematics, comparable to the rich gifts

[120] For an instructive discussion of a particular instance of perpetual criticism, see Hodges (1998).

[121] Hilbert (1931).

he had already made. From our modern point of view it is surprising that a man like Hilbert, with a gift for sweeping methods, should not have been the first model theorist instead of the first proof theorist. In the light of the exacting demands that model theory made where intuition was concerned, proof theory was indeed more accessible to traditional mathematical techniques.

So far it seems as if the *Grundlagenstreit* was some sort of personal quarrel between Hilbert and Brouwer. What were the reactions from the mathematical community? From the isolated comments in correspondence or in print, one might easily get the impression that the foundational issues, including the surrounding gossip, went largely unnoticed. It would be a mistake, however, to conclude that the conflict was a private affair between a few prominent mathematicians. Hilbert's lectures and publications always attracted a good deal of attention, and although most readers would not be in a position to judge the merits of Hilbert's program, they would certainly read Hilbert's diatribes. And on the other side Brouwer in his soft-voiced lectures could capture large audiences. His Berlin lectures had exerted a strong fascination on the audience. In contrast to Hilbert's more aggressive approach, Brouwer's lectures made their impression through a mixture of persuasive argument, wit, vigorous proof and cynical metaphor. The reports of these lectures leave no doubt that Brouwer could recruit and inspire a not inconsiderable following. Nonetheless there are few explicit references to the foundational crisis. The man who did most to enlighten his contemporaries in the confused matter of the foundations was Fraenkel, his books and papers covered the basic issues of the period. His expertise in foundational matters was widely recognised by his colleagues. Hausdorff, after reading Fraenkel's *Ten Lectures on set theory*,[122] spoke his mind without reservation, 'I still nurse hopes that you, as the best expert in this literature, will at some time launch a vigorous and witty attack at intuitionism—although it might be more advisable to let this castrates' mathematics suffocate in its own complicate obtuseness. It is indeed stupid to stick into every mathematical theorem, like the egg of an ichneumon wasp, an unknown number fabricated out of the decimal expansion of $\pi$....'[123] Study's reaction to Fraenkel's book was one of relief, 'I cannot find the outcry of Brouwer and Weyl, and also that of Hilbert, as remarkable as you and others do.'[124] The function theorist Konrad Knopp shared this sentiment; after the publication of Fraenkel's *Introduction to set theory*[125] of 1924, he confided, 'I am glad to draw from it the confirmation of my feeling, that the 'shake-up of the foundations' by Brouwer was by no means as disastrous as it seemed to be.'

Hausdorff was an implacable enemy of intuitionism, he was not prepared to relent, as appears from another letter to Fraenkel, in reply to the receipt of a copy of Fraenkel's 1928 edition of the *Introduction to set theory*, 'I hope that it will leave my heartfelt aversion to intuitionism untouched.'[126]

---

[122] *Zehn Vorlesungen über die Grundlegung der Mengenlehre*, 1927.

[123] Hausdorff to Fraenkel, 20.II.1927.

[124] Study to Fraenkel 5.III.1927.

[125] *Einleitung in die Mengenlehre*.

[126] Hausdorff to Fraenkel, 19.XI.1928.

The geometer Finsler had in his inaugural address in 1922, *Are there contradictions in mathematics?*, noted that Brouwer and Weyl rejected the principle of the excluded middle. He generously granted that 'such assumptions may in themselves lead to very interesting research', but, he continued, 'an exact science cannot be based on these, not mentioning the great complications that would thus arise; also many of the most certain results must be given up'.[127] A positive appreciation of intuitionism is to be found with Ludwig Bieberbach. He had earlier deplored the decline of the intuitive approach to mathematics as advocated and practised by Felix Klein, but the propagation of formalism by Hilbert and his school had made him aware that there was a real danger that formalist tendencies could seriously harm mathematics, in the sense that intuitive and applied mathematics could become a neglected, if not extinct, part of mathematics. He sensed in Brouwer's intuitionism a healthy antidote against the formalist epidemic. In an address for an audience of mathematics teachers he stressed the importance of Brouwer's intuitionism—including choice sequences.[128] 'There is a fresh breath of air thanks to intuitionism', he called out. In how far Bieberbach was right, is debatable. After all in Hilbert's Göttingen applied mathematics flourished, e.g. in the hands of Courant. And Hilbert occupied himself intensely with theoretical physics (albeit with formalisation in mind). But undeniably, there certainly was strong influence of Hilbert's formalist doctrines on theoretical and methodological levels.

Spectators, who observed the foundational conflict with more detachment from some distance, often saw clearer that this middle European muddle of loyalty, common sense and foundational ingenuity, was a serious matter, and not just a clash of personalities. As the American mathematician Pierpont observed

> And yet when one hears one of the greatest living mathematicians calmly telling the world that a considerable part of our analysis is devoid of proof, if not nonsense, and when one beholds the mighty efforts which the champions of Weierstrass are making to repel these attacks, it is reasonable, in view of such facts, to ask ourselves, 'Is all well?'[129]

The German philosopher Grelling described the *Grundlagenstreit* pointedly:

> If intuitionism has been characterised with a certain propriety as revolutionists who overturned the ancient régime, Hilbert might be compared with Napoleon who, without regard for considerations of legitimacy, established, through a brilliant political stroke, a new order whose success is the substitute for legitimacy.[130]

The working mathematician, more concerned with practising mathematics than with reflecting on its soundness, could not stop wondering why Brouwer, the wizard of

---

[127] Finsler (1925).

[128] About the scientific ordeal of mathematicians (*Vom Wissenschafts ideal der Mathematiker*), 1926.

[129] Pierpont (1928), p. 37.

[130] Grelling (1928).

topology, had given up the riches of traditional mathematics for a life in the arid desert of the foundations. André Weil, who attended some of Brouwer's Lectures in Berlin,[131] wrote to Fréchet:

> people here are very excited at the moment, because Brouwer has just arrived and he has started a series of lectures, not on topology, but on intuitionistic mathematics; this is very particular, as you know. I have no pretensions to understand it, but Brouwer is a very interesting man. He has declared in his first lecture that the principle of the excluded third is a superstition which is about to disappear. It is a pity that such a remarkable man devotes himself exclusively to such bizarre things.[132]

Weil's view was shared, and is still shared, by many mathematicians all over the world. The often somewhat esoteric issues of the foundations of mathematics have traditionally met with tolerance at best, and distrust at worst.

## 14.5 The Menger Conflict

The third battle was partly a clash of personalities, and partly a scholastic argument concerning dimension theory. The arguments, subterfuge and confusion resulted in a case so complicated that it is difficult to keep track of the finer details. We will stick here to a simplified account that will give a rough idea of the matter. The interested reader may find a detailed exposition in van Dalen (2005).

When the dust of the Bologna Conference had settled, Brouwer was confronted with the next conflict. In 1928 Menger dropped a bombshell that shook the usually quiet community of topologists; he had been writing a book on dimension theory[133] that was announced in a flyer of the publisher with the words 'These and numerous other questions are answered by the dimension theory, founded by Menger... and Urysohn.' This formulation was more than enough to evoke Brouwer's curiosity and wrath. He had more or less closed the books on his dimension involvement after the settlement of a misunderstanding with Urysohn, and the subsequent publishing of Urysohn's *Mémoire*, and here was something that could not pass unchallenged. So far he had every reason to believe that the matter had been settled with mutual consent of the persons involved. Urysohn had fully acknowledged Brouwer's place in the genesis of the dimension notion, and Menger might have grumbled in private, but in his publications he had not questioned Brouwer's priority. An all out attack was certainly not foreseen. And here, suddenly out of the blue, a book had appeared that boldly informed the reader that the author had submitted a paper on dimension theory in February 1922 (published in 1926 in the Amsterdam proceedings) and, a page later, that Urysohn had published his *Comptes Rendus* note in September

[131] Cf. p. 499.

[132] Weil to Fréchet, 31.I.1927.

[133] Menger (1928b).

**Fig. 14.4** Karl Menger. [Courtesy Eve Menger]

1922. The order of the events subtly suggested that Menger had a justified claim to priority, leaving it up to he reader to wonder why there was such a discrepancy between the submission of Menger's paper and the publication (elsewhere). The story of Urysohn's entrance on the stage of dimension theory has been dealt with in Chap. 12, cf. p. 397 ff., and Menger's discovery of the notion of dimension can be found in Sect. 12.3. Urysohn had discovered a mistake in Brouwer's 1913 dimension paper, but in the ensuing exchange following the first letter, he accepted Brouwer's explanation of the 'slip of the pen' (pp. 408, 415). Indeed, Brouwer's account was convincing enough, and there was even written evidence to support his point—the proof sheet of Schoenflies' *Bericht* containing the footnote with the correction to Brouwer's separation definition. Brouwer had inserted the correction in writing, which was for whatever reason not adopted by Schoenflies, unfortunately without informing Brouwer of his decision.

The initial contact between Brouwer and Menger was pleasant enough, but it appears that gradually Menger started to feel unappreciated and even aggrieved in Amsterdam (cf. Menger 1979, p. 237), and the dimension book may have been the outcome of his growing discontent—were people, and in particular Brouwer out to steal his priority of dimension theory? (cf. Menger 1928b, 1930). The statements and their formulations in the book leave little to the imagination of the readers; for example, after discussing the definition of dimension, he describes Brouwer's position as 'This notion of general degree of dimensionality already comes quite close to

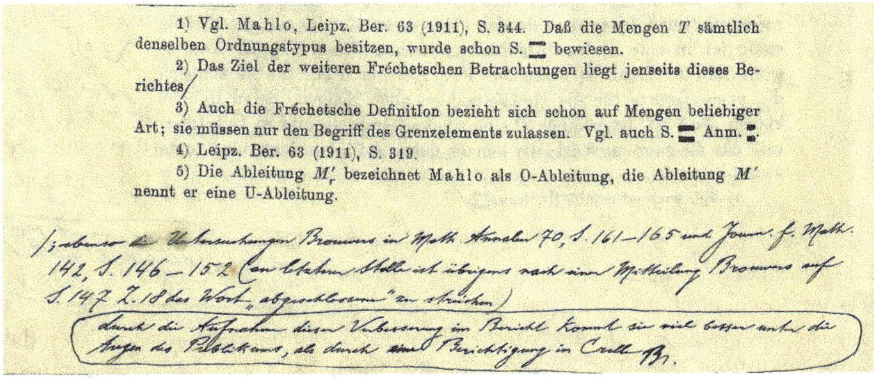

1) Vgl. Mahlo, Leipz. Ber. 63 (1911), S. 344. Daß die Mengen $T$ sämtlich denselben Ordnungstypus besitzen, wurde schon S. ▬ bewiesen.

2) Das Ziel der weiteren Fréchetschen Betrachtungen liegt jenseits dieses Berichtes/

3) Auch die Fréchetsche Definition bezieht sich schon auf Mengen beliebiger Art; sie müssen nur den Begriff des Grenzelements zulassen. Vgl. auch S. ▬ Anm. ▬.

4) Leipz. Ber. 63 (1911), S. 319.

5) Die Ableitung $M'_r$ bezeichnet Mahlo als O-Ableitung, die Ableitung $M'$ nennt er eine U-Ableitung.

/; ebenso die Untersuchungen Brouwers in Math. Annalen 70, S. 161–165 und Journ. f. Math. 142, S. 146–152 (an letzterer Stelle ist übrigens nach einer Mitteilung Brouwers auf S. 147 Z. 18 das Wort „abgeschlossen" zu streichen)

(durch die Aufnahme dieser Verbesserung im Bericht kommt sie viel besser unter die Augen des Publikums, als durch eine Berichtigung in Crelle Br.)

**Fig. 14.5** The footnote mentioning the slip of the pen, inserted by Brouwer in the proofs of Schoenflies' *Bericht*. [Brouwer archive]

the notion of dimension that underlies the theory of dimension.' Brouwer was allocated a place in dimension theory among the predecessors: Euclid, Cantor, Poincaré and Brouwer. If Menger had wanted to insult Brouwer, this passage would do very well; the patronising tone, usually reserved to pat a harmless colleague who did not make it on the shoulder, would be enough to make the blood boil of the person at the receiving end of Menger's wit. On the next page Menger characterised Brouwer's paper 'Remarks on the notion of natural dimension'[134] as a second attempt to provide a definition of dimension—this time with the right result. Even Menger's friend and student Hurewicz thought that Menger had gone too far:

> What you write about Brouwer, about his character and his 'morals', is nothing new for me. Nonetheless I find that your conduct with respect to him is wrong and inappropriate. As to the matter itself, in my opinion the historical survey in your book is not completely free from a subjective representation. [......]

'Why didn't you send the proofs to Brouwer, so that he could have reacted', Hurewicz continued; 'the matter would now only cost time and energy, and the satisfaction of the beautiful book would be spoilt'. 'A "life and death" struggle because of these foolishnesses! But that you have not for a minute given a thought to the impossible situation into which you have brought me, I very much hold against you.'[135]

Menger went, so to speak, out of his way to repaint the picture of dimension. He relegated, for example, Brouwer's theorem of the invariance of dimension to geometry or analysis, outside dimension theory. The fact that dimension is invariant under homeomorphisms is, in a way, a pleasant phenomenon, but if it were not the case, the notion of dimension would not suffer. At most it would show that

[134] Brouwer (1924e).

[135] Hurewicz to Menger, 10.X.1928.

homeomorphisms had no place in it (Menger 1928b, p. 243). Thus, one might say, Brouwer had just stumbled upon a marginal fact about dimension.

Brouwer could indeed not have found the right definition of dimension, Menger claimed, since he did not know the right notion of connectedness (Menger 1928b, p. 86). This was exactly the point that Brouwer had cleared up with Urysohn, see p. 410. One can hardly assume that Brouwer had not explained the facts to Menger during his stay in Amsterdam in 1925.

Brouwer reacted almost immediately; in the paper 'On the historiography of dimension theory'[136] he discussed his own work and the relation with the work of Urysohn, Menger and Alexandrov. He left no doubt: 'I have founded the theory of dimension in my paper 'On the natural dimension'.' Furthermore he had shown the correctness of his definition for $n$-dimensional Euclidean spaces, which was a necessary and non-trivial corroboration of the definition. Here one might remark that a definition plus a correctness proof does not constitute a whole theory, but it certainly is the starting point for a theory. Brouwer was exceptional in the sense that he had no compulsion to exploit his ideas; usually he laid down the basics of some subject and left the further development to his fellow mathematicians. This was the case with dimension theory, and even with his intuitionism. In the latter case he usually pursued the fundamental problems and left the mathematical elaboration to his students Belinfante and Heyting. In a way his new topology was an exception; that he spent in that case more time pursuing the consequences of his innovations was mainly the result of the competitive behaviour of Lebesgue and Koebe.

The bone of contention with the dimension definition was the fateful 'slip of the pen'. Brouwer maintained that in 1913 he had the right notion of separation in mind, but that somehow in the proof reading stage a restrictive term had been introduced, cf. p. 408. He was convinced that any competent reader would have noticed the slip immediately. This may seem a somewhat bold claim, but one should keep in mind that for a person like Brouwer, with a miraculous topological intuition and insight, it would be natural to assume that fellow topologists shared his acuity. Finally, one may ask why Brouwer did not produce the evidence of the proof sheet of Schoenflies' *Bericht*; it would have settled the matter once and for all. There are two reasons for not doing so, in the first place Menger could simply have refused to acknowledge the evidence, in which case Brouwer would look like a schoolboy trying to mislead the teacher, in the second place Brouwer had set strict norms for what evidence in general would be acceptable. In a letter to Hahn,[137] discussing Menger's reference to his early documents, he explained

> ... with respect to scientific historiography only such documents can be considered pure, thus can be printed by an author, which either have been written and dated by, if possible uninvolved, scholars, or if they come from the author himself, have been in the custody of other, still living, if possible uninvolved, scholars, who declare that they have been uninterruptedly from the time of

[136] Brouwer (1928f).

[137] Brouwer to Hahn, 4.VIII.1929.

> their originating to the time of publication in their possession. One's own papers, that have been, if only for a short while, in one's own possession, can never, not even a little bit, be counted as a proof, as long as there is not at the same time an official report, in which on the basis of an analysis of the ink, the age of every page and every change has been established (assuming that this is possible).

It would have been highly inconsistent to relax the standards of evidence in the case of his own documents.

Freudenthal, who knew Brouwer's handwriting better than anybody else, considered the footnote correction in the Schoenflies text genuine, and there seems to be little doubt that this is the case. Brouwer was a difficult man with a gift for clever arguments, but his integrity was beyond doubt.[138]

On another point Brouwer did score: he could supply written evidence that he had checked the relevant papers of Lennes on Blumenthal's request. But even here Menger put forward a twisted argument to the effect that Brouwer had indeed known the correct notion of connectedness, but that the Lennes paper was so little known that the reader could not be assumed to be aware of it. By now the discussion had reached a level of advanced nitpicking. Even Hahn could not get the two contestants to agree. Hahn and Brouwer had come to agree that the proper thing to do for Menger was write a rejoinder to Brouwer's historiography note that Brouwer could accept, and that would have a conciliatory statement of Brouwer attached to it. Menger did write a note, Brouwer had it type set for the Amsterdam proceedings, and Menger retracted it. In the following months Hahn tried to keep the negotiations going. He had contact with both parties, and there seemed to be a solution in sight. Menger, in the meantime, started his investigations into the editorial activities of Brouwer and Alexandrov with respect to Urysohn's mémoire for reasons that can only be guessed. On the main issue hardly any progress was made. Brouwer got ill, Hahn kept prodding him and when finally an agreement was in sight he made an unfortunate remark in a letter to Brouwer, 'I have exerted myself, since I got to know your role, to convince Menger that a calm, objective settlement were possible, and the only thing to be desired.' Brouwer, with his previous experience with Menger, thought this line an obvious consequence of an 'evidently preceding *accusation of Menger* that a calm and objective discussion with him were impossible'. This led to a hardly flattering outburst on his side, describing the behaviour of Menger in Holland.

Menger, when pressed by Hahn to produce the necessary evidence of claims made in the preceding note, resorted to a lame 'that has slipped from my memory'. Altogether Brouwer was not convinced that Menger would produce a balanced historical presentation of the facts involving Brouwer's slip of the pen, the right notion of 'connected', .... Under these circumstances he saw no possibility to deliver his conciliatory epilogue. In spite of Hahn's frantic diplomatic activity no agreement

---

[138]Moreover, a recent preliminary analysis by the Dutch forensic institute showed no deviation among the corresponding handwritten corrections.

was reached, and Menger cut without consultation the knot by submitting his text to the *Monatshefte*. When Brouwer heard this he was outraged, and even more so when the *Monatshefte* refused to accept a rejoinder from him. Here the affair essentially ended. Brouwer protested, he wrote to the editors of the Monatshefte, considering Hahn guilty of treason. But nothing could be done. Menger's Reply could hardly be considered a balanced account of dimension history. One point irked Brouwer in particular—and no wonder—Menger openly accused Brouwer of fiddling the published version of Urysohn's mémoire. This was indeed a serious accusation, based on (inconclusive) evidence of material made available to Menger by Sierpinski. Menger had conducted the investigation of Brouwer's activity as if he were a police officer on a criminal case, and he published the indictment without the intervention of a court, in this case the editorial board. The whole affair left Brouwer with bitter feelings. The mathematical community did not react in any way to the conflict, nonetheless Brouwer suffered intensely. An attack of this sort was beyond his imagination.

In the end the damage was not very serious. Those who knew the inside story of Brouwer–Menger and dimension did recognise the significance of Menger's topological work and his contributions to dimension theory, but in the matter of priority for the notion of dimension they sided with Brouwer and Urysohn. Alexandrov remarked 'In my view Brouwer is objectively right, and if I were asked for an opinion, I can only—in accordance with my information, and my conscience—do as I have done, that is, agree with Brouwer and Urysohn, and not with Menger.' Hopf, in his reply of 3 March cautiously agreed, 'Brouwer seems to be right'. The smoothing effect of time has taken the sharp edge from the matter. In the book of Hurewicz and Wallman (1948), the standard text for dimension theory for a long time, the issue is not mentioned at all, and Brouwer is simply given credit for his definition of dimension.

One particular event took place in the middle of this dimension dispute, it was mentioned in Brouwer's letter of August 9 to Hahn on his vacation address in Bellagio. It was the nightmare of any scientist; the letter opened with a dramatic statement written in a remarkable mood of resignation:

> Four days ago my briefcase (*Brieftasche*),[139] which also contained my scientific diary, was stolen from me on the front platform of a Brussels tram, by a pickpocket, and both the police and the detectives consider the case as hopeless. Since in this diary my collective scientific thoughts and ideas of the last three years, which have largely disappeared from my memory, and of which only a few have already found a registration elsewhere, had been recorded, this event means for my scientific personality a serious personal mutilation (*Verstümmelung*), in a way that is like the 'decapitation' (elimination of the central process) for a pine tree. To my amazement, I remain so far, fairly calm under this blow of fate; I believe, however, from certain phenomena, that I have nonetheless suffered a nervous collapse, the consequences of

[139] Brouwer uses *Brieftasche*; it is more likely that he was carrying a small type of briefcase that was very common at the time, than a wallet.

> which will perhaps only later become visible, together with a disorganisation of my scientific thoughts.[140]

After the exchange of some more letters and cards, Brouwer reported that there was no progress in the matter of the stolen notebook.

> For the recovery of the lost papers I mobilise everything possible—advertisements in the newspapers, police, detectives, clairvoyants—but so far no success.[141]

The loss of this notebook may shed more light on Brouwer's unexpected withdrawal from research for a considerable period. A young man would go ahead, reconstructing and remembering his ideas, but Brouwer was approaching 50, and the mental scars of his battles were visible to the discerning eye.

---

[140] Brouwer to Hahn, 9.VIII.1929.

[141] Brouwer to Hahn, 17.VIII.1929.

# Chapter 15
# The Thirties

After the emotional controversies of the *Mathematische Annalen* and dimension theory, Brouwer carried on in his characteristic way, be it that he occupied himself more with secondary activities than research. After the breakdown of the negotiations with Hahn concerning Menger's 'Reply', Brouwer had decided to withdraw the paper 'The structure of the continuum' from the *Monatshefte.* He thought it impossible to publish a paper in a journal where the editors allowed Menger to attack him without allowing him the right of rejoinder. Even when Wirtinger, Hahn's co-editor, offered to publish the rejoinder, Brouwer despaired of the possibility of giving a good presentation: 'without an unabridged publication of a series of authentic letters and documents from Urysohn's estate we would not succeed'.[1] Ehrenhaft, who had also incurred Menger's wrath, still hoped to get Brouwer's 'Continuum' paper published before Menger's 'Reply', so that at least the reader would not be prejudiced by Menger's note, but even that proved impossible. So we find Brouwer deploring in a letter to Ehrenhaft the necessity to withdraw his paper, referring to the editors of the *Monatshefte* 'who had openly and actively supported the robbery of the person Menger'.[2] Having no quarrel with the committee that had invited him to lecture, he had no objections to the publication of the manuscript as a pamphlet.

When Brouwer finally saw that Menger's 'Reply' had appeared, he angrily blamed Hahn,[3] 'Hahn has finally thrown off his mask'. In his anger Brouwer completely overlooked Hahn's efforts for a solution to the Menger conflict, and Hahn's undisputed integrity. After a long exchange of letters between Brouwer and Ehrenhaft, the Continuum paper finally appeared at the end of 1930. This really and definitely put an end to Brouwer's involvement in the Menger affair. He had with meticulous care collected an immense file containing copies of letters, originals,

[1]Brouwer to Wirtinger, 11.I.1929.

[2]Brouwer to Ehrenhaft, 1.IV.1930.

[3]Brouwer to Ehrenhaft, 19.IV.1930.

D. van Dalen, *L.E.J. Brouwer – Topologist, Intuitionist, Philosopher*,
DOI 10.1007/978-1-4471-4616-2_15, © Springer-Verlag London 2013

drafts, proof sheets, ... The file contains no documents later than 1930; in particular Menger's note on the references in Urysohn's memoir is not mentioned.[4]

## 15.1 Freudenthal Arrives

Freudenthal, the bright student from Berlin, had come to Amsterdam in 1930 after writing his dissertation under supervision of Heinz Hopf and Bieberbach.[5] Brouwer was immediately taken with the erudite young man, who was generally well-informed about mathematics, including foundational matters, and who was equally well at home in philosophy, philology and related topics. When he arrived in Amsterdam, Brouwer more or less ordered him to live in Laren. It was a matter of principle for Brouwer that his visitors and assistants lived in his direct neighbourhood. For Freudenthal this was not attractive at all, after Berlin he longed to live in a city with all the comforts of shops, culture, etc. So he asked Brouwer permission to move to Amsterdam; Brouwer grudgingly agreed, but insisted that Freudenthal should get himself a telephone connection, for 'I must be able to reach you day and night'. So Freudenthal did acquire a telephone and... Brouwer never did call him! In Freudenthal's correspondence descriptions of academic life in Amsterdam, or Holland, are to be found which are instructive because as a keen foreign observer, he saw clearer than the local mathematicians how the curriculum and the research compared to the international centres.

The University of Amsterdam, he said, was comparable to a small German university, say, Erlangen. The mathematics professors were decent mathematicians, but, with the exception of Brouwer, not of a level that could match one of the better German universities. So much, one could say, for Brouwer's Göttingen at the river Amstel. The other professor with an international reputation, the local expert in invariant theory Weitzenböck, for example, was not in touch with modern mathematics. Freudenthal recalled that at one occasion Weitzenböck asked him to explain what Galois theory was.[6] Mannoury and De Vries were popular teachers, but certainly not innovators of the mathematics curriculum. Mannoury's fame was mostly based on his role in significs, but in the international philosophical community he was certainly not seen as one of the leaders of the period. Freudenthal, anyway, was not impressed by the signific aims and achievements. Brouwer gave a variety of courses, among other things a regular course in mechanics (mathematical physics),—in fact a

[4]Part of the file was taken to Moscow after Brouwer's death by Alexandrov, and the remains that survived two fires and several transfers is in the Brouwer Archive.

[5]For more on Freudenthal's Berlin years, see Freudenthal (1987a).

[6]Freudenthal may have been a bit prejudiced; Rutherford mentions another experience: 'In the winter of 1929 Professor Weitzenböck pointed out to me that there was no complete account of the theory of modular invariants embodying the work of Dickson, Glenn and Hazlett. ... The substance of Part II is largely taken from a course of lectures entitled "Algebraïsche theorie der lichamen" which Professor Weitzenböck delivered in Amsterdam University during the session 1929–30.' Rutherford (1932).

rather old-fashioned course. Brouwer was good at traditional mechanics, he enjoyed the topic, but his active knowledge of modern physics was limited, he had not kept up with the developments. In general Brouwer appreciated physics and physicists, he would not have approved of a mathematics curriculum without physics.

Freudenthal did not find life at the university demanding; Brouwer left him free to pursue his research, although he was originally taken on to assist Brouwer in his intuitionistic program. He did, as a matter of fact, occupy himself with intuitionistic mathematics, gave alternative proofs of a number of Brouwer's results, and in 1936 published a beautiful paper on intuitionistic topology. Brouwer had been flattered that a young man from a German elite university was interested in intuitionism, and had asked sensible mathematical questions—not the routine silly ones.

When Freudenthal took up his post in Amsterdam, Withold Hurewicz had already been Brouwer's assistant for four years. Hurewicz, who had come to Amsterdam in the wake of Menger, had stayed on when Menger returned to Vienna. In the year following the publication of Menger's *Dimension Theory*, Hurewicz felt so oppressed by the atmosphere of hostility that he made up his mind to leave Amsterdam and look for a job elsewhere. This was, however, easier said than done; moreover with Menger safely in Vienna, life gradually returned to normal. The co-operation between Freudenthal and Hurewicz was exemplary; between the two of them they developed the fundamentals of homotopy theory and they definitely contributed towards the pre-war fame of Amsterdam as a research centre for topology. Hurewicz was an unusual man, the son of a rich industrialist in Lodz (Poland). In Amsterdam he made himself useful in many respects. Like Freudenthal he refereed manuscripts for Brouwer, supervised students, taught various courses, and so on. His lectures were polished and very clear. Brouwer assigned various tasks to his assistants, e.g. they conducted seminars for advanced students, refereed papers for Brouwer's new journal, presented new results in private seminars, etc.[7]

When Freudenthal had settled in his new job, he wrote two extensive reports on his new surroundings; one to Heinz Hopf, and one to the organisation for student grants in Dahlem. The letter to Hopf is of interest, as it represents the first impressions of a young mathematician in his first job. We can do no better than quote part of his letter,[8]

> At the moment I am almost exclusively occupied with dimension theory; and, as a matter of fact, as a part of my duties (*dienstlich*), I should indeed help to 'arm' for the 'Menger war'. Brouwer is planning all kinds of new battles against Menger. When he speaks of Menger, he makes an almost pathological impression. I am not as belligerently minded as Brouwer, but I can understand if somebody, who has known Urysohn personally, even if his name is not Brouwer, gets into such moods when faced with the intrigues of Menger; for having now occupied myself intensively with the papers of Urysohn and

[7]According to Freudenthal, Hurewicz presented Gödel's incompleteness theorem in a seminar. He also refereed foundational papers (including Heyting's big logic papers).

[8]Freudenthal to Hopf, 22.XII.1930.

**Fig. 15.1** Hans Freudenthal (1935). [Courtesy Mirjam Freudenthal]

> Menger, I have gained the opinion that Urysohn was far superior to Menger. Urysohn makes in all his papers the impression of a great genius, whereas Menger appears only as a talent (though a very gifted one). But whether a continuation of the conflict serves the case, is very doubtful. But with these arguments one cannot come to Brouwer!

We have seen that no new actions were undertaken by Brouwer, but the above passage illustrates how much Brouwer was obsessed by Menger's tactless, to say the least, operations.

The letter to the *Deutsche Studentwerke* contained more information on academic life; for a German the Dutch universities had some surprising aspects. Freudenthal's observations may have been a bit one-sided, but there was certainly some truth in them. Students, he said, heard lectures for some 6 to 8 hours a week, and it was no exception that a full professor lectured 2 to 4 hours a week. Nonetheless the students complained about an overloaded program. But perhaps this was exactly the consequence of the low teaching load, for it could force students to learn subjects from books. Freudenthal was not terribly impressed by the Dutch student:

> The level of the Dutch student is lower than that of the German one; that has to do with the somewhat passive national character. The Dutch student certainly works diligently, but is, however, very slow in grasping things. One hardly sees original productive work of him.

Freudenthal, with his Berlin past, saw that there was some room for an improved curriculum. He indeed was quite active in stimulating the students, and in modernising the program.

Freudenthal found to his surprise that topology was not a standard part of the mathematics program at the birthplace of modern topology; when he and Hurewicz—at Brouwer's suggestion—wanted to start a seminar on topology, it turned out that nobody knew the subject. The initiative of Alexandrov and Menger to establish a basic topology curriculum had not been followed up. And so they had to start elementary courses on topology, Hurewicz taught set theoretic topology and

Freudenthal combinatorial topology. Brouwer lectured (one hour a week) on foundations of geometry to roughly 30 students and likewise on canonical differential equations to a dozen students.[9]

## 15.2 Intuitionistic Logic

After the twenties, ironically enough, intuitionism became respectable, mainly through Heyting's formalisation, which appeared on the twentieth of February 1930 in the Proceedings of the Prussian Academy.

In 1929 the Belgian mathematician De Donder had urged Brouwer to write an exposition of intuitionistic logic.[10] The reason for this request was a series of papers on intuitionistic logic by Glivenko, Barzin, Errera, and P. Lévy, presented by De Donder to the Royal Academy of Belgium. De Donder had serious doubts about a recent paper of Barzin and Errera 'On the principle of the excluded third' (1929) and was eager to have Brouwer's comments. In June 1930 Brouwer finally got round to the request, and wrote that he was about to write an exposition, but that he had not been able to get hold of some of the relevant papers. He asked if De Donder could provide him with them.[11] De Donder answered by returning post that he would send the requested papers. And there the matter unfortunately ended, for in October Brouwer informed De Donder that 'while preparing a note on intuitionism for the Bulletin of the Royal Academy of Belgium, I was pleasantly surprised to see the publication of a note of my student Mr. Heyting which elucidates in a magisterial manner the points that I wanted to shed light upon myself. I believe that after Heyting's note little remains to be said.'[12] And so Brouwer's view never appeared in print. Indeed, Brouwer left logic entirely in the hands of Heyting. Heyting had represented the intuitionistic viewpoint at the famous Königsberg meeting, September 1930;[13] upon the receipt of Heyting's account of the meeting Brouwer wrote to him: 'Many thanks for your letter of the 16th of this month; I conclude from it that you must have been satisfied with your talk at Königsberg, and I share in this satisfaction with all my heart.'[14] In the same letter he requested Heyting to write an exposition for *Forschungen und Fortschritte*, which Heyting duly did.[15]

The logic discussion carried on by Heyting with Barzin and Errera, mentioned above, was characteristic of the confusion that intuitionism created among superficial readers. For example, Barzin and Errera were convinced that Brouwer in effect

[9]First semester of 1930/31.

[10]De Donder to Brouwer, 26.X.1929.

[11]Brouwer to De Donder, 13.VI.1930.

[12]Brouwer to De Donder, 9.X.1930.

[13]Heyting (1931b).

[14]Brouwer to Heyting, 20.IX.1930.

[15]Heyting (1931a).

introduced a 'third' into logic (namely $A \vee \neg A$ which is in their opinion neither true nor false). Heyting effectively refuted the two Belgians, and—rather exceptional for a cautious man like Heyting—indulged in a mild verbal wordplay; 'the classical logician resembles an imaginary mathematician who claims that all abstract spaces admit a metric and who reproaches those who demand a proof of that theorem for wanting to attack the liberty of science'.

In the exchange of views Barzin and Errera finally accepted that it is the interpretation of the logical connectives that makes the difference; e.g. they summed up the differences as follows, 'So we have arrived at this, the intuitionists call true what the formalists call demonstrated.' On the accepted reading of 'demonstrated' this shows a lack of understanding of Brouwer's position. Heyting, reconciliatory as ever, had stated, 'One or the other, either mathematics consists of human thoughts, or it is purely formal. Brouwer's goal is to draw all the consequences of the first alternative.'[16]

What Heyting might have added, is that Brouwer considered this the (only) right way to practise mathematics, but Heyting was in a way a better statesman than Brouwer—he knew how to sugar the pill. Barzin and Errera could thus remark: 'How meagre is this outcome, for what should have been a grand revolution.' The Barzin–Errera criticism of intuitionist logic sparked a wider discussion in which Wavre, Lévy, Glivenko, Khintchin and others took part.

This small episode illustrates the fact that there was a rather confused conception of Brouwer's enterprise. And not only minor mathematicians, but even the best did not quite grasp the full extent of the intuitionistic program. Herbrand considered intuitionism as something like finitism; he, so to speak, only took the discrete part of intuitionism into account.[17] Goldfarb in his edition of the collected works conjectures that Herbrand derived his knowledge of Brouwer's intuitionism from Hilbert, a quite plausible claim, since Herbrand was in contact with the Göttingen group and spent some time there. A similar explanation may cover the case of von Neumann, who, at one point, also identified intuitionism with some form of finitism.

## 15.3 The Sodalitas Affair

In 1927 Brouwer had invested some money in shares of a Spa in Budapest. The investment was considered almost infallible for the simple reason that the Spa belonged to the Order Sodalitas-Medicorum Oblatorum Ordinis Sancti Benedicti. The Austrian prior Robert Koch was accredited by the Cardinal Primate of Hungary with full authority with the training of doctors for the mission. The Spa The Elisabeth Salzbad, a fully fledged organisation with baths, hotel, machine room, music pavilion, etc., became the object of either clumsy or shrewd machinations. Koch got

[16]Heyting (1932). Note that the discussion actually took place in 1933. For the Barzin–Errera–Heyting discussion, see also Hesseling (2002).

[17]Cf. Herbrand (1971), pp. 273, 282 ff.

into contact with a Dutch banker, Leenes, and the two of them took the financial and real estate side in hand. It soon became apparent that under Koch's supervision the proceeds of the sale of land had vanished. When Leenes complained to the Primate of Esztergom, the latter simply denied that the Order had been approved, and therefore did not exist. After Leenes explained to him that in that case the assets and liabilities became his responsibility, the Primate calmly answered, 'then it should exist again'.

In 1930 the rumours of mismanagement, to put it mildly, reached the Dutch investors. In a number of meetings the affair was discussed and fingers were pointed to the Hungarian clergy, but also to Leenes. Brouwer took part in these meetings in such a way that his name popped up in newspaper reports. The Spa was declared bankrupt in January 1931, and a committee, which included Brouwer, founded the Dutch-Hungarian Company for the Exploitation of the Bath and Terrain. From that point on Brouwer got seriously involved in the Sodalitas affair. He repeatedly travelled to Budapest to confer with the responsible persons. From time to time it happened that a mathematics class sat waiting for professor Brouwer to give his course, only to be told that he was in Hungary. Freudenthal, in a letter to Hopf, described Brouwer's involvement in the bankruptcy of the Spa—'only after just escaping his own bankruptcy'. He observed dryly that 'Brouwer had already succeeded in getting some bishops in jail; they receive, so to speak, the punishment that Brouwer had in mind for Koebe, Hilbert, and Menger'.[18]

It is incredible how much time Brouwer wasted on the Hungarian affair; in a private note he added up his files in 1944 to a total of 67 kilo's, and he estimated the time spent on the affair at 'many thousands of hours'.[19] Students of that period reported of lectures that were not given, of classes that were waiting for Brouwer who did not show up. What brought Brouwer to this unholy activity can only be guessed, probably it was a mixture of his exaggerated sense for justice and a form of academic vanity: as one of the sharpest minds in mathematics, I should be able to clean this financial Augean stable.

Leenes and the committee did not hesitate to exchange accusations of fraud, deception, selfish spending, etc. At the meeting of October 29, 1930, another such exchange took place, but when Leenes accused the committee of forgery, the chairman, Brouwer, ordered Leenes to leave the meeting. When he refused to comply, Brouwer summoned the police to carry out the order; 'two plainclothes policemen came up, whereupon Leenes left the room, while the audience cheered'. The press considered the proceedings interesting enough to mention the event.

A financial weekly flashed the headline

**PROF. DOCTOR L.E.J. BROUWER NON-ACADEMICUS.**

Not all are chairmen that wear gowns. The most learned man as a dictator and as power-mad.

---

[18]Freudenthal to Hopf, 2.IX.1931. The reference to bishops in jail is probably a bit of irreverent gossip, the notes in the archive do not mention such a thing.

[19]He visited Budapest at least 37 times between 1930 and 1939!

Brouwer remained frantically active, but the situation had worsened to the point that he was perpetually in fear of his own impending bankruptcy. Indeed his financial situation was precarious for a long time to come. One of the principal actors, Leenes, was convicted and spent time in jail. The Sodalitas affair dragged on for years; finally in 1938 a settlement was reached, but only in 1940 could bondholders collect their money.[20]

The affair cast a dark shadow over Brouwer's life for more than a decade, his finances were in a desperate state, and his health suffered severely.

Brouwer would not have been Brouwer if he had not found an opportunity to mix some pleasure with the dreary Sodalitas matters; he befriended a beautiful Hungarian lady. One day, to the great surprise of the Van Anrooy family, he showed up at their vacation cottage in Switzerland with this new friend.

All the time that the Sodalitas affair claimed his attention, Brouwer had to carry on his normal duties. As we have seen, the university suffered most. Nonetheless we find Brouwer's name regularly in the minutes of the faculty meetings. In 1930 he initiated an action for improved salaries for faculty members and for a form of sabbatical. Needless to say that in those days of financial hardship, the proposals were rejected. In the next meeting Brouwer came back with a proposal to introduce a salary differentiation via a new function of 'chairman of a department'.[21] When that too failed, he suggested that the matter should be taken to the senate. The minutes recorded that 'Brouwer asks if we, like the curators, should agree that a raise of the assistants' salary is more urgent than that of a professor's income. We have here a nice occasion to point out the insufficient character of the present salaries.'[22] In another meeting a proposal of Brouwer is mentioned to add 1000 guilders to the faculty budget for guest speakers,—turned down by the curators. The proposal of a sabbatical every ten years was, however, 'considered favourably'.

1931 saw the publication of an intuitionistic topology paper of Brouwer: 'On free enclosures in space', it took eight years before the next mathematical paper appeared!

Freudenthal was settling in, and on Brouwer's proposal (faculty meeting of 18.III.1931) he was admitted as a *privaat docent*; on May 28 of the same year he gave his inaugural address.[23]

The University of Amsterdam celebrated in 1931 its fiftieth birthday, with a rich variety of ceremonies—a reception in the Rijksmuseum, another one in the Royal Palace, a concert by the Concertgebouw Orchestra, conducted by the famous Mengelberg, .... In the same year Korteweg was congratulated by the Senate of the University on his golden doctorate (he was in 1881 the first doctor of the University!)

The mathematics section of the faculty in Amsterdam had already lobbied for some time at the Board of the University for a separate Mathematical Institute, so

[20]However, the last correspondence concerning Sodalitas was dated 1951, and the last remittance took place in 1971!

[21]Departementshoofd.

[22]Assistants earned at that time 1000 guilders. The proposed raise was 200 guilders.

[23]The so-called 'openbare les', i.e. 'public lecture'.

far lectures had been held in the centre of Amsterdam in the Oudemanhuispoort, where the mathematicians had a reading room and a cloak room. It was felt that, following the developments abroad, a building was required to function smoothly. The Board informed Brouwer and Mannoury in 1932 that their demand could not possibly be met. A serious set back, but no reason for Brouwer to despair. He kept actively reminding the Curators of the desperate situation and finally his efforts were crowned with success, but not after many a disappointment.

When it was the faculty's turn to propose an honorary doctorate, the choice was between Levi-Civita, Féjer and Sierpinski. The latter was, thanks to Brouwer's active lobbying, awarded the doctorate in the formal session of 25 April 1932.

At the same ceremony Brouwer's uncle Poutsma, the English teacher, was awarded an honorary doctorate. Although Brouwer loved to promote the interest of members of the family, it is not likely that he influenced that particular choice.

Over the years the minor (and not so minor) duties of Brouwer started to accumulate, for example, some time during the thirties Brouwer had become a commissioner of the *North-Holland Publishing Company*. Generally speaking, the time of great achievements seemed over, Brouwer's activities present a picture of a rather mixed bag of duties, jobs and hobbies of the common professor. One should, of course, keep in mind that the Sodalitas affair was always in the background during the thirties, and that it emotionally and financially drained him. The thirties also brought him the recognition as the formidable scholar he was. Hopf, as one of the organisers of the international congress of mathematicians in Zürich, invited him as a main speaker in the philosophy section. Brouwer seems to have wavered between accepting the invitation and staying away; Freudenthal reported to Hopf that Brouwer was rather inaccessible (i.e. often absent), but that he had said not to go to Zürich—for 'he was afraid that he would lay one on Menger' (*Menger eine runterzuhauen*), but that he would send Heyting. Probably Hopf eventually talked Brouwer into visiting the conference, but just before the conference he cancelled his attendance, 'serious personal circumstances prevent me to give a talk'. Curiously enough, he did after all attend the conference,—the proceedings mention him as the chairman of section III (topology).[24] From Budapest he sent a card to Hopf, thanking him for the hospitality during the conference, 'although I am perfectly aware that I attended somewhat as an outsider'.

In Holland a certain routine had established itself for Brouwer; his student Belinfante produced research papers on intuitionistic complex analysis and Heyting, after his participation at the Königsberg meeting, was making a name for himself as a logician. Indeed, the Springer Verlag engaged Gödel and Heyting to write a book in the *Ergebnisse* series. As it turned out, Gödel never got round to write his part of the projected book, so eventually Heyting produced by himself the monograph *Foundational Research. Intuitionism and Formalism* in 1934.[25] The book gives an impartial survey of the foundational activities in mathematics, and to this day it is

[24] Speakers: Alexandrov, Borsuk, Hurewicz, Kaufmann, Knaster.

[25] *Mathematische Grundlagenforschung. Intuitionismus, Beweistheorie.*

well worth reading. One novelty was his presentation of the so-called *proof interpretation*. This interpretation went back to Brouwer's ideas on logic, based on the notion of 'construction'. Heyting had already been aware of this notion at the time of his prize essay in 1927.[26] In a way it stands to reason that he had to possess some kind of test for intuitionistic validity, after all, he went through Russell's logical axioms, weeding out the non-intuitionistic ones.

There is little doubt that the proof interpretation goes back to Brouwer. In his dissertation of 1907, Brouwer discussed logic and its role in mathematics. Indeed, he paid a good deal of attention to the problem of 'hypothetical argument', and there indications for the 'construction'-aspect of implication can be found. In Brouwer's correspondence with his Ph.D. advisor, Korteweg, the role of construction is even more explicit.[27] It does not require too much imagination to trace the construction-meaning of implication in Brouwer's famous 'jump from ends to means', as I have argued elsewhere.[28]

As Brouwer was parsimonious with elucidations in his writings, one has to see how Brouwer actually handled implication.[29] In daily practice Brouwer must have mentioned the 'construction' aspect of implication freely. There is, as a matter of fact, a set of lecture notes of his course on Intuitionistic Order in 1925 which mentions explicitly that certain axioms in the form of an implication should be read in the *intuitionistic sense*, i.e. that there should be a means of construction (*Konstruktionsmittel*) yielding the conclusion from the premises.[30] Heyting's contribution consisted of two parts—he found an axiomatisation of intuitionistic logic, and he gave the proof interpretation its elegant and correct formulation.

The mathematical landscape in Holland was nominally determined by the professors at the various universities, including the *Technische Hogeschool* (Institute for Technology) in Delft. In the late twenties and the thirties chairs in mathematics were occupied by J.A. Barrau, L.E.J. Brouwer, J.G. van der Corput, Hk. de Vries, J. Wolff, J. de Vries, J.F. Koksma, W. van der Woude, R. Weitzenböck, G. Mannoury, J.C. Kluyver, J.A. Schouten, and B.L. van der Waerden. Furthermore there were a number of lecturers, e.g. D. Kloosterman, J. Droste, H.B.A. Bockwinkel, A. Heyting, and numerous *privaat docents*. But the mathematical community was by no means exhausted by the above list; there were capable researchers, such as H.J.E. Beth, D. van Dantzig, E.J. Dijksterhuis, H. Freudenthal, W. Hurewicz, F.R. van Kampen, H. Looman, J. Ridder, and G. Schaake. Some of these names will appear in later pages. For the history of mathematics that concerns Brouwer, we turn now to an ingenious and original young man, David van Dantzig.[31] David was

---

[26] Heyting, oral communication.

[27] Brouwer to Korteweg, 23.I.1907.

[28] See e.g. van Dalen (2004).

[29] See for example Brouwer (1923d, 1925d).

[30] The student who took the notes was David van Dantzig, who was to contribute to the foundations himself much later.

[31] Most of the historical information on Van Dantzig is taken from Alberts' biography, Alberts (2000).

born in 1900 in a Jewish family. His father owned a modest factory of a chemical-pharmaceutical nature. Van Dantzig attended the HBS,[32] and enrolled at the age of seventeen in the university to study chemistry. Financial problems forced him to give up his study prematurely. Several years later he studied mathematics in the teachers' curriculum, following in Mannoury's and Korteweg's footsteps. Having obtained the desired teacher's diplomas, he entered the university again, this time to study mathematics. He was almost immediately captivated by Brouwer's topology and intuitionism, and by Mannoury's significs.

It was topology that was to make his name; he wanted to write a dissertation under Brouwer's supervision, but something went wrong, there must have been some friction of a personal nature. Brouwer, who was not eager at all to accept Ph.D. students, told Van Dantzig point blank, that 'he had no idea of mathematics, and that Van Dantzig had taken up more than enough of his and Mannoury's time with all his questions'.[33] The cause of this sharp rejection is not known. One may guess that Van Dantzig, for whom tact was not a natural asset, had in some way ruffled Brouwer's feathers, and Brouwer had after all those years little patience to foster human relations that were not his choice. Van Dantzig drew the right conclusion, and after consulting Mannoury he turned to his friend Van der Waerden, who had been appointed in Groningen in 1928. The result was more than satisfactory, he wrote a beautiful dissertation on topological algebra;[34] the term 'topological algebra', by the way, was Van Dantzig's. It is no exaggeration to consider Van Dantzig as the creator of the subject, there had already been publications on topological groups, but the general treatment of topological rings and fields was largely lacking. The actual dissertation was a very condensed version of Van Dantzig's work, the full exposition appeared subsequently in the *Mathematische Annalen* and in *Compositio Mathematica*. Van der Waerden had in the mean time accepted a call to Leipzig, where he would spend a large part of his career

Brouwer seems to have acted with a certain lack of consistency. He had turned down Van Dantzig, and yet felt led down the garden path when Van Dantzig found a Ph.D. advisor elsewhere. The fact that Van Dantzig's choice fell on Van der Waerden, and not on the advisor preferred by all students who could not make up their mind, Hk. de Vries, or on Weitzenböck as a fair second must have contributed to his discontent. Later, when Schouten was looking for a candidate for a chair in mathematics in Delft, there was a rumour that Brouwer claimed part of the results and methods of Van Dantzig's dissertation. Van der Waerden, when asked, vigorously denied the rumour. Topologists will of course recognise Brouwerian traits in Van Dantzig's topological work, but that should not come as a surprise. It simply was a fact that Brouwer's investigations had set the tone for research until the fifties.

In 1932 Van Dantzig was appointed as a lecturer in Delft. This brief episode shows a pattern that will be recognised in later relationships. Brouwer had an

---

[32]See p. 4.

[33]Van der Waerden to Schouten, 22.IV.1932.

[34]van Dantzig (1931).

enormous charisma, students naturally fell under his spell. But since he jealously guarded his privacy, conflicts between the opposing forces of attraction and protection were not unusual. Van Dantzig acknowledged Brouwer's genius and insight, without sacrificing his right to be critical. Brouwer maintained collegial relations with Van Dantzig, but the two never became close friends.

In the thirties the older generation of significists had mostly withdrawn; Van Dantzig, inspired by Mannoury, provided new blood. He published and lectured on the topic. But even an outstanding mathematician like him could not get the subject going. He was definitely interested in the foundations of mathematics; he presented, for example, Gödel's incompleteness results in the meeting of September 26, 1931 of the Dutch Mathematical Society, *Wiskundig Genootschap*. He did, however, not take part in the development of the formalisation of intuitionistic logic, that area remained exclusively Heyting's domain in Holland. Van Dantzig only published in that area after the Second World War.

Van der Waerden's departure from Groningen was a loss for Dutch mathematics. At that time, however, Groningen was often considered as a temporary sojourn, many a leading scientist moved from Groningen on to greener pastures. As a matter of fact there were forces that tried to keep Van der Waerden in Holland. It was in particular Paul Ehrenfest who made an effort to get Van der Waerden appointed in Leiden. Ehrenfest had a certain personal interest, as abstract methods, for example of group theory, were becoming prominent in theoretical physics. On 8 October 1928 he wrote to Van der Waerden that he would appreciate his assistance, as 'the group pest has broken out in our physics journals'. Two years later he was already negotiating with Van der Waerden, 'how nicely and soundly the problem of the mathematics appointments here in Leiden has developed'. He was aware that Leiden could not compete with Göttingen, 'The idea that in the fall you will start to work here, and that Leiden will develop into one of the centres of mathematics has been so much fixed in my head..., that I would be totally discouraged if you were snapped away in the last moment.'[35] How serious that option was appears from the fact that Hilbert had, at Ehrenfest's request, written a recommendation for Van der Waerden.[36]

Although Brouwer was no longer active in topology, he was still held in reverence by those who knew the notorious pitfalls of the subject. He was consulted, received reprints, etc. Alexandrov, who had personally experienced the influence of the old master, sent him a copy of his lovely monograph *The simplest fundamentals of topology*,[37] an elegant presentation of the basics of combinatorial topology. The booklet, originally intended as an appendix to Hilbert–Cohn–Vossen, *Anschauliche Geometrie*, was the outcome of lectures of Alexandrov in Göttingen. For a long time it was on the desk of every self-respecting mathematician, many learned their modern topology from it. Alexandrov carried out an admirable balancing act by keeping

[35]Ehrenfest to Van der Waerden, 6.II.1930.

[36]Courant to Ehrenfest, 9.II.1930. Courant informed Ehrenfest that Van der Waerden was number 3 on the list for Hilbert's succession.

[37]*Einfachste Grundbegriffe der Topologie*, Alexandrov (1932).

**Fig. 15.2** David van Dantzig. [CWI archive]

on the right side of both Brouwer and Hilbert. The latter wrote an appreciative introduction, and Brouwer was lavishly praised for his epochal shaping of modern topology:

'Facing these two wings[38] the monumental building of Brouwer's topology is rising up. [...] In modern topological researches there is hardly any question of substantial nature, from which no thread runs to Brouwer's papers, for which not already a—often completely ready for use—tool could be found in Brouwer's stock of topological methods. [...] With Brouwer the period of flourishing development of topology begins [...] that has led to the present 'Golden Age' of topology. ...'[39]

The gift reached Brouwer from Ascona, the centre of artists, social and spiritual guru's, health freaks, millionaires and their usual corona. It says something about Alexandrov that he found time, and had the courage to mix with that crowd. Brouwer subtly showed his appreciation for Alexandrov's gift and for his choice of location, 'Thank you for your beautiful booklet and for your card from Ascona, the Southern branch of Laren.' He added sadly: 'I am still contemplating in astonishment the process of disintegration of my life, which takes place with admirable universality and thoroughness, and I am curious if yet another season will germinate.'[40] The card underscores the general state of mind of Brouwer in the middle of the Sodalitas muddle, which would not have left the strongest untouched. Moreover, a sharp observer like Brouwer could not be misled about the stagnation in his scientific career.

Lize, for whom her husband had no secrets, viewed Brouwer's depression with deep concern. In particular at the height of the Sodalitas affair, when Brouwer had, often on short notice, to visit Budapest, Brouwer despaired of his scientific future. Would the well of his miraculous mathematical mastership ever flow again? In one

[38] Combinatorial and set theoretic topology.

[39] Alexandroff (1932), p. 26.

[40] Brouwer to Alexandrov, 20.X.1932.

of her letters to Bertus in Budapest she tried to put his mind at ease, 'In particular don't be sad about mathematics. I know that you can do it all like before, if only tranquility returns to your mind. And even if you could not do it, you have already done enough.'[41]

In the middle of all the worries and disappointments of the *Grundlagenstreit* period, marred by the legal and financial problems of the Sodalitas affair, there was one event that put balm on the scars of the recent past: it had pleased Her Majesty Queen Wilhelmina to confer the knighthood of the *Nederlandse Leeuw* (the Lion of the Netherlands) upon Brouwer. The knighthood was conferred on 27 May 1932, at the palace 'Het Loo' in Apeldoorn (the favourite palace of William and Mary). This in recognition of his outstanding contributions to science. It is not unthinkable that, when his name occurred on the honours list, Brouwer cast his mind back to his student days. As a student he had wondered if he would end up in the 'coarse mansions of society', and 'light its chandeliers and grace its doorposts' (cf. p. 80). The knighthood, pleasing as it might be—and there is no doubt that Brouwer cherished the honour—undeniably confirmed that the establishment had asserted its rights.[42]

In spite of the distractions of his financial and other projects, Brouwer carried on his intuitionistic project, albeit at a modest scale. He lectured on the subject and taught from time to time a class on intuitionistic mathematics. His activity in this direction could best be termed 'polishing'.

In 1932 Brouwer had, probably at the invitation of Van der Corput, given a semester course on intuitionism in Groningen. This course was attended by a good sized audience, and the professors Van der Corput and Schaake had taken it upon them to produce a set of notes. Van der Corput proposed to have the notes typed and published in semi-official form, he also invited Brouwer to give another series of lectures. The proceeds from the sales of the notes, he suggested, would go to the faculty student association. Brouwer answered that he was perfectly willing to oblige but that his present financial state was so desperate that—after consulting informed colleagues—he thought that it would not be unreasonable if two-thirds of the proceeds would go to him.

> Whereas I deplore that my state of emergency forces me to adopt this commercial attitude, you must on the other hand take into consideration that in the absence of this state of affairs I would not easily decide to give lectures away from my own post (nor to travel during the summer for two months as an examiner), as all this is at the expense of the continuation of my scientific researches, for which the vital conditions have been so strongly decimated since the theft, a couple of years ago, of my scientific documents.[43]

[41] Lize to Brouwer, 7.IX.1932.

[42] It should be pointed out that a knighthood in the Netherlands carried no side benefits; there is no title attached to it and it does not lend social or legal status to a person. It is no more and no less than a sign of royal appreciation.

[43] Brouwer to Van der Corput, 23.XI.1933. Cf. p. 600.

In 1935 the Groningen lecture notes were mentioned once more, but nothing came of the publication.[44]

The general acceptance of intuitionism as a part of the scientific world was a fact in Holland by 1930. This did not mean that mathematicians were contemplating a switch of allegiance to the stern Brouwer, but they came to consider it as just another school in mathematics, and there was a certain amount of quiet pride in the fact that a fellow countryman had apparently found recognition in the international world. As for the philosophers, it seems highly doubtful in how far they understood Brouwer's philosophy.

In the circle of Dutch philosophers, Brouwer's intuitionism was not a topic that reached the agenda. There is a notable exception; a philosopher from the Free University, Cornelis van Vollenhoven, wrote a dissertation *The philosophy of mathematics from a theistic point of view*[45] in which intuitionism plays an important role. The Free University was (and still is) a protestant university based in Amsterdam. Theistic philosophy was introduced by Dooyeweerd.[46] Van Vollenhoven found much in Brouwer's intuitionistic mathematics that fitted his philosophical views. Curiously enough, nothing is known about contacts between Brouwer and Van Vollenhoven.[47]

The younger men, mostly on the borderline of mathematics and philosophy, did take up the ideas and issues from intuitionism. In particular E.W. Beth, H. Freudenthal and A. Heyting played a role in the study and development of modern intuitionism.

Brouwer himself kept a rather low profile. He gave his lectures in Groningen and featured as a speaker in a series of lectures at the University of Amsterdam under the heading 'The mode of expression of science'. On 12 December 1932 he delivered an abbreviated version of his Vienna Lectures. The title of the talk was *Will, Knowledge, Speech*, a somewhat lame translation of the pithy Dutch title *Willen, Weten, Spreken*. This title has the dynamic feature of the activity rather than the act, it refers to the willing, knowing and speaking rather than the frozen concepts. We may drawn attention, in passing, that Brouwer had a penchant for three part titles—*Leven, Kunst en Mystiek*; *Wiskunde, Waarheid, Werkelijkheid*; *Mathematik, Wissenschaft und Sprache*; *Consciousness, Philosophy and Mathematics*.[48]

The series was intended as an exposition of the epistemic character of the sciences, practised at the university. Apart from Brouwer, the physicist J. Clay (a former follower of Bolland, see Sect. 2.4), the theologian A.H. de Hartog, G. Mannoury, the philosopher H.J. Pos, the psychologist G. Révész, the economist J. Tinbergen, and the physicist J. Van der Waals Jr. took part.

---

[44]Brouwer to Van der Corput, 5.X.1935.

[45]Vollenhove (1918).

[46]Herman Dooyeweerd (1884–1977) was a law professor at the Calvinist *Vrije Universiteit*. His fame rests on his philosophy, as presented in *De wijsbegeerte van de wetsidee* (1935–36).

[47]For a discussion of the role of intuitionism in Van Vollenhoven's thinking, see Blauwendraat (2004).

[48]Brouwer (1905, 1919b, 1929a, 1933a, 1949c).

Révész' participation was probably the result of Brouwer's influence. He was a Hungarian scholar, who fled from Hungary in 1920 under the regime of admiral Horty. He settled in Amsterdam, where he became a *privaat docent* in 1923. Brouwer must have met him in the twenties, a time that universities in Holland were still of a modest size, with lots of social contacts. He apparently was sufficiently impressed by the Hungarian psychologist to find it worthwhile to find him a permanent position. In 1932 the opportunity offered itself when the physics and mathematics faculty showed interest in a chair for philosophy. One can imagine that Brouwer, with his philosophical reputation, took the matter in hand and actively promoted Révész' case. This turned out to be a non-trivial matter. There were objections concerning Révész' teaching capacity. His inaugural lecture as a privaat docent had made such a poor impression, that after all those years it was still brought up as an argument against the appointment. Moreover, two serious candidates showed up: J. Clay and H.J. Pos.

At this point Brouwer decided to assert his influence. At the faculty meeting he explained that philosophy and psychology were of great importance for his faculty, 'for example for the insight in the notions of space and time, for which the psychology of the tactile sense, the sense of hearing, etc. is necessary; where for example developments in mathematics yield indeed in psychology a clarifying insight'. And Révész, he said, was a prominent figure. Here he was confronted with a web of rumours. Eventually the faculty chairman had to bring the dark rumours from Budapest up in the faculty meeting. Brouwer remained firm. It was just a matter of slander, he said. A personal enemy of Révész, professor Hekker, had misled the literary faculty. The Ministry of Justice had no negative information about Révész.

> During the white terror, which succeeded the communist period, 30% of the professors in Budapest was brought before an investigation committee. Révész also had to appear before the committee; he demanded that the minutes of the interrogation should be sent with his signature to the minister. When this demand was turned down, Révész left Budapest and went abroad.

There were more questions to be parried. Were there not certain criminal acts of which Révész was accused, asked the chairman. Brouwer, who had done his homework, replied that there were three accusations: machinations with the jewels of a Hungarian family, misappropriation of a grand piano, and unlawful moving of books. In all three cases Révész acted correctly. Few candidates for a chair will have been subjected to this kind of opposition. Justice (or Brouwer) triumphed however, and Révész got his chair. The choice turned out to be fortuitous; Révész was extremely active and he founded the Amsterdam School of psychology.

Brouwer and Révész became close friends, and it should be mentioned that during the Sodalitas affair, which was about the same time as the appointment of Révész, it was very convenient to be able to fall back on the advice of a born Hungarian. Révész, who greatly admired Brouwer, dedicated his book *The creative-personal aspect and the collective* to 'My friend L.E.J. Brouwer'.[49]

---

[49]Revész (1933).

In 1932 Brouwer's old friend Frederik van Eeden passed away. His last years had not brought the contented tranquillity that old age traditionally seems to promise, instead it visited Van Eeden with an increasing loss of mental power. He was in a way fully aware of the phenomena, which is not surprising with his trained background. Mannoury recounted an incident during a walk with Van Eeden—suddenly Van Eeden stopped and said to Mannoury, 'I know it is not true, but there is a snake here on the street and we must be careful not to tread on it.'[50] Van Eeden passed away on June 16, while Brouwer was out of the country. The widow wrote sadly: 'Yes, Bertus, I could guess that you were travelling, for I must honestly confess, that this silence astonished me and even hurt me.' Two years earlier Van Eeden had turned seventy, and this was celebrated with some ceremony. In particular a committee of intellectuals and artists from all parts of society presented a Liber Amicorum to the grand old man. Brouwer had joined the committee, but he was not one of the contributors. This may perhaps be seen as a subtle distinction between his relation with Adama van Scheltema and Van Eeden. The committee had its fair share of celebrities, e.g. Van Anrooy, Berlage, Sigmund Freud, Van Ginneken, Gutkind, Mengelberg, Henriette Roland Holst, Romain Roland, Rabindranath Tagore, and Stefan Zweig. The list of contributors also contained names of the flower of Dutch (and European) culture. The significists were represented by Mannoury, Giltay, Van Ginneken, and Henri Borel.

## 15.4 Göttingen Under the Nazi's

The year 1933 brought Hitler and his NSDAP to power in Germany.[51] The consequences belong to the darkest parts of modern western civilisation. The Third Reich was to bring terror, death, destruction, and suffering to Europe on a hitherto unknown scale. The symptoms soon became painfully visible in Germany. Almost immediately Hitler started to carry out his designs to rid Germany of Jews and left wing opponents. The civil service was one of the first targets, and the universities were soon under siege both from outside and inside. On April 7 the Law for the Restoration of the Civil Service[52] was passed. This law was the lesson drawn by Hitler from the Weimar Republic, which suffered continually from the lack of loyalty if not worse of the civil service. The law with its Arian-paragraph was the first systematic legal step in eliminating Jews from public life. Under the law non-arians were no longer eligible for civil service, but also political unreliability was a ground for dismissal. Anybody who could not be trusted to 'defend at any moment the national state without reservation' could be dismissed as well, and finally there was

[50] Oral communication, Mrs. C. Vuijsje.

[51] See Remmert (2004a, 2004b).

[52] *Gesetz zur Wiederherstellung des Berufsbeamtentums*. Cf. Schappacher and Kneser (1990), Craig (1981), p. 579.

the overall clause that for the purpose of simplification of management individuals could be pensioned off.

With such a law the government virtually had a free hand. The old president Hindenburg registered a protest, and as a compromise non-arians who had been in the civil service before 1914, or who were veterans (*alte Frontkämpfer*) of the war, were exempted from the dismissal rule. The authorities could, however, force them into early retirement. In fact, the latter often functioned as a waiting room for later actions. All universities fell victim to the 'cleaning up'-action of the new government, the '*Gleichschaltung*' as it was called euphemistically. For the losses in mathematics one may consult the reports of Pinl and Furtmüller.[53] Göttingen, for a long time the home away from home for Brouwer, was hit particularly hard. There had always been a large proportion of Jewish mathematicians, and politically the climate had been fairly liberal. Strictly speaking the Göttingen mathematics department should have had a reasonable portion of exempted professors: Felix Bernstein and Edmund Landau were pre-war appointees, and Richard Courant was a war veteran. Hermann Weyl and Gustav Herglotz were not of Jewish descent. Emmy Noether was at the time an extra-ordinary professor without a regular appointment, so the new law did not apply to her.

Nonetheless the authorities hastened to suspend Bernstein and Courant; they were informed by telegram of their immediate suspension. Emmy Noether was also suspended per telegram. As Schappacher put it, 'it was the only time that she was treated by the department of education in accordance with her scientific status'.[54]

Göttingen also lost Paul Bernays, Hertz, Hans Lewy, Neugebauer, Prague and Hermann Weyl. From the younger generation (assistants) Busemann, Fenchel, Heilbronn, F. John, K. Mahler, Steinhaus and Warschawski were dismissed.

Hermann Weyl could have stayed on, but he chose to give up his chair. Having a Jewish wife, he felt that under the new regime the circumstances were becoming unbearable. He emigrated to the USA, where he was appointed at the newly founded Institute for advanced study in Princeton. Bernstein was at the time of the political tragedy in the USA; when it became clear that he chose not to return, he was discharged. Courant, who was the managing genius of the Göttingen department, remained in Göttingen, hoping for a favourable development in the new academic policy. In the end he was disappointed, he retired voluntarily and emigrated in the summer of 1935 to the United States.[55]

The man who put the new regime, and in particular the Nazi-students, to the test was Landau. He decided to resume his teaching in the winter term of 1933. On the second of November he had taken the brave step to start his lectures, but at the

---

[53] Pinl and Furtmüller (1973).

[54] Schappacher and Kneser (1990), p. 27. Schappacher pointed out that the authorities might have been in a hurry to handle the Göttingen mathematics department, in order to forestall possible student actions against the institute or individual mathematicians. The institute was viewed as a 'bastion of Marxism'. For background information on the political landscape and the ensuring developments, see also the above mentioned exposition.

[55] See Reid (1986), Chap. 15.

institute he was met by a guard of SA members who denied him entrance. In a way it was a piece of luck that he got away without bodily harm. One of the students in charge of the boycott of Landau was the brilliant Oswald Teichmüller, a gifted algebraist.[56] Landau subsequently handed in his resignation and retired. He died in 1938, and was thus spared the fate of many of his colleagues.

Europe watched the events in Germany with horror, although in some quarters the violence was viewed with fascination. Whereas the terror of the storm troopers, and the mediaeval laws against Jews and political opponents shocked the population at large, there was here and there a certain sympathy for the authoritarian leader, who dared to take cruel and drastic steps in order to dispel real or imagined evils. Ever since the communist doctrine that you cannot make an omelette without breaking eggs, the limits of 'necessary violence' had been probed. And indeed, in most countries there were minor ultra-right or even fascist political parties, that usually made a lot of noise, but were not very successful in attracting the voter. Germany with its economic crisis, inflation and Versailles syndrome was an exceptionally receptive nation for state terrorism. But the brief transition period from Weimar Republic to Third Reich offered enough to convince right-minded citizens that Germany was reverting to a barbaric stage. The boycott of Jewish business, the banning of Jews from civil service, the institution of concentration camps, book burning, Hitler's termination of German membership of the league of nations the Reichstag-fire—all signs that spelled a collapse of civilisation.

In Holland, where in spite of various political and economic crises, war and civil strife had been unknown since the Napoleonic wars (with the minor intermezzo of the campaign against rebellious Belgium (1830)), the German events were watched with incredulity and disapproval. There were a few fascist movements, the largest of which, the *Nationaal Socialistische Beweging* (NSB, National Socialist Movement), could count on a certain sympathy in middle class circles, it was represented with varying success in parliament. That party was, however, by no means as radical and brutal as its German sister party. In fact it was in its early stage not even anti-semitic.

Dutch scientists were as a rule prepared to support their German colleagues, and once the refugees started coming, many passed through Holland or stayed there. There are no records of Brouwer's activity, as far as national politics is concerned.

The only immediate trace of these events found in Brouwer's files concerns a German female mathematician, Marie Torhorst, who had a Ph.D. in topology. Her sister, who had left Germany, had begged Mannoury to assist Marie to get permission to visit her. Mannoury, who had not shrunk from belabouring Stalin by telegram, had practised the same technique on Hitler; he cautiously concluded that therefore he was not the right person to sponsor Marie's request because 'she could thus get in trouble, as I have sent a kind telegram to Hitler, that possibly was not

[56]Teichmüller would have become one of the top mathematicians of his generation, had he survived the war. He volunteered in 1939 for military service and fell in 1943 at the eastern front. For mathematical and historical information, see Schappacher and Scholz (1992).

taken kindly'.[57] Brouwer indeed invited Marie and in July he had the pleasure of receiving her in Blaricum.

Dutch universities also acted on behalf of the German refugees; in the faculty meeting of July 7 a circular letter concerning Jewish students and scientists was tabled.

## 15.5 Bieberbach's Conversion

The Third Reich created deep rifts in the German mathematical community. On the one hand there were the anti's, the Jews and the political victims, on the other hand there were the pro's—idealists and opportunists alike. In as far as there was any discussion after 1933, it was a distorted one—freedom of the press and freedom of speech were things of the past. In Hitler's conception there was no place for dissenting opinions. '*Gleichschaltung*'[58] was carried by him and his followers to its ultimate form. The Nazi doctrine of '*Ein Volk, Ein Führer*' did not tolerate alternative political views, let alone criticism. Some mathematicians had been eagerly awaiting the hour of 'Germany's waking up', although, generally speaking, political extremism was not prominent in mathematical circles. In fact, quite a number of mathematicians had leftist views and some had even been active during the short-lived Weimar republic, Courant and Bernstein were examples. On the right wing of German politics there had been a loose confederation of '*Deutschnationalen*', not to be confused with national socialists. After the Nazi take-over these small groups were quickly and efficiently abolished by Hitler, some just disappeared, some became part of the omnipresent Party. There are two mathematicians who became the paradigm for Nazi-scientists: Theodor Vahlen and Ludwig Bieberbach.[59] Vahlen's name has already come up, he was cited by Brouwer in his dissertation, and Bieberbach was the brilliant young analyst who recognised Brouwer's topological ideas at an early stage, and who joined Brouwer in his fight against the boycott of German scientists. Since both play a role in our history, let us have a look at their political past.

Theodor Vahlen was already an established mathematician at the time that Brouwer entered the profession. He had acquired a reputation in the area of Hilbert's Foundations of Geometry through his monograph *Abstract Geometry*.[60] He initially felt drawn towards the then fashionable and innovative axiomatic theory and also

[57] Mannoury to Brouwer, 17.VI.1933.

[58] The National Socialist regime carried the political use of language to unknown heights; such terms as 'Gleichschalten' had a normal everyday meaning, but under the regime it acquired a new one: 'following the Nazidoctrine', or even cruder, 'eliminating opposition and deviant ideas and practice'.

[59] Vahlen's life and career is discussed in Siegmund-Schultze (1984). For Bieberbach, see Mehrtens (1987).

[60] Abstrakte Geometrie, Vahlen (1905b).

to set theory. Max Dehn, in a review, butchered Vahlen's *Abstract Geometry* as inexact and conceptually weak.[61] Although the book had obvious deficiencies, the review was in the eyes of some colleagues a bit too sharp; there were rumours that the Göttingen group used the occasion to prove its own superiority and to chastise an outsider. So much is certain, that it did not endear the Göttingers to Vahlen, something they would come to regret later. After his brief career in pure mathematics Vahlen turned towards applied mathematics. He worked in traditional areas and stuck to traditional methods. His publications and studies cover descriptive geometry, geodesics, technical mechanics, ballistics and the like. In 1911 he published a book *Constructions and approximations*,[62] which was well received.

During the First World War he saw active service as commander of an artillery battery. His book *Ballistics*,[63] which probably was inspired by his wartime experiences and which was dedicated to the memory of the fallen comrades in arms, was certainly not an up to date exposition, modern results and methods were often overlooked. At that time Vahlen had already made up his mind to enter politics. After Hitler's putsch he decided to throw his lot with the prisoner at Landsberg, and he became a prominent Nazi. One of his early functions was that of *Gauleiter* (chief of the district) of Pommern. As pro-rector of the University of Greifswald he demonstrated his rejection of the Weimar Republic by bringing down the flag of the republic in 1924 on the day of the Constitution. Vahlen was immediately suspended, and—quite exceptionally for a rather reactionary judiciary—eventually convicted.

After Hitler's triumph in 1933, Vahlen quickly rose in the ranks and became chief of the academic section of the Prussian department of culture. In this capacity he had a considerable influence on the nazification of the universities.

Where Vahlen was a middle of the road mathematician, respectable, but not original or deep, Bieberbach was a totally different case. He was a versatile mathematician with a lively presentation. His research interests were mainly in the area of function theory, and they bordered on many other areas. His career was fairly traditional, he studied in Heidelberg and Göttingen. In Göttingen he studied with Koebe and Klein. Koebe was his mentor in function theory and Klein influenced his view on mathematics—he became a lifelong adept of the geometrical spirit. His early research dealt with automorphic functions, in 1910 he wrote a dissertation on that topic, and in 1912 he was one of the invited speakers at the automorphic function symposium in Karlsruhe (cf. Sect. 5.3). His habilitation was on the topic of Hilbert's 18th problem, it dealt with groups of Euclidean motions. In 1910 he became a *Privat Dozent* in Zürich; the next year he moved to Königsberg.

After a brief spell as a professor in Basel he got a chair in Frankfurt, where Schoenflies was in charge of the new mathematics department. Bieberbach's star was rising fast. After Carathéodory left for Smyrna, Brouwer was offered his chair in Berlin. The offer was turned down, and subsequently also by the next candidate,

[61] Dehn (1905), Vahlen (1905a).

[62] Vahlen (1911).

[63] Vahlen (1922, 1942).

Weyl. In the following search Bieberbach emerged in the end as the most promising candidate. From then on he remained in Berlin, where he was considered a stimulating teacher, albeit a bit sloppy. His books, on a large number of topics, were popular, and went through many editions. In the mathematical community he played an active role, since 1920 he was a secretary of the DMV and editor of the *Jahresbericht*. Furthermore, as we have seen, he was one of the active members of the board of the *Mathematische Annalen*. He was generally appreciated by his students and colleagues, his tendency towards mild vanity was accepted with a forgiving smile. Einstein wrote on one occasion to Max Born's wife, 'Bieberbach's love and devotion for himself and for his muse is quite priceless. May God keep it with him, this is the best way to live.'[64] Before the advent of the third Reich Bieberbach was somewhat ambiguous about politics. On the one hand he thought '*deutschnational*', as appears from his resistance against the *Conseil* and the *Union mathématique* (cf. Chap. 9), on the other hand he freely mingled with leftists and communists. His house was open to guests of all political persuasions, and he had a reputation among the students for unorthodox open-mindedness.[65]

Almost overnight Bieberbach changed into a fervent adherent of National Socialism. An account and analysis of Bieberbach's conversion can be found in Mehrten's influential paper 'Ludwig Bieberbach and Deutsche Mathematik'.[66] Reading the story of Bieberbach's life and convictions, one cannot but conclude that Bieberbach's political extremism was not dictated by an iron fate, but rather the result of a mixture of personal preferences and ill-founded political views. Be that as it may, it remains a fact that Bieberbach joined the Nazi-movement lock-stock-and-barrel.

Now the interesting thing about Bieberbach is that he was not an opportunist who sought more power or comfort. After all, a professor of good standing in Berlin was already quite high on the social ladder. He sought to harmonise his political views with his professional expertise. In fact, Bieberbach had developed views about the nature of mathematics which he eventually tried to bring into line with the Nazi ideology. It is not easy to be categorical about Bieberbach's alleged 'lust for power'. There is no doubt that vanity came to him naturally, but in a stable society his brand would be viewed with a smile. He did hold a number of prominent positions in the mathematical community, and there were no significant complaints about his functioning. His personal and scientific history after 1933 seems to be more the result of unconditional acceptance of the Nazi doctrines, than of clever schemes to accumulate power. In how far he was driven by political correctness, or by a desire for personal advantages and honour, remains a matter of conjecture. Viewed from the perspective of the German citizen of that period, the convolutions of Bieberbach fell under the traditional 'not running away from one's responsibilities', unfortunately that phrase is, more often than not, a cover-up.

Early in his career Bieberbach took a view of mathematics that was in line with the modern trend of axiomatics. He quite explicitly endorsed formalism as

[64] Einstein to Hedwig Born, Einstein and Born (1969).

[65] Communication of H. Freudenthal.

[66] Mehrtens (1987). Our presentation makes substantial use of this paper.

the proper approach to mathematics, 'that does justice to the factual state of mathematical science [...]. We do not want to create knowledge of the truth but merely methods to gain knowledge [...]. The truth in mathematics rests solely in its logical correctness and consistency.'[67] But slowly he started to shift his allegiance. A few years later he started to take a broader view and defended in his books the significance of external problems, by pointing out that mathematics could sever the ties between mathematics proper and its applications only at its own peril.

He spoke of the 'living spirit of external stimuli which created mathematics, and to which it constantly and abundantly owes new life'. If anything, Bieberbach was developing and expressing ideas along the lines of Felix Klein's socio-cultural reflections on styles, the real world, and the like. He never actively engaged in professional, technical foundational work, nor did he display an inclination towards philosophical analysis.

One can observe, however, in Bieberbach an increasingly negative attitude towards formalism. In 1924 he spoke of the 'late rampant growth of Hilbert's axiomatics in the field of physics'.[68] Still later Bieberbach felt attracted to the ideas of Pierre Boutroux' book *The mathematicians' scientific ideal*,[69] which stressed a dichotomy: 'synthetic–analytic', or 'intuitive–discursive', or 'order of invention–order of proof'. Boutroux sided with the 'synthetic', which he claimed to be the rule rather than the exception with mathematicians. He had little sympathy for the 'algebraico-logical' artistry of mathematicians, which he considered a form of 'juggling'. In a lecture, held for the *Förderungsverein*[70] in 1926, Bieberbach expressed his views on the nature of mathematics. The title, *On the mathematician's ideal of science*,[71] already betrays the grand view the speaker was to unfold. This was not a philosophical discussion of the foundations of mathematics, a pedagogical guideline for teachers, or a historical orientation, but all of these at the same time. Bieberbach saw in Felix Klein the ideal type of a mathematician, who combined physical heuristics with mathematical intuition; Klein could be viewed as a mathematician in the old tradition of Newton, Laplace, Gauss and many others who were rooted in a global scientific idea, one in which mathematics was inspired by physics, and in which physics adopted the virtues of mathematics. On Bieberbach's view, the new exactness of the nineteenth century had separated mathematics from its external inspiration. Function theory, for example, had in the hands of Weierstrass become a showpiece of pure mathematics; Klein on the other hand found his inspiration in the tradition of Riemann—a function theory with a geometrical and physical flavour. Klein's work is indeed permeated with his intuitive geometrical spirit. It is not the logic and the formalism that determines mathematics for him, but rather the wealth of structures and their interrelations. Klein's style of non-Euclidean geometry and

[67] Bieberbach's inaugural lecture, Bieberbach (1914), the translation is Mehrtens'.

[68] Bieberbach (1924).

[69] Boutroux (1920). Bieberbach read the German translation.

[70] A society of patrons and supporters of education in the exact sciences.

[71] *Vom Wissenschaftsideal der Mathematiker*, 15.II.1926.

his group theoretical view of the notion of transformation was in the first half of the twentieth century common practice in the mathematical world.

After extolling the virtues of Klein's intuitive mathematics, Bieberbach went on to contrast it with the formalistic tendencies of his day. In the wake of Weierstrass' novel counterexamples to traditional mathematical intuitions, an understandable distrust of intuitive considerations became the rule, 'Now we have the psychological foundations for the tendency, which shows itself in the most extreme form in Hilbert's axiomatics, and that wants to have nothing to do with that what is intuitive object-oriented.' And so 'What is more plausible than the radical cure, to eliminate substance and meaning from mathematics. There are only objects of thought, which mean nothing, and with which one operates according to certain rules.' This was, in Bieberbach's opinion, Hilbert's ideal of science.

The choice for Bieberbach was not difficult, mathematics should remain a meaningful subject, its objects should have substance and meaning. Moreover, Hilbert's ideal was automatically 'hostile towards the demands of the applications'. The formal, logical approach to mathematics would be disastrous for a harmonious building of mathematics, and worst of all for the teaching of mathematics.

Weierstrass had found the uncompromising Kronecker on his path, who prophesied that 'the day would come that [mathematics] would wake up from Weierstrass' analysis and Cantor's set theory, as from a horrible dream'. 'And now', Bieberbach said, 'it is intuitionism that would aspire to herald this day'. The intuitionistic ideal of science, as characterised by Bieberbach, is in a way an amalgam of Klein's and Hilbert's ideals, it takes into account the points of Klein's philosophy, and it has a sharper notion of exactness in common with the Weierstrass–Hilbert ideal. However, where the formalist needs an extra-logical domain, which 'can only have come into their possession by unintended chance', 'the intuitionist tries, on the other hand, to do justice to the circumstances that it is people who practice mathematics, and he tries to take systematically, and everywhere in the construction of its theories, the intuitive-concrete aspects into account'. In particular, the intuitionist accepts the existence of an object satisfying certain conditions, only if a construction can be given. Considering the circumstances of this lecture for a mixed audience, Bieberbach managed to give a reasonable fair presentation of the situation in the fundaments of the mathematical building. It is quite clear that he sides with the intuitionists, Brouwer and Weyl—'Thus indeed in intuitionism the fresh air of spring is blowing', he observed enthusiastically.

One would be wrong to see Bieberbach's lecture as part of the *Grundlagenstreit*. Bieberbach was rather campaigning for the normal activity of the 'working mathematician' and the teacher, for a meaningful mathematics as opposed to the axiomatic games of the formalists. It is not unlikely that he misinterpreted some of the signs on the wall. Whereas in the early days abstract algebra *à la* Emmy Noether and Van der Waerden was seen as part of the axiomatic trend, we now know that 'modern algebra' is dealing with very concrete structures from the practice of mathematics. Views of this kind tend to change in time. We dwelt expressly on Bieberbach's views and his 1926 lecture, as it shows how intuitionism could be drawn into a political debate, where Brouwer would have been very reticent.

Politically, Bieberbach had been drifting towards the right. In the Riemann affair, the Bologna affair and the *Annalen* conflict, he had showed considerable affinity with the German nationalists. This by itself did not predestine him to become a Nazi. Not all German nationalists flocked to Hitler after the fall of the Weimar republic.

Bieberbach, however, seemed to have made up his mind in favour of the National Socialist Party. In 1933 he marched with his sons in one of the notorious SA[72] processions from Potsdam to Berlin. Soon after he joined the SA, and faithfully discharged his duties as a member. He strongly supported the new national awakening, and condoned, if not praised, the actions of nazi-student organisations against the old guard. Bieberbach's political choices and his apparent faith in the new movement surprised his colleagues and students. On the whole he had been a somewhat vain, but well-liked and friendly colleague and teacher, there was no clearly discernible cause for resentment or feeling of unfair treatment. He had been slighted by Hilbert in the Bologna affair, but certainly that would not have been enough to seek revenge.

One is inclined to view Bieberbach's alliance with the Nazi's as a deliberate, argued choice for a specific political and social system, of which he expected great things on all levels. He acted as an idealist, who, once he has chosen a particular conviction, decides to close his eyes for all negative aspects and to follow the party line right or wrong. There is no doubt that Bieberbach was totally serious about his political commitment. Mehrtens paints a vivid and disturbing picture of a man who not only follows the party line, but develops new initiatives.[73]

In fact, all of Bieberbach's activities were centred around mathematics. His conceptual views of mathematics had become more pronounced and had acquired political overtones. In particular he started to see the matter of mathematical styles in a political, and especially racial, frame. He applauded, for example, the above mentioned boycott of Landau's class, 'A few months ago differences with the Göttingen student body ended the teaching activities of Herr Landau... This should be seen as a prime example of the fact that representatives of overly different races do not mix as student and teacher.'[74] In another lecture before the *Förderverein* Bieberbach launched in 1934 his classification of mathematics and mathematicians along racial types.[75] He had borrowed these types from a psychologist, Erich Jaensch, who had introduced two basic types. The 'S-type' with unstable psychic functions internally generated synaesthetic perception, and a tendency towards disintegration. The 'J-type,' on the contrary, had stable psychic functions, and strongly integrated perceptual imagery and conceptual thinking (the *integration* type). In view of his earlier statements, it is not surprising that Bieberbach's heart was with the J-type, with great Germans like Gauss, Weierstrass and Klein as representatives. It would take us too far to discuss the types in full detail, the reader may consult Mehrtens

[72] *Schütz Abteilung*, the storm troopers of the Party.

[73] Mehrtens (1987).

[74] Ibid. p. 227.

[75] Mehrtens (1987), p. 224. Bieberbach (1934).

(1980, 1987), Lindner (1980) for further details. It may suffice to mention the proliferation of subtypes, and—of course—the classification of Jewish mathematics under the S-type. The S-type, according to its creator, led to a 'cognition liberalism', which was hostile to science, 'The cognitive-liberals basically hate truth, because it means commitment and limitation'.[76] It should not come as a surprise that Bieberbach drew his conclusions for the formalism-intuitionism clash. He saw the *Grundlagenstreit* as racially determined. Note, however, that he chose a fairly safe, academic formulation: 'By itself the J-type will tend towards intuitionism or the style of Klein, whereas formalism seems to belong to the S-type.'[77]

It would be logical therefore to classify Hilbert as an S-type, but that would go too far, even for Bieberbach. On account of Hilbert's Baltic origin, and his wide ranging publications in areas other than his formalism, 'Hilbert cannot possibly be taken as an S-type.' But, he added 'a form of J-type is known which tends to be open the influences of the S-type'. Brouwer and Hilbert would be '*bien étonné de se trouver ensemble*', but Bieberbach had no problem explaining this phenomenon, 'The difference is quite compatible with the fact that both Hilbert and Brouwer should be classified under the psychological type J3/J2. The fact that two men approach their science with an ideal norm, does not necessarily imply that it has to be the same norm in both cases.'[78]

Bieberbach was not the only person to see the formalism-intuitionism debate in the light of the politics of the day. In mathematics there were a few exponents of this trend. However, even the hard-line Nazi mathematicians were not inclined at all to have their foundational views dictated by politics. Certainly, there would be an amount of lip service to the Nazi doctrines, but as a rule that would be gratuitous formulas. It would not make much difference for handling, say, differential equations. No, the intuitionism-formalism considerations flourished best in certain marginal philosophical circles. The term 'marginal' is in a way rather misleading, as the strange mixture of psychology, mathematics, intuitionism and anti-semitism went down well with the Third Reich. If the Nazi's had on the other hand made good their claim on a thousand year domination, 'marginal' might have turned out to be the wrong adjective. As it is, we can from our point in history only express our amazement at the curious activity and arguments of the Nazi foundationalists. The names that come to mind first are those of Dingler and Steck, both philosophers with a habilitation in mathematics. These scholars launched determined attacks at decadent formalists and logicists. Not even Hilbert escaped the wrath of the national socialistic inspired philosophers, albeit that the grand old man at his Olympus was out of reach for their arrows. Brouwer, who should have been fairly safe, got his share of criticism too; Steck set himself to prove that neither Hilbert, nor Brouwer, could capture the essence of mathematics.[79]

---

[76]Lindner (1980), p. 95.

[77]Mehrtens (1987), p. 228.

[78]Mehrtens (1987), p. 228.

[79]See Menzler-Trott (2001), Chap. 4, and the literature cited in that book.

Bieberbach's excursions into the field of psychology hastened his downfall in the mathematical community. His lecture of 1934 caused some stir abroad; Hardy, for example, sent a scathing 'letter to the Editor' to *Nature*.[80]

> It is not reasonable to criticise too closely the utterances, even of men of science, in times of intense political or national excitement. There are many of us, many Englishmen and many Germans, who said things during the war which we scarcely meant and are sorry to remember now. Anxiety for one's own position, dread of falling behind the rising torrent of folly, determination at all costs not to be outdone, may be natural if not particular heroic excuses. Prof. Bieberbach's reputation excludes such explanations of his utterances; and I find myself driven to the more uncharitable conclusion that he really believes them true.

The Danish mathematician Harald Bohr, who had close ties with Germany, in particular Göttingen, strongly criticised Bieberbach in a Danish newspaper. Thereupon Bieberbach made the capital error to respond to Bohr's article, by means of an 'open letter to Harald Bohr' in the *Jahresbericht*, against the wishes of his fellow editors. In doing this he had overplayed his hand. The DMV did not want to antagonise its numerous foreign membership, and Bieberbach had more or less given the impression that he represented the views of the DMV. When the matter was put on the agenda of the Bad Pyrmont meeting of the DMV (1934), Bieberbach decided to organise a coup, or at least to put pressure on the meeting by marching into the meeting at the head of a group of students in SA uniforms. The chairman, Hecke, showed considerable courage by calling his bluff; he appealed to the statutes of the DMV to excluded all non-members, including the uniformed students, from the meeting. The Nazi-Sympathisers were, however, in so far successful, that they turned the tables on the DMV. Bohr, they claimed, had insulted and attacked the German state, which Bieberbach then had sought to defend. The rest of the assembly did not stand up to this arm twisting, and thus Bieberbach's actions were only 'regretted', whereas Bohr was 'condemned' for his attack on the German state.

But when Bieberbach subsequently proposed to introduce the 'leadership principle' (*Führerprinzip*), with Tornier as leader, he was flatly voted down. Apart from a great deal of embarrassment, the matter could have ended here with a red-faced Bieberbach. The latter, however, in his function of secretary of the DMV tried to thwart the procedure. This was indeed the last straw; his colleagues demanded his resignation. When his political patrons showed no inclination to step in, Bieberbach resigned. Having lost his foothold in the German mathematical community, he created his own journal and his own entourage. The new, heavily subsidised, journal went by the name *Deutsche Mathematik*. It never acquired the status of the journals that Bieberbach had edited in the past. The journal was intended as a platform for politically correct authors and papers, it also published news items about Nazi-sponsored activities, such as summer camps. A certain amount of propaganda was

[80] Hardy (1934). Also in *Math. Intelligencer*, 6 (1984).

printed in its columns, but one would not be justified to consider every author who published in the *Deutsche Mathematik* as a politically suspect character. Paul Koebe published for example one of his expositions, *Wesen der Kontinuitätsmethode*, in the 1936 volume, probably on request of, and to please his student Bieberbach.

## 15.6 Compositio Mathematica

Bieberbach and Brouwer had been on the same side in the affair of the Riemann volume, and in the War of the Frogs and the Mice. After that there had been little contact, if any at all. Both parties went their separate ways, Bieberbach to pick up his daily routine, and Brouwer to lick his wounds. The loss of the *Mathematische Annalen* had hurt; the plan to found his own journal, mentioned to Springer,[81] perhaps for tactical reasons, had not been forgotten. Springer himself had at the time judged this a fair solution, he may have had his doubts about the feasibility, but that clearly was not his business. Indeed Brouwer cautiously inquired with the Teubner Verlag if it was interested in founding a new journal. The reaction was predictable but disappointing. After consulting Bieberbach the publisher reached the conclusion that a new journal was a highly doubtful business proposition, and that it was questionable whether enough subscribers could be found to make the journal profitable. One must keep in mind that the world, but in particular Germany, was involved in one of its worst economic crises, so any publisher would think twice before starting a new journal. The loss of the status journal, *Mathematische Annalen*, to Springer must still have rankled, for the spokesman wrote that they would not gladly run the risk of another debacle, 'after the *Annalen* had already been wrenched from our hands'.

However discouraging this might have been, Brouwer did not give up so easily; he approached the publisher Noordhoff, a firm with some experience in mathematics, as it published the journal of the Dutch Mathematical Society, *Wiskundig Genootschap*, and it brought Dutch language mathematics text books on the market. Already in 1929 Brouwer opened negotiations with Noordhoff. On 24 October Noordhoff and Brouwer met in the house of Wijdenes, the publisher's advisor in mathematical matters. The mathematician Wijdenes, also one of Korteweg's students, was an extremely successful author of mathematics textbooks for high schools; the acquaintance with Brouwer went back to his student years. The agenda for the meeting mentioned 'the Journal and further publications'. Noordhoff was no stranger to Brouwer; his publishing company had already marketed the commercial versions of Brouwer's dissertation and his inaugural lecture, and it had published a small booklet containing the unreliability paper and both inaugural lectures.[82] In a letter of 18 March 1919 to Wijdenes Noordhoff reported about Brouwer's dissertation

---

[81]Cf. p. 562.

[82]Brouwer (1909a, 1912a, 1919b). Noordhoff listed Brouwer's publications regularly in his catalogue. In 1922, 1926, 1928, 1933: De onbetrouwbaarheid der logische principes, Het Wezen der Meetkunde, Intuïtionisme en Formalisme (collected in *Wiskunde, Waarheid, Werkelijkheid*), Over

'of which I just recently have received the stock. I had a number of copies bound in linen. There is still linen left for a small number of copies, and people usually prefer a bound copy.' Noordhoff marketed the dissertation for 4.50 guilders, with 50% royalty for Brouwer. In 1929 Noordhoff tried to convince Brouwer that a new edition of the dissertation should be published. Brouwer was, however, not too keen on the idea, 'The book is now out of date and it would have to be totally revised. In principle I am prepared to do so, but first there is a lot of other work to be done by me, among other things the publication as a book of the course on intuitionism I gave in Berlin, which I hope to submit to you soon, if in the meantime the journal has been realised.'[83]

Noordhoff was sufficiently interested in the publication of an international mathematics journal to give Brouwer the go-ahead. And so the preparations started; one of Brouwer's first decisions was the name: *Compositio Mathematica.* An important detail was the choice of editors for the journal. Brouwer decided to follow the example of the old *Annalen*, a modest board of managing editors and a large board of associate editors. In June 1930 the first letters went out to sound the prospective editors, and in October the definite letters of invitation were mailed.

The first list of mathematicians invited to join, contained the names of Alexandrov, Baer, Bieberbach, Borel, Cartan, Cech, Van der Corput, De Donder, Doetsch, Eisenhart, Feigl, Fréchet, Fubini, Fujiwara, Garnier, Hadamard, Hardy, Heegaard, Heyting, Hille, H. Hopf, Julia, Khintchine, Lefschetz, Levi-Civita, Lévy, Loewy, von Mises, Montel, von Neumann, Nörlund, Ostrowski, F. Riesz, M. Riesz, Saxer, Severi, Sierpinski, Süss, Szegö, Takagi, Tonelli, Valiron, de la Vallée-Poussin, Veblen, Wavre, Weitzenböck, Whittaker, Wilson, and Wolff.[84]

The board was in fact as international as one could possibly wish, and there was a judicious mix of the older, established generation, and the younger, coming generation.

One may well assume that most of the above, if not all, were aware of the motivation for the founding of this new journal. This is illustrated by Brouwer's old friend Hadamard. Their friendship went back to 1910, when Brouwer stayed with his brother in Paris (see p. 153). Brouwer had a very high opinion of Hadamard; he was eager to get him on the board, but Hadamard did not quite know what to make of the invitation. He wrote for advice to Einstein, saying that it was tempting to join a truly international journal, but that he was somewhat uncertain if he would in this way be used as a pawn against Hilbert.[85] Einstein replied a month later that there had indeed been a fell struggle,

de Grondslagen der Wiskunde, Luchtvaart en Photogrammetrie. In 1938, 1940,1942, 1948, De onbetrouwbaarheid der logische principes, Het Wezen der Meetkunde, Intuïtionisme en Formalisme. In 1949 only *De Uitdrukkingswijze der Wetenschap* (containing Brouwer 1933b), this item appeared for the last time in the catalogue of 1958.

[83] Brouwer to Noordhoff, 10.X.1929, cf. also p. 504.

[84] From Brouwer's letter to Veblen, 11.X.1930.

[85] Hadamard to Einstein, 16.X.1930.

> for which Hilbert, in my opinion, carried most of the blame. Brouwer, however, behaved at this occasion so excessive and obstinate, that he appears to me a man of pathological irritability.[86]

He advised Hadamard to steer clear of this new journal,

> I would unconditionally wash my hands of it, in spite of all respect for the subtleness and the honest character of Brouwer, who is not aware of the abyss of his temperament.

Einstein's dark predictions concerning Brouwer's handling of the *Compositio Mathematica* turned out to be unfounded, partly because Brouwer was a conscientious scholar, who could not sin against scientific norms, partly because his new assistant, Hans Freudenthal, first under Brouwer's guidance and gradually on his own, conducted the managing of the journal. Eventually Freudenthal just submitted each complete issue to Brouwer for his fiat, often Brouwer would not even answer, but he could also, suddenly, show interest in certain papers, and spend his time lavishly on the refereeing and the supervision of the corrections. Sometimes Brouwer noticed a particular point in a paper a year later, but by then it had already been published.

When the journal was about to be launched, the publisher sent out flyers with information. Among Freudenthal's documents there is a draft of the German text of the flyer; apparently he was asked to edit the final wording. It is interesting to read Brouwer's views on the role of a scientific journal in a time when in certain quarters the primacy of politics over science was taken for gospel. As a true internationalist Brouwer was not going to give in to new trends.

> Shortly the first issue of the mathematical journal Compositio Mathematica, edited by representatives of the mathematical sciences from 48 countries, will find itself in print with the publishing house Noordhoff. It will be the task of Compositio Mathematica, not only to encourage the development of mathematics by accepting for publication valuable mathematical papers, but also to serve the international scientific co-operation, which is at present more than ever necessary. To do justice to this aim it is not sufficient to abstain from imposing any national or language-barrier; rather, as far as possible, an international composition of the editorial board is required in order to avoid any bias with respect to national aspects. In view of the nowadays often occurring specialisation of mathematicians of specific nations on specific areas of research and methods of research, such a composition offers at the same time a guarantee against any one-sidedness with respect to the mathematical character of the published papers. ...[87]

Alexandrov was confronted with a difficult problem, he fully realised that support from the Göttingen group was more valuable than the support Brouwer could

[86]Einstein to Hadamard, 15.XI.1930.

[87]This is a somewhat free translation of the German text. There is probably an English version somewhere in some archive, but I have not found any.

give. Although Brouwer had got him a Rockefeller grant, the effective influence of Brouwer was limited and as things were in the world of mathematics, the backing of Hilbert's circle, including the publisher Springer, carried infinitely more weight than Brouwer's influence. So when he was asked to join the editorial board of the Compositio, he feared a clash of interest with the followers of Hilbert (who would, he thought, not welcome a competing journal). So he declined the invitation, Freudenthal, always a good observer of the mathematical scene and usually well informed, deplored Alexandrov's urge to ingratiate himself with the Göttinger people, '... who knows if they are so sincere. From the way they treat Noether, one might conclude that they will think twice to get him something in order not to lose him for Göttingen.'[88]

Brouwer did not take kindly to Alexandrov's refusal,[89] he was doubly disappointed as he had been using his influence attempting to get Alexandrov a chair in Groningen. As we will see, Alexandrov joined the board after all.

Compositio entered the mathematical world in a very awkward period. When Brouwer composed his first list of candidates for the editorial board, the political horizon was unclouded, but by the time real commitments had to made, the political landscape in Germany was no longer the same. Many competent mathematicians had been forced into exile, or silenced. The first list Brouwer had made contained the following German mathematicians: R. Baer, L. Bieberbach, G. Doetsch, G. Feigl, H. Hopf, A. Loewy, R. von Mises, J. von Neumann, and W. Süss.

In 1933, however, when the journal was about to be launched, this list had become rather suspect in the eyes of those Germans who followed the party line. And thus in 1934, when the first issue appeared, Bieberbach had developed strong political objections against the presence of some names on the list of editors. He wrote to Brouwer that the founding of *Compositio Mathematica* (Compositio for short) must have given satisfaction for the brutal dismissal from the board of the *Annalen*.[90] He for himself had considered his membership of the editorial board—he had indeed become a managing editor—a good thing, for it made certain that the name of a man of German spirit appeared on the cover of an international journal.

> I assumed that one would recognise this as an example that the new Germany, notwithstanding its fight with the international Jewry, would gladly co-operate with other nations, that meet us, if not with sympathy, then at least with loyalty. Instead people now see often the crucial point in the fact that Jews occur on the cover of Compositio.

'And', he continued, 'this was explained as a sign of my co-operation with Jews'.[91]

The fact that the names of Jews appeared on the cover was, for Bieberbach, a demonstration that he was prepared to tolerate their presence on the board, and this

---

[88]Freudenthal to Hopf, 22.XII.1930.

[89]*Brouwer schimpft jetzt auf Alexandrov.*

[90]Bieberbach to Brouwer, 21.VI.1934.

[91]For Harald Bohr's reaction on Bieberbach's anti-internationalist position and his co-operation with the internationalist journal Compositio, see Bohr (1934), Schappacher and Kneser (1990).

was, in view of the demonstrated willingness to join the international community, in his opinion, a defect. Other nations, he assumed, would in the end recognise the necessity of the German actions. To his disappointment he had been subjected to hostile reactions from all sides. And so 'I feel obliged to make the disappearance of the Jews from the editorial board a condition for my presence in the editorial board of Compositio'. He hoped, he wrote, that the old alliances in matters of international co-operation, would make it easier for Brouwer to carry out the necessary steps. The letter ended with the barely veiled threat that the present composition of the board would cause difficulties for the distribution of Compositio in Germany.

Bieberbach's appeal does not seem to have had much effect on Brouwer, who did not want to strike candidates from the list of editors because someone had personal or political objections. And thus in the beginning of 1935 Bieberbach resigned from the editorial board. It appears that Brouwer had tried to find a face saving formula for Bieberbach. He had sent a circular to all editors with the message that, in view of the delicate character of the present international situation, 'Science seems more than ever called to form a refuge for mutual understanding.' And hence the editors were advised that 'any editor's public participation in manifestations which could harm the mutual esteem of people and nations was incompatible with his function'.[92] This message has its own irony. Would Hilbert not have said the same thing in 1925? Responsibility apparently breeds prudence.

Bieberbach remained, however, adamant. One can see that Brouwer's message offered Bieberbach, in a sense, protection against insults, but it equally condemned his attacks on Jews and emigrants. His letter repeatedly referred to the international Jewish community and its subversive war against his fatherland. 'My national feelings forbid me to belong to an editorial board, in which so many representatives of international Jewry and in particular also emigrants are found.'[93] If Bieberbach had hoped that Brouwer would back down when threatened with his resignation, he had made a miscalculation. Brouwer did not even try to keep Bieberbach aboard. He replied: 'I hardly have to say that your decision upset me *very* painfully, but on the other hand is completely respected by me, as I know that it was dictated by your conviction and your conscience.'[94] And that was the end of a long association between two persons who had shared a mathematical interest for many years, and who had fought the *Conseil* and its boycott shoulder to shoulder.

For a Nazi the large proportion of Jews and émigrés among the German editors must have been an insult. Indeed, Baer, Loewy, Von Mises and Von Neumann were objects of the racial persecution; Heinz Hopf, although not Jewish, had emigrated to Switzerland, and hence could from Bieberbach's point of view not be relied on. Baer and Von Neumann had already left Germany, and Alfred Loewy was still in Freiburg, where he died in 1935. After Bieberbach's resignation four German editors were listed: Artin, Doetsch, Feigl and Süs. Of these four, Gustav Doetsch and

[92]Cf. Remmert (1999), p. 18.

[93]Bieberbach to Brouwer, 8.I.1935.

[94]Brouwer to Bieberbach, 15.I.1935.

Wilhelm Süss were national socialists. The fact that after all German editors belonged to a board of dubious character was in Bieberbach's eyes a serious blemish on the honour of German science. A blemish that could only be removed by the resignation of the remaining Germans. He decided to put pressure on them, at first just in the form of a suggestion, 'In agreement with *Ministerialdirektor* Vahlen, I am asking you to follow my action.'[95] Brouwer, on the other hand, did not wish to lose the editors that he had selected with so much care. He asked Doetsch to fill Bieberbach's place in the board of managing editors.[96]

Doetsch clearly found it difficult to determine his position. In July 1934 he was still on Bieberbach's side, as appears from a card to Feigl,[97] in which some scathing remarks about Jewish mathematicians and reviewers are to be found. The card ends with comments on the position of the Germans in the board:

> If Bieberbach resigns from Compositio, then it would be most desirable that we remaining German editors and editors of German descent act unanimously. Only you and SüSAs are to be considered. Szegö is a Jew, isn't he? Anyway, he is married to a Jewess. I will just wait for the result of the discussion between Bieberbach and Brouwer, but I am very much inclined to join Bieberbach. Heil Hitler

On the other hand Doetsch apparently did not want to give up a prestigious editorial post, and, as Remmert points out, he may have wanted to dissociate himself from Bieberbach, who by now had lost quite a bit of prestige.[98] He tried to bypass Bieberbach by getting himself a permission to remain on the board of Compositio. At the same time he attacked Bieberbach for his inconsistent actions in the Compositio matter, and for his bungling of the Bologna affair. The Prussian ministry of education, after some deliberation, decided that German scientists could publish in Compositio, but not join the editorial board. This clinched the matter. Doetsch, Feigl and Süss resigned from the board. Probably none of them would have objected to stay on. Feigl, who was not a Nazi, might have wished to side with Brouwer, but in the young totalitarian state one could not disobey without risking serious penalties. So in 1936 the only German on the board was Artin, who by then had emigrated.

In March 1936 Brouwer wrote to Hopf and Alexandrov, thanking them for the 'marvellous book, with which you have associated my name'[99] and informing them at the same time that all the German editors had resigned from the editorial board.[100] He asked if one of them would be prepared to join the executive board. In fact both of them did.

---

[95] Bieberbach to Doetsch, 19.I.1935.

[96] Brouwer to Doetsch, 20.III.1935.

[97] Doetsch to Feigl, 16.VII.1934.

[98] See Remmert (1999). Bieberbach had failed in his coup at the meeting of the DMV, Schappacher and Kneser (1990).

[99] Alexandroff–Hopf, *Topologie*. Dedicated to Brouwer.

[100] '*... aus der Redaktion vom Compositio Mathematica sämtliche reichsdeutsche Mitglieder ausgeschieden sind, ...*'

Compositio thus certainly had its share of difficulties at the start. But once the journal was on its way, things ran smoothly. Although Brouwer was the responsible editor, most of the work was done swiftly and competently by Freudenthal. Those who had judged Brouwer incapable of running a journal properly, turned out to be wrong. All the fears that Hilbert claimed to have for the disastrous influence of his Dutch opponent were after all ill-founded. Brouwer did not do any of the things he was suspected of; he did not stop French or Belgian authors or editors, he did not turn his journal into a vehicle for intuitionistic mathematics, he did not reject Russian Jewish authors. In short, Compositio became a normal respectable journal. Intuitionistic mathematics did not play an important role; until the Second World War only six such papers appeared, written by Belinfante, Freudenthal, Heyting and Johanson.

The scientific journal landscape, in particular in Germany, was changing dramatically in 1933. The new regime did not lose time in infiltrating existing journals; whenever possible and convenient, the *Führer* principle was enforced. This meant as a rule that political motives could, and often did, overrule scientific standards. The *Mathematische Annalen* was no exception. There is a pressing letter from Blumenthal to Hilbert in November 1933, in which he painted in vivid colours the dangers that lay ahead. The worst effect of the new times was the uncertainty that surrounded the Göttingen faculty. 'If Göttingen becomes a desert, or is populated by professors who discard tradition, then we have to open up new wells, or we come to nought.' In the light of the present threats, the founding of Compositio, which seemed so harmless at the time of the *Annalen* conflict, assumed ominous proportions: 'On the other hand the Annalen are threatened by Brouwer's newly founded Compositio Mathematica, in which, in numbers, a very large staff of international associate editors (*Mitarbeiter*) is brought together. Since Bieberbach and Feigl have joined this staff, it is clear that we cannot hope for the co-operation of the Berlin school for the Annalen. It is even more worrying that also Heinz Hopf (Zürich), with whom we always have worked well, has committed himself to this competing enterprise.'[101] Blumenthal's conclusion was that the *Annalen* urgently needed an expansion of the editorial board. He suggested Van der Waerden as a perfect candidate.

The relatively weak position of the *Annalen* at this point in time was a consequence of an over-confident decision in the past: to minimise the editorial board, and to make the *Annalen* even more exclusively a Göttingen enterprise. Nobody could have surmised that a momentous decision, taken for the wrong reason, would be regretted so soon.

## 15.7 Göttingen Reconsidered?

After the wholesale elimination of staff members that did not conform to the requirements of the new government, only a few mathematicians of the old school were left

[101] Blumenthal to Hilbert, 11.XI.1933.

to keep the Göttingen department going. Herglotz took care of the institute until a new director would be appointed. Eventually Hasse and Tornier were appointed. Hasse was a number theorist of the first rank; politically he was *deutschnational* and he apparently accepted the national socialist domination as a not unwelcome fact. Schappacher characterized Hasse's position under the Nazi regime as 'Just as he liked to identify himself during both world wars with the glamour of the German Navy, his army unit, it was for him a personal concern to contribute in the thirties to a reborn strong Germany. In the process he apparently did not hesitate to contact the more traditional Prussian side of the national socialist regime, rather than the... 'revolutionary' side.'[102] Tornier was no outstanding mathematician, his main interest became probability theory. He was an active Nazi and he followed the party line in all respects. Soon Tornier and Hasse fell out, and finally, after an internal struggle, Tornier was transferred in 1936 by Vahlen to Berlin. There he lost his job and membership of the party in 1938. Tornier must have cultivated strange company; Constance Reid reported that 'he embarrassed the mathematics faculty by being pictured in the newspaper walking on a fashionable boulevard with a notorious prostitute on his arm and a tame tortoise on a leash'.[103]

Hasse and Tornier had to staff the mathematical institute and for whatever reason they must have considered the famous Dutchman Brouwer, with his record of activities for the re-admittance of German mathematicians into the international scholarly community, a good catch.[104] One may guess that Brouwer was partly selected in order to embarrass the old Göttingen school. In view of the fact that Brouwer had for all practical purposes left topology, one can rule out that the need for a topologist motivated the choice. Neither does it seem likely that Hasse envisaged the founding of an intuitionistic tradition in Göttingen. So, in all, it seems most likely that the two directors saw in Brouwer a mathematician with universally acknowledged status, who moreover had shown himself to be an ally in the cause of rehabilitation of German science. The first letter to Brouwer mentioning the vacant chair came from Tornier, but one may be certain that he wrote to Brouwer with Hasse's approval. So there must have been some discussions in May 1934, or perhaps earlier. Tornier's letter has all the characteristics of the academic upstart. He did not bother to hide his satisfaction over the successful elimination of the Jewish mathematicians.[105]

> As you well know the flooding of Germans with alien races, in particular also in the teaching staff of the mathematical institute here, has resulted in intolerable situations.
>
> I permit myself to put the question, if you, considered by me and many German mathematicians for a long time as one of the greatest scholars of

---

[102] Schappacher (1987), p. 354.

[103] Reid (1986), p 402. See also Schappacher and Kneser (1990).

[104] Brouwer was in German circles described as *deutschfreundlich*. This term acquired after 1933 a very specific meaning: pro Nazi. But before 1933 it just meant what it said: sympathetic towards Germany and Germans. It is not unusual for commentators to confuse the two meanings.

[105] Tornier to Brouwer, 19.VI.1934.

> a typical German disposition, would be prepared to help to restore the old reputation of mathematics in Göttingen.[106] [...]
>
> I beg you to be so kind to tell me if you are in principle inclined to enter into negotiations about a call to Göttingen. I may add that my joy to see you perhaps for always in Göttingen, is shared by the responsible department of the ministry, that is Herr Ministerial-Direktor Professor Dr. Vahlen.

Considering the circumstances and the political overtone, this was not an offer to be pleased with. What would a man like Brouwer, jealous of his privacy, and a poor organiser, do in the hornet's nest of a mathematics department in Nazi Germany?

One would have expected a firm but polite rejection of the offer, but Brouwer decided to play the game. All available evidence seems to point at a financial motivation. Before Tornier's letter there had already been some correspondence with Karl Kerkhof, the man in charge of the *Reichszentrale für naturwissenschaftliche Berichterstattung*. Brouwer must finally have succeeded in finding a buyer for his house in Berlin-Zehlendorf. Considering the fact that he had been trying to sell his property for years, this was a most welcome event for Brouwer, who was at that time involved in the Hungarian Spa affair, and was threatened by nothing less than a financial disaster. But now he was confronted with interminable bureaucratic rules on housing, on sales, on the transfer of currency. Handling bureaucratic problems from a distance was (and still is) not easy, so he had enlisted the help of Kerkhof; and the latter had made his enquiries at the foreign office. On June 8 he wrote Brouwer that Brouwer's 'request was granted under certain conditions'. The vagueness of this formulation was typical for the whole procedure. Whatever Brouwer tried, and whoever he begged to intervene, each time some local government organisation or a bank would promise that the matter would be solved in a few days, and each time a new blockade would appear. Brouwer was as tenacious as a bulldog. He repeatedly asked Kerkhof and Vahlen to intervene, and each time success seemed just round the corner. Brouwer found the Foreign office, the Reichsbank, the foreign currency office (*Devisenstelle*), the Department of Finance of Berlin, the Land registry, the Dedibank, and probably a few more institutions in his way. A correspondence that ran from before June 1934 until far into 1935 pictures a Kafkaesque struggle against an impervious bureaucracy. Kerkhof and Vahlen acted a number of times on Brouwer's desperate calls for help, but more than friendly promises could not be obtained. It is not known if in the end Brouwer got his money out of Germany. Dependable information on Brouwer's strategy in the matter of the Göttingen vacancy is hard to come by. It seems highly unlikely that he seriously considered a chair in Göttingen. At the heyday of Göttingen's mathematics he had declined a call, and now a professorship under Nazi-rule was not attractive at all, to put it mildly. Apart from any other motive, his attachment to his property in Blaricum, to the pharmacy, was enough to keep him in Holland. Moreover, he would rather 'build a Göttingen'

[106] *Ich erlaube mir nun die Anfrage, ob Sie, den viele deutsche Mathematiker mit mir schon lange für einen der grössten Forscher von typische germanischer Prägung halten, bereit wären, den alten Ruf der Göttinger Mathematik neu begründen zu helfen.*

in Amsterdam than in Göttingen and a call to Göttingen, even then, came in handy in negotiations with the City Fathers of Amsterdam. Taking all arguments into consideration, one must conclude that he was using the call to Göttingen as a means of putting pressure on the various government and financial institutions that played a role.

However, he had to negotiate with the Göttingen representatives in order to keep up his credibility. Any sign of a hidden agenda might have jeopardised his financial plan. From the letters that have been preserved, the picture of a feet-dragging Brouwer appears. He wrote to both Tornier and Hasse (the former soon disappeared from the scene) in such a manner that one would be inclined to discern a sincere interest in the Göttingen chair. He postponed his visit to Göttingen, however, as long as possible. On July 27 he wrote to Tornier, telling him how despondent he felt for not having come to Göttingen so far. Tornier was elated, he felt it a great honour that the famous Dutchman would probably within a week come over and visit him, although no such promise was made by Brouwer, who just mentioned the possibility. At this point Hasse stepped in, he clearly did not want to leave Brouwer in the hands of a Nazi-adventurer. On August 11 Brouwer informed Tornier that he had been bitten by a dog, and that medical complications had arisen that meant he would be confined to his bed for some weeks. The following months Brouwer claimed that he could not leave because of threats of financial disaster.

The fact that the appointment of Brouwer in Göttingen was seriously considered did not escape the old guard. It understandably caused much anxiety among those who had known Göttingen before 1933. Harald Bohr reported Veblen about the situation in Göttingen in vivid terms.[107]

> Tornier's next idea is to get Brouwer to Göttingen. Schmidt told us that Brouwer was extremely pleased about the idea and was very much inclined to come; in this moment he discusses the financial terms with the Nazi-government. All who have had the privilege of coming in closer contact with Hilbert feel the idea of bringing Brouwer to Göttingen perhaps the dirtiest trick, especially as those people do not really believe in Brouwer's ideas of a new foundation of mathematics, but simply wish to liquidate completely the époque of Hilbert and his school.

Bohr's letter makes it quite clear that the appointment of Brouwer was seen as the ultimate insult. Apparently Bohr and his fellow Göttingers took Brouwer's words seriously, forgetting that usually in such a situation information loses part of its accuracy in its transfer from one person to the next. Bohr overlooked the fact that anybody who wants to use an offered chair for some specific benefit is likely to show some interest. Without it the leverage would not amount to much. Far away in Princeton, Hermann Weyl had also heard of the offer to Brouwer; in a report to the Rockefeller Foundation one reads '... the authorities attempted to obtain Nevanlinna [...] and the Dutchman L.E.J. Brouwer. W. considers that B. has definitely

[107] Bohr to Veblen, 11.VIII.1934, cf. Segal (1986).

passed his period of maximum productivity. The negotiation collapsed however, because B., as a good Dutchman, insisted that his salary be paid in gold.'[108] In the absence of further confirmation, we will probably have to take Weyl's statement as a tongue in cheek comment on Brouwer's preoccupation with money.

Hasse, who apparently sincerely wanted Brouwer to join the Göttingen department, was running out of time, he dearly wanted Brouwer's advice on the matter of the further vacancies in Göttingen, and Brouwer replied that he would be happy to give his views on the various candidates, but that this was more a matter for a personal discussion, so it had to wait for his visit to Göttingen, which—he said—could now take place any minute. In the end he sent a written advice. In October Hasse again prodded Brouwer into action; he desperately needed Brouwer's information on the course that he was supposed to give. From Hasse's letter (12.X.1934) it appears that in the meantime Brouwer had been in Göttingen, and promised to give a course on intuitionistic mathematics. Subsequent correspondence shows that Brouwer had managed to remove the full time appointment from the agenda, and to promise to come to Göttingen, for the time being, as a visiting professor. But even that did not work out. His attempts to get a transfer of his money (from the sale of the house) did not yield any results, and apparently even Vahlen no longer reacted to his entreaties. The last letter in the archive is resigned in tone; on February 5, 1935 Brouwer wrote to Hasse that he could not accept the visiting position as his financial situation made a prolonged absence impossible. His letter to Hasse ended with the sad words, 'Hopefully the expected rescue action from Berlin will come in time to prevent the definitive destruction of my spiritual activity. A scientific corpse will be of no use to Göttingen.'

The persecution of Jews and persons of mixed marriages could not escape anybody who read his newspapers, but for many outside the immediate sphere of influence of the Nazi's it remained an abstract injustice in a central European country. It becomes a different matter, however, if one is confronted with an individual case, with a friend or relation who has become the victim of the regime. This happened to Brouwer when he learned in 1934 that Schoenflies' son-in-law Kaemmel was threatened with dismissal. Brouwer immediately set himself to support Kaemmel by collecting testimonials of Schoenflies' importance. In a letter to Veblen[109] he wrote 'There seems to be a chance to save him, if the German Government could be convinced that Professor Schoenflies has been at the time not only a functionary of loyal national feelings, but also a scientist of importance, having played his part in the development of German science.' Veblen almost immediately complied, adding that 'I very well remember the world-wide impression that was made by the publication of Schoenflies' famous *Bericht* on point-set theory. This, as you well know, was for a long period the chief handbook of all those who worked in this field, and there is no doubt that it contributes largely to the success of German science in that direction.'

[108] Siegmund-Schultze (2001), p. 191.

[109] Brouwer to Veblen, 20.X.1934.

It is unknown if Brouwer's action made any difference. The history of Nazi-period makes us fear the worst.

## 15.8 Dutch Affairs

1935 saw the beginning of another parallel episode in Brouwer's life: local politics. In that year Brouwer was elected in the council of the town Blaricum. He had joined '*The Neutral Party*', which stood for local interests. Political or social issues played almost no role in the party program. For a beginner in politics Brouwer did very well, his party (no. 2) obtained 235 votes, of which 117 were for Brouwer. There were altogether 1372 votes, and the Neutral Party came out number 3. On 3 September Brouwer was installed as a member of the council, and right away made temporary acting alderman under the chairman of the council, mayor J.J. Klaarenbeek.

Given the multitude of projects and functions of Brouwer, he was surprisingly faithful; he attended all council meetings in 1935 and most of the meetings in later years. Looking at the minutes of the meetings—and the reports in the local newspaper *De Bel* one cannot say that Brouwer had a striking record. He did what he was supposed to do: look after local interests. So we see him pleading for the preservation of the local heath (a nature reserve), advocating better access for walkers while keeping out cars. He proposed a raise for the local police (*veldwachters*), added improvements to a plan for the construction of a cycle path (he had done his homework and quoted a report of the ANWB[110]), and pleaded for the improvement of the soccer field,—'sport has to do with national health'. Apart from the routine management of the town, incidentally a matter of principle cropped up. When, for example, the committee of action 'For God' demanded the barring of the periodical of the freethinkers from the public library, Brouwer remarked that according to the state's instructions all groups must be represented. Moreover he deemed it incorrect to act as a censor in the absence of dependable information. At one point he came into head-on collision with the mayor when defending the right of the Committee of Advice and Support to meet without the mayor and aldermen being present. The mayor was terribly upset and blamed Brouwer for displaying an unfriendly attitude in supporting the motion. The minutes record that: 'Mr. Brouwer did not wish to enter into the degree of friendliness which should be displayed by the council members. In his opinion a council member has reached the age at which he no longer needs to be educated in friendliness!' Brouwer won. For an individualist like Brouwer party lines were not the ultimate political wisdom. He was, for example, fairly close to alderman De Klerk, a shoemaker. Although they did not belong to the same party, De Klerk often sent his son with drafted proposals to Brouwer, who then went over the text and returned the documents typed and all.[111]

---

[110] Sister society of the AA and AAA.

[111] Interview, H.P. de Klerk, junior.

Fig. 15.3 Brouwer as seen by a local artist

The elections of 1939 returned Brouwer to the council with the highest number of votes,—310 out of 1601, while his party had become the second largest on the list. To run ahead of our tale, Brouwer remained councillor until the German occupation authorities disbanded all city, town, and village councils[112] (1941). After the war he did not return to politics.

Brouwer's council activities have been the subject of some speculation, 'why should a man like Brouwer stoop to enter town politics?' Evil tongues have suggested that there was something in it for Brouwer,—after all he owned large tracts of land in and around Blaricum. There is nothing on or off the record that supports these views. Probably the simplest explanation is the true one: Brouwer cared for his town and its people, so when the local politicians appealed to him, what was more natural than allow himself to be put on the list of candidates for the council? In view of Brouwer's strong ecological views (cf. *Life, Art and Mysticism*) it is not surprising that he opted for a local party, without the power structure of the established parties, without the directives from above. There is no doubt that Brouwer was, in spite of his international orientation, thoroughly part of the town. Many of his friends were from village middle class circles, one of his close friends was the shoemaker Bus (not the above mentioned alderman), with whom he could spend hours, talking about everything under the sun. When Brouwer died the shoemaker was inconsolable, 'my friend Brouwer is dead'. Brouwer was sincerely concerned with the families of the town people. At one occasion, when he heard that a child was stricken with a disease and bed-bound, he went up to the house with a book under his arm, stepped in and sat down at the child's bedside, reading aloud for an hour.

Brouwer did not feel too high and mighty to take an interest in the lives and problems of his fellow villagers. On one occasion, one of them, a local shopkeeper, by the name of Oversteegen, confided to Brouwer that he found it difficult to assess the potential of his bright son; what should the boy study? Brouwer invited the

---

[112] *gemeenteraden.*

boy over to his house and gave him a couple of private tutorials on mathematics, and concluded that no great mathematician would be lost to the world if he did not enrol in the mathematical-physical faculty. So the boy entered the university to study languages and literature, and in due time became a well-respected professor.[113]

A fair number of friends and (even) relations lived in Laren-Blaricum. His own brother Lex lived ten minutes from Brouwer's house, one of the Poutsma relatives lived across the street, and some colleagues lived in the neighbourhood, e.g. Weitzenböck. The latter had become one of the townsmen, he shared in their pastimes and was for some time the president of the chess club of Blaricum. He was a fervent chess player, so maybe it was no wonder that Max Euwe, the one-time world champion, was his Ph.D. student. Without actually going public, or trying to influence his students, he, a former Austrian military man, fostered virulent revanchist feelings. For some reason—probably the same reason that was put forward by the majority of right wing German-Austrian citizens: the treaty of Versailles—he had become violently anti-French and, at a later stage, Nazi. There is a curious manifestation of his Francophobia: in his well-known book *Invarianten Theorie*[114] the first letters of the sentences of the Introduction spelt '*Nieder mit den Franzosen*' (down with the French). Weitzenböck's book, by the way, became something like the bible for students in Amsterdam, his courses were well attended, and he had a substantial number of Ph.D. students. His students and the villagers remembered that before the Second World War he displayed no political activity at all. During the war he openly showed his political sympathies.

Brouwer got another taste of the new political masters in Germany in 1937. The story was told by Elias Balke, a German artist, who kept a diary from 1933 onwards, and published fragments in his '*Chroniknotizen*'.[115] Brouwer is mentioned in this diary. It appears that Brouwer owned some land (and probably a cottage or a cabin) in north Prussia, in the neighbourhood of Fischerkaten (now in Poland).

> In the spring of 1937 Prof. Luitzen Brouwer came one evening in the Forest house, to ask me how I saw the political developments and whether he should accept a call to Göttingen. I advised against it. Brouwer was a striking visitor. Extensively informed, travelled, world-wise. A bronzed head with a bold profile, a tropical skin like parchment, eyes that see in the distance. He had property at Java, where I just came from. As the founder of the intuitionistic school he was the publisher of an international journal for mathematics. We got into a marvellous conversation. In a tense, pointed dialogue that gave pleasure to each partner. We had common friends in Java and Germany. In the meantime we had also heard a great deal from each other, as he had, at my request, immediately given shelter to Helen,[116] when I had to bring her in 1935 to safety for the Gestapo. He was—to my joy—strongly impressed by

[113]Private communication, J.J. Oversteegen.

[114]Weitzenböck (1923).

[115]Balke (1973), p. 101.

[116]Helen Ernst, an artist.

> this talented young woman, and we knew to praise her. At the break of dawn Brouwer left. The bottles were empty, the sun was rising, the blackbirds sang and one could hear the surf. When he was already at the bottom of the stairs, he once again looked up to me and took his leave in such a charming way that this little scene remained especially dear to me.

Helen Ernst was a gifted young artist who had been imprisoned by the Gestapo in 1933, was released with great difficulty, and when she was about to be arrested again, Balke took her to Dantzig where he eventually put her aboard a Dutch ship and when she arrived in Amsterdam, Brouwer took her to Laren. She found work at the *Nieuwe Kunstschool* in Amsterdam. In 1940 she was arrested and taken to the concentration camp in Ravensbrück. Helen survived the war, and died of tuberculosis in 1948. The story of Helen has been made the subject of Hans Hübner's biography.[117] Hübner mentions 1937 as the year that Balke and Brouwer met, but that was probably taken from Balke's book. Brouwer had offered Helen one of the little cottages on the grounds,[118] before she moved to Amsterdam.

The dates mentioned in the story are rather puzzling. In 1937 Brouwer did not get a call to Göttingen. Brouwer may have told Balke about his call to Göttingen in 1934, and Balke may have mixed up the years. The fact that Helen Ernst is mentioned adds evidence for 1937 as the date of meeting. Brouwer had become an almost compulsive talker, he enjoyed the attention of an audience, and when he got an opportunity, he could easily draw from his large stock of stories, both from his own experience, and from others. So he may have told about his brother's time in Java, who worked there as a geologist, and Balke may have got the impression that he was talking about himself. The real surprise is to find Brouwer in an east Prussian forest at the Baltic sea. Would it not be natural for him to sit at home and indulge in self-pity? The urge to move around was probably stronger than his feelings of despair over financial matters. Brouwer's habit to buy real estate had probably at some time resulted in the acquisition of a piece of forest in East Prussia, and, while seeking solitude in this quiet corner of Germany, he had met Balke and his wife.[119]

Life at the Brouwer household went on along the established lines. Lize usually spent her days in Amsterdam, supervising the pharmacy, and Brouwer stayed in Blaricum or Laren, as the case might be. After the first influx of topologists had left to return to their respective home universities, the company at Brouwer's place consisted mainly of personal friends and relations. He had a particular fondness for female company and, generally speaking, the feeling was mutual. A small but devoted group of ladies visited Brouwer regularly and basked in the sun of the wit and the wisdom of their idol. In Brouwer's house tea was of great importance. There was always an ample choice of plain and exotic tea. He would, when he had visitors,

[117] Hübner (2002).

[118] *eine Hütte "mit bohème-artiger Verpflegung"*.

[119] There are some post 1945 documents in the archive that refer to land property in Poland. No details are given.

**Fig. 15.4** Irmgard Gawehn, Brouwer, Dolly Kiehl, Willem Langhout, Tine Langhout-Vermey [Brouwer archive]

pay special attention to the selection and mix of the various brands. He would look out of the window, and say, 'It is grim day, let us have Lapsang Suchong, and...' The tea ceremony was an elaborate affair with cookies and sweets. But the main attraction was the conversation.

Although the times had changed, Brouwer still kept up his practice of healthy living, including swimming, open-air 'bathing', dieting, .... The son of his friend Ru Mauve reported that he once was present at a discussion in the garden between Brouwer and Cor Jongejan, Irmgard Gawehn and Tine Langhout. The topic of the discussion was 'being afraid in the dark', a theme that appealed to the lively imagination of the ladies, and on which Brouwer could easily keep the company spellbound, being gifted with a perfect memory of important and less important bits of lore, and with a vivid imagination. Indeed, the ladies were moved to tears. Suddenly Brouwer disappeared into the house, a little later he reappeared stark naked, picking up the discussion where it was left. No comment was made, no suppressed smiles; the ladies were completely wrapped up in the conversation.

The years between the end of the *Grundlagenstreit* and the Second World War are almost devoid of creative mathematics in Brouwer's life. A great deal of time was taken up by all kinds of non-mathematical activities, e.g. the Hungarian investment adventure and the town council in Blaricum. Among the more mathematically oriented activities, the founding and organisation of the new journal *Compositio Mathematica* was far out the most prominent. Brouwer, after his initial enthusiasm, soon withdrew from the editorial tasks that he had so conscientiously carried out for the *Mathematische Annalen*. He did handle a number of papers himself, and corresponded with the referees and authors. He refereed for example the notes of Heyting and Freudenthal on intuitionistic logic and the meaning of implication. The result was a succinct approval:

> Report on the discussion Freudenthal–Heyting. Interesting discussion on the meaning of the implication of a theorem by another, when nothing is known about the correctness of the latter.

Both papers were duly published in Compositio.[120] Unfortunately that is all Brouwer said about the discussion, it would have been of historical value to know his position in the matter. Freudenthal defended an interpretation of the implication that came close to Brouwer's original views of 1907. Heyting took the modern view, known now as the 'proof interpretation'. As Brouwer's statements in his dissertation are far from precise, and leave room for various exegeses, some comments would have been most welcome. On the whole, Brouwer seems to have sided with Heyting's interpretation, but there are no explicit statements to the effect. The undesirable effects of Brouwer's overly strict view of 1907 are more likely the result of a lack of precision than of an ultra constructivistic position.[121]

Freudenthal handled the editorial matters so diligently and efficiently, that one might wonder why he had not been made an editor. Indeed, he was promised a place in the board, but it never came to anything before the war. It should be pointed out that prestigious journals insisted on prestigious editors, and the fact that Freudenthal was already making his name in mathematics did not compensate the fact that he was not a professor. Brouwer attached a good deal of value to these formal matters.[122] Indispensable as Freudenthal was, his rise in the academic levels in Amsterdam was by no means exceptionally smooth. He arrived as an assistant, and remained in that position until 1937. In 1931 he was admitted as a '*privaat docent*',[123] but that did little to boost his income. The salary was modest, and under the pressure of the economic crisis it was even reduced. In the middle of the summer vacation of 1935 Brouwer wrote him a letter as director of the Mathematical Institute, informing him that the board of the University had decided to reduce all salaries retroactively from January 1! He sadly added: 'And this will probably not be the last reduction, now that our country is slowly economically drying up, where arts and sciences are always abandoned first to whither.'[124]

In the history of Brouwer's life there are some controversies that derived from, intentional or unintentional, attempts to cheat or belittle him. We have seen the conflicts with Lebesgue, Koebe, Menger and Hilbert. These conflicts did have a ground, although it is true, generally speaking, that offence is not only given, but must also

---

[120] Heyting (1936a), Freudenthal (1936).

[121] Cf. van Dalen (2004), Kuiper (2004).

[122] The fact that Heyting was an associate editor may be explained by the distribution of the specialisms. There was already ample topological expertise in the board, but Heyting was the only foundationalist.

[123] Inaugural address, 28.V.1931.

[124] Brouwer to Freudenthal, 20.VIII.1935.

be taken. Among all the conflicts there is one, however, born out of ordinary friction, which is harder to explain on the basis of hard facts: the conflict between Brouwer and Freudenthal. It started, innocuously enough, in 1936.

In that year Hurewicz left Amsterdam and a vacancy occurred. The assistants, lately, had been topologists, and apart from Irmgard Gawehn, of good standing. Since there was no prominent candidate with a record in the foundations of mathematics in sight, it seemed not far fetched that the successor of Hurewicz should have good antecedents in the heartland of mathematics, and one might even conjecture that a topologist of some sort would be a plausible choice.

However, Brouwer made a curious choice, he offered the position to Miss Geldof, a student of nondescript merit. Freudenthal was not informed about the matter at all, although he was by far the first interested party. Just by chance the porter-caretaker of the Mathematical Institute, Koppers, told Freudenthal that Hurewicz would have a female successor.

The choice, as I said, was rather surprising, Brouwer could have easily picked stronger candidates, certainly so if he were to consider foreign mathematicians. Indeed, Freudenthal had already drawn Brouwer's attention to Erich Rothe, a friend of his and a fellow topologist, as a possible successor of Hurewicz. One might suspect that Brouwer's partiality for female company could have played a role, but there is no indication of that sort. Freudenthal conjectured that Brouwer chose Miss Geldof just because he wanted to make his own decision. Whatever the explanation, Freudenthal was incensed, and he wrote a polite letter that could not fail to annoy Brouwer:[125]

> Dear Professor, I heard from Mr. Koppers that a 'young lady'[126] has been appointed as a successor of Hurewicz; he did not know her name. If you have given up your original plan to replace Hurewicz by Erich Rothe, then that is something I immediately accept. That I was not informed about it, that my advice was not asked in this, for me so important, appointment matter, and that I only through a chance conversation with Mr. Koppers was informed—these facts have not failed to make a strong impression on me. I have till now enjoyed your trust in so many matters, that I may hope that you will find my inquiry into the motives for this treatment not indiscrete. I am sorry that I have to write to you about those things, I would have preferred a discussion with you. I hope, however, that you will grant me a moment about these things as soon as possible; I am prepared to come to Laren for this conversation. Looking forward to your reaction, I am, Yours truly,
>
> Hans Freudenthal

This was definitely not the kind of letter Brouwer enjoyed receiving—from nobody, least of all from a *privaat docent*. His answer was, predictably, testy:[127]

[125] Freudenthal to Brouwer, 12.VIII.1936.

[126] In Dutch, a '*juffrouw*'. Today it would be insulting, it was not so in the old days.

[127] Brouwer to Freudenthal, 17.VIII.1936.

> Dear Freudenthal, Since the assistant appointments regard the joint professors in mathematics, I cannot enter into answering or discussing your letter of the twelfth without consulting my colleagues.
>
> Yours,
> L.E.J. Brouwer

Freudenthal immediately reacted and tried to find out what had gone wrong. This time Brouwer answered in extenso, writing that nothing had happened until the letter of the twelfth that could have changed his opinion with respect to Freudenthal, but that he considered Freudenthal's letter an unseemly attempt at pressure, based on unjustified assumptions. Here the matter stopped. Freudenthal felt himself not taken seriously, and Brouwer had suffered from, what he considered, the impertinence of a young subaltern.

This was no reason for Brouwer, however, to think less of Freudenthal as a mathematician, and so when Mannoury and De Vries retired in 1937, he duly considered means to promote Freudenthal. In the faculty meeting of March 19, Brouwer proposed to replace the chairs of Mannoury and De Vries by a lecturer's position plus the position of '*conservator*' of the mathematical institute. It may seem strange to replace two professors in this manner, but one should bear in mind that this was the time of the economic crisis, and the university tried to balance the budget by downgrading positions, when possible. Already some time ago, in 1930, Brouwer had joined the discussion in the Senate on the topic of the introducing of the new position of conservator. For laboratories and clinics a conservator was generally accepted, but a conservator for a mathematics department was a rather surprising phenomenon. The authorities, nonetheless, went along, and Brouwer proposed to make Freudenthal conservator and Heyting, who was at that time a high school teacher in Enschede, a lecturer. Against some mild opposition of the non-mathematicians, the proposal was accepted and thus Freudenthal moved up in the hierarchy, doing exactly the same things he did before. His teaching duty (as a *privaat docent*) consisted of Analysis, the Theory of Groups and Topology.

Brouwer had talked about the matter to Freudenthal in his warm, charming manner, 'I would have made you with pleasure a lecturer too, but there are difficulties. We will first make Heyting a lecturer and then we shall see what we can do for you. But you will not suffer financially.'[128] Freudenthal, however, observed at the same time that Brouwer was taking over a few lectures from the former teaching duties of De Vries and Mannoury, for which he got some extra pay—exactly the difference between a lecturer's and a conservator's salary. 'I never talked about it, but he knew that I knew. From that moment on he had a bad conscience, and the problems between Brouwer and me began. More from his side than from mine, for I considered him a great man.'[129] Indeed, Brouwer had taken over the lectures in Mechanics. The correspondence of the Curators of the University of the University shows that the

[128] Freudenthal, oral communication.

[129] Ibid.

**Fig. 15.5** Brouwer entering his mathematical institute (1937) [courtesy C.B.J. Jonkers]

assignment was extended in 1943, and that Brouwer touched 1125 guilders a year for it.

Suddenly, in November 1937, tempers flared up again. Brouwer, as the director of the institute, had decided that 'at the request of the Van der Waals Committee' there would be no classes on November 25 and 26,[130] and he had informed the staff in writing. In the words of Brouwer: 'Thursday, December 2, Freudenthal, after my class is over, shows me that letter with the words that I 'have no right' to issue such announcements..., stating that he finds this note most unfriendly.' Freudenthal probably realised that he did not have a very strong case and wrote a soothing letter, but the harm was done, Brouwer was extremely sensitive to these little incidents that threatened his peace of mind, and questioned his authority; he did not forget them.

The above skirmishes were, objectively speaking, of no consequence. If anything, they showed that the personalities of Brouwer and Freudenthal had too much in common. Freudenthal's aggressive ambition did not differ much from that of Brouwer 30 years ago. The relation Brouwer–Freudenthal was in no way comparable to the relation Brouwer–Heyting. Freudenthal was a fair approximation to a universal mathematician, imaginative, geometrically minded, with literary, philosophical and historical talents to boot, and above all, without any hesitation to speak out; feared by his opponents, but easy company for his friends and students. Heyting, on the other hand, was certainly gifted, but lacked that exuberance that made Freudenthal stand out. He was cautious, a miracle of precision and formality. He would not expose himself to Brouwer's wrath during the latter's academic years, and so Brouwer came to see in this quiet introverted young intellectual a useful and docile follower, who could be promoted in due time.

Shortly before the war Brouwer and Freudenthal clashed again, this time there was a scientific issue. On Saturday April 24, 1939 Brouwer had presented a talk at

[130]To commemorate the centenary of Van der Waals' birth.

the monthly meeting of the Dutch Mathematical Society, his topic was, quite unexpectedly, a topological one: 'On the triangulation problem.' This problem had a long history, and can be described roughly as follows: given a surface of some kind, is it possible to subdivide it into triangles. The formulation for higher dimensional manifolds is similar.[131] The problem was one of the many fascinating challenges to topologists, and it is likely that at some time in the past, Brouwer had investigated the problem. However, by the late twenties Brouwer had expressed his loss of interest in the glamour of impressive publication lists. So he carried out his research in a somewhat eclectic manner; when his curiosity was roused by a nice problem, he sat down and applied his still powerful mind to it. Although he had exchanged his topological research for the dream of an intuitionistic mathematics, it was generally recognised that he had not lost his touch in topological matters—be it that he persistently clung to his own brand of 'geometrical' topology. Rumours had it that his drawers contained many an interesting (possibly fragmentary) topological gem. The results on the triangulation problem may well have been in one of those drawers, pulled out when a topic for a talk was required. At that particular meeting Brouwer presented a solution for the triangulation of differentiable manifolds, without the intention to pursue the matter further, let alone to publish it. The lecture was probably not well appreciated by the audience, not only because the subject had its own intrinsic difficulties, but also because Brouwer had long ago shed his classical feathers, and presented mathematics whenever possible in an intuitionistic framework. The result was a presentation that was hard to follow for traditional topologists.

If Freudenthal had not been present, that would probably have been the end of it, but he was, and he started to turn over the problem in his mind. Brouwer's idea was new to him, but no matter how he tried—he could not use it. In his own words, 'I never understood it'.[132] But his reflections bore fruit; in due time Freudenthal came up with his own solution, which he wanted to publish—of course giving proper credit to Brouwer.[133] At this point Brouwer started to raise objections, the trauma of earlier priority conflicts was so deep that he could not extricate himself from the awful dilemma—either Freudenthal could publish his paper, but then the reader had to take Freudenthal's word for the existence of Brouwer's proof,—or he could ask Freudenthal to hold back his paper for a while, so that he could publish first, but then he would have to do some work right away, for which he claimed to have no time.

The matter became urgent when Freudenthal had finished the first draft of his proof, and offered to send Brouwer a copy. Brouwer's reaction was to play for time, 'I think that after all it is more correct that I do not take notice of your proof until I have sent you the proof that I lectured about at the time. Be so good as to keep

[131]For a survey of triangulation, see Kuiper (1979), also reproduced in James (1999).

[132]Oral communication, Freudenthal.

[133]Freudenthal slightly extended the results by showing that $C^q$-manifolds allowed $C^q$-triangulations.

your proof for a short while longer.'[134] In spite of the moderate tone, Brouwer had already started to see the triangulation matter as a real priority conflict. He desperately looked for a way out of the impasse, when it occurred to him that Rosenthal was a suitable referee; he was sufficiently familiar with the problem and the techniques involved, and he was a friend of both parties. It is not fair to speak of contestants here, for Freudenthal did not wish to lay down any claim to priority, but he was more or less forced into the role of contestant by Brouwer's overprotective attitude. Rosenthal came to Holland (on account of the political situation he was planning to leave Germany anyway, so Holland served as a halfway house), and on 4 September Brouwer and Rosenthal met. Rosenthal reported that Freudenthal's paper had been finished at the end of August—beginning of September. On September 14 both met again and Brouwer launched his proposal that both parties should hand over their manuscripts at the same time to Rosenthal, who then would pass them on to the other party. On October 4 Brouwer was ill, he asked Rosenthal to change the procedure so that Freudenthal would hand over his manuscript in a closed envelope, addressed to Brouwer, whereupon Rosenthal would visit Brouwer and exchange the envelope against his own proof.[135] He wrote on the same day to Freudenthal that he was much pleased that a recent letter from Freudenthal had opened the road to a improvement of the mutual relations, damaged by a previous letter of Freudenthal. The letter then went on to outline a procedure for exchanging manuscripts, and to justify his somewhat extreme caution.

> Already the national honour forbade the acceptation of your letter of 1 September. For such an acceptation would create in the Netherlands a situation, where a chairman of the Mathematical Society, who, in an address at the General Meeting of this Society, puts on record a scientific result with an exposition in extenso of a proof, which is then not immediately published, would be obliged as a protection against theft of priority by his audience, either to record at the meeting his address by means of a dictaphone, or to depose the sealed manuscript, used at his exposition at the meeting, forthwith at a notary or at a qualified body, under penalty of having to protect later his priority at the high cost of a chemical analysis in order to determine the age of his manuscript.[136]

One sees to what length Brouwer was willing to go to draw the lessons of his painful clash with Menger. Freudenthal, who knew the ins and outs of the Menger affair, must have recognised the symptoms of that conflict, and he could not possibly be amused by such a comparison.

Furthermore Brouwer pointed out the inconsistency in Freudenthal's intentions,—in July the mailing of the manuscript to Brouwer was seen as a private matter between scholars, whereas on the first of September the manuscript was prepared for publication. To this he added an elaboration of his letter of July 8:

[134] Brouwer to Freudenthal, 8.VII.1939.

[135] Brouwer to Rosenthal, 4.X.1939.

[136] Brouwer to Freudenthal, 4.X.1939.

> I have immediately underlined that a written communication of these considerations to third parties can only be allowed posterior to a publication of my lecture, and together with a reference to such a publication or with an account of the contents of this lecture.

Freudenthal replied that he was quite willing to co-operate in the 'Rosenthal-procedure', but that he deplored the fact that in their mutual relations trust had been replaced by legal formulations. He further protested against passages in which implicit accusations had crept; indeed, certain arguments of Brouwer had a general ring, but could only be construed as referring to Freudenthal. Finally, he disagreed with the general tenor of the passage on the protection of non-written communications.

'The consequence of your view would be that one could block whole areas of research by giving out oral communications about one's research, without publishing them, and thus preventing others from publishing their researches.' Nonetheless he was quite willing to go along in this special case.

In most exchanges of the sort described above, there is usually an argument behind an argument, behind an argument, .... In this case, one may guess that old wounds had started to burn at the prospect of another priority conflict. There was, moreover, an extra ground for Brouwer's feelings of discomfort not mentioned in the discussion so far, but cropping up in Brouwer's next letter.[137]

> It is not the case that my objections in this matter came up only on 1 September at the receipt of your letter, in which I missed every similarity with the intention you announced on July 8, but already with your remarks on July 8, which I could not bring in connection with the reply which I gave you in March or April to your statement that—as you concluded from a conversation with a foreign colleague, mentioned by you—the topic of my talk in April 1937 was seen as extremely important in another country, and which therefore had begun to occupy intensively your thoughts. Wondering about this foreign appraisal, I answered at the time, that I could only interpret your information as an exhortation directed to me to publish; but that I could hardly make the fragmentary results of the lecture public without their natural completion, and that it would, alas, take many more months before I could find any time at all to accomplish this completion. I had to conclude after our discussion of 8 July that you felt, after this answer of someone, to whom you have after all been officially assigned as an assistant (*medewerker*), and whose handicapped operational potential was known to you, like undertaking yourself the completion of the results of the lecture of your superior, which were put by him on *his own* work schema. And this was the situation, which I had to observe after our conversation of July 8. And this (as I am prepared to believe, unjustifiably, but in any case understandable) brought gradually in me such memories of experiences with earlier gifted assistants out in me, that I lost any ground

[137] Brouwer to Freudenthal, 25.X.1939.

> to me to make myself familiar with your promised written communications, and also, subsequently, to send you the manuscript of April 1937....

Clearly Brouwer was thinking of the horrors of the Menger episode; he had no wish to go through such an ordeal again. This may explain why he in a sense overreacted to Freudenthal's interference in the privacy of his mathematics. What he chose to ignore was that after a public lecture the topic had become part of the public domain, so that one could not put up a sign 'Trespassers Keep Out'. We must add that the personalities of Menger and Freudenthal were not in the least similar, so that Brouwer's fears were largely imaginary.

Freudenthal was not altogether satisfied with Brouwer's account of the matter, he replied:[138]

> My reflections on the triangulation problem date back to October 1938. I actually arrived at the triangulation when I was preparing the theory of the volume of curved surfaces for my lectures and looked for a simple proof for the transformation theorem of multiple integrals. That is when I found my proof, and reconstructed *your* proof, of which not all details had become clear to me during the lecture—and as I have observed in the meantime—reconstructed it faithfully, whereby the two activities, the original and the reproductive, influenced each other to the extent that I would not be able to separate them anymore.
>
> I have not told you anything about my research at that time: but I have in chance conversations told some mathematicians about the existence of your proof, among them as you know, Lefschetz, at the end of February 1939. According to Lefschetz, Whitney has also looked into the problem, but he had not yet succeeded. Hearing this I thought it almost unavoidable that Whitney, who works at the borderline of topology and analysis, and the author of some twenty papers, and the outstanding expert of my generation, would publish in a couple of years or months, such a triangulation proof. I therefore decided to talk to you about this matter; the conversation took place—I think—half a week after my visit to Lefschetz. I cannot understand how you could have got the idea that I could have started and finished my study of the triangulation problem in 3–4 days. Lefschetz' and Whitney's interest in the problem was the cause, not of my own research, but of my attempt to get you to publish yours.

Basically, here the triangulation conflict ended; there was some more correspondence, mainly about corrections and formulations, but the dispute seemed, to all appearances, closed.

The story of the Brouwer–Freudenthal triangulation theorem got an unexpected twist, when Freudenthal received a bundle of reprints from the American topologist Cairns; these showed that he had already solved the problem in 1934/35. Freudenthal duly informed Brouwer,[139] who studied the papers of Cairns and wrote back

---

[138] Freudenthal to Brouwer, 27.X.1939.

[139] Freudenthal to Brouwer, 18.III.1940; Cairns to Freudenthal, 14.I.1940; Cairns (1935).

that he felt a 'residue of suspicion' 'whether his proofs could pass the test of a precise, detailed critique'.[140]

Brouwer's paper appeared in the *Indagationes Mathematicae* of 1939 with an apologetic footnote, mentioning the reason for its publication, ending with the sigh that the circumstances had made him publish this paper which he otherwise would have considered premature.[141]

The position of both parties can be understood. Whereas the younger party was on his way up and moved in circles where things were changing fast, the elder mathematician had no urgent drive to publish, moreover, with his status he could not risk a fiasco—either through a mistake or by publishing below his level. There was moreover Brouwer's conviction that the only good result is an intuitionistic result, so he cast the paper in his intuitionistic terminology. This by itself must have robbed him of a possible audience. Curiously enough, the paper is republished in the collected works in the intuitionistic part, whereas its main interest was of a topological nature. Even more curious: Freudenthal's paper is missing in the references.[142]

Life in the thirties did not exclusively consist of conflicts and scientific or other business, Brouwer always found time to receive and visit friends. Among his friends, Peter van Anrooy took a special place. Van Anrooy was the conductor of the Residentie Orchestra in The Hague, and Brouwer from time to time visited the Van Anrooys in their home in Scheveningen. He used to stay with them during his usual examination spree if he was assigned to schools around The Hague. His behaviour was unconventional to say the least, entering the living room, he would often make straight for the couch, lie down and conduct the conversation in the horizontal position. Nor would he shrink from admonishing Mrs. van Anrooy about her cooking.[143] Van Anrooy was not overly impressed by Brouwer's Beethoven renditions on the piano, he used to refer to it as 'hammering'. As a young girl, Van Anrooy's daughter Fien occasionally stayed with Brouwer in Blaricum.

At one particular occasion Brouwer publicly came to Van Anrooy's support; in 1937 the wedding of Princess Juliana, the daughter of Queen Wilhelmina, to Prince Bernhard von Lippe Biesterfeld took place in The Hague. The marriage was received with great enthusiasm in The Netherlands, all kinds of festivities were organised. There was, however, one detail that displeased a good many Dutchmen; since Prince Bernhard was a German subject, representatives of the German State had to be invited, and the usual honours had to be observed. This meant that at the wedding not only the Dutch national anthem, but also the *Horst Wessel* song, the official song of the Nazi-party, that had been elevated to the rank of German national anthem, had to be played. It fell to the Residentie Orchestra to play the national anthem and the *Horst Wessel* song; its conductor, Van Anrooy, however, and twenty five members of the orchestra bluntly refused and left. The Dutch authorities,

---

[140]Brouwer to Freudenthal, 30.IV.1940.

[141]Brouwer (1939).

[142]Freudenthal (1939).

[143]Oral communication Mrs. J.F. Heyting-van Anrooy, the later wife of Arend Heyting.

who had to reckon with the whims of its neighbour on the eastern border, were not pleased at all—the German Government of the day was usually quick to construe an 'insult of a friendly head of state' out of such events; Van Anrooy's step found fortunately a good measure of popular support. Brouwer also came to the defence of his friend, together with 38 other 'prominent intellectuals' he sent a telegram in support of Van Anrooy.[144]

Should one interpret Brouwer's support of Van Anrooy as a political signal? That is indeed hard to say, it seems likely that an injustice committed to his friend counted more than the political gesture. One cannot see this kind of action isolated from the mood of the times; the opinions on national socialism and its German form were the subject of heated discussions. There was a Dutch national socialist party, the *Nationaal Socialistische Beweging*, *NSB*, which was actually a rather heterogeneous collection of all kinds of political convictions, mixing socialist ideas with right wing nationalist doctrines. The rich fabric of political movements in Holland generated a certain atmosphere of hostility, which induced the more cautious elements to avoid statements or actions that could invite repercussions. Brouwer was not inclined to hide his opinion, as we have seen. That the luxury of a political opinion could have consequences is illustrated by one of Brouwer's colleagues, Bonger, a one-time fellow student of Brouwer and then the professor in sociology. Bonger was a fervent anti-Nazi; his activities were carefully watched from Germany, and radio Bremen openly threatened him with reprisals.[145]

1937 brought also the 25th anniversary of his professorial appointment; Brouwer was flooded with letters and well-wishes. He had become a well-known person in his own country, revered by some and feared by others. His former students, generally, worshipped him, he was an easy talker when he felt inclined so, and he did not easily forget the name and face of a student. Somewhat surprisingly for a man who started his career with a definite lack of care for conventions, he had acquired the respectability and admiration that go with the conscientious fulfilling of one's duties.

In the same year two of his old friends, Hendrik de Vries and Gerrit Mannoury, had reached the retirement age. Their valedictory lectures were given on the same day in the *Mathematisch Instituut* in the Roeterstraat. Brouwer, as the director of the institute, addressed them in name of the faculty, and Max Euwe spoke on behalf of the students. The newspaper *Het Volk* mentioned that their old teacher, Korteweg, was also present at the occasion.

The Brouwer of 1937 certainly was a far cry from the young idealist who was master in his own kingdom. The metamorphosis from rebellious moralist to respectable professor had been thorough—although the professor had retained the mystic and radical views of the student. The road from withdrawn student to international celebrity was a long and lonely one. Real friends were few, and two of these had already fallen by the wayside, both artists—Adama van Scheltema and Frederik van Eeden.

[144]Published in *De Tribune*, a communist journal, 16.I.1937. See also Fasseur (2001), p. 144 ff.

[145]Bonger took his life when the Germans invaded Holland.

**Fig. 15.6** Hendrik de Vries, leaving the mathematical institute after his retirement party [courtesy C.B.J. Jonkers]

The physical and mental health of his teacher, Korteweg, was slowly deteriorating; the grand old man of Amsterdam mathematics withdrew more and more from the professional scene. He was still able to celebrate his golden doctorate and to accept an honorary membership from the grateful Mathematics Society, at his 90th birthday at April 30, 1938. At the meeting of the *Wiskundig Genootschap*, Brouwer gave the presidential address with the title *Intuitionistic discontinuous functions of a real variable*.

The major event in the faculty was Zeeman's death. The consequences were hotly discussed. Zeeman's chair would automatically bestow a great measure of prestige and influence on its occupant. Brouwer vigorously campaigned for Zernike, but the majority went against him and Zernike. The general opinion was that Zernike lacked the authority and discipline to be the head of the physics department.

All this time the sky over Europe was darkening, in 1939 Poland was overrun and divided between Germany and the Soviet Union; the contacts with Polish colleagues were no longer possible. The dramatic fate of Poland did not strike everybody as necessarily a bad thing—e.g. Pontryagin wrote: 'recently the number of Soviet mathematicians has been greatly increased. In Lwow the following are present: Banach, Schauder, Knaster, Mazur and a number of others. Sierpinski and Kuratowski remained in Warsaw and are now in concentration camps.'[146]

Rosenthal was fortunate enough to leave Holland in March 1940. Not everybody was as lucky. Brouwer wrote to Freudenthal that the group theorist Remak

[146]Pontryagin to Lefschetz, 29.XII.1939.

was in imminent danger to be extradited on the grounds of misbehaviour.[147] The economistist Frijda thought that the only way to save Remak was to have him admitted into a psychiatric hospital. Remak was indeed not an easy person to get along with; he had the reputation of being difficult and rude. In spite of the support of Frijda and others, he was deported to Auschwitz, where he died in 1942.

Another unfortunate German in exile in Holland was Blumenthal.[148] He had been discharged on the ground of political unreliability from Aachen on 22 September 1933. Until then he had expected to be exempted according to the rules as a First World War veteran (*Frontkämpfer*); but apparently the rules did not particularly worry the authorities. Blumenthal had in the meantime made friends with Professor Burger from Delft. In the 'old days' they exchanged comments on literature and discussed a wide variety of topics; in one letter Blumenthal told Burgers that he had procured the best-known book of Frederik van Eeden—*The little Johannes*, a moralistic literary fairy-tale parable. In spite of the treatment he had received from the Nazi government, he would not allow nasty comments on the regime. It seemed as if he could not understand the political climate in the later years of his life, as if he refused to accept the possibility of an ultimate and systematic evil. In 1934 Blumenthal and his daughter visited Burgers in Delft; the latter managed to scratch up some money to support him, but he could not provide him with a job, and so he had to remain in Germany. From the correspondence with Burgers it appears that he was a fairly regular visitor to Holland before his final emigration from Germany.

Blumenthal had put his life in the service of Hilbert's Göttingen; he played, even from a distance, a role as the managing editor of the *Mathematische Annalen*. Surprisingly enough, he could carry on this duty until he was forced to resign in 1938. In a letter to Bernays (17.II.1938) we read that Behnke had taken over his position at the *Mathematische Annalen*. Behnke was a member of the younger generation, without hang-ups about the past of the *Annalen*. He told that he was rather surprised, and embarrassed, when, in Amsterdam in 1938, after he had given a talk, Brouwer and Weitzenböck came up to him, and asked why Blumenthal was still officially active as an editor of the *Annalen*.[149] In spite of the anti-semitic policy and atmosphere, Blumenthal could still move around freely in Germany, and had visited Göttingen in January. Hilbert, he wrote, only dreams at the sofa. His loyalty to Hilbert was not a matter of calculation, but of sincere devotion; nobody was therefore better qualified to write the biographical note for Hilbert's collected works.[150] Although his children managed to get away to England, Blumenthal had difficulty getting the required visa. In 1939 he tried to get to the United States, Weyl was prepared to intervene, but to no avail. He was able to visit his children in England, but he had to return to Holland.[151] Expecting the worst, he asked Burgers, in case of a new war, to relay his

[147] Brouwer to Freudenthal, 2.III.1940.

[148] See Behnke (1978).

[149] Behnke to Van Dalen, 27.XI.1976.

[150] Blumenthal (1935).

[151] Blumenthal's 1939–1943 diaries have been published by Volkmar Felsch, Felsch (2011).

letters to his children in England. Finally he moved to Utrecht where the war caught up with him. After moving from one place to another he was 'housed' for some time in a camp in the neighbourhood of Utrecht; Freudenthal visited him there. After the rounding up of Jews had started, he was moved to Westerbork,[152] where his wife died of pneumonia. Blumenthal was eventually transported to Theresienstadt, where he died on 12 November 1944. Various people had tried to do something for him, in particular Burgers, Wolff, Mannoury, and Van Dantzig. There is, however, no report of a renewed contact between Brouwer and his former friend and later adversary. Under the circumstances one would have thought that the agonies of the present obliterate the bitterness of the past. But generally speaking, the Dutch were so much used to their traditional neutrality, that they almost considered it a natural right to be left alone in times of war; thus they were inclined to consider the exiles from Germany as reasonably safe. Coupled to the earnest promises of its German neighbour to respect the Dutch borders, the Dutch could not easily share the fears of the German immigrants—Jewish or not.

We cannot end the account of Brouwer's pre-war actions without mentioning an event that was already at the time considered a grave faux pas. In 1939 Brouwer published a short note in *Forschungen und Fortschritte* for the 70th birthday of Theodor Vahlen. The fact that Vahlen had joined the national socialist movement with heart and soul had not endeared him to his colleagues. For most German mathematicians his name was anathema. One had to accept the reality of his rise to a high function in the Third Reich,[153] but one would avoid contact with him whenever possible. The only correspondence between Vahlen and Brouwer in the Brouwer Archive concerns financial problems at the time of the call to Göttingen. There is no evidence that the two ever met, other than possibly at some meetings of the DMV in the old days. Whatever could have induced Brouwer to write such a dedication is unclear. Freudenthal conjectured that Brouwer protected Vahlen (as a mathematician) to annoy the Göttingen crowd. It is a fact that Dehn had written an extremely negative review of Vahlen's book 'Abstract Geometry',[154] and it is said that Vahlen never forgot the humiliation. Brouwer, on the other hand, recommended Vahlen's book in the university guide right until the end of his career—and his successor Heyting kept it on the reading list. With the Vahlen dedication, Brouwer had succeeded at one stroke to amaze the party sympathisers among the German mathematicians, and to shock the rest, by paying homage to a mediocre mathematician with odious political views.

The small note opens with a philosophical introduction, which concentrates on the dichotomy 'subjective–objective', or 'man–outer world'.

> ... we must go back to the fact that man directs the operation of his impulses of will by means of subjective thought, his subjective with objective thought

[152] A transit camp from where Jews were sent to the camps in Germany.

[153] Cf. p. 623, see also Siegmund-Schultze (1984).

[154] In Brouwer's words '*sein tiefsinniges und suggestives, viel zu wenig gewürdigtes Werk über geometrische Grundlagenfragen; "Abstrakte Geometrie"*'.

> and his objective with mathematical thought, and that these impulses of will are activated by the belief in the interaction of thinking and acting man with an outer world (*Anschauungswelt*), which contains himself, his thoughts and his actions, including its laws, and which exists independent from thought.
>
> The thinking that incorporates the thinking subject into the outer world now necessarily leads to objective self-contemplation. This self-contemplation leads first to the insight that the laws of the outer world are products of the objective thought, and this amazing insight generates for the time being a paralysing doubt, first of the existence at all of an outer world, independent of thinking, next of the seriousness of the private impulses of will that presuppose the outer world. In this situation thinking can only be practised as a game and a sport, and indeed with thought as its only object.

Brouwer then went on to point a way out of these problems by using the notion of responsibility. The most notable fact about this brief discussion of Brouwer is that he does not just brush away the outer world, but looks at its inherent problems.

After describing the successive phases in the mathematician's development, Brouwer continued, 'The above mentioned norms have become especially apparent for the present generation of mathematicians, in the figure of Felix Klein, which today radiates undiminished from the past; they will again become visible in the figure of Theodor Vahlen.' This piece of traditional flattery is all the more remarkable as Vahlen cut a rather poor figure when compared to Klein, not even considering the political aspects.

The private world of Brouwer was drastically changed some years before the war. Brouwer had managed to get Cor Jongejan a position as an '*adjunct-assistant*' (the lowest rank of assistant at the university) in 1925. She was 32 years old and still a highly attractive young woman, witty and with a lively spirit. Living in the Brouwer household did not keep her from male company, there was no lack of admirers. Even Paul Alexandrov appreciated her charms, to the extent that Brouwer showed covert signs of jealousy (cf. p. 474). For Cor life at Brouwer's had the fascination of contact with this wonderful spiritual man and the added attraction of all those casual contacts with a rich variety of visitors, but it had become clear to her that the relationship with Brouwer stood in the way of the fulfilment of her own life and talents; her place was just that of a member of the court of a great man.

Although the reports picture her as an unfettered, easy going person, she worked hard enough for Brouwer. As an unofficial secretary and an official assistant at the Mathematical Institute, she had quite enough to do with all the copying of manuscripts (often by hand, later mostly typewritten)—not only Brouwer's, but at the time of his editorial activities for the *Mathematische Annalen* also of submitting authors—and the correspondence in the many affairs Brouwer was involved in. It was not unusual for Brouwer to request five or more copies of his own letters, but he also had incoming correspondence retyped by Cor.

Brouwer's place was a regular meeting point for a number of female admirers, and a good deal of intrigue was going on among the ladies. Apart from Cor the main claimant for Brouwer's attention was Tine Langhout-Vermey, an intelligent lady of good taste, who earned a living as a freelance interior decorator. Tine quite often accompanied Brouwer on his travels in Holland and abroad. At the time she met Brouwer she was married to Willem Langhout, a man of no particular merit, if one does not count his financially independent position. Willem cultivated the art of philandering to the extent that Tine divorced him. She made a successful effort to become a member of Brouwer's entourage, and fought occasional fierce battles with Cor Jongejan. Cauliflowers, it was said, went crashing through the windows. Tine was a person of good taste, she in due time became a well-respected interior designer.

None of this bothered Lize Brouwer, she was well aware of the fact that Bertus loved to bask in the sun of (in particular female) admiration, but she was equally convinced that none of these romantic episodes or flirtations would come to anything serious. When all was said and done, Lize was in actual control of the household, her invisible hand arranged the daily affairs by and large. She often used to stay in Amsterdam, taking care of the pharmacy, her main problem being her awkward position between her husband and her child. There was an unmistakable aversion on Brouwer's part to Louise, and, not surprisingly, the feeling was mutual. Lize was in the position of the unfortunate ferryman, who had to carry a goat and a cabbage, she always had to arrange things, so that Louise could come home when Brouwer was absent. This, indeed, was her real cross; she had to choose between husband and child, but whichever choice she made, she would feel guilty.

In this hectic environment, where her young life was simply running away, Cor suddenly reached the decision to leave and lead her own life. After living with Brouwer from 1915 on, she left him to live in to Zandvoort, a fashionable seaside resort. Zandvoort could not satisfy her for long, so soon she moved to Amsterdam, where at the instigation of a friend[155] she rented a big apartment at the Stadhouderskade 136 and took in a number of girls. One of the girls was Janny van Wering, a music student who later became one of the leading cembalo-players in Holland.[156] The household at the Stadhouderskade was run by Cor, who could finally display her social and organisational talents; the girls felt happy and were quite attached to Cor. Of course, the change of address did not mean a complete separation from Brouwer, for she still was an assistant at the Mathematical Institute where Brouwer was her boss, but at least she had regained a measure of independence.

The house at Stadhouderskade eventually (and inevitably) attracted a fair number of male visitors, among them was Brouwer's brother Aldert, the geologist, who at that time was at odds with Bertus. Aldert had a keen eye for female attractions, and he did not waste time in getting familiar with the renown Cor. Eventually he hardly missed an evening meal in the cosy circle of Cor and her friends.

It took quite a while, but eventually Bertus Brouwer also started to frequent the house at the Stadhouderskade. This lasted until well into the war. At one occasion Brouwer and Max Euwe (the chess master, one time world champion) were in the middle of a mathematical discussion when the air-raid-sirens sounded; they both

[155]Wim Bierens de Haan.

[156]Most of the information on this episode was provided by Mrs. Van Wering.

moved out of the room, sat themselves on the stairs and continued their discussion as if nothing had happened.

At one occasion Brouwer dropped in with Tine Vermeij and Gerda Holdert; when Cor went to the kitchen, Brouwer stretched himself on the couch and said: 'I want to be caressed.' Gerda, who was at that moment not in the mood for any of the Brouwerian ceremonies, curtly said—'Tine, why don't you do it, the way I do it is not good enough anyway.' Janny van Wering thought the exchange so silly that she burst into laughter, so that Gerda could not suppress a curt 'I would like to know why that little spook is laughing?'

Eventually Brouwer got a small room in the house, where he could have his siesta, or stay the night if he had late business in Amsterdam (e.g. senate meetings or concerts).

In the end Cor moved back to Blaricum. Janny van Wering spent one summer during the wartime in the Padox—one of the little cottages at Brouwer's property, she lived there with her cembalo and gave some house concerts for the Brouwers and the guests. She remembered one sunny afternoon in Blaricum that Cor Jongejan came on her bike home in Blaricum with a box of cakes to celebrate her assistant-pharmacist diploma. The pharmacy was indeed always in Brouwer's mind, and so he had with some foresight suggested that Cor should make herself familiar with the running of the pharmacy and get herself a diploma.

Among Brouwer's extra-academic activities there is also one that had a remote relation with his university teaching; he, like a number of his colleagues, was involved in the organisation and supervision of secondary education. The contacts between secondary and higher education were more or less automatically ensured by the existing regulations of the final examinations of high-schools and gymnasia. Brouwer always faithfully took part in the examination tours, which could take him to any town in The Netherlands. He and many of his colleagues considered this one of the self-evident duties of the academic world. The fact that there was a modest fee connected with the function of examiner might have had some influence in a time when the economical crisis was still acutely felt. On the whole, Brouwer took his responsibilities for secondary education seriously, even to the extent that in 1938 he became a member of the board of trustees (a *curator*) of the Gymnasium at Hilversum,[157] which was a municipal institution in the sense that the town of Hilversum was the responsible authority (subject to general rules of the national government).

Brouwer's appointment was renewed from year to year, also during the war years. A gymnasium was an institution with a considerable status, as a rule teachers had a university degree and quite a number had obtained a doctorate at one of the universities. The gymnasium teachers (and to a lesser degree those of the HBS) formed a pool of future professors; the rector was a man of consequence and the trustees were carefully selected from the upper strata of society. In this circle Brouwer felt perfectly at home, he always enjoyed the company of the literary minded, in particular the classicists. Even in this select board, Brouwer could not avoid conflicts. A minor

[157] Mayor of Hilversum to Brouwer, 15.VI.1938.

one took place in his second year; the board of trustees had made a short list of candidates for the vacancy of a mathematics teacher, but after some consideration the council of Hilversum changed the order, switching the numbers 2 and 3. Number 2 on the list was a young physicist, Dr. E.M. Bruins. Both the rector and the deputy rector (*conrector*) were of the opinion that Bruins lacked the experience to qualify for a full-time appointment at the gymnasium. This was seen by the trustees as an infringement of their rights, and they protested.

Brouwer, who was in a better position than most to judge the mathematical merits of the candidates, drafted an answer,[158] saying that 'the trustees recognise in Bruins such an excellent personality of special qualities of talent and character, who can play a role in the cultural life of the town, and that they feel justified to do so, as the trustee Prof. Brouwer in the first place had admired, during the years of study at the University of Amsterdam, Dr. Bruins' versatile way of thinking and his often unselfish devotion, in the second place he has seen him excel subsequently, not only as an independent scientific researcher, but also in charge of courses of the *Volksuniversiteit*[159] and as a tutor'.[160]

One might wonder why such an excellent mathematician chose to apply for a job at a gymnasium, instead of following an academic career. The answer is simple, there were hardly any tenured jobs below the rank of professor at the universities and in view of the magnitude of unemployment a position at a gymnasium was definitely to be preferred to that of ticket collector or to the fate of the unemployed. We will meet Dr. Bruins again in our history.

[158]Trustees to council, 13.VII.1939.

[159]A semi-official institution providing (mostly evening) courses in various areas.

[160]*repetitor.*

# Chapter 16
# War and Occupation

## 16.1 Occupied Holland

War came to the Netherlands on a beautiful clear day in May 1940; on the tenth the weather was exactly as any supreme command could wish for an all out attack. The German forces crossed the borders, and German paratroops descended on vital points in the west of the country. The country was not totally unprepared; the army, consisting mainly of conscripts, was mobilised in 1939, measures had been taken to protect the civilians—bomb shelters were erected in the parks and the public gardens of the cities, black-outs were ordered, air-raid wardens were appointed, the proverbially impregnable waterline was activated by means of its system of inundations, ... But still, hardly anybody had believed that the Germans would violate the neutrality of this harmless little country.

It took the Germans only five days to break the resistance, and after the atrocious and large scale bombing of Rotterdam, the army capitulated and the government, including the Royal family, fled to England.

The legal situation after the capitulation was confused, the Dutch government in exile, residing in London, considered itself the real government, but the German occupation authorities were the de facto governing party. The Dutch democratic system comprised quite a number of parties; the main parties being the Socialist Party, the Liberal Party, the Catholic and two Protestant Parties, followed by a number of smaller parties, including the Communists and the National Socialists. The latter, the National Socialist Movement (NSB), had tried unsuccessfully to imitate their German-Italian models, but remained small and rather marginal. The NSB welcomed the German invasion, and hoped to play a major role in the new constellation. This is not the place to present a history of the Dutch under German occupation, but it may suffice to say that the overwhelming majority of the Dutch resented the brutal invasion and instinctively distrusted every move and statement of the new authorities. There was also a group that, stunned by the military superiority of the German armies, took a 'realistic' view of the new situation in Europe, and were prepared to adept itself to a future under German supremacy and there were those who actively gave aid and comfort to the enemy, to the extent that the Germans succeeded

D. van Dalen, *L.E.J. Brouwer – Topologist, Intuitionist, Philosopher*,
DOI 10.1007/978-1-4471-4616-2_16, © Springer-Verlag London 2013

in attracting volunteers for civil, paramilitary and military organisations. There was a great demand for labour force in Germany, there was a Dutch division in the SS, there were places to be occupied in the civil service, the police, the secret police, etc. The Dutch, neither being better nor worse than the average nation, produced both men of principle, who refused to bow for the usurpers, and 'collaborators' or downright Nazi's, some out of political motivation and some for opportunist reasons.

After a brief period of military authority, the Netherlands were put in charge of a *Rijkscommissaris* (State commissioner) Seyss-Inquart. This new head of state was a seasoned national socialist of Austrian descent. Seyss-Inquart's plan was to keep the Dutch civil service in working order and to use it as a tool in the nazification of the country. In particular he tried to keep the civil heads of the government departments, called *secretaris-generaal* (secretary general), in function. One of his first acts was to demand from civil servants, judges and teachers a declaration of loyalty, to the effect that they would obey the instructions of the State Commissioner and his staff, and abstain from anti-German activities.

Since any promising strategy for nazification had to build on the education of the nation's youth, Seyss-Inquart almost immediately set out to put a national socialistically minded person in charge of the department of education. The most suitable man for the job was an Amsterdam professor in German philology, J. van Dam. Already before the war Van Dam had become attracted to the national socialist principles, so when he was offered the influential position of secretary general of the department of *Education, Science and Culture Protection* (the new name of the department) he had soon made up his mind. Van Dam was basically a harmless man, but with a strong dose of ambition and vanity. As soon as he was appointed, he was of course the object of intense contempt. During and after the occupation the two adjectives 'goed' and 'fout'—'right' and 'wrong', acquired a political meaning. 'Right' was patriotic and anti-Nazi and 'wrong' was pro-Nazi. These labels persisted until long after the war. Van Dam definitely was wrong. The new NSB mayor of Amsterdam, Voûte, also belonged to that class. As president curator of the University of Amsterdam he plays a role in the wartime history of the university.

When the Germans took over the civil government, they did not immediately set about revising the existing traditions and rights; the subversion of the Netherlands was a gradual, sometimes imperceptible, sometimes brutal process.

A few examples may suffice: the Germans introduced the summertime–wintertime system (now universally accepted in Europe); in the country this was seen as a pernicious German trick, people doggedly stuck to the pre-war time and referred to 'the new time' as something evil, and to 'the old time' as the right thing. Another, more serious example is the system of social care that the Germans introduced after the pattern of their own '*Winterhilfe*' (winter care). This Dutch system, *Winterhulp Nederland*, was persistently and with great tenacity boycotted by the Dutch. The average Dutchman distrusted all regulations introduced by the enemy. One did not give money to organisations with National Socialist leanings.

A major visible effect of the occupation was the ever intensifying persecution of the Jews, for whom Holland (in particular Amsterdam) had, ever since it became a free republic, been a haven.

At the university the consequences of the war were at first rather modest, and life went on much as usual. Freudenthal, Alexandrov and Hopf had started in 1939 the preparations for a *Festschrift* for Brouwer's sixtieth birthday; even after the occupation of Holland, correspondence with the Soviet Union and Switzerland was carried on as usual. Hopf had reacted positively to Freudenthal's suggestion to honour Brouwer in this way, adding '—Moreover, Brouwer's work is not sufficiently valued; it would therefore be doubly advisable to demonstrate the contemporaries how much he is appreciated.'[1] The plan to dedicate a volume of *Compositio Mathematica* to Brouwer was frustrated by the war. In the end the initiators advised the prospective authors to submit their paper for the Brouwer *Festschrift* to a journal of their choice.

The sixtieth birthday itself was a low key affair, as to be expected in occupied Holland. The newspaper *Het Handelsblad* mentioned the scholar in its columns, and that was it.

Freudenthal complained in June that it had become very difficult, not to say impossible, to reach the editors of Compositio,[2] and he asked Brouwer what to do. Should one appoint editors that could easily be reached by mail? The publication of the next issue had become problematic. Brouwer replied that the first issue of volume 8 could be published, but that in view of the difficulties it would be better not to start any new typesetting.[3] A month later Wijdenes told Freudenthal that Brouwer and he had decided to stop the publication of *Compositio Mathematica* for the time being,[4] and a couple of weeks later Brouwer wrote that no permanent closing down of Compositio was intended.[5] In the beginning of September censorship of newspapers and journals was introduced with respect to information with military significance, including a large number of civil topics, e.g. the building of roads and bridges. Even *Compositio Mathematica* received the instructions of the Military commander in the Netherlands.[6] In view of all the problems and uncertainties, Brouwer, after some deliberation, decided to end all activities of *Compositio Mathematica*.[7] Obviously, authors of already submitted and refereed papers should be completely free to resubmit their papers elsewhere. In view of the fact that the first issue of volume 8 had not yet appeared, five months after the announced date of appearance, he also decided that issue should be cancelled altogether.[8]

[1] Hopf to Freudenthal, 21.XII.1939.

[2] Freudenthal to Brouwer, 15.VI.1940.

[3] Brouwer to Freudenthal, 26.VI.1940.

[4] Wijdenes to Freudenthal, 27.VII.1940.

[5] Brouwer to Freudenthal, 9.VIII.1940.

[6] *Wehrmachtbefehlshaber in den Niederlanden. Militarische Zensurstelle.* 9.IX.1940.

[7] Freudenthal to Hopf, 10.XI.1940.

[8] Brouwer to Freudenthal, 17.X.1940.

## 16.2 Weitzenböck's Choice

In the first year of the occupation an event took place in Blaricum that so far has not been explained. This curious incident has baffled experts and the villagers alike. We recall that the Austrian mathematician Roland Weitzenböck joined the Amsterdam mathematicians in 1921, after Hermann Weyl had turned down Brouwer's offer. Weitzenböck had settled in Blaricum, where he become a fully accepted member of the community. He was a man of few words, without observable political views. Appearances are often, however, deceptive, and in this case the solid imperturbable exterior hid a considerable amount of frustration resulting from the disastrous course of the First World War. As so many German and Austrian ex-service men, Weitzenböck became a hard-core revanchist, and an implacable enemy of France. But whereas Brouwer actively campaigned for the rehabilitation of German scientists, Weitzenböck refrained from political activity. However, after the '*Anschluss*' of Austria in 1938, he started to vent his approval of Hitler's policies in private conversations.

Walter Ledderman recalled that in 1938 in St. Andrews he ran into Turnbull and Weitzenböck walking in the street. 'Turnbull tried to introduce me to Weitzenböck, who, however, refused to shake hands with me.' Shortly before the war Weitzenböck wrote to Turnbull that Britain would certainly be defeated in this war, and that the British empire would be destroyed. 'But he would like to express his condolence to Dan [Turnbull], whom he had liked since his student days in Amsterdam.'[9]

Weitzenböck's son Willy had joined the Dutch air force and was trained as a pilot at the airbase Soesterberg, but when the war broke out on May 10, 1940, his superior and his fellow pilots did not trust him well enough, so he was transferred to a non-combatant place. Whether he felt so insulted that he had to make a gesture cannot be said with certainty, but after the capitulation he joined the SS. Weitzenböck himself joined the NSB, as a result the whole village knew that the nice mathematics professor had become an enemy. In reaction he and his family were completely ignored, nobody greeted him, the chess club did not want to have any further contact. After the war, when Weitzenböck was arrested and was awaiting trial, Brouwer wrote a memorandum for the district attorney (*officier van justitie*); it is instructive to read Brouwer's view of the case. Here is a fragment:

> He began to acquire a reputation of 'greater Germany'-inclinations, when at the time of a possible war and political tension between the Third Reich and the Austria of Dollfuss and Schuschnig he applied for Dutch naturalisation; the purpose of this application was to prevent that his sons, who were approaching the age of military service, could be forced to take up arms as Austrians *against Germany*. This naturalisation must have been invalid if Weitzenböck at the time was still bound by his oath as an Austrian reserve officer. This probably has been the case, as during the occupation his naturalisation was annulled.

[9]Ledderman to Dyckhoff, 39.I.2004.

> Only after the annexation of Austria in 1938, Weitzenböck's 'greater Germany' disposition became undeniably clear when friends and colleagues offered their condolences and he, while rejecting the condolences, declared himself openly a supporter of the German empire. He has, since that time, in conversations, always frankly given expression to this disposition in such a manner that almost all his earlier relations stopped seeing him, or at least restricted their contact to the necessary professional contact. At the same time Germans and members of the NSB naturally sought his company; but he never became an active party member. On the contrary, he withdrew more and more into himself, and made, even more than before, all his working energy available for his scientific researches.

We may assume that Brouwer's view was correct and it is largely borne out by the available evidence.

In the early morning of 6 October 1940 the house of Weitzenböck was hit by a bomb, which killed his wife and son. Weitzenböck and the housekeeper survived the attack. The house was completely destroyed. The local paper wrote 'an English bomb attack, carried out in a dive, claimed two civilian victims'. The local population saw in the act the revenging hand of a villager, who had fled to England and joined the RAF. Weitzenböck himself blamed the Dutch government, and he thought it possible that one of his son's former mates had carried out the attack. An attack by the resistance must be ruled out, at that time it was not up to such a job, moreover, the police, or the German military, would have found out. The idea of a precision bombing in those days on a villa hidden, together with numerous other villa's, among trees in half dark, seems preposterous. According to the archives there was no English squadron flight over that part of Holland at the time, but of course a plane could have lost contact, and got rid of its bombs. Finally, there may have been anti-aircraft fire, and an unexploded grenade could have dropped at the house. It all adds up to a chance bomb or projectile. The bombing became a sort of legend. Stories were told about a secret meeting of high military officers, about Weitzenböck's work for military research, but none of it holds water.

## 16.3 Freudenthal Dismissed

The mathematics department had two staff members, who, as Jews, were in actual danger, Belinfante and Freudenthal. Belinfante belonged to the old Portuguese Jewish community. He was a quiet, conscientious man. Since 1924 he had been teaching intuitionistic topics on a part-time basis. In addition he was a high school teacher in mathematics. At the outbreak of the war he taught at a girl's college in Amsterdam (*Lyceum voor Meisjes*). Belinfante did not play an important role in the mathematical institute, he restricted himself to his courses and his research in intuitionistic real and complex analysis.

Freudenthal, on the other hand, was the centre of activity in the institute. He was in fact, after Hurewicz had left, the only modern mathematician, with ideas about almost all topics. His courses were notorious for their merciless use of new methods and level of abstractness. Nonetheless he was extremely popular with the (better) students. Nobody who took his courses, or enrolled in his seminars, and grasped what the young magician with his German accent was doing, would ever forget the thrill of being shown the mysteries and vistas of mathematics.

Freudenthal came to Holland in 1930 at the age of 25, but for some reason he had never acquired Dutch citizenship, although he had become completely assimilated. He was known to his mathematical colleagues all over the country, and he was at home in Dutch cultural and artistic circles. Since Jews who left Germany for a longer period automatically lost their citizenship (they were *'ausgebürgert'* (denationalised)), Freudenthal was at the outbreak of the war stateless. The Dutch authorities had in the months preceding the war, in view of the increasing tension, started to check on persons without Dutch nationality. As a consequence, Freudenthal had to report in March 1940 at the police station, where he had to show his working permit. Freudenthal, who cared more about topology than bureaucratic documents, had completely forgotten about the permit. Brouwer, when asked by Freudenthal, made some inquiries, with the result that Freudenthal had to apply for an extension at City Hall.

Freudenthal soon became the target of small bureaucratic harassment. Although Brouwer had personally approved Freudenthal's next term as a *privaat docent*, the authorities (no doubt as part of the plan to eliminate Jews from public life) had not automatically extended this appointment, and requested him to apply for a new one. Fourteen days after the capitulation he was asked to appear at some later time at the office of the head of education of the city council in order to submit a request. Freudenthal flatly refused this: 'It surpasses all reasonable limits to demand from me that I submit a new request for a *privaat docent* position, and I will inform the head of this.'[10]

Brouwer gave the request a more neutral interpretation, according to him this was a normal procedure, carried out every five years. It did not take long, however, for the Germans to start the elimination of Jews from public life. The first steps were of an administrative kind, not directly visible out on the streets, but the open discrimination followed soon, the banning of Jews from public places, parks, cinema's, pubs, swimming pools, .... One of the first steps in this direction was the so-called 'declaration of Aryan descent'; all civil servants had in October 1940 to fill out and sign a form specifying their racial descent. A month later all Jews were dismissed from the civil service, including the educational sector. On November 23 Freudenthal lost his position as a conservator, and at the same time Belinfante's permission to teach as a *privaat docent* was withdrawn. For the mathematics curriculum the loss of Freudenthal was serious, he was in charge of a number of central topics. Belinfante's teaching had always been restricted to parts of intuitionistic mathematics, with him the institute lost a conscientious teacher, albeit of somewhat marginal topics. So at one stroke Freudenthal and Belinfante had joined the class

[10]Freudenthal to Brouwer, 6.VI.1940.

of the unemployed, albeit that they received a form of payment by the state. Not only were the Jewish staff members dismissed, they were no longer allowed on the premises of their former institutions; for a scientist that was a serious restriction of his professional activity.

When a number of students approached Freudenthal with the request to conduct the examination of the parts of the curriculum that used to be his responsibility, Freudenthal asked Brouwer to take care of their interests, since he was in no position to do so.[11] He ended the letter by assuring Brouwer that he was, of course, willing to assist him in matters that were not of an official nature.

In his reply, Brouwer conjectured that the authorities would, before the end of the year, propose a solution to the problems mentioned.[12] He thought that there was no possibility for Freudenthal to continue further his duties as a *privaat docent* (as such he had not been fired), since Belinfante had already been dismissed, expressing his sympathy—'May you hold your own in life, and may new perspectives open up for you before long.'

In the eyes of the Germans, unrest had to be avoided, 'business as usual' was an important principle in occupied territory. So the universities, in as far as they had not been closed down, as was the case with Leiden and Delft, had to carry on the normal teaching duties. The Amsterdam mathematics department had to find a replacement for Freudenthal, who had been de facto responsible for the analysis-topology curriculum. The faculty, no doubt prompted by Brouwer, soon found a replacement: Evert Marie Bruins, the same man who was considered by Brouwer as a mathematics teacher in Hilversum (cf. p. 662). On 10 January 1941 Bruins was appointed to teach the analysis course. A few days later Brouwer informed Freudenthal of the new appointment, and asked him to provide Bruins with any information he might need. In his letter to Freudenthal, Brouwer spoke of a 'temporary replacement' (*waarneming*).

Why the 'temporary'? Like many Dutchmen, Brouwer probably trusted that the Third Reich would not last. In 1941 the outcome of the war was no longer a foregone matter. The Germans were seemingly invincible, but nonetheless the majority of the Dutch clung to the conviction—perhaps slightly irrational at the time—that the German superiority would eventually break down. History, so to speak, was on their side, had they not survived the King of Spain and Napoleon? The military events of 1941 carried the germ of an eventual downfall; the German army was dangerously overextending itself. The successes in North-Africa and Russia were bought at a price. Moreover, the outcome of the battle of Britain and the continuing support of the United States were reasons for cautious optimism. So Bruins' services might well have been of use for a short period only.

Under the circumstances the Dutch were cautiously optimistic about the final outcome of the war, although they suspected a great deal of misery before the Third Reich would be a thing of the past.

---

[11] Freudenthal to Brouwer, 30.XI.1940.

[12] Brouwer to Freudenthal, 27.XII.1940.

Bruins was perhaps not the optimal choice for a mathematics department; by training he was a physicist, with a taste for down-to-earth calculus, algebra and geometry, *à la* Clebsch. He had a solid acquaintance with the traditional subjects before the advance of modern algebra, topology, functional analysis and the like. But in the area of his choice he was an accomplished expert. While still a student, he had found his first results in applied mathematics at the chemistry laboratory of Büchner. Subsequently he wrote a dissertation under Jacob Clay on cosmic rays.[13] Clay was one of the physics professors in Amsterdam; he got his physics training in Leiden under Kamerlingh Onnes. In his early years he had been a follower of the philosopher Bolland, be it only temporarily. He kept a lifelong interest in the philosophy of the sciences (in particular physics). His fame was based on his work on cosmic rays (1927). In addition to the standard package of mathematics for physicists, Bruins had made himself thoroughly familiar with Weitzenböck's theory of invariants, in the style of 'symbolic notation'.

At the age of 31 he became a teacher at the mathematics department. Needless to say that he was in no way the person to bring Amsterdam to the forefront of modern mathematics, as Freudenthal did. Even in his applied mathematics he cultivated an almost nineteenth century taste, books like Courant–Hilbert's *Mathematische Methoden der Physik* were not his cup of tea. For a temporary position Bruins was not a bad choice, but soon he was upgraded to a permanent position; on 6 September 1941 he was made conservator, Freudenthal's old rank, and on 1 July 1942 he was promoted to lecturer in analysis, a position that Freudenthal had not attained after ten years in Amsterdam. On 7 July 1943 Bruins gave his inaugural lecture '*Mathematicians and Physicists*'.

The appointment and the quick rise through the ranks surprised the insiders. Although nobody denied that Bruins was an intelligent and resourceful man, he would have been placed better at the physics department or at the institute of technology. In addition it was considered 'not done' to take the position of a discharged Jew. One might help out with the teaching, but to accept the vacancy on a permanent basis, and to seal it, so to speak, with an inaugural address was a demonstration of poor taste, to say the least.

After the war Freudenthal conjectured that this promotion would have fitted Brouwer's wish to make a return for him harder,[14] for although at that time there was no animosity between Bruins and Freudenthal, it was unthinkable that Freudenthal would serve in his old rank, as a conservator under Bruins, who was mathematically no match for him. The inaugural lecture had not escaped Freudenthal, and he mentioned it as an example of Bruins' curious sense of humour. Commenting on the present circumstances, Bruins told his audience, 'And finally this, the only fact, whereby the events of war could have had an inhibiting influence on the activities of a mathematician, is in fact provided by the stories about Archimedes.'

[13] Bruins (1938).

[14] Freudenthal to Comm. of Restoration, 9.VII.1945—not sent. Freudenthal to De Groot, 17.IX.1945.

As Freudenthal viewed it, 'a phrase which contains an insult for all mathematicians of good repute, who were at that moment in camps, and for the memory of those who were murdered by the Germans, for the students too, who as a consequence of the war and the German usurpation had to interrupt their study, and finally, an insult for our whole nation, where Dr. Bruins draws a line between the misera plebs on the one side and the mathematicians on the other side, who cannot be affected by the war'.

## 16.4 University—Resistance or Survival

For the present biography the wartime history of the University of Amsterdam is of course the first subject of interest; one should, however, not forget that the universities did not act in isolation, nor were they treated by the authorities on an individual basis. Events at one university usually had their repercussion at other universities. The University at Leiden was the first to protest on a large scale against the anti-Jewish laws of the occupation authorities, in reaction it was closed down in November 1940. The other universities had avoided open action; the views of the student bodies and the boards of the universities were diametrically opposed on this issue. The history of the universities during the war is to be found in the series of books of L. de Jong.[15] For the University of Amsterdam the books of A.W. de Groot and P.J. Knegtmans should be consulted.[16]

Once the Jewish staff members were removed from the universities, it was only a matter of time before the Jewish students would be the target of new decrees. Indeed, on 5 February 1941 secretary general Van Dam, in a confidential letter, informed the universities that Seyss-Inquart had ordered a *numerus clausus* for Jewish students. At first he had ordered the total ban of all Jewish students, but after a protest of the universities he had allowed them the compromise of a *numerus clausus*. He had practised the standard tactics of the Germans in Holland: first ask too much, and then grant the other party an illusion of a reasonable deal. The Dutch had to learn the hard way that negotiating with the Nazi government was not really possible, the government would change the rules at will.

When the Amsterdam rector, a man with a name that might invite confusion: Bertus Brouwer—no family, was informed about the plans, he immediately set to work in order to prevent strikes, protest marches and the like. He was by no means pro-German, or pro-Nazi, but, like many of his fellow authorities, he thought that by avoiding open clashes, and hence the closure of the university, he would be able to salvage the main part of the Dutch educational system. The matter was almost immediately discussed in a special session of the senate of the university. After a lengthy debate, the senate decided to send a letter of protest to Van Dam. But since this letter was sent to his private address, it had hardly any public effect.

---

[15] Het Koninkrijk der Nederlanden in de tweede wereldoorlog, de Jong (1969).

[16] de Groot (1946), Knegtmans (1998).

**Fig. 16.1** Gerrit Mannoury. [Courtesy J. Mannoury]

The retired, but still active, Mannoury had considered it his moral duty to stand up for his Jewish students, in his archive there is a draft of a motion he introduced in the senate:

> The senate is of the opinion that by any discrimination against the academic rights of the Jewish students in comparison to other students, not only the interests of the University and of science is harmed, but also the higher interests of society and humanity are seriously harmed, and it resolves to bring this notion to the attention of occupation authorities of The Netherlands.

In the meeting he argued that the rejection of the announced discrimination is based on two principles: that of freedom of thought and of justice. The principle of justice carried more weight because,

> Justice is itself a limit, a demarcation and a balancing of material and cultural interests of the individual and society, of group and subgroup, of majority and minority, located in the sub- and supra consciousness of human beings. The enforcement of this principle of justice demands a high level of self-control, that prevents us from acting under unrestricted passionate emotions of anger, hate or fear; and which allows us to test our feelings with reason. And that this self-control and testing is completely lacking in the mentality of which I spoke, is shown in every line of the documents, in which this mentality is advocated.

Mannoury's fearless defence of the values of our western culture in the face of the new barbarism was a brave act indeed. After all, even the senate contained members who were not exactly opposed to the new order.

The local political activity of L.E.J. Brouwer came to an end on July 29, 1941, when the public council meetings were banned by the occupation authorities. His membership of the council had not resulted in spectacular facts that would engrave his name in the hearts of the town people, rather he had been an ordinary, dependable member, just like thousands of other council members all over the country. There was one fateful event, however, that seemed innocuous, and even beneficial at the time, that was to cost him dearly after the war.

In their effort to eliminate the old traditional political, cultural and social institutions of Holland in order to create new institutions that would be at best indifferent towards the Nazi politics, and at worst actively support the National Socialist goals, the Germans had gradually suppressed all kinds of organisations, ranging from Scouting to the trade unions, from sanatoriums to welfare organisations. The next step was the founding of new centralised institutions, that either were actively or passively supporting the German war effort and Nazi policies, or at least were easier to control than the myriads of societies, institutions, clubs, charities, ... that were part of the rich fabric of Dutch social and cultural life.

In Germany charity had been monopolised by the *Winterhilfe* organisation. This supplanted all previously existing organisations, and as such it was an example of successful *Gleichschaltung*. What seemed more logical than copying this idea in Holland? Indeed, in June 1940 *Winterhulp Nederland* was founded. In spite of large scale propaganda actions, the organisation was immediately viewed as an enemy institution, and no Dutchman was going to support the street collections of it, if he could help it. In fact, the fund raising power of *Winterhulp* was negligible.

There was, however, another way to get control of the charity market, and this was another German creation, the *Nederlandse Volksdienst*, NVD.[17] The NVD was designed after the German example, which had replaced the traditional religious and social organisations. It should provide the care and support that was of old part of the program of the churches and the labour organisations, such as mother care, childcare, health care, holiday camps for the needy, support for the unemployed, etc. With the help of the confiscated property of secular and religious institutions, such as parish youth clubs, trade unions, the NVD, founded on 28 July 1941, started to compete for the hearts and minds of the Dutch population. Whereas the *Winterhulp* was a professional organisation, the NVD was intended to involve the population in a direct way: it was an organisation of members. As might be expected, this was inviting defeat, in the first four months of its existence the NVD, it attracted nationwide no more than three thousand members, and most of these were members of the NSB.

When it became clear that the NVD would under normal circumstances never develop into a healthy organisation with enough members to demonstrate national support, the authorities decided to put pressure on bodies and on persons in prominent places to promote the NVD. Brouwer was confronted with the phenomenon in a letter from Mr. Klaarenbeek, the Mayor of Blaricum. By what means Klaarenbeek, who was not an NSB sympathiser, was ensnared in the nets of the NVD remains unclear; usually some sort of veiled threats were involved. On November 21, Klaarenbeek, who had been made the head of the local *Nederlandse Volksdienst*,[18] sent out a circular letter, informing the readers of this new organisation, which in

[17] The people's support organisation of the Netherlands.

[18] *Buurtschapshoofd van de NVD te Blaricum.*

his words was 'not connected with any political movement or group', and which solicited the support of 'all persons with a warm heart for their fellow men' and which was to 'reach out to all Dutchmen in need of its help'. After describing the (in itself laudable) social goals of the NVD, the letter continued with the request to join the organisation. Klaarenbeek (naturally, one would say) did not invent this membership campaign himself, he was 'asked' to carry out a membership drive among the leading citizens of the town. The letter ended with the words 'In the (unlikely) event that you should be of the opinion that you will not be able to join as a member, I will appreciate to be informed by you, preferably with the reasons that led you to this decision.'

Confronted with this embarrassing request, Brouwer contemplated the matter[19] and came to the conclusion that Klaarenbeek—'whose patriotic feelings were above suspicion'—should in the general interest remain as long as possible in function during the occupation. As a (suspended) council member and deputy alderman, he felt that he had to support his mayor.

To run ahead of our history, Klaarenbeek remained in function until 1944, when he was arrested by the Germans, and at his release after the liberation, reinstated as mayor.

When Brouwer observed, moreover, that a charwoman of his institute, who was a hard-core communist, and active in the resistance, had received financial support from the NVD, he concluded that the organisation did not pose a serious threat: 'I was gradually strengthened in the opinion to see in the *Blaricum Volksdienst* a charitable institution, although modelled after the German pattern, nonetheless operating according to Dutch standards, that, since all our previous charitable institutions were suspended, in spite of its noxious origin, *après tout* more or less filled a social gap.'

After ample consideration Brouwer decided to join the NVD, he duly paid his membership[20] until 1944, as he stated after the war, when he found out that the control of the NVD was completely in the hands of the National Socialists—but by then he had already paid his dues for the whole year, so he could not withdraw from the NVD before the end of 1944. The matter will come up again in the next chapter.

The war years were certainly not easy or pleasant in any sense, one tried to muddle through as well as possible. In 1941 the hardships were more of a political than of a material nature. Food and clothing were rationed, one had to stand in line, keep an eye on the sudden appearance in the shops of goods such as vegetables, fish, toilet paper, ..., but one could survive if no emergency occurred. And this was exactly what happened to Brouwer. On March 11, 1941 his cottage in Blaricum was almost totally destroyed by a fire. At 8 o'clock in the morning a curtain close to the gas heater caught fire. Brouwer quickly acted and extinguished the flames in the curtains. But a few minutes later a penetrating smell told him that the fire had already reached the thatched roof. The fire alarm was raised and within ten minutes the fire

[19]Cf. Brouwer to Comm. of Restoration, 30.VIII.1945.

[20]The dues were one and a half guilder a month, not a big deal. It was rather the symbolic value of the membership that the authorities were after.

brigade was on the spot. Brouwer was already carrying his valuable possessions, in particular his scientific archive and books, out of the cottage. The commander of the fire brigade saw the house as his first priority; he judged that he could not spare his men for the salvaging of all this paper. He ordered his men to direct their hoses at the cottage. As a result the fire was extinguished, and the cottage was reduced to the four bare walls. Brouwer bitterly remarked, trying to save what he could from the soaking remainders—'So I will leave no *Nachlass*'. Indeed, the collection of letters and papers that was eventually left after his death was far from complete, and some sheets still show traces of the fire.

The builder who inspected the remains judged them a total loss. Fortunately, Brouwer could move into the villa 'De Pimpernel' on the grounds, which he had bought in 1937. One can, however, easily imagine what a traumatic experience it must have been for a 60-year-old scientist to see his house and scientific papers become the prey of flames.

On May 10, 1941 Korteweg, the founder of the Amsterdam Mathematics Department, died at the age of 93. Brouwer spoke at his grave.[21] Korteweg had played an important role in the faculty in Amsterdam, and in mathematics in Holland. He was a well-respected man, whose judgement was equally appreciated in the Faculty, the Senate, the Royal Academy and the Mathematical Society. Of the latter, he had been president three times, and he had been on its board for almost 60 years. A whole generation of applied mathematicians was educated and shaped by him. He was already a legend before he died, and ... soon forgotten. Eventually he was only known by the first name in the Korteweg–de Vries equation and as Brouwer's Ph.D. adviser.[22] The mathematical work he did for his Ph.D. adviser J.D. van der Waals is nowadays also fully recognised.[23] Korteweg was an able and original mathematician, he was magnanimous and modest. He fully deserves his place in the gallery of great Dutch scientists.

The effects of the occupation were clearly visible for anyone who wanted to look around. It was not just a matter of a military occupation with its logical military consequences, but it was a regime that wanted to impose its political views, in particular those concerning the extermination of the Jews.

## 16.5 Freudenthal's Fortunes

One would expect Brouwer to be pleased that Freudenthal had offered his personal assistance in mathematical matters. In the past Freudenthal had carried out many duties that relieved Brouwer from a dreary routine. He had, for example, written a report on the state of topological algebra, and the result came in handy when one of

[21] Brouwer (1941).

[22] Even his own department in Amsterdam thought it safer to add 'de Vries' to the name for its 'Korteweg–de Vries mathematics institute', when a name had to be chosen in the eighties.

[23] Levelt Sengers (2002).

Brouwer's students, Frans Loonstra, asked for a topic for his dissertation. Again, in April and May 1941, Freudenthal sent Brouwer a report on Loonstra's manuscript for his dissertation.[24] The report is a rather cautious attempt to let Brouwer know that it was not a masterpiece, but more or less a competent survey and completion of known work. The last sentence, for example, leaves little doubt of Freudenthal's view, 'Loonstra succeeded in understanding the paper of Schnirelmann, which is partly written in a very condensed manner, to supplement missing proofs, clarify some obscure points, and put the text into an easily readable form'. It is doubtful how much Brouwer had seen of the manuscript; Freudenthal said that he would not be surprised if Brouwer had not read it at all.[25] The doctorate was awarded to Loonstra on 11 July 1941 for the thesis *Analytische Untersuchungen über bewertete Körper*.

The persecution of the Jews went through various stages, each time the screw was turned a bit further. On June 29, 1942 the authorities announced their decision to deport *all* Jews to the eastern part of Europe, and on July 14 the first trains rolled from the Central Station in Amsterdam to the camp at Westerbork[26] (a small place in the eastern part of the country) to deliver the Jews that had complied with the summons.

The first intuitionistic follower of Brouwer, Belinfante, was also in jeopardy. He was Jewish, and somewhat like Blumenthal, he could not imagine that any government could wish to harm him. He was a conscientious citizen, and nothing could be said against him. In a way, he stood a chance of getting a special treatment. He belonged to the group of Portuguese Jews whose fate was not yet settled. There were casuistic discussions among the German experts on Semitic affairs about this group. Were they real Jews, could they be given permission to emigrate to Portugal? In that typical Nazi thoroughness the matter was discussed by an assortment of departments and groups, and in the meantime the Portuguese Jews were granted deferment of deportation.[27] Eventually the group was put into camps anyway. As it happened, Belinfante was noted by a Dutch author, Durlacher, who was an inmate of the camp Theresienstadt at the same time as Belinfante. In his book, 'Quarantine', Durlacher recalled his encounter with Dr. Belinfante. A small man with glasses, who noted the author during a lecture of a philosopher on Hegel, and offered to teach him mathematics. Belinfante taught the young man his favourite subject, infinite series. 'In the following weeks simple series acquire meaning for me in a few evenings, in spite of the tiredness that rests on my eyelids, and in spite of the gnawing hunger. The thoughts of my teacher often wander, and his wife excuses the silences with a reference to his health. But I know that it is the disease of which we all suffer: fear and sorrow.'[28] Belinfante spent eight months in Theresienstadt, on 12 October 1944 he

[24]Freudenthal to Brouwer, 28.IV.1941, 23.V.1941.

[25]Interview, 22.XI.1988.

[26]*Durchgangslager* (transit camp).

[27]The matter is discussed in Presser (1965).

[28]Durlacher (1993).

was transported to Auschwitz, where he died two days later. In his commemoration Max Euwe wrote that Belinfante did not use the opportunities he had to go underground, distrust was not part of his world of thinking. In Euwe's words, 'A great child, but a great man.'

The fortunes of Freudenthal, one of the central figures in this period of Brouwer's history, are by no means exemplary for what happened to the Dutch Jews at large, but they illustrate well the unpredictability of events under a systematic persecution. Freudenthal had married a Dutch wife of non-Jewish descent, thus he was exempted from deportation (although a wise man did well to doubt the permanence of such laws). This did not mean that he was out of danger; anybody could be arrested at any moment, and the Germans had no moral scruples to deport a Jew against the rules. A justification could always be found.

Freudenthal was attempting to lead a life that preserved a certain degree of normality, he did mathematics, talked to colleagues and students, submitted papers; all this as a private citizen, who stayed away from the mathematical institute.

In spite of the mixed marriage, there was no certainty for Freudenthal under the German regime; in general there never was, not for Jews and not for non-Jews. This was brought home to Freudenthal when, as a foreign Jew (with a mixed marriage), he had in 1942 to register for an emigration permit. The *Obersturmscharführer* who happened to check the passports discovered that in Freudenthal's passport the stamp 'J' was missing. This was perfectly understandable, as the J's dated back to 1938, whereas Freudenthal's (German) passport was older, and living in Holland he had, of course, not felt the necessity to get a stamp somewhere in Germany. But a sin against the legal rules could not be tolerated. So Freudenthal had to present himself at the Gestapo headquarters in the Euterpestraat in Amsterdam.[29] There the officer in charge hardly looked at him, and said 'six weeks'. As a result from 17 February till 30 March Freudenthal was incarcerated in the central prison at the Weteringschans, which was partly used as an SS-prison. So, all of a sudden, Freudenthal had disappeared from the face of the earth; nobody, not even his wife, was informed. Fortunately she was a determined woman, with a perfect command of the German language. As a matter of fact she had studied German language and literature with the same professor Van Dam who was now the secretary general of the department of education. Telephoning all police stations and German agencies she could find, she finally got by chance through to the secret telephone number of the Gestapo office. Her German impressed the man at the other side of the line so much that he gave the '*gnädige Frau*' all the information, including the promise that her husband would be out after six weeks. And indeed, after six weeks of fear and uncertainty Freudenthal was a free man again.[30]

Soon Freudenthal was his usual active self again; he reported in April and May on Loonstra's manuscript, he recommended a paper of J. de Groot on extending cer-

---

[29]During the war, the name 'Euterpestraat' had become a synonym for all the Gestapo horrors, so much so that the word itself held terror for the average citizen. After the war the street was renamed after Gerrit Van der Veen, a renown member of the resistance.

[30]See Freudenthal (1987b), p. 205 ff.

tain homeomorphisms for publication in the *Indagationes*. De Groot was a student of Schaake and he worked on topology under Freudenthal's supervision. Schaake, one of the Groningen mathematics professors, was not an expert in the area, so it was a natural move for De Groot to turn to Amsterdam for support. Brouwer, as a rule, left these things to Freudenthal. As a consequence, Freudenthal, although not a professor, was de facto the Ph.D. adviser of a number of young mathematicians. Brouwer was prepared to submit the paper of De Groot to the Academy modulo some definitions and references. On July 6 he wrote Freudenthal that De Groot had not adopted all of his suggestions, and that De Groot had failed to react to Brouwer's subsequent letter. Could Freudenthal perhaps check what was the matter? Freudenthal's mediation must have been effective, for Brouwer confirmed to Freudenthal the receipt of the corrected manuscript, which he intended to submit to the Academy.[31]

For a long time no letters were exchanged, until on 24 May 1942 Freudenthal received a somewhat testy letter in which Brouwer demanded the return of the books that Freudenthal had taken out of the library of the mathematical institute *against the rules*. The library had always been Brouwer's pet, he had exerted considerable pressure to get a separate library for the mathematicians, which was, after the example of the Leiden reading room, only for consultation on the spot. Only on special request a book could be taken out. Furthermore he asked Freudenthal to instruct the students to return their books, borrowed with Freudenthal's permission, and finally to turn over the complete Compositio administration to Bruins.

After the war, when Freudenthal was fighting for his rehabilitation, he recalled the gradual erosion of his relation with Brouwer.[32] Until the spring of 1942, he wrote, the relation was with the exception of a few incidents, good. But after Freudenthal's imprisonment the situation seemed no longer the same. In spite of the fact that Brouwer was informed about Freudenthal's arrest, he did not contact or comfort Freudenthal's wife (who had three small children and was expecting the fourth). His first letter after Freudenthal's discharge from prison was the formal demand of May 24—no word of personal concern. 'After receiving that letter from professor Brouwer, it was clear to me that he had not presented my note in the April meeting of the Academy, and that he wished to provoke a quarrel by means of a sharp letter, in order to free himself from the unpleasant task to present my note.'[33] Determined to ignore the provocation, Freudenthal wrote thereupon Brouwer that he had called the Academy to inquire after De Groot's note, and that it apparently had not been submitted in the April meeting. Out of concern for De Groot (and Loonstra) he therefore asked Brouwer not to postpone the submission any longer, as 'one of the interested parties could or would possibly make other arrangements'.[34] If Freudenthal had planned to alienate Brouwer, he could not have followed a better strategy. If there was anything at all that Brouwer could not tolerate, it was interference with his

[31]Brouwer to Freudenthal, 26.VII.1941.

[32]Freudenthal to Van der Corput, 24.VIII.1945.

[33]Ibid.

[34]Freudenthal to Brouwer, 26.V.1942.

publication policy. Freudenthal's letter reached Brouwer in Roosendaal, where he was conducting the annual examinations of gymnasia and high schools; he replied, 'You and the two other gentlemen involved seem to forget that the members of the Academy constitute an editorial board for the Proceedings, and that editorial boards of journals are no printing automata!'[35] This letter was followed a day later by an envelope with the manuscripts, and a brief note, 'Hereby the manuscripts of you and the two other impatient gentlemen are returned for "making other arrangements".' Freudenthal interpreted the letter as the definite end of their relationship. In his reply to the letter of May 24, he observed that he had immediately returned the library books, but he could not recall having given students permission to take books home. As to the Compositio archive, it was all the time in a cupboard in the assistants room, but without the permission of the *Reichskommissar* he could not enter the institute to hand it over to Bruins. That, roughly, was what there was to say. Freudenthal just added that he had not passed on the message of May 29 to the other authors. 'They might perhaps feel little appreciation for the insults, that one can only bear out of respect for a great mathematician.'

After the war, when confronted with this episode, Brouwer described the events as follows: 'just before a meeting of the Academy I received some 100 pages of manuscript. I took these on my trip. A bit later I got a letter with a reprimand for my negligence. Subsequently he threatened to publish elsewhere. Then I have returned it.' The accounts are not all that different, be it that Brouwer and Freudenthal interpreted the events completely differently.

Freudenthal had in fact been supervising three Ph.D. theses; he was not a professor, so he could not officially act as an advisor, but in practice he had been the one who showed Loonstra, De Groot and Van Heemert their way through the intricacies of modern topology. In a way the style of the topology involved was very much that of Brouwer and Alexandrov–Hopf. The new, more algebraically oriented topology had not yet become familiar in Holland.

Loonstra had already obtained his doctor's degree under Brouwer, and it would have been natural enough if the two other candidates would also have chosen Brouwer as their Ph.D. adviser, he was after all the proper expert to handle topology dissertations. The difficulties with the manuscripts for the Academy had however cast enough doubt on a speedy and efficient procedure, so that De Groot had chosen to obtain his doctorate in Groningen with Schaake. On 13 October 1942 he defended his dissertation, *Topological Studies. Compactification, extension of mappings and Connectedness*.[36]

Van Heemert, a student of Brouwer, had also written his dissertation with hardly any consultation of Brouwer. He had worked with Freudenthal on a topic that was studied first by Brouwer: indecomposable continua.[37] On 1 June 1942, Freudenthal had sent a strongly positive report on Van Heemert's work to Brouwer, and it looked

[35] Brouwer to Freudenthal, 29.V.1942.

[36] de Groot (1942).

[37] Cf. p. 141.

as if the doctorate was within reach. A couple of weeks later Freudenthal wrote a disconcerted letter, in which he informed Van Heemert of the unfortunate matter of the manuscripts of De Groot, Loonstra and himself for the *Indagationes*. He advised Van Heemert to present the manuscript of his dissertation directly to Brouwer. When Van Heemert showed some hesitation, he urged him not to antagonise Brouwer, and just submit the manuscript. Brouwer, Freudenthal wrote, had developed a strong dislike for him. The reason he could not fathom. 'This incident surpasses in intensity and candour all earlier ones, and I would, when asked, have to refuse resolutely to take up my old position, although this incident has—in my opinion—nothing to do with politics, but it is one of his normal tantrums, only stronger than before.'[38]

Taking Freudenthal's advice, Van Heemert wrote to Brouwer that he would like to send him the manuscript of his dissertation, explaining that his military service and a teacher's position had caused a delay in the research as discussed in 1939.[39]

After the reminder of 13 July, Van Heemert sent a second request to Brouwer. He had hoped to finish the discussion of the content of his dissertation during the summer vacation. But in case Brouwer had no time for him, would he give him permission to take his dissertation to the Groningen faculty? Brouwer replied on 28 July that he had a large backlog of all sorts, and that he had no time to read the manuscript for the next half year. He therefore advised to send the manuscript to Loonstra, who had agreed to read it. Van Heemert had in the meantime contacted Van der Corput, and explained the situation. Apparently he had little hope that Brouwer would oblige, and thus he expressed his preference to defend his thesis in Groningen.

Brouwer's suggestion, to ask Loonstra to read the manuscript, did not appeal to Van Heemert. What could be won by involving someone who was only marginally informed rather than relying on Freudenthal's expert opinion? Van Heemert obviously wanted to keep his options open, so he called Brouwer and informed him that further deliberation was required before he could decide what to do. This clearly piqued Brouwer; four days later he wrote that he was amazed that Van Heemert had not reached a decision. 'I see now in this an argument to advise you to look for a Ph.D. adviser whose suggestions you can follow without protracted deliberation.'[40] Now Van Heemert was free to go where he wanted. When asked for advice, Van der Corput told Van Heemert that the proper advisor in Groningen was Schaake. The latter had no wish to alienate Brouwer, so he urged Van Heemert to stick with Brouwer. But eventually he accepted the dissertation, and on 5 February 1943 Van Heemert received his doctorate in Groningen.[41]

The conflict between Brouwer and Freudenthal—for a conflict we may call it—was of a totally different nature than earlier conflicts. Looking at the relation between the two, one gets the impression that here is a classical case of *incompatibilité d'humeurs*. Both men had too much in common to live in peace together; both were

---

[38]Freudenthal to Van Heemert, 26.VI.1942.

[39]Van Heemert to Brouwer, 29.VI.1942.

[40]Brouwer to Van Heemert, 3.VIII.1942.

[41]van Heemert (1943).

spiritual, highly intelligent, well read, ambitious, both had a quick wit and a sharp tongue. Brouwer, who in spite of his many public obligations, jealously guarded his private domain, and resented attempts from others to intrude on his privacy, no matter how slightly. Freudenthal came dangerously close to the impregnable citadel of Brouwer's personality. There is, by the way, no doubt of Freudenthal's sincere admiration for Brouwer, nor of Brouwer's recognition of the younger man's gifted personality. And so the classical Saul-David situation arose, more or less triggered in 1937, at the time of Freudenthal's conservator appointment (cf. p. 648). And as experience shows, these things get worse, unless one has the greatness to make a clean breast and admit one's own shortcomings. In a way Brouwer had been Hilbert's David, but the two could easily avoid personal contact. In the case of Brouwer–Freudenthal this was far more problematic. Moreover, the war had made it improbable that Freudenthal could be promoted to a professorship elsewhere, so Brouwer and Freudenthal were condemned to each other's company—even the unintended German intervention did not really separate them, as Freudenthal remained in Amsterdam and continued his activities.

There is one example of Freudenthal's persistent presence in Dutch mathematics; the Dutch mathematical society traditionally published prize questions of a purely academic nature. The crowning of a prize essay was considered a sufficient honour, no prize money or any other benefits were attached. Participants sent in their essays, which were judged blindfolded. Only after an essay was crowned was the envelope with the name of the author opened. During the war Freudenthal had submitted a solution to one of the prize problems, and his submission was inspected by the committee and awarded. This highly displeased the government department for education, science and culture protection. A disapproving letter followed;[42] the department informed the mathematical society of its displeasure in threatening terms: 'Freudenthal, who is a Jew, has participated in a Prize contest, and been awarded a prize. The Commissar-General finds it remarkable that your board lacks the perception to abstain from granting the award in such a case, and he reserves the right to investigate the case further.' In fact the matter was not followed up, but it shows that at every level in society the authorities kept a tight watch. It also showed that Freudenthal had no intention to make himself invisible.

Confronted with Freudenthal's undaunted spirit and art of survival, Brouwer probably started to think of means to transfer Freudenthal to a place where he could make a career without further clashes; later developments are consistent with that possibility. The question to ask here is, did Brouwer wish Freudenthal ill? Again, one can only guess, but the answer appears to be negative. It is more likely that he felt uncomfortable, or even threatened by the presence of a person who wanted things—including Brouwer—to get moving. In plain words, Freudenthal got on his nerves. The situation is comparable to that of the rabbi in the musical Anatevka who

[42]Department of Education to Wiskundig Genootschap, 11.I.1944. Note the stubborn way in which Nazi authorities concentrated on side issues, at a time that there were more serious (military) survival issues to be considered. Perhaps this is a universal characteristic of bureaucracies.

was perfectly willing to bless the Czar, if only the Lord 'kept him far away from here'.

In 1942 Brouwer resumed his publication activity for a brief spell. He communicated in March, April and September three short notes to the Academy.[43] The first note, *On the Free Genesis of Spreads and Functions*, was a translation of a letter to Heyting, written on 30.III.1936, which dealt with an extension of the notion of function. At various places in the early publication on intuitionistic mathematics functions were defined as 'laws', but since then Brouwer had realised that the procedure of generating points (say in the plane) as choice sequences was equally applicable to functions (think of functions as points in a function space).[44] The paper also contained an explicit formulation of the higher-order restrictions in the definitions of the notion of choice sequence.

The next paper, *The Representing Spread of the Continuous Functions of the Unit Continuum*, carried out the construction, foreshadowed in the preceding paper (this, by the way, is literally the construction of Brouwer 1918a).

Finally, the last note, *Proof that the Notion of Spread of Higher-order is not appropriate as a Fundamental Notion of Intuitionistic Mathematics*, showed that one could reduce a generalised sort of choice sequence, which allowed the chosen items of a choice sequence to be choice sequences themselves, to ordinary choice sequences. All of these publications had, so to speak, been waiting in a drawer to be called up. In this particular case the publication had an external cause, i.e. a passage in a recently published book of E.W. Beth, *Introduction to the philosophy of mathematics* (2nd edn.).[45]

Although a certain amount of international activity was taking place, the Dutch mathematicians became more and more isolated; Brouwer was no exception, his international contacts had been modest before the war, but now he maintained a self-chosen isolation. When in 1942 he was selected by the Secretary General of Education, Professor van Dam, to represent the Netherlands at a conference in Italy, he flatly refused to go.[46]

The Academy kept up its activities, albeit in a smaller scale, and the mathematics and physics section continued their monthly meetings. Brouwer, apparently, was sufficiently impressed by his most successful intuitionistic follower, Arend Heyting, to start lobbying for his election in the Academy. In 1942 these efforts were crowned with success. Heyting was not without worldly wisdom, he took no steps to occupy his place in the Academy. A judicious decision, as the academic community did not appreciate appointments by a government that was only an extension of the

---

[43]Brouwer (1942a, 1942b, 1942c).

[44]Brouwer seemed to have forgotten that he had already established basically the same facts in Brouwer (1918a) (pp. 14–15). There he showed that the set of continuous functions from $[0, 1]$ is of the same cardinality as Baire space, but overlooked the fact that the left hand side consisted of lawlike elements and the right hand side of choice sequences. The argument is correct in Brouwer (1925a), p. 253, where the lawlike part of $\mathbb{N}^{\mathbb{N}}$ is considered.

[45]Beth (1940).

[46]City Council to Secretary General, 23.IX.1942.

occupying forces. The historian of mathematics, Dijksterhuis, was not as cautious as Heyting, and his membership of the Academy was cancelled after the war.

Let us now return to the fortunes of the university during the occupation. No systematic attempts had been made to turn the universities into Nazi, or even pro-German, institutions, but the elimination of the Jews had been systematically carried out already in an early stage.

The moment of truth had arrived on November 26, 1940 in Leiden, when after the dismissal of all Jewish professors and staff members the eminent Law professor Cleveringa held his famous address in the auditorium of the University of Leiden, summing up the events and the lack of legal basis for dismissing Jewish colleagues, going even further by pointing out the violation of the Geneva rules. The students lost no time to conclude that a strike was the only possible and justified answer. The German General Commissioner for Law and Administration, Dr. Wimmer, immediately retaliated by closing down the University of Leiden. The other universities were spared that fate, but the occupation authorities had made it crystal clear that they would not tolerate student unrest. That was, certainly in the beginning of the occupation, not difficult. The Dutch were as a rule a law abiding nation and the students too were not used to political issues and actions (the Socialist or Communist Student Associations made up a small portion of the total student population); it was, however, a fact that the majority of the students was downright opposed to National Socialism and to the occupation forces.

Since our story is mostly concerned with the events at Amsterdam, we will as a rule not comment on Dutch universities in general. When the Jewish students were excluded from the student corporations, the corporations considered this unacceptable, they decided that this was a reason for their dissolution (October 1941). Apart from confiscation of the property of the corporations, no further action was undertaken by the authorities.

## 16.6 The Declaration of Loyalty

The first violence in the Amsterdam student world erupted on January 22, 1942, when a bomb exploded in the building of the NSB student organisation. The German representative in Amsterdam concluded (without proper evidence) that the attack was carried out by students, and the highest German authority, Dr. Seyss-Inquart, took the matter into his own hands. He decreed that six professors should be dismissed without pension, and sent to a concentration camp,[47] where 10 students should also be incarcerated. The University judged that such a draconian measure made it impossible to carry out its task any longer and the rector had informed the authorities that he himself and the leading professors would resign if the plans were carried out. After fervent protests and hectic conferences of various groups of authorities, the Germans reconsidered the sentence.

[47] Romein, Tielrooy, Hazewinkel-Suringa, Kollewijn, Kranenburg, and H.J. Scheltema.

The Mayor of Amsterdam, Voûte, an appointee of the occupation forces, had joined the Dutch branch of the German SS. In the mean time Voûte had been granted an audience with State Commissioner Seyss-Inquart, who had subsequently modified the sentence:

1. Nobody would be fired at this moment. Possible dismissals at a later time would not be connected with the bomb attack.
2. Fifty citizens of Amsterdam would be arrested.

In addition it was promised that a certain professor (Hazewinkel-Suringa), who had already clashed with an NSB colleague, would not be arrested.

The reaction of the State Commissioner was typical for the Nazi way of handling opposition, a tactical shift would make the sentence cosmetically more acceptable, and the opposition would lose much of its thrust and coherence. The revision of the verdict could, however, hardly be considered a victory for the university, as some later claimed it to be.

Eventually the Germans arrested a large number of citizens of Amsterdam, including four professors and three students (the arrest of three more specially designated students failed). Pannekoek, the retired professor of Astronomy, was also arrested, but he was soon released—his communist past was not sufficient reason to keep him in custody.

Soon, however, the Nazi's ordered the dismissal of nine professors. A meeting of the Academic Senate was immediately called, and long and earnest discussions followed. One faction proposed that unless the dismissals were repealed, Seyss-Inquart would be asked to close the university. The rector and secretary of the university tabled another motion: unless the dismissals were repealed, the university would issue a strong protest.

The arguments pro and contra were many and varied, but in the end the view of the radicals boiled down to: we and our students cannot and will not accept dismissals unrelated to the basic tasks of the university. Therefore we must act and shield our students. Hence we should, if necessary, discontinue the activities of the university. The other side pointed out that at the closure of the university, staff and students risked forced (or at least non-voluntary) labour in Germany. In the end the moderates won.

At the opening session of the university, looking back on the academic year 1941–42, rector Brouwer could report only misfortunes. Six students had died (he did not mention that five of them had died in concentration camps), twelve professors had been dismissed, and since August 1942 Jewish students were no longer admitted to the universities. His successor, rector Deelman, had no pleasant prospects.

The next move in the battle for the universities was made again by the Germans. On 5 February 1943 General Seyffardt, who was in charge of the Dutch Legion of the SS, was shot by the resistance. That same evening Reydon, the secretary general for Mass education and the Arts, was wounded in an attack, and his wife was killed. Almost immediately General Rauter, assuming that students had been involved in the attack, raided the Universities of Amsterdam, Delft, Utrecht and Wageningen. At half past ten on Saturday morning machine guns were mounted on the lecterns, and a large number of students were arrested and taken to the concentration camp in Vught (near 's-Hertogenbosch in the southern part of the country). In this case the rector of the Amsterdam university and his assessors took a brave and determined step: until further notice all lectures and practical instruction were to be suspended. Surprisingly enough, the Germans did not act on this provocation.

The presentation of the intricate story of the universities in general, and in particular that of Amsterdam during the occupation cannot be told here in full detail. The reader is referred to the books of De Jong, De Groot or Knegtmans. For the following episode, the reader may recall that the battle of Stalingrad had been lost in January 1943, and that the Africa campaign was in serious trouble after the battle of El Alamein. The Germans were for the continuation of the war depending on labour from the occupied countries. And so after the attacks on Seyffart and Reydon, the policy of the Germans with respect to the universities had two objectives: the prevention of unrest and obtaining extra labour force for Germany.

After rumours about German intervention had been circulating, causing considerable excitement, the Secretary General for Education Van Dam unexpectedly announced on 10 March 1943 the following *Act for the safeguarding of the order at universities and institutes of higher education*, which laid down that any student following courses or taking part in examinations had to sign what became known as the *declaration of loyalty*:

> The undersigned . . . , born at . . . at . . . living in . . . solemnly declares hereby that he will on good faith obey the laws, ordinances and other commands in force in the occupied Dutch territory, and will abstain from any action directed against the German Empire, the German army or the Dutch authorities, and also of actions and behaviour, which endanger the public order at the institutions of higher education, in view of the present circumstances.

This was a curious document indeed. It was a mixture of general civil jurisdiction with what traditionally belonged to the jurisdiction of the universities. The declaration seemed to be inspired by the wish to separate the 'friendly' students from the opposition. Furthermore, it was a reliable device to get people entangled in a web that could at any moment be made more complicated, more compromising. The registration of the Jews was a similar device—in itself harmless, but potentially lethal. Finally, and most people realised this, the relation citizen–authority had fundamentally changed. After 1940 all the advantages were at the side of the authorities, all regulations could be repealed, all civil rights could be abolished.

Almost at the same time the State Commissioner, Seyss-Inquart, issued an ordinance of even greater consequence, which supplemented and strengthened the status of the declaration of loyalty. It laid down that:

1. The Secretary General of the Department of Education can impose a *numerus clausus* for any subject (discipline) at any university.[48]

[48] Not just an entrance restriction, but applicable to any stage of an academic study.

2. Every student who finishes his studies has to enrol in the *arbeidsinzet* [an obligatory period of labour for projects determined by the authorities, which usually meant projects supporting the German war effort, be it directly in military industries or as a replacement in other projects (e.g. hospitals, technical industries, ...].

The practical consequence of this ordinance was that any student who did not sign the declaration of loyalty had to terminate his or her study, and thus, being unemployed, automatically was drafted into the *arbeidsinzet*. The latter invariably meant a period of (forced) labour in Germany.

The publication of the above ordinances created understandably considerable unrest among the students, their professors, and advisors. The reactions were of two kinds: anyone who signs the declaration gives moral support to the enemy, or: the declaration is legally meaningless, signing does no harm. Even the spiritual guides of the Dutch were in confusion, e.g. the Archbishop of Utrecht declared: 'The declaration can be signed in good conscience; I consider, however, the signing as an anti-patriotic act.'

For the senates of the universities and for the individual professors the declaration posed a serious problem. Although it did not concern them, it concerned their students and in the end the position of the Dutch universities. Up and down the country the matter was debated in all kinds of boards—senates, faculties, laboratories, ....

In Amsterdam an emergency meeting of the Senate was called on March 26, 1943. At the meeting a draft of a statement of the Senate, to be sent to the Secretary General of education, Van Dam, was presented to the members. It contained comments on the *numerus clausus* ordinance, mainly to the effect that if there had to be one, the senate should determine its size, and that no student, who had already enrolled, should be subjected to it. It remarked that no unanimous advice as to the signing of the declaration of loyalty could be agreed on. In addition it included the refusal to resume the lectures until all students had been released from the concentration camp in Vught. After pointing out that some students possibly could not sign the declaration on moral grounds, and that some members of the senate would feel it intolerable if they had to continue their teaching without these students, the draft concluded with the statement that if both ordinances were carried out in the given form, the members of the Senate had no choice but to resign.

The last sentence was a clincher, it actually amounted to the closing of the university of Amsterdam. A substantial debate ensued with arguments on all sides—which in this case basically boiled down to either a flat refusal, or a compromise of some sort. It is important to keep in mind that the meeting was attended also by persons of National Socialistic leanings such as Weitzenböck.[49] It must be pointed out that even though Weitzenböck was known to be pro-German, so far no indications or evidence have been found that he ever denounced or harmed persons. Yet, ... one never knew!

[49] In fact, one lecturer sent reports of the meetings directly to the chief of the *Sicherheitsamt* (security service). Knegtmans (1998), p. 181.

At the time of the meeting 3600 forms for the declaration of loyalty had been stored in the safe of the university. In spite of the pressure of the President Curator (i.e. the NSB mayor) none had been laid before to the students to sign.

In the discussion about the draft and its contents a number of views were defended. It was argued that the draft admitted that the Senate could not issue a unanimous reaction, and that it was unwise to advertise this.

Brouwer, who had been unusually low-key during the preceding Senate meetings, had this time decided to make himself heard. He had prepared a number of amendments, all designed to weaken the formulation of the draft and to improve the chances of a compromise. His main point, and indeed the only one that was discussed, exclusively dealt with the last line, with the threat, or rather—as the Rector said—the factual statement of a collective resignation. Brouwer clarified and defended his amendment with the words:

> My version [of the draft] differs from the one of the Rector mainly with respect to the final passage, which in the form of the Rector, strikes me as objectionable, and this for two reasons, which I would elucidate as follows: We want to try to get a favour of an authority that has an absolute power over us. This is a reality for which none of us, no matter how he assesses this situation, can close his eyes. If we really want to see our wishes granted and our expectations fulfilled, then we first have to make an effort to create the indispensable atmosphere of goodwill and understanding. And for that no worse method can be imagined, than—at least indirectly—to send a request to an authority that considers the withdrawing of one's manpower from the community as one of the most unbearable crimes, ending in an ultimatum-like threat of strike. A second ground for finding the final passage objectionable, I deem even more serious, namely that, when the threat of strike by their teachers becomes known among the students, it will inevitably lend support to possibly already existing agitation for the refusal to sign the declaration of loyalty, and which therefore would cast a serious co-responsibility on the Senate for the indescribable misery, which this refusal could bring over our students.
>
> The new version of the final passage proposed by me intends to express to some insights, wishes and expectations, just as the original one, with the deletion of the strike threat.

Reading Brouwer's arguments, one gets the strong impression that he had no illusions about the willingness of the authorities to act or think fairly and that therefore—in view of the ruthlessness of the occupation forces, as demonstrated in the brutal suppression of the so-called 'February strike'[50]—one has to take care not to antagonise them. His information about the policies of the Nazi's in Germany may have influenced his views, but he failed to take into account the absolute and calculated unreliability of the German authorities. What good does it do the mouse to please the cat?

[50] February 25 and 26, 1941.

After a long list of speakers had addressed the Senate, the Rector asked for a vote. Brouwer's amendments were rejected by a solid majority.

With the approved draft the Rector went to a meeting of the rectors of the universities to discuss a collective reaction. This meeting did not yield much; it proposed to deal with the numerus fixus matter, in the hope that this would yield a bargaining position with respect to the declaration of loyalty and its consequences. The rectors clearly misjudged the situation, they were not dealing with polite Dutch civil servants, but with the State Commissioner himself. After a discussion of their proposals with the Secretary General, the State Commissioner was informed of the views and proposals of the universities. Seyss-Inquart ignored the proposals of the rectors and simply insisted on the complete introduction of the ordinances, including the signing of the declarations of loyalty. So suddenly the universities found themselves short of time, their adversary was not so naive to play the game on their terms.

The Senate in Amsterdam reconvened on April 8 to discuss the new situation, but in the meantime the students had organised themselves and they informed the meeting that roughly 1400 students had signed a letter to the Senate to the effect that they would not sign the loyalty declaration. This meeting hardly achieved anything, the firm resolutions of the previous meeting seemed to have been forgotten. The promise of collective resignation was forgotten by all but a few. The central question, 'should the students be advised to sign the declaration of loyalty?', proved too tough for a clear 'yes' or 'no'. When no firm view on this issue emerged, Brouwer proposed to leave the matter to the discretion of the rector: [51]

> The Senate, taking into account that neither the law of higher education or any other law or ordinance, lends it authority to undertake any action in the matter of the declaration of loyalty submitted to the students, expresses its confidence in the Rector Magnificus, and passes on to the business of the day.

Eventually no decision was taken.

The above simple resumé of the crucial meetings of the senate does no justice to the sincerity and emotions of the participants. There were basically two views on the matter of 'signing or not': that of the rector and his assessors—signing is permissible and advisable, and that of the group called '*het hooglerarenverzet*' (the professors resistance group)—under no circumstances sign the declaration. Between these two opposites there was a scale of opinion. Apparently the two extremes were not able to get a majority. Brouwer's modest contribution had steered a kind of middle course. For some reason the professors resistance group had taken Brouwer's interference badly, it was explained as obstruction of their resistance. The reason is obscure; perhaps Brouwer behaved provokingly with respect to the group, perhaps his words were taken more seriously than the group considered proper. It may be remarked that the group was considered with a measure of reservation by some of the colleagues;

[51] The fact that Brouwer's role in the Senate is discussed here should not give the reader the impression that he was a prominent member. He used his right to speak and to submit amendments like everybody else. That he is mentioned so often in the context of the Senate, is because he happens to be the subject of the present biography.

the unwavering righteousness of a number of members of the group could easily be confused with arrogance.

On April 6 the Secretary General of education, Prof. van Dam, went on the air to address the nation, explaining the meaning of the declaration and exhorting the students to sign. One day later, Bolkestein, the Dutch Minister of Education in exile,[52] presented the views of the legitimate government on the declaration in a broadcast from London. Bolkestein was quite explicit in his address, 'The country (*het vaderland*) and the government also demand that the president curators and the rectors of the universities abstain from any co-operation concerning the acquisition of the declarations', and to the students he said: 'The country demands that you don't do this.' The general interpretation was that the London government went further than a fatherly advise, that it in fact ordered the rectors and students to ignore the loyalty declaration. In how far this is a legally correct view is debatable, but those who did sign the declaration probably viewed the London parole as just a patriotic advice. At this point there was at least no longer any misunderstanding, clear instructions were given, albeit by London radio and imperfectly received[53] by a part of the Dutch population.

Reading accounts of the Senate meetings discussing the declaration of loyalty, one should not get the mistaken idea that it was a topic of calm debate among the Dutch population. Every measure, be it directly proclaimed by the German authorities, or indirectly by the Dutch, acted divisively. A good patriot rejected it out of hand and if you did not, you were almost certain to be considered guilty of high treason (*landverraad*). There was good and bad, and nothing in the middle. Brouwer, the individualistic thinker, either was not aware of this wartime dichotomy, or he chose to value his intellectual integrity over the wartime national feelings.

As we have seen, Brouwer thought that it was of the greatest importance that the students should not be deported to Germany, moreover he considered the signing of the declaration an act without legal justification, and hence without consequences. In the opinion of a number of members of the Senate, the declaration could at any moment be renounced by those who had signed it. It is in this respect instructive to note that even some staunch Marxists thought the signing unobjectionable, and even good strategy.

Brouwer went so far as to go beyond the advice of the Senate, he posted on April 8 at the bulletin board of the Mathematical Institute a note informing that in his opinion and that of his colleagues, the students could sign the declaration without scruples.

> The undersigned are of the opinion that the obvious interpretation of the declaration of loyalty given to sign to the students, is such that students can sign this without any essential objection, and that their own interest and that of the country requires that they sign the declaration.

---

[52] Not to be confused with Brouwer's fellow student H. Bolkestein, the later history professor in Utrecht, cf. van Dalen (1984), Wiessing (1960).

[53] The broadcasts from England were continually jammed. Listening to the English radio was moreover strictly forbidden by the Germans.

In the hypothetical case that students who have signed, feel that they can no longer endorse it, they can always revoke their declaration and discontinue their studies.

L.E.J. Brouwer
E.M. Bruins
A. Heyting

The note was removed at Brouwer's orders on the 12th; on the 19th the note was again put on the bulletin board. On that day the Senate was convened to discuss a letter to the Secretary General of Education, dealing with the ordinances of the Commissioner of State and the declaration of loyalty. The authorities forestalled a possible protest by the Senate by simply forbidding the meeting. This did not stop the rector from dispatching the letter that had been drafted, but no discussion and no amendments had been possible. The letter stressed that since no answer had been given to the preceding letter, the Senate saw no grounds to resume teaching. Brouwer had, as he stated later in a letter to the postwar tribunal, expected that this particular senate meeting would settle the 'sign' or 'not sign' advice. So when the meeting was cancelled, he decided to stick to his old position. Apparently he was so poorly informed that he was not aware that the deadline for signing was April 13. Brouwer could not have known the contents of the letter of the rector, but he felt that the students had every right to expect advice from their mentors, and so he reissued the old advice, which represented his sincere opinion. He did not consult Bruins and Heyting on this occasion.[54]

It was his bad luck that Brouwer, although he, Bruins and Heyting had heard about the radio address of Minister Bolkestein, did not, according to a later statement of his, know the contents. Indeed, most professors who advised the students to sign, contrary to the instructions of the government in exile, later pleaded ignorance of the details of the radio speech. Given how difficult it was to listen to '*Radio Oranje*', the London broadcast of the government in exile, it was indeed quite likely that a large portion of the population did not learn the details of the address, but the general contents must have been discussed extensively, if not freely. Nonetheless, a legalistic person could (and perhaps rightly) say that in such an important matter one could not act on the grounds of mere hearsay. In how far all these professors and academic authorities were truthfully ignorant of the content of Bolkestein's address remains a matter of conjecture.

The rest of the story of the University of Amsterdam in wartime is simple. The authorities ordered the university to resume its activities in June 1943. Some 19 professors and eight lecturers resigned in protest; the Germans simply refused to accept the resignations and General Rauter, known for his ruthlessness, telephoned the university on May 12 at 11.30 that the resignations should be withdrawn before three o'clock. Otherwise summary justice would be carried out. All resignations were withdrawn.

[54]Brouwer to Comm. of Restoration, 30.VIII.1945.

Heyting had at an earlier stage informed the senate that he would under no circumstances resign. Had he foreseen the German reaction? There is a Dutch proverb '*Beter blô Jan dan dô Jan*' (Better a living coward than a dead lion), and at this stage most academics followed that adage. The reader should, however, be aware, that there was a sizeable part of the academic community that joined the resistance and took considerable risks, often to a heroic and tragic end. The skirmishes of 1943 ended in a kind of stalemate.[55] The Germans had ordered the universities to carry on as usual, be it that only part of the former student body was allowed to attend classes and work in laboratories. The London government, on the other hand, did not allow any teaching to students who had signed the declaration of loyalty, and the resistance made it clear that teaching to 'signers' was a form of treason. In the confused situation that followed, professors sabotaged their courses, student members of the resistance intimidated signers, .... When professors were put under pressure by NSB students or German authorities, they reluctantly put up a show of compliance, but no successful courses were offered. The students who had not signed were obliged to enlist in the labour service, and, as a rule, to go to Germany. Quite a number of those went underground, hiding wherever they could. Some professors organised illegal classes or private instruction for the non-signers. Brouwer and Heyting played double roles, they kept the mathematical institute going, lectured for signers, but they also instructed non-signers, and conducted illegal examinations. At a certain point, when the public transport had more or less broken down, they rode together on a tandem to Amsterdam for illegal teaching and examinations.[56] The conduct of the universities and their various governing bodies, and of individual scientists as well, was of vital interest to the student resistant movement. The discussions and decisions in the matter of the declaration of loyalty were therefore closely watched, and commented on in the underground student periodical '*De Geus*'.[57] In the issue of December 7, 1943 it critically commented on the general attitude of the universities and their reactions to the numerous violations of international law and the Geneva conventions, such as the dismissal of Jewish staff members and the declaration of loyalty. 'They think their honour is not involved when they are treated as schoolboys by a patented puppet. They are all the more sensitive to vague threats, in which he [Van Dam] has shown himself to be a master, and which they are immediately prepared to give a concrete form in the eternal "Pistol on one's chest", the only ratio to which the Dutch professor yields ...'

Summing up the situation, *De Geus* listed the names of some professors who carried on their lectures as ordered by the Secretary General. From the Mathematics and Physics faculty in Amsterdam Brouwer, Bruins and Weitzenböck were mentioned by name;[58] this was not just a piece of information for interested readers, but

[55]See Knegtmans (1998), p. 191 ff.

[56]Oral communication, A.J. Abels, L. van den Brom.

[57]'Geuzen' was the name for the group of freedom fighters (in modern terminology) who resisted the Spanish forces during the dark hours of the eighty-year war, when according to the military rules the official opposition to the Spanish had been decisively defeated.

[58]Weitzenböck had in fact already been put on a blacklist in 1941.

rather a warning, who could, or could not be trusted, and it held a message for the inevitable day of reckoning after the war. In patriotic circles there were bitter feelings indeed against those who collaborated with the Germans. The political picture was for most people a black and white one. Seen in that light the student underground press was on the whole remarkably objective, there was a dignified and earnest effort to rise to a level of serious discussion and comment. It was as if this most serious and desperate situation brought out a higher level of responsibility and moral awareness. The student resistance of the war years bears in no way a resemblance to the general atmosphere of welcome divertissement of post-war student revolts.

Knowing that *De Geus* stuck to a low key review of facts, one might assume that Brouwer (and Bruins for that matter) was sharply criticised by the Amsterdam underground student population.

It is unknown if Brouwer himself was aware of the mentioning of his name in the underground student press, if so, he had reasons to worry. The mood of the Dutch population turned more and more grim and revengeful at every new injustice, and there was a general looking forward to the '*bijltjesdag*' (hatchet day).

The faculties of law and medicine had great difficulty and little inclination to keep the teaching going, the faculty of the sciences decided to carry on as well as possible. In the mathematics department the usual routines were still carried out.

Mathematics at Amsterdam did not fare well in those years. Whereas a certain superiority over the rest of Holland was tacitly recognised before the war, the loss of Freudenthal seriously reduced the status of the department. In fact, it had lost its only modern researcher—Freudenthal, and furthermore the *privaat docent* Belinfante, who was at the time quite productive. The remaining mathematicians were not able to fill the gap. Brouwer had long since given up the publishing industry and Heyting was far too busy teaching and taking care of his large family. Weitzenböck was the only remaining productive man, but even he flagged after 1943. Moreover he practised a kind of mathematics that was on its way out. The particular Clebsch–Gordan–Weitzenböck theory of invariants was already past its heyday. Bruins was to be the main exponent of that school in the Netherlands, but resuscitation did not prove feasible.

In general, university teachers tried to survive the war as well as possible. With the exception of those who were dismissed or arrested, they, as a rule, discharged their duties when possible. Research was impossible for the laboratory scientist working at home, but the 'paper-and-pencil' scientists could at least privately pursue their research. This was far from easy, as the daily routine of obtaining food and fuel became ever more time consuming.

A number of mathematicians went underground, among them Van Dantzig. Curiously enough Freudenthal was not directly in danger. He nonetheless took the wise precaution not to expose himself. In general it was wise to avoid trouble spots and to remain invisible at times of razzia's.[59] Freudenthal practised mathematics in private, and he also tried his hand as an author. Under the pseudonym F. Sirolf he wrote a

[59]In Holland the name 'razzia' has since the war been reserved for the raids and manhunts during the occupation.

prize winning historical novel. As a Jew he could not go and collect the prize money. Fortunately Willy Bloemendal, who later became a professor in Delft, was willing to pose as the author, and to collect the prize money. Under the circumstances this required a good deal of courage. The prize, by the way, was five-thousand guilders, a considerable sum for those days (1944).[60] In the spring of 1944 Freudenthal was commandeered to Havelte, a place in the eastern part of the country; there he had to work at an airfield which was under construction. In his autobiographical book '*Write that down, Hans*' he related both the dark and light sides of the camp. One anecdote may be included here: there was a simple trick that Freudenthal applied to get a bit of rest, he picked up a spade or some tool and walked purposefully from one end of the field to the other. Any guard who saw him pass, thought 'this man is on his way to a specific job'.[61]

On *Dolle Dinsdag* (mad Tuesday) September 5, 1944, the day that a panic broke out among the Germans and their followers in the Netherlands, because of persistent rumours that the allied forces were about to liberate Holland, Freudenthal decided that he had quite enough, and he absconded. In Amsterdam where the worst was still to come during the so-called hunger-winter, he and his family waited for the liberation.

Of Brouwer's scientific activities little or nothing is known. In view of his post-war publications, one may conjecture that he returned to his old intuitionistic research and started to grapple with loose ends. In 1944, on April 28 he again fell victim to a fire in his house in the Torenlaan.

In 1943 the man who in Brouwer's words had 'added many mathematical theories, as monuments of crystalline simplicity to the spiritual property of humanity', and who regrettably had become totally estranged from his former Dutch admirer, died in a Germany that had undone his life's work. His Göttingen had been replaced by a caricature of the once so superior mathematical establishment that was largely his making. Hilbert had no longer played an active role in mathematics in the Third Reich. He had remained the chief editor of the *Mathematische Annalen*, but rather more as a symbol. Surprisingly, Blumenthal had been allowed to carry on in his function as an editor until 1939. The last big enterprise of Hilbert, the completion of the comprehensive exposition of mathematical logic and proof theory, had seen its second and last volume in 1939. The scientific-political views in Nazi Germany were not exactly positive on Hilbert's Formalism (cf. p. 628), but the support of the majority of the mathematicians, and the immense status of Hilbert, seems to have outweighed political opposition to his program.[62] This was made clear on Hilbert's eightieth birthday, when he was awarded the Goethe Prize by the Führer Adolf Hitler.[63] Süss, one of the politically prominent mathematicians in the Nazi

[60]Freudenthal (1987b), pp. 214–215.

[61]The anecdote has a follow up. When he told this story at one of the workshops in Oberwolfach, some of his German colleagues were upset that one could sink so deep as not to carry out given orders.

[62]Cf. p. 628.

[63]Thus sharing this dubious honour with Vahlen, cf. Inachin (1960).

period, only saw the negative interpretation of this gesture: according to him it illustrated the lack of status of mathematics, as 'the most famous mathematician, David Hilbert, was not honoured by the requested "Adlerschild"' (Insignia of the Eagle), but, just like 'hundreds of respectable and diligent professors, fopped off with the Goethe-medal'.[64] Knowing the pre-occupation of the Nazi party with political correctness, this event can only be interpreted as a clean bill of health for Formalism, signed by the highest authority. In view of the aggressive criticism of formalism, this Goethe Prize has its political significance, cf. p. 628. It would be totally wrong to draw any personal conclusions in this case. Hilbert was already for some time out of touch with the world around him, and one may wonder if he was conscious of the political background of the prize. After seeing his mathematics department ruined by the Party, there is little likelihood that he could be placated by an honour from the hands of the perpetrators. In scientific circles the event drew little attention, only Lietzmann mentioned it in the paper dedicated to Hilbert's 80th birthday.[65] In 1943 Hilbert died and he was buried in Göttingen.[66] The sad news travelled fast, and all over the world his death was commemorated.

## 16.7 The Brouwer Family in Wartime

If one forgets for a moment the hidden and open atrocities perpetrated against Jews and all forms of opposition, the war years presented the view of a gradually worsening economic and civil crisis. A citizen who held no political or social responsible position, and who did not read the underground press, nor listen to the London radio, and had no Jewish friends, would only see the decline in consumer goods, personal facilities and the like. In the first two years the food rationing restricted the variety in the shopping basket, but still one did not have to go hungry. It became difficult to buy shoes, clothing; vacations became a rare luxury; spare parts for almost everything became scarce, .... But in the beginning of the war years, frugality had already become a way of life to a population that had already gotten used to the hardships of the big Crisis of the thirties. Later, public goods, such as gas and electricity became scarce, and eventually were cut off altogether. Curfew restricted social contact; harassment by all sorts of military and semi-military organisations increased. And finally, in the last year, survival became the immediate concern. In addition there was the systematic confiscation by the authorities of private cars, horses, radios (it was not only a crime to listen to the BBC, but even to possess a radio), bicycles, brass objects, church bells, ... Parts of the country were made restricted military areas, Nazi's swaggered on the streets looking for victims, ....

And after the great railway strike (called by the London government in exile in September 1944, after the invasion in France) almost the entire work force of the

[64] Remmert (2004a).

[65] Lietzmann (1942).

[66] Cf. Braun (1990).

Dutch railways went on strike until the end of the war. The Germans retaliated with large scale robbery. Against this background one has to imagine daily life. And it was not different for the Brouwer family. Lize was still spending most of the week in the pharmacy and Brouwer commuted between Blaricum and Amsterdam, until finally public transport broke down. Already in 1941 the trip to Amsterdam was an uncertain and arduous adventure. Lize, in one of her letters to Louise, wrote that travelling back and forth sometimes cost Brouwer five hours a day.

The running of the pharmacy absorbed a lot of energy, finding new assistants, interviewing applicants, maintaining a reasonable stock of medicaments, it all took a lot out of Brouwer. Another urgent problem was the rebuilding of the hut left to the mercy of rain, wind and snow, there would soon be little left of the remnants of Mauve's creation. Brouwer pointed this out to the mayor. The preservation, he wrote 'anticipating a later rebuilding, may thus be considered in the interest of Dutch economy'.[67] Therefore he asked permission to install a temporary roof. Brouwer was not in luck, because the State Commissioner forbade, from July 1 onward, any private building activity.[68] Nonetheless the mayor must have considered Brouwer's cottage of economic interest, for on July 29 he got his building permit. Of course, between the permit and the rebuilding of the cottage, there was a long way to go.

During the war Brouwer senior, who was almost ninety years old, had moved in with his son. Brouwer never got on with his father, so the arrangement did little to brighten his life. The old gentleman indeed did not spread much happiness. Although a female help was engaged for him, he spent his days grumbling. Even the warm-hearted, serene Lize found it difficult to put up with the demands of the old gentleman. But her own husband was sometimes also a handful. Once, when Brouwer returned from Amsterdam, he caught a cold and ran a high temperature. But Brouwer was not the person to exercise patience and take time to recover. In letters to Louise Lize complained about the short temper of her husband, 'Dad still has a fever, but he had absolutely to go to the institute.' And a bit later 'Daddy is not yet well, but he behaves so silly. Goes out for a half hour's walk with a temperature, receives visitors, eats all sorts of things, and then flies into a temper about underpants that have not been properly mended, and then dead tired to bed and complains about his heart. He is terribly tiresome.'[69] The war and the consequences of the fire in the hut made life extremely complicated; the Brouwer couple had more than before to live in Amsterdam, as Lize wrote in a letter to her daughter. Commuting between Blaricum and Amsterdam had become so difficult now that the usual connections were scarce and uncertain.

Tensions ran high in the small family; Brouwer could not suffer his step daughter under normal circumstances, but the close contact that was the result of wartime restrictions was more than he could bear. It came to explosive scenes between him

[67] Brouwer to Klaarenbeek, 30.VI.1942.

[68] To be specific, 'Requests for exemption, which are not clearly motivated by economic or military arguments, according to art. 1 of the regulation of the Führer concerning the protection of armament industry of 29.III.1942, are under penalty of detention, and in particularly serious cases of death.'

[69] Lize to Louise, 16.XII.1941, 18.XII.1941.

and his stepdaughter, each blaming the other for all kinds of misery. A complicating factor was the financial motive: Louise's father had, when he died, left 10,000 guilders to Louise. Lize and Bertus had invested the money in an annuity, and not trusting Louise's worldly wisdom they had built in a clause against pressure from third parties to hand over (part of) the annuity. Apparently Louise had not quite understood what had happened, and she had become convinced that her stepfather had used the money for his own purposes. Lize did her utmost to calm the child, 'I am so afraid that you will get insane from whipping yourself into a frenzy all the time. For Dad has not appropriated your money, he is not using it for the hut, but he has invested it and that has always been his intention.'[70] Lize was in the unfortunate and uncomfortable position of the conflict between her feelings as a mother and as a wife. She tried, with moderate success, to intervene between Bertus and Louise; but even she could not prevent clashes. Moreover, even to her Bertus could be quite unreasonable.

Mother and daughter had a regular correspondence, albeit that Lize was the more active writer. The letters confirm that the relation between Brouwer and his stepdaughter had not improved over the years. Lize always tried to improve the relation, 'And, my dear child, try, for once, not to be so angry at Daddy. It is so bad for you and it causes you so much stress which blocks your recovery.'[71] Indeed, Louise was the permanent victim of all imaginable inflictions, the legs, the stomach, the throat, ...it was all in a state of perpetual ill health. In how far some afflictions were imaginary or exaggerated is hard to say; some of them were doubtlessly genuine. But being a martyr suited her wonderfully well.

During the war no more international trips were possible, and so in the summer of 1942 the Brouwers took their vacation in a guesthouse in South Limburg, where they made short walks. Lize had to take it easy with her seventy-two years. Brouwer accepted the limitations, trying to work in spare moments; 'Daddy is quietly doing his mathematics in the small dining room', Lize wrote to Louise.

The final years of the war brought extreme hardship to a considerable section of the population. It became quite normal for city dwellers to make long trips into the country to obtain food from the farmers either for money, or by barter. Under these circumstances not much time was left for intellectual activity. Heyting, for example, reported that the quest for food and fuel took so much of his time and energy that he simply had to give up mathematics temporarily. In an area like Blaricum trees were rather abundantly present and an inviting commodity for the villagers who had no coal to heat their houses. It is characteristic for Heyting's unparalleled honesty that he asked and obtained Brouwer's permission to chop down a few trees at his estate. Brouwer himself, of course, was also heating the various houses and cottages with wood of his own trees.

These same little houses were used for hiding persons who were at risk, be it for evading labour service in Germany, for being Jewish, for belonging to the resistance movement, or simply for being a male of the appropriate age group. Under

[70] Lize to Louise, 20.VIII.1942.

[71] Lize to Louise, 20.VIII.1943.

**Fig. 16.2** Brouwer posing with cat. [Brouwer archive]

the wooden floors hiding places were dug out and one got access to the space underneath by lifting the planking. Riet Musters, the daughter of a neighbour, recalled that during the last years of the war these hiding places were used (often at night); it fell to her to clean the houses during the day.[72] Her memories of Brouwer are those of a kind, somewhat introvert man, who walked his grounds reading a book. The children of the adjoining houses used to play on Brouwer's grounds, if they met Brouwer, he would greet them and immediately warn them to stay on the paths. There was a modest sized wood behind Brouwer's house, called 'Brouwer's copse' (*Brouwer's bosje*). Sometimes, weather permitting, Riet's father would get out a pair of folding chairs, and sit down with Brouwer in the copse to discuss whatever subject came up. Lize, as she recalls it, was rather strict with the children; Brouwer had given them permission to come to the back door, however, if Lize found them there, she invariably wanted to send them away. Lucky for them, Brouwer would often come to their rescue, and say, 'you cannot send these kids away without candy!'

Brouwer was lucky enough to have good friends and ex-students to keep an eye on him. Max Euwe gave up his job as a mathematics teacher at a Lyceum for girls and became the director of VANA, a chain of grocery stores in Amsterdam. In that

[72] Interview, R. Musters.

capacity he could from time to time provide the Brouwer household, just as he had done for the Freudenthal family, with groceries that had become hard to obtain; the goods that one was supposed to obtain on coupons were often illusory.[73]

Also in his quality of trustee of the Hilversum gymnasium, Brouwer saw the new authorities at work; Jewish students were expelled, teachers were fired or taken hostage, the building was eventually requisitioned by the German army. In the middle of the war there was even a brush with the local authorities; the council wished to reduce the power of the trustees to that of a mere advisory board. It goes without saying that Brouwer without reservation took the side of the trustees.

## 16.8 Weitzenböck in Uniform

There was one mathematics professor in Amsterdam for whom the war held a vista of redressed injustice. Weitzenböck had welcomed a reborn Germany under Hitler and the union of Germany and Austria. His political conviction, which he did not try to hide, had made him extremely unpopular in Blaricum, and at the university students and staff members had started to avoid him. There were no open conflicts, nor were there any reports of unfitting manifestation on Weitzenböck's side. He did not betray Jewish students or colleagues, nor did he report any resistance activities that came to his knowledge.

His position in the faculty had always been loyal and neutral. Although he owed his position to Brouwer, he was not owned by Brouwer. When he thought Brouwer was wrong, he did not hesitate to withhold his support in the faculty. Max Euwe, who was politically above any suspicion, and who was actively involved in the post-war 'cleansing' (*zuivering*), pointed out that the Marxist Mannoury and National Socialist Weitzenböck were friends before the war. During the war Weitzenböck used his influence to protect Mannoury.

When in May 1940 the German army invaded the Netherlands, Weitzenböck was already known as pro-German. In his own words, '...initially it was a strongly unpleasant surprise for me, but later I had to reconcile myself with the necessity of the attack as part of the campaign against France. At that time I saw the realisation of a German victory as of necessary importance for Europe'. After the capitulation in May, he joined the Dutch Nazi party, the NSB, from which he resigned again in September 1941.

In 1942 Weitzenböck acquired the German nationality. As a former officer of the '*k. und k.*'[74] Austrian army, he became a reserve officer in the German army. Although he was from that moment a German subject, he did not join the NSDAP.[75]

[73]Euwe also provided Karel van het Reve with a meal at his house every week in exchange for lessons in Russian. After the war Van het Reve became well-known in Holland as an author; from 1957 on he was the professor of Slavonic Languages in Leyden. Euwe profited from the lessons in Russian when he became the president of the FIFA, the international chess association.

[74]Imperial and Royal.

[75]National Socialist German Labourer's Party.

**Fig. 16.3** Roland Weitzenböck in uniform. [Courtesy Nationaal archief, Den Haag]

In 1942 a *Schützgruppe* (guard unit) was formed, and Weitzenböck was detailed to it. He did not take part in the various exercises of the group. In 1944 the war had reached a stage where the army started to scrape the bottom of the barrel. The unit became active, uniforms were issued and had to report in Hilversum, where he had to replace the commander of the quartering office (*Quartier Amt*). In this quality he had to requisition housing for army units. This was of course a rather painful matter; nobody enjoys being expulsed from his house for the enemy, but it amounts to adding insult to injury when the requisition officer was a fellow villager. It hardly needs to be said that Weitzenböck was hated by the local population. He did not go out of his way to provoke the villagers, but he considered it, for example, his right to put out the Swastika flag on Hitler's birthday, and, according to the records of his post-war interrogation, he once did so. It was clear at the end of 1944 that it was only a matter of time before the German army would collapse, nonetheless Weitzenböck stuck to his obligations as an Austrian-German subject. After Hitler's death, the small group of Weitzenböck's unit, consisting of the older generation (Weitzenböck was almost sixty at the time), was ordered to take the oath on Hitler's successor, admiral Dönitz, in the reading room of the public library in Hilversum on 4 May. The next day the capitulation of the German army in the Netherlands was signed, and Weitzenböck's deplorable adventure had come to an end. He had lost a son and his wife in the bomb attack of 1940, and his second son had fallen on 13 February 1944 in Tserkassie on the eastern front—a high price for a dubious political conviction.[76]

All testimonials confirm that Weitzenböck preserved a high standard of integrity. No reports of acts against persons is known. In the village and at the university he was given the cold shoulder, and so he must have felt rather desolate. He was

[76]The information on Weitzenböck's political and military activities is drawn from the file in the *Algemeen Rijks Archief* in The Hague.

probably aware of the underground activities that were going on at the mathematical institute, indeed some witnesses at the post-war trial said as much. 'He was definitely aware that in the mathematical institute things were going on that were connected with the illegal movement. From certain statements it even appeared that he knew that Jews were hidden in the institute.'[77]

He never alluded to the activities of the resistance movement in the institute; he may even have given (unintentional) protection, as his presence lent the institute a certain respectability in the eyes of the Germans.

Brouwer also ran a certain risk through the underground activities. If the Germans had found the explosives, counterfeiting material and whatever was stored and used, it is unlikely that he would be spared. The illegal activities are confirmed from various sides, the building with its nooks and crannies offered ample space for hiding. The janitor, Koppers, was the key figure in the whole organisation. It is reported that he was a bit careless where security was concerned. Eventually he was arrested, taken to Germany from where he did not return. Bruins told in his valedictory lecture the other part of the story. He was in Limburg, conducting examinations, where the news of Koppers' disappearance reached him. He hastened by train back to Amsterdam and managed with another person to remove the explosives from the institute's coal cellar. The Germans, who raided the building, found nothing. There is a private note of Mannoury on the incident: 'From Mathematical Institute ammunition removed in as far as possible. Later it appeared that some was left.'

[77]E.M. Bruins in 'Proces Verbaal', 24.VII.1947.

# Chapter 17
# Postwar Events

## 17.1 Purging the University

The winter of 1944/45 is remembered as the 'hunger winter' (*hongerwinter*); it reduced the western part of the Netherlands to a state of malnutrition unparalleled in modern Dutch history. It also brought a fiercer repression. And finally the long awaited capitulation arrived: on May 5, 1945, the German commander General Blaskowitz signed the capitulation of his army in the Netherlands, and three days later the German army as a whole laid down its arms. The liberation brought rehabilitation to the victims of wartime persecution, and retribution to the traitors, members of National Socialist organisations, and the like. Up and down the country the former resistance forces, their ranks swelled with many an opportunist, filled the jails and camps with the creatures of the defeated enemy. Tribunals were everywhere installed and the process of bringing this strange assortment of suspects to justice commenced. Indeed the variety was perplexing, among the new population of the prisons and camps there were notorious politicians, secret agents, Jew hunters, collaborationists, builders of fortifications, but also quite simple souls who had joined the NSB for naive political reasons and who lacked the degree of opportunism to leave the sinking ship in time.

The end of the war also ended the isolation that was caused by a gradual and almost complete breakdown of the postal services. All over the world people were anxious to learn about the fate of their friends and relations. There was a great deal of correspondence between German emigrants and their colleagues who had stayed behind. A large number of letters was exchanged, bringing sad news of the victims of the war or happy news of the survivors. Brouwer's contacts with the world beyond the Dutch borders were also revived; his former student De Loor from South Africa, for example, inquired after the fortunes of his old Ph.D. adviser. In his answer Brouwer wrote that a horrible period had come to an end, and that fortunately he and his family had been spared death and imprisonment. Mannoury and De Vries[1] had survived, but, he sighed, Johan Huizinga had died during the

[1] Hk. de Vries had moved to Palestine before the war.

D. van Dalen, *L.E.J. Brouwer – Topologist, Intuitionist, Philosopher*,
DOI 10.1007/978-1-4471-4616-2_17, © Springer-Verlag London 2013

war—'His death is a great loss for the Dutch nation; of which he with his three aspects of humanist, man of letters, and scholar, was one of the most representative elements.'[2]

Unfortunately, he wrote, his house had been on fire twice, and in addition part of the furniture was stolen,

> as a consequence my scientific archive and with it the manuscripts of my unpublished research have been lost. And in order to be able to reconstruct the content of those from memory (which is already gradually darkening at my age of 64), a period of leisure, and free of care, should be my share before long; there won't be any chance for that in my native country which will only very slowly raise itself from the state of material and spiritual upheaval, resulting from five years of plunder and terror. Where, as I believe, the interest of science would be served by my temporary expatriating, the plan, which we discussed in the past, to create at some time a temporary position in South Africa, is again in my mind (and this time much more vivid than at that time) and I would even greet a speedy realisation of it as a delivery.

The reaction to this letter is unknown, but little did Brouwer know what the future had in store for him. A similar sentiment was expressed in a reply to Arnold Dresden's inquiries after the war;[3] it sheds an interesting light on Brouwer's global political views:

> ... Yes, the securing of peace here also absorbs a large part of the available intellectual capacity. The first requirement seems to me the abolishing of the splitting of the earth into various areas with separate centres of military power.
>
> In anticipation of the realisation of this, the United States together with the other American States, the British Empire, Scandinavia, Switzerland, Austria, Italy, Spain, Portugal, France, Belgium and the Netherlands, should in my opinion without delay be united into a single state as far as military matters are concerned... In the meantime, I would, fearing that the unified state will take some time, at the moment (unlike before) gladly emigrate from the Netherlands. If now the possibility of a position was offered in America, I would eagerly grasp the possibility with both hands. For the confused situation during the first months after the liberation has here (also in scientific and academic circles) brought persons to power...[4]

The reader may guess the rest of the sentence. When Brouwer wrote to Dresden, he had learned the painful lesson that the status quo had in some respects completely changed. As soon as the Germans had evacuated Holland, special courts and committees started the process of 'purging' (*zuivering*, the equivalent of the German de-nazification).

---

[2]Brouwer to De Loor, 20.VI.1945.

[3]Brouwer to Dresden, May 1946, draft.

[4]The rest of the draft is missing.

The general impression of this purge leaves us with rather mixed feelings. The simple harmless followers of the National Socialist banner, whose only fault had been that they had made an objectionable political choice, complained at the severity of their punishment and members of the underground movement grumbled at the same time that the guilty often got off far too lightly. A few leading National Socialists were sentenced to death, but the clever operators escaped justice. Later historical research has fully confirmed these facts. On the whole, the practice of exploiting the confusion for personal advantages was frowned upon, but nonetheless, a certain amount of fishing in troubled waters was practised by those with a taste for it. In the process people with enemies could easily suffer. The academic world was no exception; Brouwer, in particular, felt that his treatment was largely the effect of old feuds and jealousies. Looking back at Brouwer's postwar treatment, it is indeed not so certain whether there was not a poorly veiled desire to isolate the undoubtedly great, but also inflexible, man.

The purge of the universities was energetically undertaken by so-called 'Committees of Restoration' (*College van Herstel*). At the University of Amsterdam this Committee consisted of the following persons: C.T. van Valkenburg, G. van Hall and W. Bruin. Van Valkenburg was a neurologist who had been active in the resistance, and who, during the war, had been in prison for some time. He had been passed over a few times for a professor's appointment. Academic circles entertained doubts as to his appreciation of the operation of the universities. Wieger Bruin was a well respected artist; he was appointed by the Chief of the Military Authority, Christiaan Kok, to counterbalance the hard liners Van Hall and Valkenburg.

The Committee was assisted by a committee of professors and a similar committee of students. Its task was to advise the temporary authority, the so-called 'Military Authority' (*Militair Gezag*), which for the time being exercised all military and civil authority in the Netherlands, with respect to the purging of the members of the University of Amsterdam. The legal steps, such as dismissal, reprimand, loss of pension, were in the hands of the Minister of Education, so the Military Authority in turn had to report to the minister. As we will see, the minister and the Military Authority were not always of one mind. The '*professors' resistance group*' was quick to form the advisory committee and the students followed suit. These committee's were in a fairly good position to provide information and to collect evidence on their colleagues and fellow students.

In the excited atmosphere that followed the liberation, the thought of justice and revenge dominated the minds of most Dutchmen. Having experienced five years of perversion of the citizen's rights, there was a strong tendency to give Nazi's, collaborationists, black marketeers and in general all those who had shown a lack of patriotism, a taste of their own medicine. On the whole the special courts of justice observed the law well enough, but outside the courtroom popular justice was often meted out.

The advisory committee of professors had a precarious task, it was so close to the persons it had report about, that objectivity became a precious commodity. Not being trained in legal traditions, the committee could not agree on a policy on confidentiality. Professor Derkje Hazewinkel-Suringa, a member of the Law Faculty, attached importance to it, but her view was not generally shared. Borst, a prominent professor in the medical faculty, for example, was of the opinion that the purge as an expression of the national conscience was of such an importance that each individual injustice became insignificant in comparison.[5] In fact, from a legal point of view, much about the purges, carried out in haste and in a grim mood, did not meet the standards of normal justice.

The first act of Brouwer's part in the purge opened on July 3, when the commander of the Blaricum squad in charge of arresting suspects with a questionable war-time record (the *arrestatieploeg*) dispatched the membership cards of the *Nederlandse Volksdienst*[6] of Brouwer and Weitzenböck to the Mayor of Amsterdam, in his function of president curator of the University of Amsterdam. These records only added to the complaints that were being brought against Brouwer. Both the professors and the students had weighty objections to Brouwer's wartime conduct. On July 7 the Student Contact Committee, in answer to questions posed by the Committee of Restoration, informed the Committee that 'it would greatly astonish the students of the faculties concerned if [...] H.A. Brouwer, L.E.J. Brouwer, [...] would without further consideration be put in a position to resume their teaching'. 'A large number of students', it added, 'considers these professors by definition as unacceptable (these numbers are not negligible)'.

The Committee of Restoration wished to push ahead with the investigations, a long drawn procedure would stand in the way of a speedy recovery of the university. The Committee of Restoration, once it had collected incriminating material and accusations, started to summon some twenty staff members. Brouwer's brother Aldert was interrogated on July 11 and two days later it was the turn of the mathematician Brouwer. On the seventeenth he had to appear again before the Committee; Bruins and Heyting followed on that same day.

The interrogation was a summary affair, in fact four questions and their answers were recorded. The actual conversation may have been more extensive, but the Committee apparently concentrated on these particular key points.[7]

The first question dealt with the declaration of loyalty. Brouwer stuck to the position taken in the Senate. He argued that the signing of the declaration had material advantages for the Netherlands. Not signing would have resulted in large scale forced labour (*Arbeidsinzet*), with disastrous consequences for the health of the students. Moreover, a student who signed had a better opportunity to take part in underground activities. He declared that after the bulletin board episode he had never personally advised any student to sign. He insisted that at the time he did not know the precise contents of the London broadcast.[8]

The second question concerned his position in the Senate. Had he agitated against the resistance movement of the professors? This clearly referred to the Senate

[5]Knegtmans (1998), p. 241.

[6]Cf. p. 674.

[7]The records of these interrogations are kept in the *Stads Archief Amsterdam*.

[8]Cf. p. 689.

Meetings in March and April 1943 when the declaration of loyalty was discussed. Brouwer's reply here was: 'I and many others did disagree at times with the professors' resistance movement. In March 1943 a draft of a letter to the Commissioner of State was laid before the Senate, with the final clause that a rejection of our demands would lead to the resigning of the Senate members. I opposed that, because in my opinion one could not request and threaten at the same time. The majority of the Senate agreed with me.' One might well question his argument. After all, bargaining traditionally consists of a certain degree of 'carrot and stick'. Perhaps he judged a collective resignation so unacceptable to the Germans (which it turned out to be), that one could no longer speak of bargaining. He probably felt that his argument could do with some elaboration, for in a letter he tried to spell out his reasons with more force,[9] the letter lined up all the socio-philosophical ideas from earlier days, in a defensive array:

*The declaration of loyalty 1943.*

> When a company of civilised travellers is overpowered by superstitious cannibals, their conduct, in particular their language, or gesticulation, exchange with their captors, will be exclusively aimed at their deliverance. Guile, deception and dissimulation will be their tools, both with respect to communications, and proposals and promises. Honesty, chivalry and demonstrative testimony will not only be practically improper because of its counter-effecting the goal, but also be without reasonable content: for the essential binding of the meaning of word gesture or sign, essential for honest understanding, is only possibly on the basis of tacit co-operation of the communicating parties as 'good listeners' and this co-operation can only derive its moral justification from a (in the case under consideration missing) common orientation of will.[10]
>
> A situation of that kind existed in the Netherlands during the occupation. The manner in which the enemy had descended upon us and in which he subsequently had violated the good faith and trampled on the human rights had, on the one hand, eradicated any common orientation of will or respect between the Dutch population and the occupation forces, on the other hand had directed the conduct of the Dutch population with respect to the occupation forces exclusively towards the following goals:
>
> 1. To serve the occupation authority as little as possible.
> 2. To harm the occupation authority as much as possible.

---

[9]Brouwer to Comm. of Restoration, 17.VII.1945. Either written after the interrogation, or prepared beforehand.

[10][Brouwer's footnote] Cf. my lecture '*Willen, weten en spreken*' (published in '*De uitdrukkingswijze der wetenschap*', Groningen, Noordhoff, 1933, in particular I.3).

> 3. To safeguard as well as possible our national heritage against destructive interference of the occupation authority.
>
> And in this context guile, deception and dissimulation were for the language- and gesture-exchange of the Dutch population with this enemy with respect to communications, proposals and promises proper on the grounds mentioned in the preceding paragraph; honesty, chivalry and demonstrative testimony had, on the contrary, become both practically improper and lost its rational content.
>
> In the light of this exposition, it is therefore incorrect that, as was claimed at the time, that the signing or not signing of the declaration of loyalty by Dutch students involved ethical or ideal goods of the Dutch community. On the contrary, there was the possibility that a general signing would have the following consequences:
>
> 1. Less Dutch labour would serve the enemy.
> 2. The students who worked in the underground movement would have better conditions for their work under the guise of loyalty.
> 3. The health and intellectual training of the Dutch students would on the whole be less harmed.
>
> At that time this had to imply that a general signing of the loyalty declaration would have served both the interest of the students and of the nation. And this conclusion I could not, therefore, when it pressed itself on me, keep concealed, because the venerated tradition (in particular in the Dutch community), which not only designates the expression of a sincere opinion as an inalienable right, but also as an imperative duty in matters of general interest, represents to me one of the most valuable goods, which made that I have felt lastingly bound to the Netherlands, also against personal material interests.

One wonders if the members of the Committee found time to read this document, and if they did, whether they accepted it at its face value, or whether they judged it an elaborate excuse of an old professor.

Reading Brouwer's statements at the interrogation and his written defence, one can only conclude that Brouwer had no intention of renouncing his war-time actions and statements. He did not wish to don the garb of penitence in order to mollify his interrogators. If anything, he was still convinced that at the time it was the best thing for the students to sign the declaration of loyalty, although he seemed to admit that had he known the text of the ministerial address on the topic of declaration of loyalty, he would have acted accordingly.

There is always the difficulty of analysing someone's statements in the light of past and present experience. Of course, there is a possibility that Brouwer invented the above justification after the events. It does not seem plausible, however. The arguments are too much in accord with his earlier philosophical considerations, and in the Senate Meeting of April 1943, he had already used a similar, but abbreviated argument. The true explanation is probably that Brouwer really had a convincing private social theory for the optimal strategy in the case of a formidable enemy.

The weakness of his own argument is that it allows the population to practice deception to oppose the enemy. Why, then, should the enemy not do the same—as he indeed did. In this case, supposing all students had signed the declaration, what would stop the enemy from nonetheless deporting them if the war effort made such profitable? In short, there is no reason to accuse Brouwer of insincerity, it is rather short-sightedness, or perhaps blindness, which he is to blame for.

The remaining two questions of the Committee were: 'Have you lectured for students who signed?' and 'Have you had contact with Professor Weitzenböck?' The answer to the first question was 'yes—but poorly!' Indeed, after the statement of the London Government, the students who signed the declaration had been marked as pariahs and it was considered unpatriotic to teach them. The general reaction at the universities had been mixed. Some professors gave classes for students who had signed, and at the same time conducted illegal classes for the others; some did not teach at all, but sabotaged their classes right from the beginning. Brouwer had also privately instructed students who went underground and examined them. Heyting went so far as to visit students at their home and offer his help, sometimes to the great discomfort of the parent who opened the door[11]—how was one to know if this was a real academic, and not an agent provocateur?

Brouwer's answer to the last question was—'after he turned Nazi, only professionally'. Brouwer did add that he considered Weitzenböck a great mathematician and a good person.

In early August the Committee of Restoration had sent a first tentative report to the Military Authority. It advised to dismiss a number of professors and lecturers (among them Weitzenböck and Bruins), to reprimand some of them (among them Heyting) and to suspend some for a period of further investigation (among them Révész and the brothers Brouwer).

The Military Authority acted promptly and took its decision on August 8. On the 14th the bailiff showed up on Brouwer's doorstep to serve him an order of suspension. Since nobody was at home, he left the order at the town hall. It must have been a shock for Brouwer to find out that he was still a suspect. He drew the conclusion that he was suspended from all public duties, so he also laid down his office of trustee of the Hilversum gymnasium. He rightly considered the suspension a serious blow, if not an insult. It was a painful social stigma, to be considered a Nazi-supporter or a war-time profiteer. The popular view at the time of the purge was 'guilty until proved innocent'. The general public was not going to be bothered about the finer distinctions of 'being investigated' and 'being guilty'. Another detail that would add to his worries was that the universities would have to open soon, and one did not have to be a genius to expect that a good deal of organising and negotiating would follow. Not to be included in the discussions was almost certainly to have direct consequences for an academic, let alone for Brouwer, who had run the mathematics department for years.

Brouwer, who had probably not been worried about the outcome of the review of his wartime record, now began to fear that he might become the subject of a

[11]Oral communication, J.S. Abels.

drawn-out investigation and political investigation. Most professors and lecturers, who were subjected to the post-war purge, had been informed on August 24 of the Minister of education's verdict, and subsequently reinstated. So he clearly was singled out for special treatment. He decided not to wait for future developments in subdued resignation, but to present the Committee of Restoration with additional information. On 20 August he sent a letter to the Committee motivating the teaching activities at the mathematical institute after the London government had forbidden such activities for students who had signed the loyalty declaration (the other students could not risk to attend classes). Had he known that the topic would come up during the interrogation, he wrote, he certainly would have prepared a note on the matter; therefore he had decided to send the motivation anyway.

And so Brouwer came into action in order to prevent a possible conviction and discharge. He started to send letters with additional information and arguments to the Committee of Restoration.

The first letter contained an explanation why the Mathematical Institute was not closed after May 1943, in particular why the lectures were continued (this topic, by the way, was not part of the interrogation on July 17).[12] He explained that after the ban on resignation in May 1943 the faculty members of Mathematics and Physics got together and agreed that in order to protect the instruments, books and the collections, the activities should preferably go on as usual. The argument did not quite fit the Mathematical Institute. This was problematic, for it was mainly used for lectures, which 'formed the camouflage for an underground operation, for which the director[13] had made available space and resources and to which the janitor devoted almost his whole working day, and in which a "docent",[14] as well as a female adjunct assistant,[15] took part'.

Emeritus Professor Mannoury also came to the rescue. He had probably heard that the Committee had not re-instated Brouwer and he felt that Brouwer was being punished for his frank views expressed at the Senate Meetings. He gave his views in a letter[16] which contained a psycho-significant analysis of the situation and the participants. His basic thesis was that in the early days after the announcement of declaration of loyalty the students and staff wavered between (a) an emotional and conscientious rejection of the declaration, and (b) a rational and tactical inclination to accept it. The students themselves opted for (a). He argued persuasively that free discussion and forming of opinion are an indispensable good for the sake of upholding democratic principles in the group processes that lead to decisions of the sort involving the loyalty declaration. 'However, it would be in flagrant contradiction with this principle of unhindered opinion forming to prosecute in retrospect, now that the power of the oppressor has been broken, a number of those who in the

[12]Brouwer to CvH, 20.VIII.1945.

[13]i.e. Brouwer.

[14]Bruins.

[15]Cor Jongejan.

[16]Mannoury to CvH, 18.VIII.1945.

stage of deliberation have given their views in good faith and to the best of their knowledge.' He added that 'the advice to sign, in the conviction that the declaration could be withdrawn at any moment, now seems odious to many, is understandable, but it may not be concluded that any sympathy with the enemy was involved'.

Brouwer must have realised by now that the Committee meant business, in view of the kind of advice it had been giving (and which certainly must have been known to Brouwer), everything was possible—ranging from rehabilitation to dismissal. Apart from the shame and humiliation ('there goes that professor who collaborated (was *fout* (bad)) during the war'), there were the ominous professional consequences. So to speed up matters and to clear his name, Brouwer started to collect evidence and witnesses of his innocence.

On August 30, he sent a letter with enclosures to the Committee of Restoration that served to explain the circumstances and background of his membership of the *Nederlandse Volksdienst*. These have already been related in the preceding chapter (cf. p. 674). Mayor Klaarenbeek confirmed the account that Brouwer had given of the *Nederlandse Volksdienst* episode; a copy of this letter was duly made and dispatched to the Committee of Restoration.

It is probably no exaggeration to use the word 'hostile' to describe the views of the Committee in Brouwer's case. Brouwer was not the person to feign deference for persons he did not respect. His icy politeness could, however, be more forbidding than harsh words, and it is likely that no love was lost between the Committee and the defendant.

That the mathematical institute was used for resistance purposes is an established fact. Apart from the oral tradition there is in this case even some written evidence. The janitor Koppers, who had been arrested on June 7, 1944, was sent to Germany, where he died in a hospital in Hamburg on 7 December 1944. This was reported to Brouwer by another former political prisoner. Brouwer informed the responsible alderman of Koppers' fate, and urged him to grant the widow of Koppers a pension.[17] Koppers had, he wrote, from the beginning of the occupation to his arrest been active in the resistance, 'and in that way saved the lives of hundreds of Dutchmen'. He enclosed the testimony of two former resistance members. One of them, Van der Hurk,[18] had declared in a letter to the Committee of Restoration that 'Professor Brouwer, by making the mathematical institute available for illegal work in the widest sense of the word, has created the opportunity for various acts of sabotage directed against the occupation forces, and for the saving of life and health of a great many fellow citizens'.[19] The obvious corollary was that Brouwer, the director of the institute and the direct superior of Koppers, could not reasonably be suspected of pro-German feelings or actions, since the institute was used with his tacit approval.

After the letters to the Committee of September, there was a long silence. It took its time in the case of the few remaining difficult cases, the brothers Brouwer, Révész and five more members of the teaching staff.

---

[17] Brouwer to City Council, 12.9.1945.

[18] A former associate of the famous resistance fighter Van der Veen.

[19] Van der Hurk to Committee of Restoration, 12.IX.1945.

The advice of the Committee was submitted to the minister. The minister had no time for long deliberations, the new term was about to start and teachers could not easily be missed. And so at the end of August he announced that a number of individuals would be *reprimanded*.[20]

Bruins and Heyting belonged to that group. On 3 September the suspension of Bruins and Heyting was lifted. The typical formulation of the reprimand was 'In view of [incriminating facts], the Minister of Education, Arts and Sciences condemns the position taken by...' In serious cases the minister could write 'strongly condemns', and in mild cases such as Heyting's, he would add a sentence 'there is no doubt of the correct patriotic disposition'.

Heyting and Bruins were reprimanded in the light form; Heyting, said the reprimand, had acted wrongly, he had given guidance, but in the wrong direction. In fact, when Heyting was eventually considered for a royal honour, this reprimand was not seen as an obstacle. The official who handled the investigation into Heyting's antecedents informed the mayor that the Committee of Restoration had allowed Heyting to conduct examinations, while suspended. 'This may illustrate that one did not attach too much importance to it.' Heyting took a very serious view of the purge. He did not have the poise of Brouwer, nor the brashness of his fellow lecturer, Bruins.

As Freudenthal wrote in a letter to Fraenkel, Heyting was not exactly a paradigm of courage during the war. After the liberation he came to Freudenthal, probably to keep him from negative comments. But Freudenthal had, as he wrote 'not for a moment contemplated to register a complaint'. Fraenkel, who was extremely sensitive to Nazi-tendencies, was apparently trying to find out which scientists were still acceptable. He must have asked Freudenthal for advice about Heyting. 'There are no objections to resume your relationship with him', was Freudenthal's comment.[21]

Bruins was an altogether different fry. When the Committee of Restoration had summoned him, he was not in the least impressed; on the contrary, he went to the attack right away. He denied all charges, explained them as ill-founded or misrepresented, and he presented the Committee with an impressive illegal record. His own testimony shows him as an undaunted freedom fighter.

There is no doubt that he had the making of an enterprising and courageous man. He was very much the master of his own mind and actions; mathematically he chose his own style and topics without caring for the fashion of the times, socially he displayed considerable talents for useful relations, and later for making enemies. He was an easy talker, with a fund of amusing and interesting (although not always true) stories. Fear was unknown to him, and a role in the resistance appealed to him. His wartime connections proved useful in the period of the purge; his sources at the Military Authority (for some time the only real authority in Holland) provided him with valuable information on the activities and plans of the Committee of Restoration. And so, when he was summoned to be interrogated, he was in possession of all the facts. By and large he was in the same situation as Heyting, he had taught

[20]The official term for misbehaviour of civil servants is 'reprimand' (*berisping*) which in the civil service casts a serious blemish on a person's record.

[21]Freudenthal to Fraenkel, 15.XI.1945.

students who had signed the loyalty declaration, and he had endorsed the note that was posted after the discussions of the loyalty declaration (cf. p. 690). But there was something else and worse; Bruins had been appointed as Freudenthal's successor. In itself there was nothing wrong with this, after all courses had to be given. The real problem was that he held the position of a dismissed Jew, and the Committee of Restoration had decreed that all appointments of that sort would be reconsidered.

Any Jew who returned after the war should be restored in his old position. In fact most wartime appointees gave up their position without being forced to do so. But Bruins was not prepared to step down, he felt no obligation whatsoever to give up his lectureship. His argument was that his present position was not Freudenthal's old position, and he had a case. After all, when Freudenthal was dismissed, he was a conservator, and Bruins had been promoted to lecturer. And so Bruins saw no reason to give up his position. In a way his argument only served to change the status of his case, for there was a general rule that *all* wartime appointments should be terminated, and only after further investigation the university might decide to continue the appointment. His information on the line of action of the Committee of Restoration was so accurate, that he had even prepared typed replies to the questions that were going to be asked.[22]

Although the Committee failed to impress Bruins, or to solicit interesting information, it was firmly convinced that he should be dismissed as a lecturer. So it advised the minister accordingly; the Military Authority, however, advised the minister not to adopt the Committee's proposal. When the Committee found out that the minister was not going to dismiss Bruins, it made a last attempt to outmanoeuvre the latter. It advised Bruins to step down voluntarily (August 22).

The Committee took the matter of Bruins' appointment in the position left by a discharged Jew [Freudenthal] extremely seriously. When it asked if he considered laying down his lectureship, he did not even deign it an answer; instead he declared that he was well aware that Mr. Freudenthal was spreading nasty rumours about him. As a matter of fact, the university had immediately after the liberation adopted the policy that persons appointed during the occupation should be dismissed, with a possibility to re-apply for the job. The historian Dijksterhuis lost for that reason. For some reason Bruins weathered the storm: he had found out that the case against him was not strong enough to dismiss him, and so, when the Committee advised him to step down out of free will, he knew that the Minister would not endorse his dismissal, and therefore he simply ignored the advice. Two days later he got off with the lightest reprimand and a lecturer's post.

Most academic teachers who were appointed during the war did in fact give up their positions, and those who were 'clean', fairly easily got (new or their old) jobs. Bruins' reasons for not doing the right thing can only be conjectured, but one may safely assume that Brouwer had a vested interest in preventing the return of Freudenthal, so he would certainly not have encouraged Bruins to step down.

A curious case worth mentioning here was that of Professor Révész, founder of the Amsterdam School of Psychology. After leaving Hungary, he had settled in

[22]Bruins (1982).

Amsterdam, where he was strongly supported by Brouwer (see p. 618). During the war Révész, in the interest of the resistance, cultivated certain contacts with the Germans, and after the liberation he belonged to the small group of suspended professors. After some time the suspension was lifted without further arguments. He understandably insisted that his reputation should be restored; the faculty sent a supporting note to the Military Authority and the Committee of Restoration. Révész had in the meantime started some private investigations into the curious circumstances of his suspension. It turned out that a girl who consorted with a member of the Gestapo had sometime during the war told Bruins that Révész worked for the Gestapo. The inveterate gossip Bruins had passed the rumour on, so that the Committee of Restoration and the organisation responsible for enforcing the political order (*Politieke Orde Dienst*) picked up the information after the war. It soon turned out to be unfounded. Révész was indeed known to have rescued people in trouble, as in the case of the daughter of the well-known history scholars, Jan and Annie Romein. In 1944 he managed to get her out of the notorious headquarters of the Gestapo in the Euterpestraat.[23]

Eventually Bruins himself admitted that the accusation lacked substance.[24] Révész had managed to survive the hardships of the war reasonably well. In spite of his evidently non-Dutch name, and his heavy accent, he had been left in peace. His relationship with Brouwer had always been very close, Brouwer over the years became a good friend of the family, dropping in whenever he felt like it. During the war the frequency dropped and Freudenthal told that after the war he had a long talk with Révész. They (naturally) eventually came to discuss Brouwer. Not surprisingly, Freudenthal related the treatment he had received in the hands of Brouwer, but even Révész complained about Brouwer's behaviour during the war. According to Freudenthal, Brouwer had all but cut Révész dead. Révész was at a loss for an explanation, and conjectured that Brouwer might have thought that Révész was Jewish, and hence dissociated himself from an embarrassing relation. This story conflicts with the testimony of Révész's daughter Judith, who resolutely denied any such estrangement. She had recalled quite clearly that not only during the war, but also after the war, Brouwer visited the Révész family regularly.[25] The conflicting reports may be due to the stress of the post-war situation. One should also keep in mind that during the war it became harder to move around, and so the frequency of Brouwer's visits may indeed have dropped on the grounds of factual difficulties (e.g. curfew, break down of the transport system).

No progress was made in Brouwer's case; it was apparently no use to try to hurry the Committee. In the beginning of December Brouwer was still in the dark, and so he once more approached the Committee again with a number of facts.[26] He informed it that: 1. he had stopped the publication of *Compositio Mathematica*,

[23]Romein-Verschoor (1970), vol. 2, p. 52.

[24]Faculty meeting, 1.XII.1945.

[25]Oral communication, August 1994.

[26]Brouwer to Committee of Restoration, 3.XII.1945.

'because I could not possibly have my editorial policy controlled by the occupation authorities'; 2. he had tried to stop the creation of a Corporation of Scientific Organisations in the Netherlands[27] and when that turned out to be impossible, had tried to exercise his influence to make the statutes as harmless as possible. When this corporation decided to publish a survey of Research in the Netherlands, he first tried to talk the Mathematical Society out of it, and finally boycotted it himself; 3. he did not attend mathematical conferences in enemy territory, and bluntly told the authorities so;[28] 4. he refused to ask for a reconnection of his telephone line in 1944, when virtually the whole network was closed down. It is a touching detail that Brouwer mentioned the fact that he was approached by someone who tried to persuade him to submit an application for the re-establishment of his telephone connection, and all that was needed was a sort of pledge of loyalty. Brouwer indignantly turned the advice down, he even convinced the man that such things were 'only allowed for the benefit of the community at large and never for the advantage of individuals'. After some discussion the person decided to drop his own request; and 5. 'during the occupation I have adhered to the view that persons or groups, that had no direct or official relation to the authorities and did not possess any means of power (e.g. miners and medical men), should direct neither requests, nor admonitions to the occupation authorities. For such approaches could only derive their meaning and content from a basis of understanding between the parties, and indicated thus implicitly the existence of such a basis of understanding, which would have an encouraging effect on the ever present pursuit of penetration, and could furnish opportunities or pretexts for further measures of nazification. This view was the reason that during the occupation I have only been in personal contact with the occupation authorities, when under force, such as checks, search of the premises and police interrogations.'

Again, Brouwer had felt compelled to offer a long theoretical view. It was perfectly sound, but it is doubtful if the Committee would attach much importance value to it.

The letter also mentioned, and contained as evidence a newspaper cutting, Brouwer's involvement in a protest in support of the conductor Van Anrooy in the conflict over the Horst Wessel song (cf. p. 654). This, together with the fact that all editors living in Germany had left the board of *Compositio Mathematica* in the thirties, should lend support to the claim of Brouwer's aversion to the Nazi-spirit.

While the officials of the ministry were still pondering the remaining cases of suspended scientists, the student bodies had their own deliberations. The association of students of the exact sciences (*Natuur Filosofische Faculteitsvereniging*) had paid a great deal of attention to the purge of their teachers. Although most cases in Amsterdam had been decided, there were strong emotions as far as some of their professors were concerned. Two chemistry professors were strongly criticised, and their rehabilitation was violently opposed. But Brouwer also came under fire. In the student weekly one could read in the report about the meeting of 13 December that

[27] Werkgemeenschap van wetenschappelijke organisaties in Nederland.

[28] e.g. the Italian Conference in Rome, 1942.

'About Professor L.E.J. Brouwer little was known, but some indisputable facts were for the meeting sufficient to pronounce the "unacceptable". A resolution was passed to inform the person under consideration, as he had agreed to offer his resignation in this case.' No written evidence has been found to confirm this statement, on the contrary, in his letter to Wibaut of 12 January, Brouwer refers to the above report in Propria Cures as a 'total fabrication'. True or not, to offer his resignation would have been a clever ploy. There was a fairly effective anti-Brouwer lobby among the students, which dominated the meeting. Those who remember the post-war years know that the meetings were not always the theatre of rationality. It is not clear what exactly motivated the anti-Brouwer mood. It certainly was not helpful for his case that Brouwer strongly resented fools and pretentious persons. He could be merciless in his reactions. Somehow he managed to inspire either love and admiration, or hatred and repugnance. Apparently there were enough students in the first category to tip the scales in his favour in the end. The weekly *Propria Cures* of 4 January 1946, which reported about the meeting in December, wrote on the same page: 'The accusations brought against Prof. L.E.J. Brouwer have for the time being not been sufficiently confirmed, as a consequence of contradicting testimonials and a lack of witnesses, to justify further action in the sense of the resolution of the meeting.' The students showed, after all, to have a commendable respect for justice and procedure.

The City Council had in the meantime run out of patience with the Minister, it asked him to make up his mind in the matter of the eight suspended professors.[29] If in some cases there could as yet be no decision taken, could he at least give the names? The regular teaching required decisions.

A month later the Minister gave his verdict.[30] It apparently took some time to inform Brouwer of the decision, as he was still sending corrections and additional material to the Committee of Restoration on 30 December.[31] Brouwer was found at fault at three points:

(1) The posting of the advice that students could sign the loyalty declaration 'without essential scruple' and re-posting it after the statement of the exiled government in London.
(2) The opposition in the Senate to the resistance movements of the professors and of the students.
(3) The financial assistance to the Nederlandse Volksdienst.

And thus, the statement ran, the Minister 'disapproves of the attitude adopted by Professor L.E.J. Brouwer'. The matter ended here, there was no possibility of appeal against the reprimand.

[29]Mayor to Minister of Education Van der Leeuw, 13.XI.1945.

[30]Minister to Committee of Restoration, 11.XII.1945.

[31]In at least one memorandum to the Committee of Restoration (undated) Brouwer returned to the accusations that were made against him, referring to the letter of December 11 of the minister. He argued that the Committee had not represented his case faithfully to the minister, and that the same accusations that were raised against him, were overlooked in the case of colleagues. In short, that the principle of equality of rights had not been upheld properly. He thus carried on the defence when the authorities had already closed the case.

This was the end of the procedure, the University was free to re-instate Brouwer—which it did. Although one might consider this a light punishment, it was not considered so in Holland. A reprimand was an almost visible mark, like a 'scarlet letter'. It certainly cast a shadow on Brouwer's authority for the rest of his academic career.

The most problematic case of post-war justice, to put it mildly, was undoubtedly Weitzenböck. He had been arrested in March 1946,[32] and taken to the detention centre 'Oude Molen' in Naarden (not far from Blaricum). There his wartime record was thoroughly checked. Witnesses were produced, statements were compared, but apparently no actions against persons or institutions could be ascertained. All that could be ascertained was that he was pro-Nazi, that he had been called up for active army duty towards the end of the war, and that in his capacity of billet-master he had requisitioned housing in 't Gooi. After his interrogations he was interned in Camp Vught, the former Nazi camp. In March 1948 the Attorney General decided that Weitzenböck was to be released. His case was further handled by the courts in Amsterdam and Hilversum. The eventual outcome of the legal procedure was that his German nationality was annulled, part of his personal property and savings were confiscated. He lost his position and rights of pension. He was indeed barred from professional positions for life. In addition he was denied a passport, so that he could no longer travel abroad. Being a free man again, he settled in the village of Zelhem, not far from the German border, mainly because his daughter who at time was a nurse in Germany, could visit him more easily. In his exile in Zelhem, he occupied himself until his death in 1955 with mathematics; some papers were published, and the republication of his book on four-dimensional space was prepared—it appeared posthumously in 1956 with Birkhäuser. He was also a reviewer for the Zentralblatt. In 1949 the Free University in Berlin put out feelers to offer him a chair. The Dutch government, however, did not give Weitzenböck permission to leave the country. And so Weitzenböck, who gravely erred in his political choices, and who out of traditional military loyalty joined the German army (albeit as a member of a protection group (*Schützgruppe*)), had to bear the punishments fixed by the post-war courts. Was his fate inhuman or unfair? Certainly not when compared to the treatment of Jews and of the opposition in wartime; what may count in favour of him is that he did not go beyond the support of an odious political movement, there are no reports of personal actions of his against persons or institutions.

As to be expected, the rumours of Brouwer's fate under the post-war purge also travelled abroad. When Fraenkel inquired with Courant what the attitude should be with respect to people like Bieberbach, he also expressed his interest in Brouwer and Weitzenböck. Courant replied[33] that he had come to doubt the notions of collective guilt and the like, but 'As to individuals, I don't think that anyone behaved as crazily as Bieberbach. I cannot help feeling that he is and always was just crazy but not really dangerous. Much more dangerous are people such as Brouwer, who has been

[32]According to Euwe he was no longer in Holland at the end of the war, after the arrest he was returned to Holland for trial. Interview, 25.VI.1977.

[33]Fraenkel to Courant, 5.IX.1945; Courant to Fraenkel, 19.X.1945.

an ardent collaborationist and has been deposed.' Clearly, the grapevine did not get it right, but it certainly helped to further the rumour that Brouwer had been a Nazi, or at least had supported the Nazis. On the basis of the facts, one has to reject that idea; but also Brouwer's personality would make him quite unfit for either subscribing to the Nazi doctrines, with its *Blut und Boden*, anti-semitism, and the like, or for lending pragmatic support. Brouwer at any time despised rabble-rousers and cheap talk; even Bieberbach's political creed he could not take seriously, he mockingly characterised him as 'the white Negro'.[34] The utter vulgarity of National Socialist practice and its supporters were sufficient to put a man like Brouwer off.[35] His philosophical justification of his views and actions during the occupation are too true to type, and too much a consequence of his earlier philosophical thoughts, to dispose of as an argument of convenience.

There is no doubt that Brouwer suffered under the post-war accusations,—he, one of the leading Dutch scientists, who had always come to the defence of the underdog, was classified as a spineless collaborationist!

There are few documents that shed light on his plight; at the end of 1945 he re-established contact with his old friend Gutkind, who had moved to the States long ago. The letter is no longer extant, but we can guess from Gutkind's letter to Einstein[36] that it dealt with war and post-war problems. Indeed, Gutkind gave Brouwer's kindest regards to Einstein, and told Einstein that Brouwer had expressed a strong wish to settle in America. After mentioning Brouwer's support of illegal activities ('that Brouwer had hidden some Jews in his institute and took care of them', and 'Moreover, I recall that Brouwer travelled a couple of times to Germany to get German mathematicians out of German concentration camps'),[37] he asked Einstein to help Brouwer:

> Brouwer writes that he would prefer a research position over a teaching job. He believes that he would fit well in the Princeton Institute for Advanced

[34]The curious combination of Bieberbach's political views and the fact that he did not in the least fit the ideal of the German blond warrior often elicited smiles.

[35]There are reports that Brouwer, with his curious sense of humour, sometimes answered professional mail from Germany which ended with the obligatory 'Heil Hitler' by ending his reply with 'Greetings from Queen Wilhelmina'. Oral communication, Mrs. Sapir. I have, however, not found written evidence.

[36]Gutkind to Einstein, 3.I.1946.

[37]The first refers to the role of the institute in the resistance movement. Since Brouwer's letter is no longer extant, it is hard to say what exactly its content was. As to the second statement, Gutkind's wording suggests that he had independent information on this matter. It is not wholly impossible, cf. the book of Balke, but then it probably took place before the war. In the absence of further information, there is little to be said. Balke's book contains a reference to Brouwer's support of a Nazi victim (cf. p. 643), there may have been more instances. In fact in some interviews Brouwer's role in helping and hiding Jews and other victims of the regime have been mentioned; as no further confirmation could be obtained, it is no more than a rumour. There is similar, but more concrete, evidence that Brouwer's cottages were used by people who had gone in hiding (*onderduikers*), cf. p. 697. Brouwer did not mention facts of this sort in his defence; his standard of argumentation was so high that it could easily turn against himself. Even here he practised *Methodenreinheit*.

> Studies. He would like equally well to go to one of the Californian Universities, where the mild climate would suit him better. Both Brouwer and his wife have, in spite of their great toughness, a tender constitution...
>
> Surely, you, my dearest Professor Einstein, will understand my initiative to approach you, to save this great man and to deliver him from his present difficult situation.

Einstein did not dismiss the matter without some serious consideration. He wrote that he had discussed the matter with a colleague, but that, although everybody agreed that Brouwer had outstanding scientific merits, little could be done for him.

> It has become known that he has been suspended by the University of Amsterdam on political grounds, and it is clear that with such an important man one would not have decided lightly upon such a move. Therefore it seems to me that it is rather hopeless to take steps for his permanent appointment in this country, until the matter has been settled.[38]

There is ample evidence that the climate in Amsterdam had become too unpleasant, not to say hostile, for Brouwer, to be considered an acceptable place for spending the last few years of his professional career. Undoubtedly, the purge and its consequences were felt as unbearable insults by Brouwer.

The end of the war had raised Freudenthal's hope for a happy end to his involuntary exile. There was every reason for optimism. After the defeat of the invaders one could reasonably expect the rehabilitation of the victims of the occupation. Freudenthal and his family had survived the war as well as one could hope. There had been little contact with his colleagues during the last years. Koksma later told that at the entry of the Canadian forces in Amsterdam he was watching the long line of vehicles when he suddenly saw Freudenthal sitting on a Canadian tank, 'When I saw Freudenthal on that tank I was really convinced that the war was over.'[39]

What seemed more natural in those hectic days, full of optimism, than a speedy reinstatement of Freudenthal at his university? Of course, he expected more than just a re-appointment in his old function; the person who had made use of the opportunity provided by the Nazi policy not only had succeeded Freudenthal, but also obtained the promotion that was promised to Freudenthal. In spite of the glaring injustice, nothing happened. In a letter to the mayor[40] he drew attention to his treatment and expressed the hope that his 15 years of service to the University of Amsterdam and his international status would result in his re-instatement. Actually, Freudenthal as an unemployed citizen, living under dangerous circumstances, had contributed more to the wartime mathematical activities than all the official staff members of the Mathematical Institute together. He had, for example, supervised three Ph.D. theses: of J. de Groot, A. van Heemert, and F. Loonstra. His appeal to the mayor asked for recognition, 'after the disqualification, which any outsider

[38]Einstein to Gutkind, 10.I.1946.

[39]Interview in Alberts et al., p. 116.

[40]Freudenthal to Mayor, 11.VI.1945.

could discern in the career of my successor, who, as a physicist was a stranger in the realm of mathematics, and remained so, and who nonetheless was already after one year appointed as a lecturer, a position which was not granted to me after ten years, in spite of numerous promises'.

In spite of the treatment he had received from Brouwer, and of Brouwer's disputed war record, Freudenthal had expected some token from Brouwer, 'and I must say, I am not sufficiently vengeful not to have entered into a reconciliation',[41] but nothing happened. Brouwer and Bruins insisted that the most Freudenthal could expect was the position of conservator under Bruins.

The conflicts between Brouwer and Freudenthal had left Brouwer with the conviction that co-operation had become impossible. He wished Freudenthal no harm, but he could not bear to have him around, and so he pursued a double policy. He strongly opposed Freudenthal's reinstatement and promotion, and at the same time used his influence to get him a chair at another university. For Freudenthal that solution was unacceptable, to be denied the lecturer's position would be a grave insult.

Since Brouwer was under investigation, Freudenthal hoped that his rehabilitation would not be difficult. In order to show Brouwer that the personal conflicts of the past did not enter into his request for re-instatement, Freudenthal had sent Brouwer a copy of the letter to the Mayor. And Brouwer had passed the letter on to... Bruins, a not uninterested party. Bruins immediately set out to refute Freudenthal's claims and statements in a letter to the mayor.[42] Apart from a rather artificial objection—Bruins denied the fact that he was Freudenthal's successor by pointing out that duties were not identical. He simply stuck to the argument that a lecturer could not be the successor of a conservator. Bruins' letter consisted mostly of attempts to show that he was indeed a mathematician, combined with hardly veiled insults addressed to Freudenthal. A couple of letters followed, repeating arguments, providing new details.[43] Freudenthal's style was by far the more dignified one; he carefully avoided getting drawn into a vulgar match of abuse. Bruins' evidence for his strong mathematical background was easily dealt with. Most of his articles at that time were clever applications of (fairly elementary) parts of mathematics to problems in physics, chemistry and medicine, but they could hardly support his claim to his status as a mathematician. Likewise, Freudenthal added, not much weight could be attached to a prize essay (which was advanced by Bruins as a serious argument) that did not merit an award. In short 'If Dr. Bruins thinks that, on the grounds of a prize essay that was not successful, and some other minor papers, he enjoys a reputation in the mathematical world beyond the limits of the University of Amsterdam, he is grossly mistaken.'

The only new fact that Bruins mentioned in a subsequent letter was that his position was already agreed upon by Brouwer and his colleagues in 1938. On the whole

[41] Freudenthal to Van der Corput, 25.VIII.1945.

[42] Bruins to Mayor, 20.VI.1945.

[43] Freudenthal to Mayor, 22.VI.1945; Bruins to Mayor, 1.VII.1945; Freudenthal to Mayor, 6.I.1945, Freudenthal to CvH, 9.VII.1945.

this would not be impossible, but highly unlikely, and it was never mentioned by Brouwer. The subsequent letters of Freudenthal repeated and further underpinned his case against Bruins.

Both Bruins and Freudenthal appealed to the mayor in a number of letters, in which the given arguments were further elaborated. Freudenthal's arguments can be summarised as 'in essence Bruins has taken over my position, and he is not a proper mathematician'; Bruins on the other hand maintained that 'legally my position is different from Freudenthal's last position, and I am a good and proper mathematician'. On both counts Freudenthal's side was the stronger one. Anybody who compared Freudenthal's mathematical record with Bruins will agree that Freudenthal was already at the top of his profession, whereas Bruins was respectable middle class. And where the claim on the position is concerned, Bruins' argument appears as a legal trick, possibly useful in the hands of a slick lawyer in court, but definitely weak on the moral side. Freudenthal, as a persecuted Jew, clearly had the moral right on his side.

Bruins, bent on the defence of his position, apparently used curious means to thwart Freudenthal's claim. Freudenthal told Van der Corput that Bruins, with his taste for the bizarre, 'had me lured to an address in Amsterdam-South where he had a simple minded acquaintance of his make a proposal to withdraw my claims on a restoration, for which he would compensate me financially; Bruins, by the way, was hiding behind the door when this proposal was made'.[44] Nonetheless, Freudenthal was optimistic that justice would be done, and most colleagues would have shared this optimism.

The matter of Freudenthal's rehabilitation was, as most people at the time agreed, and as we nowadays see it, clear enough for the university to make a decision; it would have meant some reshuffling of the budget, but it could be done. That, however, is not the way big institutions solve their problems. One could not take such steps without proper advice, and whose advice is one to ask? Of course that of the Director of the Mathematical Institute, suspended or not.

There is no doubt that the sympathy of the University officials and the mathematical world was on the side of Freudenthal; Bruins' legal position was, however, strong enough for him to brave all threats, but his greatest asset was the support of Brouwer.

The latter was invited to give his views on Freudenthal. On August 28 he had an interview with the official in charge. According to the note that Brouwer composed after the interview, the young, promising German mathematician Dr. Freudenthal had come to Amsterdam in 1930 to conduct research in areas relating to and supporting Brouwer's research. Since Freudenthal (and Hurewicz, for that matter) carried on their own research and did a bit of teaching as *privaat docent*, and since this could not fully justify their presence, Brouwer (according to this note) founded

---

[44]Ibid.

*Compositio Mathematica*, in order to supervise the editing 'with the help of both assistants'.[45]

At the same time he had fostered hopes that both would find suitable jobs elsewhere. Hurewicz did, but Freudenthal stayed. As we have seen, at the retirement of De Vries, Freudenthal was entrusted with the analysis courses and was appointed conservator. 'The professors at the Mathematical Institute [Brouwer, Mannoury, de Vries, Weitzenböck] expected, I think unanimously—and did not hide this from Freudenthal—that the natural course of things would automatically result in his inclusion in the ranks of the official teaching staff (*docentenkorps*), if his request for naturalisation, submitted already a long time ago, was granted, if it should turn out that he was well-suited for teaching other than advanced and specially gifted audiences, and if an adequate understanding and co-operation between him and the other mathematical staff members should develop.'

Then the grim conclusion followed:

> However, the fulfillment of this threefold condition until now has not taken place. Moreover, Freudenthal's attitude of impertinent irritation with respect to me, of which signs had been manifest already before 1937, has since then become more outspoken and had taken the sharper form of slander and aggression. Only the hope that his activity in cultivating relations here and abroad would after the war soon bring him an appointment elsewhere has enabled me to maintain so far my patience with respect to him.

This letter confirms an earlier diagnosis of the conflict. Brouwer could simply not bear to have somebody around who criticised him, no matter how discretely. Brouwer was no fool at any time, and he obviously realised that more than a decade of his scientific career had been wasted, but he could not bear to look into the mirror that Freudenthal held up for him. All this talk about aggression was a clumsy but nasty device to counter Freudenthal's unspoken criticism.

After all these years Brouwer was as highly strung as ever and his urge to get rid of Freudenthal assumed alarming proportions. At the hearing he defended himself against Freudenthal's reproach that he had refused the papers of De Groot, Loonstra and Freudenthal, saying that in 1942, shortly before a meeting of the Academy, he got more than 100 pages of manuscripts from Freudenthal.

'I took them along on a trip. Shortly afterwards I got a reproving letter. Then he threatened to publish them elsewhere, and I returned the papers. I have not approved a dissertation of a student of Freudenthal just like that, as he demanded.'[46] And when asked if Freudenthal could be charged with the running of the institute,[47]

---

[45]The argument is a curious mixture of fact and fiction, clearly designed to add plausibility to Brouwer's views. Hurewicz and Freudenthal were indeed appointed in Amsterdam before Compositio began to appear, so their presence was convenient for the running of the journal, but nobody would maintain that Compositio was designed to keep Brouwer's assistants occupied.

[46]Notes of the Secretary of the Comm. of Restoration.

[47]i.e. act as temporary director.

Brouwer exclaimed, 'Not necessary, I will yield only for bayonets!' Brouwer probably managed to get his message across, he could argue very persuasively, and charm or repulse at will. It is a fact indeed, that Freudenthal felt later that the chairman of the Committee of Restoration, Van Valkenburg, was rather cold towards him.[48]

Almost immediately after the interview with the Committee of Restoration, Brouwer must have learnt something that, in his opinion, confirmed his worst fears. On the thirtieth of August Brouwer reported a violation of the Mathematical Institute;[49] he had (with the permission of the chairman of the Committee of Restoration) gone to the institute, as a part of his private library and archive was in his room. On entering his room he perceived that his papers were disturbed and on closer examination a large number of the keys of various drawers and filing cabinets had disappeared. These keys were hidden between archive folders at spots only known to Brouwer and Cor Jongejan. The cleaning lady, Mrs. Peters, told him that she had found an unknown person in Brouwer's room who was inspecting papers, and who told her that from now on she had to deal with him as a Director of the Institute, and that furthermore only one of the members of the institute would be allowed to return.

On inquiry, the temporary supervisor Mr. Pater declared (according to Brouwer's report) that on Monday August 20, the chairman of the faculty, Professor Clay, asked him to give the key to the Mathematical Institute to Freudenthal. Pater objected that he could not do so without the permission of Brouwer, or his deputy, Bruins. On Clay's assurance that he would take full responsibility, Pater accompanied Freudenthal to the institute, who said that he was going to teach. After some time Freudenthal sent Pater away and when Freudenthal left, Pater saw that the briefcase of Freudenthal had considerably grown in size between arrival and departure. The matter was not only known to Brouwer, but also to Bruins, who immediately saw the value of the incident, either as interesting gossip or as convenient smear.[50]

For Bruins (and Brouwer) this was a welcome argument in the battle against Freudenthal. Brouwer left it at the above letter, but Bruins started to spread accusations against Freudenthal.[51] When Freudenthal got wind of this, he immediately asked the authorities to investigate the matter. The Alderman duly started an investigation and asked Clay to report to him. Clay checked the facts and interviewed the persons involved. His findings were:[52]

- Freudenthal was authorised by an official at City Hall[53] to enter the Mathematical Institute, which he had to run for the time being.

---

[48]Freudenthal to Van der Corput, 17.IX.1945.

[49]Brouwer to Alderman for Education, 30.VIII.1945.

[50]Freudenthal to Van der Corput, 17.IX.1945.

[51]Freudenthal to Mayor, 21.IX.1945.

[52]Clay to Alderman, 15.X.1945.

[53]Ottens.

- Clay had asked him to look for the examination administration of Weitzenböck (which was vital for the final exams). This was to be found in the room which Weitzenböck shared with Brouwer.
- Mr. Pater had left Freudenthal alone on his own accord.
- Freudenthal had only opened unlocked drawers and cabinets.
- The cleaning lady stated that the majority of the keys were missing.

Clay ended the report by noting that nothing could be concluded against or in favour of Freudenthal. However, he continued,

> As it became clear to me that there existed previous grievances between Prof. Brouwer and Dr. Freudenthal, it is unpleasant for Prof. Brouwer to learn that Dr. Freudenthal has been in his room without his knowledge; on the other hand it is most unpleasant for Dr. Freudenthal to be suspected of malicious intent, which is not confirmed by any of the facts that have been brought to my attention, and for which I cannot find any indication.
>
> There has been here an unfortunate coincidence, of which both gentlemen have become victims.
>
> A better understanding in general, or should this be impossible, a definitive separation of their areas of activity, shall have to bring the solution to the incident.

This incident is related here at length, to illustrate that the relation had really deteriorated to the point of cloak and dagger stories.

Indeed Freudenthal had already asked Brouwer in July to be admitted to the institute, without success. Freudenthal was indeed authorised to run the institute, but this authorisation was mysteriously withdrawn. He bitterly complained to the chairman of the Committee of Restoration that it was surprising to say the least, that he—the only 'clean' staff member—could not enter the institute and go about his business, while all the suspended members (Brouwer, Bruins, Heyting) were in the possession of a key.[54] Partial justice was done to Freudenthal, when on September 14 he was re-appointed as conservator, while 'the possibility of honouring your request for appointment to lecturer is further investigated'.

For Freudenthal this was not enough, he saw the wartime appointment of Bruins as a trick to pre-empt any claims of his and he viewed the lectureship as his due. The fact that he should work under Bruins is in retrospect indeed preposterous, and Freudenthal completely ruled out this option. Roughly at that time various actions were taken to support Freudenthal. His students and former students sent an address to the Committee of Restoration, and Van der Corput alarmed the Minister.

All of this was to no avail: the Faculty, although on the whole sympathetic towards Freudenthal, could not make up its mind to support him. There were too many conflicting interests, and when Van der Corput, Van der Waerden and Van Dantzig all wanted to be on the Amsterdam faculty, Freudenthal was no longer a priority.

---

[54] Freudenthal to CvH, 23.VIII.1945.

Furthermore, the faculty felt understandably that, even though Brouwer was suspended, no major decision on mathematics should be taken without consulting him. And so Brouwer could even at this time delay the decision process.

At the first post-war faculty meeting,[55] Clay suggested that Van der Waerden would be a good candidate for Weitzenböck's vacancy. The chairman (Weevers, appointed by the Committee of Restoration) objected on the grounds that Van der Waerden had been in Leipzig during the war. He suggested that in Brouwer's absence Freudenthal, who was close to Brouwer, should conduct Brouwer's exams.

## 17.2 Faculty Politics

It was no coincidence that Van der Waerden was mentioned in the faculty meeting. This man, who was the brightest Dutch mathematician after Brouwer, had made a shining career. After his study in Amsterdam he had gone to Göttingen to study and work with Emmy Noether. In 1926 he got his Ph.D. in Amsterdam under Hk. de Vries, and in 1929, at 26 years old, he became a professor in Groningen. Three years later he accepted a chair in Leipzig, and became the successor of Otto Hölder. During his Leipzig years he taught a large variety of courses and published on a wealth of topics. In that period he was considered the leading expert on algebraic geometry.

The advent of the Nazi regime brought him in conflicts that were almost predictable in the case of an intelligent, original man with a talent for speaking his mind. In 1934 he was reprimanded by the local government for joining Heisenberg in a protest in the faculty against the dismissal of four Jews,—a foreigner should stay out of German politics (unless he supported the party, of course)! He never compromised himself by supporting the Party; when the Germans invaded Holland, he was arrested, released after the capitulation and ordered not to leave Germany. For all practical purposes he had become a high-class forced labourer. In 1943 the Faculty in Utrecht offered him a chair at the retirement of Barrau, a Dutch geometer, who like Brouwer wrote his dissertation under Korteweg. In the same year Van der Waerden's house and property was destroyed by bombing. After a difficult odyssey he returned after the war to Holland and found that the post-war government did not wish to endorse the appointment that the Utrecht University had arranged for him.[56] Freudenthal brought him in contact with Professor W.J.D. van Dijk, the director of the research department of Shell Oil, who was at that moment looking for mathematical expertise for his laboratory.[57] As a result Van der Waerden got a research position at the Shell establishment in the Hague, so that he at least had a decent income.

[55] 15.VIII.1945.

[56] The Committee of Restoration and the 'Van der Corput'-committee played a somewhat dubious role in the matter, see below.

[57] Alberts et al. (1987), p. 134.

The reader may recall that Brouwer could not get along with the young Van der Waerden, who used to speak his mind no matter where and when. He had, however, supported him in going to Emmy Noether in Göttingen. But the fact that Van der Waerden became a fervent admirer of Hilbert certainly contributed to a further alienation between the two. As long as Van der Waerden was in Leipzig, things were fine, but a Van der Waerden in Holland represented a threat to Brouwer's emotional stability. Sooner or later Van der Waerden was going to play a role in Dutch mathematics, and worse—he might be interested in a position in Amsterdam!

Indeed, from Brouwer's point of view the situation was fraught with dangers and unacceptable consequences. He would have to fight a war at two fronts at least, and as we will see, the number of fronts quickly multiplied. When the faculty met again, the new chairman Clay had faithfully consulted Brouwer on the matter of Weitzenböck's successor. Brouwer had, he said, told him that the chair of Weitzenböck was a piece of luxury the faculty could dispense with—no successor was needed. Instead Brouwer pushed Heyting's promotion to full professor. The emeritus Pannekoek (an advisory member of the faculty) drew attention to Van der Waerden as a successor of Weitzenböck; Clay answered that in view of the well-known aversion of Brouwer to Van der Waerden, he had not undertaken anything in this direction. Nonetheless he insisted that the vacancy should be filled. There was an external reason for doing so, as we will see.

The situation in the faculty (and in all of Dutch Mathematics, I may add) was indeed complicated by a new initiative of the Minister of Education; he had installed a committee to design plans for a reorganisation of mathematics in Holland. The records of the committee are somewhat elusive; the ministry has not been able to locate any minutes or recommendations in their archives, but from bits of correspondence and from various archives some information could be collected.[58] The committee, called the 'Committee for co-ordinating and reorganising higher education in mathematics in the Netherlands' was created by Minister G. van der Leeuw on 26 October 1945; it was disbanded in the beginning of 1947. There were two well-defined tasks of the committee: 1. to co-ordinate the appointments in the then existing vacancies in mathematics; and 2. to investigate the possibility and desirability of a centre for scientific mathematical activities, and to consider means to make closer contact between pure mathematics and its applications in other areas.

The committee consisted of four mathematicians, one physicist and an astronomer: J.G. van der Corput (Groningen), J.F. Koksma (Amsterdam, Free University), J.A. Schouten (retired), D. van Dantzig (Delft), H.A. Kramers (Leiden), and M.G.J. Minnaert (Utrecht). Three of the four mathematicians were not friendly inclined towards Brouwer, to say the least. The plain fact that Brouwer, the greatest Dutch mathematician of his day, was not a member is telling. Formally speaking, Brouwer could not very well have been a member, after all, at the time of the establishing of the committee, Brouwer was suspended, and one cannot blame a minister for disregarding a candidate with a problematic political profile. Factually, however,

[58] I am indebted to G. Alberts for putting the following facts at my disposal.

leaving out Brouwer could only be seen as a calculated insult, not so much from the side of the minister, as from the side of his colleagues. In a sense it was a good thing that Brouwer was not a member of the committee; at least he could not be held responsible for the havoc that the committee created. The committee did not contain a member from the Amsterdam faculty. This was in a way surprising as the University of Amsterdam had for a long time been at the top of the profession where mathematics was concerned. Counting out Brouwer, Freudenthal or Heyting might very well have qualified; both had a good reputation in the field. Technically it would perhaps have been a bit awkward, as neither was a professor at the time, so they would have to discuss their own promotion. There is little documentation about the committee left (I have actually seen no original material), so most information is based on recollections of mathematicians who at that time were junior members of the community. The general impression was that Van der Corput had made a judicious choice of (mathematical) members of his committee (to which I will refer as the 'Van der Corput Committee'), and that to all appearances he had a hidden agenda.

The committee took its task very seriously; it went so far as to dictate faculties the choice of candidates for chairs. For example Van der Woude, who was a lecturer at Leiden, was commandeered to a chair in Utrecht—completely against his wishes. He wisely ignored the instructions of the committee and remained in Leiden. In Delft the committee ran into a vigorous opposition. The mathematician Bottema bluntly refused to continue the discussion with the committee, almost threatening Van der Corput with bodily harm.[59] The older generation of Dutch mathematicians could recite instances of undesirable interference and nepotism. Van der Corput and Van Dantzig were for example reported to have used the committee to get appointed in Amsterdam. Both of them had earlier offers from Utrecht, but Amsterdam apparently appealed more in those days. Indeed, Utrecht had no mathematics professor at the time. Julius Wolff was deported during the war; he died in a German camp. Barrau had retired in 1943, and the vacancy had not yet been filled. The faculty and the board of the university had not neglected the matter. As a matter of fact, they had sounded Van der Waerden, who was at that time still at the Leipzig university. Van der Waerden was in a difficult position, as a professor in Germany he was not free to accept offers without the government's permission. He had already received calls to Munich and Göttingen, but permission to accept had been denied by the authorities. It is highly doubtful if a call from Utrecht would fare any better. He replied to the Utrecht faculty that he could not be certain to accept the offer under the present circumstances.[60] It should be added that his German colleagues would be very sorry to see him go. Carathéodory, for example, asked Van der Waerden if

[59]Oral communication, N.G. de Bruijn. Van der Corput also wanted to transfer Blumenthal's reprint collection from Delft to the Mathematical Centre in Amsterdam. The request was turned down. In retrospect, it would have been better that his request had been granted, for eventually the valuable collection was lost, and not so long ago items of the collection turned up in antiquarian bookshops.

[60]Utrecht faculty to Van der Waerden, 27.VII.1943, Van der Waerden to Utrecht faculty, 19.IX.1943.

he could not temporise in the Utrecht matter. Van der Waerden later claimed that he did not want to accept a chair under state secretary Van Dam; a wise decision indeed, he was already a suspect person for his teaching in Germany, and to accept a position in wartime Holland would have been a real blot on his career. Once the war was over, Van der Waerden was repatriated to Holland, and the Utrecht job came in sight again. On 14 August 1945 the Utrecht University sent a list to its Committee of Restoration with names of candidates for the mathematics vacancy: 1. Van der Waerden, 2. Van Dantzig, 3. Struik, 4. Bottema.[61] These were the candidates for the geometry chair, simultaneously candidates were offered for the chair in 'analytic mathematics'. There Van der Corput headed the list. The faculty did its best to convince the Committee that Van der Waerden was politically clean.

In the meantime Van der Corput's committee had considered the options for a national mathematics institute. Two issues were of immediate importance: the place and the staff. In The Hague at the ministry there were voices that favoured Utrecht. With Van der Waerden (and at least one more professor) it would have an extremely strong and competent mathematician. And Utrecht was in the geographical centre. However, the Van der Corput committee considered Amsterdam the better choice. When the rumours about the proposed centre started to circulate, the Utrecht faculty again approached the Committee of Restoration.[62] It summed up the prospective distribution of mathematicians: Van der Corput opted for Amsterdam, Koksma stayed at the Free University, and Kloosterman in Leiden. Van der Waerden now also had an offer from Amsterdam, but he would prefer Utrecht if the appointment could be realised soon.[63] The faculty had no objection to Amsterdam as the seat of the new institute, but with Van der Corput and Van der Waerden both in Amsterdam, the Utrecht students would probably defect to Amsterdam. With Van der Waerden in Utrecht the balance would be restored. The Utrecht hopes were dashed when the minister announced that persons who worked voluntarily in Germany during the war were not eligible for appointments (in any of the departments).

Exit Van der Waerden.[64] There were persistent rumours (neither confirmed nor denied) that Queen Wilhelmina personally blocked Van der Waerden's appointment, and that the above formulation was an official arabesque.[65]

The requested advice about a research institute for mathematics eventually led to the founding of the Mathematical Centre in Amsterdam.

---

[61] Van Dantzig and Bottema were professors in Delft, Struik had been at MIT since the twenties.

[62] Faculty Utrecht to Committee of Restoration, 13.XII.1945.

[63] In fact the Amsterdam faculty used a similar argument. Van der Waerden had apparently decided to sit on the fence.

[64] Van der Waerden's life has recently been the subject of a series of papers by A. Soifer—Soifer (2005a, 2005b, 2005c). On account of the advanced stage of preparation of the present book, no use could be made of this material.

[65] Until 1981 professors and lecturers at the state universities were appointed by the Queen on the recommendation of the Minister of Education. The Minister in turn was presented with an ordered pair of two candidates by the university. Usually the Queen followed the recommendation of the faculties, but even in ordinary times, politics could play a role and a candidate could be 'returned' to the faculty.

**Fig. 17.1** B.L.G. van der Corput. [Photographer Aart Klein, copyright Centrum Wiskunde & Informatica (CWI)]

For the proper understanding of some of the delicate workings of Dutch politics in this particular case, the reader should know a bit more about the major actor, namely Professor Johannes Gualtherus van der Corput, a mathematician of good standing. He had a fine reputation in number theory and his publications had appeared in the leading journals. We have already met his name in connection with Brouwer—it was on his instigation that Brouwer gave a course on intuitionism in Groningen in 1932. Van der Corput's prominence in post-war Dutch mathematics was partially the consequence of his friendship with the first Minister of Education after the war. This Minister, Van der Leeuw, had been a professor in theology in Groningen; he was respected and known for his views and publications on the philosophy of culture. In the exciting period following the liberation, there was enough turmoil to admit people into the political arena who had not made a lifelong career in politics. One of those persons was Van der Leeuw. Van der Corput, as a Groningen professor, happened to be on good terms with Van der Leeuw, and when the latter became the Minister of Education, Van der Corput had the minister's ear. Consequently, Van der Corput soon found himself chairman of the most important committee for Dutch mathematics. The idea for an independent Mathematical Institute had in fact been born during the war. Mathematicians in the forced period of inactivity said to each other, wouldn't it be nice if we had an institute where we could sit and talk mathematics, and work?[66]

Clay, the faculty chairman in Amsterdam, mentioned the plan for a national mathematical institute in the third post-war faculty meeting. In his view the plan envisaged a new Mathematical Institute in Amsterdam, provided the leading mathematicians of the University of Amsterdam should (at least partly) belong to the institute. He had already consulted Brouwer, and the latter agreed to the appointment of Van der Corput in Amsterdam. Clay added that one should not wait too long, for otherwise the new institute might go to Utrecht.

[66]Quoth Koksma and Van Dantzig (oral communication, N.G. de Bruijn).

The discussions at the first post war faculty meetings were confused; it is almost as if its members did not dare to make decisions 'while father was away'. Heyting's promotion was discussed, but there was little enthusiasm for a copy of Brouwer; moreover, one of the members thought it improper to put a recently suspended colleague up for promotion. The chairman reported that Brouwer had expressed the wish to work with Van der Corput, while at the same time making it clear that the appointment of Van Dantzig and Van der Waerden would strongly displease him. He had even mentioned to some members of the faculty that the appointment of Van der Waerden would be a reason for him to resign.

The discussions at the faculty meetings later that year were dominated by the prospect of securing the new institute for Amsterdam. The offers of Utrecht to Van der Corput and Van der Waerden were a serious setback, but if the faculty managed to outbid Utrecht, there was a chance. Van der Corput was willing to favour Amsterdam, provided the new institute came to Amsterdam, and that Van der Waerden was to join him there. In view of this fact, Clay motioned that Van der Waerden should be put forward as a candidate at the next City Council Meeting, the more so as the Minister and the responsible councilman had in principle agreed to share the financial responsibility for the new institute. Although Brouwer was known to have personal objections to Van der Waerden, the chairman said that apparently Brouwer was willing to give in if the new institute were to be situated in Amsterdam and if Van der Waerden were to be attached to the new institute. When the meeting continued to discuss the aversion of Brouwer to Van der Waerden, the physicist Michels remarked dryly that if Brouwer didn't like it, he should get out. After all, he was 65.[67]

The consequence of all these new prospects was that suddenly Freudenthal had become a side issue. The appointment of at least two new professors, it was argued, made it inopportune to promote Freudenthal at this moment to a lectureship, as the new professors should not be presented with a fait accompli.

Although the faculty deplored the injustice of the situation—Bruins sticking to his chair and Freudenthal left out in the cold,—it did not go beyond a truly academic 'on the one hand..., on the other hand...' message to the Committee of Restoration. The Committee urgently wanted a bit of practical advice, but the faculty did not give any.

Freudenthal was certainly not mistaken when he wrote to De Groot,[68] 'how powerful Brouwer is, although he is suspended'.

In spite of numerous letters to City Hall, the Rector, the Committee of Restoration, the faculty, ..., Freudenthal did not succeed in getting justice done. Even worse, he shared the fate of so many men who simply wanted justice—when in May 1946 Freudenthal implored Clay, the faculty chairman, to see that he was finally rehabilitated, and made it clear that the sympathy of the faculty was no substitute,

[67]Brouwer had thwarted Michels' appointment in the thirties. Not much love seems to have been lost between the two, Maas (2001), p. 180.

[68]Freudenthal to De Groot, 17.IX.1945.

Clay replied testily, 'It is not clear to me why you keep speaking about rehabilitation. You have been *re-instated* in the position you occupied before the war... The feelings of distrust that you ventilate with respect to the faculty are not helpful to reach a just solution.'[69]

When Brouwer returned to the faculty after the suspension was lifted, he immediately set out to recover lost ground. His most urgent goals were (a) to remove Freudenthal, and (b) to prevent an appointment of Van der Waerden.

He had already personally, and through Heyting, made it clear that the faculty could do very well without a successor of Weitzenböck. The chairman, Clay, faithfully informed the board of the university,[70]

> In addition to my letter of October 19, I inform you at the request of colleague Brouwer, that he and the lecturers Dr. Heyting and Dr. Bruins find that in Amsterdam, also after Weitzenböck's departure, there are enough mathematical teachers. They are of the opinion that the appointment of a successor of Weitzenböck—who (in order to meet a condition of L.E.J. Brouwer for turning down the offer of a chair abroad) was appointed as an extra-ordinary professor at the time when there was a strong influx of interested foreigners,—is under the present drastically changed circumstances not justified.[71]

Having complied with Brouwer's wishes, he added that the faculty, in view of the prospective institute, did not share Brouwer's views.

After the resumption of his regular duties, the first thing Brouwer did was to call a special meeting of the faculty to discuss 'proposals for teaching mathematics'.[72] No sooner had the chairman given the floor to Brouwer, than he embarked on a lengthy harangue. He had, he said, been absent against his will and on returning he had found undesirable persons in his scientific home. The three mathematical staff members,[73] according to him, had wished to promote Heyting to a professorship. That being taken care of, he had further proposals in mind. He was surprised by the plan to appoint Van der Waerden. After making it quite explicit that co-operation with Van der Waerden would be most painful to him, Brouwer cleverly argued that it would be the proper thing to appoint Van der Waerden at the new institute. The chairman, Clay, had no difficulty in grasping the immediate consequences of Brouwer's proposal; he vigorously pleaded that no time should be lost. The minister had been at the point of appointing Van der Waerden in Utrecht, he said, when some members of the Mathematical Committee begged him not to do so. Van der Waerden himself, it was said, was willing to accept an offer from Amsterdam provided it was made before the Utrecht offer. The founding of the new institute in Amsterdam was vigorous pleaded by Clay.

---

[69] Clay to Freudenthal, 15.V.1946.

[70] Clay to rector, 22.X.1945.

[71] The teaching staff consisted at that moment of Brouwer, Bruins, Freudenthal (who had been reinstated as a conservator on September 14), and Heyting.

[72] 24.I.1946.

[73] Brouwer, Bruins and Heyting.

The discussion did not yield any particular results. Brouwer tried to push Heyting's promotion, but some members raised the objection that Heyting had broken the unanimity of the professors at a critical period of the occupation. In Brouwer's eyes this was unfair, taking into account Van der Waerden's past, etc. It was clear, however, that Brouwer was again casting his spell over the faculty. His arguments against an immediate offer to Van der Waerden—namely that no commitment could be made to Van der Waerden until the funds for the new institute had been granted—were so successful that the chairman finally declined to send any letters about the appointment to the university authorities.

Brouwer had not waited, however, for a faculty meeting to pay attention to Freudenthal's lecturer's position. He had found out in December that two lecturers in mathematics were proposed to the City Council, and it seemed plausible that one of these was for Freudenthal. He complained that neither he nor Bruins or Heyting were informed of these proposals.[74] In discussions with members of the Van der Corput committee for mathematics, he had already pointed out that co-operation with Freudenthal had become extremely difficult for him, and that it 'has had such a paralysing influence on his energy, that anybody who can end this co-operation without personal disadvantage to Mr. Freudenthal should do this in the general interest. Thus, in my opinion, the above mentioned Committee is faced with the duty to see to it as far as it can that the various vacancies at the Dutch Universities are filled in such a way that a fitting position outside Amsterdam is allocated either to Freudenthal, if he turns out to be acceptable in Holland, or to me.' A member of the committee had, however, conjectured to Brouwer that Freudenthal might refuse to leave Amsterdam. If that were the case, Brouwer argued, Freudenthal's inflexibility would only be reinforced by an offer of the position of lecturer at the present moment. In the same letter to the City Council Brouwer maintained that Bruins should be made a lecturer in applied and introductory mathematics (*propaedeutische wiskunde*).

The subsequent developments show that Brouwer's power had eroded. The chairman of the faculty, Clay, received as a matter of routine, a copy of Brouwer's letter, and he firmly took Brouwer to task:[75] he assured that no proposals for lecturer appointments had been made; but, he said, even if Freudenthal were offered the position of lecturer in Amsterdam, it would make it easier for him to get offers from elsewhere. The letter closed with a telling remark: nothing was decided with respect to Bruins, but 'it makes me sad that at this moment he is active in ruining his own position by a smear campaign'.

Power is concentrated in decision making and advisory bodies, and the single most interesting body at the time of Brouwer's suspension was the Ministerial Committee for Mathematics. So what could be more natural for Brouwer than to request a seat on it? On the same day Clay sent his firm letter, Brouwer wrote to Van der Corput, 'Now that I have been re-instated in my position, don't you think it right

[74] Brouwer to B & W (Mayor and Aldermen), 7.I.1946.

[75] Clay to Brouwer, 10.I.1946.

and proper, and in the interest of further developments, that I am after all given a seat in the co-ordination committee, installed by the minister and presided over by you?' The answer is not known, but Brouwer's demand had no effect.

It must have dawned on Brouwer that the new colleague, Van der Corput, was no push-over. The domination of Brouwer had been challenged, and the decline of his empire had set in. Whatever his actions and protests were, he could no longer make the law, or keep City Hall under his control.

A demonstration was to come earlier than he expected; on January 16 Van der Waerden's appointment was discussed in the City Council. Brouwer was furious; the chairman of the faculty had kept him informed of the progress of the new institute, and he had managed to postpone Van der Waerden's appointment till later (indefinitely, he hoped), at the price of accepting Van der Corput as the successor of Weitzenböck.

He indignantly complained to the chairman that he was the victim of repeated breach of faith.[76] First, he claimed, the letters of advice re Van der Corput had been sent with a text not approved by him, at a time not determined by him—contrary to recent promises. Next, Clay and Van der Corput had made plans to guarantee the participation of Van der Waerden in the new institute. And now the Van der Waerden matter had reached the City council. 'This cannot fail to fill me with mistrust with respect to the persons involved in this matter.'

The procedure of Van der Waerden's appointment did not make any progress that year, the urgency had also somewhat diminished as he accepted a visiting professorship at the Johns Hopkins University in Baltimore. Nonetheless, Brouwer felt he had to be vigilant, for he started to see enemies and conspiracies everywhere (indeed, by now people started to look for means to get around this formidable obstacle). It goes without saying that the suspension with its humiliation, including some attacks from the side of the students, did not leave Brouwer unaffected, apart from little scraps of paper with remarks on his persecutors, there is little known about his reactions. When the chemistry Professor Wibaut, who had been, on the whole, a good rector in the last year of the university's war years, and who was to a far greater degree the target of student actions, wished Brouwer happy New Year and said that he would not attend the Dies (the birthday of the university) as long as certain colleagues were treated so shabbily, Brouwer for a moment could not withhold his bitter comments:[77]

> I see in the actions of the aggressive groups in the Senate such a striking similarity with the Jacobin and Fascist clubs of the past in their as yet bloodless initial period, that I can neither believe in their early disintegration, nor in their early humanisation. Averting the dangers that threaten from that side will only have a chance of success, as I fear, through a militant union of the colleagues of good-will, ... For this reason alone, the defeat of the fascist states had to be paid with so much blood and tears and such an endless aftermath of

[76]Brouwer to Clay, 23.I.1946.

[77]Brouwer to Wibaut, 12.I.1946.

desperation, and brought so much desperation, because the militant union of the 'united nations' (in particular of the safe outsiders) has been realised so slowly.

This team forming is not at all in my line, but it looks as if, after the collapse of the liberal state structures there will no longer be a place for isolated persons, and that one can only obtain the right and the means for existence through uniting.

Although Brouwer kept a stiff upper lip in his dealings with colleagues and with the faculty, he was privately extremely bitter about the treatment he had received at the hands of the new men in power.

Among the drafts in the Brouwer-archive there are some drafts of letters to (presumably) Gutkind, in which Brouwer complained about his humiliations. The letters must have been written roughly at the time of the battles in the faculty. The reader will recall that in his early years, Brouwer reacted extremely emotionally to obstruction and unfairness. His brushes with, for example, Reiman (in the matter of the International School of Philosophy), Denjoy and Menger, almost produced mental breakdowns.

A few samples from the drafts will illustrate Brouwer's feelings:

—I am still flabbergasted by your message that I have to make a "statement" in order to enter the USA.

—In the vacuum after the liberation, when democracy was temporarily suspended, a take-over has taken place under cries of the hysterical usurpers, who, in order to maintain themselves at the place where they don't belong, have to exterminate the liberal, and as long as they have not succeeded to do this, to detain them in order to prevent that they write their 'I chose freedom'.[78] As long as they cannot kill or imprison him, they have to slander him abroad, so that the flight abroad is prevented.

—Here too, at the time of the military authority and the suspension of democracy after the liberation, a take-over by a section of the population consisting of the ambitious mediocre has taken place, who had no chance under a normal democracy. And who are now forced, in order to remain in their usurped positions, in the first place, to persecute by means of fraud, slander, insult and maltreatment the 'independent' persons, in the second place to erode democracy...

—The usurpers hate most, not the helpers and collaborators of the enemy, but those who do not admire, and never have admired, their so-called resistance style. And among these they carry out their persecution, and in this they are not checked, even though the persecuted have obstructed and hindered the enemy with any considerable risk *in another style*.

—When a band of ambitious mediocrities, who do not have a chance in times of a solidly united democracy, have penetrated, in times of a damaged social structure, with a hysteric device in the centre of power, for which

[78] A popular book written by the Soviet defector, Victor Kravchenko.

> they have no competence, the structure cannot regenerate itself. And thus the usurpers are more and more forced, in the first place, to persecute the independent (i.e. those who think for themselves, who are honest and candid and do not belong to a clique), and to hinder their emigration; in the second place to extend state interference to dumb automatisms, because they are not fit for intelligent official positions, and to suppress the instinctive spiritual opposition (or criticism, at least) of the non-thinking masses, by propaganda (schools for politics, state information service), and curtail the freedom of speech, . . . .

It is not difficult to guess what Brouwer was referring to. His personal position was thoroughly undermined by the purge procedure, but he was equally upset about the changes in the university in general, and in mathematics in particular. Moreover, like any citizen he was confronted by the new elite, political or otherwise, a certain portion of which consisted of opportunists and incompetent bunglers with a fair dose of political flair.

In his young years Brouwer, incidentally, was a victim of spells of nervous outburst; later in life there were fewer signs of those, and the glowing anger often made place for utter despair. The above quotations illustrate a mixture of anger and despair that was to mar his remaining years.

In the beginning of January 1946 Brouwer collected some items that he had entrusted to the safe of the *Nederlandse Handelmaatschappij*. With the uncertainties of the war in mind, he had deposited his most precious notes. One would think that a man like Brouwer would give priority to the essential research notes in as far as they were salvaged from the fire in his cottage, but in fact the safe held a curious collection:

| | |
|---|---|
| Package 1 | first part main course |
| collected 11.6.45 | notes main course 1940 and additions |
| | notes main course 43/44. |
| Package 2. | Lecture notes mechanics |
| collected 12.14.46 | id. canonical equations, + id. classical vibrations |
| | + enclosures |
| Package 3 | Basis notes main course + higher algebraic curves |
| collected 11.6.45 | list of examinations with lecture reports. |

and furthermore

> Foreign lecture series on intuitionistic mathematics in folder, 16 quarto sheets (with 5 additional enclosures) on measure and one-parameter continuous groups in intuitionistic precision (in cover).
>
> Annotated academy documents with enclosures (in cover), de la Vallee Poussin : 'Integrales de Lebesgue' etc. (with notes); Lecture 30 April 1938 (in one package)
>
> Debet accounts of Sodalitas committee. Advances Sodalitas.

It is surprising that Brouwer first and foremost thought of the security of the above material. It may well reflect his feeling of responsibility for his duties as a teacher.

For the time being Brouwer's first priority was the appointment of Freudenthal elsewhere. There seemed to be hope for him, for Utrecht was interested in filling the vacancies left by Barrau and Wolff. In fact, Utrecht was left with one lecturer, Bockwinkel, a man with a curious reputation; at one time he had shown interest in constructive mathematics, but he actually paid more attention to perfecting his lecture notes than to research; needless to say that his teaching was conservative. So far the Utrecht faculty had not been very fortunate. Van der Corput and Van der Waerden had opted for Amsterdam. In the post-war negotiations Utrecht won at least one prize: the prominent mathematician David van Dantzig was appointed. When, however, Amsterdam made him an offer, he almost immediately left; he did, however, complete his one year course in Utrecht.[79] So the University of Utrecht began to look like the poor bride who was twice left before the altar, and once deserted in haste.

Brouwer, who was of course aware of the vacancy at Utrecht, probably sent a glowing recommendation of Freudenthal to the chairman of the Utrecht faculty, the astronomer Minnaert. Minnaert was not fooled for a minute, the goings on in Amsterdam were not kept a secret from the rest of Holland. He was, however, sensible enough to realise that Freudenthal would be an asset to the Utrecht faculty.[80]

For Freudenthal this was a perplexing situation; he did not object to a chair in Utrecht, but he felt that by accepting the Utrecht chair, he would also accept the injustice inflicted on him in Amsterdam. Moreover his students more-or-less appealed to him not to give up Amsterdam. The choice was difficult and painful. Freudenthal's students admirably summed up the argument: after such intensive campaigns for rehabilitation a sudden transfer to another university could give support to the idea that there was a reason after all for denying Freudenthal the rehabilitation. This was indeed entirely Freudenthal's sentiment; it was in this spirit that he wrote to Minnaert.[81] He could not leave Amsterdam without at least an effective rehabilitation. Giving up would mean deserting all those who fought for his rehabilitation, or that of others.

> I would betray my patriotic duty, and would be labelled as a deserter from the ranks of thousands, who now fight for their rehabilitation or that of others, if I would acquiesce without very compelling reasons, (now in liberated Holland), for the second time in the orders of the enemy, i.e. my dismissal, with all its consequences. I could only have done so if I had been able to learn the motives of Professor Brouwer, and if I could have convinced myself and my supporters, that by giving in I would not serve the interest of my successor, but that of a great mathematician. Professor Brouwer has to my deepest disappointment never made an appeal to the filial respect and devotion with which I have served him since 1930...

[79] Van Dantzig was appointed in Amsterdam on 22 May 1946. Cf. Alberts (1998).

[80] In those days one did not apply for a chair. One was recommended by colleagues and sister faculties. After the traditional bargaining the University submitted the name of the candidate to the Minister for approval.

[81] Freudenthal to Minnaert, 6.V.1946.

> Rather, by appealing to my hypothetical fear for a public discussion of my appointment, Professor Brouwer had made it impossible for me to withdraw my candidacy with honour.
>
> I don't reject hereby the possibility to convenience Professor Brouwer after my appointment in Amsterdam, under reasonable conditions, e.g. by accepting a possible appointment in Utrecht.

The letter was dispatched two days before the City Council of Amsterdam was going to discuss the appointment of a lecturer in analysis, group theory and topology.

Brouwer, who by now really feared that the City Council might appoint Freudenthal, had on May 1 again addressed the Mayor, praising Freudenthal's capacities: 'a mathematician of exceptional decisiveness, universality and productivity, to whom we thank important results in a diversity of fields, and who as a consequence, considering his scientific merits, is fully entitled to an independent position at a university'. But, he repeated, 'I for myself have to avoid any contact with Dr. Freudenthal for reasons of self-preservation... As, furthermore, my relation to Dr. Freudenthal has already paralysed my scientific production for years, the continuation of an official connection between him and me will probably destroy the remaining chances on a resuming and finishing of my lifework.'

The above two letters illustrate the tragic quality of the conflict, sought by none of the participants, but after the fashion of a classical drama, ordained by the Gods. Here *Schicksal*, to use the once fashionable German term, had brought together the young man, admiring and loving Brouwer as a father, only to find himself rejected, and the older spiritual giant, who has to free himself of the younger budding intellect in order to protect his vulnerable inner life. It is a tragic aspect indeed that these emotions can be confessed to third parties but seldom to the object of love-hate.

Be that as it may, the struggle went on. Brouwer even went so far as to write to all the chairmen of the parties in the City Council, begging them to postpone the decision.[82] At the same time he drafted a letter from the faculty to Freudenthal, saying that 'The faculty of Mathematics and Physics would like to inform you, that it unanimously judges your position in the higher education [system] completely unsatisfactory. The faculty will thus remain alert, in order to bring this position in line with your scientific merits. For a speedy realisation of an optimal solution, the faculty counts on your personal understanding and co-operation.' This letter was duly signed and dispatched by Clay.[83]

The letter, coupled to the unsatisfactory outcome of the council meeting, incensed Freudenthal, who pointed out to Clay that the letter did not explain why he still had not been rehabilitated one year after the liberation, that his scientific merits had never been doubted, that the appeal to 'personal understanding and co-operation' rather seemed to indicate a prelude to the abandoning of the rehabilitation, and finally that the factual content of the letter was nil. He said that he had heard some rumours that in the near future he would have to make certain choices which would

[82] Brouwer to political parties, 6.V.1946.

[83] Clay to Freudenthal, 8.V.1946 (cc. Brouwer).

solve all the problems for the faculty. The least the faculty should do is outline these choices.

Of course, Freudenthal was perfectly aware that the choices involved the chair in Utrecht, but he objected to the strategy that made him the sole actor in this matter. The faculty, he said, was in honour obliged to acknowledge the fact. This was more than Clay could stand; he got so annoyed that he more or less gave up Freudenthal's cause.[84] Freudenthal was, understandably, disgusted about the treatment he had received at the hands of the authorities and his colleagues. Apart from the loyal support of the students little had been done to further his cause. Even his close allies offered scarcely more than verbal sympathy. In an interview in 1987[85] Freudenthal looked back at the painful events after the war: 'At this time, when my fight for rehabilitation was going on, I received neither from Van der Corput, nor from Van Dantzig any support. In words yes, but not in deeds, for they wanted a position in Amsterdam themselves.'

The whole episode is a classic example of the pressure a system can bring to bear on an individual who simply wants his grievances redressed. It is the more a tragic example, as a friendly solution was perfectly possible and as a minority of two could apparently dictate the actions of a, on the whole reasonable and fair, faculty. The affair ended with Freudenthal's acceptance of the chair in Utrecht. To Freudenthal's immense surprise, Brouwer, entering the Mathematical Institute in Amsterdam, when Freudenthal was just leaving, grasped his hand and warmly congratulated him with his appointment. On 9 December 1946, Freudenthal held his inaugural address in Utrecht and from that time on he was the driving power behind the mathematical activity in Utrecht. Sometimes his conceptions were taxing the imagination of his company too heavily. At the traditional interview between the Curators and the new professor, he was asked what in his opinion the size of the staff of the Utrecht Mathematical Institute should be. His answer was '11 professors'. This man must be out of his mind, was the general reaction. However, before he retired he had built an institute that surpassed the demands of his early days. He introduced, for that time, modern methods and topics in the mathematics curriculum, and since Freudenthal's appointment, mathematics at Utrecht has obtained a status that made it stand out in the Netherlands. He remained a great man till the end, he never tried or wished to take his revenge on Brouwer. As a matter of fact, his edition of the Collected Topological Works of Brouwer is a fitting monument to the man who gave topology a new lease on life. Freudenthal has played an important role in Dutch mathematics and education in general. He was a man overflowing with ideas, always ready for new initiatives. Yet he never became part of the 'old-boys-network', his outspoken opinions and his habit of voicing unpleasant truths (e.g. his criticism of the Mathematical Centre, which soon after its founding developed into an Amsterdam instead of a national institution) gave him the reputation of being 'difficult'.

[84] Clay to Freudenthal, 15.V.1946.

[85] Alberts et al., p. 118.

## 17.3 Back to Research

With Freudenthal safely in Utrecht, Van der Waerden on his way to, or in Baltimore, Brouwer could have relaxed and enjoyed the last few years of his professorship. He had five more years to go, and according to himself, he had an ambitious program for finishing and redoing his mathematics. There was even a return to his old research activity; from 1946 onwards papers started to appear with some regularity. The first post-war paper was a retrospection, *Synopsis of the signific movement in the Netherlands. Prospects of the signific movement.* The paper was the result of a lecture at the 'Second International Summer Conference of the Signific movement', held from 24 through 31 August. In it, he summed up the participants and aims of the *Signific Circle* and the *International Academy for Philosophy*, as discussed in Chap. 10. He also added some remarks on the contemporary use of the same methods. After all, World War II had brought even more 'abuse of false slogans for the satisfying of dark instincts'. He listed the following desiderata 'serving self-realisation of the individual': public safety, public welfare, mental freedom, and as much as possible freedom of action for the individual. These desiderata were not surprising in a world that had been subjected to the Nazi domination, but even today they would stand unchallenged. Furthermore he indicated two necessary conditions: '1. The utmost moderation of the state influence over the individual and the utmost reduction of the possibility of domination of the individuals over each other; 2. the existence of a relatively harmless and innocuous mode of diverting ineradicable dark and frivolous instincts such as lust of power, sadism and gambling.' The role of language surfaced finally in the closing passages:

> ..., it is my opinion that in any case in a happy humanity, state intervention will have to be prudently handled, and the state will have to use a language strictly indicative. If it deviates from this duty and admits into its language vaguely spiritually tinged terms, such as *principles, attitude, character, moulding, firmness of character, resoluteness, leader, qualities, heroisms* (for continuation of the list see Göbbels), then inquisition, denunciation and man hunting will still have their chance, man will oppress man and man will mistrust man.
>
> The fight against abuse of hysterical devices, the fight for unmasking them in private, and for removing them from public life will in future remain a preponderant part of the business of significs.

All this is perfectly reasonable, and it summed up the subtle and less than subtle tricks of the totalitarian regimes that had then been toppled (and the remaining one that was to haunt mankind for another fifty years), but we know that Brouwer was also referring to the post war powers in the Netherlands, no doubt in the light of his own recent experiences.

Significs also figured prominently in the major first post-war event in the mathematical circle in Amsterdam: the honouring of the senior mathematician Mannoury. On September 11, 1946 Brouwer awarded in the familiar surroundings of the auditorium of the University of Amsterdam an honorary doctorate to his former teacher and lifelong friend. Knowing how reluctant universities are to present honorary doctorates to their own scientists, one may be certain that Mannoury had strong supporters in the Senate. Brouwer honoured Mannoury not in the first place as a mathematician, but most of all as significist. Mannoury was indeed revered by all his students and followers, and the honorary doctorate was not only well-deserved, but also widely applauded. Those who have known Mannoury treasured for always the memory of a kind, wise enthusiastic friend and teacher.

Brouwer's mathematical papers that appeared after the war were mainly reports on earlier work. They were in part triggered by publications of Griss and Van Dantzig. Griss, a student of Weitzenböck, was a mathematician with pronounced philosophic interests. His analysis of mathematical intuitionism led him one step beyond Brouwer: he denied the meaningfulness of negation (and falsity, e.g. $0 = 1$). The argument being (basically) that one cannot have a mental picture of constructions of such things.

Brouwer, in his first paper in the series dealing with negative properties,[86] introduced the so-called *creating subject*,[87] a notion already present in his Berlin Lectures, cf. pp. 505, 517. He used it to show that there are real numbers for which one could not possibly show that they are not positive, but for which one (so far) has no proof that they are positive. In Brouwer's terminology 'order is stronger than virtual order'. His result can also be stated as 'apartness is stronger than inequality'. That is, 'inequality is essentially negative'. The result is still not optimal, as it depends on the state of the creating subject's knowledge (or our knowledge, if you like); it is still a weak counter example.[88] In a subsequent paper Brouwer really exploited the creating subject in full force, by showing that the identity of apartness and inequality is even contradictory, in symbols $\neg\forall xy(x \neq y \rightarrow x\#y)$.[89]

This was his ultimate evidence that $\neq$ is not a positive notion. Here again, Brouwer had demonstrated his ingenuity and insight. He also formulated this new result in a geometrical context in a somewhat provoking manner: Euclidean plane geometry is contradictory.[90] Of course, this way of describing the result is somewhat misleading. What he meant was that a particular classical theorem turned out to be false in the intuitionistic setting. But that he had already established in the twenties, when he showed that for the intuitionist continuum the dichotomy property was false.

---

[86]Brouwer (1948a).

[87]Later called 'creative subject' by Kreisel.

[88]Brouwer (1949a).

[89]There is a thin line separating the weak from the strong refutations. Here the matter is particularly delicate, as one would like to exhibit a real that was distinct from 0, but not apart. However, $\neg(a\#0 \rightarrow a = 0)$, so such a straightforward example is not available. The weak counterexamples are usually of the form 'we have no evidence for...', whereas the strong counterexamples are plain contradictions, 'it is not the case that ...'. Brouwer's new strong counterexamples used the full strength of his theory of choice sequences (including the continuity principle).

[90]Brouwer (1949b).

The use of the creating subject remained mysterious for some time.[91] Later metamathematical researches have shown it to be consistent with most of the traditional intuitionistic principles.[92] To be sure, allowing sequences to depend on the mental activity of the creating subject (or the *idealised mathematician*), comes to a revision of mathematical universe, and hence it is by no means trivial (or even true) that all the old principles still hold.

No mathematical activity, unfortunately, could keep Brouwer from academic politics. He felt that he had made Amsterdam mathematics what it was, and to a large extent that was true. In the twenties and early thirties Amsterdam had even been a prominent centre for topology, but later developments, partly caused by the economic crisis, partly by Brouwer's disinterest in, and emotional unsuitability for, active leadership. Nonetheless, he still clung to the old promise of 1920, when Amsterdam had bought his continued presence and activity. He did not realise that administrative promises have a hidden date of expiration, in particular administrations are not prepared to honour promises to parties that have not fulfilled their part of the bargain. In Brouwer's case one could argue that after the golden decade of the twenties, Brouwer had done little to keep his centre in perfect shape. Weitzenböck's solid routine did little to enhance the prestige of Amsterdam mathematics; the succession of Hurewicz and Brouwer's own withdrawal into his private world of research were not helpful to boost its status—not to mention the replacement of Freudenthal by Bruins.

His choice of candidates for the vacancies had not always been a happy one. The choice of Weitzenböck had been a mystery right from the beginning, the choice of the assistants Gawehn and Geldof was difficult to justify and finally his swapping Freudenthal for Bruins did nothing to raise the status of the mathematical establishment in Amsterdam. And so his claim to an Amsterdam equivalent of Göttingen had lost much of its force.

He must have been vaguely worried when he was informed that the minister contemplated the founding of a solid, real research institute for mathematics. Was it going to be in Amsterdam, and if so, what would the consequences be for his mathematical institute? In the beginning of 1946 he still was not worried about the new institute (which eventually got the name *Mathematical Centre*,[93] so we will, somewhat anachronistically, adopt that name from now on, although at this point in our history it was as yet nameless). It could even serve as a means to keep Van der Waerden at bay. The driving power behind the Mathematical Centre was Van der Corput, who was the chairman of the ministerial committee for mathematics, and who was to advise the Minister on the desirability and feasibility of the centre.

In December 1945 Brouwer had asked Van der Corput to bring him up to date as to the centre—he was still suspended and hence could not take part in the deliberations. We have seen that Brouwer considered himself entitled to a seat in the committee, and he understandably felt neglected.

---

[91] Heyting, in his monograph (Heyting 1956), put the topic in the chapter 'Controversial subjects'.

[92] It is however incompatible with $\forall\alpha\exists\beta$-continuity, Myhill (1966).

[93] Nowadays *Centrum voor Wiskunde en Informatica, CWI.*

A major step towards this Mathematical Centre was taken on 11 February 1946, when a special foundation '*Het Mathematisch Centrum*' was created in Amsterdam.[94] The board of the foundation consisted of the members of the Van der Corput committee, J.G. van der Corput, D. van Dantzig, J.F. Koksma, H.A. Kramers, M.G.M. Minnaert, and J. Schouten, supplemented with the two chairmen of the science faculties of the University of Amsterdam and the Free University, J. Clay and G.J. Sizoo. The active kernel of this group was made up of Van der Corput, Van Dantzig, and Koksma—the most powerful men in postwar mathematics. In October they formed, together with Van de Waerden, the Administrative Council of the Centre. Simultaneously the board was replaced by a Curatorium, where the leading industries, the City Council of Amsterdam, the Institute for Technology at Delft and both Amsterdam Universities were represented. Schouten assumed the central position of secretary.

In September 1946 there is a puzzling bit of correspondence between Brouwer and Van der Corput:[95]

> Amice, In a telephone conversation with Heyting (who had called me about something else) I recently let slip something about the consequences which could possibly follow from deceptions which I have experienced lately. Since these deceptions have by no means as yet resulted in a definite plan, it would have been better to practice complete silence about possible consequences, even towards a person whom I consider in the matter more-or-less a fellow.
>
> Having learned—somewhat to my surprise—that the content of my telephonic impulse has been communicated by Heyting to you, the above explanation had to be given.

Somewhat later he returned to the deceptions,[96] saying that as long as the deceptions had not turned into hard facts ('which could easily not just hurt me, but also go against the general interest'), he should try to prevent those facts.

Cryptic language! One does not have to be *clairvoyant* to guess what made Brouwer write these notes. It is almost impossible that rumours of the wheeling and dealing around the Mathematical Centre should not have reached Brouwer.

The conflict at the faculty, the threatening developments around the Mathematical Centre, could not escape Lize. She realised that her husband had become an isolated person in the Amsterdam mathematical community. In a letter to her daughter, Louise, she saw early retirement as the plausible solution.[97]

> We seriously think that Dad, now that he has reached the age of 65, should resign. That miserable hurrying, and the hasting back and forth should be ended now. He is now entitled to his full pension, and that is indeed quite a

[94] An account of the history of the Mathematical Centre can be found in Alberts' dissertation, Alberts (1998).

[95] Brouwer to Van der Corput, 27.IX.1946.

[96] Brouwer to Van der Corput, 8.X.1946.

[97] Lize to Louise, 6.III 1946.

> difference with his salary. But we should then live more economically. Dad can spend more time on the administration of the pharmacy, and pick up his own mathematics again. If he carries on until his 70th year, he will be totally exhausted, and no longer able to get something done. So that will probably happen before long.

It is surprising that a clever operator like Van der Corput had overlooked the consequences of withholding information from a close associé. In an open discussion some form of compromise could have been reached, but when Brouwer learned that the man whom he had welcomed as a likeminded colleague was practically sidetracking him, he saw that Van der Corput was not after co-operation but rather capitulation. Desperate situations ask for desperate measures, so he decided to turn directly to the mayor and the alderman for education. On 8 October he explained, in a long letter, how a centre for mathematics, if it were to be established in Amsterdam, was rightfully his.

In the summer of 1946 the City Council had reserved a budget of 25,000 guilders with the motivation, 'It is proposed to grant for the year 1946 to the Foundation Mathematical Institute a subsidy to the sum of D.fl. 25.000,–, for the Mathematical Institute that will take the place of the European Centre for Mathematics at Göttingen.' As there was only one Mathematical Institute in Amsterdam, which was in 1920 designated by the City Council as to be organised, as soon as the city finances allow this, on the same footing as the Mathematical Institute in Göttingen, the council must therefore have been under the impression that the old promise was now going to be fulfilled.

In Brouwer's words:

> One of the two promises has at the time been fulfilled, the realisation of the second one was between 1920 and 1934 repeatedly initiated, but these initial steps led each time, through causes unknown to me, to nothing. In that period there was a pronounced influx of foreign mathematicians to Amsterdam, which I have welcomed for some time. But because in the absence of an institute and an appropriate equipment for directing a group of studying foreigners, the necessary personal, spiritual and financial sacrifices eventually became too much for me (in particular after an indispensable source of income, which was discussed as such in the 1920 negotiations, was strongly reduced after an expropriation by the City),[98] this hospitality had to terminate. Thus the board of the city had through its temporising not only duped me personally, but it had also nipped the international centre, which was evolving by itself in Amsterdam, in the bud.
>
> It is possible that the fact that the fulfilment of the promises made to me in Amsterdam, and the vanishing of my instruction of foreigners in Amsterdam, had attracted international attention; once again a chair in Göttingen was offered to me in 1934. This was, in spite of the high remuneration and

[98]Brouwer refers here to the pharmacy, see p. 512.

> ample equipment, already because the established form of government then in Germany was ab initio unacceptable for me. Only after again explaining in extenso to the president curator the disappointments that the City Council had caused me, and after hearing from his mouth that he seriously deplored the course matters had taken, and that from the side of the board of the city everything that could be done by way of redress, also *would* be done, I formally turned down the offer. [...]
>
> On the basis of the above historical exposition, in combination with the passage of the concept budget for 1946, it is difficult, in my opinion, to interpret the granting of the mentioned budget item other than that the amount mentioned—whether or not via a foundation—should be spent for the benefit of the mathematical institute of the University of Amsterdam, for purposes to be determined in agreement with the director of this institute.
>
> Moreover, on the basis of the above account, it would in my opinion be inadmissible in whatever way, to take the authority out of my hands, after I had raised it to the present level under difficult circumstances and under the sacrifice of a variety of personal interests. Alas, there are indications that in certain circles there is a design for such a step; also that the preparatory actions in the direction already took place under the protection of the smokescreen of confusion of the Liberation. If this scheme should succeed (quod consules avertant) I believe that a page in the history of science would be written that will not fail to arouse astonished interest with future generations.

The above document apparently caused quite a stir; City Hall warned Van der Corput that things threatened to take a turn for the worse, and Van der Corput was so alarmed that he had the whole letter read to a secretary, who took a stenographic report. A copy of Brouwer's letter was forwarded to the ministry of education. One gets the impression that the fathers of the Mathematical Centre were not so certain of their case as they professed to be.

Following the above mentioned telephone conversation with Heyting, which inadvertently was conveyed to Van der Corput, Brouwer wrote to the latter that he was not quite ready to discuss the matter with him, but a few days later he spelled out his discontent to Van der Corput.[99] It was better, he wrote, 'if to begin with I give you a written exposition of the thoughts with respect to the new mathematical institute, which so depress me, and that you subsequently answer in writing. Thus we can at least lay a basis for later personal discussions.'

The promised exposition followed three days later, it was a bitter complaint about the steps that had already been taken in the matter of the founding of the Mathematical Centre. He was, he wrote, unpleasantly surprised when he found out that the State and the City of Amsterdam had taken steps to open a 'Göttingen' Mathematical Institute in Amsterdam, without consulting him—an institute which had been promised to him in 1920, as soon as the financial situation allowed such. But, he

[99] Brouwer to Van der Corput, 8.X.1946, 14.X.1946.

guessed, the 'insult of being ignored' was probably a consequence of the 'political persecution'.

Since Van der Corput had a large say in the preparations, Brouwer had felt assured that if the preparations were successful, the leading position to which he was entitled would not be withheld from him. Indeed, at the end of October 1945 he was even invited for a discussion with colleagues who had already been informed. When, subsequently, Van der Corput had asked him if he would object to a possible co-operation with Van der Waerden within the framework of the new institute, he had concluded that the question could only be meaningfully posed under the assumption of his own participation in the board. That he did not hear of further developments did not overly worry him, but suddenly in July 'I had to hear from a member of the ruling group that dealt with the organisation of the new institute, that there was no intention of giving me a place in the managing bodies'. The reason given was 'to spare me the strenuous work that was connected with the leading position'. Brouwer bitingly remarked that he did not care for such a respect for his peace. After that he heard no more, until he happened to find out that the leading positions had been filled.[100]

> Now I have to count to my great disappointment with the possibility that people dare indeed to expel me from the leading positions, since long entrusted to me, of the mathematical organisation in Amsterdam. I say 'dare', because the history of my thirty-seven years of activity at the University of Amsterdam will make it nothing less than an impropriety if from now on a mathematical institute, modelled after Göttingen, will be supported, where I am excluded from a leading position.
>
> But you, with your influence, can still stop my being passed over, and eventually it will appear that you have served the interest of our country both scientifically and morally.

In a post script he added a final private concern:

> An additional circumstance is the danger that the dark figures, who had already earlier discredited me abroad, will now make my exclusion from a leading position of the new institute public and argue that it is a token of my reduced status here.

He felt insulted; after being the leader of Dutch mathematics for more than 30 years, he was quietly pushed aside. The purge and Brouwer's reprimand cannot have played more than a marginal role; some Dutch mathematicians may have borne Brouwer a grudge, but there may well have been a general feeling that Brouwer had a disproportionate influence on Dutch mathematics. Kloosterman, the mathematician from Leiden, took a commonsensical view of the anti-Brouwer feelings: 'Brouwer is a better mathematician, and the gentlemen cannot stand that, but it does not worry me.'[101]

---

[100] Brouwer to Van der Corput, 17.X.1946.

[101] Oral communication, T.A. Springer.

The letter had its effect, Van der Corput sent Brouwer a short note[102]

> I will use my influence to see that you get a position that you deserve on the grounds of your capacity and personality at the Mathematical Centre.

The first assault at the bastion of the Mathematical Centre had created a visible breach. The two men discussed the problems the next day, apparently in complete harmony, for Brouwer wrote, 'I have the feeling that our conversation yesterday might have been most clarifying, both on account of the new facts we heard from each other, and of the impression of our mutual disposition and opinions we have exchanged.'[103]

But Brouwer was too wily to let it go at that, he knew the game too well; so he added that he considered that no actions were promised, as long as they had not been given in writing. Van der Corput immediately sat down to compose an answer; this was a delicate matter, for Brouwer had a reputation for getting the most out of any written statement. The first draft was written on 4 November and the final letter was dated 13 November.

After agreeing with Brouwer that the conversation of 1 November had a clarifying effect, Van der Corput went on to enumerate 'the points on which we, as I think, have reached an agreement'.

> 1. There will be a friendly co-operation between the Mathematical Institute and the Mathematical Centre.
>
> 2. In the spirit of the team of the Mathematical Centre, personal sympathies and antipathies will have no influence where the interest of the Centre, or of mathematics, is concerned. Let me mention as an example the possible appointment of Van der Waerden at the Municipal University, which appointment should be not only of great importance for mathematics in Holland, but also be financially profitable for the Mathematical Centre.
>
> 3. You give a description of the work you think you will do for the Mathematical Centre.
>
> 4. Your letter of October 8 to B. & W. [mayor and aldermen] of Amsterdam will be withdrawn.
>
> 5. I have to add that the board of trustees considers it self-evident that no one of its members will send on his own initiative complaints, advice or counter-advice to authorities such as the City Hall, the Government, industry or the Curators. A new member of the Board of Trustees is asked to commit himself to this line of conduct.
>
> As soon as I have your endorsement in writing of the above five points, without exception, expressis verbis, it will be a pleasure to propose you as a member of the Board of Trustees.

One wonders if Van der Corput was quite serious about appointing Brouwer in any position at all. He could not very well have expected Brouwer to agree—in partic-

[102] Van der Corput to Brouwer, 31.X.1946.

[103] Brouwer to Van der Corput, 2.XI.1946.

ular to points 2, 4 and 5. Whether one agrees with Brouwer, yes or no, the passage concerning Van der Waerden is plainly in contradiction with the results of the faculty discussions, and point 5 would make Brouwer (and any member) a hostage of the rest of the company. As if pushing Brouwer yet a bit further over the brink, Van der Corput told him that on acceptance of the conditions, and hence on a possible appointment, the stipulation that only written agreements were valid, was to be abandoned in the interest of an efficient management.

Brouwer must have been stunned,—what Van der Corput asked him was nothing else than an unconditional surrender; he had, as the expression has it, been adding insult to injury. Brouwer took his time and answered coolly a month later;[104] he pointed out that the conversation of 1 November had a purely informative character; nothing was agreed, ideas and views were exchanged. He wrote that he had drawn attention to the fact that the activities of the Mathematical Centre seemed largely of a purely academic nature and hence should be practised within the university, and further that 'the previous history of the mathematical activity in Amsterdam does not allow that I am included in an Amsterdam mathematical organisation in a way that places other officials above me'.

The letter of 13 November had completely perplexed him, he said, for no agreement was reached on any point, with the exception of the desirability of further consultations.

> —How you could write 'the points, on which I think we reached an agreement' instead of 'which I had expressed as my wishes and proposals' escapes me completely. Approximative equivalence of these two utterances exists only for very egocentric or very dominating natures, but for the time being I have no right to classify you as such.

The letter illustrated that an altogether written communication would have been preferable; now all misunderstandings had to be repaired. 'In October you twice declared yourself unable to meet my wish for written negotiations. However, how much less time would a written formulation of your proposals have asked at the time, than the amount I have to devote now to correcting conclusions that you have drawn from oral communications!' Brouwer's letter is one big demonstration of his dissatisfaction with Van der Corput's manner of operating. Van der Corput was for example implying in his letter that personal discussions with Brouwer were a waste of time: 'Furthermore you impute me the tendency to consider given promises valid "when they are laid down in writing". I will consider these words as a slip of the pen. I have to bear myself too much grudge as a consequence of oral agreements that have been broken against me. However, it is something totally different to distinguish between negotiating and informative discussions, and likewise the stipulating that a certain discussion will have the character of the one or the other.' In order to answer Van der Corput's letter in extenso, he wrote, he would have to see the relevant files first. The final passage, too, bore the first signs of pessimism and resignation:

[104]Brouwer to Van der Corput, 14.XII.1946.

> So much is clear to me that if the supervision of the mathematical enterprise in Amsterdam should be taken away from me, only the parties inflicting the injustice should be responsible for the resulting situation and that therefore the suffering party should under no circumstances be subjected to conditions for the purpose of a possible repair. And also that the position of the mathematical sciences founded in the past at the University of Amsterdam by the authorities, should be protected by those same authorities.

As far as the correspondence shows, this is the end of a possible accommodation between Brouwer and the Mathematical Centre. Brouwer could not expect any compromise from the leading proponents. Van der Corput did take Brouwer's opposition seriously, not in the sense that he would reconsider the position, but rather he wished to forestall any possible moves of Brouwer against the Centre. Before Brouwer's reply he had already shared some of his worries with G. Bolkestein, the former London Minister, who had put his mind at ease—no action of Brouwer against the Mathematical Centre had reached the ministry. Nonetheless, Bolkestein advised against further dealings with Brouwer:

> You and Professor Clay know better how to assess Professor Brouwer's attitude than I do, but in my opinion the M.C. should no longer seek contacts with him; whatever the consequences. There must after all be a certain feeling of self-respect on our side (sic), that should prevent us from negotiating with someone, who, like Mr. Brouwer, agitates against the M.C.[105]

From this point onwards the roads of Brouwer and the Mathematical Centre separated, Brouwer considered it an odious creation, which had stolen his 'Göttingen in Amsterdam'. There was a half-hearted attempt to mollify Brouwer by offering him the title of 'honorary chairman' of the *Mathematisch Centrum*. Needless to say that the suggestion was ignored.

One might get the impression that the post-war years offered nothing but misery to Brouwer, but fortunately there were also some brighter spots. In 1946 Brouwer had received an invitation from the University of Cambridge to give a course on intuitionism; as one can easily imagine, this appealed very much to Brouwer.

The (spiritual) sponsors on the Cambridge side of Brouwer's lecture series could very well have been the resident logician Steen and the topologist Newman. We find his course listed for the Michaelmas term of 1947 as 'Intuitionistic Mathematics'. Among the persons attending Brouwer's lectures was a man who was in due time to give a new impetus to the study of intuitionism, Georg Kreisel. He reported later that he asked Brouwer after a lecture, if he meant all he said. On Brouwer's reply that he followed Shaw's dictum, 'you have to exaggerate in order to make an impression', Kreisel innocently pointed out that, 'nobody promised you that it would be a good impression. Incidentally, Brouwer was not amused. Apparently he did not like to be interrupted anyway; fittingly, for a good solipsist.'[106] For Brouwer the course was

[105]Bolkestein to Van der Corput, 21.XII.1946.

[106]See Kreisel (1987), p. 147.

not only a welcome escape from what he considered a hostile environment, but it also gave him an opportunity to prepare an up to date account of his intuitionistic theories. After all he was lecturing to bright young mathematicians, and he was casting his lecture notes in English, the new lingua franca of science. The course notes were destined to be published by the Cambridge University Press, and Brouwer seriously worked and reworked his notes in order to get a polished exposition. He even made an appointment with Steen, for 'consultations on the intuitionistic vocabulary for the forthcoming book'.[107] But the Cambridge Lectures fared no better than the Berlin ones.

In the middle of all these conflicts and machinations, there was at least one event that assured Brouwer that not the whole world had turned against him. In 1947 it was forty years ago that Brouwer got his doctorate. Friends, students and colleagues who had not deserted the old master, organised a jubilee symposium on 19 February. The ceremony took place in the lecture hall of the Geological Institute of Brouwer's brother Aldert. Mannoury addressed his famous student and colleague; he sketched how Bolyai, Riemann and Peirce had broken the hegemony of the Euclidean and Archimedean axioms, how Einstein had 'widened the visual field of physical science', but how Aristotelian logic had survived all disturbances until Brouwer 'emancipated human thought from the authority of the logical principles and so ran down that stronghold itself'. He mentioned Weyl and Dingler who had recognised in Brouwer's ideas a revolution and chaos, without appreciating the positive element in Brouwer's program.[108] He went on to appraise the fundamental contributions in the words of the paper *La question vitale: 'A ou B'*.[109]

Faculty politics did not wholly absorb Brouwer's scarce time, he gradually picked up his international contacts. In the summer vacation he took part in a symposium of the '*Institut des Sciences Théoriques*' in Brussels. There, in the Palace of the Academy, from 8 through 13 September a small select company got together to discuss the modern developments in the exact sciences. The Netherlands were represented by Beth, Brouwer, Heyting, and Pos. For Brouwer this meeting offered an extra bonus: after all those years he got together with his friend Hermann Weyl. Both parties were extremely pleased to find that they could pick up the threads of their friendship. Beth was one of the speakers, but Brouwer and Weyl only took part in the discussions.[110]

The story of Brouwer and the faculty now is speeding towards its end. Although Brouwer remained active till the last day, both in his teaching and as a Director of the Institute, his influence was waning. In fact, Van der Corput was laying down the

---

[107] Steen to Brouwer, 27.XI.1947.

[108] Mannoury did Weyl an injustice. In fact Weyl was one of the first who fully appreciated Brouwer's ideas, cf. Weyl (1921), van Dalen (1995).

[109] Mannoury (1943), 'insiders knew that 'A or B' meant 'Aristotle or Brouwer'.'

[110] Recorded in the proceedings, *Problèmes de philosophie des sciences: premier symposium, Bruxelles*, 1947. Hermann, 1948–1950. 7 vols. *Serie Archives de l'Institut International des Sciences Théoriques. Ser.A. Bulletin de l'Académie Internationale de Philosophie de Sciences.*

**Fig. 17.2** Celebration of the 40th anniversary of Brouwer's doctorate. Front row: Clay, Brouwer, Mannoury. Two persons to the left of Clay—Van Dantzig; to the right of Brouwer—Lize; right behind Mannoury—Heyting, next to him Louise; third row from the right—Evert and Cor Bruins. [Courtesy R.A.F. Guasco]

law in Amsterdam mathematics, and all Brouwer could do was carry out rear guard actions.

At the end of 1947 the faculty resumed its activity to secure Van der Waerden for Amsterdam. The faculty had called a meeting on December 10, knowing that Brouwer would be in Cambridge at that time and making sure that the invitation would not reach his home address until after the meeting.[111] At the meeting a plan to extend the mathematics section vigorously in the 'technical direction' was accepted. Brouwer found out what was going on when in Cambridge, he was—with good reason—furious, he immediately sent a telegram to Bruins,

> Please send stencilled circular to faculty members containing my protest against faculty adopting proposal without previous duly convoked discussion by first section.

A letter to the faculty was dispatched the same day, it suggested that 'the question should be faced, if such extensions [in a technical direction] are not in conflict with the calling of the university, and that therefore the action cannot be allowed to be carried out in haste and without written preparation'.

---

[111]The faculty files show that Brouwer had duly requested the Minister's permission to obtain foreign currency at the rate of £.5,– a day, 3.XII.1947.

Upon returning, he called a meeting of the lecturers Bruins, Heyting, de Groot and the assistant Loonstra to discuss the matter and to present his views. The result was a joint letter to the Central Committee of the Faculty;[112] it stated that at the request of City Hall the wishes for an extension of the faculty had been registered, that some sections had expressed interest in an extension in a technical direction, but that the undersigned, not being involved in extra-universitary activities for the sake of their mathematical practice, 'Asked the committee to pass their desiderata on to City Hall.' It hardly needs mentioning that the desiderata coincided with Brouwer's views: Pure mathematics (including rational and stellar mechanics) should be brought up to a fairly complete strength, 'as already promised to the first undersigned in 1920'. This required two full professors. Furthermore it contained the hardly veiled recommendation that Bruins should get a chair for applied mathematics; the second chair was reserved for those areas of mathematical research out of which the rational and stellar mechanics had developed and remained interwoven with. However (strange switch of subject), the first priority was a renovation of the Mathematical Institute, including the addition of a coffee room for the students. The letter made one thing abundantly clear: the undersigned could not agree with the change in a technical direction.

The chairman, Clay, did not care one bit for Brouwer's objections, and sent a list of desiderata to the Senate, mentioning Brouwer's personal wishes, without mentioning the support that Brouwer had found. The faculty wished, he wrote, a chair for applied mathematics, to be filled by a scientist of the first rank (exit Bruins, enter Van der Waerden), and a position for 'modern computation methods and computing machinery'—this could be shared with the Mathematical Centre on a part-time basis.

It may seem curious, from the viewpoint of fair representation, that the preferences of one mathematician (Van der Corput) in the faculty prevailed over those of the remaining four mathematicians. One should keep in mind, however, that the Mathematical Centre was a fact since February 1946, and that Clay and Van der Corput had leading positions in it.

In February J. de Groot (the topologist) was offered a chair at the Institute for Technology (*Technische Hogeschool*, TH) at Delft. Clay confronted the Curators with the fact that the TH had eight professors in mathematics and was to get two more in the course of 1948, whereas Amsterdam, the Dutch capital of mathematics, had to offer a complete curriculum with two professors and three lecturers. 'It does not become the faculty to compare the position of a mathematics professor in Delft with that of a university lecturer in the same subject, but it should not be forgotten that the latter has the task to educate and train mathematicians such that his students will be able to conduct independent and original research.'[113] Now that De Groot had an offer from Delft, steps had to be taken to secure the mathematical education in Amsterdam. The faculty did not think it necessary at this moment to promote all

[112]Brouwer, Bruins, Heyting, De Groot, Loonstra to Faculty, 27.XII.1947.

[113]Clay to Curators, 27.II.1948.

lecturers to professors, but at least De Groot and Heyting should be made extra-ordinary professors. The choice for De Groot and Heyting was not a difficult one. Both had a good reputation, and Brouwer had composed a flowery recommendation for Heyting. 'Science', he wrote, 'owes some fundamental discoveries to Dr. Heyting, which have also influenced philosophy, and in particular have thoroughly renewed epistemology'. After extolling Heyting's significance for non-Aristotelian geometry and for the formalisation of the Heyting-logic, originating from intuitionism, he concluded: 'Heyting is now 50 years old. It is more or less poignant that at this age, a man of his merits and his fame, still has the academic position of a lecturer, which previously had always been occupied by professors.'[114]

In March the battle over Van der Waerden's appointment was renewed. The faculty chairman pleaded in a letter to the Rector for the admission of a special chair in applied mathematics, to be administered by the 'Foundation for Higher Education in Applied Mathematics'.[115] A word of explanation may be in order. The Dutch academic system basically knew three kinds of professors (chairs), the full professor, the extra-ordinary professor and the special professor.[116] The second one is for all purposes a regular appointee of the university, but in a lower rank, with a lower salary and possibly with a part-time appointment. The special professor is appointed and paid by an extra-academic institution, society, foundation, etc. As a result there are special chairs for a wide range of topics, ranging from parapsychology to theoretical physics.

The request of the above mentioned foundation was therefore nothing special, but in this particular case, it was but another attempt to attach Van der Waerden to the University of Amsterdam. Indeed the Foundation was just a derivative of the Mathematical Centre, with Clay and Van der Corput in the driver's seat. The Foundation did not beat about the bush, it straightforwardly proposed to appoint Van der Waerden, who had a visiting chair in the USA, and was prepared to resist some attractive American offers, provided he could pursue his research in Holland, combined with a special chair in Amsterdam, in order to train students and researchers for applied mathematics.

Brouwer immediately countered the proposal; he presented his objections in a five-point note to the Senate. For one thing, he said, we don't need any more applied mathematics than is already provided by Dr. Bruins; for another, this is a device to get around the earlier objection of the Minister to the appointment of Van der Waerden (in Utrecht). Nonetheless the appointment of Van der Waerden is presented as so urgent, 'that it is not too high a price to pay for the definite insolvency of the City of Amsterdam with respect to promises made to the undersigned in 1920, trusting in which he therefore remained in the Netherlands'.

At the subsequent Senate Meeting there was a lively discussion. When Brouwer was given the floor, he began by remarking that there was a small group of experts

[114] Brouwer to faculty, 27.II.1948.

[115] Clay to Rector, 13.III.1948.

[116] At present there are full, part-time and extra-ordinary professors. Apart from the change in title, the positions are the same as the old ones.

in the Senate, of which he was the oldest. He—to everyone's surprise—supported the faculty's request and warmly recommended Van der Waerden. There was just, he said, a difference of opinion between himself and the Faculty. It was his personal opinion that applied mathematics had no place at a university; it were not the merits of Van der Waerden in that area that had inspired the proposal, Van der Waerden was known for completely different reasons. He then suggested that he should formulate his views and present them to the Rector—Van der Corput could add a rejoinder if he wished to do so.

This move, clearly, had not been expected by Clay; he remarked that he was glad that Brouwer supported the candidacy of Van der Waerden, but he did not see how Brouwer's proposal could be carried out. The secretary of the Senate thought there was no problem:—let Brouwer's views, as received by the Faculty, and Van der Corput's, be sent to the Rector. As to be expected, there was also opposition to the proposal of the faculty for a different reason. It was pointed out that not too long ago the Minister had refused to confirm Van der Waerden's appointment for political reasons. Could one support such a questionable candidate? But now, Clay countered, 'Van der Waerden is at Johns Hopkins and we will sound the Minister on the acceptability.' These arguments, coupled with Brouwer's support, were for the Senate good enough to go along with the faculty's proposal.

Brouwer held his part of the bargain, he composed a memorandum for the Curators, in which he gave an exposition of his views.[117] Since the arguments have a definite Brouwerian philosophical flavour, it is worthwhile to reproduce them here:

> 1. Mathematics is an introvert science, as such it merges with philosophy, theology and reflexive psychology, but is to a higher degree than these, constructive. And the mathematical urge to create is therefore directed not only at inner clarification, but also at beauty, a beauty related to that of architecture and music, but more immaterial.
>
> 2. On these grounds the mathematical state of mind is usually indifferent with respect to natural science and definitely unfavourably inclined towards the promoting of the exploitation of nature and towards technique, which creates possibilities for it.
>
> 3. This does not detract from the generally known fact, that *and* technique *and* natural science *and* a manifold of other extrovert sciences have been able to reach their present status and size only because they 'reckoned' (arithmetically or graphically), i.e. they operated mathematically on the mathematical systems which were 'projected' on their activity, or their domain.
>
> 4. Although, in this manner, in the first place the technical, but furthermore all other extrovert sciences belong more or less to 'applied mathematics', they are in their essence fundamentally different from that of the introvert mathematics.
>
> 5. Where applied mathematics is mixed with the activity of a university as an all-permeating incidental circumstance, and almost totally represents the

[117] Brouwer to Curators, 2.IV.1948.

> substance itself of the activity at an institute of technology, there is, exactly on the ground of this ubiquity, no place for a special curriculum at either of the two institutions. On the contrary, every extrovert science is interwoven with its own applied mathematics, and should remain inseparably connected to it in its teaching.

The reader will recognise a Shakespearean flavour in the argument, it recalls Anthony's great oratory in 'Julius Caesar'. After thus determining the place of applied mathematics in the general curriculum, he went on to argue that one could imagine a sort of education in applied mathematics, namely the very elementary part which belongs to the 'propaedeuse'. This, he added, is already taken care of by Bruins. The rest of the arguments (counting up to 11) are a repetition of his earlier ones—abuse of power by the Faculty, procedural mistakes, the violation of the 1920 promise, Van der Waerden's expertise in pure mathematics.

Not content with presenting his views on the proposed appointment of Van der Waerden to the board of the university, Brouwer went one step further and sent his arguments to the minister of education.[118] He repeated the above arguments, but added one that reflected on Van der Waerden's career in Germany:

> From a researcher like Professor Van der Waerden, who is only theoretically, but not experimentally active, the scientific influence is almost independent of personal presence. Thus, as soon as a materially and scientifically favourable position has been secured, the question of his presence here in the country loses all scientific and national importance, and it becomes almost exclusively a matter of national prestige. From a viewpoint of national prestige the motivation of his appointment here in the country seems however extremely weak to the undersigned. For if it is claimed that by the presence of Professor Van der Waerden in Amsterdam the strength of our nation is enhanced, the reply is forced upon us that in that case the national strength of the German empire has been enhanced during the whole period of the Hitler regime by the presence of Professor Van der Waerden in Leipzig. And if it is argued that if Professor Van der Waerden is not offered a suitable position by the Netherlands, this will be done by America, the reply is forced upon us that if at the moment there are positions open to Professor Van der Waerden in America, this should not have been less the case between 1933 and 1940, when many prominent and right-minded German scholars and artists were welcomed with open arms in America, and that therefore one has to assume that Professor Van der Waerden had not felt the desire to turn his back on the Hitler regime.

There are no indications that the ministry reacted to Brouwer's letter. It is not unlikely that Brouwer intended to use Van de Waerden's past in the attempt to keep him away from Amsterdam—after all, the minister (and ultimately the Queen) had the last word on appointments, but the feeling expressed in this passage perfectly

[118] Brouwer to Minister of Education, 15.IV1948.

reflected the general opinion of the Dutch, and in particular the students, in this matter.

Clay and Van der Corput had once more been put on the defensive. They wrote a rather unconvincing letter to the Rector and Van der Corput was given the task to dissociate the three lecturers from Brouwer.[119] Van der Corput announced that the faculty was going to interview the lecturers, because 'In my opinion the enclosure added to the letter of April 2nd does not completely correctly represent the views of the gentlemen Heyting, Bruins, De Groot and Loonstra.'[120] In short, Brouwer was a fraud or the four gentlemen had signed something without reading it! Even a milder man than Brouwer would have resented such an imputation. A short note was the reaction:[121]

> Prof. Dr. J.G. van der Corput, Amsterdam. Your message of the ninth, received by me on the twelfth and your attitude, expressed in it, have deeply offended me and have changed things for me.
>
> Yours truly
> L.E.J. Brouwer

No 'amice' or 'friendly greetings'. The friendly relations between Brouwer and Van der Corput were a thing of the past. Did Van der Corput provoke this parting of ways on purpose, or did he not quite know how to formulate his thoughts? There is little to go on, he was a kind man with a good reputation with students and colleagues, but a man who knew to get what he wanted. In a mixture of post-war emotions and academic power play, there was a regrettable tendency to assume that in a confrontation with Brouwer one was allowed to bend the rules and the mores. This attitude even lingered years after his death.

Once Brouwer had accepted the inevitable, he stopped taking Van der Corput seriously. One way of coming to terms with persons he considered despicable, was to provide them with nicknames. And Brouwer was a master at the game. There are a few nicknames that have been handed down from generation to generation. Van der Corput's nickname was particular apt (in Brouwer eyes), he called him *Van der Corrupt*. There is an anecdote about the Brouwer–Van der Corput relationship that has been vouched for by students and colleagues of the period. It had not escaped Mrs. Van der Corput that her husband and Brouwer were no longer on the friendly footing of the beginning. She therefore took the bold decision to go and see Brouwer about the matter. She hoped to accomplish what the men could not: a reconciliation. So she spent an afternoon with Brouwer, who turned on his charm at full strength. The matter of the deplorable deterioration of the relations got all the attention it deserved, but no conciliation was accomplished. When she returned home, and ran into her husband, she gazed at him, and half seriously, half mockingly asked 'are you really so corrupt?'

---

[119]Clay and Van der Corput to Rector, 9.IV.1948.

[120]Van der Corput to Brouwer, 9.IV.1948.

[121]Brouwer to Van der Corput, 14.IV.1948.

One gets the impression from the surviving correspondence that at this point Brouwer had given up fighting the opposition any longer. There are some isolated letters, for example a curious one from Clay, who claimed not to understand how one could prevent in 1948 the fulfilment of a promise made in 1920.

The Secretary of the faculty, the physicist J. de Boer, tried to repair the rift between Brouwer and Clay–van der Corput; but Brouwer simply ignored the attempt and he refused to attend the faculty hearing of the lecturers Heyting, Bruins, De Groot and the assistant Loonstra. Being convinced that the Faculty was conspiring against him, he saw in the fact that the convocation, dated April 10, reached him on April 13, one day before the meeting, another indication that Clay c.s. intended to make it difficult for him to attend.

De Boer wrote Brouwer right after the Faculty Meeting, that three of the four who had signed the letter of 27.XII.1947 had declared that they did not wish their letter to be used at this moment against an appointment of Van der Waerden.[122] This, by itself, meant that Van der Corput's insinuation that Brouwer had made them sign something that they did not mean to say, had not been substantiated. In the same letter, De Boer informed Brouwer that the Minister no longer objected to Van der Waerden's appointment as a special professor.

At the Senate Meeting of July 5 Brouwer conducted a minor rear guard action, consisting of a correction of the minutes, so that his view point was properly recorded.

The events in the faculty might give the reader the impression that Brouwer was living in a state of permanent siege, with the ever-present prospect of utter destruction. In Brouwer's mind this may have been so to a certain extent, but there were also heart-warming signals of recognition. One such spot of sunlight was his election to Foreign Member of the Royal Society. J.H.C. Whitehead had written a brief but forceful recommendation:[123]

> Brouwer revolutionised the foundations of mathematics by his critique of the notion of 'existence' in the mathematical sense and by his constructive theory of 'Intuitionism', which arises from it. Also he was the outstanding figure in topology during the twenty odd years which followed the publication of Poincaré's papers on the subject. His work on fixed points, on the degree of a mapping and on the concept of dimension opened up some of the most fruitful fields of research in the subject. It is not unlikely that he will subsequently be considered the most original mathematician now alive.

Brouwer was elected on May 27, 1948 together with a number of other worthies (among them Linus Pauling, Kurt Mahler (the number theorist) and J.M. Whittaker).

Brouwer's gradual return to the other fascination of his youth, philosophy, found an eloquent expression in 1948, when he took part in the Tenth International Congress of Philosophy at Amsterdam, which was held from 11–18 August.

---

[122] De Boer to Brouwer, 15.IV.1948.

[123] 5.II.1948.

Brouwer used the occasion to spell out his views in the address *Consciousness, Philosophy and Mathematics*.[124] In a way this was an update of his Vienna lectures, be it that there was a notable shift in style and presentation. Where the Vienna lectures were the credo of a conqueror of the world in the strength of his life, compact, to the point, with little consideration for his readers, the Amsterdam lecture was more reflective and resigned in nature. The opening sentences faithfully expressed Brouwer's feelings,

> First of all an account should be rendered of the phases consciousness has to pass through in its transition from its deepest home to the exterior world in which we co-operate and seek mutual understanding. This account does not imply mutual understanding and in some way may remain a soliloquy. This can be said of other parts of this lecture too.

The first part of the lecture can be viewed as an elaborate justification and elucidation of the ur-intuition of the dissertation. The key notion, of course, is the subject, experiencing sensations. Through the *move of time*, already (namelessly) occurring in the Vienna Lectures, the whole world of the subject is constituted—both the egoic and the exterior. The jump from end to means (cf. pp. 66, 'leap from end to means', 516) is upheld in its full extent, here under the name 'cunning act' ('mathematical act' in the Vienna Lectures). The exposition expressly mentions an aspect that, although obviously part of Brouwer's views, was not mentioned before:

> In this connection there is a phenomenon of *play*, occurring when conative activity or causal thinking or acting is performed *playfully*, i.e. without inducement of either desire or apprehension or vocation or inspiration or compulsion.

Brouwer recognised three phases of the 'exodus of consciousness from its deepest home':

- the *naive* phase, of 'the creation of the world of sensation';
- the *isolated causal* phase, where the causal acts take place; and
- the *social* phase, in which 'co-operation with the individuals' finds its place.

Given the phases, 'the question arises, whether and where, on and after the exodus of consciousness, *beauty, mutual understanding, wisdom* and *truth* can be found'. In the last two phases 'there is beauty in remembrance of the miracle of bygone naivety, remembrance evoked either by reverie through a haze of wistfulness and nostalgia, or by (self-created or encountered) works of art, or by certain kinds of science. Such science evoking beauty reveals or playfully mathematises naively perceptible forms and laws of nature, after having approached them with attentive reverence, and with a minimum of tools.' Here, and in later passages, play and playfulness rank highly in the intellectual activity, in particular where non-pragmatic acts are concerned.

The subject recognises in the 'outer world' object individuals, who behave with a certain similarity to the subject; this, according to Brouwer, cannot be derived

[124] Brouwer (1949c).

from 'mind', as it would induce the subject 'to place in each individual a mind with free-will dependent on this individual, thus elevating itself to a mind of the second order experiencing incognisable alien consciousnesses as sensations. Quod non est! '*Plurality of mind*' is thus firmly rejected by Brouwer, and 'in default of a plurality of mind, *there is no exchange of thought either*'. In view of these passages there can be little doubt of Brouwer's solipsistic inclinations at that stage of his life.

Although Brouwer explicitly mentions the social phase, one should not be misled by this term; fellow (human) beings are not independent beings in a pre-existing outer world, they are, like the rest of the universe, creations of the subject. For logicians and philosophers the interesting topic is of course that of truth and its place in logic. '*Truth* is only in *reality*', quoth Brouwer, 'i.e. in the present and past experiences of consciousness'. The purpose of logic being to preserve truth, a certain reliance on logic in this respect developed; however, the conclusions of logic do not 'convey *truths* before these truths have been experienced', nor is it certain that 'these truths always can be experienced'. Brouwer's conclusion is that 'logic is not a reliable instrument to discover truths, and cannot deduce truths which would not be accessible in another way as well'. In other words 'there are no non-experienced truths'.

The lecture also contains creating subject arguments, albeit not in the strong form of his 1949 paper.[125] A trace of the defence against Griss' negationless criticism can be discovered in the modest claim 'there seems to be little hope for reducing irrationality of a real number $a \dots$ to a constructive property'.

A number of more or less expository papers that followed spelled out the same messages; but none with the clarity and explicitness of this philosophical lecture.

## 17.4 The Loss of Compositio Mathematica

The fight for *Compositio Mathematica* is one more dramatic episode in Brouwer's life—the last big one. It took place at the end of his academic career, and it is a vivid illustration of the erosion of his position in Dutch mathematics, his inability to build and maintain a sufficient support in mathematical circles. For the lone operator Brouwer it was no longer possible to defend his position. Even his considerable command of argumentation and persuasion had lost its magic power. As Kreisel put it, in the obituary of Brouwer for the Royal Society: '..., while, ..., solipsism seems an excellent first approximation for an analysis of mathematical reasoning, it would not be expected to be equally sound in public relations.'[126]

Most of the documents of the Compositio affair are to be found in the Brouwer archive. Unfortunately the publisher Noordhoff has not preserved the correspondence and documents pertaining to the matter.[127]

---

[125] Brouwer (1949a).

[126] Kreisel and Newman (1969), p. 46.

[127] In fact Noordhoff merged with Wolters, and it is no longer an independent company. In the transition the relevant material was probably discarded.

When life resumed its course after the war, many threads had to be picked up which were either dropped at the outbreak of the war, or which had become entangled in a number of ways during the war. In almost all organisations and companies there was a, sometimes subtle, sometimes not so subtle, power struggle between the forces of renewal and those of restoration. Next to the old political parties, new parties sprung up with new names and new programs. In art young men eagerly waited for the fall of the establishment. New dailies and weeklies appeared, most of them the legal successors of the underground papers published by the various resistance movements.

In the universities one could also observe a mild echo of the social-political changes in the Netherlands. By and large the most significant phenomenon was a temporary speed up of appointments of professors. The war and the purge had left vacancies to be filled. On the whole one could speak in the case of the post-war developments in academia more of a restoration than of a revolution. A disruption like that of the sixties and its democratisation was out of the question. The scientific organisations, as a rule, resumed their activities, their regular meetings and publications.

The publishing houses could not immediately join the upsurge of economic and cultural life, hampered as they were by the shortage of paper. This had consequences in particular for scientific publications; for a long time libraries, professors and students alike had to make due with second-hand pre-war copies and with books donated by (mostly American) universities.

As the man in charge, Brouwer had to consider the future of his *Compositio Mathematica*. The journal had been discontinued in 1940, when it was confronted with serious difficulties. Freudenthal, who had run the journal almost single-handedly, was the first to bring up the matter of re-issuing Compositio. In a letter to Hopf he gave an account of the situation:[128]

> Concerning Compositio, the matter is that I have officially no business with Compositio. I am simply not a member of the editorial board. Compositio can probably not appear legally with the editorial board as it was on May 14, 1940. For here everything is 'purged', the civil service, the professions, associations, editorial boards, etc. If an editorial board has not itself been infected, it can of its own proceed to purge itself. How this is done with editorial boards in which foreigners are also present, I do not know. In the case of Compositio the matter is especially unpleasant; Weitzenböck is stepping down anyway. The purging of Brouwer is yet open—I mean his purging as a professor, and the result will have its consequence for his further membership of the editorial board. [...] If Brouwer returns as a professor, he will certainly claim his right to sit on the editorial board. But the remaining Dutch mathematicians (apart from Heyting) probably have no wish to work with him. This can be said with certainty of Van der Corput. [...] I don't see at the moment any possibility but the founding of another journal under a similar name. I will discuss the

[128]Freudenthal to Hopf, 9.X.1945.

matter with Van der Corput. Perhaps he can say something. What would be your position with respect to a Compositio without Brouwer or with Brouwer thrown out?

In fact, nothing happened at all. Brouwer did not even consider a quick re-animation of Compositio. It would have been rather unlikely that the authorities would have allotted the required amount of paper for the journal.

The activities around Compositio during the first years after the war are somewhat obscure. On the one hand, Brouwer started to explore the possibilities of a re-issuing of the journal, on the other hand, a number of Brouwer's opponents would rather see a Compositio without Brouwer. It seems that Brouwer was approached by Noordhoff with the request to resume the publication of Compositio.[129] Apparently the efforts of Noordhoff were not very satisfactory, for Brouwer was in 1947 cautiously shopping around for a new publisher. In January 1947 he inquired with Father Van Breda, professor of philosophy in Leuven, renowned for his founding of the Husserl Archive, if there were printers in Belgium who could handle an international mathematics journal. Van Breda supplied the information.[130] In the same letter he invited Brouwer to Leuven for a series of lectures. In view of the awkward financial situation, he could not promise Brouwer a suitable fee, 3500 Belgian francs was all he could offer, Brouwer would have to pay for his stay in Leuven out of his own pocket, but 'undoubtedly you will repeatedly be invited by various professors for lunch and dinner'. In spite of the scant fee, Brouwer accepted the invitation and lectured six times at the end of March.[131]

On February 3, 1948 the difficulties had been sufficiently overcome that Brouwer informed the Committee of Administration (editorial board) of Compositio (de Donder, Hopf, Julia, Whittaker) of his plans to send out a circular letter to all editors.[132] In this letter the editors were asked to stay on and to publish their own papers and 'those originating from your school' in Compositio.

Noordhoff set itself to produce a first post-war issue, but it discovered that the printer had lost patience, and re-used the lead of the type of the 1940 issue.[133] Having some doubts as to the wisdom of leaving the daily affairs of Compositio to Brouwer, Noordhoff casually asked Freudenthal's opinion on the future of Compositio under Brouwer. Freudenthal expressed his willingness to give his opinion, but declined to do so in writing.[134] He urgently counselled Noordhoff to clean up the editorial board—no more than one third of the old board ever actively took part in the editing. One should, in his opinion, attract some 20 young mathematicians 'who

[129] Brouwer to Ed. board Comp. Math., 10.VII.1949, 27.I.1950.

[130] Van Breda to Brouwer, 25.I.1947.

[131] 17, 19, 21, 24, 26 and 27 March 1947.

[132] Strangely enough 'to the editors belonging to the United Nations'. What had happened to his internationalist convictions of 1919?

[133] Noordhoff to Freudenthal, 1.XI.1948.

[134] Freudenthal to Noordhoff, 1.XII.1948.

are at the peak of their creative power, and who are not yet members of other editorial boards'. Moreover Noordhoff should attract a young, active mathematician with a broad interest and good qualifications for the position of secretary. The person should have relations with the top circles in mathematics, should have enough personal courage to reject mediocre work, etc. He ended with the harsh words: 'if one wishes to salvage anything at all of the goodwill of Compositio Mathematica, one should take action promptly and energetically. A journal that keeps plodding on or that degenerates into a rubbish dump would do considerable harm to the international reputation of Dutch mathematics.'

Noordhoff did not act solely on Freudenthal's advice, it even went so far as to poll the mathematics professors in the Netherlands. In January 1949 the publisher sent a letter to the Dutch mathematics professors, asking them for their support, announcing the re-issuing of Compositio under the temporary secretarial care of Brouwer, who had taken the initiative. The letter contained the seemingly harmless sentence: 'As we are of the opinion that the journal with its good reputation should appear at the same level as before, we would appreciate if the journal in addition to the support of its foreign contributors, would also receive the total support of the Dutch mathematicians.'

The letter elicited quite a number of reactions, one of which was provided by Van der Corput, who was one of the old editors. Brouwer apparently had not included Van der Corput in his list of recipients of the announcement of the re-animation of Compositio. One can easily imagine why: a man who had deftly outmanœuvred Brouwer in the faculty and in the Mathematical Centre affair, was not to be trusted in an editorial board.

Van der Corput did not accept Brouwer's move without protest, he complained to Noordhoff that he was, to his surprise, unaware of the plans concerning Compositio.[135] Noordhoff cleverly made use of Van der Corput's dismay, expressing their surprise that one of the co-founders of Compositio, with the same rights as Brouwer, had not been informed by Brouwer.[136] The plan for resuscitating Compositio was greeted with applause by most Dutch mathematicians, wrote Noordhoff to Van der Corput; perhaps one should ask Brouwer how (and why) he happened to overlook Van der Corput. 'Is there any objection on your side, that we show your letter to Professor Brouwer?', he subtly inquired. This was not quite what Van der Corput had in mind; he immediately replied that 'Some of the mathematicians consulted by you have expressed themselves very cautiously. It seems to me that my answer should rather not be passed on to Professor L.E.J. Brouwer.'[137]

As one could expect, Brouwer did not react kindly to the Noordhoff circular letter. He interpreted it as an attempt to import more Dutchmen into the editorial board; worse, he viewed it (according to Schouten)[138] as 'an action (by some person or persons unknown) to throw him out, and he took the whole thing as a personal affront'.

---

[135] Van der Corput to Noordhoff, 26.I.1949.

[136] Noordhoff to Van der Corput, 29.I.1949.

[137] Van der Corput to Noordhoff, 31.I.1949 (draft).

[138] Schouten to Hopf, 8.XI.1949.

On these grounds he refused to work any longer with Noordhoff, and the preparations came to a complete halt. That did not mean that Brouwer had put Compositio out of his mind altogether. He actively looked for new editors; one of the persons approached was Paul Bernays; in order to get a better representation of the subject of mathematical logic in Compositio, Brouwer invited him to join the board of editors, asking at the same time his advice as to another editor from the logical corner of mathematics. Hopf had suggested MacLane, but Brouwer thought that Kleene might be a good candidate. Bernays apparently advocated Kleene's membership, for Brouwer wrote to Kleene 'I have the pleasure to invite you, *firstly* to enter the editorial staff of Compositio Mathematica, *secondly* to favour this periodical with some work of your own.'[139] From the letter it also appears that Brouwer was keenly interested in what was going on in recursion theory. He thanked Kleene for some reprints, and asked for a specific paper. Kleene was at the time occupied with finishing his monumental *Introduction to Meta-Mathematics*, so he accepted the invitation provided he could finish the book first. As a matter of fact he spent a term in Amsterdam in 1950, where he could get first-hand information on intuitionism.[140]

In order to get the journal under way again, Noordhoff and some mathematical colleagues called in the help of Schouten, who was the Dutch mathematician following Brouwer in seniority. By now Schouten was one of the more prominent Dutch mathematicians. He had in 1943 resigned from his Delft chair, and withdrawn himself to a quiet part of the country, but his influence was still considerable and Noordhoff must have seen in him a valuable ally in the attempt to edge out Brouwer. Although Brouwer and Schouten had their differences in the early twenties—patched up in 1929 after the mediation of Weitzenböck (cf. Sect. 8.3)—animosity was certainly not Schouten's motivation to take Noordhoff's side. Schouten was one of the editors of Compositio of the first hour; it was probably a sincere wish to restore Compositio to its old glory that made him an actor in the Compositio affair.

Schouten met Brouwer on May 28 and discussed the matter. According to Schouten, Brouwer agreed to enlarge the Committee of Administration with Kloosterman, Heyting, and Gerretsen[141] as a secretary.[142] When this agreement was reached Schouten immediately informed Noordhoff and a meeting with Brouwer was scheduled for July 5. To the general disappointment Brouwer asked for postponement of the meeting,[143] and subsequently did not respond to any letters. In all fairness it must be said that no attacks at Brouwer were envisaged, Van der Corput at one point argued forcibly that the combination 'Brouwer–Compositio' was from an international point of view the strongest possible, and that Noordhoff should really try to keep Brouwer in charge. Neither was Schouten out for Brouwer's removal, but he clearly wanted to reduce him to 'one of the editors'. Unfortunately Schouten did

[139]Brouwer to Kleene, 12.IV.1949.

[140]Kleene to Brouwer, 19.IV.1949.

[141]Professor in Groningen, a function theorist.

[142]ibid. The information is mostly based on Schouten's letter.

[143]Telegram, 30.VI.1949.

not possess the tact needed to handle a mercurial person like Brouwer. His letters, obviously well meant, were of the half-patronising, half-schoolmasterly kind that goes against the grain. Brouwer in particular had no wish to be lectured. In the end it must have been a mixture of exasperation and genuine worry about the future of Compositio that drove Brouwer to desperate steps.

Brouwer clearly had given up hope to reach an agreement with the Noordhoff faction. Why is not quite clear. Maybe it was the old story of a personal consultation interpreted differently by the parties. Brouwer had learned a lesson in his relation with Van der Corput: never rely on verbal agreements. Whatever caused the final disruption of connections with Schouten and Noordhoff, Brouwer lost no time in taking counter measures. On July 10 he sent a letter to the members of the Committee of Administration, proposing to sever all ties with Noordhoff.

> Dear Colleagues,
>
> When the House of Noordhoff Groningen, which had functioned from 1934 to 1940 as bookseller-publisher-agent of Compositio Mathematica, offered us to resume from 1945 its old function, there was no reason to refuse it the opportunity to prove its claim to be up to that task. However, having taken up this task, it had started by working so miserably, be it through a lack of equipment, be it through a lack of zeal, be it through a lack of good will, and it finally demanded, before continuing its work, a reorganisation of the editorial board which would change completely the character, and in particular the international character, of our journal.

Under the circumstances, he went on, I propose to take our business to another publisher, 'I have good hopes to find for that purpose a house of renown, well directed and equipped, which will serve us better than the one that has deceived us.'

Brouwer must have thought of the North-Holland Publishing Company (which also printed for the Academy, and of which Brouwer was a member of the board of commissioners), for he had already approached that firm in early July. Unfortunately for Brouwer that particular plan fell flat. When North-Holland was informed by Noordhoff of its purported rights, it lost interest in the acquisition of a journal that might bring a string of lawsuits.

A majority of the Committee of Administration agreed with Brouwer: de Donder, Julia and Saxer sent their approval for further action. Schouten in the meantime tried his best to get Brouwer back to the negotiating table; he asked Van der Corput to talk Brouwer round. Noordhoff wrote a conciliatory letter and Heyting tried to influence Brouwer. None of this was of any avail.[144]

With nothing to lose, Schouten decided to take to the offensive. His first object was to make the members of Committee of Administration see the Compositio problem his way. In a long letter to Hopf, Schouten set out to justify his and Noordhoff's cause and to prove Brouwer wrong.[145] Apart from a recount of the events

---

[144] Schouten to Hopf, 8.XI.1949.

[145] Freudenthal guessed that similar letters went to the other members.

the letter contained a list of refutations of Brouwer's claims or suspicions. Some of Schouten's arguments and claims had a degree of plausibility, but if they contained some truth, certainly not the whole truth. In particular his protestation that he did not attempt to remove Brouwer seems a bit lame, unless one supplements the claim that the action was not being directed against Brouwer, by the clause 'as long as Brouwer does not interfere with the journal'. As Schouten saw it, the situation held a grave risk for Brouwer:

> Up till now as a mediator I was able to prevent legal action from the side of the publisher. But after the strict refusal of Mr. Brouwer this will not be possible any longer. So if nothing is done, there will be a legal action and the Committee of Administration, especially Mr. Brouwer, will be made responsible for further delay. As I see it now, it would have been better if I had written to the other members of the Committee of Administration at an earlier time. But my intention was to be very careful and to make things for Mr. Brouwer, as little disagreeable as possible, and this held me back till now from this action. Up till now the most influential Dutch mathematicians agreed with me that we must aim at a solution giving Mr. Brouwer the place and the honour that are naturally due to the man who founded the 'Compositio Mathematica'. But with Mr. Brouwer now turning down any compromise, a solution has to be found in whatever way. For Mr. Brouwer this would lead to a very serious defeat and I think we ought to try, if possible, to avoid such a defeat for a man of his age and fame.

On the whole, Schouten's action should be taken at its face value. He was not the evil man Brouwer had thought him to be. It is more likely that he had taken the role of mediator in a sincere wish to solve the problem without hurting Brouwer. He would probably have preferred to solve the Compositio conflict without damage to the parties concerned. Nonetheless he had to play the game for Noordhoff, and in that role he cleverly bent the facts to his advantage. One should not forget that at the same time Schouten, Van der Corput et al. were the subject of Brouwer's guerrilla warfare in the faculty. So, if Schouten showed some exasperation, he was entitled to it.

In order to minimise the damage to all concerned, Schouten launched a proposal for salvaging Compositio. A Temporary Committee of Reorganisation should be installed, consisting of four Dutch members (one distinguished man from each of the four Dutch Universities)[146] with the following tasks:

1. Start the editing of Compositio.
2. Arrange the election of a new Committee of Administration.
3. Draft rules for Compositio and a contract with the publisher.
4. Submit the rules and contract to the vote of the general committee.
5. Dissolve itself.

---

[146] Schouten, for unknown reasons, counted only four universities. There were five at the time.

Schouten did not wish to become a member of the Committee of Administration, 'At my age it is a big mistake to do things or to go on doing things that younger people can do so much better. A wise man has to know the time at which he has to withdraw', he wrote, but he was willing to act as the 'central man who has to constitute the Temporary Committee and to work as its Chairman, and to mediate the parties concerned. He would make it his special duty to ensure that Mr. Brouwer got the honour and the place due to the founder of the *Compositio Mathematica*.

The secretary designate, Gerretsen, had the task to get legal advise, for it was not unthinkable that Brouwer would take the publisher or the new board to court. After weighing the possible actions, the legal adviser deemed it safe to proceed along the lines indicated by Schouten.[147]

Schouten set to work without delay, and soon he could present the General Committee with his Temporary Committee of Reorganisation, Freudenthal, Gerretsen, Kloosterman, and Koksma.[148]

He had ascertained, as he stated, that the greatest possible majority of the Committee of Administration had agreed to his proposals. Brouwer was blessedly unaware of all the goings on (which shows how a once central person may get isolated).

Noordhoff had reacted with an invitation for further talks, an unacceptable proposition for Brouwer, who insisted that the work should be resumed first. Van der Corput, in a letter of January 20 to Brouwer, completely ignored the Compositio-matter. Probably for a good reason, for the Temporary Committee had drafted a rather tactless letter to Brouwer, informing him of the existence of the committee and demanding peremptorily that he hand over the administration of Compositio. This letter reached Brouwer not before February. Brouwer only learned about the activities of the Noordhoff party through a chance remark of a French mathematician, probably P. Levy, who told Brouwer that he was informed by Gerretsen that his paper would soon appear in Compositio.

From correspondence of Lize it appears that in the fall of 1949 (and perhaps earlier) Brouwer had been suffering from a mixture of complaints. In particular he feared that his heart and lungs were in a bad shape. On 27 November Lize wrote her daughter that Brouwer had been thoroughly examined by one of his colleagues in the medical faculty, professor Formijne. The examination showed that his heart and lungs were in good order, and that the stomach was the problem. Brouwer was immediately put on a diet. In any case he was greatly relieved, he was eying the future with some more optimism, all the more as he had been invited to lecture in Paris in December 1949–January 1950. He had started his lectures on December 13, returned for Christmas to Blaricum, and he taught the second part of his course in January. He had duly reported to the board of the university that his work as a 'professeur d'échange' at the Sorbonne was not yet finished, and he had asked

[147]Kluyver to Gerretsen, 11.XII.1949.

[148]Schouten to General Committee, 2.I.1950. Cf. Brouwer to eds., 1.I.1950.

permission to return to Paris for another month of teaching.[149] His absence suited the Compositio conspirators wonderfully.

On January 9, the day before he returned to Paris in order to resume his series of lectures on intuitionism, he had written to Van der Corput that the latter's intervention with Noordhoff had been a failure. In October Brouwer had asked Van der Corput to inform Noordhoff that he agreed to enlarge the editorial board as soon as Noordhoff resumed the work at the forthcoming issue of Compositio.

Brouwer was furious when he found out what was going on. He immediately wrote a long letter to the Committee of Administration.[150] In his function of (temporary) secretary of Compositio he was responsible for the manuscripts of the authors, he wrote, and now he found that Noordhoff had betrayed his confidence by giving Gerretsen, the 'self-styled secretary', access to the manuscripts. After a short résumé of the history of Compositio, Brouwer concluded that the latest démarches of Noordhoff left him no choice but to go to court. He asked the members of the committee to authorise him to start the legal procedures. 'I don't expect an easy law suit. The manœuvre of Mr. Gerretsen made me think that there is a well meditated machination, by which the House of Noordhoff hopes to get Compositio in its possession, and for which purpose it has succeeded to ally itself to some mathematicians in my country who wish to enter into the Committee of Administration of the journal.'

Returning home, he had found the new issue of Compositio waiting for him. He again turned to the members of the Committee of Administration.[151] The publication of the issue under the name *Compositio Mathematica* with the traditional cover constituted such a colossal fraud, that the impudence of 'our adversary should hopefully facilitate our law suit'.

The letter had hardly been written when Brouwer received, with more than a month's delay, Schouten's letter of January 2. This was the first indication that Schouten had managed to turn the (or at least some) members of the Committee of Administration. Brouwer again addressed the Committee of Administration, explaining what had been discussed between him and Schouten.[152]

Hopf, who was well aware of the difficult sides of Brouwer's personality, felt that he could not take Schouten's side. The years of friendship with Brouwer could not be erased that easily. In reply to a card from Brouwer with a picture of *Le Penseur*—a reference to the past: 'Does Hopf still remember how he, with Neugebauer, found me here in the Louvre?',[153] the reply was 'Of course I remember very well how Neugebauer and I found you in the summer of 1926, sitting in front of the Mona Lisa.'[154]—he lamented the recent developments. In spite of the advanced stage of

[149] Brouwer to Mayor and Aldermen, 3.I.1950.

[150] Brouwer to Comm. of Adm., 27.I.1950.

[151] Brouwer to Comm. of Adm., 3.II.1950.

[152] Brouwer to Comm. of Adm., 7.II.1950.

[153] Brouwer to Hopf, 28.I.1950.

[154] Hopf to Brouwer, 12.II.1950.

the fight, he implored Brouwer to accept a compromise. He argued persuasively that one could not blame a publisher for clinging to a journal. Publishers need editors who are prepared for minor compromises, he said, it would not be a shame at all for Brouwer to accept a younger Dutchman at his side in the board. There would be enough foreigners to guarantee the international character.

That same day Hopf sent a letter to Schouten, informing him of his letter to Brouwer. Hopf also castigated Schouten for not informing Brouwer of the setting up of the Temporary Committee. He disapprovingly commented on Freudenthal's membership of the committee; in this way, he wrote, the committee lost its neutrality right at the beginning. Schouten answered with an extensive justification of his policy.[155] Since Brouwer had cut all communication, it had been impossible to consult him. Apparently it had not occurred to Schouten that in order to inform a person, one does not have to consult him.

The choice of Freudenthal, Schouten argued, was not motivated by anti-Brouwer feelings. One simply had to look for four prominent mathematicians from the four universities. Freudenthal, as a Utrecht professor, was a natural choice; the other professor, Popken, was a young man who, moreover, was the son-in-law of Van der Corput. In view of the situation in Amsterdam, Schouten did not wish to aggravate the differences between Brouwer and Van der Corput. Freudenthal's long experience with Compositio made him a valuable addition to the committee, and, he added, 'I note with pleasure that Freudenthal is extremely correct, and in no way hostile to Brouwer.' 'Of course, I was aware of the frictions between Brouwer and Freudenthal', he wrote, 'but as Brouwer permanently gets into a fight, then with one person and then with another, one could not seriously take this into account. Where would the world be if everybody could insist on keeping company only with those they could love, and by whom they are loved? This is ridiculous child's play, and it is most regrettable that an excellent scholar possesses so little wisdom, that he stoops to playing cowboys and indians.'

Schouten left no doubt that he saw no place for Brouwer in the affairs of Compositio.

> Personally I hope for officially a reconciliation with Mr. Br. He had 'terminated the friendship' on 28.I.1950 from Paris, apparently in response to some correspondence of Gerretsen with a French mathematician on overdue proofs, without having read the letters of 1 and 2.I.1950 and not being aware of the appearance of the Compositio Mathematica, which by then had already taken place, nor of the real state of affairs. Moreover he had written that he did not wish to receive letters from me and that he would not read any. Those are, however, children's games, which will not induce me to withdraw my friendship from him. From one day to another this could change, as I have experienced before. As you see from the new statutes, the possibility has been created to appoint an especially excellent man, who has been of great importance for the Compositio Mathematica, to honorary president of the editorial

[155]Schouten to Hopf, 14.II.1950.

board (*Hauptredaktion*). Perhaps the solution is to be found there. At his age something like that would be exactly the right place for Mr. Br., and the honour would be saved!

In spite of Schouten's attempt to treat Brouwer 'en bagatelle', one has no difficulty in seeing through the rather obvious 'propaganda'. Schouten had succeeded in reducing Brouwer's position in the considerations of the committee to that of a capricious child that could be placated with a shining toy. The last sentence, in particular, gives away Schouten's opinion of Brouwer—an elderly dodderer, only good for some honorary position.

History seemed to repeat itself; exactly as in the case of the *Mathematische Annalen*, Brouwer was completely ignored by the new rulers. The reorganisation was carried out according to plan, and on May 5 Schouten could announce to the members of the General Committee of Compositio that the new regulations had been accepted by a majority vote. In a letter to Veblen,[156] Schouten explained the reasons for certain formulations of the rules; he added that 'As matters are now, it is all a bit disagreeable for Mr. Brouwer. It is his own fault, but personally I should like to make things as pleasant as possible for him. I have in mind to propose him for 'Honorary President' of the General Committee. It is quite impossible to make him a member because then the difficulties would begin all over again, he being what he is. Best if he is so wise to accept this honorary Presidentship, the honour is in this way saved and in this position he is quite unable to do any mischief.' It could not be expressed more clearly!

On May 31 a letter was sent to the members of the General Committee (formerly the editorial board) containing the list of candidates for the General Committee (Cartan, Gerretsen, Kamke, Kloosterman and Koksma), the Special Committee (formerly the Committee of Administration) (Cartan, Kamke, Kloosterman, Koksma, Saxer and Whittaker) and finally the name of L.E.J. Brouwer for honorary President of the Special Committee.

In fact, when volume 8 of Compositio was completed (1951), the cover carried more new members than the above listed: Ancochea, Bompiani, Eilenberg, Freudenthal, Kleene, MacLane and Picone.

Here the Compositio affair ended. Brouwer had not only lost a battle, he had lost for the second time a journal—this time it was his own journal, expropriated by his colleagues and supposed friends.

The cover of Compositio did not list Brouwer as a honorary president, so either the members did not support Schouten's proposal, or Brouwer refused to accept the honour. Brouwer refused to resign himself to the inevitable, as late as December 1950 he was still corresponding with his lawyer.[157]

He had not been able to carry on the matter, he wrote, because of two disastrous developments:

[156]Schouten to Veblen, 9.V.1950.

[157]Brouwer to Baron van Haersolte, 22.XII.1950.

1. Shortly after I wrote you for the last time, I had to observe that my foreign confrères in the Comm. of Adm., who were in July 1949 without any reserve on my side, have abandoned me as a consequence of communications and promises of my adversaries, which have remained a secret for me.
2. The physical shock, inflicted on me by this bewildering observation, left me, after a heart attack, weakened in mind and body to such a degree, that I have been put out of action for a longer period, with respect to the mentioned aggression, and that every sojourn in the realm of thought of this conflict was forbidden for some time.

Nonetheless he planned further action against Noordhoff because it 'not only reflected on my honour, but also on the honour of my country'.

Apparently these plans did not materialise, the conclusion to the Compositio conflict is lost in vagueness. Brouwer remained very bitter, however, about the whole affair, he used to refer to 'the theft of my journal'. In spite of the insulting treatment, however, he remained a member of the editorial board of Compositio until his death.

Looking at the Compositio affair from a distance, one can see what the parties wished and feared. Schouten and his followers thought that Brouwer would stand in the way of a recovery of Compositio, and they were more inclined to be loyal to the publisher than to the founding father. Brouwer, on the other hand, was confident that he could maintain the pre-war quality even without Freudenthal's help. He was firmly convinced that his Dutch colleagues together with the publisher would reduce Compositio to a provincial periodical. Justified or not, he feared that the lesser gifted would finally see a possibility to enhance their status by joining an editorial board, rather on the basis of their nationality than on the basis of competence. And Noordhoff had a straightforward commercial interest. The publisher successfully played on the secret dreams of the mathematical community—to be an editor of this prestigious journal!

Brouwer probably did not fully realise to what extent the success of Compositio was due to Freudenthal, but it is not impossible that with a suitable substitute for Freudenthal, he could have made Compositio a success. His complaints about Noordhoff were, one would guess, the usual complaints of editors, aggravated by the post-war shortage. In short the conflict seemed to based as much on personalities as on facts.

The standard of conduct in the conflict is perhaps best characterised by the motto 'dealing with Brouwer, anything goes'. If one judges by the standards of 'obeying the rules', Brouwer definitely cut a better figure than Schouten. The episode does not do credit to the Dutch mathematical community, it is at best an interesting topic for psychologists.

## 17.5 Rearguard Actions

The Compositio conflict, for the time being, overshadowed most of the other matters that called for Brouwer's attention. There was little news from the faculty, Brouwer was content to discharge his duties as a Director of the Mathematical Institute; only in March 1949 he suddenly came to life when Bruins was offered a chair at the Agricultural University at Wageningen.[158]

Brouwer happened to be in Switzerland when he alarmed the faculty with the news that it was at risk to lose Bruins. Brouwer was suffering from a bout of asthma, which had already prevented him from attending Schouten's inaugural address 'On the interaction between Mathematics and Physics during the last 40 years', which marked his entry into the Amsterdam Faculty.[159]

The faculty had felt that it would profit from the co-operation of the leading Dutch differential geometer, not only because of the prestige that his appointment would lend to Amsterdam, but also in view of his usefulness as an organiser for running the new Mathematical Centre. The Centre indeed benefitted from Schouten's expertise, but his contribution to the mathematics curriculum in Amsterdam was negligible. As Schouten, called back from his self-chosen early retirement (he resigned during the occupation in 1943), was due for mandatory retirement in 1953, the appointment did not carry serious risks for the faculty.

Inaugural addresses traditionally ended by thanking all and sundry, from the Queen down to the assistants; Schouten addressed the (absent) Brouwer with the kind words, 'Dear Brouwer, that serious health problems prevent you to be present here, is a source of serious regret for me. May your cure abroad bring you a complete recovery, so that you may return soon completely recuperated in our midst'. The Compositio conflict had at that point not yet reached its climax, and at least superficially everything seemed peaceful.

We read about Brouwer's visit to Switzerland in a letter to Hopf; Brouwer rejoiced that he was finally back in his beloved Tessin, where he hoped to recover from bronchial asthma, which had been troubling him since August 1948.[160] Unfortunately, he wrote, he would have to return soon, since the currency that he had finally—after months of patience—obtained, was disappearing faster than he expected.

From the Kurhaus in Cademario (near Lugano) he called Van der Corput's attention to the possibility of Bruins' exit from the Amsterdam stage, in order to accept a chair in Wageningen. Isn't it time, he wondered, that we should do something to keep this mathematician, who had a 'lively and easily roused interest in other sciences than mathematics, which was found before him perhaps only in Professor Korteweg' in Amsterdam. Bruins' flair for tackling problems in chemistry and physics was recognised in Amsterdam; in Brouwer's words these activities 'have to a high degree stimulated in Dr. Bruins the gift to provide scientists in a wide range of non-mathematical areas, who call in his assistance, now and then with simple (but therefore not obvious) mathematical methods, which enable them to conquer difficulties in their field and to open new vistas'. These talents, in Brouwer's view, could

[158] *Landbouw Hogeschool.*

[159] 21.I.1949.

[160] Brouwer to Hopf, 10.III.1949.

possibly be saved for Amsterdam by offering Bruins an extra-ordinary chair (if necessary without cost to the City of Amsterdam). Bruins did not accept the chair in Wageningen, but he was not promoted in Amsterdam either. Although Freudenthal's appointment in Utrecht had removed the immediate threat to Bruins' Amsterdam career, the position of Bruins had not improved. Hardly anybody in the mathematics department supported him. His role in the treatment of Freudenthal, the paltry handling of the lecturer's position, had left an unpleasant taste. And if that was not bad enough, Bruins' mathematical profile did not really fit the Amsterdam department, where, definitely after Van Dantzig, Beth, De Groot, Van der Waerden and De Bruijn had joined, innovation and progress became the order of the day.

In 1949 Brouwer received a letter from a young Swiss mathematician, Ernst Specker, who had just introduced the analogue of Brouwer's methods in recursive mathematics. Specker's letter contained a beautiful and short argument that a monotone, bounded sequence of recursive reals need not have a recursive real as a limit. This was the first step in applying the young theory of recursive functions to the foundations of everyday mathematics, in this case analysis of real functions. No reaction of Brouwer is known—unfortunately.

Specker was personally acquainted with Brouwer, for during one of Brouwer's visits to Zürich Specker was present when Bernays had invited Brouwer. Brouwer was always interested in young mathematicians and their work, so he soon got into a lively conversation with Specker. In the course of the discussion (or monologue?) Specker inquired if Brouwer had any intuitionistic followers in Holland. The answer was rather revealing, 'Yes, there are some, but they cannot be a lot of good, or they would not tag along after me.'[161]

Although the situation in Amsterdam was, from Brouwer's viewpoint, far from satisfactory, his old urge for travelling became stronger than his desire for the preservation of his position in the faculty. In April 1949 the authorities gave him permission to travel to Spain—and more important, granted him permission to obtain the necessary currency. He was invited to represent the Netherlands at the Centenary Celebration of the Royal Academy in Madrid. This was his first visit to Spain, and he greatly enjoyed it.[162]

On his return to Amsterdam, Brouwer took a decision that must have been difficult, he handed the supervision of the Mathematics Library over to Schouten. This library had been his pet project since the early days of his career, when the mathematical literature was still a part of the general University Collection. Under Brouwer's management the private mathematics library had slowly expanded into an up-to-date collection (the reader may recall that one of Brouwer's conditions for staying in Amsterdam was a considerable sum for enlarging the book and journal collection). It was a so-called '*hand bibliotheek*', not a lending library, but a collection to be consulted by staff and students at the premises.

[161] Oral communication, E. Specker.

[162] Brouwer to Hopf, 3.V.1949 (from Madrid).

In July 1949 Brouwer again travelled abroad, this time he permitted himself the luxury to fly to Belgium. Evert Beth, the new professor in logic in Amsterdam, picked him up at the airport in Hulsbroeck.

Later that year a series of lectures in Paris at the Sorbonne followed; in the month of December he presented an overview of intuitionism. It was during this second part of his stay in Paris that the Compositio affair exploded.

The return to Paris after years of absence was a memorable occasion for Brouwer, who, after all, started his international career in Paris at the apartment of his brother during the Christmas holiday 1909–10, where the new topology was born.[163] In his opening address of the lecture course he recalled the old days with a certain measure of nostalgia.

> It is for me a great honour and a great satisfaction to have the opportunity to give an exposition of some of my ideas under the aegis of the Sorbonne and in the institute that bears the name of the French scholar[164] who on the one hand was the greatest mathematician of his time, and who on the other hand, gave the philosophy of Science one of the strongest impulses man has known.
>
> Addressing you here, I also feel an emotion of recollection. It is now exactly forty years ago, it was in December 1909 and January 1910, that I inhaled for the first time the scientific air of Paris. And I recall vividly and with a deep gratitude the encouraging benevolence, with which I, a young beginner, was received by the grand old men of that time, whose names had been tied to the grandiose evolution of the mathematical sciences which was taking place, and to whom, through their published courses and their monographs, my generation owes the greater part of its knowledge and a considerable part of its inspiration. It was the five classical memoirs on analysis situs of Poincaré and the reflections on dimensionality of Poincaré that were to open the perspectives in which my thoughts on topology developed themselves, and it was the studies of Poincaré and Borel, in particular the manner in which the latter had introduced the notion of measure, that made me glimpse the direction in which I had to seek the primordial origin of mathematical thought. It is a very special sensation to find of these great old men of the past, after a so considerable lapse, some who survive full of vigour.

This tribute to Borel and Poincaré was a proper demonstration of his intellectual debt to the two great Frenchmen. It unambiguously mentions the role of Poincaré's topological memoirs; the influence of Poincaré's algebraic approach, or rather the lack of it, seems therefore to have been rather a matter of personal taste (*esprit géometrique*), than of ignorance. To be precise, it was Hadamard who drew Brouwer's attention to Poincaré's topological work. Brouwer, being largely a self-educated man in the more advanced parts of mathematics, had read Cantor and Schoenflies, but missed Poincaré.

---

[163]Cf. p. 153.

[164]Poincaré.

During his stay in Paris he wrote two notes for the Comptes Rendus of the Académie des Sciences, presented by Emile Borel, both of them dealing with the notion of order. In the last one he published the proof that the continuum can not be 'classically ordered', i.e. that an ordering satisfying the dichotomy principle is not possible. This was a general result; that the usual ordering did not satisfy the trichotomy was not so surprising, but the paper showed that no suitable binary relation of that sort could be found at all. This result had already been presented (as a number of the post-war publications) in the Berlin Lectures of 1927.

Evidently, Brouwer and Borel met during Brouwer's stay in Paris, and it is most likely that they had met in the past. The strange thing is that, although they had strikingly similar ideas on foundational matters, little of mutual influence was acknowledged in their writings. And even at this point Brouwer explicitly mentioned the Borel measure, but not Borel's remarks on choice sequences. Of course, from Brouwer's point of view, Borel contributed no more than heuristic considerations, which in the absence of a systematic revision of the underlying logic, could not compete with Brouwer's approach. Nonetheless Borel quite explicitly spotted the phenomenon, and made appropriate philosophical distinctions. At the same time it is surprising, to say the least, that Borel from his part ignored Brouwer's ideas.

In 1949 Arend Heyting started to make plans for the publication of Brouwer's collected works. He was cut short by Brouwer, who was of the opinion that the time for such an enterprise had not yet come:[165]

> Dear Heyting,
>
> In my opinion, an author who is still alive and in a state of scientific responsibility, and who now brings out again works that have appeared earlier, is obliged to justify for himself for each of those works, both the significance it had at the time of the first publication for the state of science at that time, and the significance it could have for the present state of science if it had appeared now for the first time. And he ought to consider on the basis of this justification, to which comments on the re-printed text the present reader is entitled.
>
> Therefore a re-publication of my collected works would impose on me an amount of work of such a size, that the required time will lack me for several years to come.
>
> To a lesser extent this objection holds also for the planned re-publication of my dissertation with connected publications in English. For this too, I will not have time for the efforts involved for me, as long as not in the first place Compositio Mathematica will permanently be back in operation, subsequently my Cambridge Lectures have appeared, and finally the manuscript of my intuitionistic function theory has been completed. But that time is, I believe (if my working power is at least not completely paralysed by the consolidation of the nazification of Dutch mathematics) in a fairly near future, ...

[165] Brouwer to Heyting, 28.X.1949.

It is clear that Brouwer did not want a simple reprinting of his papers, he indeed wanted to put all his new insights and a host of corrections and addenda into an edition of his collected works. The ambitious plans that were to be carried out first, all came to naught. The Compositio enterprise was successfully taken over by Noordhoff and Schouten, the Cambridge Lectures were not finished,[166] and a post-war manuscript of a theory of real functions has not even been found.

The bitter reference to the 'consolidation of the nazification' expresses Brouwer's views on the take-over of mathematics by the new bosses. The clash between Brouwer and the other Dutch mathematicians was the result of a mixture of causes. For one thing, he had dominated Dutch mathematics for a (too) long period; ever since his meteoric rise in international mathematics, none of the Dutch mathematicians matched him in status. The only person who made a comparable rise was Van der Waerden, but he hardly played a role in Dutch mathematics since he left Holland. The older mathematicians, like Barrau, Van der Corput, Weitzenböck, Schouten, Van der Woude and Kloosterman went their separate ways and left Brouwer alone. The younger ones, like Freudenthal, Van Dantzig, Loonstra, De Groot and Beth, played only a minor role before the war, and for them Brouwer was one of those great men of the past—a man to admire, but not to obey, as their predecessors had done. So when the war was over and mathematicians all over the world were looking forward to a period of new life and new ideas, it was natural that the Dutch mathematicians wanted to strike out on their own. That they should find an un-cooperative Brouwer on their path is not surprising. He had long aspired to make Amsterdam a mathematical centre comparable to Göttingen, and when efforts were made after the war to create a mathematical research centre in Holland, Brouwer automatically assumed that it should be situated in Amsterdam, if not actually under his supervision, then at least in harmony with his ideas and goals.

It seems highly unlikely that the post-war Czar of Dutch mathematics, Van der Corput, would have allowed Brouwer much influence anyway, but the political purge was in this case a convenient gift—Brouwer had not been admitted to the inner circle of planners which was going to determine the future Dutch mathematics. Looking back, one gets the impression that Brouwer's political *faux pas* was no serious reason for pulling the rug out from under him. Of course, there were some who considered him guilty of high treason, but that was a small minority, a minority that dwindled after the first excitement was over. The major factor that contributed to Brouwer's downfall was an old-fashioned generation conflict between a new generation and a dominating father. Similar features were certainly also present in the case of the world-famous conductor of the Concertgebouw Orchestra, Willem Mengelberg. Although there were more substantial accusations against Mengelberg, the desire of the younger generation to let fresh air into the repertoire of musical life in Holland cannot be neglected in the history of Mengelberg's downfall.

Brouwer did not view the developments with composure. He saw in the démarches of Van der Corput, Schouten, etc. the continuation of a practice that he

[166] Posthumously published as Brouwer (1981).

had witnessed first in Germany and then in Holland. For him it simply boiled down to a repetition of earlier take-overs, new actors—same play.

The post-war events in Amsterdam had Brouwer firmly convinced that he was surrounded by enemies, and he felt betrayed by colleagues who, generally speaking, were actually kindly disposed to him. His sharp mind analysed all that had happened and came to the conclusion that there was a large scale conspiracy to get him out of the way. The take-over of the faculty and the 'theft' of his journal were two points that rankled so much that he more and more avoided contact with his colleagues. He had gradually cultivated his grievance and anger to such an extent that when his seventieth birthday was coming up, he refused and forbade any attention. In 1951 Holland was still an orderly, neat country, which paid attention to its great scholars, and as a rule important dates of respected scientists were observed in the national newspapers. For example, at the occasion of the fortieth anniversary of Brouwer's doctorate in 1947, the *Algemeen Handelsblad* published a short note of Mannoury. This time Brouwer took no risk, he sent a letter to the national press agency ANP, with the request to inform all newspapers that he did not wish that any attention should be paid to his seventieth birthday.[167] He expressly notified friends, colleagues and editors of journals that any attention would strike him as painful. When he discovered 'to his embarrassment' that the proceedings of the Academy had published a number of papers dedicated to him on his 70th birthday, he poured out his heart.[168] He did not wish mathematical circles to pay any attention to his seventieth birthday as:

> at all events, which have made me scientifically homeless these last years, and have destroyed my health, working power and joy of life, such as
>
> – my persecution by the Military Authority (*Militair Gezag*),
> – the deloyally realised founding of Mathematical Centre,
> – the breach of faith of the authorities towards me, which was interwoven with the above,
> – the thus displayed violation of fair play in the arena of Dutch mathematics,
> – the destruction of the mathematical education at the University of Amsterdam which resulted from this,
> – and finally, the theft of my journal effected with the help of deceit, which had almost turned into robbery with murder,
>
> the professional community has, insofar as it did not actively participate, watched doing nothing, and has refrained from any help against the aggressors, whenever and wherever there was an occasion to do so.
>
> If there is thus no occasion for festivity at my 70th birthday, but only for the reflection on a macabre position, this holds in particular with respect to the professional community.

[167] Brouwer to ANP (General Dutch Press Office), 9.II.1951.

[168] Brouwer to Heyting, 23.II.1951.

Mannoury, who had submitted a paper to the home journal of the Dutch Mathematical Society, *Het Nieuw Archief voor Wiskunde*,—dedicated to Brouwer on his 70th birthday—received a more melancholy letter, begging him to withdraw the paper from this journal, published by the 'deceitful' Noordhoff, and to submit it elsewhere. He closed the letter on a faint note of apology and self-reproach, 'Forgive me my necessity of observing that nowadays I am in a somewhat sorry condition. It is done in a slightly ashamed resignation and in the light of the silent fixed star, in which your person is too.'[169]

The request was easily made, but Mannoury felt that the cancelling of this particular dedication, with his own hands, was more than he could reconcile with his conscience and sense of justice, 'For this dedication has a wider intent than the expression of a personal feeling towards you, it is a recognition of your part in the insights that I have acquired, *but which I do not at all consider my property.*'[170] This plea from his former teacher and long-time friend could not fail to impress Brouwer, but he remained nonetheless adamant. A dedication that would even faintly be connected with Noordhoff or his seventieth birthday ('which has been for me in the res publica scientiarum a day of embittered mourning') was too painful to even consider. He even went so far as to offer to bear the cost of having Mannoury's paper printed elsewhere!

In 1950 the appointment of Van der Waerden, long thwarted by Brouwer, was finally about to be realised. There are indications that one of the reasons for the failing of the attempts to get him a chair (both in Utrecht and Amsterdam) was to be found in the royal palace. Queen Wilhelmina was determined not to accept proposals for appointing individuals who had voluntarily served the enemy. But after her abdication in 1948, her daughter, Queen Juliana, took a more moderate view. Van der Corput had this time taken all precautions, he had asked the Minister of Education in December 1949 if he would object if Van der Waerden would be proposed for a chair in applied mathematics. When the Minister answered that he would not raise any objections,[171] the faculty, now chaired by Professor Cornelia van Arkel, sprang into action. It presented a plan for the number of chairs in mathematics in Amsterdam; it listed four full-time professors, one lecturer and one extra-ordinary professor. The faculty defended the plan as a modest extension of the staff compared to 1937. Van der Waerden was proposed for a new chair in applied mathematics; as usual the obligatory couple of other candidates were added, but the faculty made it abundantly clear that it would not settle for less than Van der Waerden. It pointed out that such an exceptionally gifted man should not be lost for the Netherlands, let alone Amsterdam.

---

[169] Brouwer to Mannoury, 25.III.1951. Mannoury's paper with dedication appeared in Synthese (Mannoury 1951).

[170] Mannoury to Brouwer, 28.III.1951.

[171] Secretary General to Van der Corput, 31.I.1950.

Van der Corput informed Brouwer in Paris about the faculty's plans and told him that Van der Waerden would be put number 1 on the list for the chair that was going to replace Loonstra's lectureship.[172]

Brouwer had already resigned himself to the inevitable, he confined himself to sending a letter to the Curators of the University drawing their attention to the fact that the tasks of Van der Waerden and Bruins overlapped.[173] He suggested that Van der Waerden could equally well be made a professor in 'Mathematics' or 'Pure and Applied Mathematics'. The letter contained moreover a cautious dose of praise of Dr. Bruins, 'As I have already explained in a letter in 1948, it is difficult to attach a clear meaning to the term 'applied mathematics', but insofar as this might succeed, the gift and the fondness for applied mathematics has not since Korteweg been found so evidently in a Dutch mathematician as in Dr. Bruins.' Before a month had gone by he had changed his mind, he sent after all a letter to the Faculty pointing out that the procedure for the appointment of Van der Waerden had been extremely sloppy, which it indeed had been. And there the matter ended.

Van der Waerden was appointed in the Academic year of 1950.[174] Van der Waerden's activity in Amsterdam did not last long, already in September 1950 the University of Zürich had made him an offer that was hard to resist. And so after one year he left the University that finally had succeeded in creating the chair that both parties wanted. The Rector of the University, in his annual address in 1951, remarked that Van der Waerden had presented the mathematical novelty of coinciding inaugural and farewell addresses.

No sooner had Van der Waerden's departure been announced, than Brouwer tried to regain the initiative in the shaping of the mathematics department. He suggested that this was the perfect occasion to reduce the mathematical staff to a size fitting the 'reduction of the number of positions at the universities, which took place after the war, and the post-war decline of receptive and investigative intelligence potential of the mathematics students, and the urgency of the discontinuation of any consumptive liberality by the authorities'.[175]

He added that such a 'deflation of the mathematical staff', if combined with (1) a strict specification of the professorial tasks, (2) giving up the practice of leaving the teaching of courses to assistants, (3) an abandoning by the authorities of the 'extra university institution which diminishes and tarnishes the mathematical milieu at the university, and at present drains and sequesters the pre-war Amsterdam Mathematical School, which for a time had some importance in the world and drew some attention' could possibly help the Amsterdam School, 'which is down at the moment and only barely breathes, to get up and show itself again'.

---

[172] Van der Corput to Brouwer, 20.I.1950.

[173] Brouwer to Curators, 5.II.1950.

[174] I count myself lucky to have heard his lectures in that year, Van der Waerden was a wonderful teacher.

[175] Brouwer to the Chairman of the Faculty, 7.III.1951. The letter consisted of two sentences of 12 and 15 lines!

The chairman, Professor van Arkel, probably considered this as a letter to her personally, and so no faculty reactions appeared to have been formulated. But a week later Brouwer felt obliged to act again, this time he had found on the agenda of the Faculty Meeting of May 14 an item called 'vacancy mathematics'. It required no superior intelligence to see that this was the vacancy resulting from Brouwer's imminent retirement. In Brouwer's eyes this was the last occasion to undo all the harm that was inflicted upon mathematics in Amsterdam, as he viewed it. He immediately addressed the Faculty,[176] pointing out that it could have to count with the possibility that

1. The Trustees might have understood before September 17, 1951[177] that the manner in which the position of mathematics at the University of Amsterdam was gradually changed after the war, increasingly hampered the discharge of his duties,[178] and had made the discharge of these duties in accordance with the intention, at the time of their establishment, completely impossible.
2. The Trustees might therefore consider to make it possible that I will retire on 17 September, without appointing a successor for me, or that I may be asked to occupy my position a bit longer, with the temporary restoration of the leadership of the mathematical activity at the University of Amsterdam, which was taken away from me after the war, and with the mandate to prepare and initiate the required measures for the re-integration of the pre-war efficiency of the activities.

When no reaction was forthcoming, he again addressed the faculty[179] repeating the arguments, but this time he specifically pointed out that by just adding 'analysis' to the teaching duties of Dr. Bruins—and by promoting him to full professor—the mathematics curriculum would be taken care of.

'My own position', he wrote, 'had been eroded by the founding and operating of the Mathematical Centre since 1946, and it has lost its raison d'être'.

Referring to his letter of October 8, 1946, he remarked that the page of history mentioned there was almost written, but

> a conciliating last paragraph could still be added, if the City Council recognised at the eleventh hour the hollowness of the argumentation, which had persuaded it to subsidise the Mathematical Centre instead of the University of Amsterdam, and would return to me during a brief extension of my employment, the supervision of the mathematical activities in Amsterdam, with the commission to strip it of all expensive glamour and of all unjustified privileges, and to concentrate again on the discipline that in the past gave the mathematics in Amsterdam some importance in the world, which is still in Amsterdam better represented than elsewhere in the world, and which increasingly attracts the attention of the mathematical world.

---

[176] Brouwer to faculty, 14.III.1951.

[177] Brouwer's official day of retirement.

[178] As a professor and as a director.

[179] Brouwer to Faculty, 1.V.1951.

The letter contained a number of proposals for the reorganisation of the Mathematics Department, including one that went further than anything that Brouwer had ever proposed: to uniformise the advanced part of the curriculum[180] and to make for all candidates an examination 'of some depth' in intuitionism and symbolic logic mandatory.

This was the first time that Brouwer had asked for a privileged position for his intuitionistic school. The reason is not easily guessed; did he fear that intuitionism was about to disappear, or had he observed (in Leuven, Cambridge and Paris) signs of a new blossoming of intuitionism and did he want Amsterdam to be in a leading position by the time the world would ask for more constructive mathematics?

The above letter finally drew fire; Schouten wrote an agitated letter to Van der Corput and Van Dantzig, using terms such as 'we cannot leave it at that', 'give me weapons', 'these scandalous, nonsensical scribblings', 'give me the ammunition, the shooting will be taken care of', ...[181]

A flurry of letters and memoranda followed. Schouten set out to refute Brouwer's points one by one in a first note (undated),—a draft was prepared by the Faculty Secretary (5.V.1951), Brouwer replied (29.V.1951), Schouten sent a letter to the Faculty (31.V.1951), accompanied by a letter to the Chairman, and it is more than likely that this was only part of the correspondence.

Apparently, Schouten c.s. viewed Brouwer's proposals with great apprehension; the arguments were not devastating in themselves, but Brouwer's view that the mathematical institute could be run on a much smaller budget could present authorities with more temptation than they could reasonably be expected to resist.

Schouten's note on the whole was rather the document of a politician than of a scientist, he mostly misrepresented Brouwer's argument and carried out a shadow fight—which he then won on points. For example, where Brouwer complained that the raison d'être of his position has disappeared, eroded as it was by the Mathematical Centre, he simply stated that 'It is correct that Professor Brouwer has not given many courses the last few years. The reason is partly the fact that Professor Heyting at the time unofficially has taken over some classes from him, in order to enable him to devote the last years of his professorship to the research and teaching of his special field. By causes beyond the control of the faculty, Professor Brouwer has, however, made only little use of the offered opportunity.' The message is clear enough: Brouwer is himself to blame for his ineffectiveness. While there is a great deal of truth in this, it was not what Brouwer meant by 'erosion of task' and 'loss of raison d'être!'

The draft of the faculty was much more balanced and tempered then Schouten's memorandum. It argued that the present teaching load was excessive—9 hours per week against 8 hours per week in Delft.

---

180 The first part was already strictly uniform, the second part (after the candidaats exam) was rather flexible.

181 Schouten to Van der Corput/Van Dantzig, 7.V.1951.

After a consideration of various factors, the faculty settled on the following main subjects to be taught and studied in the Mathematical Institute:

1. Number Theory (Van der Corput)
2. Philosophy of Mathematics (Heyting)
3. Set theory (Brouwer)
4. Applied mathematics (Van der Waerden)[182]
5. One or more parts of Analysis, Algebra or Geometry.

The Faculty proposed to split Brouwer's chair (set theory, function theory and axiomatics) into two chairs and to redistribute some topics, so that the following chairs resulted:

I. Analysis, algebra and applied mathematics.
II. Geometry, analysis and set theory.

Running ahead of our story, we can already tell that the two new chairs were approved and that N.G. de Bruijn (I) and J. de Groot (II) were eventually appointed.

In a separate letter Schouten (probably as a Trustee of the Mathematical Centre) informed the Faculty of his (or the M.C.'s) views. He persuasively argued that close ties between the University and the Mathematical Centre were to the advantage of both parties, but he firmly denied all claims of Brouwer with respect to possible negative effects of the Mathematical Centre on the Mathematical Institute. He plainly stated that as Brouwer had difficulties in separating personal and objective aspects, he would be an unacceptable risk in a young organisation, where tact was an indispensable commodity. The founders would gladly have given a position in 'accordance with his scientific past', but could not risk doing so.

The accompanying letter to the Chairman was less guarded, Schouten asked her to enclose his letter in the file to be sent to the Trustees, not, as he said, to reinforce the arguments of the Faculty, but also because Brouwer had insisted that the faculty should enclose his letters. He bluntly told the chairman that Brouwer's letters bristled with incorrect statements and misconceptions, but he went on to say that he felt himself in a difficult situation; '... I have always maintained that one should not demand from a man like Mr. Brouwer, whose critical sharp mind once managed to revolutionise the foundations of mathematics, that he should on top of that also be moderate, tactful and easy going.' He advocated a liberal attitude towards Brouwer, because he had been in difficulties in 1945, and suffered setbacks and severe disappointments. The state of misery and bitterness of Brouwer, although largely self-inflicted, he added, was deplored by all mathematicians, but when the interests of mathematical education was at stake, Brouwer should be addressed in the same manner as all other staff members.

[182]The draft mentioned that although Bruins taught applied mathematics, the enormous interest in the courses of Van der Waerden had shown that advanced courses were also in demand.

In the matter of the destiny of Brouwer's chair, not even Heyting could influence Brouwer. Heyting had not hoped, but at least tried, to change Brouwer's views;[183] Brouwer returned the letter unopened, he kindly but determinedly explained to Heyting[184] that 'As I have to make a most economical use of the little time and strength that is probably left to me for my struggle to liberate the Amsterdam Mathematical School from the occupational powers of the Mathematical Centre, I am faced with the necessity to carry on this struggle only in discussion with the authorities involved, and thus with exclusion of any private discussion.' He suggested that Heyting, if he really wanted to ventilate his views, should send them to those same authorities.

Although time was running out fast—only a few months separated Brouwer from the date of retirement—he launched a new campaign. He tried at the last minute to construe a case for a chair for Bruins. He asked Bruins to prepare a list of publications and he collected the opinion of a number of scientists on Bruins' work, e.g. Rutherford (St. Andrews), Turnbull (London), de Broglie (Paris), Destouches (Paris) and Chatelet (Paris).

The mathematicians in Amsterdam were not particularly impressed by Bruins' mathematics, no support for a promotion to full professor could be found. Indeed they had either asked or found reviews and comments on the work of Bruins, which did little to shore up his status. Brouwer sent these comments to Bruins to give him the opportunity to exercise his *droit de réponse.*[185] Apparently all this activity never led to a discussion or proposal in the faculty.

In July 1951 Brouwer acted for the last time as '*promotor*' (i.e. Ph.D. adviser) of Miss J.A. Geldof, the same lady that caused a clash between Brouwer and Freudenthal in 1936. Her thesis, 'On the arithmetisation of axiomatic geometries', was a fairly traditional affair, with little originality. Freudenthal had despaired of her ever getting a doctor's degree and the last assistant of Brouwer, the topologist Daan Kan, spent a great deal of time in straightening out the details.[186] Mannoury, who took part in the opposition (i.e. the traditional public examining of the candidate), said that Brouwer awarded the degree with such an elaborate speech, that it almost had the character of a farewell address.

Kan was Brouwer's assistant after the war and he pursued his career further in Jerusalem, and later at MIT in Cambridge, Massachusetts. He recalled that Brouwer warmly recommended him to Fraenkel (who at that time had very mixed feelings about Brouwer, to put it mildly), and it struck him as amusing that those two old gentlemen indulged in an almost ritual criticism of each other.

Shortly before the end of the academic year 1951, Brouwer participated for the last time in a conference as an active professor. It was the St. Andrews Mathematical Colloquium of the Edingburgh Mathematical Society. Brouwer was one of the

[183] Heyting to Brouwer, 10.II.1951.

[184] Brouwer to Heyting, 14.III.1951.

[185] Brouwer to Bruins, 13.VII.1951.

[186] Oral communication D. Kan.

invited speakers; at the invitation of D.E. Rutherford, the Secretary of the Society, he presented a talk on July 25 from 5.30 to 6.30.

At that time, travelling was still something of an achievement, one had to get a visa and to obtain permission to take money out of the country. The university had no funds to provide its staff members with the necessary cash, and so a prospective traveller had to scrounge money from scarce sources. Brouwer carried on a lengthy correspondence with the representative of the British Council, Mr. R.P. Hinks, on the topic of a possible financial support. The efforts of Hinks on Brouwer's behalf were finally crowned with success: both the St. Andrews Conference and the stay at Cambridge would be supported.[187] The choice of mode of travel was left to Brouwer; the boat services between Holland and Scotland not having been reopened yet, he travelled by plane from Schiphol to Prestwick and then by train from London to St. Andrews. In view of his age and state of health he asked permission to travel first class, a request that was granted.

The colloquium was, in Brouwer's words, 'on a high level, both successful and delightful'. The colloquium brought him together with the Canadian geometer Coxeter. The basis for a later invitation to the Canadian Mathematical Conference in 1953 was probably laid in St. Andrews.

Brouwer ended his academic career at the University of Amsterdam, which he had served with an unwavering loyalty for forty-two years, without any official events, no farewell address, no symposium. The Department that he had loved and raised to a status comparable to that of physics in Leiden or Amsterdam let him go without so much as a 'thank you'.

Even the harmless symbolic gesture of asking him to continue some courses until a successor had been appointed was not made. Heyting, who had succeeded Brouwer as the Director of the Mathematical Institute, wrote a polite letter that barely veiled the beginnings of impatience, which expressed his opinion that Brouwer himself was the cause of the 'misunderstanding'. On Heyting's view, Brouwer had announced that he wished to be put in charge of a reorganisation of the mathematics curriculum in accordance with his views, and he professed not to be aware of Brouwer's willingness to accept the humble role of a substitute. 'After all', he wrote, 'it would not have been possible to make an arrangement so that you could take care of the courses, as you are going in October to England. Moreover, the fundamental conflicts of opinion, which have surfaced during the last months, would doubtlessly have led to problems.'[188] It is one of those quirks of history that Heyting, who had to thank his career to Brouwer, so soon after Brouwer's retirement showed his former benefactor politely but resolutely the door. It could not have been a pleasant task for Heyting to disappoint (and actually offend) the man to whom he owed his place in the world of scholars, and who had time and again recommended him. In spite of his indebtedness to Brouwer, Heyting was enough of a realist to have reached the conclusion that the future of the faculty was in better hands with

[187] Hinks to Brouwer, 4.VII.1951.

[188] Heyting to Brouwer, 25.IX.1951.

Van der Corput c.s. than with Brouwer. Brouwer had no difficulty in recognising the mood in the Mathematics Department, where nobody (except Bruins) was any longer on his side. He replied coolly that he could not discuss the matter of the interim arrangements for his courses with Heyting; this only concerned the Chairman of the Faculty and the emeritus professor.[189]

[189]Brouwer to Heyting, 27.X.1951.

# Chapter 18
# The Restless Emeritus

## 18.1 The Traveller

Having parted with his creation, the Mathematical Institute, Brouwer had no intention of retiring to the life of a tranquil pensioner. He almost immediately departed for his beloved Cambridge to discuss with Steen the publication of his lecture notes. While in England, he submitted a paper to the Royal Society, of which he was a foreign member since 1948; it was in the vein of his Berlin lectures, *An intuitionist correction of the fixed point theorem on the sphere*.[1] Brouwer had observed long ago that the fixed point theorem in its classical generality failed (already for continuous functions mapping the unit interval into itself). In this paper he gave a counterexample to the existence of a fixed point for continuous maps of the sphere that preserved the indicatrix (orientation), and showed that although there is not necessarily a point that remains in its place under such a mapping, there are points that move less than any given distance. This is a basic phenomenon in the constructivisation of classical theorems, precise results are replaced by results 'up to an $\varepsilon$'.

The British Council had generously agreed to pay Brouwer's stay in Cambridge and the return fare; this being negotiated and approved in July, Brouwer planned to visit Cambridge in 1951 from October 10 to November 10. However, a serious attack of bronchitis forced him to postpone his arrival until November 29. The exchange of letters with the British Council official Hinks is of interest as Brouwer explicitly states his grounds for this (and earlier) visit to Cambridge:

> May I take the liberty to recapitulate the object and aim of my planned visit to Cambridge? I am writing an English book on intuitionism, a branch of science initiated by me, but, as I think, rather universally recognised today. The book is to be published by the Cambridge University Press.[2] Up to now a textbook on intuitionism does not exist in the English language. There is not even an English vocabulary. Having little training in English expressions, and,

[1] Cf. Sect. 4.4 and p. 503.

[2] Cf. Brouwer (1981).

D. van Dalen, *L.E.J. Brouwer – Topologist, Intuitionist, Philosopher*,
DOI 10.1007/978-1-4471-4616-2_18, © Springer-Verlag London 2013 

> moreover, having to coin a lot of English terms for notions introduced in my theories, with the help of my English friends I found the following way to get the book ready: I was given the opportunity to deliver an introductory course in intuitionism in the University of Cambridge. During this course there were regular consultative meetings with some of my hearers, in which intuitionistic notions newly introduced in the last lectures got their definitive nomenclature, whilst the paragraphs of the book relative to the last lectures were discussed and rewritten to get them as understandable as possible for English readers.[3]

In 1947 and 1948 Brouwer had spent three terms in Cambridge, and five of the six planned chapters were finished in this way. But in 1949 the university ran out of money, so that the project came to a halt. The planned visit in November was in fact part of the publication project. Apparently the effect of this visit was not satisfactory—the book was never finished. After Brouwer's death the 'final' version of the first five chapters was found among the material that was collected by the Dutch Mathematical Society, but chapter six was, as far as we know, never written. The book, eventually published by the Cambridge University Press,[4] indeed shows traces of intense terminological activity; Brouwer introduced a large number of colourful terms for concepts of intuitionistic set theory, but few of those names have survived. One must admit that Brouwer's terminology does justice to the geometrical motivation, but the introduction of a large number of novel names is a rather hazardous didactic policy. Brouwer's paper 'Points and Spaces' uses the same terminology, the foundational literature however, eventually adopted a more conventional vocabulary, mainly under influence of Brouwer's German papers and of Kleene's papers.

Although Brouwer was downright bitter about the treacherous treatment he had suffered at the hands of his mathematical colleagues, one should not get the impression that he sat moping at home. He had made an imaginary partition between his mathematical world and the rest. He continued life much as in the old days. People would drop in, his female admirers would sit at the feet of the great man under the spell of his eloquence. He would gladly set forth on any topic and display an almost unparalleled knowledge of literature, art, politics, medicine, agriculture, history, .... The philosopher Frits Staal,[5] who was in Amsterdam at that time, recalled an instance of Brouwer's extraordinary conversational talent. A visiting mathematician from India had said to him that it was his greatest wish to meet Brouwer. Staal made a telephone call, and Brouwer immediately agreed to receive the visitor. The next day Staal took his colleague to Blaricum. After the usual introductory phrases, the conversation got on its way, and Brouwer soon saw that a mathematical discussion would not be very rewarding. He immediately changed the subject, and embarked on a lengthy and precise discussion of Indian politics; he questioned and lectured the visitor on the Hyderabad affair, demonstrating a complete familiarity

[3] Brouwer to Hinks, 15.III.1951.

[4] Brouwer (1981).

[5] Oral communication, see also Noorda and Staal (1999), p. 31.

with all the facts, persons, and backgrounds. The Indian was dumbfounded. This was a demonstration of Brouwer's phenomenal memory and command of facts; a lesser man might be tempted to show off by boning up on the facts the night before, not so Brouwer, he simply had a natural talent for storing and retrieving information.

In the village Brouwer was well-known for his volubility. Maarten Mauve, the son of Brouwer's old friend, recalled that from time to time, Brouwer would, without a warning, drop in, flatten his foldable hat, and embark on endless stories. Any topic would do.

Scientists from all over the world visited Brouwer in Blaricum and enjoyed his erudite conversation. Mrs. Beth gave an amusing account of one such visit by Tarski. Tarski was a close friend of Beth and whenever he came to Holland the two met. At one of those occasions Tarski expressed the wish to see Brouwer. As Beth was on a friendly footing with Brouwer, an appointment was readily made, and Mrs. Beth drove Tarski and Beth to Blaricum. During the ride Tarski was briefed on the proper conduct during the ceremonies and proceedings which were part of Brouwer's way of life—much as a visiting dignitary was instructed before entering the court of an illustrious monarch. Mrs. Beth judged this a wise precaution in view of the characters of the two men, who each had a strong opinion of their place in science. The party was a great success, Brouwer displayed a magnificent hospitality, and the conversation was spirited. Now it was the case that Brouwer used to invite female company to add lustre to his parties, and the importance of the visitor could be measured by the number of ladies present. Tarski had been informed of this significant detail, and so when the time had come to go home, Tarski and the Beths took leave of Brouwer and the company, and they got into the car. No sooner was the door closed or Tarski asked: 'How many ladies did Brouwer invite for Carnap?'

Mrs. Beth also told that once, when she visited Brouwer, she was much impressed by Brouwer's white tropical suit with gilded buttons. She complimented him how well he looked in it, and she could see that he was very pleased with her remark. Next visit, Mrs. Beth told, he wore the same suit, evidently expecting the same praise, but this time 'I did not pay him a compliment'. In general, Brouwer loved to wear white suits, a reason for the locals to call him jokingly the ice-cream man.

Brouwer's resentment against the post-war bosses of Dutch mathematics, who had 'stolen his journal' and snatched his 'Göttingen at the Amstel', did not stop him from attending the monthly meetings of the mathematical society. He is still remembered by the junior members of that period as the sharp-witted old gentleman, who did not feel too high and mighty to be interested in their ideas and work.

At the mathematics department two important vacancies were to be filled in 1951. They resulted from Brouwer's retirement and van der Waerden's appointment in Zürich. Brouwer had, understandably, showed a keen interest in the matter. As we have seen, Brouwer had suggested that his chair should not be filled for the time being (cf. p. 776). The faculty had wisely ignored Brouwer's desideratum, and submitted a request for the appointment of two new professors, both for Brouwer's chair, N.G. de Bruijn and J. de Groot. The two held chairs in Delft. De Groot was a topologist of the set-theoretic kind, he practised a mixture of Dutch and Polish topology. De Bruijn was a young, versatile mathematician, at that time mostly known for his number theory and algebra; he had started his career as an assistant of Van Dantzig in Delft. When the proposals were submitted to the city council, which had the final say in these matters, a heated discussion followed. Part of the council supported the proposal of the University, but a number of councillors was of the opinion that one chair was enough and that Bruins was the proper person to fill it. It is probably not far-fetched to see the hand of Brouwer in this move. On the other hand, a persistent rumour went around that Bruins had in person been visiting the individual councillors to present his case, and that his efforts had been rather counter-productive. Whatever may have been the case, in the meeting of 10 October 1951 the university proposal got a majority of the votes; De Bruijn later told that it was the support of the communists that made the difference.

Although Brouwer would have preferred another outcome, he did not begrudge the two newcomers their success. When both were about to present their inaugural lecture, he wrote to them that, 'At the time I have hesitated to congratulate you with your appointment in Amsterdam, and in the end I have refrained from doing so, even though, from my side, I could not imagine a more felicitous extension of our academic mathematics staff, than the joining of you both.' The reason being that 'I could not convince myself that there is a sufficiently favourable spiritual climate in the mathematics section of the University of Amsterdam, to warrant a congratulation at your reception by it.'[6] He ended by asking a favour: not to mention his name in their inaugural addresses, for, 'if there were to remain any chance at all to bring my life's work to a satisfactory end, I will not only have to forget my connections in the past with the mathematics section at the University of Amsterdam, but also to make, as far as in my power, others forget it'.

In the last years of his career Brouwer resumed his old travelling habit; he had always been an enthusiastic traveller, be it for pleasure, health or professionally. In his final professional years he had a number of gratifying invitations for mathematical events. So did he revisit his beloved Göttingen in 1951 when he represented the Royal Academy at the bicentenary of the Academy of Sciences in Göttingen.

When Brouwer was making preparations for his visit to the Colloquium in St. Andrews,[7] the next invitation reached him.[8] The provost of University College, London, R. Pye, invited him to give the Shearman Memorial Lecture, in memory of Dr. A.T. Shearman, who had bequeathed a sum to the College for a course of lectures on Symbolic Logic and Methodology. The status of the lectures may be judged by Brouwer's eminent predecessors, to wit: Russell[9] (on 'Scientific Inference'), Schrödinger (on 'The origin and Nature of Scientific Thought'), and Tarski (on 'Fundamental Ideas and Problems in Meta-Mathematics').

Brouwer was delighted to accept the invitation, provided he could deliver the lectures in the Spring. In the course of the ensuing correspondence, Brouwer, when

---

[6]Brouwer to De Bruijn and De Groot, 22.VI.1952.

[7]See p. 780.

[8]Dr. R. Pye to Brouwer, 16.V.1951.

[9]Weyl (1946), p. 275.

asked whom of his friends he would like to be invited for the traditional tea party of the Provost in honour of the guest, provided an interesting list: K. Popper, H. Dingle, Mr. and Mrs. Haynal Conyi, Miss Winifred Gordon Fraser, Whitehead, Kneale, Waisman, Steen, Braithwaite, Routledge, Newman, Polanyi and Turing. He added, 'If some of them would have the opportunity to attend my first lecture, I should be happy to see them at the tea party.' It is not known whom of the above actually turned up.

The series of lectures was scheduled for May 12, 16, 19 and 23 of 1952, and the title was 'Outline of Intuitionism'.

During this visit to London[10] Brouwer was briefly confronted with the past. During and after World War I Brouwer had become acquainted with esoteric groups, leftovers from the pre-war designs of intellectual circles. Gutkind was one of those high minded mystical philosophers that had set out to bring a happier world about. After the war Brouwer met Gutkind and his wife, and he was immediately taken with them.[11] After Gutkind found a safe haven in the USA, Brouwer lost contact with him. And that, basically, was the case with all members of the old circles. A particular flamboyant representative of the old generation was Dimitri Mitrinović. He had been a student in Herzegovina in the old Austro-Hungarian empire; like most of his comrades, he was opposed to the Danube monarchy. After an intense period of political-literary activity, he decided to start a new life of sober cultural reflection. In 1914 he moved to Munich, and after a long odyssey he landed in London. In the meantime he had made friends with the members of the 'blood-brotherhood', the men of a royal spirit, including Gutkind, Van Eeden and Brouwer.[12]

In England Mitronović made his name as a writer for the influential magazine '*New Age*'. Until his death he was the centre of a varying group of thinkers and seekers, idealistic mystics, and political idealists. After World War II the circle had shrunk to a size that did not permit large scale projects and actions. The group met in the 'Renaissance Club', also called the 'Anti-Barbarus Renaissance Club of the New Atlantic'. It became a meeting point for exalted spirits. Martin Buber, one of the great thinkers of the age, dropped in, and Brouwer did so too. He was probably informed about the existence of the New Atlantic group by Winifred Gordon Fraser. She belonged to the group, and she must have recognised a kindred spirit in Brouwer. Doctor Ralph Twentyman described Brouwer's visit, 'I remember vividly one evening when a number of us were at Charlotte Street and Brouwer came in, tall and with the most brilliant eyes I have ever seen. They seemed to light up the whole room. George Adams was present, and he and Brouwer soon established a warm relationship, and became engrossed in higher mathematical discussions together. George Adams was a fine mathematician and a pupil of Rudolph Steiner. He was engaged in trying to work out indications of Steiner about 'Ethereal' space

---

[10]The time of the following events is a matter of plausibility, since there are no documents providing the exact information.

[11]Cf. p. 243.

[12]Mitrinovic's life is described in Rigby (1984). See also p. 243 ff.

and projective geometry, and their relevance particularly to plant morphology. He worked with the polarity of point and plane, seeing the planar space as helpful in approaching Steiner's concept of Etheric.'[13] Twentyman never forgot Brouwer's glorious entrance, 'the high spiritedness Brouwer brought into any room is one of my most treasured memories. Everything moved up a floor or two when he appeared.' We may safely assume that Brouwer and Mitrinović met at the time, although there is no direct information on that point. Twentyman noted that, 'Certainly Mitrinović had a very high regard for him, both as a man and for his orientation in mathematical higher philosophy.'

That same year Brouwer wandered further afield. He finally fulfilled an old promise, to visit his former Ph.D. student Barend de Loor in South Africa.[14] There are no letters in Brouwer's archive that show all that much contact between the two, be it that in the summer of 1945 Brouwer seriously discussed with De Loor the possibility of (a temporary) emigration to South Africa. Apparently De Loor had managed to arrange a lecture tour for Brouwer in 1952. In July Brouwer addressed the South African Association for the Advancement of Science in Cape Town. The lecture was published as 'Historical Background, Principles and Methods of Intuitionism',[15] and a parallel *Zuid-Afrikaanse* version was also published. The paper is interesting because it signals a retreat from earlier positions on choice sequences; Brouwer asserted that higher-order restrictions on choice sequences were not 'justified by close introspection'. This mental reserve is stronger than an earlier one in his Cambridge lectures: 'But at present the author is inclined to think this admission superfluous and perhaps leading to needless complications.'[16]

In Cape Town he had a minor accident with the result that one of his fingers was incapacitated for some time. Brouwer gratefully thanked the rector of the University of the Witwatersrand for paying the doctor's bills. He also used the occasion to see a bit more of the country, staying with South African friends and colleagues. Back home, he kept up a correspondence with them. A substantial part of the letters is concerned with the correction of the proofs of his paper for the proceedings; Brouwer was assisted by the South African mathematician Welters.

The next year brought two more major trips. The first one was to Finland, where he was Von Wright's guest in Helsinki. He gave two lectures for the Philosophical Society of Finland, *Turning points in the relations between logic and mathematics* (20.V.1953) and *Order on the Continuum* (22.V.1953).[17]

The years had not lessened Brouwer's love of nature, nor his unconventionality. When he stayed with Von Wright, he asked his host one night, 'would you mind if I slept out on the ground tonight?' Von Wright may have had his doubts about his

---

[13] Twentyman to Van Dalen, 24.III.1992.

[14] See also p. 702.

[15] Brouwer (1952b).

[16] How right he was can be seen in e.g. van Atten and van Dalen (2002).

[17] *Wendungen in der Beziehungen zwischen Logik und Mathematik. Die Ordnung im Kontinuum.*

aged guest, but he saw no reason to refuse the request. And so Brouwer once more was one with nature under the sky, as he loved to be.[18]

It is hard to say that Brouwer had mellowed with age, he could fly off the handle like the next man, but generally speaking he was no longer in the line of fire. Old injustices could still provoke him into sharp and sometimes rambling harangues, but the fixation on the misfortunes of his life was slowly giving way to the recognition of the love and appreciation of his friends and students. In particular, Brouwer more and more cherished his love and admiration for his old teacher and friend Mannoury. As it happened, his visit to Helsinki coincided with Mannoury's birthday. As an attentive friend he wrote to Mannoury before leaving. The letter is a touching document, a mixture of friendship and self-pity.

> An early warm congratulation on your eighty sixth birthday, which I will celebrate in Finland in my thoughts. And a belated thank-you for your sign of life at the occasion of my birthday. I really hope that you have now completely recovered. I too have been in bed for months during this winter at a relapse of a heart attack, which I suffered in 1950 being suddenly informed of the theft of my journal and the unbelievable means of deceit and fraud used. It is, I believe, through this event which I on the one hand out of self-preservation must try to forget and to which I can, on the other hand, not resign myself, that I have definitely become misanthropic within our national borders. Remarkable, how little philosophy sometimes protects against psycho-somatic reflexes and reactions.
>
> Have a pleasant day on the seventeenth and do greet the members of your family.

Brouwer's post-war lecturing did not take him to Germany, the country that had shown him hospitality and admiration before the essentially scientific discussion, known as the '*Grundlagenstreit*', was ended so abruptly by a non-scientific move. Apart from a few official visits, he had after the war no dealings with German mathematics. This may have been as much his choice as a defensive pro-Hilbert atmosphere lingered in Germany for years after the *Grundlagenstreit*, and even long after Hilbert's death. His lectures, as well as his publications, were now mostly in English; the lectures in Finland were exceptions, and probably a courtesy towards a country where the mathematics was still practised in the German scientific tradition. Nonetheless, even the affronts of the *Mathematische Annalen* affair could not erase from Brouwer's memory the true and sincere friendships of happier days. The man who did most for Brouwer in the dark days of the *Annalen* conflict, Carathéodory, had died in 1950 and the two had not met since the tragic events of 1928; the loss of such an old friend, even after a long separation, deeply moved Brouwer. In his letter of condolence, he wrote:

> The tiding of the death of your father, which arrived only just now, has deeply shaken me. His friendship, and the awareness of his great importance

[18]Personal communication, Von Wright.

> as a thinker and as a humane character, were for me since a number of decades something really essential. By his departure from life, the world has become poorer for me. How I have looked for an occasion to see him again. It was not to be so and hence it only rests me to express my sincerest sympathy to you and the rest of the family and the other relatives at the terrible loss which you suffered, and I assure you that I will keep the memory of Constantin Carathéodory in the highest honour.[19]

Brouwer's last big paper 'Points and Spaces' resulted from his lectures at the *Canadian Mathematical Congress* at Kingston, Ontario in August 1953. Probably, the geometer Coxeter, who had met Brouwer at the St. Andrews conference in 1952, had arranged an invitation for Brouwer as a speaker at this congress. Brouwer immediately accepted the invitation, and proposed a series of talks on intuitionistic mathematics.

Once having accepted the invitation, Brouwer set out to obtain funds for the trip. This was not a simple operation, Coxeter had informed Brouwer in his letter of invitation that the Dutch Government had promised to pay the travelling expenses and that the Congress would take care of the housing plus a fee of $300,–. Brouwer was immensely pleased to get the opportunity to visit Canada; indeed, when he accepted the invitation, he added that he would enjoy a lecture tour after the conference. Coxeter made some inquiries and he soon came up with Queens University at Kingston and McGill in Montreal as possible hosts.

The Conference actually consisted of two parts, the conference proper—in conjunction with the Fifty-eighth Summer Meeting of the American Mathematical Society, and the Summer Seminar. The topic of the Summer Seminar was Topology and Geometry. Brouwer's talks were part of the four Lecture Series on Research Topics. The other speakers being Henri Cartan—*Homologie des Groupes discrèts*; Newman—*Theory of fixed points, particularly of periodic functions*, and Beniamino Segre—*Algebraic Geometry*.

Flattering as the invitation might be, it had its problems, mostly of a financial nature; travelling was expensive in those days and a retired professor was certainly not in a position to throw his money around. And so even during his stay in Helsinki Brouwer carried on his correspondence with the Ministry of Education and the Canadian organisers in order to arrange a proper funding. An unfortunate illness had stopped him from answering the mail from the ministry, which had been in his possession since January. The Ministry's letter thoroughly dampened Brouwer's enthusiasm for the trip to Canada. The Minister had politely inquired if Brouwer was planning to accept the Canadian invitation, about which it has been informed by the Foreign Office. If so, it was willing to contribute towards the cost of travelling by boat.[20] In those days flying was still more expensive than sailing, a flight to Canada at that time cost about 2,750 guilders. Brouwer expressed his disappointment at this announcement, he pointed out that the invitation was accompanied with

[19]Brouwer to Stephanos Carathéodory, 28.II.1950.

[20]Minister to Brouwer, 16.I.1953.

a promise that the Dutch Government would furnish the travel expenses. 'In particular', he wrote, 'taking into account the personalities that were invited together with me (Cartan from Paris, Newman from Manchester and Segre from Rome), I would be sorry to withdraw in the end, after I had made my promises'.[21] After a good many letters, the financial problems were solved. Even the excess luggage was the subject of tactful but tenacious negotiating. Brouwer felt he had to remind the Minister that as an emeritus he earned a paltry 5,050 guilders a year, and he added bitterly 'which in my case has the consequence that I support my family by labour, which has nothing to do with science and which thousands can do better than I can'.[22] Eventually Brouwer arrived at August 7 at Montreal Airport, from where he travelled on to Kingston. Brouwer not only lectured on intuitionism, but at the request of the organisers, also on 'High School Teaching in Holland'. The lectures on intuitionism, with the title 'Points and Spaces', delivered from 8 through 31 of August, were divided into six sections:

1. History of the gradual disengagement of mathematics from logic.
2. Spreads and fans.
3. Generation of Cartesian and topological spaces.
4. The virtual order of the continuum.
5. Precision analysis of the continuum.
6. Intuitionistic recasting of some classical geometric theorems.

At the request of the organiser, Williams, he had provided a list of reading material; this list by itself showed that intuitionism had lacked resonance in the mathematical community: apart from *Consciousness, Philosophy and Mathematics*, all the material was from the twenties:

Weyl—*Über die neue Grundlagenkrise der Mathematik* 1921;
Dresden—*Brouwer's contributions to the foundations of mathematics* 1924;
Wavre—*Y a-t-il une crise des mathématiques?* 1924, *Sur le principe du tiers exclu, Sur les propositions indémontrables* 1926/28;
Lévy—*Sur le principe du tiers exclu, Critique de la logique empirique, Logique classique, logique Brouwerienne et logique mixte* 1926/27;
Borel—*A propos de la discussion entre M. R. Wavre et M. P. Lévy* 1926/27;
Brouwer—*Intuitionistische Betrachtungen über den Formalismus* 1928.

The conference itself was a joyful experience for Brouwer, he met Coxeter again, talked with Mrs. Coxeter, who was of Dutch descent, and also ran into Paul Gilmore, a young Canadian logician who had studied with Beth in Amsterdam, and who had finished his dissertation on Griss' negationless mathematics in 1953.[23] Jim Lambek, who was also present, told that he had some talks with Brouwer.[24] At one occasion

[21] Brouwer to Minister, 27.V.1953.

[22] Brouwer to Minister, 27.VII.1953, and to Williams, 24.VII.1953.

[23] Gilmore (1953).

[24] Oral communication of J. Lambek.

Lambek happened to mention Wittgenstein, whereupon Brouwer politely inquired what Wittgenstein had done for logic.

– He has invented the truth tables.

– And, pray, what are the truth tables?

Lambek proceeded to explain what truth tables were, and how they worked. Only later did he realise that the question had been an ironic one. Lambek also reported how, during a conference picnic of the participants, Brouwer, who had joined the party, had suddenly disappeared. After some excited searching it turned out that the septuagenarian had not been able to resist the urge to climb one of the trees, and he was discovered somewhere above the heads of the assembled mathematicians.[25]

The mathematician Bob Wisner reported about his encounter with Brouwer when he visited the University of British Columbia (Vancouver).[26] Together with a fellow staff member, Bob Stanley, he was more or less assigned to Brouwer as a guide and assistant. This included meals, coffee, visits to the town and the like. Being somewhat familiar with Brouwer's career and subject, he discussed all kinds of topics. After inquiring into the status of Brouwer's fixed point theorems—with the predictable outcome: "not valid"—the two brought up the status and prospects of intuitionism.

> On yet another day, we inquired about what he thought was the future of constructive mathematics. His response was really an embarrassment, for he told that the Netherlands Government had promised that there would be established a Mathematics Institute in his honour, and he had since become convinced that there would never be such. In telling this, he became very emotional, and tears were streaming down his cheeks. This was such a poignant scene, being as it was in a public place. We felt so sorry for him and didn't really know how to act under the circumstances. So we did the best we could to change the subject, and I am not now sure that we succeeded very well. I do not recall how this session ended.

Brouwer had more surprises in store for his young companion; his blithe disregard for traffic was most discomforting for his fellow pedestrians.

> At busy cross streets with traffic lights, he would just ignore the lights and begin to walk across, and I had to hold him back physically, which was not at all easy. I had the feeling that I saved his life more than once before we got to our target address, a building of maybe ten or more stories. Once there, I looked at the directory and found that the office we wanted was on the fifth or sixth floor, as I recall. We went to the elevator and pushed on the button. The elevator did not arrive at once, and Brouwer did not want to wait, so he went to the stairs nearby, striding up them two or three steps at a time. He was much taller than my height at the time of a bit over six feet, so it was difficult

[25]This fact was also mentioned in a letter of Professor Williams (Williams to Brouwer, 28.VIII.1959) who recalled 'Your prowess in climbing trees'.

[26]R.J. Wisner, *Reminiscences About L.E.J. Brouwer*, 1998, unpublished.

for me to keep up with him. He got some papers to fill out, and we left after making an appointment for the next day. Returning to the car, I had to save his life a few more times at intersections.

After the Conference Brouwer set out on a long lecture tour through the United States and Canada. It turned out to be not so trivial to enter the USA, Brouwer had to wait patiently for a visa, but finally he got one. While still in Canada, he had received an invitation from Cairns, the topologist who had proved the triangulation property of differentiable manifolds five years before Brouwer. Brouwer reacted late but enthusiastically from his first stop in the States.[27] He wished, however, first to visit a relative of his wife, Jan de Holl in Birmingham, Alabama. The lecture tour through the US was largely if not completely organised by Stephen Kleene, the recursion theorist, who knew Brouwer from his stay in Amsterdam in 1950. It was in fact Kleene who had informed Cairns of Brouwer's presence. The precise details of the lecture tour have not been preserved, but we find Brouwer on November 22 in Los Angeles, from where he arranged his lecture in Urbana, the University where Cairns taught. He talked on November 30 on 'The intuitionistic continuum' and moved on to Purdue University, where he gave a talk on December 2. According to Brouwer's family doctor, he also visited the National Laboratory in Los Alamos, for this purpose he obtained a special pass.[28]

The content of Brouwer's lectures is not recorded, but we may safely assume that it closely resembled the Canadian lecture, and perhaps the Finnish one. That is to say, that it contained a good deal of, at that time still rather esoteric, not to say hermetic, material explaining the basics required for a proof of the bar theorem and the fan theorem.

For non-experts this certainly was no easy subject. The success of the lectures may thus be questioned. In particular since Brouwer made no concessions to his audience; he lectured in much the same way he conducted his classes in Amsterdam, addressing the blackboard rather than the audience, while developing his ideas in a soft voice. When Brouwer had lectured in Berkeley at the mathematical colloquium, Tarski spoke at the end of the lecture, 'It is very good that Brouwer has delivered his lecture, then you mathematicians can see that there are also complicated things in the foundations of mathematics.'[29]

Anil Nerode, who was at that time a young logician in Chicago, had a more positive recollection of Brouwer's visit. He had already privately studied Brouwer's papers beforehand, and so he was, as a student, the only person in the department who could discuss the content of the talk with the speaker.[30]

The Canadian-American trip gave great pleasure to Brouwer, who enjoyed travelling abroad. And in America and Canada there was quite a number of émigrés from Germany and Austria, many of those were known to Brouwer from the happy

[27] Wilton, Connecticut. Brouwer to Cairns, 30.IX.1953.

[28] Interview, C. Emmer.

[29] Oral communication, Dana Scott.

[30] Oral communication, Anil Nerode. (One wonders if Menger was present at Brouwer's lecture.)

years of scientific progress in the years before the *Third Reich*. It had moreover one particular side-benefit that carried for Brouwer probably more weight than all his mathematical lectures together; he hoped and planned to revive the old friendship with the Gutkinds. After their departure from Europe they had not met, and it is even doubtful if there was much contact. The only trace so far is the letter Gutkind wrote to Einstein, relaying Brouwer's wish to emigrate. This trip provided a unique occasion to re-establish the old relations Brouwer eagerly looked forward to a reunion—as a card from Kingston shows:[31]

> Please send me a sign of life by return of post. Badly longing to see you again. Why have we been so silent to each other for so many years, and did I only get your last book from Mitrinović?[32] Love, love
>
> Bertus (Papagei)[33]

On November 28 he wrote to Lucia Gutkind that he had completed the larger part of his tour; the talks, he said, had taken more time than he expected, but 'between now and my returning home to you, only Urbana, la Fayette, Toronto and Ithaca are left as stages'. The ties of friendship and harmony between Brouwer and Erich and Lucia Gutkind had been very special indeed; there was that mysterious affinity of souls that never failed to manifest itself with Brouwer. The days which he spent with the Gutkinds in New York were afterwards remembered with deep emotion by Brouwer.

A curious incident took place during Brouwer's stay in New York. Roy Finch, who was in close contact with the Gutkinds, and who witnessed Brouwer's visit to New York, described it in a letter:[34]

> Brouwer did not stay with the Gutkinds, but evidently spent most of a day and an evening with them. (I was there for the evening.) The Gutkinds told me this story. They lived in the Master Institute and Hotel on Rivers Drive. Downstairs in the same building there were music studios. One of the musicians who had a studio there, in which he taught classes, was Paul Wittgenstein, the brother of Ludwig, who was a pianist. Paul had lost an arm in World War I, but continued to play the piano with his left hand (a number of famous composers wrote compositions for him, including Ravel). Paul had a serious falling-out with his brother, evidently over the question of inheritance, Ludwig having given his share (the family was very wealthy) to his sister. This is all preliminary to what happened in the elevator at the Masters Institute when Gutkind, Brouwer and Paul Wittgenstein all happened to be in it at the same time. According to Mrs. Gutkind, who was also present and told me, Eric introduced Brouwer to Paul Wittgenstein with the words 'This is a friend

[31] Brouwer to Gutkind, 31.VIII.1953.

[32] Cf. p. 787.

[33] Parrot.

[34] Finch to Van Dalen, 11.II.1991.

**Fig. 18.1** Brouwer lecturing in Princeton (1953) [Brouwer archive]

> of your brother's.' To this Paul Wittgenstein replied: 'I do not wish to have anything to do with anybody who is a friend of my brother's.' He turned his back and left the elevator at the next stop.

Finch had met Brouwer at Gutkind's apartment, after Brouwer had been to Princeton. In Princeton the contact between Brouwer and Einstein, which was broken off after the *Mathematische Annalen* affair, was re-established. He told the company that he had a lengthy conversation with Einstein, and commented 'Einstein's mind never leaves the subject he is interested in, for one minute. He does not want to discuss anything else.'[35] The meeting with Einstein seemed to have healed the old wounds; Casper Emmer was told by Lize that Einstein and Brouwer 'had made peace'.[36]

Finch had asked Brouwer what his present view was of infinity, in a non-technical sense; Brouwer's reply was simply 'It just means that you can go on and on.' When the conversation got round to Wittgenstein, Brouwer told that he and Wittgenstein had met privately for an all-day meeting on an island, during which they discussed Brouwer's Vienna Lecture. It is unfortunate that Brouwer did not volunteer any more specific information on this meeting.[37]

---

[35]Finch to Van Dalen, 10.XII.1990.

[36]Oral communication, C. Emmer. Brouwer had resented Einstein's neutrality in the *Annalen* affair. In his book there was no neutrality in the conflict between good and evil.

[37]R. Finch gathered that the island was somewhere off the coast of Holland.

Gödel's comments on Brouwer's visit to Princeton can be found in a letter to his mother.[38] After reporting that the faculty now took up more of his time, he went on to discuss Brouwer's visit:

> Moreover a professor from Holland came to visit for 14 days, a famous man who was strongly involved in my field. I had therefore to invite him now and then at our place (once for lunch and once for tea). He is already 72 years old and no longer quite up to date. His lectures have not found much response (and rightly so). He apparently only came to America to earn money here. There one can see once more the difference between here and over there. The pensions in Holland have generally not been raised (since 1939), although the prices have been tripled, so that this famous man has to work as a bookkeeper in a drugstore just to survive with his wife and daughter. It is really very sad when something like that happens to distinguished men.

Apparently Brouwer had painted a dramatic picture of his economic position, which was indeed his own perception of the situation. Unfortunately, nothing more is known about the discussions between Brouwer and Gödel.

The lecture tour included more than the above mentioned addresses, for example he also visited the University in Boulder, Colorado; although it is not confirmed, it seems most likely that he met there his old friend Rosenthal. Brouwer moreover lectured at MIT; Hartley Rogers, who attended the talk, recalled its unusual nature.[39] Brouwer's lecture had the appearance of a soliloquy, the audience was evidently of secondary interest. The lecture itself went on longer than usual, lectures traditionally ended punctually after one hour, but Brouwer did not seem aware of the time, he just went on. And so after one and a half hour, a great number of prominent scientists, who otherwise would not tolerate such a breach of decorum, started to sneak out and call their wife that they would be late for dinner.

The visit to Cambridge was a wonderful opportunity to renew contact with Dirk Jan Struik, who had attended Brouwer's lectures in 1916/17. Struik, like Brouwer an inveterate hiker, took Brouwer on a walk in the surroundings of Boston. The reader will not be surprised to learn that Dirk and Ruth Struik took Brouwer also to Walden Pond. There is a picture of Brouwer contemplating the spot of Thoreau's former cabin. The contact with the Struiks was kept up, and when Struik was a visiting professor in Utrecht in 1962, Brouwer found the couple a nice cottage in walking distance from his own house.

Kleene, who had mostly organised Brouwer's itinerary, hosted Brouwer during his stay in Madison (where he also lectured); he told the following anecdote:

> When Brouwer arrived at one of the places on his lecture tour, his sponsor apologised that there had been a misunderstanding, the funds for his honorarium were not forthcoming. Brouwer told his sponsor not to be concerned

[38] Gödel to Mrs. Gödel, 31.X.1953.

[39] Personal communication.

> about it; he then gave them instead of his one-hour lecture his two-hour lecture.[40]

The tour ended where it had begun; on 31 December Brouwer was back in Montreal, from where he sent a New Year's card to his friend Mannoury. On January 3 he found himself again in Blaricum, full of memories and in particular with a rekindled and loving feeling for Lucia and Eric Gutkind. He immediately sent a card to them, signed by Lize, Tine Vermey, Cor Jongejan and himself, followed by another card, with the words 'In my thoughts I am still with you.'[41]

Brouwer was enthusiastic about America, in particular where nature and hygiene was concerned. He had since his childhood been sensitive to dust; he in particular suffered from respiration afflictions and inflammation of the nose, and here he had found a country 'where one is not troubled by dust'. For a man like Brouwer, with a strong craving for clean living, the appeal of the United States was quite understandable.

Once back in Holland, he submitted in a quick succession the papers '*Addenda and Corrigenda on the role of the principium tertii exclusi in mathematics*', '*Ordnungswechsel in Bezug auf eine coupierbare geschlossene stetige Kurve*', '*Further addenda and corrigenda on the role of the principium tertii exclusi in mathematics*', '*Intuitionistic Differentiability*', '*An example of Contradictority in classical theory of functions*'.[42] All of those papers consisted of intuitionistic refinements of old results and notions, and of corrections of old papers. None of these papers are notable for new revolutionary ideas, but they show that Brouwer had not lost his mastery of the technical parts of intuitionistic mathematics.

The inevitable loss of old companions and friends that is the fate of the aged, saddened Brouwer's life too; on March 3, 1954 Hendrik de Vries, his fellow mathematician in the faculty, with whom he used to share jokes, and who in the old days came over to play the violin with Brouwer, died in Benjamina (Israel). In December of the same year, Peter van Anrooy, the conductor, passed away. The musician from the Hague had been a true friend, and he was sorely missed by Brouwer. He wrote to the widow,

> Dear Freddy,
>
> Thus Peter has fought the struggle of his life to the end. A life that, directed by a great and indomitable talent, and a tempestuous richness of thoughts and feelings, found the predestined triumphs *and* conflicts on its way. My sincere condolences and my assurance that I, together with the many who are better qualified, will highly cherish his memory.
>
> Your Bertus Brouwer

[40]Kleene to Van Dalen, 3.II.1977.

[41]*Fühle mich noch immer dort bei Euch.* Brouwer to Gutkind, 12.I.1954.

[42]Submitted resp. 30.1, 27.2, 20.3, 24.4, 24.4 1954.

**Fig. 18.2** Queen Juliana with a selected group of participants of the International Mathematics Congress 1954 (Amsterdam) at the palace in Soestdijk. Left in front—Von Neumann; the Queen is flanked by the Field medal winners—Serre and Kodaira; behind Serre—Hermann Weyl; behind Weyl—Schouten; to the right behind the Queen—Heinz Hopf; second to the right of the last row—Veblen; partly visible in front of Veblen—Alexandrov. [Photographer Aart Klein, copyright Centrum Wiskunde & Informatica (CWI)]

A year after Brouwer's American lecture tour the international mathematics conference was held in Amsterdam. It brought the flower of mathematicians to Holland. Many of Brouwer's old friends attended, but he apparently stayed away. However, a number of his friends used the occasion to visit him in Blaricum. Alexandrov, who was happy to revisit Holland after all those years, was invited with a number of prominent participants to the royal residence. At the tea party he told Queen Juliana that in the thirties they were more or less neighbours in Katwijk at the seaside in Holland.

Although he had reached the age of 73 years, Brouwer did not intend to give up his involvement in scientific matters and events. And 1954 indeed offered some occasion for renewing old contacts and making new ones. In March he was invited by Felix Hackett to participate in the Commemoration of Boole's *Laws of thought* of 1854. Brouwer gladly accepted, and on 24 May he found himself in Trinity College Dublin, where he delivered a talk *The effect of intuitionism on classical algebra of logic*. He used the occasion to stress the basic differences between classical and intuitionistic logic. The lecture presented in a nutshell the salient points. 'Classical algebra of logic', he said, 'furnishes a formal image of common-sensical thoughts'.

**Fig. 18.3** Max Euwe, playing a simultaneous chess game at the International Mathematics Congress 1954. [Photographer Aart Klein, copyright Centrum Wiskunde & Informatica (CWI)]

As such it is based on a threefold belief: (i) the existence of a *truth*, independent of human thought; (ii) obtaining new truths from old ones by logical reasoning; and (iii) the principle of the excluded middle. He went on to sketch the status of logic:

> Until not long ago this threefold belief of common-sensical thought was shared by scientific, also by mathematical, thought. As long as mathematics was considered as the science of space and time, it was a beloved field of activity of logical reasoning, not only in the days when space and time were believed to exist independently of human experience, but also after they had been taken for innate forms of conscious exterior human experience. There continued to reign some conviction that a mathematical assertion is either false or true, whether we know it or not, and that after the extinction of humanity mathematical truths, just as laws of nature, will survive.
>
> Only after intuitionism had recognised mathematics as an autonomic interior constructional mental activity, which although it has found extremely useful linguistic expression and can be applied to an exterior world, nevertheless neither in its origin nor in the essence of its method has anything to do with language or an exterior world, on the one hand axioms became illusory, on the other hand the criterion of truth or falsehood of a mathematical assertion was confined to mathematical activity itself, without appeal either to logic or to a hypothetical omniscient being. An immediate consequences was that in mathematics no truths could be recognised which had not been experienced, ...

After giving a number of counter-examples to classical theorems, Brouwer concluded on a positive note,

> Fortunately classical algebra of logic has its merits quite apart from the question of its applicability to mathematics. Not only as a formal image of the technique of common sensible thinking has it reached a high degree of perfection, but also in itself, as an edifice of thought, it is a thing of exceptional harmony and beauty. Indeed, its successor, the sumptuous symbolic logic of the twentieth century, which at present is continually raising the most captivating problems and making the most surprising and penetrating discoveries, likewise is for a great part cultivated for its own sake. Don't let us forget that it is Boole who has been the originator of all this.

The lecture appeared in print in 1955, and it was Brouwer's final paper. The paper is worth reading, more because of its reflections on the nature of logic, than because of the logical content. After all those years Brouwer continued to forego the benefits of modern predicate logic; he used the time honoured notion of Boole and the school of algebraic logic. For his purpose that was indeed good enough. In a way the paper was his farewell to mathematics; in a modest way he remained active in the subject, but nothing substantial resulted after 1955. He planned to revise his old papers, but that did not come to more than isolated private notes.

Three months later Brouwer again represented Dutch mathematics. This time he was asked to represent the Academy in Paris at the centenary of Poincaré's birth. A better delegate could not have been found by the Academy; Brouwer was in a double sense a successor of Poincaré. Both in his topological research and in his foundational innovations, he carried Poincaré's torch.

It is good to keep in mind that at almost the same time Heyting was finishing the last parts of his monograph *Intuitionism. An Introduction*. This book finally provided a larger public with an easier access to the mysteries of intuitionism. Its role should not be underestimated, for a large number of logicians it was their first encounter with intuitionism; its readability has always been praised. Brouwer's views on this Introduction are not known, one may safely assume that he approved of its publication, because it appeared in the series *Studies in Logic*, founded by Brouwer himself and of which he was an editor until his death. On the other hand, after his retirement Brouwer got more and more estranged from his colleagues and even Heyting had not escaped the fate of falling into disgrace. The fact that Heyting deserted him in his conflict with the faculty must have hurt Brouwer more than most desertions.

In 1955 Brouwer travelled once more as a delegate of the Royal Academy to Göttingen. This time for the centenary of Gauss' death. It was a memorable occasion to be back in the old capital of the mathematical world. Not only did the town and the university awaken old memories, but some of the participants were witnesses of the old days; Hopf, Radon, Denjoy, Julia, Perron, Reidemeister, Nöbeling, Süss, A. Schmidt and Courant were present, and at one time or another in the past, they had been more or less close to Brouwer. There may have been some awkward moments when former antagonists came face to face, but by now Brouwer had more or less accepted the past as a fact of life, and, unless provoked, he would behave as a perfect man of the world.

A well-deserved honour befell Brouwer in June of the same year; the University of Cambridge granted him an honorary doctorate. Although he had been rather out of sorts lately, he gathered all his courage and, accompanied by Tine Vermey,[43] he set out for Cambridge. Lize had not felt like travelling to England, and it was a load of her mind that Tine was prepared to take care of Bertus. Bertus had been in a poor state already for some time, and it was a bit of relief to have a few days for herself. She decided at once to have Louise over in Blaricum.

On the second Thursday of June, Brouwer strode in the colourful procession along King's Parade to the Great Hall. He was in the company of fellow honorary doctors, men of renown, among others the famous architect sir Giles Gilbert Scott, the lawyer sir Charles Bruce, Locker Tennyson, the grandson of the poet, and the Right Honourable David Robert Alexander Earl of Crawford and Balcarres.

The virtues of the aged revolutionary, who even at this occasion managed to look eccentric, were aptly summed up by the Chancellor, lord Tedder, Marshal of the Royal Air Force, in his laudation,

> 'Everyone knows', said Cicero, 'that those who are called mathematicians deal in matters of incomprehensible complexity and subtlety'. Copernicus put it more briefly: 'Mathematics is written for mathematicians.' Hence my quandary. I shall prove a fool if I try to expound the merits of a top-grade mathematician, a coward if I shirk my appointed task. What then? I proceed and beg your indulgence. In two ways Professor Brouwer has earned an outstanding reputation. First, he has so advanced the science of topology that with Henri Poincaré and Georg Cantor he may be reckoned among the founders of this study in modern times. Secondly he has invented a new kind of mathematics, which he calls intuitional and which has revolutionised the whole foundations of the subject. It contains two acts (as he calls them). In the first, mathematics is defined as a languageless activity of the mind, having its origin in the conception of time. From this beginning, he demonstrates the fallibility of those who regard the logical principle of 'tertium non datur' as a reliable instrument for the advancement of knowledge and as for the argument of 'reductio ad absurdum', he exorcises it with a kind of magical incantation: 'The absurdity of the absurdity of the absurdity is absurd.' The second act—but before the second act my courage fails me. Like Socrates on another occasion, I fear I may fall into a great pit of nonsense. Your applause must make amends for the deficiencies of my speech.[44]

Before leaving the eminent seat of learning that had so graciously recognised his role in science, Brouwer gave a talk on foundational matters. After that he returned to Holland, where he and Cor Jongejan withdrew to Zandvoort for a holiday on the seaside.

---

[43] Tine had been married to Willem Langhout, the marriage had ended in divorce.

[44] Quoted from *The Times*. Brouwer had himself made a translation from the Latin text into Dutch. A note in his handwriting has been preserved. Cf. van Dalen (2001a), p. 477.

**Fig. 18.4** Brouwer in the procession in Cambridge at the occasion of his honorary doctorate [Brouwer archive]

Hermann Weyl had in 1951 exchanged the United States for his beloved Switzerland, be it that he spent each year a few months in Princeton. He had renewed his contacts with European colleagues and friends, and attended for example the International mathematics conference in Amsterdam in 1954. He died on 9 December 1955 in Zürich.

Brouwer visited his old friend and fellow revolutionary when Weyl was already seriously ill. Considering how close they had been and how they shared their views on philosophical and foundational matters, there must have been an abundance of material for them to discuss. Brouwer later told that the conversation naturally had led to the foundations of mathematics, to the continuum, choice sequences, laws, continuity, and of course to the underlying philosophy. No man could cast his spell better than Brouwer if he wished; his softly spoken, beautifully formulated ideas and visions must have evoked in Weyl the memories of the carefree days of their meeting in 1919 in the Engadin.

– '*Ach Brouwer*', Weyl had sighed, '*es ist alles wieder schwankend geworden*'.[45] And so Weyl ended where he had begun his foundational activities.[46]

[45] Everything has again become vacillating.

[46] There is no doubt that, whatever Weyl's technical mathematical practice may have been, philosophically he remained on Brouwer's side. Freudenthal, who knew Weyl very well, and who was

In 1955 Brouwer was once more party in a conflict in Blaricum; a small business in his neighbourhood was asking permission to mix explosive fluids, this was for Brouwer and some fellow residents a reason to take legal action to prevent this. When the city council made up its mind to grant the permission, Brouwer went all the way to fight the decision, he even appealed to the Crown. Legal battles like this one took a more and more important place in Brouwer's life; they cost him dearly, both in fees and in emotional stress. Although the arguments were always cool and ingenious, the man behind the arguments was emotionally involved with every fibre of his person. In this particular case, he even travelled to the Hague in order to attend the ruling of the State Court (*Raad van State*) personally (17.XII.1956). The case was eventually won. There were more cases, and some of them had an operetta-like quality.

On January 30, 1956 Gerrit Mannoury died; Brouwer grieved for his long-time companion, in a letter to his colleague Stomps he wrote:[47]

> I feel deeply grieved by the loss of Mannoury. At a very early age I have already learnt from him and many decades I have warmed myself at his personality and received fruitful suggestions for my work from him. His character was of a sincerity and detachedness which commanded respect, and his realm of thought was of an almost infinite wealth [...]. Should at some time his writings be translated into a world language, then I cannot but expect, as far as I can see, that the judgement of the world will grant him a place in the Pantheon of civilisation among the great thinkers.

The treatment at the hands of his former colleagues had filled him with bitterness, which even showed some mild signs of paranoia. He never forgave, for example, Clay for his role as a chairman of the faculty, and it is said that he refused to sit next to Clay at dinners, fearing that Clay could poison his wine. It is hard to say in how far he really meant this; he was no stranger to a certain provocative eccentricity, and such claims could perfectly serve to produce an effect of shock in an otherwise pleasant conversation. Brouwer's mind remained as brilliant and sharp as ever but his taste for the bizarre may easily have tempted him to shock his visitors. Being well-acquainted with the more cruel sides of history and the practices of (for example) the Borgias, he liked to warn that one had to be careful, because 'many a sugar pot contains ground glass'. In spite of his traumatic experiences, he kept up his habit of visiting the meetings of the Mathematical Society, which still met at that time the last Saturday of the month in the hotel Krasnapolski in Amsterdam; there he mostly talked to the younger visitors. Furthermore he remained a regular visitor of the Royal Academy; at reaching the age of 70, at which an 'active member'

certainly qualified in foundational matters, wrote in his obituary (Freudenthal 1955), '*Er entscheidet sich für Brouwers intuitionistische Deutung der Mathematik, aber dem Systembauen abhold, verschmäht er Brouwers aufs Allgemeine zielende Methode. Das ist kein Verrat, denn das Wesen seiner Mathematik war und blieb intuitionistisch.*'

[47] Brouwer to Stomps, 17.II.1956.

becomes a member in retirement,[48] he informed the Academy that he had no wish to have his membership of the Section Physics extended, instead, he wrote, he was willing to join the 'Free Section' if the subject 'exact introspection' was added to its domain.

In due time Brouwer got in mathematical circles the reputation of a quarrelsome, unreasonable man. Heinz Hopf used to say jokingly that he was the only mathematician who never got into a fight with Brouwer, and in particular in the Netherlands, Brouwer's sharp tongue became something of a legend. The following account of De Groot may illustrate this. Once De Groot gave Brouwer a ride home after a meeting of the mathematical society; the conversation was lively, and Brouwer started to list all the fights he had with mathematicians. Suddenly he sat up and said to De Groot, 'but we also have a fight'. De Groot smiled and said, 'well, let us forget about it'. Brouwer possessed enough sense of humour to see that this quarrel at least had lost its grounds for continuation.

It was not only his discontent with the academic world and his colleagues that bothered him. His health, which was, in spite of the above mentioned heart problems, largely in reasonable condition, started to fail him on the smaller points, something that comes with the process of ageing. In 1956 he suffered from complaints of the urinary tract. Far from seeking help from surgical quarters, he stuck to the recipes of his wife: extract of celery and leek. Eventually the pains were alleviated. Some of his afflictions were most likely of a partially nervous origin. In January 1956, Brouwer feared once more for his life. As Lize reported to Louise,[49]

> Dad[50] had in the night of Saturday again a poor night. He thought it was the heart, but that was not the case. It was rather the stomach. He was violently sick during the night, and his heart reacted in a disturbing manner. He has been through this before, e.g. in Almelo in the examination committee; that time he had eaten strawberries, and a herring and a glass of cold water to top it of! He was then taken to the hospital next door.

Lize had hastened to Almelo and with the help of a few bandages cured her husband. A week later she could reassure Louise that,

> Fortunately Dad is well again. It is not the heart, but rather the stomach, which is always fed cookies by Cor. I have cured him with the triangular compress, and warned him not to eat sweets and to eat and drink all kinds of things between meals. Now everything is well for the time being.
>
> Now and then he asked me if I thought that he would die soon, because it was his heart after all. I told him that it was up to him.

With most of their money tied up in the pharmacy and in real estate, the Brouwers led a sober life. Indeed the only striking extravaganza on Brouwer's side was his love

[48]Nowadays at 65.

[49]Lize to Louise, 15.I.1956.

[50]In her letters to Louise, Lize used the term 'Paps' for Bertus, often 'de Paps'.

**Fig. 18.5** The two L.E.J.'s—Jan Brouwer and his uncle at a meeting of the Shell Company in London (1960) [Brouwer archive]

for travelling, in particular abroad. They had to watch their budget very carefully. Under the circumstances, Louise presented a real problem.

Brouwer's financial worries were not wholly imaginary. The pharmacy, in particular, drained his sources. On the other hand he owned large tracts of land and some real estate that could have been turned into money. Brouwer had a naive hope and conviction that he might find a miracle investment; at one time he invested a small sum in an instrument to stretch the hand of a piano player. The gadget did of course not sell at all, and the contraptions laid rusting in his garage for years. His nephew Jan (son of Aldert, named L.E.J. after his famous uncle) was the subject of Brouwer's admiration. Jan, who was like his uncle, a paragon of intelligence, turned his capacities to good use in the service of the Royal Shell. After serving the company in almost every corner of the earth, he ended his career as a very successful president director of Shell.

The relation with Louise had at the best of times been touchy, apart from the personal likes and dislikes there was an ongoing concern for her financial management. The allowance, paid from her father's estate, had not kept pace with the rising cost of living. In the late fifties the burden of a semi-invalid stepdaughter, who had by then already left her 65th birthday behind her, became a serious problem. Replying to Louise's congratulations on the occasion of the fifty-fifth wedding anniversary of her mother and stepfather, Brouwer wrote in September 1959:[51]

> Dear Louise,
>
> Many thanks for your sympathy at our 55th anniversary, and for your fascinating book on Balzac, who appears to have had in his life almost as many difficulties as the undersigned.
>
> As to the financial circumstances these have, alas, deteriorated so much for your mother and me during the last months, as rapidly at the health of both of us, that we can no longer afford you the same standard of living as so far.

[51] Brouwer to Louise Peijpers, 7.IX.1959.

> In the first place we can no longer allow you to scorn the services of the excellent National Health organisation A.Z.A.[52] offers for free both for your teeth and your internal health, and to waste pointlessly money by having you treated privately for the same things.
>
> In the second place we can no longer afford you to continue the payment of a rent which is far above the average, for a length of time ...
>
> Should your mother and I have a bit longer to live and should our financial circumstances then take a fortunate turn, then you may rest assured that you will immediately profit from that favourable turn. But you should also reckon with the possibility of a sad event in a very near future.

And indeed that sad event took place sooner than expected.

Lize, more than ever tiny and fragile, had tenaciously run the pharmacies and the Blaricum household with a velvet hand; in spite of the demanding and often grumpy nature of Brouwer, she had managed to protect the interests of Bertus—and also of her daughter Louise. The latter had left the house long ago, but she never managed to earn a living, so that she was an extra financial burden to the somewhat chaotic economy of the Brouwer household. The surviving letters show us Lize perpetually in a hardly enviable position between the Scylla and Charybdis. On the one hand she stuck to her loyalty to her husband, and on the other hand she was visited by feelings of guilt for neglecting her only child.

In the course of time Lize, who had already in 1923 become a Catholic, had grown more and more pious, she sought the assistance of the church in all kinds of matters, for example she asked in November 1955 Father Superior Drehmann to pray for a provisor for the pharmacy when it was almost impossible to find one, and when one had turned up, she was sincerely grateful for the prayers of the Father Superior; in view of the success, she did not hesitate to asking for prayers for shop assistants, who were also very scarce at the time.

In her piety she put unlimited trust in the healing powers of faith, later in life she expressed her admiration for the Christian Scientists, who had, in her opinion, taken over the monopoly of 'faith healing' from the church. She urged Louise to look into the Christian Scientist principles of faith, remarking that 'Max Euwe, the genius chess master and his wife had been healed' by the Christian Scientist faith and practice.[53]

Louise had a semi-independent existence in Amsterdam. She lived in one of the apartments over the pharmacy at the Amstelveenseweg. Lize used to praise her for 'taking care of the shop so well', but that seems rather an attempt to placate her difficult daughter. There was not much sympathy lost between the staff of the pharmacies and Louise, the assistants viewed her, rightly or wrongly, as the lazy, sulking daughter, who reported on the daily affairs when Lize was away.

---

[52] An Amsterdam organisation, which provided medical help for the lower income classes for a moderate fee. A subscription to the National Health Organisation was to a certain degree a social stigma. For private patients it was generally considered degrading to join the National Health.

[53] Lize Brouwer to Louise Peijpers, 11.II.1957. Later Mrs. Euwe became head of the Christian Science in the Netherlands.

**Fig. 18.6** Tea on the lawn of the Pimpernel. Left to right: Cor Jongejan, Non Emmer (wife of Brouwer's family doctor), Brouwer, Lize, Tine Vermey. [Courtesy Casper Emmer]

Lize was not blind to her daughter's shortcomings, but her prevailing attitude was one of protectiveness. Brouwer's dislike for Louise was no secret, and in the interest of domestic peace Lize did her utmost to keep the two apart. Whenever something was going on in Blaricum, a jubilee, a birthday, Christmas, she advised Louise to stay away—'nothing that would interest you'. And when the coast was clear, she would invite Louise, or visit her in Amsterdam.

Louise on her side, was perfectly aware of Brouwer's feelings, and those of her mother. Usually she stayed in one of the houses that belonged to the pharmacy and kept up some pretension of taking care of the premises. She had a certain shrewd insight into the feelings of her mother, and she exploited these without much scruples. Some of the surviving letters display that entreating style parents use with difficult or even aggressive children.

Lize had made a will in 1921, which left Louise enough for a comfortable life, the other beneficiaries were Bertus and Cor Jongejan. The proportions were precisely indicated, with—depending at the circumstances—a quarter to Brouwer, a quarter to Cor and a half to Louise. Of course Brouwer had agreed to the will, he was one of the witnesses, but neither Lize nor he had foreseen what problems would be in store for the settling of the estate.

In retrospect one easily discerns a definite pattern in the relations of Lize; she had to deal not only with Bertus and Louise, but with the whole entourage of Brouwer and the pharmacies. The impression one gets is that of a resigned and even relaxed

**Fig. 18.7** The Brouwer couple in the late fifties [Brouwer archive]

attitude towards Brouwer and his numerous girlfriends. None of these ladies seriously threatened her position, she clearly filled the role of the caring and loving mother and spouse, and less that of a passionate, let alone jealous, wife. She once remarked that the sexual aspects of marriage (and by extrapolation, of the relations with the weaker sex) were not prominent in Brouwer's life. Lize had gradually accepted Brouwer's way of forming attachments, in particular she had accepted Cor Jongejan—to whom Brouwer always turned in the long run—much as one accepts a sometimes cheerful, sometimes obnoxious child. Whatever the situation was, there was no sign of jealousy or a hidden struggle for power. With a superior resignation she knew that in the end her views would carry the day.

On the whole their marriage had been a harmonious one, it had started out with the dreams and idealism of the young; Brouwer's meteor-like career had surprised and delighted them both. Brouwer had consistently taken a keen interest in Lize's affairs, in particular he ruled the business matters of the pharmacy as if he were the pharmacist and not his wife. She from her side had guarded his physical and spiritual well-being. Both of them were ardent health adepts, Lize was a master of herb lore and she practised all kinds of cures that avoided the blessings of the pharmaceutical industries. But not only the physical well-being of Bertus claimed her interest, she fervently hoped to bring Brouwer also into the fold of the Catholic Church; the difficulty was that the object of her desires was flatly anti-catholic, so that the only way left open for a conversion was that of constant and intense prayer. Whether Brouwer was aware of the attempts at his soul is unknown. It is a fact that he remained allergic to the Catholic Church. When a self-styled philosopher, Dr. A.J. Reesink from Laren died in the St. Jans hospital, Casper Emmer told Brouwer that he happened to enter the room where the deceased was lying and saw that he had a rosary in his hands. Brouwer flew up and cried 'Do you hear that Kleine Moek,[54] when my time has come, you keep them out of the house, those body snatchers.'

[54] The Brouwer couple used to refer to each other as '*klein Moek*' (little Mom), and to '*lief Man*' (sweet husband), in letters abbreviated as kM and lM.

Kleene observed during a visit in 1950 how strongly Van Eeden's conversion to Catholicism still rankled after all those years, when Brouwer was still upset and furious about what he called 'Van Eeden's treason'.

In 1959 Brouwer severed the connections with Bruins. The cause was a paper of Bruins that he had asked Brouwer submit to the Proceedings of the Academy. The paper dealt with invariant theory. Brouwer had checked the contents and sent a number of comments and corrections, for example he suggested that a suitable title would be *On pathological metrisation elucidating theory of invariants*. He also pointed out that the references were in need of correcting.[55] Somehow Bruins was dragging his feet, and when he returned the revision, Brouwer was not satisfied, so that he waited for further corrections. This led to an exchange of letters, and when Brouwer had still not submitted the paper in August, Bruins made the mistake to hint that he wished to take his paper elsewhere. When that did not work, he took a high tone and demanded his manuscript back. As we know from past incidents, this was exactly the kind of thing Brouwer could not tolerate. For all practical purposes that was the end of the relationship between Brouwer and Bruins. The impression left by this episode is that Bruins had come to take Brouwer for granted. Brouwer's repeated attempts to get Bruins promoted, combined with Brouwer's isolation in Dutch mathematics, probably gave Bruins ideas.

## 18.2 The Pharmacy

Among all the changing projects and affairs, there was one element that remained constant: the pharmacy. It is not easy to guess what its attraction was for Brouwer; was it a material tribute to his wife, or was it a palpable testimony to his importance and success? A man like Brouwer could not have overlooked the fact that, say, the fixed point theorem was a much better investment in eternal fame than the ownership of a pharmacy. Yet his attachment to the pharmacies was genuine and no doubt dictated by emotion rather than by calculation. Probably his fascination with the shops was the result of a sincere devotion to his wife, mixed with a feeling of family obligation, and the conviction that a clever man like himself could make a success of this commercial activity. Whatever the motivation for Brouwer was, it had started innocuously enough with the old shop at the Overtoom. This was the place where Bertus and Lize started their married life, that is to say, the Amsterdam part of it. Here Brouwer did part of his research, took care of his correspondence and met with his friends of the signific circle. If the city had not chosen to redevelop that part of the Overtoom, it is likely that the pharmacy had lasted as a calm centre for dispensing medicine and for homeopathic specialities. Fate, however, decided otherwise. In 1929 the shop had to be moved elsewhere. Brouwer found a suitable place, roughly a mile to the south at the Amstelveenseweg (cf. 523). Somewhat later Brouwer acquired two more locations, not far from the shop at the Amstelveenseweg, he bought

[55] Brouwer to Bruins, 4.VII.1959.

a location at the Surinameplein and rented a wooden temporary building at Surinamestraat number 12.[56] The shops at the Amstelveenseweg and those at the Surinamestraat and -plein (square, place) were separated by a canal; after a bridge was built, it took no more than 10 minutes to get from one place to the other.

In the middle of the war Brouwer started to make plans for moving the pharmacy of the Surinamestraat (which was only a temporary location) to the shop at the Surinameplein. Since the house at Surinameplein did not satisfy the requirements for a pharmacy, a great deal of reconstructing was required. Plans were made by a local contractor, but understandably, no actual work was done before the end of the war. The situation during the war was, as a matter of fact, far from pleasant, as the German army had constructed a concrete platform in the back garden, right behind the pharmacy, and installed an anti-aircraft battery.[57] After the war Brouwer managed to get permission to make large scale alterations to the house at the Surinameplein. In fact, he planned to have a basement added to the pharmacy, necessary for safe storage of pharmaceutical material. This was an unconventional plan; the newer houses in Amsterdam had as a rule no cellars or basements, the soft underground and the technique of building houses on piles made such extras far too expensive. Finally in 1955 the new basement was added. As a kind of constructional tour de force, the building was jacked up, so that work under the house could be carried out. As it was a four story brick building, this was at the time a major operation. The building did survive it rather well, although some unavoidable cracks showed up. The tenants, understandably, blamed the construction activity for it. Brouwer, on the other hand, blamed it on shortcomings of the original building (1936). We must keep in mind that Brouwer was already 74 years old when he undertook the basement construction, an age at which most people prefer a life of leisure and reflection. His wife, for whom he was creating the perfect pharmacy, had passed her eighty fifth birthday! It seems clear that retirement was not an option for her, but considering her age, it was unlikely that she could enjoy the new shop for long. Financially speaking, the extension was also an irrational project. It was highly unlikely that a sale would make up for the investments. Perhaps one should admire the tenacity with which Brouwer extended his pharmacy, but his business instincts seem to have played a secondary role.

Brouwer carried on lengthy and cunning negotiations with surrounding tenants and owners to obtain a larger share of the central court. Surprisingly enough, he was successful; he indeed acquired enough space to connect the pharmacies of the Surinamestraat and the Surinameplein by way of his own backyard. And, to put the finishing touch on the new pharmacy, he commissioned the building of a laboratory for Lize on this ground behind the pharmacies. This was a simple, one story structure, which had all the facilities one could wish for a laboratory. Like the pharmacy at the Amstelveenseweg, the laboratory was more than a utilitarian construction,

---

[56]The transaction was concluded on 2 April 1930.

[57]The pre-war building style in Amsterdam was that of (mostly) four story housing blocks with an enclosed court, split up among the tenants of the ground floor apartments.

its interior was drawn up in style, with nice panelling and stained glass, decorated windows.

The shop itself was furnished in the traditional style of a chemist's shop. Of course in the process of perfecting the building, misfortunes occurred, for example, most of the year 1951 was taken up by a dispute with the firm that installed the central heating. The heating did not work at all, and finally Brouwer had to call in another firm for the reparations, which cost him an extra 1384 guilders.

Brouwer's relation with the pharmacy was somewhat ambiguous. On the one hand he could paint it as a cruel master, for which he had to slave as a free-lance administrator. There are a number of private notes in Brouwer's archive on various topics; the following refers to the pharmacy: 'For several years I have been forced to earn my living by doing unskilled labour of the lowest level uninterruptedly, all day long, Sundays and days of illness not excepted.' Gödel's letter to his mother in which he reported on the visit of a great man from the past (i.e. Brouwer) also mentions Brouwer's complaints, see p. 796. There is a letter from Brouwer to his solicitor, Hardenberg, in which he expresses similar grievances; he wrote that he 'has lived in poverty ever since, to save the estate. Works 10 hours a day at it, must count with a sudden death.'[58] It was not just a matter of keeping the books in order and of keeping an eye on the staff. Brouwer spent a considerable amount of time and energy on negotiations with the city office for building and planning, with contractors, with third parties who made all kinds of absurd demands, ... In Lize's letters there is often mention of the state of exhaustion Brouwer was in, after spending whole days on business. True, the bleakest period in Brouwer's relation to the pharmacy came after his wife's death, when his stepdaughter enlisted legal assistance to get the most out of the inheritance, but it is equally true that Brouwer spent much time at all kinds of chores in the pharmacy.

On the other hand he considered the pharmacy his private pet. He devoted much time and thought to the well-being of the pharmacy and its staff. His relation to the assistants was that of a benevolent feudal lord, he cared about the 'girls' and he kept them on until a ripe old age. One of the assistants had joined the pharmacy as a very young girl, she worked her way up to the managing of the shop at the Amstelveenseweg, and when she retired at the age of eighty, it turned out that many customers had taken her for Mrs. Brouwer. Brouwer had a liking for all the assistants, but some more than others, he was for example very much impressed by a blond beauty, Greetje. None of this worried Cor Jongejan, she just made fun of it, as she made fun of almost everything, including herself. This was possibly her way of protecting herself.

When Brouwer bought the building (a temporary wooden construction) in the Surinamestraat, he obtained a long lease on the ground (a standard procedure in Amsterdam) with the condition that the city could at any time give notice when it needed the ground for development. In 1953 the city plans, which had been postponed on account of the war, had reached a phase in which the ground of the Surinamestraat

[58] Brouwer to Hardenberg, 11.II.1963. He is referring to Lize's estate and the lawsuits of Louise.

was required. On March 13, 1953 the official notice of the city, announcing the termination of the lease, dropped in Brouwer's letter box. In 1954 the City of Amsterdam officially cancelled the rent for the lot of the premises at the Surinamestraat, in view of plans for developing the street. And so Brouwer had to move out of the temporary pharmacy.

The actual moving of the pharmacy to the Surinameplein presented a serious problem, as the pharmacy, including its laboratory and office did not fit into the ground floor of the premises at Surinameplein 8. An obvious solution to the problem was to give notice to the tenant of the first floor, and to add that part of the house to the pharmacy. However, the general housing shortage, caused by the demolitions during the war and the post-war baby boom, did not allow for such steps. There was a strict rent protection. By regulation, no space should be converted from living- to office- or business-space. Unless, of course, one could get permission from the official housing office to do so. Brouwer set immediately to work on the problem. He engaged his solicitor and spent his own time to convince the present tenant that she should find alternative quarters, while fighting the City of Amsterdam at the same time. The tenant was not inclined to co-operate, and the City of Amsterdam was adamant. In November, 1955, Brouwer lost the lawsuit and was ordered to evacuate the temporary shop before December 1. He appealed, and managed to win time. In June 1956 the shop in the Surinamestraat was finally and definitely pulled down.

From then on the business was conducted at the Amstelveenseweg and the Surinameplein, be it that the Surinameplein shop became the more important of the two. The following example may illustrate how seriously Brouwer took the commercial aspects of the pharmacies. When the City, after pulling down the house in the Surinamestraat, reconstructed the Surinameplein, it designed a small public garden in front of the shops. The plan also implied the moving of the big red letterbox from the front of the pharmacy to this public garden. Immediately Brouwer started a protest: it would reduce the number of customers, he argued. And when the City did not give up its plan, he took it to court and won.

Since there were two chemist shops to be manned, again a 'provisor' was required. This provisor was to take care of the shop at the Amstelveenseweg, while Lize supervised the shop at the Surinameplein. The engagement of new provisors was always left to Brouwer. One of the later provisors, Mrs. Paulssen, remembered clearly how one day in 1952, Brouwer dropped in at her place in the Jan van der Heydestraat. This was in the older part of Amsterdam, notorious (or famous) for its almost infinite stairs that led from the street all the way to the top floor. Mrs. Paulssen was pregnant and her husband was away working for his Ph.D. One day the bell rang, and when she had pulled the rope that opened the street door, a tall, lean man with long white hair stormed up the stairs, taking two steps at the time. He had reached her floor before she could take of her apron, he looked round and said 'not very suggestive, the way you live here'. This vital elderly gentleman turned out to be professor Brouwer, the owner of a pharmacy; he invited her to become the provisor in his pharmacy, and she gladly accepted the offer. To her astonishment there were virtually no customers at the pharmacy that was assigned to her, yet the shop was open all day. Thus she simply went over to the shop at the Surinameplein, where she helped out. There she found a free-and-easy group; the customers bossed over the staff, and it took her some time to straighten things out. Mrs. Brouwer was completely out of touch with professional matters. Brouwer was very tender and patient with her, he took his time, kindly explaining things to her when she happened to be puzzled. Lize was equally concerned about Bertus and his health. At one occasion Brouwer was asked to go out and do some shopping. It was very cold, and Lize insisted that Bertus could not go out without a hat. Unfortunately he had forgotten his own hat. After a long exchange of arguments, Brouwer finally left the shop with one of Lize's hats on his head, blissfully unaware of his rather unusual look.

Mrs. Paulssen's impression of Brouwer was that of a remarkable person, a bit unworldly, and sometimes inclined to talk about mysticism. When her child was born, she got a bed jacket with thirty moth holes. After four years she moved to Rotterdam, remembering the pharmacy Brouwer–de Holl as quite an exceptional place.

On 11 October, 1959 Lize died at the age of 89 years in the hospital in Bussum. In spite of her advanced years, she had insisted on regularly visiting the pharmacy in Amsterdam and on carrying out the small duties that she considered hers. In the autumn of 1959 she fell down in the shop and broke her hip. After being laid up for several weeks in her apartment over the pharmacy, she was transported to the hospital in Bussum.[59]

Lize's death was a bitter blow to Brouwer, life without her would never be the same. Although Cor Jongejan from now on ran the household and took charge of the pharmacy, one cannot say that she filled the place Lize had left. With the death of Lize the duties of the executor of the will fell upon Brouwer. In itself the handling of the estate should not have been an overly taxing job, but there were aggravating circumstances. In the first place, the estate was rather more complicated than the ordinary 'family with some savings' estate. Not only did the couple, which was married without a settlement, own land and real estate, but the pharmacy posed real problems. The value of a pharmacy depended on a lot of factors, such as its house, the circle of clients, the management, etc. Nonetheless Brouwer succeeded in getting all the required figures right, but that was not enough; his stepdaughter, who heartily returned the dislike that Brouwer felt for her, finally saw her opportunity to get even with her stepfather. She got herself a lawyer, who like Louise was an ultra catholic, bordering on bigotry.[60] With his help she managed to delay the settling of the estate for some five years. This plunged Brouwer once more into a quagmire of legal and financial problems.

On 24 August 1960, Brouwer had completed the registration of the credit and debit of his and his deceased wife's joint possessions. The document drawn up by the notary lists an impressive array of real estate: two houses in the Amstelveenseweg

[59] She was buried on 13 October in Blaricum in a private grave, from where her body was moved on 21 October to a double grave. Brouwer insisted on two separate graves for him and his wife.

[60] Much later in life Louise had acquired enough critical sense to separate her worldly interests from her religious convictions—her adviser at the end of her life was a shrewd and impartial Jewish lawyer.

(Amsterdam), two houses on the Surinameplein (one of which housed the pharmacy) (Amsterdam), two villas and a wooden cottage (the hut) in Blaricum and Laren, furthermore an assorted collection of holdings in 't Gooi, mostly heather and wooded land, but also agricultural land. The estate also comprised a fair (but not large) collection of shares (roughly 86,000 guilders), the capital of the pharmacy (28,497 guilders) and some fifteen hundred guilders at various banks. On the debit side there was something like D.fl. 80,000,–. (We note that Brouwer was so conscientious as to even list a claim of D.fl. 15,000,– on the Sodalitas Bath Company; a dubious claim, as he expressed it.)

On 5 July 1961, Brouwer, Louise, and Cor reached under the watchful eye of the notary a settlement, it was agreed that the joint property would not be split up for the next five years. Louise immediately received an amount of ten thousand guilders (to be deducted from her share). Brouwer was put in charge of the administration of the estate, from which he had to pay out 400 guilders monthly (and 600 in December) to Louise—again to be deducted from her share. As Brouwer had to manage the pharmacies, he had to be able to raise the required funds for running the shops. For this purpose he was allowed to mortgage the property for no more than fifty thousand guilders. Once an agreement was reached, Louise's solicitor started to raise problems, after long negotiations he was prepared to concur with a mortgage of thirty thousand guilders on Brouwer's property Torenlaan 68, on the condition that Brouwer surrender his mandate for a mortgage of twenty five thousand guilders at the 'Nationale'.[61] 'Miss A.L.E. Peijpers acquiesces—for the time being— in the situation, now that her stepfather is prepared to allow her five thousand guilders, to be deducted from her legitimate inheritance, where he wants to draw 30,000 guilders from the estate', as the solicitor of Louise wrote to the notary.

In spite of the formal arrangement, a substantial residue of distrust remained on both sides. Louise was in spite of everything a shrewd operator. She was determined not only to get the maximum out of the estate, but also to put a spoke in her stepfather's wheel whenever she could. Her lawyer apparently had no compunctions carrying out her wishes, at the same time serving his own interest. There is an extremely bitter letter from Brouwer to his notary, Mr. W. van der Ploeg, which shows that Brouwer was at the end of his tether.[62]

> Now that the position Mr. Groos has taken with respect to the settlement of the estate Brouwer–de Holl seems to be beyond the power of the judiciary, on the basis of this position Mr. Groos assigns himself, in the matter of the settling and its duration, the exclusive authority to charge me with the complete management of the estate, which demands from me complete workdays of at least 12 to 14 hours, under refusal of remuneration, to forbid me to have the mentally deranged co-heiress Miss Peijpers placed under legal constraint, to forbid me to charge the cost of the administration to this administration, and to

[61] Insurance company.

[62] Brouwer to van der Ploeg, 12.VIII.1963.

> use the revenues of the estate, which are by far exceeded by the expenditures, exclusively for fees, determined by himself, for himself;
>
> now that this attitude of Mr. Groos has finally, after almost four years, caused me bitter poverty, serious illness, and a harmed international reputation;
>
> now that any prospect of rescue from this situation is lacking,
>
> logically speaking the consequence of this situation would for me be suicide, were it not that the consequences for others which follow from it, prevent me from it.

The letter is no doubt a dramatic protest against mostly real injustice; this time Brouwer was up against a formidable opponent, a slick lawyer with a keen eye for weak spots. In cases like this logic and common sense are no match for the legal mind, sharpened by ages of casuistry and sophistry. Although Brouwer could not produce hard evidence, he was convinced that even his stepdaughter was in the end the victim of her councillor; there is a private note of Brouwer in the archive in which he mentions that he was told that Louise had said to the informant, ‘How did dad find out that Mr. Groos had given me no more than one half of the ten thousand guilders?’ Not that this information was very useful to Brouwer, but it did little to make him more accommodating towards his stepdaughter.

Under the management of Cor Jongejan the shop went gradually further down the road to friendly anarchy. She was easygoing to the extreme; at the end of the day she would often, without even counting, empty the cash register into her shopping bag or a plastic bag. When, as happened from time to time, she forgot her bag, one of the assistants would find it the next day. The stock was not systematically checked, old and obsolete medicines packed the cupboards and the replenishing of the stocks was left to the wisdom of the visiting salesman. When the pharmacy had night duty, instead of the single assistant required, there were often four assistants on duty. It was not unusual that cash was taken from the till and used to buy delicacies at the night shop round the corner, to pass the night. By and large Brouwer lost a good deal of money on the pharmacy. He often dropped in at the Surinameplein, carrying out his administrative duties, surrounded by the small army of assistants (all female). Brouwer enjoyed these visits, it was a special treat for him to retire to the bathroom at the first floor and to conduct his business from the bath. Covered by a towel he would sometimes call in an assistant to give instructions; lying in the bathtub he read books, recited poems, did the administration and felt happy.

In 1965 Brouwer finally sold the pharmacy; this was not a simple business transaction, dealing with Brouwer required patience and subtlety. The new owner, Mrs. Hanny van Lakwijk, negotiated for a full year before the sale was concluded. Brouwer expected the new owner to come up to his expectations, just any buyer would not do. Hanny visited Brouwer repeatedly in Blaricum to discuss business, and to be inspected. The negotiations were more similar to a courtship than to a business transaction—the pharmacy had to be won. Hanny was thoroughly familiar with the strong and weak points of the pharmacy, as she had served as a ‘provisor’, so she could not be fooled about the business details. But dealing with Brouwer was not just a matter of looking at figures and real estate. She would find Brouwer in Blaricum reclining on his old wicker chair, in his white tropical suit, under a straw hat with holes. Brouwer opened the discussions innocently enough and acted as though he was a stranger to business, leaving the floor to the other party. Then suddenly he would draw a sheaf of paper with all the relevant data from under his chair, and argue sharp and to the point. Brouwer, moreover, had simple but non-negotiable conditions, for example Cor Jongejan should stay on as an assistant, and the pharmacy should continue to operate under the name 'Brouwer–de Holl'.

The negotiations were finally successful, partly because Brouwer had developed a strong sympathy for Hanny, without however allowing her to get too close to him. He kept a certain distance by corresponding with her in English. Cor said to her 'the professor is quite fond of you, but he wants to keep his distance'. He viewed his relation to the prospective owner more or less as a grandfather would look at a promising grandchild. He could in a paternal manner be quite strict with Hanny's correspondence. It happened, for instance, that Cor Jongejan handed her back a letter of hers with the words, 'The Professor has corrected your letter, he wants you to rewrite it.' He had said to Cor, 'It looks like the letter of a little girl; she has studied, she must be able to do better.'

Brouwer's reputation as a difficult man was made clear to Mrs. van Lakwijk by her former professors; when it became known that she was interested in buying the pharmacy, she suddenly got a telephone call from Prof. Cornelia van Arkel, 'would she immediately come and see her'. Van Arkel warned her in no uncertain words that Brouwer could not be trusted, illustrating this with many examples. If there was any person that could mislead you, it was Brouwer! A couple of days later the warning was repeated by Professor Kok, he summoned her to his house—what was she thinking of? Kok lectured her in even stronger terms on Brouwer's evil reputation, his ruining the pharmacy, etc.

In spite of this undoubtedly well-meant advice, Hanny carried on the negotiations, and finally in 1965 she could call the pharmacy her own. Brouwer was proud of her because she had conquered the pharmacy, as he put it, and left him the honour of graciously handing over the property.

An amusing detail of the transaction was that Brouwer insisted on payment in cash; she travelled to Blaricum at the appointed day with a suitcase full of small denominations! At the opening of the pharmacy under the new owner, Brouwer presented Hanny with a beautiful antique mortar filled with flowers.

Although the pharmacy had changed hands, Brouwer kept a vivid interest in it; he often visited the Surinameplein, where the assistants saw him, watching the shop from a distance.

—Oh, Mrs. van Lakwijk, there is Professor Brouwer watching us.

—Well, then he will shortly be in.

After some time, Brouwer would then move off in the direction of the shop, crossing the heavy traffic and walking along the track of the trams. The traffic then miraculously parted, like the waters the Red Sea. On this point all witnesses agree that Brouwer managed to ignore the traffic in his path. Students and colleagues who saw him making his way to the mathematical institute, or some other place, were horrified to see him plunge absentmindedly into the flow of cars and cyclists.

One thing Brouwer was loath to give up; owner or not; he would insist to take his customary bath in the bathroom belonging to the pharmacy, if Hanny happened to be absent. The collected assistants tried to stop him—'Oh no professor, you can't do that, you are no longer the owner', and when they failed to do so, they started to talk him out of the bath. Finally he would give in and say, 'All right, then give me a glass of warm milk.'

## 18.3 The Last Years

The accumulated pressure of the legal problems, the loss of his wife and the never forgotten or forgiven injuries of the post-war years increased Brouwer's tendency to see enemies all around him. His last years were not free from forms of paranoia, he felt threatened by criminal conspiracies. In the most literal sense he was convinced that 'they' were after him. He was under the impression that criminal parties broke into his house and spirited away incriminating material. Often the police were called in after a reported theft; usually the missing files or briefcase were discovered somewhere else in the house. The legal battles, of which he fought a number simultaneously, gradually transformed the living room. The floor was occupied with stacks of documents of case 1, case 2, ...Like the commander of a beleaguered fortress Brouwer would, before going to bed, make the rounds, and bolt all the doors and windows, only to have them reopened by Cor Jongejan, who quietly let in a friend when Brouwer had gone to bed. After the ladies had spent the rest of the evening with animated discussions, she would re-lock the doors and Brouwer was none the wiser.

His paranoia did not improve over the years. There is a (unconfirmed) story that he wrote to Dag Hammarskjöld, the Secretary General of the United Nations, telling him about his situation, and the criminal designs at his life and property. According to the story, Hammarskjöld send a kind reply guaranteeing him protection within 30 kilometres of his home.

Although Brouwer had taken his leave from the Science section of the Academy, he kept an interest in the matters of the Academy. He visited the meetings of the Science section when he felt like it. In the early sixties he was asked to present a revised set of voting rules. A considerable amount of time and correspondence was taken up by that project in 1961, and well into 1962.

The fact that Brouwer considered himself the victim of machinations of some of his colleagues, and deplored the lack of support from the remaining mathematicians, did not stop him from attending the Academy meetings. He was a frequent and outspoken visitor of the meetings of the Physics section (which he had left to join the Free Section). He could comment on the past with a cynical detachment. One such case was recalled by N.G. de Bruijn: after a meeting of the Academy Schouten came upon Brouwer, who was talking with a small group of members. Schouten, in whose opinion the conflict with Brouwer had lost its edge long ago, judged this a good opportunity to restore peace.

**Fig. 18.8** Brouwer posing for the sculptor Wertheim [Brouwer archive]

Schouten—let us forget about the old differences.

Brouwer—do I then get my journal back?

In 1960 Brouwer suddenly became active in the Academy; at the extra-ordinary meeting of the section of the sciences the appointment of a new foreign member had been tabled. A number of members had put forward Kuratowski as a candidate, but Brouwer had proposed his own candidate, Sierpinski. Brouwer conducted his campaign with great display of words. Apparently one of the members had remarked that if Sierpinski—a man well advanced in years— were appointed, it might be for only a year or so. This provoked a spirited rejoinder of Brouwer, who was actually only one year older. He praised Sierpinski's qualities with gusto,

> During his whole life so far, Sierpinski has continuously broadened, deepened and rejuvenated his thoughts; recently his researches took him into number theory, and he has started with fervour a seminar. His leadership on the areas which he opened up has remained undisputed—no matter how famous, astute and original some of his collaborators may have been. Without claiming completeness, I mention among them Borel, Baire, Lebesgue, Hausdorff, Young, Hobson, Alexandroff, Kuratowski, Tarski and Carnap. Among them Sierpinski shines with undiminished luminosity in the epistemological sky as a star of the first magnitude.

The nomination of Sierpinski was put into a flowery language that clearly betrayed Brouwer's hand,

> When round the latest turn of the century the physiology of the real functions had drawn the general attention, and thereby a vast domain of new problems had emerged, the treatment of which demanded a drastically deeper foundation of epistemology, and questions such as the justification of existence and the scope of the axiom of choice and the notion of continuity could no longer be ignored, the international mass of researchers, which was thereby inspired, has right from the beginning been led by Sierpinski, who, through his mental power and originality, obtained results that mark him as a grandmaster, and by means of his inspiring suggestiveness shaped the Polish mathematical

> school which finds its expression in the renown journal Fundamenta Mathematicae. [......]
>
> With Chopin, Paderewski and Mrs. Curie, Sierpinski belongs to the group of admirable figures which Poland has given to the world through the ages.[63]

Freudenthal had been one of the supporters of Kuratowski's appointment, he sharply criticised the decision of the Academy to reject Kuratowski and to appoint Sierpinski instead. He remarked that it would be hard to substantiate the mere claim of Sierpinski's fame as an equal of the above mentioned Poles. 'It would not have been an easy task to unearth from the possibly thousand mathematical papers of Sierpinski (mostly of 1 or 2 pages) enough material to justify this comparison.' Brouwer reacted with more praise for the old master—and he poured his indignation out over 'a member' who had remarked that on account of Sierpinski's advanced years, he would probably serve probably for no more than a year. In a first indignant reaction Brouwer quoted a member[64] describing Sierpinski as a senile old gentleman, well into his eighties. In the end Brouwer won the battle and Sierpinski became a foreign member of the Academy.

In spite of his advancing age Brouwer still considered emigration as a way to start a new life abroad, leaving his beloved, but, in his eyes, corrupt and degraded native country behind. In a letter to Marston Morse, he fondly recalled his stay in Princeton, and added that he was 'toying with the idea of being called back to the Institute some day for a longer stay. Which might also be to the profit of science, my circumstances in the Netherlands being absolutely prohibitive for scientific research.'[65]

In 1959 B.N. Moyls (acting head of the Mathematics Department of the University of British Columbia) approached Brouwer, asking whether it was true that he was interested in a position;[66] Brouwer eagerly replied, 'As a matter of fact I want to emigrate from the Netherlands, find a field of activity on your continent, and change my nationality as quickly as possible.'[67] When this did not work out, he asked in desperation if the 'joint universities could rescue me?' Apparently there was little that tied him to Holland after the death of Lize, nonetheless he remained active in his own characteristic way.

In spite of his advanced age he kept up his travels. At the third centenary of the Royal Society in 1960, he together with the astronomer Oort represented the Netherlands.

Brouwer had lost none of his fierceness and his sympathy for the underdog. His last action in the Academy is a worthy illustration of the spirit that never left him. In 1962 Brouwer put A.D. de Groot up as a candidate for a vacancy in the 'Free

[63] There are a number of versions of this text in the archive, which differ in formulation.

[64] Not Freudenthal, probably Schouten.

[65] Brouwer to Morse, 4.I.1955.

[66] Moyls to Brouwer, 7.VIII.1959.

[67] Brouwer to Moyls, 5.IX.1959.

**Fig. 18.9** The three brothers Brouwer [Brouwer archive]

Section' of the Academy. De Groot was a psychologist who had as a student enrolled in the mathematics faculty and subsequently had switched to psychology. He was a student of Révész, the father of the Amsterdam school of psychology. He wrote a dissertation on 'The thinking of the chess player', a work that more or less established his reputation (and which has been rediscovered now by the cognitive artificial intelligence people). Brouwer devoted much time and ingenuity to De Groot's candidacy, but the opposing forces this time got the better of him.

Among the smaller actions in which Brouwer took part one must mention his support of the candidacy of the Dutch author Simon Vestdijk for the Nobel prize for literature (1964).[68] This action was not successful either.

Brouwer still regularly visited the meetings of the Academy, where he had a fierce reputation. It was, for example, not unusual to see him quarrel there with his brother Aldert, the geologist.

By now his old friends had died or were no longer active. His brother Lex had died in November 1963, and Brouwer spoke at his grave:

> In Izaak Alexander Brouwer there was an experience of beauty and a capacity for expressing beauty, in painting, music and literature, which seemed boundless.
>
> Further a rare erudition in the area of technology, critical, inventive and constructive.
>
> Next to this talent, which was also visible in his personal appearance, there was no place for a social fighting spirit, not for a regular and hygienic way of life, not for finishing, displaying in public and delivering own creations, not for efficient measures against setbacks.
>
> Izaak Alexander Brouwer has been surrounded in his inexorable isolation by the admiration, love and friendship of many. They will cherish an unforgettable memory of him.

[68] Vestdijk was not only an outstanding but also an extremely prolific author, one of his fellow authors wrote 'he who writes faster than the gods can read'.

True and compassionate words for a brother who had great dreams, but did not know how to realise them. Lex Brouwer lives on in the memory of his pupils of the Baarn's Lyceum, where he taught French. The eccentric, scholarly man with his motor bike was one of the more striking personalities at school.

Where the oldest and the youngest Brouwer brothers had making a lightning career, the one in the middle had not chosen to follow the wishes of his parent. Lex had started to study chemistry (and according to Aldert's son Aldie, he was very successful in the subject), but he soon gave up a prospective academic career in order to become a painter. He ran away to Paris, and started to learn the basics of the art. He also found an attractive girlfriend, Winnie, whom he married. His parents were not happy at all with this unexpected demonstration of independence, they hastened to Paris, and carried the young man and his wife back to Holland. There he, rather reluctantly, enrolled in the study of French to become a teacher. In the end he lived between the two worlds of Art and Teaching. Life had not brought him what he dearly wished, and he remained a rather disenchanted man to his last days. He settled in Blaricum, where he became the owner of a charming little house with a studio. There is a legend that he was so gifted in the French language that he wrote a brilliant dissertation, but never defended or published it; instead, it is told, he buried the manuscript in his backyard (it has never been found).

In the same year Tine Vermey, one of his intimate friends, closed her eyes; at the funeral Brouwer could not master his emotions. Ru Mauve, the last of the small band of Brouwer's fellow students, also passed away in 1964. When his son Maarten went to Brouwer's house to bring the sad news, Brouwer was silent for a while, and turned to Cor, sighing ‘Ru was a good man, he learned easily, but he never in his whole life had an original idea. However, if you talked to him, he was so sharp, that he could always catch you at smaller points.’ Thus, as it is with true friends, the memories of the weak and strong points had blended into a tender remembrance. Maarten recalled vividly how Brouwer came to the house to pay his respects. Ru was lying in state at home, as was the custom. There, in a coffin lined with satin, he was in a suit with silver buttons. When Brouwer arrived, he went straight up to the room where the coffin was placed and kneeled before the coffin. There he remained in thoughts in total silence for at least twenty minutes. Returning to the family gathered in the sitting room, he immediately embarked on a lively conversation. His disregard for the conventional atmosphere of mournfulness did by no means reflect on the sincerity of his grief.

The aged Brouwer had not lost his eagerness for human company. He regularly entertained a small circle of female admirers, who loved to follow his lengthy and often fantastic tales. He had, ever since he shed off the reserved habits of his student years, been an almost compulsive talker. He could expand any subject, his phenomenal memory allowed him to use details that he had picked up, nobody knew how long ago. In addition to the, say, rational part of his conversation, he loved to add little mystifications and fantasies.

Colleagues from abroad, who visited him after his retirement, recall his conversation, from which one had to pick up little bits of historical information, as one picks gold nuggets from a mass of mud and stones.

**Fig. 18.10** Brouwer and Cor Jongejan [Brouwer archive]

When Kleene visited him, he learned the story of Brouwer's walking tour to Italy in 1902; Brouwer illustrated how he wrapped himself in his cape for sleeping in the open air.

Although he avoided contact with his Dutch mathematical colleagues, he was hospitable to visitors, and there often were old friends or students who dropped in. A regular visitor was the chess master Euwe, who could get along very well with Brouwer. Euwe, after his chess career (which coincided with his teaching of mathematics at a girl's school in Amsterdam), had become a computer scientist. He got a chair in Rotterdam, and in addition he was an advisor to Remmington. One might wonder how Brouwer viewed the advent of the computer and computer mathematics, which was already taking place during his life time. Not much is known beyond the little that Euwe told. Brouwer attended in 1962 Euwe's congress 'Man and Robot', and Euwe related how Brouwer had shown a clear grasp of the principles behind the (then) modern computers, when he was given a guided tour by Euwe along the computing machines which were under his care. The attending staff at the occasion showed surprise at the understanding of 'the old gentleman'. He had, however, enough self-knowledge to take a realistic view of the offer of an adviser's position with the firm, which would have brought him 10,000 guilders. Just his name was worth the money, Euwe told him. Brouwer declined the offer, he had to finish more urgent things first.

After the death of Lize, Cor Jongejan and Tine Vermeij had become, if not 'next of kin', at least 'next of relationship'. They played a major role in the remaining years of Brouwer's life. The relationship with Louise was a disaster, to put it mildly, a situation of mutual dislike and distrust, interrupted by moments of open warfare. With Cor and Tine Brouwer could get along well enough; where love had been a feeling of loyalty and admiration remained on the side of the ladies. In the past they had fought fierce battles for Brouwer's favour, but at a riper age—Cor was 66 and Tine 72 when Lize died, 89 years of age—the competition had petered out. On many occasions Cor or Tine accompanied Brouwer on his trips abroad. Brouwer felt that his obligations were really to Cor, who had sacrificed her personal future to that of Brouwer. Consequently, he made her a marriage proposal, which she turned down. She remained his universal support, managed the pharmacy, ran the house, did the garden etc. Hanny van Lakwijk described the quiet leisurely atmosphere in the garden. Brouwer was lying in a hammock, and from above instructed Cor how to plant what where. An old-fashioned scene of a man with a straw hat and a pretty middle aged woman in the mixture of light and shadow, among the luxurious trees and shrubs.

The housework was done by a local help, Ali. One would have thought that by now, Brouwer turning eighty, it was time to settle into a comfortable routine. One of slippers, good books, a warm fire, music, ....

Not so for Brouwer, he kept a vivid interest in the university; he became a familiar figure at Ph.D. examinations, where he took part in the questioning of a rich variety of candidates, not only in the mathematics and physics sections, but also in the literary faculty. He displayed a remarkable mastery of the classic languages, examining candidates in Latin or Greek if the occasion arose. As his knowledge of a considerable part of the subjects in mathematics and in physics was often based on first hand information from the grand old men of the subjects, he liked to interlace his questions with long and often intriguing references to his experiences in the far past. When Hans Mooij defended his dissertation on the work of Poincaré, Brouwer could not let the occasion slip by without presenting an extensive exposition of the views of Poincaré, as observed by him first hand. The excursions in the literary faculty may have had something to do with a new development: he fell once more violently in love with a most remarkable lady, a colleague in the literary faculty. The object of his love was Emily Haspels. She was a professor in archaeology, with a great expertise in Greek vases. Emily, who was thirteen years Brouwer's junior, had entered Brouwer's life through his interest in languages.

His love for the classic languages had been the factor that brought them together. They had regular sessions, during which they read together the ancient-literature. And soon this mutual interest turned into something more personal. This new person in his life was a paragon of intelligence, determination and tenacity. She had in the thirties travelled in China and worked in Turkey. It was in Turkey, where she carried out her research on vases, that the Second World War surprised her. When she learned about the fate of Holland, it was too late to return, the Turkish government gave her no permission to leave the country. After working as a cleaning lady

**Fig. 18.11** Brouwer taking part in Jan Hilgevoord's examination for the doctor's degree (1960). [Courtesy Jan Hilgevoord]

in schools, she eventually got a teaching job at the university, as the authorities had discovered her expertise. After the war she got in 1946 a chair at the University of Amsterdam.

The new relationship differed in an important respect from the loves of his younger years. This time Brouwer's relationship was of a different nature. He was the party who stayed behind while Emily travelled around the world and carried out her research. The letters that Emily wrote to Brouwer were eagerly read, inscribed with the dates of reception and other information. There is no doubt that the affection was mutual, this was a case of two strong characters, who met in a mixture of love and admiration. Both were undoubtedly serious about their affections; there is a large number of letters and jottings, probably given to Emily by Cor Jongejan after Brouwer's death, and they show a picture of sincere interest and affection. Emily travelled all over the world for her research, there are letters from Lyons, Venice, Oxford, Stanford, .... In Oxford she met the ladies Deneke, who were descendants of Maurice Philip Deneke, a musicologist, who left money for a series of lectures, called the Deneke lectures. Staying with Margaret Deneke, she found out that Brouwer had in 1949 given a lecture in the series, with the title: 'The influence of mathematics on logic'. A lecture that was not only appreciated by the ladies Deneke, but also by the scholarly community. Emily wrote that she had been in Lady Margaret Hall (in Oxford), where she was told of the impression Brouwer made, 'all the mathematicians' faces in the audience lighted up, because they got something worth their while'.[69] Suddenly, in the middle of the correspondence, there is a letter from Brouwer, mentioning an infarct, 'My infarct a few times very dangerous but already the awareness of a wide variety of indispensability keeps me alive.'[70]

In 1966 Emily visited the United States for a lecture tour, combined with negotiations with various publishers concerning her new book. The tour, from May till August, also took her to Princeton, where, clearly on Brouwer's request, she tried to meet Oppenheim and Marston Morse. Both appeared rather elusive, Oppenheim because of his terminal illness, and Morse because he made few appearances. Finally, in July, she happened to run into him. After Morse introduced himself, she immediately asked him if he knew Brouwer. 'Yes, he has been here for a few weeks once; he has a very famous name, internationally, at home he has a rather stormy career, I gather.' 'Yes', she replied, 'apparently others cannot look ahead the way he does'.[71]

The fact that Brouwer's travelling range had become smaller did not mean that he had lost his appetite for company and for social and scientific meetings. One of those meetings took him in 1962 to Middelburg in Zeeland. Brouwer was attending some meeting in Middelburg when he decided to look up Karen, the daughter of his friend Ru Mauve, who lived in Veere, in a roughly three hundred and fifty year old warehouse. Since the buses were running only once every one and a half hours, the eighty one year old Brouwer set out on foot for a walk of some ten kilometres. Agnes van den Noort-van Gelder, Karen's daughter, described her meeting with Brouwer:[72]

> I was 16 at the time, and I was with Karen in the kitchen (on the first floor, we lived upstairs, downstairs was too cold). Then there was a ring at the door, and Karen looked out of the window—'Say, there is Brouwer!', she said astonished and surprised, and hastened down the stairs.
>
> Curious, I also looked from the window and saw a piece of what appeared to be a black cape. Down there the greeting took time, but finally Brouwer mounted the steep stairs, and there he appeared from the curtain of the stair well—with his sharp pale face and the white spiky hairs of such a length that they curled into his collar. I did not know how to react—what shoes. And the cape. And how funny he smelt, dusty, not very pleasant. [...]
>
> Pretty soon I was sent to the green grocer for mushrooms. Mushrooms! Not so long ago these could only be bought to order, and then only on Saturdays! Well, Brouwer stayed for dinner: mushrooms and rice and lettuce. [...]
>
> Karen told me that Brouwer was ve-ge-ta-rian. A new word! Hence the fancy mushrooms. It was with reluctance that I sat next to him at the table—he

[69] Haspels to Brouwer, 25.VI.1964.

[70] Brouwer to Haspels, 30.VIII.1965.

[71] Haspels to Brouwer, 31.VII.1966.

[72] Agnes van den Noort-van Gelder to Van Dalen, 14.III.2002. In the same letter Agnes provided some extra information on Brouwer's famous walking tour to Italy in 1901 (Cf. p. 31). Old family documents that had been found showed that Brouwer had stayed with Ru and his wife in Florence. The young couple had rented the Villa Belvedere at the Via Dante da Castiglione.

sat at my sister's place, who had already 'left the house'. I remember that the conversation was difficult to follow. Brouwer's use of language was curious for me, and his sentences were rather complicated. But he taught me that one should *always* eat with little bites, and then *always* chew 32 times. I did not dare to think this nonsense (but I found out later that 32 times is not so much after all).

After my parents had taken him to the bus, my father said: 'Now you have seen a genius!'

A severe loss was the death of Erich Gutkind in September 1965. The Gutkinds and Brouwer had renewed their friendship in 1953, and Brouwer immediately had resumed the old relationship. The death of Erich left him 'crushed like you, and with you', as he wrote to Lucia,[73] and a couple of weeks later he begged her to join him,

> How Eka continually floats around me and looks at me! With respect to travel abroad I am for the time being a prisoner. But couldn't you come to me, for a short time or as long as you like? Eye in eye, Bertus[74]

In March 1966 his old friend Zernike died, he attended the funeral in Groningen, but when he was begged not to speak at the grave, he angrily marched out of the room.[75]

The estate of Lize was not yet settled and Brouwer (one would guess, correctly) surmised ill-will or worse on the side of Louise and her representative, Mr. Groos; on April 19, 1966 he filed a formal complaint against Groos. 'Criminal actions of private persons and official authorities', he said, 'led to his bankruptcy'. The battle with the tax collector also went on. On October 31, 1966 he got a distress warrant ordering him to pay the overdue taxes within two days, on penalty of seizure of Brouwer's possessions by the bailiff.

The year 1966 was almost over when Brouwer finally fell victim to motorised traffic, as so many of his friends had feared and predicted. Cor Jongejan was celebrating with the staff of the pharmacy Saint Nicholas' eve (*Sinterklaas*) on the second of December (3 days early), and in order not to be interrupted, she had proposed to turn off the ring of the telephone. Hanny van Lakwijk then observed that the light of the telephone was continually flashing. Finally she picked up the telephone and received the stunning news that Brouwer had died as the result of an accident. It appeared that he had been asked by Cor to deliver some presents for Saint Nicholas' eve to friends across the road; he had picked up a grey horse blanket and wrapped it around him. When crossing the Torenlaan in front of his house, almost invisible in the ill lit lane, he was hit by a car, which knocked him into the other lane. There, two more cars ran over him, and he died almost immediately. Cor was heartbroken, she felt that she had been responsible. If only ....

[73]Brouwer to Lucia Gutkind, 6.IX.1965.

[74]Brouwer to Lucia Gutkind, 17.IX.1965.

[75]Oral communication, N. van Kampen.

A few days later Brouwer was buried in the local cemetery, next to the grave of Lize. He had always expressed his objections to a shared grave. The funeral was, at Brouwer's request, attended by a handful of close friends; Max Euwe held the funeral oration.

Emily was in Oxford when all this took place. There was a small scrap of paper among her notes and letters that said 'and then when you call his house, suddenly—"he is dead—will be buried in one and a half hour". You do not survive something like that—you live on, but you don't get back to normal.'

## 18.4 Epilogue

Brouwer's will appointed Cor Jongejan as the sole beneficiary. Cor thus came into the possession of a substantial fortune, at least on paper—most of it was tied up in real estate and land. The negative side of the inheritance was the debt to the tax collector. No love was lost between Brouwer and the Internal Revenue Service, his relation with the tax collector consisted of a long series of requests for postponement of payment, appeals, arguments, personal interviews, etc. Later in life, in particular after his retirement, he regularly got into serious problems with the Internal Revenue Service. In 1964 the inspector turned down Brouwer's claim for a reduction in tax, and in 1965 the bailiff was at his door, giving Brouwer ten days time to pay his taxes, under threat of a considerable penalty. Brouwer did not cheat on his taxes, but he certainly looked for legal means to minimise the tax assessment. He had quite a reputation with the tax inspectors, he could talk until the inspector's head was spinning. Louise told that, speaking about taxes, Brouwer had said, 'When I die, I will leave a mess that nobody will be able to sort out.' Poor Cor, as the sole heiress, had to take the rap; three days after Brouwer's death she was summoned to the police station to surrender her passport. No foreign travel until the tax debt was cleared. Through Brouwer's, accidental or intended, mismanagement of his tax returns, a considerable debt had accumulated. Moreover, being no kin of the deceased, she had to pay the maximal death duty of 67%.

Brouwer's estate contained a large number of documents, of a diverse nature. There were legal documents, scientific ones, and of course personal correspondence. Cor conscientiously sifted the whole mass of paper and put the scientific *Nachlass* into the hands of the Dutch Mathematical Society. The chairman, Professor Dijkman (an intuitionist and a former student of Brouwer), collected the material and stored it at the Technical University of Delft. Freudenthal and Heyting had access to the material, which they used for their edition of the Collected Works.[76] Before the transfer of the material, Alexandrov visited Cor Jongejan when he was in Holland to give a lecture at the occasion of the opening of Freudenthal's new mathematical institute in Utrecht (1967). He asked for, and obtained, the remaining documents that related to Urysohn's and his own connections with Brouwer. As an editor of

[76] In 1976 Dijkman handed the material over to the Brouwer archive in Utrecht.

Urysohn's collected works[77] he had an understandable interest in the documents. The material was not returned, but fortunately it was preserved in Alexandrov's own archive, and Professor Shiryaev made it kindly available for the present biography. Alexandrov's talk in Utrecht was an interesting historical survey of topology in Holland, in particular in Brouwer's school, it has been published in the *Nieuw Archief voor Wiskunde*.[78] Brouwer's private library was sold to a well-known Amsterdam bookseller-antiquarian. It would have been a valuable source of historical and biographical information, as many books and monographs carried Brouwer's notes and comments in the margin. Alas, the sale of the books could not be traced in the files of the firm.[79]

Brouwer had prudently protected his correspondence with Lize from curious eyes; he had instructed the notary to burn the whole bundle. In fact, much more correspondence was destroyed at various occasions. For example the correspondence of Tine Vermey was turned over to Brouwer by her niece, Mrs. Versteegh-Vermey, and subsequently destroyed by Brouwer. Cor Jongejan sifted the correspondence and notes of Brouwer, and the more intimate items were burned. After the death of Brouwer's brother Aldert, the son Jan made a huge bonfire of his papers and letters, remarking that the world would better be spared the uncountable fights and conflicts for which his father had a reputation. In the process many letters of Bertus were lost too.

Brouwer's brother Aldert almost immediately laid a claim on Cor. He was in poor health, and he avoided spending his money if he could get someone else to pay for him. Together with her friend Mrs. Volmer, Cor went three times a week to Aldert's house with homemade vegetarian meals. She drove him around in her little car; she and Mrs. Volmer drove him, for example, several times to the Spa in Bad Pyrmont. Cor did not survive Brouwer for long; she died two years after Brouwer, leaving the estate to a son of Aldert, the teacher Aldie, stipulating that Aldert had the usefruct during his life. This transfer of the estate once more was taxed at the maximum rate.

Cor was a generous person who did not begrudge others a possible intimate friendship with Brouwer. Where female 'admirers' were concerned, she philosophically shrugged her shoulders, 'when I was young I did the same thing', she said. Her relationship with Emily Haspels was friendly and uncomplicated. After Brouwer's death she made her a present of Brouwer's own copy of '*Life, Art and Mysticism*'; a copy with Brouwer's notes in the margin.[80]

The Pharmacy *Brouwer–de Holl* was successfully run by the new owner, Hanny van Lakwijk. But when in 1995 the time had come for her to retire, the pharmacy fell victim to the reorganisation of the system of pharmacies in Amsterdam. The

[77] Urysohn (1951).

[78] Alexandrov (1969).

[79] It seems that the library was sold in its entirety to some university library in Canada or the United States.

[80] The copy came after her death into the hands of a nephew, and after his death it could not be traced.

pharmacy was taken over and subsequently closed down, so there is nothing left that reminds of the joint enterprise of Lize and Bertus.

After Brouwer's retirement, intuitionistic mathematics was taught by his student Heyting. It is hard to think of a greater contrast between teacher and student; Brouwer had the flamboyance of the artist-scholar, the man of the world, whereas Heyting was the cautious, precise scholar, who avoided the theatrical gestures and bold formulations, who felt safety in symbols. Brouwer felt no scruples about being right and telling people so, Heyting on the other hand was a man of compromise, who recognised other foundational views as legitimate possibilities, even though he did not share them. The only condition for him was exactness. For him the *Grundlagenstreit* was a historical accident, a blatant failure of common sense and of communication. As Heyting wrote to Bernays after Brouwer's death, 'Brouwer's expectation that intuitionistic mathematics would take the place of classical mathematics will remain an illusion (fortunately), but I am pleased to say that intuitionism nowadays gets more recognition and acceptance, than was the case only a few years ago.'[81]

Heyting was successful in making intuitionism respectable, he wrote and lectured in a style that was better adapted to the taste and customs of the newer generations. He had little or no affinity with Brouwer's more extravagant views and aims, a work like *Life, Art and Mysticism* was viewed by him as something that might jeopardise the recognition of intuitionistic mathematics, and its content was alien to him. Indeed the spirit of the times had changed, the fighting mood had become a thing of the past. A man like Paul Lorenzen, who did not hide his views on right and wrong in science, strongly disapproved of Heyting's lack of fighting spirit, but he was one of the last of the foundational scholars with a moral conviction where science was concerned.

Under Heyting's supervision a steady supply of dissertations (almost all on intuitionistic topics) appeared. These were basically contributions to the program of rebuilding mathematics on intuitionistic principles. The foundational research into intuitionistic mathematics and logic was at the same time being picked up in the United States. The two leading scholars in this area were doubtlessly Stephen Kleene and George Kreisel. They, and Myhill and Kripke, gave a powerful boost to the study of systems of intuitionistic analysis. This short look at Brouwer's scientific legacy would not be complete without mentioning the man who took Brouwer's pragmatic program really seriously: Erret Bishop. Bishop launched a rebuilding of mathematics on a constructive basis, this time using all the experience gained in the past century, without being hampered by philosophical hangups, likes or dislikes. Bishop, with his considerable experience in modern analysis, set the standard for a new practice of constructive mathematics; his book *Foundations of Constructive Analysis* became the starting point for a wealth of interesting and successful research.[82]

[81] Heyting to Bernays, 25.V.1968. Indeed, at the Buffalo congress of 1968, Intuitionism and Proof Theory were given equal time in the program, cf. Myhill et al. (1970).

[82] See Bishop (1967), Bishop and Bridges (1985).

In Holland intuitionism was carried on by the next generation, A.S. Troelstra (Amsterdam) and D. van Dalen (Utrecht). The style and content was no longer along the lines of the pre-war tradition, but fitted into the international foundational activity. A substantial number of dissertations were written and the scope of the research was considerably extended.[83] In the new century Brouwer's home base, the University of Amsterdam, abandoned intuitionism and brought a historical and fruitful era to a close.

Those who had played a role in Brouwer's life were getting on in years; Van der Corput and Schouten, the two bugbears of Brouwer's later years, had left the stage soon after Brouwer's retirement. Van der Corput had spent only four years in Amsterdam before he accepted a visitor's chair in Berkeley. When he returned two years later to Amsterdam, he had lost his enthusiasm for the local activities. He soon returned to California, where he spent another twelve years.[84]

Schouten had retired in 1953; he took an active part in the management of Dutch mathematics, in particular he remained on the board of the Mathematical Centre until 1968. He died in 1971 in Epe, in the middle of the Netherlands, where he had settled in 1943.[85]

The younger generation, Beth, Freudenthal and Heyting, took over the foundations of mathematics. Beth, a man with a precarious health, had died in 1964, leaving a rich inheritance of logical ideas and techniques. Freudenthal, the man who had suffered most under Brouwer's rejection, had without a grudge acknowledged the greatness of the man who had been his inspiration at the turning point of his career.

Van der Waerden had left Amsterdam in 1951 for Zürich. To his many interests in mathematics, he had in the meantime added the history of mathematics; this earned him the lasting enmity and fierce attacks of Bruins, who used every opportunity to spell out the low opinion he had of Van der Waerden's historical expertise. Bruins' wrath was in particular reserved for Freudenthal. It may not be uncommon to see this inversion of guilt feeling; but when one realises that Bruins hated Freudenthal because he himself failed to do justice to the man whose position he had appropriated, one cannot suppress a feeling of astonishment. Indeed, Bruins went so far as to send Freudenthal a copy of his farewell lecture, with the message 'this is the first attack' on the cover. Bruins never left the University of Amsterdam; over the years he had become a specialist in Babylonian mathematics. In the fifties he had been a visiting professor in Baghdad. There he had the opportunity to study historic material and sites. Returning to Amsterdam he resumed his lecturer's position. In the following years Bruins' career was plagued by a long string of conflicts with his colleagues and the university. Eventually he was made extraordinary professor in the history of mathematics. Looking back at his career, one can but feel sad that such a gifted man generated so many conflicts—making enemies came natural to him. It is idle speculation to ask what would have become of him if there had not

[83]See, for example, *Constructivism in Mathematics, I, II* (Troelstra and van Dalen 1988a, 1988b).

[84]See Korevaar (1975), Duparc and Korevaar (1982).

[85]Nijenhuis (1972), Struik (1971).

**Fig. 18.12** Alexandrov and Freudenthal at the International Mathematics Congress 1954. [Courtesy Nationaal archief, Noord-Holland]

been a war. His role in the matter of Freudenthal's position in Amsterdam was never forgotten nor forgiven. The fact that Brouwer used him can only be seen as a partial excuse. In all likelihood he would have been a difficult man at all places and at all times.

Freudenthal, the Cinderella of the Amsterdam institute, was after all the most successful of Brouwer's entourage. He created a strong mathematical department in Utrecht, of which he was for years the unchallenged master. With a firm hand he established a powerful tradition in the subjects of his expertise: geometry, algebra, Lie groups, topology; at his retirement he left a department behind that had earned a prominent place in the international community. He never felt tempted to take his revenge on Brouwer for his dubious treatment during and after the war. On the contrary, he honoured him in word and writing. Freudenthal's obituaries fully did justice to the greatness of the man who revolutionised topology, and mathematics as a whole. Freudenthal's edition of Brouwer's topological work stands out as a precise analysis of Brouwer's ideas and as an homage to the man who surpassed all his contemporaries in the extraordinary force of his geometrical intuition. When Freudenthal retired he had become a major force in the didactics of mathematics.

Louise's history is easily told. Financial problems were no longer her worry. She was already, partially under her mother's exhortations, an example of piety, and when her mother and Brouwer had gone, she became even more deeply religious. In due time she choose to follow the anti-pope. In her house she had a rich variety of statues and icons. In her simple apartment she celebrated the mass. She gradually developed a measure of religious mania. Her mother, Cor Jongejan, and her stepfather peopled her dreams. Brouwer would rudely keep her awake at night. Only after Brouwer's centenary, when she got to read a brief story of his life,[86] she suddenly realised that the man was not crazy after all. The nightly disturbances stopped immediately, 'Brouwer feels that now justice is done' she remarked. She was bedridden

[86]van Dalen (1981).

**Fig. 18.13** Brouwer's grave at the Blaricum cemetery [Photo Dokie van Dalen]

for many years, but that did not stop her from ruling the world her way. She died at the ripe old age of eighty seven, a woman with a strong will, who had not found a world adjusted to her views.

Of Brouwer's property, the villa 'de Pimpernel' still exists, the hut and the other small houses and buildings have been replaced by run-of-the-mill estate housing for the affluent. Fortunately the town of Blaricum has generously taken over the care for the graves of Bertus and Lize; in 2003 the *Koninlijk Wiskundig Genootschap*,[87] the University of Amsterdam, and a generous private sponsor placed a tasteful commemoration glass monument at the grave.

[87]The Queen had in 2002 granted the predicate 'Royal' to the Dutch Mathematical Society.

# References


Aczel, P.: An introduction to inductive definitions. In: Barwise, J. (ed.) The Handbook of Mathematical Logic, pp. 739–782. North-Holland, Amsterdam (1977)

Alberts, G.: Jaren van berekening. Toepassingsgerichte initiatieven in de Nederlandse wiskundebeoefening 1945–1960. Amsterdam University Press, Amsterdam (1998)

Alberts, G.: Twee geesten van de wiskunde. Biografie van David van Dantzig. CWI, Amsterdam (2000)

Alberts, G., van der Blij, F., Nuis, J.: Zij mogen uiteraard daarbij de zuivere wiskunde niet verwaarloozen. CWI, Amsterdam (1987)

Alexander, J.W.: A proof and extension of the Jordan–Brouwer separation theorem. Trans. Am. Math. Soc. **23**, 333–349 (1922)

Alexander, J.W.: An example of a simply connected surface bounding a region which is not simply connected. Proc. Natl. Acad. Sci. USA **10**, 8–10 (1924)

Alexandroff, P.: Einfachste Grundbegriffe der Topologie. Springer, Berlin (1932)

Alexandroff, P., Hopf, H.: Topologie I. Springer, Berlin (1935)

Alexandroff, P., Urysohn, P.: Mémoire sur les espaces topologiques compacts. Verh. K. Akad. Wet. Afd. Natuurkd., Eerste Sect. **14**, 1–96 (1929)

Alexandrov, P.S.: Dimensionstheorie. Ein Beitrag zur Geometrie der abgeschlossenen Mengen. Math. Ann. **106**, 161–238 (1932)

Alexandrov, P.S.: Die Topologie in und um Holland in den Jahren 1920–1930. Nieuw Arch. Wiskd. **17**, 109–127 (1969)

Alexandrov, P.S.: Poincaré and topology. Russ. Math. Surv. **27**, 157–166 (1972)

Alexandrov, P.S.: Pages from an autobiography. Russ. Math. Surv. **34**, 267–302 (1979)

Alexandrov, P.S.: Pages from an autobiography. Russ. Math. Surv. **35**, 315–358 (1980)

Antoine, L.: Sur l'homéomorphie de deux figures et de leurs voisinages. J. Math. Pures Appl. **4**, 221–325 (1921)

Arzela, C.: Sull'essistenza degli'integrali nelle equazioni differenziali ordinarie. Mem. Accad. Sci. Ist. Bologna **5**, 131–140 (1896)

Baire, R.: Sur la non-applicabilité de deux continus à $n$ et $n + p$ dimensions. C. R. Math. Acad. Sci. Paris **144**, 318–321 (1907a)

Baire, R.: Sur la non-applicabilité de deux continus à $n$ et $n + p$ dimensions. Bull. Sci. Math. **31**, 94–99 (1907b)

Balke, E.: Chroniknotizen. Aus den Errinnerungen des Freundes (1902–1989). L. Mundschenk KG, Soltau (1973)

Bashkirtseff, M.: Journal de Marie Bashkirtseff. Charpentier et Cie, Paris (1888)

Beardon, A.F.: A Primer on Riemann Surfaces. Cambridge University Press, Cambridge (1984)

Behnke, H.: Semesterberichte. VandenHoeck und Ruprecht, Göttingen (1978)

Behnke, H., Sommer, F.: Theorie der analytische Funktionen. Springer, Berlin (1955)



D. van Dalen, *L.E.J. Brouwer – Topologist, Intuitionist, Philosopher*,
DOI 10.1007/978-1-4471-4616-2, © Springer-Verlag London 2013

Belinfante, M.J.: Over oneindige reeksen. Ph.D. thesis, Amsterdam (1923)

Bellaar-Spruyt, C.: Leerboek der Formeele Logica. Bewerkt naar de dictaten van wijlen Prof. Dr. C.B. Spruyt door M. Honigh. Vincent Loosjes, Haarlem (1903)

Bernays, P.: Über Hilbert's Gedanken zur Grundlegung der Arithmetik. Jahresber. Dtsch. Math.-Ver. **31**, 10–19 (1922)

Bernstein, F.: Die Mengenlehre George Cantors und der Finitismus. Jahresber. Dtsch. Math.-Ver. **28**, 63–78 (1919)

Beth, E.W.: Inleiding tot de Wijsbegeerte der Wiskunde, 2nd, revised edn. Dekker en van der Vegt, Antwerpen (1940)

Bieberbach, L.: Über die Grundlagen der modernen Mathematik. Die Geisteswissenschaften **1**, 896–901 (1914)

Bieberbach, L.: Ganesh Prasad. Mathematical research in the last 20 years (review). Dtsch. Literaturztg. **45**, 725–727 (1924)

Bieberbach, L.: Persöhnlichkeitsstruktur und Mathematisches Schaffen. Unterrichtsblätter Math. Naturwiss. **40**, 236–243 (1934)

Bishop, E.: Foundations of Constructive Analysis. McGraw-Hill, New York (1967)

Bishop, E., Bridges, D.: Constructive Analysis. Springer, Berlin (1985)

Blauwendraat, H.: Worsteling naar waarheid. De opkomst van Wiskunde en Informatica aan de VU. Meinema, Zoetermeer (2004)

Blumenthal, O.: Lebensgeschichte von David Hilbert. In: David Hilbert, Gesammelte Abhandlungen. III, vol. 3, pp. 388–429. Springer, Berlin (1935)

Bockstaele, P.: Het intuïtionisme bij de Franse wiskundigen. Verh. K. Vlaam. Acad. Wet. **XI**(32), 999 (1949)

Boersen, M.W.J.L.: De Kolonie van de Internationale Broederschap te Blaricum. Historische Kring Blaricum, Blaricum (1987)

Bohr, H.: Ny Matematik i Tyskland. Berlinske Aften (1934)

Bolland, G.J.P.J.: Aanschouwing en Verstand. Gedachten over Continua & Discreta in Wiskunde en Bewegingsleer. Adriani, Leiden (1897)

Bolland, G.J.P.J.: Zuivere Rede en Hare Werkelijkheid. Adriani, Leiden (1904)

Bonger, W.A.: Scheltema en het Socialisme (persoonlijke herinneringen). In: Ter herdenking van C.S. Adama van Scheltema, pp. 46–68. Querido, Amsterdam (1929)

Borel, É.: Leçons sur la théorie des fonctions. Gauthier-Villars, Paris (1898)

Borel, É.: Les "paradoxes" de la théorie des ensembles. Ann. Sci. Éc. Norm. Super. **25**, 443–448 (1908a)

Borel, É.: Sur les principes de la théorie des ensembles. In: Castelnuovo, G. (ed.) Atti IV Congr. Intern. Mat. Roma, vol. 1, pp. 15–17. Accad. Naz. Lincei, Roma (1908b)

Borel, É.: Leçons sur la théorie des fonctions, 3rd edn. Gauthier-Villars, Paris (1928)

Borwein, J.M.: Brouwer–Heyting sequences converge. Math. Intell. **20**, 14–15 (1998)

Boutroux, P.: L'idéal scientifique des Mathématiciens dans l'antiquité et dans les temps modernes. Felix Alcan, Paris (1920)

Braun, H.: Eine Frau und die Mathematik 1933–1940. Der Beginn einer wissenschaftlichen Laufbahn. Koechler, M. (ed.). Springer, Berlin (1990)

Brouwer, L.E.J.: Algebraic deduction of the decomposability of the continuous motion about a fixed point of $S_4$ into those of two $S_3$'s. K. Ned. Akad. Wet. Proc. Sect. Sci. **6**, 832–838 (1904)

Brouwer, L.E.J.: Leven, Kunst en Mystiek. Waltman, Delft (1905). Translation by W.P. van Stigt in Notre Dame J. Form. Log. **37**, 381–429 (1966)

Brouwer, L.E.J.: Polydimensional vectordistributions. K. Ned. Akad. Wet. Proc. Sect. Sci. **9**, 66–78 (1906a)

Brouwer, L.E.J.: The force field of the non-Euclidean spaces with negative curvature. K. Ned. Akad. Wet. Proc. Sect. Sci. **9**, 116–133 (1906b)

Brouwer, L.E.J.: The force field of the non-Euclidean spaces with positive curvature. K. Ned. Akad. Wet. Proc. Sect. Sci. **9**, 250–266 (1906c). Corr. in Brouwer (1909g)

Brouwer, L.E.J.: Over de grondslagen der wiskunde. Ph.D. thesis, Amsterdam (1907)

Brouwer, L.E.J.: Die mögliche Mächtigkeiten. In: Castelnuovo, G. (ed.) Atti IV Congr. Intern. Mat. Roma, vol. 3, pp. 569–571. Accad. Naz. Lincei, Roma (1908a)
Brouwer, L.E.J.: De onbetrouwbaarheid der logische principes. Tijdschr. Wijsb. **2**, 152–158 (1908b)
Brouwer, L.E.J.: About difference quotients and differential quotients. K. Ned. Akad. Wet. Proc. Sect. Sci. **11**, 59–66 (1908c)
Brouwer, L.E.J.: Het wezen der meetkunde. Clausen, Amsterdam (1909a). Inaugural address privaat docent, 12.10.1909. Also in Brouwer (1919b)
Brouwer, L.E.J.: Die Theorie der endlichen continuierlichen Gruppen unabhängig von den Axiomen von Lie. In: Castelnuovo, G. (ed.) Atti IV Congr. Intern. Mat. Roma, vol. 2, pp. 296–303. Acad. Naz. Lincei, Roma (1909b)
Brouwer, L.E.J.: Die Theorie der endlichen kontinuierlichen Gruppen, unabhängig von den Axiomen von Lie, I. Math. Ann. **67**, 246–267 (1909c)
Brouwer, L.E.J.: Karakteriseering der Euclidische en niet-Euclidische bewegingsgroepen in $R_n$. In: Handelingen van het Nederlandsch Natuur- en Geneeskundig Congres 12, vol. 12, pp. 189–199. J.L.E.I. Kleynenberg, Haarlem (1909d)
Brouwer, L.E.J.: Continuous one-one transformations of surfaces in themselves. K. Ned. Akad. Wet. Proc. Sect. Sci. **11**, 788–798 (1909e)
Brouwer, L.E.J.: On continuous vector distributions on surfaces. K. Ned. Akad. Wet. Proc. Sect. Sci. **11**, 850–858 (1909f). Corr. in Brouwer (1910f)
Brouwer, L.E.J.: Continuous one-one transformations of surfaces in themselves, II. K. Ned. Akad. Wet. Proc. Sect. Sci. **12**, 286–297 (1909g). Corr. in Brouwer (1911g)
Brouwer, L.E.J.: Beweis des Jordanschen Kurvensatzes. Math. Ann. **69**, 169–175 (1910a). Corr. in Brouwer (1910i, 1919o)
Brouwer, L.E.J.: Die Theorie der endlichen kontinuierlichen Gruppen, unabhängig von den Axiomen von Lie, II. Math. Ann. **69**, 181–203 (1910b)
Brouwer, L.E.J.: On continuous vectordistributions on surfaces, II. K. Ned. Akad. Wet. Proc. Sect. Sci. **12**, 716–734 (1910c)
Brouwer, L.E.J.: On the structure of perfect sets of points. K. Ned. Akad. Wet. Proc. Sect. Sci. **12**, 785–794 (1910d)
Brouwer, L.E.J.: Zur Analysis Situs. Math. Ann. **68**, 422–434 (1910e)
Brouwer, L.E.J.: On continuous vectordistributions on surfaces, III. K. Ned. Akad. Wet. Proc. Sect. Sci. **12**, 171–186 (1910f)
Brouwer, L.E.J.: Über eineindeutige, stetige Transformationen von Flächen in sich. Math. Ann. **69**, 176–180 (1910g)
Brouwer, L.E.J.: G. Mannoury, Methodologisches und Philosophisches zur Elementarmathematik. Nieuw Arch. Wiskd. **9**, 199–201 (1910h)
Brouwer, L.E.J.: Berichtigung. Math. Ann. **69**, 592 (1910i)
Brouwer, L.E.J.: On the structure of perfect sets of points, II. K. Ned. Akad. Wet. Proc. Sect. Sci. **14**, 137–147 (1911a)
Brouwer, L.E.J.: Beweis der Invarianz der Dimensionenzahl. Math. Ann. **70**, 161–165 (1911b)
Brouwer, L.E.J.: Über Abbildung von Mannigfaltigkeiten. Math. Ann. **71**, 97–115 (1911c). Corr. in Brouwer (1911h, 1921c)
Brouwer, L.E.J.: Beweis der Invarianz des $n$-dimensionalen Gebiets. Math. Ann. **71**, 305–313 (1911d)
Brouwer, L.E.J.: Beweis des Jordanschen Satzes für den $n$-dimensionalen Raum. Math. Ann. **71**, 314–319 (1911e)
Brouwer, L.E.J.: Sur une théorie de la mesure. A propos d'un article de M.G. Combebiac. Enseign. Math. **13**, 377–380 (1911f)
Brouwer, L.E.J.: Continuous one-one transformations of surfaces in themselves III. K. Ned. Akad. Proc. **13**, 767–777 (1911g)
Brouwer, L.E.J.: Berichtigung. Math. Ann. **71**, 598 (1911h)
Brouwer, L.E.J.: Intuïtionisme en Formalisme. Clausen, Amsterdam (1912a). Inaugural address, professor

Brouwer, L.E.J.: Beweis des ebenen Translationssatzes. Math. Ann. **72**, 37–54 (1912b)

Brouwer, L.E.J.: Zur Invarianz des $n$-dimensionalen Gebiets. Math. Ann. **72**, 55–56 (1912c)

Brouwer, L.E.J.: Über die topologischen Schwierigkeiten des Kontinuitätsbeweises der Existenztheoreme eindeutig umkehrbarer polymorpher Funktionen auf Riemannschen Flächen (Auszug aus einem Brief an R. Fricke). Nachr. Akad. Wiss. Gött. Math.-Phys. Kl., 2B **999**, 603–606 (1912d)

Brouwer, L.E.J.: Sur l'invariance de la courbe fermée. C. R. Math. Acad. Sci. Paris **154**, 862–863 (1912e)

Brouwer, L.E.J.: Über die Singularitätenfreiheit der Modulmannigfaltigkeit. Nachr. Akad. Wiss. Gött. Math.-Phys. Kl., 2B **999**, 803–806 (1912f)

Brouwer, L.E.J.: Über den Kontinuitätsbeweis für das Fundamentaltheorem der automorphen Funktionen im Grenzkreisfall. Jahresber. Dtsch. Math.-Ver. **21**, 154–157 (1912g)

Brouwer, L.E.J.: Continuous one-one transformations of surfaces in themselves, V. K. Ned. Akad. Wet. Proc. Sect. Sci. **15**, 352–360 (1912h)

Brouwer, L.E.J.: Beweis der Invarianz der geschlossenen Kurve. Math. Ann. **72**, 422–425 (1912i)

Brouwer, L.E.J.: On looping coefficients. K. Ned. Akad. Proc. **15**, 113–122 (1912j)

Brouwer, L.E.J.: Über den natürlichen Dimensionsbegriff. J. Reine Angew. Math. **142**, 146–152 (1913a). Corr. in Brouwer (1924h)

Brouwer, L.E.J.: Intuitionism and formalism. Bull. Am. Math. Soc. **20**, 81–96 (1913b)

Brouwer, L.E.J.: A. Schoenflies und H. Hahn. Die Entwickelung der Mengenlehre und ihrer Anwendungen, Leipzig und Berlin 1913. Jahresber. Dtsch. Math.-Ver. **23**, 78–83 (1914)

Brouwer, L.E.J.: Over de loodrechte trajectoriën der baankrommen eener vlakke eenledige projectieve groep. Nieuw Arch. Wiskd. **11**, 265–290 (1915)

Brouwer, L.E.J.: Luchtvaart en photogrammetrie. Avia **6**, 29–30 (1916a)

Brouwer, L.E.J.: Luchtvaart en photogrammetrie. Avia **6**, 122–124 (1916b)

Brouwer, L.E.J.: Luchtvaart en photogrammetrie. Avia **6**, 223–225 (1916c)

Brouwer, L.E.J.: Addenda en corrigenda over de grondslagen der wiskunde. K. Ned. Akad. Wet. Versl. Gewone Vergad. Afd. Natuurkd. **25**, 1418–1423 (1917a). Separate sheet with corrections inserted

Brouwer, L.E.J.: Luchtvaart en photogrammetrie. Het Vliegveld, 142–144, 165–167 (1917b)

Brouwer, L.E.J.: Begründung der Mengenlehre unabhängig vom logischen Satz vom ausgeschlossenen Dritten. Erster Teil, Allgemeine Mengenlehre. Verh. K. Akad. Wet. Amst. **5**, 1–43 (1918a)

Brouwer, L.E.J.: Über die Erweiterung des Definitionsbereich einer stetigen Funktion. Math. Ann. **79**, 209–211 (1918b). See Brouwer (1919c)

Brouwer, L.E.J.: Lebesguesches Mass und Analysis Situs. Math. Ann. **79**, 212–222 (1918c)

Brouwer, L.E.J.: Begründung der Mengenlehre unabhängig vom logischen Satz vom ausgeschlossenen Dritten. Zweiter Teil, Theorie der Punktmengen. Verh. K. Akad. Wet. Amst. **7**, 1–33 (1919a)

Brouwer, L.E.J.: Wiskunde, Waarheid, Werkelijkheid. Noordhoff, Groningen (1919b)

Brouwer, L.E.J.: Nachträgliche Bemerkung über die Erweiterung des Definitionsbereiches einer stetigen Funktion. Math. Ann. **78**, 403 (1919c). Re Brouwer (1918b)

Brouwer, L.E.J.: Énumération des surfaces de Riemann regulières de genre un. C. R. Math. Acad. Sci. Paris **168**, 677–678 (1919d)

Brouwer, L.E.J.: Énumération des surfaces de Riemann regulières de genre un. C. R. Math. Acad. Sci. Paris **168**, 832 (1919e)

Brouwer, L.E.J.: Énumération des groupes finis de transformations topologiques du tore. C. R. Math. Acad. Sci. Paris **168**, 845–848 (1919f)

Brouwer, L.E.J.: Énumération des groupes finis de transformations topologiques du tore. C. R. Math. Acad. Sci. Paris **168**, 1168 (1919g)

Brouwer, L.E.J.: Intuitionistische Mengenlehre. Jahresber. Dtsch. Math.-Ver. **28**, 203–208 (1919h). Appeared in 1920

Brouwer, L.E.J.: Über eineindeutige stetige Transformationen von Flächen in sich, VI. K. Ned. Akad. Wet. Proc. Sect. Sci. **21**, 707–710 (1919i)

Brouwer, L.E.J.: Ueber topologische Involutionen. K. Ned. Akad. Wet. Proc. Sect. Sci. **21**, 1143–1145 (1919j)

Brouwer, L.E.J.: Remark on multiple integrals. K. Ned. Akad. Wet. Proc. Sect. Sci. **22**, 150–154 (1919k)

Brouwer, L.E.J.: Luchtvaart en photogrammetrie, I. Nieuw Arch. Wiskd. **7**, 311–331 (1919l)

Brouwer, L.E.J.: Über die periodischen Transformationen der Kugel. Math. Ann. **80**, 39–41 (1919m)

Brouwer, L.E.J.: Remark on the plane translation theorem. K. Ned. Akad. Proc. **21**, 935–936 (1919n)

Brouwer, L.E.J.: Berichtigung. Math. Ann. **79**, 403 (1919o)

Brouwer, L.E.J.: Ueber eineindeutige, stetige Transformationen von Flächen in sich, VII. K. Ned. Akad. Proc. **22**, 811–814 (1920)

Brouwer, L.E.J.: Besitzt jede reelle Zahl eine Dezimalbruch-Entwickelung? Math. Ann. **83**, 201–210 (1921a)

Brouwer, L.E.J.: Aufzählung der Abbildungsklassen endlichfach zusammenhängender Flächen. Math. Ann. **82**, 280–286 (1921b)

Brouwer, L.E.J.: Berichtigung. Math. Ann. **82**, 286 (1921c)

Brouwer, L.E.J.: Intuitionistische Mengenlehre. K. Ned. Akad. Wet. Proc. Sect. Sci. **23**, 949–954 (1922a)

Brouwer, L.E.J.: Wis- en Natuurkunde en Wijsbegeerte. Nieuwe Kron. **2**, 3–5 (1922b)

Brouwer, L.E.J.: Begründung der Funktionenlehre unabhängig vom logischen Satz vom ausgeschlossenen Dritten. Erster Teil, Stetigkeit, Messbarkeit, Derivierbarkeit. Verh. K. Akad. Wet. Amst. **2**, 1–24 (1923a)

Brouwer, L.E.J.: Over de rol van het principium tertiï exclusi in the wiskunde, in het bijzonder in de functietheorie. Wis- Natuurkd. Tijdschr. **2**, 1–7 (1923b)

Brouwer, L.E.J.: Die Rolle des Satzes vom ausgeschlossenen Dritten in der Mathematik. Jahresber. Dtsch. Math.-Ver. **32**, 67 (1923c). Italics

Brouwer, L.E.J.: Intuïtionistische splitsing van mathematische grondbegrippen. K. Ned. Akad. Wet. Versl. Gewone Vergad. Afd. Natuurkd. **32**, 877–880 (1923d)

Brouwer, L.E.J.: Über den natürlichen Dimensionsbegriff. K. Ned. Akad. Wet. Proc. Sect. Sci. **26**, 795–800 (1923e)

Brouwer, L.E.J.: Über die Bedeutung des Satzes vom ausgeschlossenen Dritten in der Mathematik insbesondere in der Funktionentheorie. J. Reine Angew. Math. **154**, 1–8 (1923f)

Brouwer, L.E.J.: Beweis dass jede volle Funktion gleichmässig stetig ist. K. Ned. Akad. Wet. Proc. Sect. Sci. **27**, 189–193 (1924a)

Brouwer, L.E.J.: Intuitionistische Ergänzung des Fundamentalsatzes der Algebra. K. Ned. Akad. Wet. Proc. Sect. Sci. **27**, 631–634 (1924b)

Brouwer, L.E.J.: Bewijs van de onafhankelijkheid van de onttrekkingsrelatie van de versmeltingsrelatie. K. Ned. Akad. Wet. Versl. Gewone Vergad. Afd. Natuurkd. **33**, 479–480 (1924c)

Brouwer, L.E.J.: Bemerkungen zum Beweise der gleichmässigen Stetigkeit voller Funktionen. K. Ned. Akad. Wet. Proc. Sect. Sci. **27**, 644–646 (1924d)

Brouwer, L.E.J.: Bemerkungen zum natürlichen Dimensionsbegriff. K. Ned. Akad. Wet. Proc. Sect. Sci. **27**, 635–638 (1924e)

Brouwer, L.E.J.: On the $n$-dimensional simplex star in $R_n$. K. Ned. Akad. Wet. Proc. Sect. Sci. **27**, 778–780 (1924f)

Brouwer, L.E.J.: Zum natürlichen Dimensionsbegriff. Math. Z. **21**, 312–314 (1924g)

Brouwer, L.E.J.: Berichtigung. J. Reine Angew. Math. **153**, 253 (1924h). Re Brouwer (1913a)

Brouwer, L.E.J.: Zur Begründung der intuitionistischen Mathematik I. Math. Ann. **93**, 244–257 (1925a). Corr. in Brouwer (1926a)

Brouwer, L.E.J.: Intuitionistischer Beweis des Jordanschen Kurvensatzes. K. Ned. Akad. Wet. Proc. Sect. Sci. **28**, 503–508 (1925b)

Brouwer, L.E.J.: Zuschrift an dem Herausgeber. Jahresber. Dtsch. Math.-Ver. **33**, 124 (1925c)

Brouwer, L.E.J.: Intuitionistische Zerlegung mathematischer Grundbegriffe. Jahresber. Dtsch. Math.-Ver. **33**, 251–256 (1925d)

Brouwer, L.E.J.: Zur Begründung der intuitionistischen Mathematik II. Math. Ann. **95**, 453–472 (1926a)

Brouwer, L.E.J.: Intuitionistische Einführung des Dimensionsbegriffes. K. Ned. Akad. Wet. Proc. Sect. Sci. **29**, 855–873 (1926b)

Brouwer, L.E.J.: Die intuitionistische Form des Heine Borel Theorems. K. Ned. Akad. Proc. **29**, 866–867 (1926c)

Brouwer, L.E.J.: Zur Begründung der intuitionistischen Mathematik III. Math. Ann. **96**, 451–488 (1927a)

Brouwer, L.E.J.: Über Definitionsbereiche von Funktionen. Math. Ann. **97**, 60–75 (1927b)

Brouwer, L.E.J.: Virtuelle Ordnung und unerweiterbare Ordnung. J. Reine Angew. Math. **157**, 255–257 (1927c)

Brouwer, L.E.J.: Zur intuitionistischen Zerlegung mathematischer Grundbegriffe. Jahresber. Dtsch. Math.-Ver. **36**, 127–129 (1927d)

Brouwer, L.E.J.: Intuitionistische Betrachtungen über den Formalismus. K. Ned. Akad. Wet. Versl. Gewone Vergad. Afd. Natuurkd. **36**, 1189 (1928a). Dutch summary of Brouwer (1928b)

Brouwer, L.E.J.: Intuitionistische Betrachtungen über den Formalismus. K. Ned. Akad. Wet. Proc. Sect. Sci. **31**, 374–379 (1928b)

Brouwer, L.E.J.: Intuitionistische Betrachtungen über den Formalismus. Sitz.ber. Preuss. Akad. Wiss. Phys.-Math. Kl., 48–52 (1928c)

Brouwer, L.E.J.: Beweis dass jede Menge in einer individualisierten Menge enthalten ist. K. Ned. Akad. Wet. Proc. Sect. Sci. **31**, 380–381 (1928d)

Brouwer, L.E.J.: Zur Geschichtsschreibung der Dimensionstheorie. K. Ned. Akad. Wet. Versl. Gewone Vergad. Afd. Natuurkd. **37**, 626 (1928e)

Brouwer, L.E.J.: Zur Geschichtsschreibung der Dimensionstheorie. K. Ned. Akad. Wet. Proc. Sect. Sci. **31**, 953–957 (1928f). Corr. in KNAW Proc. 32, p. 1022

Brouwer, L.E.J.: Mathematik, Wissenschaft und Sprache. Monatshefte Math. Phys. **36**, 153–164 (1929a)

Brouwer, L.E.J.: Herinnering aan C.S. Aama van Scheltema door L.E.J. Brouwer, p. 69. Querido, Amsterdam (1929b)

Brouwer, L.E.J.: Die Struktur des Kontinuums [Sonderabdruck] (1930)

Brouwer, L.E.J.: Willen, Weten, Spreken. Euclides **9**, 177–193 (1933a)

Brouwer, L.E.J.: Willen, Weten, Spreken. In: Brouwer, L.E.J., Clay, J., et al. (eds.) De uitdrukkingswijze der wetenschap, pp. 43–63. Noordhoff, Groningen (1933b)

Brouwer, L.E.J.: Zum Triangulationsproblem. Indag. Math. **1**, 248–253 (1939)

Brouwer, L.E.J.: D. J. Korteweg [Obituary]. Verh. K. Akad. Wet. Amst. **999**, 266–267 (1941)

Brouwer, L.E.J.: Zum freien Werden von Mengen und Funktionen. Indag. Math. **4**, 107–108 (1942a)

Brouwer, L.E.J.: Die repräsentierende Menge der stetigen Funktionen des Einheitskontinuums. Indag. Math. **4**, 154 (1942b)

Brouwer, L.E.J.: Beweis dass der Begriff der Mengen höherer Ordnung nicht als Grundbegriff der intuitionistischen Mathematik in Betracht kommt. Indag. Math. **4**, 274–276 (1942c)

Brouwer, L.E.J.: Address delivered on September 16th, 1946, on the conferment upon Professor G. Mannoury of the honorary degree of Doctor of Science. Jaarboek der Universiteit van Amsterdam 1946–1947 (1946)

Brouwer, L.E.J.: Essentieel negatieve eigenschappen. Indag. Math. **10**, 322–323 (1948a). Transl. "Essentially negative properties" in CW 1, p. 478

Brouwer, L.E.J.: Discussion. (Following "Les conceptions mathématiques et le reel" of Gonseth). In: Symposium de l'Institut des Sciences théoriques. Bruxelles 1947. Actualités scientifiques et Industrielles, pp. 31–60. Hermann, Paris (1948b)

Brouwer, L.E.J.: De non-aequivalentie van de constructieve en negatieve orderelatie in het continuum. Indag. Math. **11**, 37–39 (1949a). Transl. "The non-equivalence of the constructive and the negative order relation in the continuum" in CW 1, pp. 495–496

Brouwer, L.E.J.: Contradictoriteit der elementaire meetkunde. Indag. Math. **11**, 89–90 (1949b). Transl. "Contradictority of elementary geometry" in CW 1, pp. 497–498

Brouwer, L.E.J.: Consciousness, philosophy and mathematics. In: Proceedings of the 10th International Congress of Philosophy, Amsterdam, 1948, vol. 3, pp. 1235–1249 (1949c)

Brouwer, L.E.J.: Sur la possibilité d'ordonner le continu. C. R. Math. Acad. Sci. Paris **230**, 349–350 (1950)

Brouwer, L.E.J.: An intuitionist correction of the fixed-point theorem on the sphere. Proc. R. Soc. Lond. Ser. A **213**, 1–2 (1952a)

Brouwer, L.E.J.: Historical background, principles and methods of intuitionism. South Afr. J. Sci. **49**, 139–146 (1952b)

Brouwer, L.E.J.: Point and spaces. Can. J. Math. **6**, 1–17 (1954a)

Brouwer, L.E.J.: Addenda en corrigenda over de rol van het principium tertii exclusi in de wiskunde. Indag. Math. **16**, 104–105 (1954b). Transl. "Addenda and corrigenda on the role of the principium tertii exclusi in mathematics" in CW 1, pp. 539–540

Brouwer, L.E.J.: Collected Works 1. Philosophy and Foundations of Mathematics. Heyting, A. (ed.). North-Holland, Amsterdam (1975)

Brouwer, L.E.J.: Collected Works 2. Geometry, Analysis Topology and Mechanics. Freudenthal, H. (ed.). North-Holland, Amsterdam (1976)

Brouwer, L.E.J.: Brouwer's Cambridge Lectures on Intuitionism. van Dalen, D. (ed.). Cambridge University Press, Cambridge (1981)

Brouwer, L.E.J.: Intuitionismus. van Dalen, D. (ed.). Bibliographisches Institut Wissenschaftsverlag, Mannheim (1992)

Brouwer, L.E.J.: The Selected Correspondence of L.E.J. Brouwer. van Dalen, D. (ed.). Springer, London (2011)

Brouwer, L.E.J., de Loor, B.: Intuitionistischer Beweis des Fundamentalsatzes der Algebra. K. Ned. Akad. Wet. Proc. Sect. Sci. **27**, 186–188 (1924)

Brouwer, L.E.J., et al.: Voorbereidend Manifest. Meded. Int. Inst. Wijsb. **1**, 3–12 (1918)

Brouwer, L.E.J., et al.: Signifische dialogen. Synthese **2**, 168–174 (1937a)

Brouwer, L.E.J., et al.: Signifische dialogen. Synthese **2**, 261–268 (1937b)

Brouwer, L.E.J., et al.: Signifische dialogen. Synthese **2**, 316–324 (1937c)

Brouwer, L.E.J., et al.: Signifische Dialogen. J. Bijleveld, Utrecht (1939)

Browder, F.I. (ed.): Hilbert's Problems. Proc. of Symposia in Pure Mathematics. Am. Math. Soc., Providence (1976). 2 vols.

Bruins, E.M.: Cosmische stralen in het aardmagnetisch veld. Ph.D. thesis, Amsterdam (1938)

Bruins, E.M.: ΑΝΑΓΚΗ. Tekst van het College van den 15-den October 1982, terafsluiting van regulier onderwijs aan de Universiteit gegeven door Evert Marie Bruins. Ex Malis Bona (1982)

Cairns, S.S.: Triangulation of the manifold of class one. Bull. Am. Math. Soc. **41**, 549–552 (1935)

Cantor, G.: Ueber eine elementare Frage der Mannigfaltigkeitslehre. Jahresber. Dtsch. Math.-Ver. **1**, 75–78 (1892)

Cartan, E.: Les groupes réels simples finis et continus. Ann. Sci. Éc. Norm. Super. **31**, 263–355 (1914)

Caspary, F.: Zur theorie der Thetafunctionen mit zwei argumenten. J. Reine Angew. Math. **94**, 74–86 (1883)

Castelnuovo, G. (ed.): Atti IV Congr. Intern. Mat. Roma. Accad. Naz. Lincei, Roma (1909)

Church, A.: On the law of the excluded middle. Bull. Am. Math. Soc. **34**, 75–78 (1928)

Courant, R.: Reminiscences from Hilbert's Göttingen. Math. Intell. **3**, 154–164 (1981)

Craig, G.A.: Germany 1866–1945. Oxford University Press, Oxford (1981)

Dawson, J.W.: Logical Dilemmas. The Life and Work of Kurt Gödel. AK Peters, Wellesley (1997)

de Groot, J.: Topologische Studieën. Compactificatie, Voortzetting van Afbeeldingen en Samenhang. Ph.D. thesis, Rijks Universiteit Groningen (1942)

de Groot, A.W.: De Universiteit van Amsterdam in oorlogstijd. H.J.W. Becht, Amsterdam (1946)

de Jong, L.: Het Koninkrijk der Nederlanden in de Tweede Wereldoorlog. Staatsuitgeverij, Den Haag (1969). 13 vols.

de Vries, Hk.: Inleiding tot de studie der meetkunde van het aantal. Noordhoff, Groningen (1936)

Dehn, M.: K.Th. Vahlen, Abstrakte Geometrie (review). Jahresber. Dtsch. Math.-Ver. **14**, 535–537 (1905)

Delvigne, R.: L.E.J. Brouwer over C.S. Adama van Scheltema. Juffr. Ida **11**, 41–44 (1985)

Dieudonné, J.: A History of Algebraic and Differential Topology, 1900–1960. Birkhäuser, Basel (1989)

Dijkstra, B.: Idols of Perversity. Fantasies of Feminine Evil in Fin-de-siècle Culture. Cambridge University Press, Cambridge (1986)

Drost, F.: Carel Steven Adama van Scheltema. Ph.D. thesis, Rijksuniversiteit Groningen (1952)

Duparc, H.J.A., Korevaar, J.: Johannes Gualtherus van der Corput. 4 September 1890–13 September 1975. Nieuw Arch. Wiskd. **30**, 1–40 (1982)

Durlacher, G.L.: Quarantaine. Meulenhof, Amsterdam (1993)

Einstein, A., Born, M.: Briefwechsel 1916–1955. Nymphenburger Verlagshandlung, Munich (1969)

Euwe, M.: Mengentheoretische Betrachtungen über das Schachspiel. Proc. K. Ned. Akad. Wet. **32**, 633–644 (1929)

Fasseur, C.: Wilhelmina. Krijgshaftig in een vormeloze jas. Balans, Amsterdam (2001)

Feigl, G.: Geschichtliche Entwicklung der Topologie. Jahresber. Dtsch. Math.-Ver. **37**, 273–286 (1928)

Felsch, V.: Otto Blumenthals Tagebücher. Ein Aachener Mathematik Professor erleidet die NS-Diktatur in Deutschland, den Niederlanden und Theresienstadt. Hartung Gorre Verlag, Konstanz (2011)

Finsler, P.: Gibt es Widersprüche in der Mathematik? Jahresber. Dtsch. Math.-Ver. **34**, 143–155 (1925)

Fontijn, J.: Tweespalt. Het leven van Frederik van Eeden tot 1901. Querido, Amsterdam (1990)

Fontijn, J.: Trots verbrijzeld. Het leven van Frederik van Eeden vanaf 1901. Querido, Amsterdam (1996)

Forman, P.: Il Naturforscherversammlung a Nauheim del settembre 1920: una introduzione alla vita scientifica nella Republica de Weimar. In: Rossi, A., Battimelli, G., de Maria, M. (eds.) La ristrutturazione delle scienze tra le due guerre mondiali, vol. 1, pp. 59–78. La Goliardica, Rome (1986)

Fourman, M.P., Hyland, J.M.E.: Sheaf models for analysis. In: Applications of Sheaves, pp. 280–301. Springer, Berlin (1979)

Fraenkel, A.: Einleitung in die Mengenlehre. Springer, Berlin (1919)

Fraenkel, A.: Einleitung in die Mengenlehre, 2nd edn. Springer, Berlin, (1923)

Fraenkel, A.: Zehn Vorlesungen über die Grundlegung der Mengenlehre. Teubner, Leipzig (1927). Reprinted by the Wissenschaftliche Buchgesellschaft Darmstadt (1972)

Franks, J.: A new proof of the Brouwer plane translation theorem. Ergod. Theory Dyn. Syst. **12**, 217–226 (1992)

Freudenthal, H.: Zur intuitionistischen Deutung logischer Formeln. Compos. Math. **4**, 112–116 (1936)

Freudenthal, H.: Die Triangulation der differentierbaren Mannigfaltigkeiten. Indag. Math. **1**, 311–332 (1939). Nachtrag, Ind. Math. **2**, 249 (1940)

Freudenthal, H.: Leibniz und die Analysis Situs. In: Homenaje a Millias–Vallicrosa, vol. 1, pp. 611–621. Consejo Superior de Investicaciones Cientificas, Barcelona (1954)

Freudenthal, H.: Hermann Weyl. Der Dolmetscher zwischen Mathematikern und Physikern um die moderne Interpretation von Raum, Zeit und Materie. In: Schwerte, H., Spengler, W. (eds.) Forscher und Wissenschaftler im heutigen Europa – Weltall und Erde, vol. 999, pp. 357–366. Gerh, Stalling Verlag, Oldenbourg (1955)

Freudenthal, H.: Zur Geschichte der Grundlagen der Geometrie. Zugleich eine Besprechung der 8. Aufl. von Hilberts "Grundlagen der Geometrie". Nieuw Arch. Wiskd. **5**, 105–142 (1957)

Freudenthal, H.: The cradle of modern topology, according to Brouwer's inedita. Hist. Math. **2**, 495–502 (1975)

Freudenthal, H.: Een manuscript van Brouwer. In: Tweehonderd Jaar Onvermoeide Arbeid. Tentoonstellingscatalogus, vol. 2, pp. 43–55. Mathematisch Centrum, Amsterdam (1979)

Freudenthal, H.: A bit of gossip: Koebe. Math. Intell. **6**, 77 (1984)

Freudenthal, H.: Berlin 1923–1930. Studienerinnerungen von Hans Freudenthal. de Gruyter, Berlin (1987a)
Freudenthal, H.: Schrijf dat op, Hans. Knipsels uit een leven. Meulenhoff, Amsterdam (1987b)
Frey, G., Stammbach, U.: Hermann Weyl und die Mathematik an der ETH, Zürich, 1913–1930. Birkhäuser, Basel (1992)
Fricke, R., Klein, F.: Vorlesungen über die Theorie der automorphen Functionen I. Teubner, Leipzig (1897)
Fricke, R., Klein, F.: Vorlesungen über die Theorie der automorphen Functionen II. Teubner, Leipzig (1912)
Gentzen, G.: The consistency of elementary number theory. In: Szabo, M.E. (ed.) The Collected Papers of Gerhard Gentzen, pp. 132–213. North-Holland, Amsterdam (1969)
Georgiadou, M.: Constantin Carathéodory. Mathematics and Politics in Turbulent Times. Springer, Berlin (2004)
Gilmore, P.: The effect of Griss' Criticism of the intuitionistic logic on deductive theories formalized within the intuitionistic logic. Ph.D. thesis, Amsterdam (1953)
Graf von Krockow, C.: Die Deutschen in ihrem Jahrhundert, 1890–1990. Rowohlt, Hamburg (1990)
Gray, J.J.: The Hilbert Challenge. Oxford University Press, Oxford (2000)
Grelling, K.: Philosophy of the exact sciences: its present status in Germany. Monist **38**, 97–119 (1928)
Gutkind, E.: Einwand von Erich Gutkind. Meded. Int. Inst. Wijsb. **2**, 33 (1919)
Gutkind, E.: Von Freundschaft. In: Liber Amicorum Dr. Frederik van Eeden aangeboden ter gelegenheid van zijn 70ste verjaardag 3 april 1930, pp. 68–69. Maatschappij tot verspreiding van goede en goedkope literatuur, Amsterdam (1930)
Haalmeijer, B.P., Schogt, J.H.: Inleiding tot de leer der verzamelingen. Noordhoff, Groningen (1927)
Hadamard, J.: Sur quelques applications de l'indice de Kronecker. In: Tannéry, J. (ed.) Introduction à la théorie des fonctions, 2nd edn., vol. 2, pp. 437–477 (1910)
Hahn, H.: Theorie der reellen Funktionen, I. Springer, Berlin (1921)
Hardy, G.H.: The J-type and the S-type among mathematicians. Nature **134**, 250 (1934)
Heijerman, E., van der Hoeven, M.J.: Filosofie in Nederland. De Internationale School voor Wijsbegeerte als ontmoetingsplaats 1916–1986. Boom, Meppel (1986)
Herbrand, J.: Logical Writings. Goldfarb, W. (ed.). Harvard University Press, Cambridge (1971)
Hesseling, D.E.: Gnomes in the fog. The reception of Brouwer's intuitionism in the 1920s. Ph.D. thesis, Utrecht (1999)
Hesseling, D.E.: Gnomes in the Fog. The Reception of Brouwer's Intuitionism in the 1920s. Birkhäuser, Basel (2002)
Heyting, A.: Die Intuitionistische Mathematik. Forsch. Fortschr. **7**, 38–39 (1931a)
Heyting, A.: Philosphische Grundlegung der Mathematik. Blätter für Deutsche Philosophie (4), 1930 (review). Jahresber. Dtsch. Math.-Ver. **40**, 50–52 (1931b)
Heyting, A.: Sur la logique intuitionniste. A propos d'un article de MM. Barzin et Errera. Enseign. Math. **31**, 121–122 (1932)
Heyting, A.: Bemerkung zu dem Aufsatz von Hern Freudenthal "Zur intuitionistischen Deutung logischer Formeln". Compos. Math. **4**, 117–118 (1936a)
Heyting, A.: Intuïtionistische Wiskunde. Math. B **5**, 105–112 (1936b)
Heyting, A.: Untersuchtungen über intuitionistische Algebra. Verh. K. Akad. Wet. Afd. Natuurkd., Eerste Sect. **18**(2), 36 pp. (1941)
Heyting, A.: Intuitionism, an Introduction. North-Holland, Amsterdam (1956)
Heyting, L.: De wereld in een dorp. Meulenhof, Amsterdam (1994)
Hilbert, D.: Mathematische Probleme. Nachr. Ges. Wiss. Gött., Math.-Phys. Kl. **999**, 253–297 (1900)
Hilbert, D.: Ueber die Grundlagen der Geometrie. Math. Ann. **56**, 381–422 (1902)
Hilbert, D.: Über die Grundlagen der Logik und der Arithmetik. In: Verhandlungen des Dritten Internationalen Mathematiker-Kongresses in Heidelberg vom 8. bis 13 August 1904, pp. 174–

185. Teubner, Leipzig (1905)
Hilbert, D.: Grundlagen der Geometrie. Teubner, Leipzig (1909)
Hilbert, D.: Neubegründung der Mathematik (Erste Mitteilung). Abh. Math. Semin. Univ. Hamb. **1**, 157–177 (1922)
Hilbert, D.: Die Logischen Grundlagen der Mathematik. Math. Ann. **88**, 151–165 (1923)
Hilbert, D.: Über das Unendliche. Math. Ann. **95**, 161–190 (1926)
Hilbert, D.: Die Grundlagen der Mathematik. Abh. Math. Semin. Univ. Hamb. **6**, 65–92 (1928)
Hilbert, D.: Die Grundlegung der elementaren Zahlenlehre. Math. Ann. **104**, 485–494 (1930)
Hilbert, D.: Beweis des Tertium non datur. Nachr. Gött., 120–125 (1931)
Hilbert, D., Bernays, P.: Grundlagen der Mathematik I. Springer, Berlin (1934)
Hilbert, D., Bernays, P.: Grundlagen der Mathematik II. Springer, Berlin (1939)
Hodges, W.: An editor recalls some hopeless papers. Bull. Symb. Log. **4**, 1–16 (1998)
Hölder, O.: Die mathematische Methode. Logisch erkenntnisstheoretische Untersuchungen im Gebiete der Mathematik, Mechanik und Physik. Springer, Berlin (1924)
Hopf, H.: Ein Abschnitt aus der Entwicklung der Topologie. Jahresber. Dtsch. Math.-Ver. **68**, 182–192 (1966)
Hübner, H.: Ein zerbrechliches Menschenskind – Helen Ernst (1904–1948). Biographie einer antifaschistischen Künstlerin zwischen Athen, Zürich, Berlin, Amsterdam, Ravensbrück und Schwerin. Trafo Verlag, Berlin (2002)
Hurewicz, W., Wallman, H.: Dimension Theory. Princeton University Press, Princeton (1948)
Hurwitz, A.: Über algebraische Gebilde mit eindeutigen Transformationen in sich. Math. Ann. **41**, 403–442 (1892)
Husserl, E.: Edmund Husserl: Briefwechsel. Schuhmann, K. (ed.). Springer, Dordrecht (1994)
Husserl, E.: Husserliana, vol. 6. Palmer, R.E., Sheehan, T. (eds.). Kluwer, Dordrecht (1997)
Inachin, K.T.: Märtyrer mit einem kleinen Häuflein Getreuer. Der erste Gauleiter der NSDAP in Pommern Karl Theodor Vahlen. (Schriftreihe der Vierteljahrhefte für Zeitgeschichte). Deutsche Verlags-Anstalt, Stuttgart (1960)
Jahnke, E.: Bemerkung zu der am 27. Februar 1904 vorgelegten Notiz von Herrn Brouwer "Over een splitsing van de continue beweging om een punt $O$ van $R_4$ in twee continue bewegingen om $O$ van $R_3$'s". K. Ned. Akad. Wet. Versl. **12**, 940–941 (1904)
James, I.M. (ed.): The History of Topology. Elsevier, Amsterdam (1999)
Johnson, D.M.: The problem of the invariance of dimension in the growth of modern topology, part I. Arch. Hist. Exact Sci. **20**, 97–188 (1979)
Johnson, D.M.: The problem of the invariance of dimension in the growth of modern topology, part II. Arch. Hist. Exact Sci. **25**, 85–267 (1981)
Jordan, C.: Cours d'Analyse de l'École Polytechnique I. Gauthiers-Villars, Paris (1893)
Karo, G.: Der geistige Krieg gegen Deutschland. Z. Völkerpsychol. Soziol. **2** (1926), 22 pp.
Kellermann, H. (ed.): Der Krieg der Geister: eine Auslese deutscher und ausländischer Stimmen. Alexander Dunker Verlag, Weimar (1915)
Kleene, S.C., Vesley, R.E.: The Foundations of Intuitionistic Mathematics Especially in Relation to Recursive Functions. North-Holland, Amsterdam (1965)
Klein, F.: Zur Nicht-Euklidische Geometrie. Math. Ann. **37**, 544–572 (1890)
Klein, F.: Lectures on Mathematics. Delivered From Aug. 28 to Sept. 9, 1893. Macmillan & Co., New York (1893). The so-called "Evanstone lectures"
Klein, F.: Zum Kontinuitätsbeweise des Fundamentaltheorems. In: Bessel-Hagen, E., Fricke, R., Vermeil, H. (eds.) Gesammelte Mathematische Abhandlungen III, vol. 3, pp. 731–741. Springer, Berlin (1923)
Klein, F.: Vorlesungen über die Entwicklung der Mathematik im 19. Jahrhundert II. Springer, Berlin (1927)
Knegtmans, P.J.: Een kwetsbaar centrum van de geest. Amsterdam University Press, Amsterdam (1998)
Kneser, A.: Leopold Kronecker. Jahresber. Dtsch. Math.-Ver. **33**, 210–228 (1925)
Koebe, P.: Zur theorie der konforme Abbildung und Uniformisierung (Voranzeige). Sitzungsber. Sächs. Akad. Wiss. Leipz., Math.-Nat. Wiss. Kl. **66**, 67–75 (1914)

Koebe, P.: Wesen der Kontinuitätsmethode. Dtsch. Math. **1**, 859–879 (1936)

Köhler, E.: Gödel und der Wiener Kreis. In: Kruntorad, P. (ed.) Jour Fixe der Vernunft. Der Wiener Kreis und die Folgen, pp. 127–158. Verlag Hölder–Pichler–Temsky, Vienna (1991)

König, D.: Sur les correspondances multivoque des ensembles. Fundam. Math. **8**, 114–134 (1926)

Korevaar, J.: Johannes Gualtherus van der Corput (4 September 1890–13 September 1975). Jaarboek KNAW, 1–6 (1975)

Korteweg, D.J.: Huygens' sympathische uurwerken en verwante verschijnselen in verband met de principale en samengestelde slingeringen die zich voordoen, als aan een mechanisme met één enkele vrijheidsgraad twee slingers bevestigd worden. K. Ned. Akad. Wet. Versl. **13**, 413–432 (1905)

Kreisel, G.: Lawless sequences of natural numbers. Compos. Math. **20**, 222–248 (1968)

Kreisel, G.: Gödel's excursion into intuitionistic logic. In: Weingartner, P., Schmetterer, L. (eds.) Gödel Remembered, pp. 67–186. Bibliopolis, Napoli (1987)

Kreisel, G., Newman, M.H.A.: Luitzen Egbertus Jan Brouwer 1881–1966. Elected For. Mem R.S. 1948. Biogr. Mem. Fellows R. Soc. **15**, 38–68 (1969)

Kreisel, G., Troelstra, A.S.: Formal systems for some branches of intuitionistic analysis. Ann. Math. Log. **1**, 229–387 (1970)

Kühnau, R.: Paul Koebe und die Funktionentheorie. In: Schumann, H., Beckert, H. (eds.) 100 Jahre Mathematisches Seminar der Karl-Marx-Universität Leipzig, pp. 183–194. VEB Deutscher Verlag der Wissenschaften, Leipzig (1981)

Kuiper, N.H.: A short history of triangulation and related matters. In: Baayen, P.C., van Dulst, D., Oosterhoff, J. (eds.) Proceedings Bicentennial Congress Wiskundig Genootschap, vol. 1, pp. 61–79. Mathematisch Centrum, Amsterdam (1979)

Kuiper, J.J.C.: Ideas and explorations. Brouwer's road to intuitionism. Ph.D. thesis, Utrecht University (2004)

Lebesgue, H.: Sur la non-applicabilité de deux domaines appartenant respectivement des espaces à $n$ et $n + p$ dimensions (Extrait d'une lettre à M.O. Blumenthal). Math. Ann. **70**, 166–168 (1911a)

Lebesgue, H.: Sur l'invariance du nombre de dimensions d'un espace et sur le théorème de M. Jordan relatif aux variété fermées. C. R. Math. Acad. Sci. Paris **152**, 841–843 (1911b)

Lebesgue, H.: Sur les correspondances entre les points de deux espaces. Fundam. Math. **2**, 256–285 (1921)

Lennes, N.J.: Curves in non-metrical analysis situs with an application in the calculus of variations. Am. J. Math. **33**, 287–326 (1911)

Levelt Sengers, J.: How Fluids Unmix. Discoveries by the School of Van der Waals and Kamerlingh Onnes. Edita KNAW, Amsterdam (2002)

Lietzmann, W.: David Hilbert zum 80. Geburtsstage. Arch. f. Landes- und Volkskunde von Nieder-Sachsen, 203–217 (1942)

Lindner, H.: Deutsche und gegentypische Mathematik. Zur Begründung einer arteigenen Mathematik im Dritten Reich durch Ludwig Bieberbach. Suhrkamp, Frankfurt am Main (1980)

Loewner, C.: Charles Loewner. Collected Papers. Bers, L. (ed.). Birkhäuser, Basel (1988)

Maas, A.J.P.: Atomisme en individualisme. Ph.D. thesis, University of Amsterdam (2001)

MacLane, S.: Mathematics at the University of Göttingen, 1931–1933. In: Brewer, J., Smith, M. (eds.) Emmy Noether: A Tribute to Her Life and Work, pp. 65–78. Marcel Dekker, New York (1981)

Mancosu, P.: From Brouwer to Hilbert. The Debate on the Foundations of Mathematics in the 1920s. Oxford University Press, Oxford (1998). Collection of papers

Mannoury, G.: Lois cyclomatiques. Nieuw Arch. Wiskd. **4**, 126–152 (1898a)

Mannoury, G.: Spheres de seconde espece. Nieuw Arch. Wiskd. **4**, 83–89 (1898b)

Mannoury, G.: Surface-images. Nieuw Arch. Wiskd. **4**, 112–129 (1900)

Mannoury, G.: Review. Over de Grondslagen van de Wiskunde. De Beweging **3**, 241–249 (1907)

Mannoury, G.: Heden is het keerpunt (1930)

Mannoury, G.: De Schoonheid der wiskunde als significh probleem. Synthese **2**, 197–201 (1937)

Mannoury, G.: La question vitale "A ou B". Nieuw Arch. Wiskd. **821**, 161–167 (1943)

Mannoury, G.: Handboek der Analytische Significa. Geschiedenis der Begripskritiek. Kroonder, Bussum (1947)
Mannoury, G.: Handboek der Analytische Significa. Hoofdbegrippen en Methoden der Significa. Ontogenese en Fylogenese van het verstandhoudingsapparaat. Kroonder, Bussum (1948)
Mannoury, G.: De wetenschap van de mens. Ned. Tijdschr. Psychol. Haar Grensgeb. **6**, 208–211 (1951). Dedicated to Brouwer at his seventieth birthday. "Opgedragen aan Prof. Dr. L.E.J. Brouwer, ter gelegenheid van zijn zeventigste verjaardag, als erkentenis van het vele, dat hij tot de gedachterijping zijner tijdgenoten heeft bijgedragen"
Martino, E.: Brouwer's equivalence between virtual and inextensible order. Hist. Philos. Logic **9**, 57–66 (1988)
Mauthner, F.: Beiträge zu einer Kritik der Sprache. J.G. Cotta, Stuttgart (1906). 3 vols.
Mayrhofer, K.: Hans Hahn. Monatshefte Math. Phys. **41**, 221–238 (1934)
Mehrtens, H.: Das "Dritte Reich" in der Naturwissenschaftsgeschichte: Literaturbericht und Problemskizze. In: Mehrtens, H., Richter, S. (eds.) Naturwissenschaft, Technik und NS-Ideologie. Suhrkamp, Frankfurt (1980)
Mehrtens, H.: Anschauungswelt versus Papierwelt – Zur Interpretation der Grundlagenkrise der Mathematik. In: Poser, H., Schütt, H.-W. (eds.) Ontologie und Wissenschaft. Philosophische und wissenschafts-historische Untersuchungen zur Frage der Objektkonstitution, vol. 19, pp. 231–276. Technische Universität, Berlin (1984)
Mehrtens, H.: Ludwig Bieberbach and Deutsche Mathematik. In: Philips, E.R. (ed.) Studies in History of Mathematics, pp. 195–241. Math. Assoc. of America, Washington (1987)
Menger, K.: Über die Dimensionalität von Punktmengen. I. Monatshefte Math. Phys. **33**, 148–160 (1923)
Menger, K.: Zur Entstehung meiner Arbeiten über Dimensions- und Kurventheorie. Proc. K. Ned. Akad. Wet. **29**, 1122–1124 (1926). Subm. 29.5.1926
Menger, K.: Bemerkungen zu Grundlagenfragen. Jahresber. Dtsch. Math.-Ver. **37**, 213–226 (1928a). On the analogy Spreads-Analytic sets
Menger, K.: Dimensionstheorie. Teubner, Leipzig (1928b)
Menger, K.: Zur Dimensions- und Kurventheorie. Monatshefte Math. Phys. **36**, 411–432 (1929)
Menger, K.: Antwort auf eine Note von Brouwer. Monatshefte Math. Phys. **37**, 175–182 (1930)
Menger, K.: Selected Papers in Logic and Foundations, Didactics, Economics. Reidel, Dordrecht (1979)
Menger, K.: Reminiscences of the Vienna Circle and the Mathematical Colloquium. Golland, L., McGuinness, B., Sklar, A. (eds.). Kluwer, Dordrecht (1994)
Menger, K.: Selecta Mathematica. Schweizer, B., Sklar, A., Sigmund, K., Gruber, P., Hlawka, E., Reich, L., Schmetterer, L. (eds.). Springer, Vienna (2003)
Menzler-Trott, E.: Gentzens Problem. Mathematische Logik im nationalsozialistischen Deutschland. Birkhäuser, Basel (2001)
Mulder, P.: Kirkman-Systemen. Ph.D. thesis, Groningen (1917)
Myhill, J.: Notes towards an axiomatization of intuitionistic analysis. Log. Anal. **9**, 280–297 (1966)
Myhill, J., Kino, A., Vesley, R.E.: Intuitionism and Proof Theory. Proceedings of the Summer Conference at Buffalo, N.Y., 1968. North-Holland, Amsterdam (1970)
Nevanlinna, R.: Uniformisierung. Springer, Berlin, (1953)
Nielsen, J.: Über fixpunktfreie topologische Abbildungen geschlossener Flächen. Math. Ann. **81**, 94–96 (1920)
Nijenhuis, A.: J.A. Schouten: a master at tensors (28 August 1883–20 January 1971). Nieuw Arch. Wiskd. **20**, 1–19 (1972)
Noether, E., Cavaillès, J.: Briefwechsel Cantor–Dedekind. Hermann, Paris (1937)
Noorda, S., Staal, F.: Varen onder eigen vlag – Later op een maandagmiddag. Vossiuspers, Amsterdam (1999)
Otterspeer, W.: Bolland. Een Biografie. Bert Bakker, Amsterdam (1995)
Peano, G.: Démonstration de l'intégrabilité des équations différentielles ordinaires. Math. Ann. **3**, 182–228 (1890)

Peano, G.: Formulaire de mathématiques. Bocca frères, Turin (1895). Last volume (4) appeared in 1903

Phragmén, E.: Über die Begrenzung von Continua. Acta Math. **7**, 43–48 (1885)

Picard, E.: Discours et Mélanges. Gauthier-Villar, Paris (1922)

Pierpont, J.: Mathematical rigor, past and present. Bull. Am. Math. Soc. **34**, 23–53 (1928). AMS lecture 1927

Pinl, M., Furtmüller, L.: Mathematicians under Hitler. In: Year Book XVIII. Publications of the Leo Baeck Institute, vol. 18, pp. 129–184. Secker & Warburg, London (1973)

Plisko, V.E.: Letters of A.N. Kolmogorov to A. Heyting. Russ. Math. Surv. **43**, 89–93 (1988a). Usp. Math. Nauk **43**, 75–77 (1988)

Plisko, V.E.: The Kolmogorov calculus as a part of minimal calculus. Russ. Math. Surv. **43**, 95–110 (1988b). Usp. Math. Nauk **43**, 79–91 (1988)

Poincaré, H.: Mémoire sur les courbes définies par une équation differentielle. J. Math. Pures Appl. **7**, 375–442 (1881)

Poincaré, H.: Sur les résidus des integrales doubles. Acta Math. **9**, 321–380 (1887)

Poincaré, H.: Analysis situs. J. Éc. Polytech. **1**, 1–123 (1895)

Poincaré, H.: Les méthodes nouvelles de la mécanique céleste. III. Gauthier-Villars, Paris (1899)

Poincaré, H.: Science et Méthode. Flammarion, Paris (1905)

Poincaré, H.: L'Avenir des mathématiques. In: Castelnuovo, G. (ed.) Atti IV Congr. Intern. Mat. Roma, vol. 1, pp. 167–182. Accad. Naz. Lincei, Roma (1908)

Poincaré, H.: Sechs Vorträge über ausgewählte Gegenstände aus der reinen Mathematik und mathematische Physik. Teubner, Leipzig (1910)

Poincaré, H.: Sur un théorème de géométrie. Rend. Circ. Mat. Palermo **33**, 375–407 (1912)

Polya, G.: Eine Erinnerung an Hermann Weyl. Math. Z. **126**, 296–298 (1972)

Presser, J.: Ondergang. De Vervolging en Verdelging van het Nederlandse Jodendom. Martinus Nijhoff, Den Haag (1965). 2 vols.

Reid, C.: Hilbert. Springer, Berlin (1970)

Reid, C.: Hilbert–Courant. Springer, Berlin (1986)

Remmert, V.R.: Mathematicians at war. Power struggles in Nazi Germany's mathematical community: Gustav Doetsch and Wilhelm Süss. Rev. Hist. Math. **5**, 7–59 (1999)

Remmert, V.R.: Die Deutsche Mathematiker-Vereinigung im "Dritten Reich": Fach- und Parteipolitik. DMV-Mitt. **12**, 223–245 (2004a)

Remmert, V.R.: Die Deutsche Mathematiker-Vereinigung im "Dritten Reich": Krisenjahre und Konsolidierung. DMV-Mitt. **12**, 159–177 (2004b)

Revész, G.: Das Schöpferisch-Persönliche und das Kollektive in ihrem kulturhistorischen Zusammenhang. J.C.B. Mohr (Paul Siebeck), Tübingen (1933)

Richard, J.: Les principes des mathématiques et le probleme des ensembles. Rev. Gén. Sci. Pures Appl. **16**, 541 (1905). Transl. in van Heijenoort (1967)

Riesz, F.: Die Genesis des Raumbegriffs. Gesammelte Arb. **1**, 67–161 (1960). Original in Hungarian, 1906

Rigby, A.: Initiation and Initiative. An Exploration of the Life and Ideas of Dimitrije Mitrinović. East European Monographs. Columbia University Press, Boulder (1984)

Ritter, P.H. (ed.): Eene Halve Eeuw. 1848–1898. Nederland onder de Regeering van Koning Willem den Derde en het regentschap van Koningin Emma door Nederlanders beschreven. I & II. J.L. Beijers, J. Funke, Amsterdam (1898)

Rogers, H. Jr.: Theory of Recursive Functions and Effective Computability. McGraw-Hill, New York (1967)

Romein-Verschoor, A.: Omzien in verwondering. Arbeiderspers, Amsterdam (1970). 2 vols.

Rowe, D.E.: Interview with Dirk Jan Struik. Math. Intell. **11**, 14–26 (1989)

Rutherford, D.: Modular Invariants. Cambridge University Press, Cambridge (1932)

Schappacher, N.: Das Mathematische Institut der Universität Göttingen 1929–1950. In: Becker, H., Dahms, H.-J., Wegeler, C. (eds.) Die Universität Göttingen unter dem National Sozialismus, pp. 345–373. K.G. Saur, München (1987)

Schappacher, N., Kneser, M.: Fachverband – Institut – Staat. Streiflichter auf das Verhältnis von Mathematik zu Gesellswchaft und Politik in Deutschland sei 1890 unter besonderer Berücksichtigung der Zeit des Nationalsozialismus. In: Fischer, G., Hirzebruch, F., Scharlau, W., Törnig, W. (eds.) Festschrift zur Jubiläum der DMV. Vieweg, Braunschweig (1990)

Schappacher, N., Scholz, E.: Oswald Teichmüller – Leben und Werk. Jahresber. Dtsch. Math.-Ver. **94**, 1–39 (1992)

Schmidt-Ott, F.: Erlebtes und Erstrebtes. 1860–1950. Franz Steiner Verlag, Wiesbaden (1952)

Schmitz, H.W.: Tönnies' Zeichentheorie zwischen Signifik und Wiener Kreis. In: Clausen, L., Borries, V. (eds.) Tönnies heute. Zur Aktualität von Ferdinand Tönnies, vol. 999, pp. 73–93. Mülau Verlag, Kiel (1985)

Schmitz, H.W.: Frederik van Eeden and the introduction of significs into the Netherlands: from Lady Welby to Mannoury. In: Schmitz, H.W. (ed.) Essays on Significs. Papers Presented on the Occasion of the 150th Birtday of Victoria Lady Welby (1837–1912), vol. 23, pp. 219–246. Benjamins, Philadelphia (1990a)

Schmitz, W.H.: De Hollandse Significa. Van Gorcum, Assen (1990b)

Schoenflies, A.: Über den Beweis eines Haupttheorems aus der Theorie der Punktmengen. Nachr. Ges. Wiss. Gött., Math.-Phys. Kl. **999**, 21–31 (1903)

Schoenflies, A.: Beiträge zur Theorie der Punktmengen. II. Math. Ann. **59**, 129–160 (1904)

Schoenflies, A.: Über wohlgeordnete Mengen. Math. Ann. **60**, 181–186 (1905)

Schoenflies, A.: Beiträge zur Theorie der Punktmengen III. Math. Ann. **62**, 286–328 (1906)

Schoenflies, A.: Die Entwicklung der Lehre von den Punktmannigfaltigkeiten. II. Teubner, Leipzig (1908)

Schroeder-Gudehus, B.: Deutsche Wissenschaft und internationale Zusammenarbeit, 1914–1928. Ein Beitrag zum Studium kultureller Beziehungen in politischen Krisenzeiten. Imprimerie Dumaret & Golay, Geneve (1966)

Schroeder-Gudehus, B.: Les Scientifiques et la Paix. La communauté scientifique internationale au cours des années 20. Les Presses de l'Université de Montréal, Montréal (1978)

Schwabe, K.: Wissenschaft und Kriegsmoral. Die deutschen Hochschullehrer und die politischen Grundfragen des ersten Weltkrieges. Musterschmidt Verlag, Göttingen (1969)

Segal, S.L.: Mathematics and German politics: the National Socialist experience. Hist. Math. **13**, 118–135 (1986)

Sieg, W.: Hilbert's programs: 1917–1922. Bull. Symb. Log. **5**, 1–44 (1999)

Sieg, W.: Towards finitist proof theory. In: Jørgensen, K.F., Hendricks, V., Pedersen, S.A. (eds.) Proof Theory. History and Philosophical Significance. Synthese Library, vol. 292, pp. 95–116. Kluwer, Dordrecht (2000)

Siegmund-Schultze, R.: Theodor Vahlen zum Schuldanteil eines deutschen Mathematikers am faschistischen Miszbrauch der Wissenschaft. NTM Schriftenr. Gesch. Naturwiss. Tech. Med. **21**, 17–32 (1984)

Siegmund-Schultze, R.: Rockefeller and the Internationalization of Mathematics Between the Two World Wars. Birkhäuser, Basel (2001)

Soifer, A.: In search for van der Waerden, Leipzig and Amsterdam, 1931–1945. Part I: Leipzig. Geombinatorics **14**, 21–40 (2005a)

Soifer, A.: In search for van der Waerden, Leipzig and Amsterdam, 1931–1945. Part II: Amsterdam. Geombinatorics **14**, 72–102 (2005b)

Soifer, A.: In search for van der Waerden, Leipzig and Amsterdam, 1931–1945. Part III: Amsterdam. Geombinatorics **14**, 124–161 (2005c)

Springer, T.A.: B.L. van der Waerden. Levensber. Herdenk. **999**, 45–50 (1997)

Struik, D.J.: Levensbericht van Jan Arnoldus Schouten (28 januari 1883–20 januari 1971). Jaarb. KNAW **71**, 94–100 (1971)

Stuurman, F., Krijgsman, H.: Family Business. On Dictionary Projects of H. Poutsma (1856–1937) and L.E.J. Brouwer (1881–1966). Stichting Neerlandistiek, VU Amsterdam, Amsterdam (1995)

Sundholm, G., van Atten, M.: The proper interpretation of intuitionistic logic: on Brouwer's demonstration of the Bar Theorem. In: van Atten, M., Boldini, P., Bourdeau, M., Heinzmann, G.

(eds.) One Hundred Years of Intuitionism (1907–2007), pp. 60–70. Birkhäuser, Basel (2008)
Tietze, H.: Über Funktionen die auf einer abgeschlossenen Menge stetig sind. J. Reine Angew. Math. **145**, 9–14 (1914)
Tönnies, F.: Philosophical terminology (I). Mind **8**, 289–332 (1899a)
Tönnies, F.: Philosophical terminology (II). Mind **8**, 467–491 (1899b)
Tönnies, F.: Philosophical terminology (III). Mind **9**, 46–61 (1900)
Tönnies, F.: Philosophische Terminologie in psychologisch-soziologischer Ansicht. Thomas, Leipzig (1906)
Troelstra, A.S.: A. Heyting on the formalization of intuitionistic logic. In: Grootendorst, A.W., Bertin, E.M.J., Bos, H.M.J. (eds.) Two Decades of Mathematics in the Netherlands, vol. 1, pp. 153–175. Mathematical Centre, Amsterdam (1978)
Troelstra, A.S.: On the origin and development of Brouwer's concept of choice sequence. In: Troelstra, A.S., van Dalen, D. (eds.) The L.E.J. Brouwer Centenary Symposium, vol. 999, pp. 465–486. North-Holland, Amsterdam (1982)
Troelstra, A.S., van Dalen, D.: Constructivism in Mathematics, vol. 1. North-Holland, Amsterdam (1988a)
Troelstra, A.S.: van Dalen, D.: Constructivism in Mathematics, vol. 2. North-Holland, Amsterdam (1988b)
Tuchman, B.W.: The Guns of August. Bantam Books, New York (1962)
Tumarkin, L.: Nouvelle démonstration d'un théorème de Paul Urysohn. Fundam. Math. **8**, 360–361 (1926)
Ular, A.: Le Livre de La Voie et la ligne-droite de LAO-TSË. Éditions de la Revue Blanche, Paris (1902)
Urysohn, P.: Les multiplicités Cantoriennes. C. R. Math. Acad. Sci. Paris **175**, 440–442 (1922)
Urysohn, P.: Sur une fonction analytique partout continue. Fundam. Math. **4**, 144–150 (1923)
Urysohn, P.: Works on Topology and Other Areas of Mathematics 1, 2. Alexandrov, P. (ed.). State Publ. of Technical and Theoretical Literature, Moscow (1951) (Russian)
Vahlen, K.Th.: Max Dehns Besprechung meiner "Abstrakte Geometrie". Jahresber. Dtsch. Math.-Ver. **14**, 591–595 (1905a)
Vahlen, Th.: Abstrakte Geometrie. Untersuchungen über die Grundlagen der Euklischen und nicht-Euklidischen Geometrie. Teubner, Leipzig (1905b). Second edition appeared as supplement to Deutsche Mathematik. Hirzel, Leipzig (1940)
Vahlen, Th.: Konstruktionen und Approximationen in systematischer Darstellung: eine Vorstufe zur höheren Geometrie. Teubner, Leipzig (1911)
Vahlen, Th.: Ballistik. de Gruyter, Berlin (1922)
Vahlen, Th.: Ballistik, 2nd edn. de Gruyter, Berlin (1942)
van Atten, M.: On Brouwer. Wadsworth, London (2003)
van Atten, M.: The hypothetical judgement in the history of intuitionistic logic. In: Glymour, C., Wang, W., Westerståhl, D. (eds.) Logic, Methodology, and Philosophy of Science XIII: Proceedings of the 2007 International Congress in Beijing, vol. 13. King's College Publications, London (2008)
van Atten, M., van Dalen, D.: Arguments for the continuity principle. Bull. Symb. Log. **8**, 329–347 (2002)
van Dalen, D.: L.E.J. Brouwer en de eenzaamheid van het gelijk. Vrij Nederland, 21.2.1981, 3–23 (1981)
van Dalen, D. (ed.): Droeve snaar, vriend van mij. De correspondentie tussen Brouwer en Adama van Scheltema. De Arbeiderspers, Amsterdam (1984)
van Dalen, D.: Eine Bemerkung zum Aufsatz "Der Fundamentalsatz der Algebra und der Intuitionismus" von H. Kneser. Arch. Math. Log. **25**, 43–44 (1985)
van Dalen, D.: Hermann Weyl's intuitionistic mathematics. Bull. Symb. Log. **1**, 145–169 (1995)
van Dalen, D.: From Brouwerian counter examples to the creating subject. Stud. Log. **62**, 305–314 (1999a)
van Dalen, D.: Luitzen Egbertus Jan Brouwer. In: James, I. (ed.) History of Topology, pp. 947–964. Elsevier, Amsterdam (1999b)

van Dalen, D.: The role of language and logic in Brouwer's work. In: Orlowska, E. (ed.) Logic in Action, pp. 3–14. Springer, Vienna (1999c)
van Dalen, D.: Brouwer and Fraenkel on intuitionism. Bull. Symb. Log. **6**, 284–310 (2000)
van Dalen, D.: L.E.J. Brouwer 1881–1966. Een Biografie. Het heldere licht der wiskunde. Bert Bakker, Amsterdam (2001a)
van Dalen, D.: L.E.J. Brouwer en De Grondslagen van de wiskunde. Epsilon, Utrecht (2001b)
van Dalen, D.: Kolmogorov and Brouwer on constructive implication and the Ex Falso rule. Russ. Math. Surv. **59**, 247–257 (2004)
van Dalen, D.: Mystic, Geometer, and Intuitionist: The Life of L.E.J. Brouwer. Volume 2: Hope and Disillusion. Oxford University Press, Oxford (2005)
van Dantzig, D.: Studien over topologische algebra. Ph.D. thesis, Rijksuniversiteit Groningen (1931)
van Dantzig, D.: Gerrit Mannoury's significance for mathematics and its foundations. Nieuw Arch. Wiskd. **5**, 1–18 (1957)
van der Waerden, B.L.: De algebraiese grondslagen der meetkunde van het aantal. Ph.D. thesis, University of Amsterdam (1926)
van der Waerden, B.L.: De strijd om de abstraktie. Noordhoff, Groningen (1928). Inaugural address
van Eeden, F.: Het rode lampje. Versluys, Amsterdam (1921). 2 vols.
van Eeden, F.: Dagboek 1878–1923. van Tricht, H.W. (ed.). Tjeenk Willink, Culemborg (1971). 4 vols.
van Eeden, F., Gutkind, E.: Welt–Eroberung durch Helden–Liebe. Schuster & Loeffler, Berlin (1911)
van Emmerik, E.P.: J.J. van Laar (1860–1938). A mathematical chemist. Ph.D. thesis, Universiteit van Amsterdam (1991)
van Everdingen, E.: Zestig Jaar Internationale School van Wijsbergeerte. Van Gorcum, Assen (1976)
van Heemert, A.: De $R_n$-adische voortbrenging van algemeen–topologische ruimten met toepassing op de constructie van niet splitsbare continua. Ph.D. thesis, Rijksuniversiteit Groningen (1943)
van Heijenoort, J.: From Frege to Gödel. A Source Book in Mathematical Logic, 1879–1931. Harvard University Press, Cambridge (1967)
van Stigt, W.P.: The rejected parts of Brouwer's dissertation on the foundations of mathematics. Hist. Math. **6**, 385–404 (1979)
van Stigt, W.P.: Brouwer's Intuitionism. North-Holland, Amsterdam (1990)
Volker (pseud. E. Gutkind). Siderische Geburt. Seraphische Wanderungen vom Tode der Welt zur Taufe der Tat. Karl Schnabel, Berlin (1910)
Vollenhove, D.H.T.: De wijsbegeerte van de wiskunde van theïstisch standpunt. Ph.D. thesis, Vrije Universiteit, Amsterdam (1918)
Wang, H.: Reflections on Kurt Gödel. MIT Press, Cambridge (1987)
Weil, A.: Soevenirs d'apprentisage. Birkhäuser, Zürich (1991)
Weitzenböck, R.: Invarianten-Theorie. Noordhoff, Groningen (1923)
Weyl, H.: Das Kontinuum. Kritische Untersuchungen über die Grundlagen der Analysis. Veit, Leipzig (1918). Translation: The Continuum: A Critical Examination of the Foundations of Analysis. Dover Publications, reprint edn. (April 1994)
Weyl, H.: Über die neue Grundlagenkrise der Mathematik. Math. Z. **10**, 39–79 (1921)
Weyl, H.: Randbemerkungen zu Hauptproblemen der Mathematik. Math. Z. **20**, 131–150 (1924)
Weyl, H.: Diskussionsbemerkungen zu dem zweiten Hilbertschen Vortrag über die Grundlagen der Mathematik. Abh. Math. Semin. Univ. Hamb. **6**, 86–88 (1928)
Weyl, H.: The Open World. Three Lectures on the Metaphysical Implications of Science. Yale University Press, New Haven (1932)
Weyl, H.: David Hilbert and his mathematical work. Bull. Am. Math. Soc. **50**, 612–654 (1944)
Weyl, H.: Mathematics and logic. A brief survey serving as a preface to a review of "The Philosophy of Bertrand Russell". Am. Math. Mon. **53**, 2–13 (1946)

Weyl, H.: Erkenntnis und Besinnung (Ein Lebensrückblick). Studia Philosophica, Jahrbuch der Schweizerischen Philosopischen Gesellschaft (1954)

Wiessing, H.: Bewegend Portret. Moussault, Amsterdam (1960)

Willink, B.: De Tweede Gouden Eeuw. Nederland en de Nobelprijzen voor natuurwetenschappen, 1970–1940. Prometheus, Amsterdam (1998)

Wilson, W.: Afbeeldingen van Ruimten. Ph.D. thesis, Amsterdam (1928)

Wittgenstein, L.: Philosophische Untersuchungen. Surkamp, Frankfurt (1984). First ed. Blackwell, Oxford (1964). Manuscript (1930)

Yoneyama, K.: Theory of continuous set of points. Tohoku Math. J. **12**, 43–158 (1917)

Young, G.C., Young, W.H.: The Theory of Sets of Points. Cambridge University Press, Cambridge (1906)

Zermelo, E.: Beweis dasz jede Mengen wohlgeordnet werden kann. Math. Ann. **59**, 514–516 (1904)

Zermelo, E.: Neuer Beweis für die Möglichkeit einer Wohlordnung. Math. Ann. **65**, 107–128 (1908)

Zermelo, E.: Über den Begriff der Definitheit in der Axiomatik. Fundam. Math. **14**, 339–344 (1929)

Zoretti, L.: Review of A. Schoenflies' "Entwicklung der Lehre von den Punktmannigfaltigkeiten. II". Bull. Soc. Math. Fr. **35**, 283–288 (1911)

Zorin, V.K.: On Poincaré's letter to Brouwer. Russ. Math. Surv. **27**, 166–168 (1972)

# Index




DOI 10.1007/978-1-4471-4616-2, 

**S**

GPSR Compliance
The European Union's (EU) General Product Safety Regulation (GPSR) is a set of rules that requires consumer products to be safe and our obligations to ensure this.

If you have any concerns about our products, you can contact us on

ProductSafety@springernature.com

In case Publisher is established outside the EU, the EU authorized representative is:

Springer Nature Customer Service Center GmbH
Europaplatz 3
69115 Heidelberg, Germany

www.ingramcontent.com/pod-product-compliance
Lightning Source LLC
LaVergne TN
LVHW011535200926
848484LV00001B/267
* 9 7 8 1 4 4 7 1 4 6 1 5 5 *